Understanding Process Dynamics and Control

Understanding Process Dynamics and Control presents a fresh look at process control, with a state-space approach presented in parallel with the traditional approach to explain the strategies used in industry today.

Modern time-domain and traditional transform-domain methods are integrated throughout and the advantages and limitations of each approach are explained; the fundamental theoretical concepts and methods of process control are applied to practical problems.

To ensure understanding of the mathematical calculations involved, MATLAB is included for numeric calculations and Maple for symbolic calculations, with the math behind every method carefully explained so that students develop a clear understanding of how and why the software tools work.

Written for a one-semester course with optional advanced-level material, features include solved examples, cases including a variety of chemical process examples, chapter summaries, key terms and concepts, as well as over 240 end-of-chapter problems, including focused computational exercises.

Costas Kravaris is Professor of Chemical Engineering at Texas A&M University, USA. He has over 35 years of teaching experience in process dynamics and control classes at both undergraduate and graduate level. He is an active researcher in nonlinear control, nonlinear state estimation and nonlinear model reduction, with applications to chemical processes.

Ioannis K. Kookos is Professor of Process Systems Engineering, in the Department Chemical Engineering at the University of Patras, Greece. He received his BSc in Chemical Engineering from the National Technical University of Athens, Greece and his MSc (1994) and PhD (2001) in Process Systems Engineering from Imperial College, London (Centre for PSE). He then worked as a lecturer at the University of Manchester, Department of Chemical Engineering.

Cambridge Series in Chemical Engineering

Series Editor

Arvind Varma, *Purdue University*

Editorial Board

Juan de Pablo, *University of Chicago*
Michael Doherty, *University of California-Santa Barbara*
Ignacio Grossman, *Carnegie Mellon University*
Jim Yang Lee, *National University of Singapore*
Antonios Mikos, *Rice University*

Books in the Series

Baldea and Daoutidis, *Dynamics and Nonlinear Control of Integrated Process Systems*

Chamberlin, *Radioactive Aerosols*

Chau, *Process Control: A First Course with Matlab*

Cussler, *Diffusion: Mass Transfer in Fluid Systems, Third Edition*

Cussler and Moggridge, *Chemical Product Design, Second Edition*

De Pablo and Schieber, *Molecular Engineering Thermodynamics*

Deen, *Introduction to Chemical Engineering Fluid Mechanics*

Denn, *Chemical Engineering: An Introduction*

Denn, *Polymer Melt Processing: Foundations in Fluid Mechanics and Heat Transfer*

Dorfman and Daoutidis, *Numerical Methods with Chemical Engineering Applications*

Duncan and Reimer, *Chemical Engineering Design and Analysis: An Introduction, Second Edition*

Fan, *Chemical Looping Partial Oxidation Gasification, Reforming, and Chemical Syntheses*

Fan and Zhu, *Principles of Gas-Solid Flows*

Fox, *Computational Models for Turbulent Reacting Flows*

Franses, *Thermodynamics with Chemical Engineering Applications*

Leal, *Advanced Transport Phenomena: Fluid Mechanics and Convective Transport Processes*

Lim and Shin, *Fed-Batch Cultures: Principles and Applications of Semi-Batch Bioreactors*

Litster, *Design and Processing of Particulate Products*

Marchisio and Fox, *Computational Models for Polydisperse Particulate and Multiphase Systems*

Mewis and Wagner, *Colloidal Suspension Rheology*

Morbidelli, Gavriilidis, and Varma, *Catalyst Design: Optimal Distribution of Catalyst in Pellets, Reactors, and Membranes*

Nicoud, *Chromatographic Processes*

Noble and Terry, *Principles of Chemical Separations with Environmental Applications*

Orbey and Sandler, *Modeling Vapor-Liquid Equilibria: Cubic Equations of State and their Mixing Rules*

Pfister, Nicoud, and Morbidelli, *Continuous Biopharmaceutical Processes: Chromatography, Bioconjugation, and Protein Stability*

Petyluk, *Distillation Theory and its Applications to Optimal Design of Separation Units*

Ramkrishna and Song, *Cybernetic Modeling for Bioreaction Engineering*

Rao and Nott, *An Introduction to Granular Flow*

Russell, Robinson, and Wagner, *Mass and Heat Transfer: Analysis of Mass Contactors and Heat Exchangers*

Schobert, *Chemistry of Fossil Fuels and Biofuels*

Shell, *Thermodynamics and Statistical Mechanics*

Sirkar, *Separation of Molecules, Macromolecules and Particles: Principles, Phenomena and Processes*

Slattery, *Advanced Transport Phenomena*

Varma, Morbidelli, and Wu, *Parametric Sensitivity in Chemical Systems*

Wolf, Bielser, and Morbidelli, *Perfusion Cell Culture Processes for Biopharmaceuticals*

Understanding Process Dynamics and Control

Costas Kravaris
Texas A & M University

Ioannis K. Kookos
University of Patras, Greece

CAMBRIDGE
UNIVERSITY PRESS

University Printing House, Cambridge CB2 8BS, United Kingdom

One Liberty Plaza, 20th Floor, New York, NY 10006, USA

477 Williamstown Road, Port Melbourne, VIC 3207, Australia

314–321, 3rd Floor, Plot 3, Splendor Forum, Jasola District Centre, New Delhi – 110025, India

79 Anson Road, #06–04/06, Singapore 079906

Cambridge University Press is part of the University of Cambridge.

It furthers the University's mission by disseminating knowledge in the pursuit of education, learning, and research at the highest international levels of excellence.

www.cambridge.org
Information on this title: www.cambridge.org/9781107035584
DOI: 10.1017/9781139565080

© Costas Kravaris and Ioannis K. Kookos 2021

This publication is in copyright. Subject to statutory exception and to the provisions of relevant collective licensing agreements, no reproduction of any part may take place without the written permission of Cambridge University Press.

First published 2021

Printed in the United Kingdom by TJ Books Limited, Padstow Cornwall

A catalogue record for this publication is available from the British Library.

ISBN 978-1-107-03558-4 Hardback

Cambridge University Press has no responsibility for the persistence or accuracy of URLs for external or third-party internet websites referred to in this publication and does not guarantee that any content on such websites is, or will remain, accurate or appropriate.

Dedicated to our families,

Irene, Michael, Evangeline and Cosmas
C. Kravaris

and

Natasa, Kostas and Georgia
I.K. Kookos

Contents

Preface	page xvii

1 INTRODUCTION — 1

Study Objectives — 1
1.1 What is Process Control? — 1
1.2 Feedback Control System: Key Ideas, Concepts and Terminology — 2
1.3 Process Control Notation and Control Loop Representation — 8
1.4 Understanding Process Dynamics is a Prerequisite for Learning Process Control — 9
1.5 Some Historical Notes — 11
Learning Summary — 15
Terms and Concepts — 15
Further Reading — 16
Problems — 17

2 DYNAMIC MODELS FOR CHEMICAL PROCESS SYSTEMS — 18

Study Objectives — 18
2.1 Introduction — 18
2.2 Conservation Laws — 20
2.3 Modeling Examples of Nonreacting Systems — 23
2.4 Modeling of Reacting Systems — 28
2.5 Modeling of Equilibrium Separation Systems — 37
2.6 Modeling of Simple Electrical and Mechanical Systems — 39
2.7 Software Tools — 43
Learning Summary — 45
Terms and Concepts — 46
Further Reading — 46
Problems — 47

3 FIRST-ORDER SYSTEMS — 55

Study Objectives — 55
3.1 Examples of First-Order Systems — 55
3.2 Deviation Variables — 58
3.3 Solution of Linear First-Order Differential Equations with Constant Coefficients — 59

3.4	The Choice of Reference Steady State Affects the Mathematical Form of the Dynamics Problem	62
3.5	Unforced Response: Effect of Initial Condition under Zero Input	63
3.6	Forced Response: Effect of Nonzero Input under Zero Initial Condition	63
3.7	Standard Idealized Input Variations	65
3.8	Response of a First-Order System to a Step Input	68
3.9	Response of a First-Order System to a Pulse Input	73
3.10	Response of a First-Order System to a Ramp Input	75
3.11	Response of a First-Order System to a Sinusoidal Input	77
3.12	Response of a First-Order System to an Arbitrary Input – Time Discretization of the First-Order System	82
3.13	Another Example of a First-Order System: Liquid Storage Tank	88
3.14	Nonlinear First-Order Systems and their Linearization	94
3.15	Liquid Storage Tank with Input Bypass	97
3.16	General Form of a First-Order System	99
3.17	Software Tools	102
	Learning Summary	106
	Terms and Concepts	107
	Further Reading	108
	Problems	108

4 CONNECTIONS OF FIRST-ORDER SYSTEMS — 115

	Study Objectives	115
4.1	First-Order Systems Connected in Series	115
4.2	First-Order Systems Connected in Parallel	119
4.3	Interacting First-Order Systems	122
4.4	Response of First-Order Systems Connected in Series or in Parallel	123
4.5	Software Tools	132
	Learning Summary	134
	Terms and Concepts	136
	Further Reading	136
	Problems	137

5 SECOND-ORDER SYSTEMS — 144

	Study Objectives	144
5.1	A Classical Example of a Second-Order System	145
5.2	A Second-Order System can be Described by Either a Set of Two First-Order ODEs or a Single Second-Order ODE	147
5.3	Calculating the Response of a Second-Order System – Step Response of a Second-Order System	148
5.4	Qualitative and Quantitative Characteristics of the Step Response of a Second-Order System	154

		Contents	xi

	5.5	Frequency Response and Bode Diagrams of Second-Order Systems with $\zeta > 0$	159
	5.6	The General Form of a Linear Second-Order System	161
	5.7	Software Tools	163
	Learning Summary		166
	Terms and Concepts		166
	Further Reading		167
	Problems		168

6 LINEAR HIGHER-ORDER SYSTEMS 171

	Study Objectives		171
	6.1	Representative Examples of Higher-Order Systems – Using Vectors and Matrices to Describe a Linear System	171
	6.2	Steady State of a Linear System – Deviation Variables	175
	6.3	Using the Laplace-Transform Method to Solve the Linear Vector Differential Equation and Calculate the Response – Transfer Function of a Linear System	177
	6.4	The Matrix Exponential Function	179
	6.5	Solution of the Linear Vector Differential Equation using the Matrix Exponential Function	182
	6.6	Dynamic Response of a Linear System	187
	6.7	Response to an Arbitrary Input – Time Discretization of a Linear System	191
	6.8	Calculating the Response of a Second-Order System via the Matrix Exponential Function	195
	6.9	Multi-Input–Multi-Output Linear Systems	197
	6.10	Software Tools	202
	Learning Summary		206
	Terms and Concepts		206
	Further Reading		206
	Problems		207

7 EIGENVALUE ANALYSIS – ASYMPTOTIC STABILITY 215

	Study Objectives		215
	7.1	Introduction	215
	7.2	The Role of System Eigenvalues on the Characteristics of the Response of a Linear System	216
	7.3	Asymptotic Stability of Linear Systems	220
	7.4	Properties of the Forced Response of Asymptotically Stable Linear Systems	224
	7.5	The Role of Eigenvalues in Time Discretization of Linear Systems – Stability Test on a Discretized Linear System	225
	7.6	Nonlinear Systems and their Linearization	228
	7.7	Software Tools	240

	Learning Summary	244
	Terms and Concepts	245
	Further Reading	245
	Problems	245

8　TRANSFER-FUNCTION ANALYSIS OF THE INPUT–OUTPUT BEHAVIOR　　251

Study Objectives　　251
8.1　Introduction　　251
8.2　A Transfer Function is a Higher-Order Differential Equation in Disguise　　252
8.3　Proper and Improper Transfer Functions – Relative Order　　257
8.4　Poles, Zeros and Static Gain of a Transfer Function　　259
8.5　Calculating the Output Response to Common Inputs from the Transfer Function – the Role of Poles in the Response　　261
8.6　Effect of Zeros on the Step Response　　268
8.7　Bounded-Input–Bounded-Output (BIBO) Stability　　273
8.8　Asymptotic Response of BIBO-Stable Linear Systems　　275
8.9　Software Tools　　279
Learning Summary　　287
Terms and Concepts　　287
Further Reading　　288
Problems　　288

9　FREQUENCY RESPONSE　　297

Study Objectives　　297
9.1　Introduction　　297
9.2　Frequency Response and Bode Diagrams　　298
9.3　Straight-Line Approximation Method for Sketching Bode Diagrams　　303
9.4　Low-Frequency and High-Frequency Response　　311
9.5　Nyquist Plots　　312
9.6　Software Tools　　319
Learning Summary　　321
Terms and Concepts　　321
Further Reading　　322
Problems　　322

10　THE FEEDBACK CONTROL SYSTEM　　327

Study Objectives　　327
10.1　Heating Tank Process Example　　327
10.2　Common Sensors and Final Control Elements　　329
10.3　Block-Diagram Representation of the Heating Tank Process Example　　332

10.4	Further Examples of Process Control Loops	335
10.5	Commonly Used Control Laws	338
	Learning Summary	345
	Terms and Concepts	345
	Further Reading	346
	Problems	346

11 BLOCK-DIAGRAM REDUCTION AND TRANSIENT-RESPONSE CALCULATION IN A FEEDBACK CONTROL SYSTEM — 350

	Study Objectives	350
11.1	Calculation of the Overall Closed-Loop Transfer Functions in a Standard Feedback Control Loop	350
11.2	Calculation of Overall Transfer Functions in a Multi-Loop Feedback Control System	356
11.3	Stirred Tank Heater under Negligible Sensor Dynamics: Closed-Loop Response Calculation under P or PI Control	359
11.4	Software Tools	366
	Learning Summary	372
	Terms and Concepts	373
	Further Reading	373
	Problems	374

12 STEADY-STATE AND STABILITY ANALYSIS OF THE CLOSED-LOOP SYSTEM — 377

	Study Objectives	377
12.1	Steady-State Analysis of a Feedback Control System	377
12.2	Closed-Loop Stability, Characteristic Polynomial and Characteristic Equation	385
12.3	The Routh Criterion	389
12.4	Calculating Stability Limits via the Substitution $s = i\omega$	394
12.5	Some Remarks about the Role of Proportional, Integral and Derivative Actions	395
12.6	Software Tools	399
	Learning Summary	404
	Terms and Concepts	405
	Further Reading	405
	Problems	405

13 STATE-SPACE DESCRIPTION AND ANALYSIS OF THE CLOSED-LOOP SYSTEM — 409

	Study Objectives	409
13.1	State-Space Description and Analysis of the Heating Tank	409
13.2	State-Space Analysis of Closed-Loop Systems	415

13.3 Time Discretization of the Closed-Loop System	422
13.4 State-Space Description of Nonlinear Closed-Loop Systems	426
13.5 Software Tools	428
Learning Summary	434
Further Reading	435
Problems	435

14 SYSTEMS WITH DEAD TIME — 437

Study Objectives	437
14.1 Introduction	437
14.2 Approximation of Dead Time by Rational Transfer Functions	446
14.3 Parameter Estimation for FOPDT Systems	456
14.4 Feedback Control of Systems with Dead Time – Closed-Loop Stability Analysis	460
14.5 Calculation of Closed-Loop Response for Systems involving Dead Time	467
14.6 Software Tools	473
Learning Summary	475
Terms and Concepts	476
Further Reading	476
Problems	476

15 PARAMETRIC ANALYSIS OF CLOSED-LOOP DYNAMICS – ROOT-LOCUS DIAGRAMS — 484

Study Objectives	484
15.1 What is a Root-Locus Diagram? Some Examples	484
15.2 Basic Properties of the Root Locus – Basic Rules for Sketching Root-Locus Diagrams	502
15.3 Further Properties of the Root Locus – Additional Rules for Sketching Root-Locus Diagrams	508
15.4 Calculation of the Points of Intersection of the Root Locus with the Imaginary Axis	524
15.5 Root Locus with Respect to Other Controller Parameters	527
15.6 Software Tools	531
Learning Summary	536
Terms and Concepts	537
Further Reading	537
Problems	537

16 OPTIMAL SELECTION OF CONTROLLER PARAMETERS — 541

Study Objectives	541
16.1 Control Performance Criteria	541
16.2 Analytic Calculation of Quadratic Criteria for a Stable System and a Step Input	549

16.3	Calculation of Optimal Controller Parameters for Quadratic Criteria	557
16.4	Software Tools	563
	Learning Summary	570
	Terms and Concepts	571
	Further Reading	571
	Problems	572

17 BODE AND NYQUIST STABILITY CRITERIA – GAIN AND PHASE MARGINS — 575

	Study Objectives	575
17.1	Introduction	575
17.2	The Bode Stability Criterion	576
17.3	The Nyquist Stability Criterion	594
17.4	Example Applications of the Nyquist Criterion	597
17.5	Software Tools	604
	Learning Summary	607
	Terms and Concepts	607
	Further Reading	608
	Problems	608

18 MULTI-INPUT–MULTI-OUTPUT SYSTEMS — 613

	Study Objectives	613
18.1	Introduction	613
18.2	Dynamic Response of MIMO Linear Systems	620
18.3	Feedback Control of MIMO Systems: State-Space versus Transfer-Function Description of the Closed-Loop System	623
18.4	Interaction in MIMO Systems	627
18.5	Decoupling in MIMO Systems	632
18.6	Software Tools	634
	Learning Summary	638
	Terms and Concepts	639
	Further Reading	639
	Problems	639

19 SYNTHESIS OF MODEL-BASED FEEDBACK CONTROLLERS — 641

	Study Objectives	641
19.1	Introduction	641
19.2	Nearly Optimal Model-Based Controller Synthesis	648
19.3	Controller Synthesis for Low-Order Models	650
19.4	The Smith Predictor for Processes with Large Dead Time	657
19.5	Effect of Modeling Error	660
19.6	State-Space Form of the Model-Based Controller	668

19.7	Model-Based Controller Synthesis for MIMO Systems	674
	Learning Summary	678
	Terms and Concepts	678
	Further Reading	679
	Problems	679

20 CASCADE, RATIO AND FEEDFORWARD CONTROL 683

	Study Objectives	683
20.1	Introduction	683
20.2	Cascade Control	684
20.3	Ratio Control	694
20.4	Feedforward Control	695
20.5.	Model-Based Feedforward Control	700
	Learning Summary	714
	Terms and Concepts	715
	Further Reading	715
	Problems	715

APPENDIX A LAPLACE TRANSFORM 719

A.1	Definition of the Laplace Transform	719
A.2	Laplace Transforms of Elementary Functions	720
A.3	Properties of Laplace Transforms	721
A.4	Inverse Laplace Transform	725
A.5	Calculation of the Inverse Laplace Transform of Rational Functions via Partial Fraction Expansion	725
A.6	Solution of Linear Ordinary Differential Equations using the Laplace Transform	732
A.7	Software Tools	735
	Problems	739

APPENDIX B BASIC MATRIX THEORY 743

B.1	Basic Notations and Definitions	743
B.2	Determinant of a Square Matrix	747
B.3	Matrix Inversion	749
B.4	Eigenvalues	750
B.5	The Cayley–Hamilton Theorem and the Resolvent Identity	752
B.6	Differentiation and Integration of Matrices	755
B.7	Software Tools	756

Index 760

Preface

Scope of the Book

When we took undergraduate process dynamics and control in the 1970s and the 1980s, the entire course was built around the Laplace transform and the transfer function. This conceptual and methodological approach has been in place in undergraduate chemical engineering education since the 1960s and even the 1950s, and it reflected the development and widespread use of electronic PID control systems, for which it provided a very adequate background for the chemical engineering graduates. Today, the vast majority of undergraduate chemical process dynamics and control courses still follow exactly the same conceptual approach, revolving around the Laplace transform and the transfer function. But control technology has changed a lot during the past 60 years. Even though PID controllers are still used, model-predictive control has evolved into an industrial standard for advanced applications. But model-predictive control is formulated in state space and in discrete time, whereas the standard control course is in the transform domain and in continuous time. There is a big conceptual gap between what is taught in the classroom and the industrial state of the art. This gap is well recognized within the chemical process control community, as is the need to bridge this gap. It is aim of this book to propose a realistic solution on how to bridge this gap, so that chemical engineering graduates are better prepared in using modern control technology. This book has evolved after many years of teaching experimentation at Texas A&M University and the University of Patras.

The main feature of this book is the introduction of state-space methods at the undergraduate level, not at the end of the book, but from day one. There are two main reasons that this is feasible. The first is that state-space concepts and methods are easy to grasp and comprehend, since they are in the time domain. The second is the availability of powerful computational tools that emerge from the state-space methods and can be implemented through user-friendly software packages. Once the student is given the key ideas and concepts in the time domain, he/she can painlessly apply them computationally.

Of course, one should not downplay the significance of manual calculations in developing an understanding of dynamic behavior in open loop and in closed loop. To this end, Laplace-transform methods offer a distinct advantage over time-domain methods. Even though industrial practitioners keep telling us that "there is no Laplace domain in their plant," there is no question about its educational value. The concept of the transfer function is also an invaluable educational tool for the student to understand connections of dynamic systems, including the feedback loop, and also to calculate and appreciate frequency response characteristics. For this reason, Laplace-domain methods are used in this book, and they are used in parallel with state-space methods. Whenever a quick manual calculation is feasible, the student should be able to go to the "Laplace planet" and come back, whenever calculations are very involved or simulation is needed, the student should be able to handle it computationally using software.

This book offers a strong state-space component, both conceptually and computationally, and this is blended with the traditional analytical framework, in order to maximize the students' understanding. But there is also an additional advantage. Because of its state-space component, this book brings the process dynamics and control course closer to other chemical engineering courses, such as the chemical reactor course. A chemical reactor course introduces local asymptotic stability in a state-space setting and tests it through eigenvalues, whereas a traditional control course defines stability in an input–output sense and tests it through the poles of the transfer function. This gap is nonexistent in the present book: asymptotic stability is defined and explained in a state-space context, input–output stability is defined and explained in a transfer function or convolution integral context, and the relationship of the two notions of stability is discussed. Moreover, there are a number of chemical reactor examples throughout the book that link the two courses in a synergistic manner.

A final comment should be made about the word "understanding" in the title of this book. It is our firm belief that engineers must have a thorough understanding of how their tools work, when do they work and why they work. If they treat a software package as a magic black box, without understanding what's inside the box, they have not learned anything. For this reason, special care is taken in this book to explain the math that is behind every method presented, so that the student develops a clear understanding of how, when and why.

Organization of the Book

A general introduction is given in Chapter 1. A review of unsteady state material and energy balances is given in Chapter 2. Reviews of the Laplace transform and of basic matrix algebra are separate from the chapters, and are given in Appendices A and B.

Chapters 3–9 and the first half of Chapter 14 cover process dynamics. The approach taken is to start from the simplest dynamic systems (first-order systems) in Chapter 3, and then progressively generalize. Both time domain (including discrete time) and transfer function (including frequency response) start from Chapter 3 and are pursued in parallel in the subsequent chapters. Chapters 4 and 5 are generalizations, studying connections of first-order systems and inherently second-order systems. Chapters 6–9 cover the dynamic analysis of higher-order systems in both state space (Chapters 6 and 7) and transform domain (Chapters 8 and 9), including asymptotic stability and input–output stability. Dead time is postponed to Chapter 14. All the dynamics chapters are to be covered; the only part that is optional is the second part of Chapter 9 on Nyquist diagrams, which is only needed in the second part of Chapter 17.

The rest of the chapters are on process control. Chapters 10–14 cover the basic feedback control concepts and analysis methods. Chapter 10 gives a general introduction to feedback control, and also defines the PID controller in both state-space and transfer function form. Chapters 11 and 12 do transfer function analysis of the feedback control loop, whereas in Chapter 13 the same analysis is done in state space. Chapter 14 discusses systems with deadtime, both open loop dynamics and feedback control. Deadtime is treated separately because of its distinct mathematical characteristics. Chapters 10–14 provide an absolute minimum for the feedback control part of the course. From that point, the instructor can choose what design methods he/she wants to put emphasis on, root locus (Chapter 15), optimization (Chapter 16), gain and phase margins (Chapter 17) or model-based (Chapter 19). Also, the instructor has the choice to discuss issues in multivariable control (Chapter 18) or stay SISO throughout the course. The last chapter (Chapter 20) discusses cascade, ratio and feedforward control. These control structures are discussed first at a conceptual level, and then model-based design for cascade and feedforward control is derived. The conceptual part is, in a sense, a continuation of Chapter 10 and it is essential to be taught; the model-based part is a continuation of Chapter 19.

The last section of each chapter is about software tools. The use of software for the application of the theory of the chapter is explained through simple examples. Two alternative software packages are used: MATLAB and its control systems toolbox is chosen because of its strength in numerical calculations, and Maple and its libraries (LinearAlgebra, inttrans, etc.) because of its strength in symbolic calculations.

The following table gives a sample syllabus for the process dynamics and control course at Texas A&M University, as it has been taught in the past three semesters. It reflects the personal choices of the instructor on (i) the design methods for the control part of the course (optimization and model-based are emphasized) and (ii) the pace of covering the material (slower at the beginning, faster at the end). Of course, there are many other options, depending on instructor priorities and students' background.

Topic	From the book	Hours
Introduction	Chapter 1	1
Review of unsteady-state material and energy balances	Chapter 2	1
Review of the Laplace transform	Appendix A	2
First-order systems	Chapter 3	5
Connections of first-order systems	Chapter 4	2
Second-order systems	Chapter 5	2
Higher-order systems	Chapter 6 and Appendix B (first half)	4 ½
Eigenvalue analysis, asymptotic stability	Chapter 7 and Appendix B (second half)	2 ½
Transfer-function analysis	Chapter 8	2
Bode diagrams	Chapter 9 – Bode part	1

Topic	From the book	Hours
The feedback control system	Chapter 10	1
Block-diagram simplification, closed-loop responses	Chapter 11	2
Steady-state analysis, stability analysis	Chapter 12	2 ½
State-space analysis of the closed-loop system	Chapter 13	1 ½
Optimization of feedback controllers	Chapter 16	2
Systems with dead time	Chapter 14	2
Bode stability criterion, gain and phase margins	Chapter 17 – Bode part	1
Model-based control	Chapter 19, excluding MIMO	2
Cascade, ratio and feedforward control	Chapter 20	2
Total lecture hours		39

Costas Kravaris and Ioannis K. Kookos, October 2020

1 Introduction

STUDY OBJECTIVES

After studying this chapter, you should be able to do the following.

- Identify the underlying reasons for the need of control systems.
- Identify the main parts of a feedback control system.
- Identify the main terminology and notation used in process control.
- Discuss recent history of process control.

1.1 What is Process Control?

Automatic control is a discipline which studies the design of man-made systems with the aim to "shape" purposefully their response. Scientists and engineers who work in this field of study, depending on their background or their area of interest, may give a more specific or more abstract definition. Automatic control is an interdisciplinary science and plays a key role in most engineering disciplines including electrical, mechanical and chemical engineering. There is a common theoretical basis that can be applied to all these systems, despite the major differences in their physical characteristics.

Process control is the branch of automatic control concerned with production plants in the chemical, petrochemical, food and related industries. Process control plays a critical role in ensuring proper operation of the plant, in terms of safety, product quality and profitability. Even though chemical processes are of different physical nature when compared to robots, unmanned vehicles and aircrafts, missiles and spacecrafts, the underlying principles of automatic control are the same.

Automatic control is a part of our everyday life. Cars, refrigerators, washing machines, public buildings and homes have numerous automatic control systems installed. What is equally impressive is that these control systems operate and function so efficiently that we hardly ever take notice of their existence. They deliver, they are reliable, they make our everyday life better and safer, and they are really cheap. They are the result of the hard work of numerous ingenious scientists and engineers who devoted their life to make our world a better place to be.

Introduction

The amount of knowledge that has been generated in the past 100 years in the field of automatic control is vast. Using this knowledge to design and operate control systems in practice is vital for maintaining the same pace of development in the years to come. This book aims at explaining the fundamental principles of process control in a way that makes it easier for future chemical engineers to comprehend past developments and to develop new tools that advance engineering practice. We also hope that learning process control methods and concepts will help future chemical engineers to interact and collaborate with control engineers of other disciplines.

1.2 Feedback Control System: Key Ideas, Concepts and Terminology

The idea of feedback control will be introduced in the present section, along with some pertinent key concepts and terminology.

Consider the primitive control system shown in Figure 1.1. A liquid stream is fed to a buffer tank (*process*) and an operator (*controller*) tries to keep the liquid level in the tank (*measured* and *controlled variable*) at the desired value (*set point*) by using a logical procedure (*control algorithm*) based on his/her training and experience. The means to accomplish this task is the opening or closing of a valve (*final control element*) that adjusts the flowrate (*manipulated variable*) of the exit stream.

A number of questions are immediately raised.

- Why does the liquid level vary during everyday operation?
- What is the "desired level" of the liquid in the tank and on what grounds is it determined?

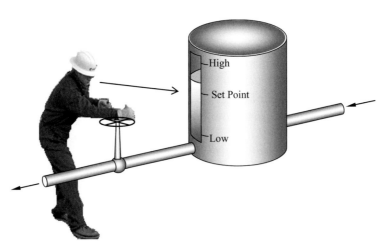

Figure 1.1 A "primitive" level control system.

- What is the best sequence of actions that the operator needs to take when the liquid level deviates from the set point?
- Do we need a human operator to control the level? Why not build an "automated system"?
- Would an "automated system" be more efficient and reliable than the human in controlling the level of the liquid?
- Are there any other measurements that could be used, in combination with the level of the liquid in the tank, in order to more effectively maintain the liquid level at the desired value?

Some of these types of questions have an easy answer, but some will need further thinking and elaboration throughout your process control course and even throughout your professional career.

Most undergraduate courses in chemical engineering consider processes that operate at steady state. This is a logical and well-documented simplification that allows chemical engineers to design fairly complex processes in a reasonable amount of time. However, in actuality, processes operate in a dynamic environment. Imagine that you design a heat exchanger that uses sea water as cooling water to cool down a process stream from 100 °C to 50 °C. At the design stage you have to make an assumption about the temperature of the cooling water (a unique value) and suppose that you have selected a temperature of 20 °C. Now think about the chances of the cooling water temperature being exactly 20 °C. Will the system fail if the actual temperature of the water is 15 °C or 10 °C? The answer is yes, the system will fail to keep the temperature at the desired value of 50 °C, unless a valve is installed, which can appropriately adjust the flowrate of the cooling water. In addition, a temperature sensor needs to be installed, to measure the temperature of the process stream exiting the heat exchanger. Then, using the measured and recorded temperature, an operator can check if the temperature is at the proper value, and appropriately adjust the cooling water flowrate to correct any discrepancies, as shown in Figure 1.2. The sea-water temperature can vary

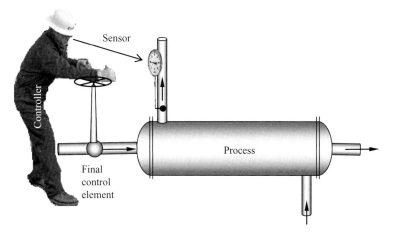

Figure 1.2 A "primitive" temperature-control system.

throughout the day, so the operator will need to perform frequent adjustment of the valve opening to keep the temperature of the process stream close to the desired temperature. In addition, the operator can implement changes in the desired temperature, if there are reasons related to the operation of downstream processes.

The basic elements of the temperature-control system shown in Figure 1.2 are also shown in the block diagram of Figure 1.3. The blocks are used as a means of representing the components of the system and the arrows denote a signal or information flow. The measurement (the line exiting the sensor and entering the controller) is not, in the case of the control system of Figure 1.2, an actual signal but an information flow and denotes the reading of the temperature indication by the operator. The operator/controller is a necessary element of the loop that processes (using a control algorithm) the information and decides on the appropriate action to be taken (opening or closing of the valve). The opening or closing of the valve determines the flowrate of the cooling medium (sea water) and thus the rate of heat transfer in the heat exchanger. Finally, the temperature of the product stream is measured by the sensor (operator's eyes) and the loop is closed. In most cases, the controller is a computer-based system that receives a signal from the sensor, executes the control algorithm and sends a signal that sets the valve position, as indicated in Figure 1.4. Computers can perform very complex calculations in a very short time, can handle more than one control system simultaneously and work continuously and, in most cases, without human intervention.

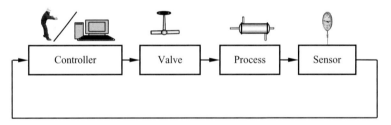

Figure 1.3 Elements of a "primitive" temperature-control system.

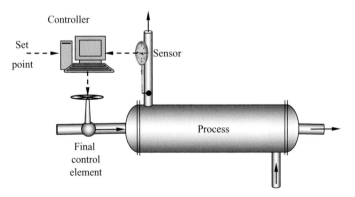

Figure 1.4 A computer-based temperature-control system.

1.2 Feedback Control System: Key Ideas, Concepts and Terminology

Let's now summarize the types of process variables encountered in the primitive control systems that we have seen so far.

Disturbance variables. Any process is affected by several external influences, and many of them vary in an uncontrollable and unpredictable manner. These are the disturbances that cause the operation of the process to deviate from the desired steady state. In the case of the heat exchanger, potential disturbances are the sea-water temperature, the temperature and the flowrate of the incoming process stream, or equipment aging known as fouling (which increases the resistance to heat transfer). Some of the disturbances could be measured in real time but others are difficult, expensive or even impossible to measure.

Manipulated variables. The manipulated variables are those process variables that are adjusted by the controller in order to achieve the control objectives. A manipulated variable is also called control input, to signify that it represents the control action that "feeds" the process. The most frequent manipulated variable in the chemical process industries is the flowrate through the installation of a control valve or a pump.

Measured variables. Measured variables are all the variables for which we have installed a sensor or measuring device that continuously measures and transmits the current value of the variable. Of course, sensors cost money and need frequent maintenance, therefore their installation should be well justified. The most common measured variables in the chemical industries are temperature, pressure, flow and level. Others, such as composition, are more costly and less frequently used.

A measured variable that the controller is maintaining at a particular desired value is called a **controlled variable**. The desired value is called the **set point** of the controlled variable. The set point is usually kept constant for a long time, but sometimes a need may arise to change the set point, and this should be handled by the controller.

When the value of the controlled variable agrees with its set point, it is "in control," otherwise there is an error. The **error** is defined as the difference between the set point and the value of the controlled variable, and the job of the controller is to make it equal to zero.

Figure 1.5 depicts a generic feedback control system. It shows all the types of variables that come into play, as well as the basic elements of the control system and how they are connected with each other.

The final control element (usually a control valve) together with the process and the sensor comprise the **physical system** or **open-loop system**. We see from Figure 1.5 that, when the sensor is connected to the controller, and the controller is acting on the final control element, the overall system has a circular structure, like a ring or a loop, and it is called the **closed-loop system**. It is also called a **feedback control system**. The idea of feedback control involves continuous monitoring of the controlled variable and "feeding back" the information, to make changes and adjustments in the process, through changes in the manipulated variable. The controller's action is usually based on the error, i.e. the discrepancy between the set point and the measurement of the controlled output. Depending on the error (its current value, its history and its trend), the controller takes corrective action. In simple terms, one can describe the operation of a feedback control system as: monitor, detect and correct.

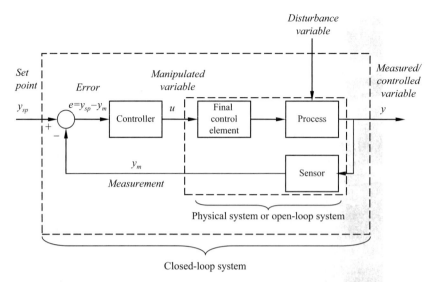

Figure 1.5 Basic elements of a feedback control system and their interconnections.

Sensors play a critical role is the proper operation of a feedback control system. Sensors use an electrical or mechanical phenomenon in order to determine the temperature, pressure, level or flowrate. Temperature sensors are based on the expansion of a liquid or gas (thermometers), on the Seebeck or thermoelectric effect, the creation of voltage between two junctions at different temperatures (thermocouples), the variation of electrical resistance of several materials with temperature (resistance temperature detectors and thermistors) and the thermal radiation emitted (pyrometers). Most pressure sensors are based on measuring the deflection or strain caused by the pressure when applied to an area (strain-gauge, electromagnetic, piezoelectric, etc.). Pressure sensors are used in conjunction with an orifice or a Venturi tube to measure flow, as differential pressure across the orifice or between two segments of a Venturi tube (with different aperture) is strongly related to flow. Pressure sensors are also used to calculate the level of a liquid in a tank as the pressure difference between the top and the bottom of a tank is directly proportional to the height of the liquid. The transmitter is used to convert the primary measurement by the sensor to a pneumatic or electrical signal. The combination of the sensor and the transmitter is called a transducer.

In a chemical plant, there may be hundreds or thousands of feedback control loops like the one depicted in Figure 1.5. The need to transmit all information and functionality to a central "control room" (see Figure 1.6) to achieve continuous monitoring and reduce drastically the manpower required was quickly identified and implemented in the 1960s. This centralization was really effective in improving the operation of the plant. At that time, the controllers (one controller for each control loop) were behind the control room panels, and all control signals were transmitted back to plant. Gradually the structure was modified as all functionalities were assigned to a network of input/output racks with their own control processors which could be distributed locally in the plant (and could communicate with the

Figure 1.6 A control room of the 1960s and a more recent DCS control room (from Wikimedia Commons, the free media repository).

control room). The distributed control system or DCS was thus born, in which the controllers were placed close to the processing units but transmitted all information to a central location through a central network to minimize cabling runs. Monitoring, interconnection, reconfiguration and expansion of plant controls were finally easy. Local control algorithms could be executed by the central units in the case of system failure and thus reliability was greatly enhanced. Recent advantages such as wireless technology and Internet of Things as well as mobile interfaces might have a real impact in the near future.

1.3 Process Control Notation and Control Loop Representation

The standard notation used in process control is also indicated in Figure 1.5. The actual value of the measured and controlled variable is denoted by y. The measurement is denoted by y_m, and this may not match y in a transient event, as the sensor signal may be lagging behind in the changes of the physical variables that it measures. The desired or set-point value of the controlled variable is denoted by y_{sp}. The error signal $e = y_{sp} - y$ is also indicated in the diagram. (The small circle with the two inward arrows with appropriate signs and one outward arrow indicates the subtraction operation.) The error signal e drives the controller, which determines the appropriate adjustments, in order to correct the error and eventually bring it back to zero. The signal u from the controller sets the value of the manipulated variable of the process, which is actually implemented by the final control element. Finally, the sensor detects the change in the response of the system and the loop is closed.

Process engineers use standard symbols to denote process units such as vessels, heat exchangers and towers when constructing the Process Flow Diagram (PFD) of a production facility. The same holds true for control and instrumentation engineers. The standards for documenting the details of control and instrumentation have been defined by the Instrumentation, Systems, and Automation Society (ISA) and are known as Standard ISA-S5. There are several publications by the ISA that document in great detail the construction of Process and Instrumentation Diagrams (P&ID) that are routinely used by process engineers during the construction, commission and operation phases. The reader is referred to these publications as they are outside the scope of this book.

The notation that will be used in this book is presented in Figure 1.7 through an example. The standard representation of a control system is shown in this figure. A process stream

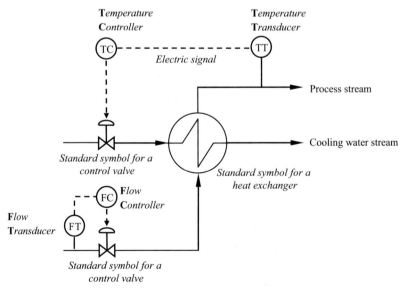

Figure 1.7 Common control loop representation.

that is flow-controlled, is cooled using a heat exchanger and cooling water. The temperature of the outgoing process steam is measured by a temperature sensor and the measurement is indicated by the temperature transducer (TT) symbol in Figure 1.7. Bubble or circle symbols are used to indicate instrumentation (measurement) or control function. Inside the circle symbol a two-letter coding system is used to denote the specific functionality of the block. The first letter in the two-letter naming system refers to the variable controlled or measured and the letters commonly used are the following.

T: temperature
F: flow, flowrate
L: level
P: pressure
C (or A): composition

The second letter indicates whether this is a measuring device or transducer (T) or a control device (C). TC is therefore used to indicate a temperature controller while FC indicates a flow controller in Figure 1.7. The two circle symbols denoted as TT (temperature transducer) and TC (temperature controller) are connected through a dashed line (– – – –) which indicates an electrical signal (4–20 mA, 1–5 V or 0–10 V). Other common conventions are the following: a pneumatic signal is denoted by —//—//— (normally in the range of 3–15 psig) and a data-transfer signal is denoted by —o—o— (usually binary signal). We will not try to indicate explicitly whether a signal is electrical or pneumatic, as it adds a complexity that is unnecessary within this book, and we will be using a dashed line to indicate exchange of information between a sensor, a controller and a final control element, as shown in Figure 1.7.

1.4 Understanding Process Dynamics is a Prerequisite for Learning Process Control

The action of a controller is not static: it is dynamic in nature. As external disturbances vary with time, the controller must take action, in a continuously changing environment. And the controller is not isolated: it keeps interacting with the sensor and the final control element, which in turn interact with the process, and all these interactions are transient in nature. To be able to understand what is happening inside the feedback control loop, we must first have a thorough understanding of transient behavior.

The process, the final control element, the sensor and the controller are all dynamic systems, whose behavior changes with time due to a changing environment (such as varying feed composition or temperature), changing process specifications (such as changing product purity) or equipment aging (such as fouling). The mathematical tools normally used to describe process dynamics are ordinary and partial differential equations accompanied, in some cases, by algebraic equations.

In the first part of this book we will study the dynamics of an isolated system, to try to understand its transient behavior. We will see different kinds of transient behavior, and we will explain the behavior and characterize it in a systematic way. We will see how to calculate these transient responses, analytically and numerically. One of the key concepts that we will discuss is the concept of stability, and we will derive tests to determine if a system has stable behavior. We will also introduce the necessary software tools to calculate routinely the dynamic response of common process systems.

Equipped with these concepts and tools, we will study interconnected dynamic systems, in a feedback control loop. We will see how the dynamic behavior of all the elements of the loop can be combined, and we will derive the dynamic behavior of the overall system, and calculate its transient response.

Typical process systems' dynamic responses are presented in Figure 1.8. The response can be fast or relatively slow as shown in Figures 1.8a and b, respectively. A characteristic commonly encountered in process systems is that of delayed response as shown in Figure 1.8c. These three general responses are the ones usually obtained by chemical processes like distillation and absorption columns, evaporators and heat-transfer equipment. There are

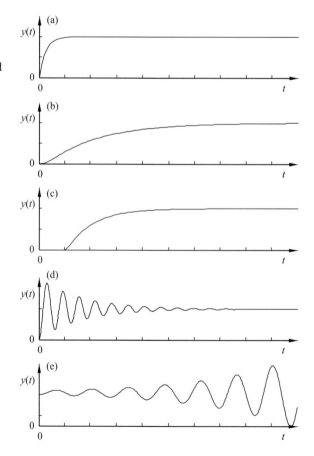

Figure 1.8 Representative cases of process system transients: (a) fast transient, (b) slow transient, (c) delayed transient, (d) oscillatory transient and (e) unstable transient.

processes, like chemical reactors, that can exhibit oscillatory (Figure 1.8d) or even unstable (Figure 1.8e) response.

1.5 Some Historical Notes

A comprehensive presentation of the history of feedback control has been written by O. Mayr (*The Origins of Feedback Control*, The MIT Press, 1970), who traces the control of mechanisms to antiquity with references to the work of Ktesibios, Philon and Heron as the main representatives of Hellenistic technology. The interested reader is referred to Mayr's book for a detailed discussion of this early work. The modern form of control systems technology is believed to have started in the middle of the eighteenth century. A famous problem at that time was the search for a means to control the rotation speed of a shaft, used for instance in the grinding stone in a wind-driven flour mill. A promising method was based on the use of a conical pendulum, or flyball governor (also called a centrifugal governor), to measure the speed of the mill. The adaptation of the flyball governor to the steam engine in the laboratories of James Watt around 1788 made the flyball governor famous. The action of the flyball governor is shown in Figure 1.9 and was simple: the balls (which spin around a central shaft) of the governor generate an angle with the horizontal that is analogous to the speed of the rotating shaft (and the speed of the engine, which is connected to the shaft). Increasing the speed increases the angle and decreasing the speed decreases the angle. The angle formed can therefore be used to sense the speed of the engine and then proper control action can be achieved (by mechanical means) by manipulating the flowrate of the steam (in the case of the steam engine). When the engine speed drops (due to an increase in the engine's load), the ball angle decreases and more steam will be admitted by mechanical means, restoring most of the lost speed. The reverse action will be achieved in the case of decreased load. A basic form of feedback control is thus realized. Apart from the famous flyball governor,

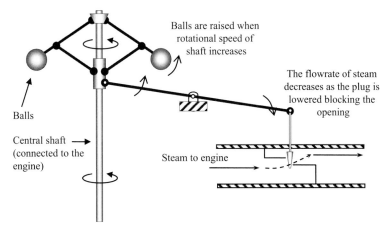

Figure 1.9 Basic operation of the flyball governor (engine speed control system).

slow progress took place in the field in the nineteenth century. Things, however, made a dramatic change in the beginning of the twentieth century.

Things were very different in the chemical industry just 100 years ago. We present the following extract from one of the most significant journals in chemical engineering (the *Journal of Industrial & Engineering Chemistry*, published by the American Chemical Society or ACS) that was published in 1918 (*I&ECR*, p. 133, February 1918, by R. P. Brown from The Brown Instrument Company, Philadelphia):

> Probably no employee causes the average works manager so many sleepless nights as does the furnace man, on whose shoulders rests the responsibility for the accurate heat treatment of the steel and the uniformity of the product. This is not only true of a steel plant, but is also equally true in the chemical industry, where the temperature of numerous processes must be accurately controlled; … The old furnace man, through years of practice, will endeavor to gauge the temperature of the furnace with his eye. Providing he has not been up all the previous night and his eye is clear, he will probably judge the temperature fairly accurately. …
>
> But we can pardon the works manager or director for asking, "Suppose John dies, gets sick or quits his job, how am I to handle the output of these furnaces?" He would like to have an understudy for the old furnace man, but the latter does not like the idea. So he wonders why someone does not develop a device to automatically control the temperature of the furnaces, so that he can cease worrying about them.
>
> This is one reason why a great amount of study has been given, not only to perfection of pyrometers, but also to the automatic control of temperature. It has, however, been only recently that real results have been accomplished in automatic temperature control.

Some primitive control systems are described only a few years later. One of these is shown in Figure 1.10. The so-called vapor-tension system consisted of a metallic bulb partly filled with a liquid, located in the container under control, and a metallic capillary leading from the bulb to an expansion or capsular chamber in the regulator case. Any change in the temperature of the space surrounding the metallic bulb will cause a change in the vapor pressure of the enclosed liquid and therefore in the pressure below the diaphragm. As the pressure in the capillary tube and diaphragm changes, this causes the stem and the plug to move up (when the temperature increases) or down (when the temperature decreases). If the plug is used in a valve to restrict or release the opening through

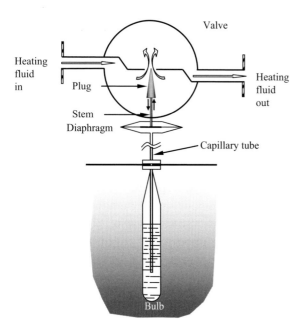

Figure 1.10 A primitive control system (adapted with modifications from *I&ECR*, **14**(11), 1016, November 1922).

which a thermal fluid is supplied to the system then a rudimentary control action can be achieved. When the temperature increases the vapor pressure of the enclosed liquid also increases, raising the stem and the plug and blocking the flow of the heating fluid. This causes less heating fluid to be supplied to the system and the temperature drops. This is a continuous control system that was chosen here purposefully despite the fact that most early control systems were simple on–off devices.

A number of interesting descriptions of early, basic control systems appeared in the literature in the late 1920s (see, for instance, I. Ginsberg, Automatic control in the chemical industries, *I&ECR*, **21**(5), 410, 1929). Real progress was achieved in the 1930s and mainly in the early 1940s with the publication of the seminal papers by Ziegler and Nichols (Ziegler, J. G. and Nichols, N. B., *Trans. Am. Soc. Mech. Eng.*, **64**, 759–68, 1942, and **65**, 433–44, 1943). These papers presented a systematic method for selecting the parameters of process controllers with minimal process upsetting and were based solely on the response of the open-loop system to a step change in the process input. The controller parameters were then determined as a function of the characteristics of the step response. The method proved to be particularly successful and gained widespread acceptance from the community. The Ziegler–Nichols method is still selected by many researchers, 80 years later, as a basis for establishing the advantages of their controller tuning techniques.

The numerous military applications of automatic control, developed and perfected during the Second World War, created a strong interest in the chemical process industries. Just before the beginning of the Second World War, a significant change started to materialize in the chemical plants as most of the large enterprises, such as E. I. du Pont de Nemours and Company, realized the benefits of continuous operation as compared with batch operation. However, it all had to wait until the introduction of microcomputers in the late 1950s and it was during the 1960s that the process control became an integral part of the operation of the chemical plants.

In the 1950s, process control theory witnessed significant progress as particular attention was placed in the asymptotic response of process systems to sinusoidal input variation. This analysis technique, known as frequency response, was particularly mature at that time in other areas such as telecommunication systems. At this time the first book in process control was published by N. Ceaglske (*Automatic Process Control for Chemical Engineers*, New York: Wiley, 1956) from the Department of Chemical Engineering, University of Minnesota, Minneapolis. Significant progress was also reported in using electrical analogs to simulate the response of common process units such as distillation columns and heat exchangers (see, for instance, work reported in *I&ECR*, p. 1035, June 1956, by J. M. Mozley from E. I. du Pont de Nemours and Company). Based on a technique that used electrical components to simulate the response of process systems using electrical analog circuits, this work offered the opportunity to simulate relatively complex dynamics efficiently when computers were expensive and relatively awkward to use (judged from today's point of view). The side effect of this development was that time-domain analysis techniques were studied by chemical engineers in parallel with frequency-response techniques.

The theory of chemical process control reached maturity in the 1960s and early 1970s with numerous applications and theoretical developments and the publication of several books that facilitated wider understanding of the field:

- P. S. Buckley, *Techniques of Process Control*, New York: John Wiley & Sons, 1964
- P. Harriott, *Process Control*, New York: McGraw-Hill, 1964
- D. R. Coughanowr and L. B. Koppel, *Process Systems Analysis and Control*, New York: McGraw-Hill Company, 1965
- P. W. Murrill, *Automatic Control of Processes*, Scranton, PA: Intext Educational Publishers, 1967
- E. F. Johnson, *Automatic Process Control*, New York: McGraw Hill Company, 1967
- J. M. Douglas, *Process Dynamics and Control*, volumes 1 & 2, Englewood Cliffs, NJ: Prentice-Hall Inc., 1972
- W. L. Luyben, *Process Modeling Simulation and Control for Chemical Engineers*, New York: McGraw-Hill Company, 1973.

All these books are mainly concentrated around what is now called classical automatic control theory, which is mainly characterized by the use of Laplace and frequency-domain analysis of control systems. These particular methods of control systems analysis and design were used extensively in electrical engineering and communication systems in the 1940s and 1950s and offered, even in the 1960s, a strong and comprehensive set of tools that could be used efficiently to analyze and design process control systems. The time-domain approach, which is based on using sets of first-order differential equations to describe system dynamics (including feedback control systems), is an alternative method of formulating control systems theory, which has become known as the modern or state-space approach. Both classical and modern approaches will be presented in this book in a parallel way and the reader will soon realize that both approaches have their advantages and disadvantages. In any case, a process control engineer must be able to understand and use system analysis and controller synthesis tools from both approaches.

In the past 40 years, process control theory is considered a well-established field of study and research in chemical engineering and has expanded its areas of application. Automatic process control systems are in extensive use in the chemical process industries. Several books were published in the 1980s and 1990s (a list of some of the most influential textbooks is presented at the end of the chapter) and the field can now be considered as a fairly mature field. In addition, computer software tools that are now available make application of the theory for controller synthesis, closed-loop system simulation and controller prototyping an easy task. Technologies such as digitalization and the Internet of Things (IoT) are expected to have a real impact in the years to come. Nonlinear control, model-predictive control and real-time optimization are now routinely installed in advanced control systems. These developments have shifted the focus and the educational needs in the field of process control. Future chemical engineers must have a solid understanding of the underlying theory but more importantly they must be able to combine this knowledge with software tools in order effectively to design controllers for large-scale, interacting and integrated plants.

LEARNING SUMMARY

Automatic control is an interdisciplinary field of study with many common elements between different engineering disciplines such as electrical, mechanical and chemical engineering. Chemical process control is a well-established field in chemical engineering and it involves the application of automatic control in the chemical, petrochemical, food and related industries. Automatic operation is achieved mainly through feedback control systems, whose main constituting elements are shown in Figure 1.3.

The controlled variable is measured continuously through a sensor and is fed back to the controller, which can be an operator or a computer system. Controller actions are implemented to the process through a final control element, which, in the case of chemical process industries, is usually a control valve.

Chemical process control has a history of roughly one century but is advancing at a fast pace by taking advantage of innovations in other fields, mainly related to computer software and hardware. Advanced control algorithms are routinely implemented at several levels of the control system. This book presents both approaches to process control theory: the classical approach that is based on the frequency response and the Laplace domain, and the modern approach that is based on the state-space approach in the time domain. It also introduces basic software tools in commercial computer software necessary for the efficient implementation of theory into practice.

TERMS AND CONCEPTS

Block diagram. A diagram that indicates the flow of information around the system where each block denotes a component and the arrows the interactions among the different components

Closed-loop system. A system with components connected in a circular (loop) structure. The term closed-loop system is used in the context of feedback control, where a controller is connected with the final control element, the process and the sensor in a circular pattern.

Controlled variable. The process variable that we want to maintain at a particular desired value.

Controller. A device (or a human operator) that corrects any mismatch between the set point and the controlled variable by adjusting the manipulated variable.

Disturbance variable. Any external variable that can affect the process but is not under our control.

Error. The difference between the value of the set point and the value of the measured variable.

Manipulated variable. Process variable that is adjusted to bring the controlled variable back or close to the set point.

Set point. The desired value of the controlled variable.

FURTHER READING

The books that follow are the most frequently used textbooks in chemical process control and presented according to the date of publication of their most recent edition.

Stephanopoulos, G., *Chemical Process Control*. Englewood Cliffs, NJ: Prentice Hall, 1984.

Luyben, W. L., *Process Modeling Simulation and Control for Chemical Engineers*, 2nd edn. New York: McGraw-Hill, 1990.

Ogunnaike, B. A. and Ray, W. H., *Process Dynamics, Modeling and Control*. New York: Oxford, 1994.

Shinskey, F. G., *Process Control Systems: Application, Design, and Tuning*. New York: McGraw Hill, 1996.

Luyben, M. L. and Luyben, W. L., *Essentials of Process Control*. New York: McGraw-Hill, 1997.

Marlin, T. E., *Process Control: Designing Processes and Control Systems for Dynamic Performance*, 2nd edn. New York: McGraw Hill, 2000.

Chau, P. C., *Process Control, A First Course with MATLAB*. Cambridge, UK: Cambridge University Press, 2002.

Brosilow C. and Joseph, B., *Techniques of Model-based Control*. Englewood Cliffs, NJ: Prentice Hall, 2002.

Bequette, B. W., *Process Control: Modeling, Design and Simulation*. Englewood Cliffs, NJ: Prentice Hall, 2003.

Smith, C. A. and Corripio, A. B., *Principles and Practice of Automatic Process Control*, 3rd edn. New York: Wiley, 2005.

Riggs, J. B. and Karim, M. N., *Chemical and Bio-Process Control*, 3rd edn. Boston, MA: Pearson International Edition, 2006.

Coughanowr, D. R. and LeBlanc, S., *Process Systems Analysis and Control*, 3rd edn. New York: McGraw-Hill Education, 2009.

Svrcek, W. Y., Mahoney, D. P. and Young, B. R., *A Real-Time Approach to Process Control*, 3rd edn. Chichester, UK: Wiley, 2013.

Seborg, D. E., Edgar, T. F., Mellichamp, D. A. and Doyle III, F. J., *Process Dynamics and Control*, 4th edn. New York: Wiley, 2016.

Rohani, S. (ed.), *Coulson and Richardson's Chemical Engineering Volume 3B: Process Control*, 4th edn. Cambridge, MA: Butterworth-Heinemann, 2017.

Information about standard symbols and convections used in P&IDs and sensor and actuator technology can be found in the following books.

Meier, F. A. & Meier, C. A., *Instrumentation and Control Systems Documentation*. Research Triangle Park, NC: International Society of Automation (ISA), 2011.

Hughes, T. A., *Measurement and Control Basics*, 5th edn. Research Triangle Park, NC: International Society of Automation (ISA), 2015.

Dunn, W. C., *Fundamentals of Industrial Instrumentation and Control*. New York: Mc-Graw Hill Education, 2018.

Toghraei, M., *Piping and Instrumentation Diagram Development*. New Jersey: John Wiley & Sons, Inc., 2019.

PROBLEMS

1.1 Draw a block diagram for the control system generated when you drive an automobile. What are the "sensors" that you use and what are the final control elements? What are the disturbances?

1.2 Draw a block diagram for an automobile cruise control system. Select some information from the internet about autonomous driving and list the sensors that you think you need to achieve such a task.

1.3 Draw a block diagram for the control system that maintains the temperature in a home refrigerator. Identify the potential disturbances.

1.4 Draw a block diagram for the control system for a home air-conditioning system. Identify the potential disturbances.

1.5 Collect information from the internet concerning the basic principles of temperature sensors.

1.6 Collect information from the internet concerning the basic principles of pressure and differential pressure sensors.

1.7 Collect information from the internet concerning the basic principles of flow metering devices.

1.8 Suppose you were to develop the first sensor for measuring the liquid level in a tank or the pressure in a closed vessel for gas storage. Think about the general principles on which we can base the development of such a sensor. Search the internet to collect information about the most used sensors for liquid-level or pressure measurement.

1.9 Suppose you have been asked to develop a system that follows the Sun and positions a surface always to face the Sun (so that the Sun beam direction is normal to the capturing surface). Think about the characteristics of the sensor that detects the position of the Sun and the necessary actuators to move the surface. Search the internet, collect information about the systems used in solar tracking control systems for photovoltaic panels and prepare a short presentation.

1.10 Compare Figures 1.9 and 1.10 and discuss the common elements.

2 Dynamic Models for Chemical Process Systems

In this chapter we discuss the main principles of modeling the dynamics of chemical processes. We start by stating the general inventory rate equation. The conservation of mass and energy, which are special applications of the inventory rate equation, are then presented and a number of representative examples are analyzed. The development of an appropriate dynamic process model is the first step in understanding the underlying phenomena that result in the transient behavior of a process system. Modeling will be essential in developing appropriate control laws and in understanding the effect that a controller has on process transients.

STUDY OBJECTIVES

After studying this chapter you should be able to:

- Possess basic understanding and skills in dynamic model development.
- Apply the appropriate inventory equations to model the dynamics of representative chemical processes.
- Apply differential equation solvers of software packages like MATLAB or Maple, to simulate the dynamic behavior of chemical process systems.

2.1 Introduction

A mathematical model is a representation of our knowledge about a physical system, which is "translated" into a set of mathematical equations. The aim is to use the model to increase our understanding of the real system's behavior (model simulation can provide insights), and also to design and optimize the operation of the process.

Dynamic models provide a quantitative description of the transient behavior of a process, in addition to its steady-state characteristics. This is very useful in order to select proper operating conditions for the process, so that undesirable transients are avoided. It is also very important for the design of controllers, because, as we will see in subsequent chapters, a controller can modify the dynamic behavior of a process, for better or for worse.

Mathematical models can be classified as lumped or distributed. Lumped dynamic models are the ones in which variables change only with time and not with spatial position. A continuous stirred tank is a representative example of a lumped system; because of aggressive stirring, uniform properties are achieved throughout the tank. Ordinary differential equations (ODEs) are commonly employed to describe lumped systems, with time being the independent variable of the differential equation. Distributed models are the ones in which variables change with time and spatial position and therefore partial differential equations are employed. Plug flow reactors are representative examples of distributed systems, where variables such as composition and temperature vary as a function of time, axial as well as radial position within the reactor.

Mathematical models can also be classified as mechanistic or empirical. Mechanistic (or "white-box") models are based on first principles such as the laws of conservation of mass, energy and momentum, augmented with equilibrium or rate expressions that relate the main process variables. The main advantage of the mechanistic models is the fact that they can be used to describe process dynamics over a wide range of conditions as they are based on universally applicable principles of physics, chemistry and biology. Empirical (or "black-box") models are based on experimental data and fitting of appropriate, but otherwise arbitrary, mathematical expressions that can capture the observed behavior of a process. Extrapolation of empirical models outside the region in which they have been developed can result in erroneous conclusions about the behavior of a system. The main advantage of empirical models stems from the fact that they are, in most cases, particularly simple and require no prior knowledge or physical understanding of the system under investigation.

In the past 50 years, a lot of effort has been expended to systematize model development methods, but with limited success. Engineering experience and intuition play a key role in successful model development and, in this sense, modeling is "more art than science." Model development can be time consuming, and the developed models can be costly to maintain and/or to use. Therefore, it is imperative that the engineer identifies the level of model complexity that is necessary for a particular application. Simple models are developed first, and model complexity can be increased, if necessary, in a stepwise manner, as understanding of the underlying phenomena is improved.

The main steps in the development of mechanistic models are the following:

(a) identification of aims and objectives of model building,
(b) creation of a schematic diagram on which the main process variables are clearly identified,
(c) statement of the main assumptions and their implications,
(d) application of material and energy balances,
(e) development of solution strategies,
(f) model validation and possible revision of assumptions.

As real-world processes can be extremely complex, it is important that the engineer identifies clearly the reason(s) for developing a model and avoids incorporation of any unnecessary complexities. The engineer needs to identify the main process variables that will be used later on, for the purpose of analyzing process performance or validating the process model. The development of any model is based on well-understood and well-justified assumptions that do not oversimplify the mathematical model of the system under study. One must take any effort to make the model as simple as possible, but not "simpler than that," as this may result in false conclusions. Material and energy balances are then applied to the system under study and then simplified (using the assumptions). At this point, some of the assumptions are revised and the equations modified accordingly. Finally, model solution strategies are developed and implemented. Simulation results are compared to experimental data (if available) and further assumptions are introduced or existing assumptions are modified to improve the balance between predictive accuracy and complexity.

In this chapter, representative examples of dynamic models of common processes will be presented. The models that will be derived are based on the classical laws of conservation of mass and energy and therefore a short review will be presented first. The last section of the chapter will give examples of dynamic models for some simple mechanical and electrical systems.

2.2 Conservation Laws

The inventory rate equation for an extensive quantity S, such as mass or energy, states that

$$\begin{bmatrix} \text{rate of} \\ \textbf{accumulation} \\ \text{of } S \end{bmatrix} = \begin{bmatrix} \text{rate of} \\ \text{flow of } S \textbf{ in} \\ \text{the system} \end{bmatrix} - \begin{bmatrix} \text{rate of} \\ \text{flow of } S \textbf{ out} \\ \text{of the system} \end{bmatrix} + \begin{bmatrix} \text{rate of} \\ \textbf{generation} \\ \text{of } S \end{bmatrix} \qquad (2.2.1)$$

The inventory rate equation applies to a well-defined area which is usually called control volume. In chemical engineering, the area of application of the inventory rate equation can be, for instance, a process unit or part of it. This is the system under study and anything outside the system is the environment.

Conserved quantities commonly used in chemical engineering are mass, energy and momentum. The term "conserved" is loosely used to denote something that cannot be destroyed or generated from nothing. The advantage in dealing with conserved quantities is that the generation term (last term in Eq. (2.2.1)) is zero. This is the case, for instance, when Eq. (2.2.1) is used for the total mass (conserved quantity) in a system. On the other hand, when Eq. (2.2.1) is used for a nonconserved quantity, such as the mass of a reactant in a chemical reactor, a rate expression for the generation/consumption is needed. This rate term may involve a number of additional variables (temperature, catalyst activity and internal/geometrical characteristics of the system and catalyst) and significant additional information may be necessary to complete the mathematical model.

2.2.1 Conservation of Total Mass

Conservation of total mass, which is a conserved quantity, is easily expressed mathematically by direct application of Eq. (2.2.1)

$$\begin{bmatrix} \text{rate of} \\ \textbf{accumulation} \\ \text{of mass} \end{bmatrix} = \begin{bmatrix} \text{rate of} \\ \text{flow of mass } \textbf{in} \\ \text{the system} \end{bmatrix} - \begin{bmatrix} \text{rate of} \\ \text{flow of mass } \textbf{out} \\ \text{of the system} \end{bmatrix} \quad (2.2.2)$$

The rate of accumulation of any quantity is the time rate of change of that particular quantity within the control volume of a system. If $M(t)$ is the total mass within the system boundaries at time t, then the accumulation term is simply its derivative with respect to time

$$\begin{bmatrix} \text{rate of} \\ \textbf{accumulation} \\ \text{of mass} \end{bmatrix} = \frac{d}{dt}(M(t)) \quad (2.2.3)$$

Using the notation shown in Figure 2.1, where n incoming and m outgoing streams to an arbitrary system are shown, we obtain

$$\frac{dM(t)}{dt} = \sum_{i=1}^{n} \dot{M}_{in,i}(t) - \sum_{j=1}^{m} \dot{M}_{out,j}(t) \quad (2.2.4)$$

where $\dot{M}_{in,i}$ and $\dot{M}_{out,j}$ are the mass flowrates of the incoming and outgoing streams, respectively. These flowrates may vary over time and therefore their time variation must be known in order to be able to solve Eq. (2.2.4) and predict the variation of the total mass within the system $M(t)$.

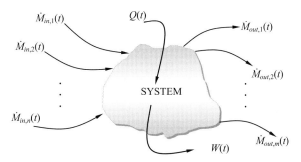

Figure 2.1 A general system and its interactions with the surroundings.

2.2.2 Conservation of Component Mass

The conservation of mass equation may also be applied to a specific component of a mixture. In the absence of chemical reaction, the conservation of component mass equation may be written, similarly to Eq. (2.2.2), as follows:

$$\begin{bmatrix} \text{rate of} \\ \textbf{accumulation} \\ \text{of mass of component} \end{bmatrix} = \begin{bmatrix} \text{rate of} \\ \text{flow of mass of component } \textbf{in} \\ \text{the system} \end{bmatrix} - \begin{bmatrix} \text{rate of} \\ \text{flow of mass of component } \textbf{out} \\ \text{of the system} \end{bmatrix} \quad (2.2.5)$$

The situation is different when a chemical reaction takes place. A particular component may be produced or consumed by the chemical reaction, therefore the component inventory rate equation includes an additional term that accounts for the rate of generation or consumption of the particular component. Also, it is convenient to express the inventory rate equation in terms of moles rather than masses. The rate of generation or consumption of a component k can be calculated using the reaction rate r, which is usually expressed in kmol/(m³·s), multiplied by the stoichiometric number v_k of component k in the reaction. It should be noted that stoichiometric numbers are negative for the reactants and positive for the products. Component mass balances are commonly written in the following form:

$$\frac{dn_k(t)}{dt} = \sum_{i=1}^{n} \dot{n}_{k,in,i}(t) - \sum_{j=1}^{m} \dot{n}_{k,out,j}(t) + V \sum_{p=1}^{\ell} v_{k,p} r_p \quad (2.2.6)$$

where n_k is the molar holdup of component k within the system, $\dot{n}_{k,in,i}$ is the incoming molar flowrate of component k at stream i and $\dot{n}_{k,out,j}$ is the outgoing molar flowrate of component k at stream j. $v_{k,p}$ is the stoichiometric number of component k in reaction p and r_p is the rate of reaction p. V is the total volume of the mixture within the system. The component mass balance (2.2.6) involves the reaction rate term and an appropriate rate expression is necessary to complete the process model.

2.2.3 Conservation of Energy

Energy is a conserved quantity and consists of internal energy, kinetic energy and potential energy. In process applications, kinetic and potential energies are usually negligible. If $\mathcal{U}$ is the total internal energy within the system, then the conservation equation for energy can be written as

$$\frac{d\mathcal{U}(t)}{dt} = \begin{bmatrix} \text{rate at} \\ \text{which energy enters} \\ \text{the system} \end{bmatrix} - \begin{bmatrix} \text{rate at} \\ \text{which energy leaves} \\ \text{the system} \end{bmatrix} \quad (2.2.7)$$

As a fluid element enters a system, it carries its internal energy, hence the incoming and outgoing flow streams bring net internal energy flux $\sum_{i=1}^{n} \dot{M}_{in,i} u_i - \sum_{j=1}^{m} \dot{M}_{out,j} u_j$, where u denotes specific internal energy (internal energy per unit mass).

But for systems open to the flow of mass, additional terms must be included in the energy balance equation. This is the energy flow due to the fact that when an element of fluid moves, it does work on the fluid ahead of it, so the incoming and outgoing flow streams' contribution is actually $\sum_{i=1}^{n} \dot{M}_{in,i}(u_i + p_i v_i) - \sum_{j=1}^{m} \dot{M}_{out,j}(u_j + p_j v_j)$, where p is the pressure and v the specific molar volume of the corresponding stream, or, defining the specific enthalpy $h = u + pv$, it is $\sum_{i=1}^{n} \dot{M}_{in,i} h_i - \sum_{j=1}^{m} \dot{M}_{out,j} h_j$.

To complete the right-hand side of the energy equation (2.2.7), we need to also include two more terms:

- a rate of heat transfer term Q that includes heat transfer by conduction, convection or radiation through the system boundaries (assumed positive when supplied by the surroundings to the system),
- a work term W that stands for the sum of the shaft work (W_s) done by the system to the surroundings and of the work associated with the deformation of the system boundaries. The latter is usually negligible, so only shaft work is included.

Thus, the energy conservation equation has the following form:

$$\frac{dU}{dt} = \sum_{i=1}^{n} \dot{M}_{in,i} h_i - \sum_{j=1}^{m} \dot{M}_{out,j} h_j + Q - W_s \qquad (2.2.8)$$

It should be noted that the energy conservation equation may be alternatively written in molar form, with molar flowrates and specific molar enthalpies.

In what follows, a number of representative examples of chemical process systems will be presented.

2.3 Modeling Examples of Nonreacting Systems

2.3.1 Liquid Storage Tank

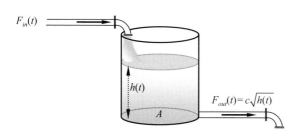

Figure 2.2 A liquid storage tank.

A representative liquid storage tank is shown in Figure 2.2. The tank is cylindrical with a uniform cross-sectional area A and the liquid level is h (the liquid volume is $V = Ah$). The incoming stream has volumetric flowrate F_{in}, hence mass flowrate $\dot{M}_{in} = \rho F_{in}$ and the outgoing stream has volumetric flowrate F_{out}, hence mass flowrate $\dot{M}_{out} = \rho F_{out}$, where ρ is the density of the liquid. Application of Eqs. (2.2.2) or (2.2.4) yields

$$\frac{dM(t)}{dt} = \dot{M}_{in}(t) - \dot{M}_{out}(t) \tag{2.3.1}$$

or

$$\frac{d(\rho V(t))}{dt} = \rho F_{in}(t) - \rho F_{out}(t)$$

For a pure component at constant temperature, the density ρ is constant and thus the mass balance simplifies to

$$A\frac{dh(t)}{dt} = F_{in}(t) - F_{out}(t) \tag{2.3.2}$$

The volumetric flowrate of the outlet stream F_{out} is a function of the liquid level h in the tank, and is usually expressed through a correlation of the form $F_{out} = f(h)$. In most applications, F_{out} is approximately proportional to the square root of liquid level h:

$$F_{out}(t) = c\sqrt{h(t)} \tag{2.3.3}$$

where c is a constant with appropriate units. Using Eq. (2.3.3), Eq. (2.3.2) can be written as

$$A\frac{dh(t)}{dt} = F_{in}(t) - c\sqrt{h(t)} \tag{2.3.4}$$

Using the last equation the liquid level in the tank under steady-state conditions (denoted by the subscript s) can easily be established

$$\left.\frac{dh}{dt}\right|_s = 0 \Rightarrow h_s = \left(\frac{F_{in,s}}{c}\right)^2 \tag{2.3.5}$$

The ODE given by Eq. (2.3.4) can be solved for any known variation of the incoming stream flow rate $F_{in}(t)$ and used to determine the evolution of the liquid level in the tank. It is important to note that there is only one unknown ($h(t)$) and one equation is available that can be used to determine the time evolution of the unknown variable.

At this point, it will be useful to introduce some terminology that will be consistently used in dynamics. The time variation of the inlet stream's flow rate $F_{in}(t)$ accounts for the external influences on the tank and it is the *cause* of changes in the operation of the tank. F_{in} is called an *input variable*. On the other hand, the liquid level $h(t)$ describes the *effect* of the input variable on the operation of the tank. h is called an *output variable*. At the same time $h(t)$, being the solution of the mathematical model (2.3.4), gives complete information on the state of the tank at every point in time; h is also called a *state variable*. The cross-sectional area A and the empirical coefficient c are constant *parameters*.

2.3.2 A Blending Process

In Figure 2.3 a stirred tank blending system is shown where two streams with different concentrations of an active component are mixed. Stream 1 has a volumetric flowrate $F_{in,1}(t)$ and a concentration of the active component $C_{in,1}(t)$. Stream 2 has a volumetric flowrate $F_{in,2}(t)$ and a concentration of the active component $C_{in,2}(t)$. The volume of the liquid in the tank is $V(t)$, the cross-sectional area A and the concentration of the active component $C(t)$. As the liquid in the tank is continuously stirred its properties are uniform and depend on time only (and not on the position inside the tank). The volumetric flowrate of the product stream is $F_{out}(t)$ and the concentration of the active component is equal to its concentration inside the tank, i.e. $C(t)$. We first develop the overall mass balance for the tank by applying directly Eq. (2.2.4):

$$\frac{dM}{dt} = \dot{M}_{in,1} + \dot{M}_{in,2} - \dot{M}_{out} \tag{2.3.6}$$

If the density of the liquid is constant (independent of the concentration of the active component) then we obtain:

$$\frac{d(\rho V)}{dt} = \rho F_{in,1} + \rho F_{in,2} - \rho F_{out}$$

or

$$\frac{dV}{dt} = F_{in,1} + F_{in,2} - F_{out} \tag{2.3.7}$$

The mass balance for the active component may be obtained by directly applying Eq. (2.2.5):

$$\frac{d(VC)}{dt} = F_{in,1}C_{in,1} + F_{in,2}C_{in,2} - F_{out}C \tag{2.3.8}$$

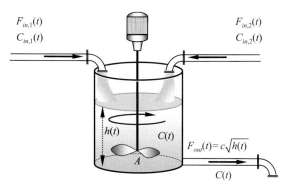

Figure 2.3 A blending process.

We may also use Eq. (2.3.7) to simplify Eq. (2.3.8)

$$V\frac{dC}{dt} + C(F_{in,1} + F_{in,2} - F_{out}) = F_{in,1}C_{in,1} + F_{in,2}C_{in,2} - F_{out}C$$

or

$$V\frac{dC}{dt} = F_{in,1}C_{in,1} + F_{in,2}C_{in,2} - (F_{in,1} + F_{in,2})C \tag{2.3.9}$$

Finally, using the fact that the volume in the tank is $V = Ah$ and assuming that the volumetric flowrate of the product stream is $F_{out} = c\sqrt{h}$, the total mass balance (2.3.7) and the component mass balance (2.3.9) take the form:

$$A\frac{dh}{dt} = F_{in,1} + F_{in,2} - c\sqrt{h} \tag{2.3.10}$$

$$Ah\frac{dC}{dt} = F_{in,1}C_{in,1} + F_{in,2}C_{in,2} - (F_{in,1} + F_{in,2})C \tag{2.3.11}$$

The differential equations (2.3.10) and (2.3.11) constitute a dynamic model for the blending process.

The inlet streams' flowrates $F_{in,1}(t)$ and $F_{in,2}(t)$ and concentrations $C_{in,1}(t)$ and $C_{in,2}(t)$ represent external influences on the blending process. If they are all time varying, then they are the *input variables* of the system. If the concentrations $C_{in,1}$ and $C_{in,2}$ are constant (e.g. if stream 1 is pure water and stream 2 is pure active component) but $F_{in,1}(t)$ and $F_{in,2}(t)$ are time varying, then the two inlet flowrates are the input variables and the two inlet concentrations are constant parameters. Depending on the situation, the blending process could have a larger or smaller number of input variables.

Considering now the dynamic model (2.3.10)–(2.3.11), we see that the dependent variables of the differential equations are h and C. If we compute the solution, then at every point in time, we have complete information on what is happening in the blending process. The variables $h(t)$ and $C(t)$ are the *state variables* of the process.

Moreover, because the purpose of blending is to have the exit stream with desirable concentration, the state variable $C(t)$ is of critical significance. It captures the effect of changes in the process inputs, therefore it is the *output variable* of the process.

2.3.3 Heating Tank

In Figure 2.4, a heating tank with an external jacket is shown. A process stream of temperature T_{in} enters with volumetric flow rate F. A thermal fluid, such as steam or a liquid at a high temperature T_J, circulates in the jacket and is used to raise the temperature of the tank's liquid content. An energy balance can be written for the tank by applying Eq. (2.2.8) with $W_s = 0$:

$$\frac{d(Mu)}{dt} = \dot{M}_{in}\hat{h}_{in} - \dot{M}_{out}\hat{h}_{out} + Q \tag{2.3.12}$$

2.3 Modeling Examples of Nonreacting Systems

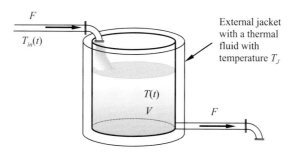

Figure 2.4 A heating tank.

Before we proceed, we make the following assumptions:

(i) the density ρ of the liquid in the tank is constant, independent of temperature;
(ii) the inlet and outlet volumetric flow rates are constant and equal to each other, and consequently, the volume V of the liquid in the tank remains constant;
(iii) the enthalpy of the liquid is given by $\hat{h}(T) = c_p(T - T_{ref})$, where T_{ref} is a reference temperature at which the enthalpy of the liquid is set to zero and c_p is the specific heat capacity of the liquid, assumed to be constant.

Finally, we can use the fact, that for liquids, internal energy and enthalpy are almost identical.

With the foregoing assumptions and approximations, and keeping in mind that the total mass in the tank is $M = \rho V$ and the mass flowrates $\dot{M}_{in} = \dot{M}_{out} = \rho F$, the energy balance equation (2.3.12) takes the following form:

$$\frac{d\left(V \rho \, c_p (T - T_{ref})\right)}{dt} = F \rho c_p (T_{in} - T_{ref}) - F \rho c_p (T - T_{ref}) + Q \tag{2.3.13}$$

or, simplifying the equation

$$V \rho c_p \frac{dT}{dt} = F \rho c_p (T_{in} - T) + Q \tag{2.3.14}$$

To complete the model, we need a heat transfer correlation for the heating rate Q, depending on the design of the heating jacket and the type of thermal fluid.

If the thermal fluid is saturated steam that completely condenses in the jacket, Q may be approximated as

$$Q = \dot{M}_{st} \Delta \hat{h}_{vap,st} \tag{2.3.15}$$

where $\dot{M}_{st}$ is the mass flow rate of steam and $\Delta \hat{h}_{vap,st}$ is the latent heat. Substituting into Eq. (2.3.14), we obtain the dynamic model

$$V \rho c_p \frac{dT}{dt} = F \rho c_p (T_{in} - T) + \dot{M}_{st} \Delta \hat{h}_{vap,st} \tag{2.3.16}$$

with $T_{in}(t)$ and $\dot{M}_{st}(t)$ as input variables, $T(t)$ as state variable and also output variable, and V, F, ρ, c_p and $\Delta\hat{h}_{vap,st}$ constant parameters.

In many cases, the heat transfer rate is fairly accurately approximated by the classic correlation

$$Q = A_H U_H (T_J - T) \tag{2.3.17}$$

where A_H is the area that is available to heat transfer and U_H is the overall heat transfer coefficient, in which case substituting into Eq. (2.3.14), we obtain the energy balance for the tank:

$$V \rho c_p \frac{dT}{dt} = F \rho c_p (T_{in} - T) + A_H U_H (T_J - T) \tag{2.3.18}$$

If the jacket temperature T_J and the heat transfer coefficient U_H remain approximately constant during the operation of the process, Eq. (2.3.18) provides a dynamic model for the jacketed heater, with $T_{in}(t)$ as input variable, $T(t)$ as state and also output variable, and V, F, ρ, c_p, A_H, U_H and T_J as constant parameters.

If T_J can vary significantly during the operation of the process, an additional equation (energy balance of the thermal fluid in the jacket) is needed in order to have a complete dynamic model. Assuming that the thermal fluid is a liquid (e.g. hot water or oil) with constant density ρ_J and heat capacity $c_{p,J}$ while the volume of the jacket is constant and equal to V_J, then the energy balance equation for the thermal fluid in the jacket can be written in the same way as the energy balance of the liquid in the tank:

$$V_J \rho_J c_{p,J} \frac{dT_J}{dt} = F_J \rho_J c_{p,J} (T_{in,J} - T_J) - Q \tag{2.3.19}$$

and, using the heat transfer correlation (2.3.17)

$$V_J \rho_J c_{p,J} \frac{dT_J}{dt} = F_J \rho_J c_{p,J} (T_{in,J} - T_J) - A_H U_H (T_J - T) \tag{2.3.20}$$

where F_J is the volumetric flowrate of the thermal fluid and $T_{in,J}$ is the inlet temperature to the jacket. Now the dynamic model of the jacketed heater consists of two coupled differential equations, (2.3.18) and (2.3.20). $T_{in}(t)$ and $T_{in,J}(t)$ are the input variables, $T(t)$ and $T_J(t)$ are the state variables, $T(t)$ is the output variable and V, F, ρ, c_p, A_H, U_H, V_J, F_J, ρ_J and $c_{p,J}$ are constant parameters.

2.4 Modeling of Reacting Systems

2.4.1 Continuous Isothermal Stirred Tank Reactor

In Figure 2.5 a continuous stirred tank reactor is shown where a reaction is taking place in the liquid phase under constant pressure and temperature. The reaction taking place is represented by the following general equation

$$|v_a| A + |v_b| B + \ldots \rightleftharpoons |v_c| C + |v_d| D + \ldots \tag{2.4.1}$$

2.4 Modeling of Reacting Systems

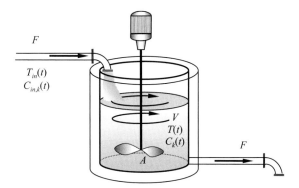

Figure 2.5 A continuous stirred tank reactor.

where A, B, ... are the reactants, C, D, ... are the products and v_a, v_b, ... and v_c, v_d, ... are the corresponding stoichiometric numbers. The reaction rate is given by the following equation

$$r = k_f C_A^{n_A} C_B^{n_B} \ldots - k_b C_C^{n_C} C_D^{n_D} \ldots \tag{2.4.2}$$

where r is the reaction rate in kmol/(m³·s), C_i is the concentration of component i, n_i is the reaction order with respect to component i, k_f is the forward reaction rate constant and k_b is the backward reaction rate constant.

Reaction-rate constants are generally functions of the temperature and therefore the reaction rate $r = r(T, C_A, C_B, \ldots, C_C, C_D, \ldots)$. When the temperature remains constant (isothermal operation), then the reaction rate depends on the concentrations only.

In the present example, we assume constant reactor temperature at all times. Additionally, we assume that the density of the liquid in the reactor is constant, that the inlet and outlet volumetric flow rates are constant and equal to each other and, consequently, the volume of the reactor contents remains constant. A component mass balance for species k can be derived starting from Eq. (2.2.6) and noting that $\dot{n}_{k,in} = FC_{in,k}$, $\dot{n}_{k,out} = FC_k$ and $n_k = VC_k$:

$$\frac{d}{dt}(VC_k) = FC_{in,k} - FC_k + Vv_k r \tag{2.4.3}$$

where F is the volumetric flowrate of feed and product streams and V is the volume of the reacting mixture. Equation (2.4.3) holds for all components k = A, B, ..., C, D, ...

Because the volume is constant, this can be written equivalently as

$$V\frac{dC_k}{dt} = F(C_{in,k} - C_k) + Vv_k r \tag{2.4.4}$$

We now restrict our attention to the simplest case of one reactant R, one product P and an irreversible reaction R $\xrightarrow{r}$ P with first-order kinetics, i.e. reaction rate proportional to the reactant concentration: $r = kC_R$. In this case, the component mass balance for the reactant R is written as (the stoichiometric number of R is $v_R = -1$)

$$V\frac{dC_R}{dt} = F(C_{in,R} - C_R) - VkC_R \tag{2.4.5}$$

and in a similar way for the product P (the stoichiometric number of P is $\nu_P = +1$)

$$V\frac{dC_P}{dt} = F(C_{in,P} - C_P) + VkC_R \qquad (2.4.6)$$

We also consider the case of consecutive reactions R $\xrightarrow{r_1}$ P $\xrightarrow{r_2}$ B, where R is the reactant, P is the desirable product and B is an undesirable byproduct. Assume that both reactions have first-order kinetics, i.e. $r_1 = k_1 C_R$ and $r_2 = k_2 C_P$. Then, the component mass balances for each species are written the same way, but with the sum of the reaction rates for all the reactions in which the species is involved (see Eq. (2.2.6)). For the reactant R, the component mass balance is:

$$V\frac{dC_R}{dt} = F(C_{in,R} - C_R) - Vk_1 C_R \qquad (2.4.7)$$

with only one reaction rate term in the equation, but for the product P it is

$$V\frac{dC_P}{dt} = F(C_{in,P} - C_P) + Vk_1 C_R - Vk_2 C_P \qquad (2.4.8)$$

where there are two reaction terms (with appropriate signs), since P is involved in two reactions.

Suppose now that we are interested in calculating the product concentration C_P. To do so, we need to solve the system of differential equations (2.4.7) and (2.4.8), whose dependent variables are C_R and C_P; these are the state variables of the system. C_P is also the output variable of the system, since this is what we want to calculate, whereas C_R is just a state variable. The input variables of the system are $C_{in,R}(t)$ and $C_{in,P}(t)$ (if they are time varying). V, F, k_1 and k_2 are constant parameters.

2.4.2 Continuous Nonisothermal Stirred Tank Reactor

When the temperature of the reacting mixture varies, we need to develop a differential equation that describes the dynamics of the temperature. For this purpose, a natural starting point would be the energy conservation equation (2.2.8) in molar form, with $W_s = 0$. This is the starting point of the derivations in reaction engineering books (see Fogler, 5th edn, chapter 13).

A simpler and more intuitive formulation is to write down an inventory rate equation for the enthalpy (instead of energy), which includes an enthalpy production term (enthalpy is not a conserved quantity), to account for the heat generated by the chemical reaction (Q_{rxn})

$$\frac{d\mathcal{H}}{dt} = \dot{\mathcal{H}}_{in} - \dot{\mathcal{H}}_{out} + Q_{rxn} + Q \qquad (2.4.9)$$

where $\mathcal{H}$ is the overall enthalpy within the system boundary, $\dot{\mathcal{H}}_{in}$ is the rate at which enthalpy enters the system, $\dot{\mathcal{H}}_{out}$ is the rate at which enthalpy leaves the system and Q is the heat transfer rate through the system boundary.

Assuming: (i) constant density ρ of the reacting mixture, independent of temperature and composition, and (ii) that specific enthalpy of the mixture depends only on temperature and not composition, according to

$$\hat{h}(T) = c_p(T - T_{ref}) \qquad (2.4.10)$$

the accumulation term is written as

$$\frac{d\mathcal{H}}{dt} = \frac{d}{dt}\left[V\rho c_p(T - T_{ref})\right] = V\rho c_p \frac{dT}{dt} \qquad (2.4.11)$$

whereas the "in" and "out" terms for the enthalpy balance are written as

$$\dot{\mathcal{H}}_{in} = F\rho c_p(T_{in} - T_{ref}) \qquad (2.4.12)$$

$$\dot{\mathcal{H}}_{out} = F\rho c_p(T - T_{ref}) \qquad (2.4.13)$$

The heat of reaction $(-\Delta \hat{h}_{rxn})$ is the heat released by the chemical reaction per mole of a reference reactant consumed. For the case of a single reaction, the generation term in the enthalpy balance is:

$$Q_{rxn} = Vr\left(-\Delta \hat{h}_{rxn}\right) \qquad (2.4.14)$$

where the product Vr stands for the total moles of the reference reactant consumed by the reaction per unit time. It should be noted that $\Delta \hat{h}_{rxn}$ is negative for an exothermic reaction and the negative sign used in Eq. (2.4.14) makes Q_{rxn} positive. Likewise, $\Delta \hat{h}_{rxn}$ is positive for an endothermic reaction and this makes Q_{rxn} negative.

Substituting Eqs. (2.4.11)–(2.4.14) into the enthalpy balance (2.4.9) we obtain

$$\rho c_p V \frac{dT}{dt} = \rho F c_p (T_{in} - T) + Vr\left(-\Delta \hat{h}_{rxn}\right) + Q \qquad (2.4.15)$$

where Q is given by some heat transfer correlation e.g. Eq. (2.3.15) or (2.3.17). Notice that the reaction rate r in (2.4.15) is a function of temperature and concentrations, therefore (2.4.15) must be solved simultaneously with the component mass balance equation(s), which in turn depend on the temperature through the rate constant(s).

For the reaction R $\xrightarrow{r}$ P, the dynamic model of the nonisothermal continuous stirred tank reactor (CSTR) involves the coupled differential equations (2.4.5) and (2.4.15), possibly further coupled with a jacket energy balance of the form (2.3.19).

When multiple reactions are taking place in the reactor, the term Q_{rxn} in the enthalpy balance is the sum of the contributions from each reaction. In the case of consecutive reactions R $\xrightarrow{r_1}$ P $\xrightarrow{r_2}$ B considered in the previous subsection, with heats of reaction $(-\Delta \hat{h}_{rxn,1})$ and $(-\Delta \hat{h}_{rxn,2})$, respectively,

$$Q_{rxn} = V r_1 \left(-\Delta \hat{h}_{rxn,1}\right) + V r_2 \left(-\Delta \hat{h}_{rxn,2}\right) \tag{2.4.16}$$

and the energy balance equation becomes

$$\rho c_p V \frac{dT}{dt} = \rho F c_p \left(T_{in} - T\right) + V r_1 \left(-\Delta \hat{h}_{rxn,1}\right) + V r_2 \left(-\Delta \hat{h}_{rxn,2}\right) + Q \tag{2.4.17}$$

Again, this needs to be combined with a heat transfer correlation for Q, and is coupled with the component mass balances of Eqs. (2.4.7) and (2.4.8) and possibly further coupled with a jacket energy balance.

2.4.3 Continuous Bioreactors

A continuous bioreactor is similar to a continuous stirred tank (chemical) reactor. The feed to a continuous bioreactor contains all nutrients (carbon source, nitrogen source, vitamins and minerals, etc.) that are necessary for the microorganism used. The cells of the microorganism are usually in suspension in the bioreactor and, in order to grow and produce products, consume a carbon-containing substrate together with other sources of nutrients and energy. The material balance equations are written for the growth-limiting component (substrate), cell mass and extracellular or intracellular products formed. To this end, the simplified view of Figure 2.6 is commonly used. According to this view, cells (with concentration X in g/L) consume substrate (with concentration S in g/L) for maintenance, production of more cells and extracellular product(s) (with concentration P in g/L).

The rate of cell growth r_X (in g/(L·s)) is defined with respect to this oversimplified view to denote the volumetric rate of increase of cell concentration (cell mass). A specific rate μ $(=r_X/X)$ is also defined (the term specific is used to denote the fact that this is the rate per unit cell concentration). There is a key assumption involved in the development of the bioreactor model: all cells, irrespective of their individual characteristics (such as age, for instance), exhibit the same behavior.

The rate of substrate consumption (r_S) is defined with respect to the rate of cell growth (r_X) and product formation (r_P) by defining appropriate yield coefficients. More specifically, a yield coefficient $Y_{X/S}$ is defined as the ratio of the cell mass produced over the mass of the substrate consumed. In a similar way, $Y_{P/S}$ is defined as the ratio of the product mass produced over the mass of the substrate consumed.

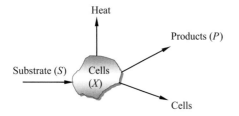

Figure 2.6 A view of a cell as an open system.

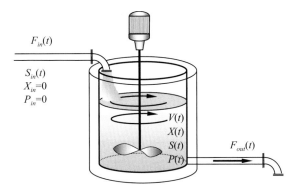

Figure 2.7 A continuous bioreactor.

To derive the mass balance equation for a representative continuous bioreactor (system and notation used are shown in Figure 2.7) we first apply the inventory rate equation for the cell mass which, in the case of sterile feed ($X_{in}=0$), can be written as

$$\frac{d(VX)}{dt} = -F_{out}X + r_X V \qquad (2.4.18)$$

The inventory rate equation written for the product takes the following form

$$\frac{d(VP)}{dt} = -F_{out}P + Vr_P \qquad (2.4.19)$$

where r_P is the rate of product formation (mass of product per unit time, per unit volume). Finally, the inventory equation is written for the substrate

$$\frac{d(VS)}{dt} = F_{in}S_{in} - F_{out}S - V\left(\frac{r_X}{Y_{X/S}} + \frac{r_P}{Y_{P/S}} + m_s X\right) \qquad (2.4.20)$$

where m_s is the (specific) maintenance coefficient that denotes the rate of substrate consumption due to maintenance processes (in g of substrate/g of biomass) which is commonly considered constant. The first term in the parentheses corresponds to substrate consumption for cell mass production, the second term corresponds to substrate consumption for product formation and the third term corresponds to substrate consumption for cell maintenance.

In order to complete the model we need to introduce expressions for the specific rate of cell growth and the rate of product formation. The most well-known equation for the specific cell growth rate is the Monod equation

$$r_X = \mu(S)X = \mu_{max}\left(\frac{S}{S+K_S}\right)X \qquad (2.4.21)$$

where μ_{max} is the maximum specific growth rate and K_S is the saturation constant (defined as the value of the substrate concentration for which the growth rate is half the maximum value). It is interesting to note that the Monod expression predicts $\mu = \mu_{max}$ for $S \gg K_S$ and

$\mu = (\mu_{max}/K_S)S$ for $S \ll K_S$, i.e. a zero-order kinetics for large substrate concentrations and a first-order kinetics for small substrate concentrations. As far as the product formation rate is concerned, the Luedeking and Piret expression is commonly employed

$$r_p(S,X) = \alpha r_X + \beta X = (\alpha\mu + \beta)X \tag{2.4.22}$$

where the first term corresponds to growth-associated production and the second term corresponds to nongrowth-associated production, and α, β are constants.

2.4.4 A Simple Model for an Isothermal Solid Oxide Fuel Cell

A fuel cell is an electrochemical device that converts chemical energy directly into electrical energy, i.e. without thermal and mechanical steps, and therefore its efficiency is not restricted by the Carnot cycle limits on efficiency. Furthermore, fuel cells can deliver the most environmentally friendly conversion of chemical energy into electrical energy because the main byproduct of their operation is water. There are several types of fuel cells operating at different temperatures and they are named after the type of material used for the electrolyte. High-temperature solid-oxide fuel cells (SOFC) consist of an anode, at which the oxidation reaction occurs, an ion-conducting electrolyte and a cathode, at which the reduction reaction occurs. A SOFC is constructed using a ceramic oxide ion-conducting electrolyte, which is sandwiched between the porous electron-conducting anode and the porous cathode. This is shown in Figure 2.8. The oxygen ions are transported through the electrolyte, but the electrons are not allowed to pass through the electrolyte. The electrochemical reactions that are taking place are shown in Figure 2.8. The overall reaction that is taking place is

$$H_2 + \tfrac{1}{2}O_2 \rightarrow H_2O \tag{2.4.23}$$

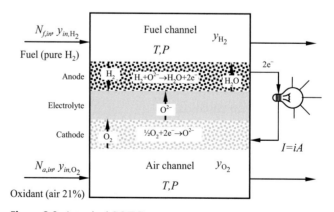

Figure 2.8 A typical SOFC.

2.4 Modeling of Reacting Systems

According to Faraday's law, if the current through the external circuit is I then the rate of the electrochemical reaction is given by

$$r_e = \frac{I}{2F} = \frac{iA}{2F} \qquad (2.4.24)$$

where F is the Faraday constant (which is equal to Avogadro's number times the elementary charge of an electron), and I is the current through the external load (that equals the current density i in A/m² times the cell area A).

In the fuel channel we can write the following component inventory rate equation for hydrogen that is a direct application of Eq. (2.2.6)

$$\frac{dn_{H_2}}{dt} = n_{in,H_2} - n_{out,H_2} - r_e \qquad (2.4.25)$$

where n_{H_2} are the total moles of hydrogen in the fuel channel, and n_{in,H_2} (n_{out,H_2}) is the incoming (outgoing) molar flowrate of hydrogen. Since the total molar flowrate of the fuel is $N_{f,in}$ and remains constant (as one mole of water is produced for each mole of hydrogen consumed, see Eq. (2.4.23)), it follows that

$$\frac{dn_{H_2}}{dt} = N_{f,in}\left(y_{in,H_2} - y_{H_2}\right) - r_e \qquad (2.4.26)$$

where y denotes mole fraction. The molar holdup of hydrogen in the fuel channel, which has a volume V_f, pressure P and temperature T, for ideal gas mixtures is

$$n_{H_2} = \frac{PV_f}{RT} y_{H_2} \qquad (2.4.27)$$

where R is the ideal gas constant. Using Eqs. (2.4.24) and (2.4.27), Eq. (2.4.26) can be written as

$$\left(\frac{PV_f}{RT}\right)\frac{dy_{H_2}}{dt} = N_{f,in}\left(y_{in,H_2} - y_{H_2}\right) - \frac{iA}{2F} \qquad (2.4.28)$$

Following the same approach we can write the oxygen material balance as follows (note that the stoichiometric number of oxygen is –½)

$$\left(\frac{PV_a}{RT}\right)\frac{dy_{O_2}}{dt} = N_{a,in}\left(y_{in,O_2} - y_{O_2}\right) - \frac{iA}{4F} \qquad (2.4.29)$$

where V_a is the volume of the air channel and $N_{a,in}$ the air molar flowrate.

In Figure 2.9 the experimentally observed voltage across an external load is given as a function of the current

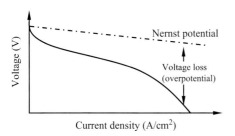

Figure 2.9 Representative plot of voltage versus current density for a SOFC.

density. At zero current (open circuit) the voltage is the maximum possible and equals the open circuit voltage (OCV). As the current is increased, irreversibilities (losses) appear and the voltage drops below OCV. The voltage is described by the following equation

$$V = V_{Nernst}(i) - \eta(i) \qquad (2.4.30)$$

where V_{Nernst} is the Nernst potential and corresponds to the maximum potential that can be achieved under ideal conditions (absence of irreversibilities) for a given temperature and composition of an ideal reacting gas mixture at atmospheric pressure:

$$V_{Nernst}(i) = E^0(T) - \frac{RT}{2F} \ln\left(\frac{y_{H_2O}}{y_{H_2}\sqrt{y_{O_2}}}\right) \qquad (2.4.31)$$

where E^0 (= $1.253 - 2.4516 \cdot 10^{-4}T$ in V) is the reversible cell voltage (which is a function of temperature only). $\eta(i)$ is the voltage loss or overpotential that is attributed to
(1) ohmic losses

$$\eta_\Omega(i) = R_\Omega i \qquad (2.4.32)$$

where R_Ω is the area specific ohmic resistance of the cell (in $\Omega \cdot m^2$);
(2) activation losses due to the barriers that prevent the electrochemical reactions occurring

$$\eta_{act}(i) = \frac{2RT}{F}\sinh^{-1}\left(\frac{i}{2i_0}\right) = \frac{2RT}{F}\ln\left[\left(\frac{i}{2i_0}\right) + \sqrt{1+\left(\frac{i}{2i_0}\right)^2}\right] \qquad (2.4.33)$$

where i_0 is an experimentally determined constant called the exchange current density;
(3) concentration losses due to the depletion of the reactant concentration, which is expressed through the following empirical equation

$$\eta_{conc}(i) = \frac{RT}{2F}\ln\left(\frac{i_L}{i_L - i}\right) = -\frac{RT}{2F}\ln\left(1 - \frac{i}{i_L}\right) \qquad (2.4.34)$$

where i_L is an experimentally determined constant called limiting current density.
Using Eqs. (2.4.31)–(2.4.34), Eq. (2.4.30) can be written as

$$V(i) = E^0(T) - \frac{f}{2}\ln\left(\frac{y_{H_2O}}{y_{H_2}\sqrt{y_{O_2}}}\right) - R_\Omega i - 2f\sinh^{-1}\left(\frac{i}{2i_0}\right) + \frac{f}{2}\ln\left(1 - \frac{i}{i_L}\right) \qquad (2.4.35)$$

where $f = RT/F$. The last equation needs to be solved for the current density i, to express i as a function of V, y_{H_2} and y_{O_2}, and the result substituted into the differential equations (2.4.28) and (2.4.29). In this way, Eqs. (2.4.28), (2.4.29) and (2.4.35) constitute a mathematical model for the SOFC. The mole fractions of hydrogen and oxygen (y_{H_2}, y_{O_2}) are the state variables, whereas the voltage V is the input variable as it is set by external means.

2.5 Modeling of Equilibrium Separation Systems

2.5.1 A Simple Model for a Constant Holdup Flash Distillation Unit

In Figure 2.10 a simple flash distillation unit is shown. A binary mixture is fed to the unit with molar flowrate F and composition z, which is the mole fraction of the more volatile component. The mixture is fed through a valve and a heat exchanger with the aim of changing the pressure and/or the temperature to produce a two-phase (vapour–liquid) mixture. A vapour stream with molar flowrate V and mole fraction y and a liquid stream with molar flowrate L and mole fraction x are produced.

The molar holdups of the vapour and liquid within the flash unit are N_V and N_L, correspondingly. These holdups are assumed constant to simplify the model. This assumption also implies that the following equation (overall mass balance) always holds true (even under dynamic conditions):

$$F = L + V \qquad (2.5.1)$$

The component mass balance for the more volatile component can be written as

$$\frac{d}{dt}(N_L x + N_V y) \approx N_L \frac{dx}{dt} = Fz - Lx - Vy \qquad (2.5.2)$$

where the fact that the vapour holdup is usually negligible (when compared with the liquid holdup) has been taken into account.

The mole fraction in the vapor phase y and in the liquid phase x are related through a vapor–liquid equilibrium (VLE) model which is of the form

$$y = K(P, T, x, y) x \qquad (2.5.3)$$

where K is the equilibrium constant that is, in the most general case, a function of pressure P, temperature T and mole fractions in the liquid (x) and vapor phase (y). When a constant relative volatility α is assumed the following VLE model applies

$$K = K(x) = \frac{\alpha}{1 + (\alpha - 1)x} \qquad (2.5.4)$$

Using Eqs. (2.5.3) and (2.5.4), Eq. (2.5.2) takes the form

$$N_L \frac{dx}{dt} = Fz - (L + VK(x)) x \qquad (2.5.5)$$

Note that, due to Eq. (2.5.1), among V and L only one is an independent variable.

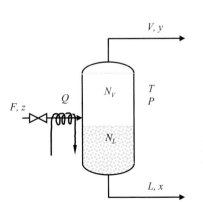

Figure 2.10 A simple flash distillation unit.

2.5.2 A Simple Model for an Ideal Three-Stage Distillation Unit

In Figure 2.11 an ideal, three-stage fractional distillation column is shown. The feed is a binary mixture and has a molar flowrate F and composition z (mole fraction of the more volatile component) and is a saturated liquid. The vapor produced in the feed tray is totally condensed and collected in the reflux drum. There is a top product stream with molar flowrate D and composition x_D. Another stream with the same composition and molar flowrate R is fed back to the top of the column (reflux). The liquid that leaves the feed tray is collected to the base of the column. There is a bottom product stream with flowrate B and composition x_B. Heat is added at the bottom of the column and a vapor stream is produced that is at equilibrium with the liquid at the bottom of the column.

Constant molar overflow, constant relative volatility, constant liquid holdup and negligible vapor holdup on all trays are assumed. Based on these assumptions, the following material balances for the more-volatile component can be written.

Condenser and reflux drum:

$$N_D \frac{dx_D}{dt} = -(R+D)x_D + VK(x_F)x_F \qquad (2.5.6)$$

Feed tray:

$$N_F \frac{dx_F}{dt} = Rx_D - \left[(R+F)+VK(x_F)\right]x_F + Fz \qquad (2.5.7)$$

Column base:

$$N_B \frac{dx_B}{dt} = (R+F)x_F - \left[B+VK(x_B)\right]x_B \qquad (2.5.8)$$

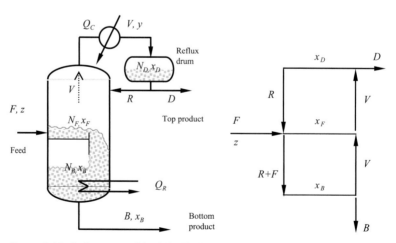

Figure 2.11 A three-stage ideal distillation unit.

The equilibrium constants K are calculated using Eq. (2.5.4) as we have assumed that the relative volatility α is constant. It is interesting to note that from the four variables B, V, R and D only two are independent as they must satisfy the overall material balances

$$R + F = B + V \qquad (2.5.9)$$

$$R + D = V \qquad (2.5.10)$$

If, for example, V and R are used as independent variables then D can be calculated from (2.5.10) ($D = V - R$) and B can be calculated from (2.5.9) ($B = R + F - V = F - D$). The model involves four constants (N_D, N_F, N_B, α) and two variables (F, z) that are determined exogenously by conditions upstream of the distillation unit (disturbances). Therefore, Eqs. (2.5.6)–(2.5.10) consist of five equations in seven unknowns (x_D, x_F, x_B, R, D, V, B). In order to be able to solve the equations and determine the response of the system to any variation of the disturbances we need two additional independent equations.

2.6 Modeling of Simple Electrical and Mechanical Systems

Electric circuits consist of interconnections of sources of electric voltage with other electronic elements. The most classical elements of electrical circuits are resistors, capacitors and inductors, and their symbols, notation and mathematical description are summarized in Table 2.1. Electric and electronic components play a central role in (among others) analog control devices, electrical sensors, signal conditioning and energy conversion devices. The basic equations of electric circuits, called Kirchhoff's laws, are also summarized in Table

Table 2.1 Ideal electrical system components

Resistor (resistance R in ohm Ω)	Capacitor (capacitance C in farad F)	Inductor (inductance L in henry H)
$v(t) = i(t) R$ $i(t) = v(t) \dfrac{1}{R}$	$v(t) = \dfrac{1}{C} \int i(t)\, dt$ $i(t) = C \dfrac{dv(t)}{dt}$	$i(t) = \dfrac{1}{L} \int v(t)\, dt$ $v(t) = L \dfrac{di(t)}{dt}$
Kirchhoff's voltage law The algebraic sum of all voltages taken around a closed path in a circuit is zero.	*Kirchhoff's current law* The algebraic sum of currents leaving a node equals the algebraic sum of currents entering that node.	

2.1. When relatively complex circuits are studied it is vital to apply Kirchhoff's laws in a systematic way. It is reminded that in a series circuit the same current ($i(t)$ in A) passes through all elements connected in series and the total voltage (in V) is the sum of the voltage drops across each element of the series circuit. In a parallel circuit it is the voltage drop that is common to all elements connected in parallel, and the total current is the sum of the currents that pass through each of the elements connected in parallel. We will demonstrate the methodology briefly through two simple examples.

Three of the simplest electrical circuits are shown in Figure 2.12. The RC circuit shown in Figure 2.12a consists of a voltage source connected in series with a resistor with resistance R and a capacitor with capacitance C. According to Kirchhoff's voltage law (see Table 2.1) the

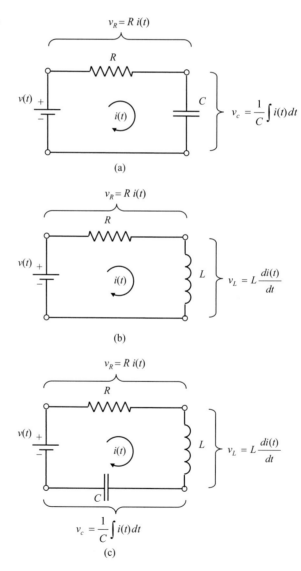

Figure 2.12 The simplest series circuits: (a) the RC, (b) the RL and (c) the RLC electrical circuits.

voltage across the source must be equal to the voltage drop across the elements connected in a serial manner:

$$v(t) = v_R(t) + v_C(t) = R\,i(t) + \frac{1}{C}\int i(t)\,dt \qquad (2.6.1)$$

We then make use of the equation:

$$C\frac{dv_C(t)}{dt} = i(t) \qquad (2.6.2)$$

to finally obtain:

$$RC\frac{dv_C(t)}{dt} + v_C(t) = v(t) \qquad (2.6.3)$$

Working in a similar way for the RL circuit shown in Figure 2.12b we obtain:

$$v(t) = v_R(t) + v_L(t) = Ri(t) + L\frac{di(t)}{dt} \qquad (2.6.4)$$

For the more general RLC circuit of Figure 2.12c we apply Kirchhoff's voltage law:

$$v(t) = v_R(t) + v_L(t) + v_C(t) = Ri(t) + L\frac{di(t)}{dt} + \frac{1}{C}\int i(t)\,dt \qquad (2.6.5)$$

and then using Eq. (2.6.2):

$$LC\frac{d^2 v_C(t)}{dt^2} + RC\frac{dv_C(t)}{dt} + v_C(t) = v(t) \qquad (2.6.6)$$

The basic elements of the translational mechanical systems modeling are summarized in Table 2.2. The spring is also used to model stiffness and expresses the linear dependence between force and deformation, while the damper (also used for viscous friction) is used to model resistance to relative motion. The mechanical systems are modeled using Newton's second law according to which the (vector) sum of forces acting on an object is equal to the mass m of that object multiplied by the acceleration a of the object (force and acceleration are vectors).

Table 2.2 Ideal mechanical system components

Spring (spring constant k in N/m)	Damper (friction coefficient b N/(m/s))	Mass and inertia (mass m in kg)
$F(t) = kx(t)$	$F(t) = -b\,v(t) = -b\dfrac{dx(t)}{dt}$	$F(t) = m\dfrac{dv(t)}{dt} = m\dfrac{d^2 x(t)}{dt^2}$

Newton's law: The mass of the system times its acceleration equals the sum of forces acting on the system (a vector equation).

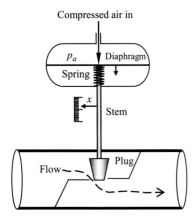

Figure 2.13 A classical pneumatic valve.

A mechanical system that is of paramount importance in process control is the pneumatic control valve shown in Figure 2.13. In a pneumatic control valve the pneumatic signal from the I/P transducer (an element that converts electrical signal to air pressure) applies directly to the actuator. Compressed air enters above the diaphragm and pushes against the spring to close the valve. The valve fully closes when the plug seats tightly against the seat ring. When air pressure decreases, the spring pressure causes the diaphragm, stem and plug to move upward, opening the valve (as this means a loss of pressure would cause the valve to open, such a valve is called a fail-open valve). Needless to say that different configurations of air inlet, spring location and valve seat arrangement result in different fail positions and determine the state of the valve when air pressure is lost.

If the moving parts of the valve have an overall mass m then applying Newton's second law results in the following equation:

$$m\frac{d^2x}{dt^2} = \text{sum of forces acting} \qquad (2.6.7)$$

where x is the valve stem position (relative to the position at which the spring is at rest). The acting forces are due to the air pressure p_a acting on the diaphragm (which has an area A), friction forces and the spring force opposing the downward movement of the diaphragm:

$$m\frac{d^2x}{dt^2} = p_a A - kx - b\frac{dx}{dt} \qquad (2.6.8)$$

If inertia force is negligible then the following model for a control valve is obtained:

$$\frac{b}{k}\frac{dx}{dt} + x = \frac{A}{k}p_a \qquad (2.6.9)$$

From the last equation, an algebraic expression is obtained when friction is also negligible:

$$x = \frac{A}{k}p_a \qquad (2.6.10)$$

If the flow through the valve is proportional to the valve stem position (linear valve) then the volumetric flowrate through the valve F can be expressed as

$$F = \alpha x \qquad (2.6.11)$$

where α is a constant, and Eq. (2.6.10) can be written as

$$\frac{m}{k}\frac{d^2F}{dt^2} + \frac{b}{k}\frac{dF}{dt} + F = \frac{\alpha A}{k}p_a \qquad (2.6.12)$$

or

$$\frac{b}{k}\frac{dF}{dt} + F = \frac{\alpha A}{k} p_a \qquad (2.6.13)$$

for negligible inertia force.

2.7 Software Tools

2.7.1 Numerical Solution of an Ordinary Differential Equation using MATLAB

In order to solve ODEs in MATLAB we need first to define the ODE function in an m-file (a text file with an .m extension stored in the current folder or in a folder on the MATLAB search path, as MATLAB looks for programs in these specific locations). The m-file (`odefilename.m`) should have the following form:

```
function dfdt = odefilename(t,y,parameters)
dfdt = A function of t and y;
```

It begins with "`function`" and `odefilename` can be any name but should match the name of the m-file (`odefilename.m`). In order to solve the differential equation of the heating tank, examined in Section 2.3.3, and described by the following differential equation

$$\rho c_p V \frac{dT}{dt} = F_{in} \rho c_p (T_{in} - T) + A_H U_H (T_J - T) \qquad (2.7.1)$$

we create the following m-file (`heat_tank.m`)

```
function dTdt=heat_tank(t,T)
% Heating tank example
% model parameters
Fin = 0.1; % m^3/h
V = 0.1; % m^3
rhoCp = 1000; % kJ / (m^3 C)
AU = 100; % kJ/(h C)
TJ = 75; % C
Tin= 25 ; % C
% equation
dTdt = (Fin*rhoCp*(Tin-T)+AU*(TJ-T)) / (V*rhoCp);
```

Note that the numerical values of the constants appearing in the model are defined in the m-file. The numerical solution of the model of the heating tank can be achieved by using any of the available ODE solvers in MATLAB. One of these solvers is `ode45`. By typing "`help ode45`", you can find out that the syntax for `ode45` is:

```
[t,y] = ode45(odefilename,tspan,y0)
```

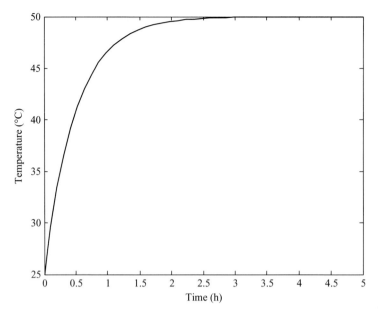

Figure 2.14 Temperature of the heating tank during starting up.

where

    `t` is the independent variable, i.e. time in most cases
    `y` is the dependent variable(s)
    `tspan = [t0 tf]` is the range of `t` (`t0` is the initial value and `tf` is the final value of `t`) and
    `y0` is a initial value of the dependent variable.

We can simulate the starting up by setting the initial condition y0=25, i.e. the temperature in the tank to be equal to the temperature of the feed, we can type

```
» [t,T]=ode45(@heat_tank,[0 5],25);
» plot(t,T)
» xlabel('Time (h)')
» ylabel('Temperature (deg C)')
```

and obtain the result shown in Figure. 2.14.

2.7.2 Numerical Solution of an Ordinary Differential Equation using Maple

The numerical solution of the nonlinear differential equation of the heating tank can be obtained with Maple as follows.

First, define the ordinary differential equation to be solved (the derivative $dT(t)/dt$ can be written as `diff(T(t),t)`, using the `diff` command):

```
> Fin:=0.1:V:=0.1:rhoCp:=1000:AU:=100:TJ:=75:Tin:=25:
> ode:=diff(T(t),t)= (Fin*rhoCp*(Tin-T(t))+AU*(TJ-T(t)))/
  (V*rhoCp):
```

Solve the ordinary differential equation numerically, using the `dsolve` command, specifying the initial condition and the range of the independent variable over which the solution should be computed.

```
> p:=dsolve({ode,T(0)=25},numeric,range=0..5):
```

Plot the result, using the `odeplot` command of the `plots` package.

```
> with(plots): odeplot(p);
```

In this way, we obtain the result shown in Figure 2.14.

LEARNING SUMMARY

- The inventory rate equation for a quantity S can be expressed as

$$\begin{bmatrix} \text{rate of} \\ \text{accumulation} \\ \text{of } S \end{bmatrix} = \begin{bmatrix} \text{rate of} \\ \text{flow of } S \text{ in} \\ \text{the system} \end{bmatrix} - \begin{bmatrix} \text{rate of} \\ \text{flow of } S \text{ out} \\ \text{of the system} \end{bmatrix} + \begin{bmatrix} \text{rate of} \\ \text{generation} \\ \text{of } S \end{bmatrix}$$

If S is a conserved quantity (total mass, energy or momentum) then there is no generation term.

- The total mass balance can be expressed as

$$\frac{dM(t)}{dt} = \sum_{i=1}^{n} \dot{M}_{in,i}(t) - \sum_{j=1}^{m} \dot{M}_{out,j}(t)$$

- The component mass balance can take into account the occurrence of chemical reaction(s) and is most conveniently expressed in molar units as

$$\frac{dn_k(t)}{dt} = \sum_{i=1}^{n} \dot{n}_{k,in,i}(t) - \sum_{j=1}^{m} \dot{n}_{k,out,j}(t) + V \sum_{p=1}^{\ell} v_{k,p} r_p$$

- The energy balance can be expressed as

$$\frac{dU}{dt} = \sum_{i=1}^{n} \dot{M}_{in,i} h_i - \sum_{j=1}^{m} \dot{M}_{out,j} h_j + Q - W_s$$

The application of the energy balance to reacting systems can be quite involved. To simplify derivations, it is possible to use an enthalpy balance:

$$\frac{d\mathcal{H}}{dt} = \dot{\mathcal{H}}_{in} - \dot{\mathcal{H}}_{out} + Q_{rxn} + Q$$

TERMS AND CONCEPTS

Classification of Process Variables

> **Input variables.** Those variables that account for the external influences on a process, causing changes in the process.
> **State variables.** The set of process variables that provide complete information on the state of the process; they are the dependent variables of the differential equations of the dynamic process model.
> **Output variables.** State variables or functions of state variables that are of major significance in assessing process performance; they measure the effect of input changes on the process.

Conservation law. A law according to which a particular property of an isolated system does not change as the system evolves.

Conserved quantities. Quantities that can be neither created nor destroyed and are (among others) the total mass, energy and momentum. The mass of a component is not a conserved quantity.

Distributed model. A mathematical model with more than one independent variables, such as time and spatial position, and consists of partial differential equations and possibly algebraic equations.

Dynamic model. Any mathematical model that describes the time evolution of a dynamical system.

Empirical model. Any mathematical model that is not based on fundamental principles of physics, chemistry or biology.

Inventory rate equation. An equation stating that the accumulation of any extensive quantity within a control volume is equal to the rate of net efflux of the quantity through the control volume surface plus the rate of its generation within the control volume. The equation cannot be applied to intensive variables (such as temperature, pressure, etc.)

Lumped model. A mathematical model where there is only one independent variable (time) and consists of a set of ordinary differential equations and possibly algebraic equations.

Mathematical model. A set of mathematical equations or expressions that describe the behavior of a system.

Mechanistic model or first-principles model. A mathematical model that is based on fundamental principles of physics, chemistry or biology.

System. A group of interacting or interrelated entities that form a unified whole.

FURTHER READING

Additional information of the model development process can be found in the following books and journal publications.

A clear and in depth presentation of the model development principles with a large number of examples is offered by the book

Ingham, J., Dunn, I. J., Heinzle, E. and Prenosil, J. E., *Chemical Engineering Dynamics, An Introduction to Modelling and Computer Simulation*, 2nd edn. Weinheim: Wiley-VCH, 2000.

An industrial approach to modeling that also includes a detailed derivation of many models used in industrial applications is presented in the book

Thomas, P. J., *Simulation of Industrial Processes for Control Engineers*, Elsevier Science & Technology Books, 1999.

The following two books present an advance treatment to modeling of chemical processes

Hangos, K. and Cameron, I., *Process Modelling and Model Analysis, Process Systems Engineering*, volume 4. London: Academic Press, 2001.

Rice, R. G. and Do, D. D., *Applied Mathematics and Modeling for Chemical Engineers*. New York: John Wiley & Sons, Inc., 1995.

Current industrial practice of process modeling is given in the journal publications

Cox, R. K., Smith, J. F. and Dimitratos, Y., Can simulation technology enable a paradigm shift in process control?: Modeling for the rest of us, *Computers & Chemical Engineering*, **30** (10–12), 1542–52, 2006.

Pantelides, C. C. and Renfro, J. G., The online use of first-principles models in process operations: Review, current status and future needs, *Computers & Chemical Engineering*, **51** (5), 136–48, 2013.

On the modeling of chemical reactors, detailed information can be found in chemical reaction engineering books, e.g.

Fogler, H. S., *Elements of Chemical Reaction Engineering*, 5th edn. Prentice Hall, 2016.

PROBLEMS

2.1 The liquid storage tank shown in Figure P2.1 contains 1000 kg of liquid. At time $t = 0$ liquid is fed to the tank with mass flowrate of $\dot{M}_{in} = 100$ kg/min which lasts for 10 min. What is the equation that describes the mass of liquid in the tank? Solve the equation to obtain the mass in the tank as a function of time.

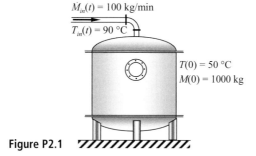

Figure P2.1

2.2 The liquid storage tank shown in Figure P2.1 contains 1000 kg of liquid at temperature of $T(0) = 50\ °C$. At time $t = 0$, liquid is fed to the tank with mass flowrate of $\dot{M}_{in} = 100$ kg/min which lasts for 10 min. The temperature of the liquid fed is $T_{in} = 90\ °C$. What is the equation that describes the temperature of the liquid in the tank? Solve the equation to obtain the evolution of the temperature of the liquid in the tank.

2.3 In Figure P2.3 a heat exchanger is shown where a process stream is heated from temperature T_{in} to T_{out} using steam at constant temperature T_{st}. Derive the energy balance equation for the process assuming a lumped system (i.e. that system properties do not depend on the spatial position).

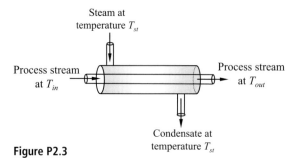

Figure P2.3

2.4 In Figure P2.4 a cone-shaped liquid storage tank is depicted. Apply a total mass balance to derive a differential equation for the liquid level in the tank.

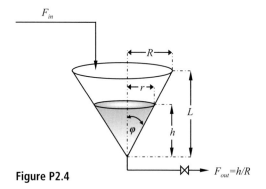

Figure P2.4

2.5 A system of two liquid storage tanks in series is shown in Figure P2.5. Use material balance equations to derive a dynamic model for the system of tanks. Assume that the volumetric flowrate of the outgoing stream from each tank is proportional to the square root of the corresponding height.

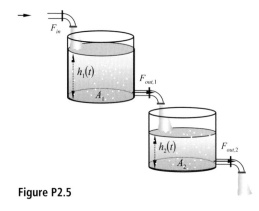

Figure P2.5

2.6 Consider a constant-volume continuous stirred tank reactor, as shown in Figure P2.6, in which the following reactions

$$R \underset{r_{-1}}{\overset{r_1}{\rightleftharpoons}} P \xrightarrow{r_2} B$$

take place isothermally. The intermediate product P is the desired product, whereas B is an undesirable byproduct. The kinetics of the first reaction is first order in both directions, with corresponding kinetic rate expressions $r_1 = k_1 C_R$, $r_{-1} = k_{-1} C_P$, but the kinetics of the second reaction is second order: $r_2 = k_2 C_P^2$. The feed is a solution of the reactant R of concentration $C_{in,R}$, that contains no P or B, and has a constant volumetric flowrate F. Derive a dynamic model for this system, consisting of differential equations, that is capable of predicting the product concentration $C_P(t)$. Identify input variables, state variables, output variables and parameters of the system.

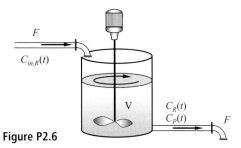

Figure P2.6

2.7 In Figure P2.7 a mixing process is shown. Pure water with volumetric flowrate $F_{in,1}$ and temperature $T_{in,1}$ is mixed with an aqueous solution of a salt with volumetric flowrate $F_{in,2}$, temperature $T_{in,2}$ and salt concentration $w_{in,2}$ (in kg/m³). Heat is added at a rate Q through an electric heater. The flowrate F_{out} of the outlet stream is approximately proportional to the liquid level h in the tank.

Derive a dynamic model of the mixing process in the form of a system of differential equations with dependent variables the liquid level h, the salt concentration w and the temperature T in the tank.

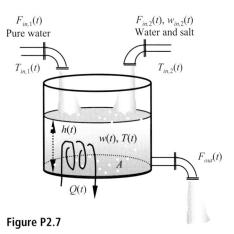

Figure P2.7

2.8 In Figure P2.8 a tray of a distillation column is shown. The liquid molar holdup on the j-th tray is $N_{L,j}$ while the vapor holdup is normally assumed negligible. The volumetric flowrate of the outflow liquid stream is given by

$$F_{L,j} = c(h_j - h_w)^{3/2} \tag{P2.8.1}$$

where c is a constant, h_j is the liquid level on the j-th tray and h_w is the height of the weir. Derive the dynamic model of the tray using the following assumptions:

(1) binary mixture
(2) negligible vapor holdup
(3) constant molar overflow
(4) constant pressure.

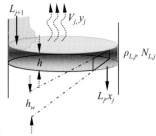

Figure P2.8

2.9 In Figure P2.9 a two-phase reactor is shown. The reaction that is taking place is A + B → C and the reaction rate is first order with respect to the partial pressure of each reactant. Feed stream 1 is a mixture of the two reactants and also contains an inert component I. Feed 2 is pure A. Both feeds are gaseous. The reaction product C is practically nonvolatile while A, B and I are very volatile at the reactor conditions. The flows through the valves are linear functions of the valve openings and are given by $F_i = c_i u_i$, where c_i is a constant and u_i is the fractional opening of the corresponding valve. A temperature-control system has been installed that achieves almost perfect control of the reactor temperature. Using the assumptions that only the product is present in the liquid phase and that the reaction rate is a function of the partial pressures of the reactants in the vapor phase, derive the mathematical model that describes the composition dynamics of the liquid and vapor phases.

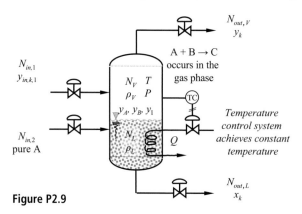

Figure P2.9

2.10 In the fed-batch reactor shown in Figure P2.10, reactant A in pure form is fed as gas to the reactor where the reaction $nA \rightarrow B$ is taking place. Assume that isothermal operation of the reactor at temperature T is achieved by other means (not shown in Figure P2.10). The reaction rate is first order with respect to the partial pressure of reactant A. The reactor is initially filled with an inert component I at pressure P_0. The feed to the reactor has a volumetric flowrate which varies linearly with the pressure difference $P_{in} - P$, where P_{in} is the upstream pressure, and the valve fractional opening x ($F_{in} = c_V x (P_{in} - P)$, where c_V is the valve constant). Assume that the gaseous mixture in the reactor is ideal and derive the dynamic material balances of components A, B and I.

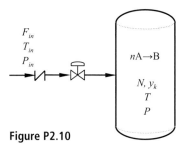

Figure P2.10

2.11 In Figure P2.11 a single-component vaporizer is shown. The feed is a pure liquid that is fed at molar flowrate N_{in} and temperature T_{in} to the vaporizer that operates at pressure P and temperature T. Heat is added (Q) to the vaporizer that causes the generation of a vapor stream with molar flowrate N_{out}. Derive the mathematical model of the system.

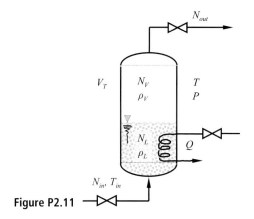

Figure P2.11

COMPUTATIONAL PROBLEMS

2.12 A microorganism is grown in the continuous bioreactor shown in Figure 2.7. The bioreactor has a volume of $V = 1$ m³ while the sterile feed has a volumetric flowrate of $F_{in} = 0.2$ m³/h with a substrate concentration of $S_{in} = 50$ kg/m³. The specific growth rate of the microorganism is given by the Monod's kinetics (Eq. (2.4.21)) with $K_s = 2$ kg/m³ and $\mu_{max} = 1$ h^{-1}. The corresponding yield coefficient is $Y_{X/S} = 0.5$ kg/kg. Determine the substrate concentration and cell mass concentration under steady-state conditions. The feed is changed from $F_{in} = 0.2$ m^3/h to $F_{in} = 0.4$ m^3/h and after steady state is achieved is changed again to $F_{in} = 0.8$ m^3/h. Assume that $F_{out} = F_{in} = F$ and therefore reactor volume is always constant. Use an ODE solver in MATLAB to obtain the bioreactor response. What is the time elapsed until a new steady state is achieved?

2.13 The parameters of the SOFC model presented in Section 2.4.4 are given in Table P2.13.
  (a) Use the data in Table P2.13 to generate the V–i characteristic plot of the cell.
  (b) Determine the power density generated by the cell defined as $P = Vi$ and find the conditions at which the maximum is achieved.
  (c) For the steady-state solution that corresponds to $i = 3$ A/cm^2 study in MATLAB the response of the system to a step change in the current density from 3 A/cm^2 to 4 A/cm^2. To achieve that use a solver of differential and algebraic equations in MATLAB such as ode15s.
  (d) Repeat (c) for a step change in the voltage from its steady-state value at $i = 3$ A/cm^2 to 0.9 V.

Table P2.13

Parameter	Symbol	Value	Units
Faraday constant	F	96485	C·mol^{-1}
Ideal gas constant	R	8.3144	V·C·mol^{-1} K^{-1}
	R'	83.144	cm^3 bar·mol^{-1} K^{-1}
Operating temperature	T	1200	K
Operating pressure	P	1	bar
Cell area	A	1	cm^2
Fuel channel volume	V_f	1	cm^3
Air channel volume	V_a	1	cm^3
Fuel molar flow	$N_{f,in}$	0.0001	mol·s^{-1}
Mole fraction of H$_2$	y_{in,H_2}	1	–
Air molar flow	$N_{a,in}$	0.0010	mol·s^{-1}
Oxygen mole fraction	y_{in,O_2}	0.21	–
Reversible cell voltage	E^0	0.94	V
Spec. ohmic resistance	R_Ω	0.05	Ω·cm^2
Exchange current dens.	i_0	0.5	A·cm^{-2}
Limiting current dens.	i_L	5	A·cm^{-2}

2.14 A binary distillation column which has a feed molar flowrate F, composition z and thermal quality (fraction of vapor in the feed) q_F has NT equilibrium trays as shown in Figure P2.14a. The fresh feed is fed on tray NF. The column has a total condenser and a partial reboiler. Liquid-level control systems are installed that vary the top-product molar flowrate to achieve perfect level control of the reflux drum and also vary the bottom-product flowrate to achieve perfect liquid-level control at the bottom of the column. The liquid outflow from each tray is given by the Francis weir formula

$$L_i = 6000 \rho_L l_w h_{ow,i}^{3/2}$$

where L_i is the liquid molar flowrate from tray i (kmol/h), ρ_L is the molar density of the liquid (kmol/m³), l_w is the length of the weir (normally $0.77 D_C$, where D_C is the column diameter) (m) and $h_{ow,i}$ is the height of the liquid above the height of the weir h_w on tray i. Taking the fact that the downcomer covers approximately 88% of the tray cross-sectional area it follows that h_{ow} can be calculated from the equation

$$0.88 \frac{\pi D_C^2}{4} \left(h_w + h_{ow,i} \right) \rho_L = M_{L,i}$$

where $M_{L,i}$ is the molar liquid holdup on tray i. Develop a dynamic model for the column assuming ideal behavior, constant relative volatility, negligible vapor holdup and constant molar overflow (CMO). Develop a MATLAB m-file and study the response of the column to a step change of the feed composition from $z = 0.5$ to $z = 0.45$. Model parameters are given in Table P2.14. Use the notation shown in Figures P2.14a and P2.14b, where x (y) is the mole fraction of the volatile component in the liquid (vapor) and V is the molar flowrate of the vapor.

Table P2.14

Parameter	Symbol	Value	Units
Feed molar flowrate	F	100	kmol·h⁻¹
Feed composition	z	0.5	–
Thermal quality	q_F	0	–
Number of trays	NT	10	
Feed tray location	NF	5	
Reflux flowrate	R	150	kmol·h⁻¹
Vapor boilup	V	200	kmol·h⁻¹
Relative volatility	a	2	–
Liquid density	ρ_L	50	kmol·m⁻³
Reflux drum holdup	M_D	10	kmol
Column base holdup	M_D	10	kmol
Column diameter	D_C	1	m
Weir height	h_w	0.0254	m

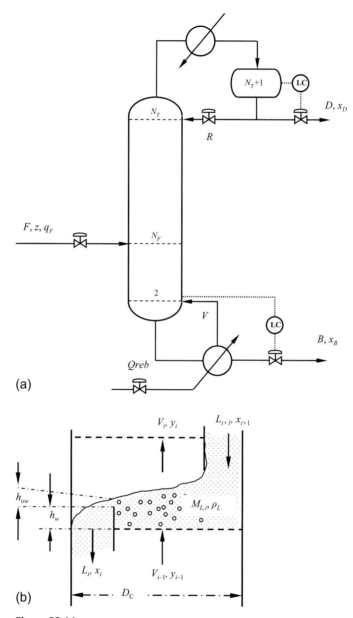

Figure P2.14

3 First-Order Systems

First-order systems are dynamical systems that are described by first-order differential equations, which are derived, for example, from material or energy balances. First-order systems are the simplest form of dynamical systems and will be used in this chapter in order to introduce the key characteristics of the response of dynamical systems. Higher-order dynamical systems are often decomposed into interconnecting first-order systems, and this will be presented in the chapters that follow. Understanding the basic principles of first-order system dynamics is essential in analysing the complex dynamics often exhibited by chemical process systems.

STUDY OBJECTIVES

After studying this chapter you should be able to do the following.

- Predict the qualitative characteristics of first-order system response to several standard input variations.
- Explain the physical meaning of the process time constant and static gain.
- Perform systematically linearization of a nonlinear first-order differential equation.
- Calculate and analyze the response of first-order systems using software tools, like MATLAB or Maple.

3.1 Examples of First-Order Systems

First-order systems are dynamical systems that are described by first-order differential equations, which are derived, for instance, from mass or energy balances. There are a large number of chemical process systems that belong to this class and a classical example of a mercury thermometer will be presented next.

3.1.1 The Mercury Thermometer

A mercury thermometer shown in Figure 3.1 is a dynamical system that has the temperature T of the surrounding fluid as the input variable and the thermometer reading y as the output variable. In order to study the dynamic behavior of the mercury thermometer we need to

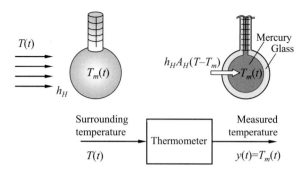

Figure 3.1 A mercury thermometer.

describe mathematically the transfer of energy between the surrounding fluid and the thermometer. To this end, we assume that

- the temperature of mercury in the thermometer is uniform,
- the heat capacity of the sealing glass is negligible compared to that of the mercury and they both always have the same temperature,
- the resistance to heat transfer between the surrounding fluid and the glass is the dominant resistance to heat transfer,
- the contraction or expansion of the sealing glass is negligible,
- all physical properties of mercury are constant and independent of temperature.

The energy balance of the system can be expressed mathematically as follows: the rate of change of the energy stored in mercury must be equal to the heat transfer rate between mercury and the environment. In mathematical terms this can be expressed as follows:

$$\frac{d}{dt}(mc_p T_m) = h_H A_H (T - T_m) \tag{3.1.1}$$

where m is the mass of mercury, c_p is the heat capacity of mercury, T_m is the temperature of mercury, h_H is the heat transfer coefficient and A_H is the available heat transfer area. Equation (3.1.1) can be written as follows

$$\left(\frac{mc_p}{h_H A_H}\right)\frac{dy}{dt} + y = T \tag{3.1.2}$$

where $y = T_m$ is the temperature measurement. The constant $\tau = mc_p/h_H A_H$, which has units of time, is called the time constant of the thermometer and it is a quantitative measure of the speed of response of the system. In summary, under the assumptions made previously, the dynamic behavior of the thermometer is described by the first-order differential equation

$$\tau \frac{dy}{dt} + y = T \tag{3.1.3}$$

where T, the temperature of the surrounding fluid, is the input variable and y, the thermometer reading, is the output variable. Under steady-state conditions, the thermometer is in thermal equilibrium with its surrounding fluid and from Eq. (3.1.3) it follows that

$$y_s = T_s \tag{3.1.4}$$

where the subscript s is used to denote steady-state conditions.

3.1.2 A Continuous Stirred Tank Reactor

In Figure 3.2 a constant volume (V) continuous stirred tank reactor is shown, where the first-order reaction R $\to$ P is taking place under constant temperature. If the volumetric flowrates of the feed and product streams are constant and equal (F) then the material balance for the reactant R is as follows

$$\frac{d}{dt}(VC_R) = FC_{in,R} - FC_R - Vk_R C_R \tag{3.1.5}$$

where C_R is the concentration of the reactant in the reactor, $C_{in,R}$ is the concentration of the reactant in the feed and k_R is the reaction rate constant. Since V and F are constants, Eq. (3.1.5) can be rearranged to give

$$\left(\frac{V}{F+Vk_R}\right)\frac{dC_R}{dt} + C_R = \left(\frac{F}{F+Vk_R}\right)C_{in,R} \tag{3.1.6}$$

and defining

$$\tau = \left(\frac{V}{F+Vk_R}\right), \quad k = \left(\frac{F}{F+Vk_R}\right) \tag{3.1.7}$$

it is written further as

$$\tau \frac{dC_R}{dt} + C_R = kC_{in,R} \tag{3.1.8}$$

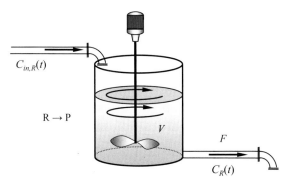

Figure 3.2 An isothermal continuous stirred tank reactor.

58 First-Order Systems

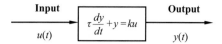

Figure 3.3 A linear first-order system.

Equation (3.1.8) relates the concentration of the reactant in the feed $C_{in,R}$ (input variable) to the concentration of the reactant in the reactor C_R (output variable). Under steady-state conditions, the input and output variables are related via

$$C_{R,s} = k C_{in,R,s} \tag{3.1.9}$$

where, as before, the subscript s is used to denote steady-state conditions. The parameter k, which relates the values of input and output at steady state, is called the steady-state gain or static gain of the system.

3.1.3 Linear First-Order System with Time Constant τ and Static Gain k

Even though the thermometer and continuous stirred tank reactor examples describe fundamentally different processes, they follow the same linear differential equation:

$$\tau \frac{dy}{dt} + y = ku \tag{3.1.10}$$

where $u(t)$ is the input variable, $y(t)$ is the output variable, $\tau > 0$ is the time constant of the system (constant parameter) and $k \neq 0$ is the steady-state gain or static gain of the system (constant parameter).

A pictorial representation of a first-order system is given in Figure 3.3, explicitly indicating the input (inward arrow) and the output (outward arrow).

3.2 Deviation Variables

In the energy balance developed for the mercury thermometer, no particular assumption was made regarding the zero of the temperature scale used. Different temperature scales have their zero placed at different reference points. In addition, one could arbitrarily set the reference point to any other plausible temperature such as

- the room temperature
- the temperature of the human body
- the normal operating temperature of a unit operation.

It is, however, important to note that the differential equation that describes the system dynamics does not depend on the reference temperature selected. If we select $y_s =$

T_s to be the reference temperature, then by subtracting Eq. (3.1.4) from Eq. (3.1.3) we obtain

$$\tau \frac{d(y-y_s)}{dt} + y - y_s = T - T_s \tag{3.2.1}$$

We then define the following variables, which are called deviation variables

$$\bar{y} = y - y_s$$
$$\bar{T} = T - T_s \tag{3.2.2}$$

and the differential equation (3.2.1) becomes

$$\tau \frac{d\bar{y}}{dt} + \bar{y} = \bar{T} \tag{3.2.3}$$

It is important to observe that the mathematical form of Eqs. (3.1.3) and (3.2.3) is exactly the same.

A similar observation can be made for the isothermal, continuous stirred tank reactor example. Selecting a reference steady-state value for the reactant concentration and subtracting Eq. (3.1.9) from Eq. (3.1.8), we obtain

$$\tau \frac{d(C_R - C_{R,s})}{dt} + (C_R - C_{R,s}) = k(C_{in,R} - C_{in,R,s}) \tag{3.2.4}$$

By defining the following deviation variables

$$\bar{C}_R = C_R - C_{R,s}, \quad \bar{C}_{in,R} = C_{in,R} - C_{in,R,s} \tag{3.2.5}$$

we can write Eq. (3.1.8) in deviation form as follows

$$\tau \frac{d\bar{C}_R}{dt} + \bar{C}_R = k\bar{C}_{in,R} \tag{3.2.6}$$

Comparing Eq. (3.2.6) to Eq. (3.1.8), we observe that the differential equation that relates deviation variables is exactly the same as the original differential equation.

3.3 Solution of Linear First-Order Differential Equations with Constant Coefficients

The differential equation that describes the dynamics of the systems considered in the previous sections has the following general form

$$\tau \frac{dy(t)}{dt} + y(t) = k u(t), \quad y(0) = \text{known} \tag{3.3.1}$$

This is a linear first-order differential equation with constant coefficients, where y denotes the dependent variable of the differential equation or output variable of the system, and u is the forcing function or the input variable. This is the simplest possible form of an ordinary differential equation and a number of alternative solution techniques are available. In this chapter, two solution methodologies will be presented.

3.3.1 Solution of Linear First-Order Ordinary Differential Equations using the Method of Integrating Factor

If we multiply both sides of the differential equation (3.3.1) by $\frac{1}{\tau}e^{t/\tau}$, we obtain

$$\frac{1}{\tau}e^{t/\tau}\left[\tau\frac{dy(t)}{dt}+y(t)\right]=\frac{1}{\tau}e^{t/\tau}\,ku(t)\Rightarrow e^{t/\tau}\frac{dy(t)}{dt}+y(t)\frac{d(e^{t/\tau})}{dt}=\frac{1}{\tau}e^{t/\tau}\,ku(t)$$

or

$$\frac{d(e^{t/\tau}y(t))}{dt}=\frac{1}{\tau}e^{t/\tau}\,ku(t) \qquad (3.3.2)$$

Simple integration from 0 to t gives

$$e^{t/\tau}y(t)-y(0)=\frac{k}{\tau}\int_0^t e^{t'/\tau}\,u(t')\,dt'$$

or

$$y(t)=e^{-t/\tau}y(0)+\frac{k}{\tau}e^{-t/\tau}\int_0^t e^{t'/\tau}u(t')dt' \qquad (3.3.3)$$

Inserting the function $(k/\tau)e^{-t/\tau}$ inside the integral, we obtain the alternative expression

$$y(t)=e^{-t/\tau}y(0)+\int_0^t \frac{k}{\tau}e^{-(t-t')/\tau}\,u(t')\,dt' \qquad (3.3.4)$$

We observe that the integral in the above expression is of convolution type (see Appendix A) and in particular

$$\int_0^t \frac{k}{\tau}e^{-(t-t')/\tau}\,u(t')\,dt'=\frac{k}{\tau}e^{-t/\tau}*u(t) \qquad (3.3.5)$$

where the symbol * denotes convolution. It should be noted that the function $g(t)=(k/\tau)\,e^{-t/\tau}$ characterizes the effect of the input on the response of the first-order system, as its convolution with the input function gives the corresponding term in the solution formula.

3.3.2 Solution of Linear First-Order Ordinary Differential Equations using the Method of Laplace Transform

If we take the Laplace transform of both sides of differential equation (3.3.1) we obtain

$$\tau(sY(s) - y(0)) + Y(s) = k\,U(s) \tag{3.3.6}$$

where $U(s)$ is the Laplace transform of $u(t)$ and $Y(s)$ is the Laplace transform of $y(t)$. Equation (3.3.6) is then solved for $Y(s)$ to obtain

$$Y(s) = \frac{\tau}{\tau s + 1} y(0) + \frac{k}{\tau s + 1} U(s) \tag{3.3.7}$$

Finally, by using the properties of the inverse Laplace transform (see Appendix A) it follows that

$$y(t) = e^{-t/\tau} y(0) + \mathcal{L}^{-1}\left(\frac{k}{\tau s + 1} U(s)\right) \tag{3.3.8}$$

where $\mathcal{L}^{-1}$ denotes the inverse Laplace transform.

Comparison of the two methodologies: The results obtained by using the two alternative methodologies are exactly the same, as expected. The first term on the right-hand sides of Eqs. (3.3.4) and (3.3.8) is the same while the same holds true and for the second term through the convolution theorem (see Appendix A), i.e.

$$\mathcal{L}^{-1}\left(\frac{k}{\tau s + 1} U(s)\right) = \int_0^t \frac{k}{\tau} e^{-(t-t')/\tau}\, u(t')\, dt' = \frac{k}{\tau} e^{-t/\tau} * u(t) \tag{3.3.9}$$

In order to calculate analytically this term for a given input signal $u(t)$ we can follow one of the two alternative approaches:

- the exact form of $u(t)$ is substituted in Eq. (3.3.4) and the integral is calculated analytically (or numerically),
- the Laplace transform of $u(t)$ is multiplied by $k/(\tau s + 1)$ and the result is inverted using the properties of the inverse Laplace transform.

The second approach is the more commonly used as it is the simpler of the two.

Useful conclusions that can be drawn from the analytic solution: The output $y(t)$ depends upon (i) the initial condition and (ii) the input $u(t)$. The initial condition influences the output through the first term of Eqs. (3.3.4) and (3.3.8). The input affects the output through the second term of the same equations. The two terms are then added together, since the differential equation is linear, and therefore, the superposition principle holds true. In the study of linear system dynamics one often distinguishes two subproblems:

- effect of initial condition(s)
- effect of the input.

This subdivision is permissible since these two effects are additive.

Figure 3.4 A thermometer is immersed into a high-temperature bath.

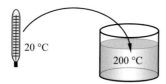

3.4 The Choice of Reference Steady State Affects the Mathematical Form of the Dynamics Problem

Consider a mercury thermometer, which has been on a table for some time, recording the room temperature of 20 °C. Suddenly, it is placed in a 200 °C oil bath (Figure 3.4). We wish to calculate the thermometer reading as a function of time. We already know that the dynamics of the mercury thermometer is governed by the first-order differential equation (3.1.3). There are many alternative choices of reference temperature in defining the deviation variables. Let's consider the following cases:

Case 1: reference temperature of 0 °C (zero on the Celsius scale)
Case 2: reference temperature of 20 °C (room temperature)
Case 3: reference temperature of 200 °C (bath temperature)

and examine what is involved in the calculation of the thermometer reading.

Case 1: reference temperature of 0 °C

Initial condition: $\bar{y}(0) = y(0) - 0 \ °C = 20 \ °C$
Input: $\bar{T}(t) = T(t) - 0 \ °C = 200 \ °C$ (constant for $t > 0$)
Here we have both nonzero initial condition and nonzero input.

Case 2: reference temperature of 20 °C (this is the initial steady-state value)

Initial condition: $\bar{y}(0) = y(0) - 20°C = 0 \ °C$
Input: $\bar{T}(t) = T(t) - 20 \ °C = 180 \ °C$
Here we have zero initial condition and the dynamic response involves only the effect of the nonzero input (which is constant for $t > 0$).

Case 3: reference temperature of 200 °C (this is the final steady-state value)

Initial condition: $\bar{y}(0) = y(0) - 200°C = -180 \ °C$
Input: $\bar{T}(t) = T(t) - 200 \ °C = 0 \ °C$
Here we have zero input for $t > 0$, and the dynamic response involves only the effect of the nonzero initial condition.

In dynamic response calculations, the following types of problems are encountered:

- superimposed effect of nonzero initial conditions and nonzero input (see case 1 in the example);
- effect of input under zero initial conditions (see case 2 in the example, where the deviation variables have been defined with respect to the initial steady state);
- effect of initial conditions under zero input (see case 3 in the example, where the deviation variables have been defined with respect to the final steady state).

As is clearly shown in the example, the type of problem can be dependent upon the choice of reference steady state in defining the deviation variables. When a process is initially at steady state and this steady state is used as the reference, the initial condition becomes zero in deviation form. Likewise, when a process reaches a final steady state under the influence of a constant input, and the final steady state is used to define deviation variables, the input becomes zero in deviation form. In engineering calculations, we are free to choose the reference steady state at our convenience, in terms of simplicity of calculations and/or interpretation of the results.

3.5 Unforced Response: Effect of Initial Condition under Zero Input

The response of a dynamical system under zero input is called an unforced response or natural response or free response. For the case of a first-order system with $u(t)=0$, $\forall t > 0$, the output of the system is given by (see Eqs. (3.3.4) or (3.3.8)):

$$y(t) = e^{-t/\tau} y(0) \qquad (3.5.1)$$

The unforced response of a first-order system is shown in Figure 3.5.

It is important to note that when time equals the time constant τ, the system output is $y(\tau) = 0.368 y(0)$ while for $t = 5\tau$, $y(5\tau) = 0.007 y(0)$. We see that at time greater than four to five time constants, the system output has become practically zero. The time constant τ is therefore a measure of the speed of the unforced response of a first-order system.

3.6 Forced Response: Effect of Nonzero Input under Zero Initial Condition

The response of a dynamical system to a nonzero input is called a forced response. The forced response of a first-order system, under zero initial condition $y(0) = 0$, is given by the convolution integral (see Eqs. (3.3.4) and (3.3.5))

$$y(t) = \int_0^t \frac{k}{\tau} e^{-(t-t')/\tau} u(t') \, dt' \qquad (3.6.1)$$

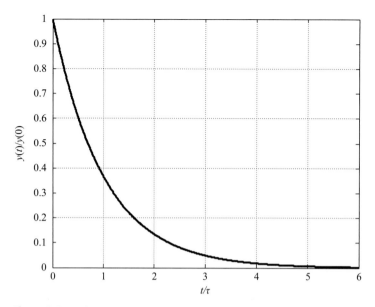

Figure 3.5 Unforced response of a first-order system.

Alternatively, using Eq. (3.3.7) we obtain the following result in the Laplace domain

$$Y(s) = \frac{k}{\tau s + 1} U(s) \qquad (3.6.2)$$

which, when compared to Eq. (3.6.1), appears to be much simpler. The function

$$G(s) = \frac{k}{\tau s + 1} \qquad (3.6.3)$$

which, when multiplied by the Laplace transform of the input gives the Laplace transform of the output, is called the transfer function.

Figure 3.6 gives a diagrammatic representation of the first-order system in terms of its Laplace domain description (3.6.2). The transfer function is placed inside the box, the Laplace transform of the input is over an arrow pointing towards the box, and the Laplace transform of the output over an arrow pointing outwards. This type of diagram is called a block diagram and is used extensively in dynamics and control.

It should also be noted that the inverse Laplace transform $g(t)$ of the transfer function

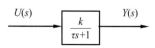

Figure 3.6 Block-diagram representation of a first-order system.

$$g(t) = \frac{k}{\tau} e^{-t/\tau} = \mathcal{L}^{-1}\left(\frac{k}{\tau s + 1}\right) \qquad (3.6.4)$$

is exactly the function that is convoluted with the input $u(t)$ to give the output $y(t)$ in Eq. (3.6.1).

Even though the Laplace transform method can simplify considerably the algebra involved in calculating the response, the relation between the system's input and output is still not at all transparent. For this reason, it makes sense to consider some representative scenarios of input changes and study their effect on the output. This gives rise to a number of idealized input variations that approximately capture those representative scenarios. In the next sections we will study the forced response of linear first-order systems to four such inputs, namely:

- the step input – idealized function for a sudden and sustained change
- the pulse input – idealized function for a sudden but nonsustained change
- the ramp input – idealized function for a change at a nearly constant rate
- the sinusoidal input – idealized function for a periodic change.

3.7 Standard Idealized Input Variations

In this section we will give the mathematical description of four idealized input variations before studying the response of linear first-order systems to these inputs. Of course, these idealized functions will not exactly appear in the real world, but they approximately capture real-life situations. The use of idealized situations is also common in other disciplines such as thermodynamics (the ideal gas law is the most striking example). Engineering intuition has developed methodologies for studying and analyzing real-world problems based on these highly idealized input variations.

3.7.1 Step Input

The mathematical description of the step change in the input of a dynamical system, which is shown in Figure 3.7a, is the following

$$u(t) = \begin{cases} 0, & t < 0 \\ M, & t \geq 0 \end{cases} \qquad (3.7.1)$$

or equivalently, $u(t) = M\mathcal{H}(t)$, where $\mathcal{H}(t)$ is the Heaviside unit step function, and M is the magnitude of the step. The step change has the following Laplace transform

$$\mathcal{L}[u(t)] = \frac{M}{s} \qquad (3.7.2)$$

The unit step input corresponds to the case of $M = 1$. The opening or the closing of a switch or a control valve are classical examples of sudden and sustained changes that can be approximated by a step change.

3.7.2 Pulse Input

A pulse change in an input signal, shown in Figure 3.7b, is described by

$$u(t) = \begin{cases} 0, & t < 0 \\ \dfrac{M}{\varepsilon}, & 0 \leq t < \varepsilon \\ 0, & t \geq \varepsilon \end{cases} \qquad (3.7.3)$$

which has the following Laplace transform (see Appendix A)

$$\mathcal{L}[u(t)] = M \dfrac{(1 - e^{-\varepsilon s})}{\varepsilon s} \qquad (3.7.4)$$

M/ε is the height of the pulse, ε is its time duration and M is the magnitude of the pulse (the area under the curve). A classical example of a sudden but not sustained input variation is the opening and subsequent closing of a switch or valve. It is important to note that a pulse input is equivalent to a step input with size M/ε applied at time $t = 0$ followed by a step change of size $-M/\varepsilon$ applied at $t = \varepsilon$:

$$u(t) = \dfrac{M}{\varepsilon}\big(\mathcal{H}(t) - \mathcal{H}(t - \varepsilon)\big) \qquad (3.7.5)$$

where $\mathcal{H}(t)$ is the Heaviside unit step function, and this property immediately leads to the Laplace transform (3.7.4).

It is interesting to examine the limiting case of a pulse with very small duration $\varepsilon \to 0$ and therefore with very large height $M/\varepsilon \to \infty$. When $M = 1$, this limiting pulse function is often referred to as the unit impulse function or Dirac delta function and is denoted with the symbol $\delta(t)$. The Laplace transform of $\delta(t)$ is obtained as the limit of the Laplace transform of a pulse of unit magnitude:

$$\mathcal{L}[\delta(t)] = \lim_{\varepsilon \to 0} \mathcal{L}[\text{unit pulse}] = \lim_{\varepsilon \to 0} \dfrac{1 - e^{-\varepsilon s}}{\varepsilon s} = 1 \qquad (3.7.6)$$

where l'Hôpital's rule has been used to evaluate the limit.

It should be pointed out that the Dirac delta function represents a highly idealized and highly unrealistic situation of a function going from zero to infinity and then back to zero in no time, with the area under the curve being equal to 1. Thus $\delta(t)$ is not a function in a usual mathematical sense; it is a generalized function. However, as we will see, it leads to useful approximations in case a thin and tall pulse is applied to a system.

3.7.3 Ramp Input

The mathematical description of a linear input variation, shown in Figure 3.7c, is the following

$$u(t) = \begin{cases} 0, & t < 0 \\ Mt, & t \geq 0 \end{cases} \qquad (3.7.7)$$

M is the rate of change of the input signal ($M = \Delta u/\Delta t$). The ramp input signal has the following Laplace transform

$$\mathcal{L}[u(t)] = \frac{M}{s^2} \qquad (3.7.8)$$

The case of $M = 1$ corresponds to the unit ramp input. The linear variation of an input is usually used in practice to implement a change in a conservative (less aggressive) manner, instead of an abrupt (step-like) change.

3.7.4 Sinusoidal Input

The mathematical description of a sinusoidal input variation, shown in Figure 3.7d, is the following

$$u(t) = \begin{cases} 0, & t < 0 \\ M \sin \omega t, & t \geq 0 \end{cases} \qquad (3.7.9)$$

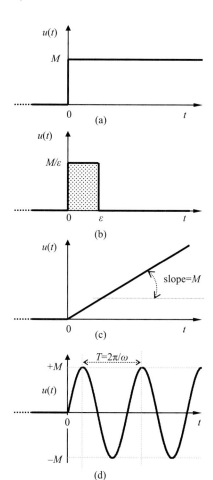

Figure 3.7 Standard inputs: (a) step, (b) pulse, (c) ramp and (d) sinusoidal.

which has the Laplace transform (see Appendix A)

$$\mathcal{L}[u(t)] = M \frac{\omega}{s^2 + \omega^2} \tag{3.7.10}$$

M is the amplitude of the sinusoidal signal and ω is the radian frequency (the period T is $T = 2\pi/\omega$). An example of an approximate sinusoidal variation is the evolution of the environmental temperature during a day.

In the sections that follow the response of a first-order system to these idealized input variations under zero initial conditions will be analyzed. The aim is to analyze the main characteristics of the dynamic response and to elucidate the role of the time constant τ and the static gain k on the response.

3.8 Response of a First-Order System to a Step Input

Consider a first-order linear system described by Eq. (3.3.1) with zero initial condition. The input of the system undergoes a step change with magnitude M as shown in Figure 3.7a. In order to calculate the output response of the system, one could either use the convolution integral of Eq. (3.6.1) or the Laplace transform relation (3.6.2).

Calculating the step response by using Eq. (3.6.1): We substitute the step function into Eq. (3.6.1). Since the step function equals M for all $t \geq 0$, we obtain

$$y(t) = \int_0^t \frac{k}{\tau} e^{-(t-t')/\tau} u(t') \, dt' = \int_0^t \frac{k}{\tau} e^{-(t-t')/\tau} M \, dt' = \frac{k}{\tau} e^{-t/\tau} M \int_0^t e^{t'/\tau} \, dt' = \frac{k}{\tau} e^{-t/\tau} M \left(\tau e^{t/\tau} - \tau \right)$$

or

$$y(t) = kM \left(1 - e^{-t/\tau}\right) \tag{3.8.1}$$

Calculating the step response using the Laplace transform: Using Eq. (3.6.2) it follows that

$$Y(s) = \frac{k}{\tau s + 1} U(s) = \frac{k}{\tau s + 1} \frac{M}{s}$$

This can be expanded by partial fractions to give

$$Y(s) = \frac{k}{\tau s + 1} \frac{M}{s} = kM \left(\frac{1}{s} - \frac{\tau}{\tau s + 1} \right) \tag{3.8.2}$$

Finally, using the Laplace transform tables available in Appendix A, we obtain Eq. (3.8.1). Figure 3.8 gives the response of a linear first-order system to a step change in the input. From this figure we observe that

- the output reaches 63.2% of its final value when the time elapsed is one time constant τ. The response is essentially complete in four to five time constants;
- the slope of the response curve at $t = 0$ is given by

3.8 Response of a First-Order System to a Step Input

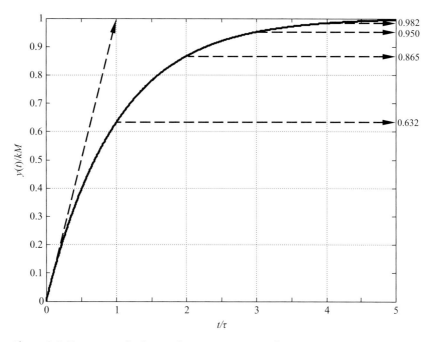

Figure 3.8 Response of a first-order system to a step input.

$$\left.\frac{dy(t)}{dt}\right|_{t=0} = \frac{kM}{\tau} \tag{3.8.3}$$

i.e. if the initial slope were maintained, the response would be complete in one time constant. From these observations it follows that the time constant τ is a measure of the speed of response.

We have already mentioned that the constant k is called the static or steady-state gain of the system. For a step change in the input it follows that

$$y(\infty) = kM \tag{3.8.4}$$

or, by using the fact that zero initial conditions are assumed,

$$k = \frac{y(\infty)}{M} = \frac{y(\infty) - 0}{M - 0} = \frac{\Delta y}{\Delta u} \tag{3.8.5}$$

where Δ denotes the final minus the initial value of a signal. We observe that the static gain relates the steady-state changes of the input and output signals. Also, note that Eq. (3.8.1), using Eq. (3.8.4), can be written as

$$1 - \frac{y(t)}{y(\infty)} = e^{-t/\tau}$$

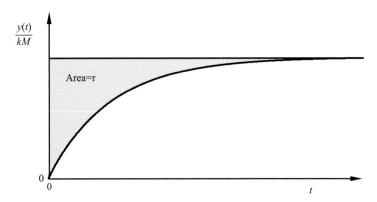

Figure 3.9 Integral of incomplete step response.

or

$$\ln\left[\frac{y(\infty)-y(t)}{y(\infty)-y(0)}\right]=-\frac{t}{\tau} \qquad (3.8.6)$$

which is, in semilogarithmic coordinates, a linear function with slope $-(1/\tau)$. This can be used to validate the hypothesis that a system behaves as a first-order system and to determine the time constant. Finally, it is interesting to observe that

$$I=\int_0^\infty \left[y(\infty)-y(t')\right]dt' = \int_0^\infty \left[kM-kM\left(1-e^{-t'/\tau}\right)\right]dt' = -kM\,\tau e^{-t'/\tau}\Big|_0^\infty = kM\tau \qquad (3.8.7)$$

This result is also shown in Figure 3.9. The shaded area of the figure equals the time constant of the first-order system. By calculating this integral from experimental data we can obtain an estimation of the system time constant.

Example 3.1 Estimation of the parameters of a first-order system

Consider the mercury thermometer shown in Figure 3.4, which has been on a table for some time, recording the room temperature of 20 °C, and suddenly, at $t = 0$, it is placed in a 200 °C oil bath. The temperature measurements as a function of time are given in Table 3.1 and in Figure 3.10. Use these data to estimate the time constant and the static gain of the thermometer.

Solution

The static gain of the thermometer can be estimated using Eq. (3.8.5)

$$k=\frac{\Delta y}{\Delta u}=\frac{y(\infty)-y(0)}{u(t)-u(0)}=\frac{200-20}{200-20}=1\ °C/°C$$

3.8 Response of a First-Order System to a Step Input

Table 3.1 Step response experiment data for the mercury thermometer of Figure 3.4

i	Time (s)	Temperature reading
1	0	20
2	5	91
3	10	134
4	15	160
5	20	176
6	25	185
7	30	191
8	35	195
9	40	197
10	45	198
11	50	199
12	∞	200

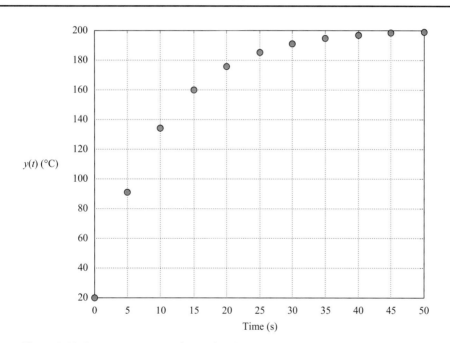

Figure 3.10 Step response experiment for the mercury thermometer of Figure 3.4.

The time constant can be estimated from Eq. (3.8.1), which can be written as

$$y(\tau) - y(0) = [y(\infty) - y(0)](1 - e^{-1})$$

or

$$y(\tau) = y(0) + [y(\infty) - y(0)](1 - e^{-1}) = 20 + 180 \cdot 0.632$$
$$= 133.8 \ °C \approx 134 \ °C$$

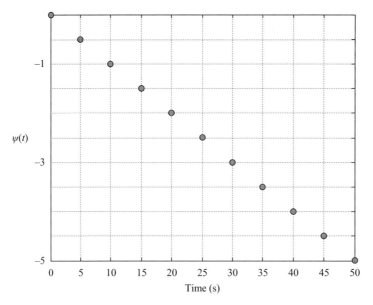

Figure 3.11 Graphical estimation of the time constant of the thermometer.

From Table 3.1 we observe that the temperature becomes 134 °C at time $t = \tau = 10$ s which is, therefore, the time constant of the thermometer. Alternatively, we can use Eq. (3.8.6) and plot the quantity

$$\psi(t_i) = \ln\left[\frac{y(\infty) - y(t_i)}{y(\infty) - y(0)}\right]$$

as a function of time. The result is shown in Figure 3.11 and it follows that

$$\text{slope} = \frac{\Delta \psi}{\Delta t} = \frac{-5}{50} = -0.1 \text{ and } -\frac{1}{\tau} = -0.1 \Rightarrow \tau = 10$$

Finally, the time constant can be estimated using Eq. (3.8.7) by calculating numerically the integral involved. If, for example, the trapezoidal rule is used, then

$$I = \frac{5}{2}\{[y(t_{12}) - y(t_1)] + [y(t_{12}) - y(t_2)]\} + \cdots + \frac{5}{2}\{[y(t_{12}) - y(t_{10})] + [y(t_{12}) - y(t_{11})]\}$$

$$= \frac{5}{2}\left\{[y(t_{12}) - y(t_1)] + 2\sum_{i=2}^{10}[y(t_{12}) - y(t_i)] + [y(t_{12}) - y(t_{11})]\right\} = 1817.5$$

from which we calculate that $\tau = \dfrac{I}{kM} = \dfrac{1817.5}{1 \cdot 180} \approx 10.1$. The transfer function of the thermometer is, therefore, the following

$$G(s) = \frac{1}{10s + 1}$$

3.9 Response of a First-Order System to a Pulse Input

Pulse response is the response of a system to a pulse input signal under zero initial conditions. Let's consider the case where the input signal is a pulse of magnitude M and time duration ε, as shown in Figure 3.7b. The output of the first-order system (3.3.1) can be calculated as in the case of the step response. The easiest way of deriving the pulse response is through the following observations: for $0 \leq t < \varepsilon$ the pulse input is identical to a step of magnitude M/ε, hence from (3.8.1)

$$y(t) = k\frac{M}{\varepsilon}\left(1 - e^{-t/\tau}\right), \quad 0 \leq t < \varepsilon \tag{3.9.1}$$

When $t = \varepsilon$, the value of the output is

$$y(\varepsilon) = k\frac{M}{\varepsilon}\left(1 - e^{-\varepsilon/\tau}\right) \tag{3.9.2}$$

For $t \geq \varepsilon$ the input becomes zero and the response is unforced, starting at $t = \varepsilon$ from the nonzero initial condition ($y(\varepsilon)$). Thus, applying (3.5.1), time-shifted, leads to

$$y(t) = e^{-(t-\varepsilon)/\tau} y(\varepsilon) = k\frac{M}{\varepsilon}\left(1 - e^{-\varepsilon/\tau}\right)e^{-(t-\varepsilon)/\tau}, \quad t \geq \varepsilon \tag{3.9.3}$$

The same result can be obtained by substituting the pulse function (3.7.3) into (3.6.1):

$$y(t) = \begin{cases} \displaystyle\int_0^t \frac{k}{\tau} e^{-(t-t')/\tau} \frac{M}{\varepsilon} \, dt', & 0 \leq t < \varepsilon \\[2ex] \displaystyle\int_0^\varepsilon \frac{k}{\tau} e^{-(t-t')/\tau} \frac{M}{\varepsilon} \, dt', & t \geq \varepsilon \end{cases}$$

and evaluating the integrals.

Calculating the pulse response by using the Laplace transform: from Eq. (3.6.2) we obtain

$$Y(s) = \frac{k}{\tau s + 1} U(s) = \frac{k}{\tau s + 1} \frac{M(1 - e^{-\varepsilon s})}{\varepsilon s} = k\left[\frac{1}{(\tau s + 1)} \cdot \frac{M/\varepsilon}{s} - e^{-\varepsilon s} \frac{1}{(\tau s + 1)} \cdot \frac{M/\varepsilon}{s}\right] \tag{3.9.4}$$

From Eq. (3.9.4) we observe that the pulse response consists of two parts: the response to a step input of magnitude M/ε applied at $t = 0$ and the response to a step input of magnitude $-M/\varepsilon$ applied at $t = \varepsilon$ (the factor of $e^{-\varepsilon s}$ in the second term signifies time variable translation by ε time units, as a result of the time variable translation theorem of the Laplace transform). Therefore using (3.8.1) we can write the following

$$y(t) = \begin{cases} \dfrac{kM}{\varepsilon}\left(1 - e^{-t/\tau}\right), & 0 \leq t < \varepsilon \\[2ex] k\dfrac{M}{\varepsilon}\left(1 - e^{-t/\tau}\right) + k\left(\dfrac{-M}{\varepsilon}\right)\left(1 - e^{-(t-\varepsilon)/\tau}\right), & t \geq \varepsilon \end{cases} \tag{3.9.5}$$

which is exactly the same as Eqs. (3.9.1) and (3.9.3). The pulse response of a linear first-order system is shown in Figure 3.12a (where the duration of the pulse is taken to be smaller than the time constant, for demonstration purposes).

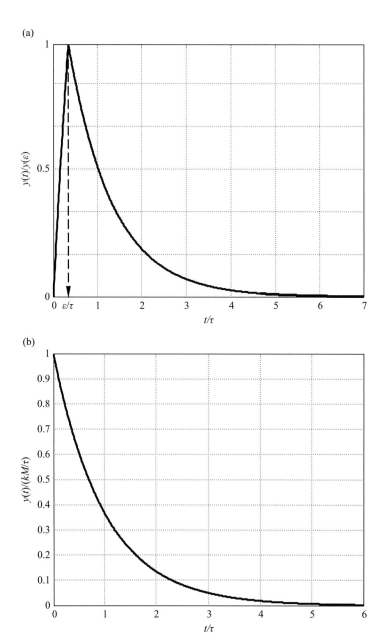

Figure 3.12 Pulse (a) and impulse (b) response of a first-order system.

3.9.1 Impulse Response

We will now consider the limiting case where the duration of the pulse becomes very small, i.e. $\varepsilon \to 0$, in which case

- the interval $[0, \varepsilon]$ shrinks to the point 0,
- the peak value of the response is given by $y(\varepsilon) = (kM/\varepsilon)(1 - e^{-\varepsilon/\tau}) \to kM/\tau$,
- the response for $t > \varepsilon$ is identical to the unforced response with initial condition $y(0) = y(\varepsilon) = kM/\tau$, i.e.

$$y(t) = \frac{kM}{\tau} e^{-t/\tau} \tag{3.9.6}$$

This limiting response is called the impulse response. It is shown in Figure 3.12b.

Alternatively, the impulse response could be thought of as the response of the system to the input $u(t) = M\delta(t)$, where $\delta(t)$ is the Dirac delta function. Because from (3.7.6) the Laplace transform $\delta(t)$ is equal to 1, we have $U(s) = M$, therefore $Y(s) = kM/(\tau s + 1)$, and inverting the Laplace transform gives (3.9.6).

3.10 Response of a First-Order System to a Ramp Input

If the input to a first-order system with zero initial conditions is a ramp input of the form $u(t) = Mt$, as shown in Figure 3.7c, then its output can be calculated as follows.

Calculating the ramp response by using Eq. (3.6.1):

$$y(t) = \int_0^t \frac{k}{\tau} e^{-(t-t')/\tau} u(t') \, dt' = \frac{kM}{\tau} e^{-t/\tau} \int_0^t e^{t'/\tau} t' \, dt' = kMe^{-t/\tau} \left(te^{t/\tau} - \int_0^t e^{t'/\tau} dt' \right)$$

from which

$$y(t) = kM \left(t - \tau + \tau e^{-t/\tau} \right) \tag{3.10.1}$$

This result is shown in Figure 3.13.

Calculating the ramp response by using the Laplace transform: Using Eq. (3.6.2) and the the fact that $U(s) = \mathcal{L}[Mt] = M/s^2$ we obtain

$$Y(s) = \frac{k}{\tau s + 1} U(s) = \frac{k}{\tau s + 1} \frac{M}{s^2}$$

or, by expanding in partial fractions

$$Y(s) = \frac{k}{\tau s + 1} \frac{M}{s^2} = kM \left(\frac{1}{s^2} - \frac{\tau}{s} + \frac{\tau}{s + \frac{1}{\tau}} \right) \tag{3.10.2}$$

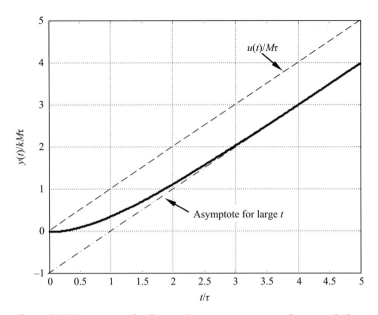

Figure 3.13 Response of a first-order system to a ramp input variation.

and by using the tables of the inverse Laplace transform from Appendix A we can obtain Eq. (3.10.1).

From Eq. (3.10.1) and Figure 3.13, we observe that for $t \gg \tau$,

- the response approaches $y(t) = kM(t - \tau)$, which is a linear function of time,
- the output follows the input with a delay of one time constant,
- the static gain is equal to $k =$ (slope of the output)/(slope of the input), or

$$\left. \frac{dy(t)/dt}{du(t)/dt} \right|_{t \gg \tau} = \frac{kM}{M} = k \tag{3.10.3}$$

Many measuring devices, such as thermometers, exhibit first-order dynamics with a static gain of 1. If the measured variable varies linearly, then the measurement signal lags behind the actual physical variable, with a delay of one time constant. The difference between the actual and the measured variable is

$$u(t) - y(t) = Mt - M(t - \tau) = M\tau \tag{3.10.4}$$

and this is called the dynamic error of the measuring device. It is important to observe that the dynamic error is proportional to the rate of change of the measured variable and to the time constant of the system. This is another reason why the time constant is considered to be a measure of the speed or response.

3.11 Response of a First-Order System to a Sinusoidal Input

If the input to a first-order system is a sinusoidal signal of amplitude M and radian frequency ω, then the output, for zero initial conditions, can be calculated as follows:

Calculating the sinusoidal response by using Eq. (3.6.1):

$$y(t) = \int_0^t \frac{k}{\tau} e^{-(t-t')/\tau} u(t') \, dt' = \int_0^t \frac{k}{\tau} e^{-(t-t')/\tau} M \sin(\omega t') \, dt' = \frac{kM}{\tau} e^{-t/\tau} \int_0^t e^{t'/\tau} \sin(\omega t') \, dt'$$

From tables of integrals,

$$\int e^{ax} \sin(\beta x) \, dx = e^{ax} \frac{a \sin(\beta x) - \beta \cos(\beta x)}{a^2 + \beta^2}$$

and thus we can calculate

$$\int_0^t e^{t'/\tau} \sin(\omega t') \, dt' = e^{t/\tau} \frac{\tau \sin(\omega t) - \tau^2 \omega \cos(\omega t)}{1 + \tau^2 \omega^2} + \frac{\tau^2 \omega}{1 + \tau^2 \omega^2}$$

and finally

$$y(t) = kM \left(\frac{\omega \tau}{1 + (\omega \tau)^2} e^{-t/\tau} - \frac{\omega \tau}{1 + (\omega \tau)^2} \cos(\omega t) + \frac{1}{1 + (\omega \tau)^2} \sin(\omega t) \right) \quad (3.11.1)$$

Calculating the sinusoidal response by using the Laplace transform: From Eq. (3.6.2) we obtain

$$Y(s) = \frac{k}{\tau s + 1} U(s) = \frac{k}{\tau s + 1} \frac{M\omega}{s^2 + \omega^2}$$

The above expression can be expanded in partial fractions in the form

$$Y(s) = \frac{k}{\tau s + 1} \cdot \frac{M\omega}{s^2 + \omega^2} = kM \left(\frac{c_1}{s + (1/\tau)} + \frac{c_2}{s + i\omega} + \frac{c_3}{s - i\omega} \right)$$

where the constants c_1, c_2 and c_3 are calculated following the methods of Appendix A. We find:

$$c_1 = \frac{\omega \tau}{1 + (\omega \tau)^2}, \quad c_2 = \frac{-\frac{\omega \tau}{2} + \frac{1}{2} i}{1 + (\omega \tau)^2}, \quad c_3 = \frac{-\frac{\omega \tau}{2} - \frac{1}{2} i}{1 + (\omega \tau)^2}$$

Then, inverting the Laplace transform we find

$$y(t) = kM \left(\frac{\omega \tau}{1 + (\omega \tau)^2} e^{-t/\tau} + \frac{-\frac{\omega \tau}{2} + \frac{1}{2} i}{1 + (\omega \tau)^2} e^{-i\omega t} + \frac{-\frac{\omega \tau}{2} - \frac{1}{2} i}{1 + (\omega \tau)^2} e^{i\omega t} \right)$$

and, converting the complex exponentials into sines and cosines via Euler's formula, we obtain Eq. (3.11.1).

3.11.1 Characteristics of Sinusoidal Response

From Eq. (3.11.1) we observe that the response to a sinusoidal input consists of two terms that involve the $\sin(\omega t)$ and $\cos(\omega t)$ functions as well as an additional exponential term. The sine and cosine functions can be combined using the identity

$$p\cos(\omega t)+q\sin(\omega t)=\sqrt{p^2+q^2}\,\sin(\omega t+\varphi),\quad \varphi=\tan^{-1}\left(\frac{p}{q}\right)$$

and thus Eq. (3.11.1) can be written as

$$y(t)=kM\left(\frac{\omega\tau}{1+(\omega\tau)^2}e^{-t/\tau}+\frac{1}{\sqrt{1+(\omega\tau)^2}}\sin(\omega t+\varphi)\right),\quad \varphi=-\tan^{-1}(\omega\tau) \qquad (3.11.2)$$

Observe that the exponential term dies out for large t, and we obtain

$$y(t)=\frac{kM}{\sqrt{1+(\omega\tau)^2}}\sin(\omega t+\varphi),\quad \varphi=-\tan^{-1}(\omega\tau) \qquad (3.11.3)$$

which is called the frequency response.

We can now summarize the characteristics of frequency response. For simplicity, let's assume $k > 0$. We observe that for large t,

- the output of the first-order system varies in a sinusoidal manner with the same frequency as the input signal,
- the amplitude ratio (AR), defined as the ratio of the amplitude of the output signal to that of the input signal is given by

$$AR=\frac{\text{amplitude of the output}}{\text{amplitude of the intput}}=\frac{k}{\sqrt{1+(\omega\tau)^2}}, \qquad (3.11.4)$$

- the phase difference between output and input is given by

$$\varphi=-\tan^{-1}(\omega\tau) \qquad (3.11.5)$$

Both AR and φ are frequency dependent. Furthermore, the output lags behind the input since the angle φ is negative. These conclusions are shown also in Figure 3.14.

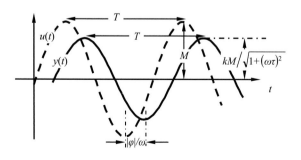

Figure 3.14 Asymptotic ($t\to\infty$) response of a first-order system to a sinusoidal input.

3.11.2 Bode Diagrams of a First-Order System

The Bode diagrams are plots of the amplitude ratio AR and the phase φ as a function of the frequency ω. The AR plot is usually in log × log scale, the φ plot in lin × log scale. Figures 3.15a and b depict the Bode diagrams of a first-order system with positive steady-state gain k. From these figures as well as Eq. (3.11.4), we can observe that:

- at low frequencies ($\omega\tau \ll 1$) $AR \approx k$ and $\varphi \approx 0°$
- at high frequencies ($\omega\tau \gg 1$) $AR \approx k/(\omega\tau)$ and $\varphi \approx -90°$.

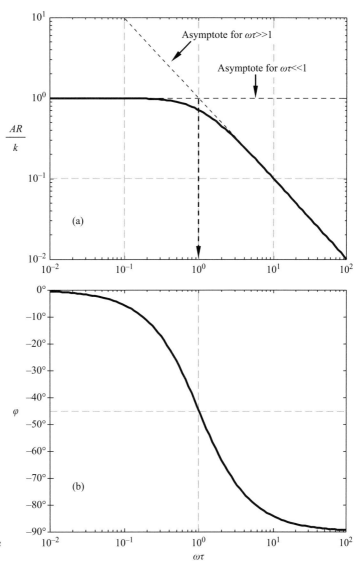

Figure 3.15 Bode diagrams of a first-order system: (a) amplitude ratio and (b) phase angle.

At low frequencies ($\omega\tau \ll 1$) the Bode diagram of the amplitude ratio in logarithmic scales is a straight line parallel to the frequency axis, while at high frequencies ($\omega\tau \gg 1$) is again a straight line with slope that is equal to -1. These two asymptotes intersect each other at the frequency $\omega_c = 1/\tau$. This is called the corner frequency.

Example 3.2 Response of a first-order system to a superposition of sinusoidal inputs

A first-order system, initially at zero, is subject to a superposition of three sinusoidal inputs of different frequencies:

$$u(t) = \begin{cases} 0 & , t < 0 \\ M_1 \sin\omega_1 t + M_2 \sin\omega_2 t + M_3 \sin\omega_3 t, & t \geq 0 \end{cases}$$

Calculate the output response $y(t)$. Plot the input and the output on a common graph when $k = 1$, $\tau = 1/10$, $M_1 = 1$, $\omega_1 = 1$, $M_2 = 1/50$, $\omega_2 = 60$, $M_3 = 1/10$, $\omega_3 = 120$. Interpret your result in terms of the AR values of each frequency component. What do you observe?

Solution

We see from Eq. (3.6.1) that the output $y(t)$ depends linearly on the input $u(t)$, therefore the effect of the sum of three different inputs is superimposed. Thus, in the present case where all three inputs are sinusoidal, the output is the superposition of sinusoidal responses of the form of Eq. (3.11.1):

$$y(t) = k \sum_{i=1}^{3} M_i \left(\frac{\omega_i \tau}{1 + (\omega_i \tau)^2} e^{-t/\tau} - \frac{\omega_i \tau}{1 + (\omega_i \tau)^2} \cos(\omega_i t) + \frac{1}{1 + (\omega_i \tau)^2} \sin(\omega_i t) \right)$$

For the given parameter values, the input and output are depicted in Figure 3.16. The low-frequency component of the signal corresponding to the frequency of $\omega_1 = 1$ (period $= 2\pi$) is the "base" of the signal, the other two components are low-amplitude high-frequency perturbations. Calculating the amplitude ratio for each component using Eq. (3.11.4) and the system parameters $k = 1$, $\tau = 1/10$, we find: $AR(\omega_1) = 0.995$, $AR(\omega_2) = 0.164$, $AR(\omega_3) = 0.083$ (see also the Bode diagram, Figure 3.15a). The low frequency $\omega_1 = 1$ and is exactly one order of magnitude smaller than $1/\tau = 10$, and this is why it has an AR of about 1. The higher frequencies $\omega_2 = 60$, $\omega_3 = 120$ are significantly higher than $1/\tau = 10$, and this is why their AR is much less than 1. When these are superimposed, the overall effect is that the low frequency "passes," whereas the higher frequencies are "cut." This is clearly visible from the output response shown in Figure 3.16.

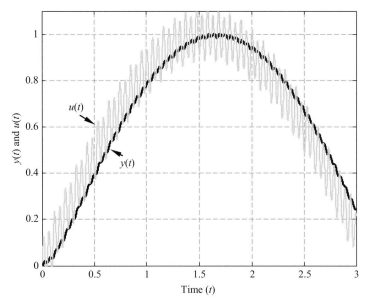

Figure 3.16 Input and output for the first-order system studied in Example 3.2.

3.11.3 First-Order Low-Pass Filter

As seen in the previous subsection and Example 3.2, a key property of a first-order system is that it cuts out high frequencies $\omega \gg 1/\tau$, but lets low frequencies $\omega \ll 1/\tau$ pass. This is the property of a low-pass filter. In electrical engineering, a filter is a device that removes some unwanted components or features from a signal, such as certain frequencies. Common types of filter include:

- *low-pass filter*, which lets low frequencies pass, but high frequencies are cut;
- *high-pass filter*, which lets high frequencies pass, but low frequencies are cut;
- *band-pass filter*, which allows only frequencies from a certain range to pass;
- *band-stop filter*, which cuts only frequencies from a certain range;
- *all-pass filter*, which lets all frequencies pass.

Figure 3.17 shows two examples of first-order RC filters, consisting of a resistor and a capacitor; one is low-pass and the other is high-pass. For the one on the left (Figure 3.17a), which is low-pass, one can apply principles of electrical engineering and show that it follows

$$RC\frac{dV_{out}}{dt} + V_{out} = V_{in} \qquad (3.11.6)$$

where V_{in} is the inlet voltage (input variable), V_{out} is the outlet voltage (output variable), R is the resistance and C is the capacitance (constant parameters). In electrical engineering

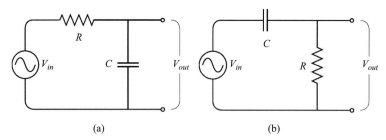

Figure 3.17 RC filters: (a) low-pass and (b) high-pass.

books, one can find a variety of electrical and electronic circuits that realize different kinds of filters. The design of filter circuitry is, of course, outside the field of chemical engineering; however, the concept of a low-pass filter, the idea of cutting out the high-frequency content of a signal, is very useful in chemical process control practice.

The general equation of a first-order low-pass filter is:

$$\tau_f \frac{d\sigma_f}{dt} + \sigma_f = \sigma \qquad (3.11.7)$$

with input σ being a signal to be processed (typically a measurement signal from a sensor), output σ_f, called the "filtered signal," and parameter τ_f, the filter time constant. The first-order low-pass filter is just a first-order system with unity static gain ($k = 1$).

3.12 Response of a First-Order System to an Arbitrary Input – Time Discretization of the First-Order System

3.12.1 Discrete-Time First-Order Systems

The study of the effect of idealized inputs of the previous sections is of great value because these idealized inputs capture (at least approximately) common situations encountered in practice. However, there are also situations where the input variation does not even resemble an idealized input and the output response needs to be accurately calculated. One will then need to apply some kind of numerical algorithm to calculate the output response.

Consider a first-order system that is subject to an arbitrary input $u(t)$. Its output can then be directly calculated from Eq. (3.3.4). Suppose now that we know the input values as a set of numbers (say the values at every second) that we read from a data file. We will then need to write a simple code to use these numbers and simulate a time-discretized form of the system. There are many ways of time-discretizing a first-order system, including numerical methods for the solution of differential equations. However, the easiest and the most practical approach is to directly discretize the convolution integral in Eq. (3.3.4). Specifically, consider

3.12 Response of a First-Order System to an Arbitrary Input

values of $u(t)$ being given in the form of a sequence of numbers, at times $t = 0, T_s, 2T_s, \ldots, jT_s, \ldots$, where T_s is the *sampling period*, i.e.

$$\begin{aligned} u[0] &= u(0) \\ u[1] &= u(T_s) \\ u[2] &= u(2T_s) \\ &\vdots \\ u[j] &= u(jT_s) \\ &\vdots \end{aligned} \tag{3.12.1}$$

We would like to calculate the corresponding values of the output $y[j]$ at times $t = T_s, 2T_s, \ldots, jT_s, \ldots$, again in the form of a sequence of numbers. The understanding is, of course, that the sampling period T_s is small enough, so that the input data provide sufficiently detailed information on the input variation. In chemical plants, this is always the case, since process variables change in a time scale of minutes or hours, or even in a slower time scale, whereas process variable data are available every second or fraction of a second.

Because the sampling period is in practice very small, we can use any numerical approximation scheme for the convolution integral, e.g. rectangle, trapezoid, Simpson, etc. As the rectangle approximation is the simplest, it is almost always the preferred choice. In particular, one can do a piecewise constant approximation of $u(t)$ as follows (see Figure 3.18a):

$$u(t) = \begin{cases} u(0), & \text{for } 0 \leq t < T_s \\ u(T_s), & \text{for } T_s \leq t < 2T_s \\ u(2T_s), & \text{for } 2T_s \leq t < 3T_s \\ \vdots \\ u(jT_s), & \text{for } jT_s \leq t < (j+1)T_s \\ \vdots \end{cases} \tag{3.12.2}$$

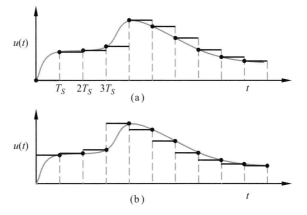

Figure 3.18 (a) Left endpoint rectangle approximation and (b) right endpoint rectangle approximation.

In this approximation, the input is held constant over each time interval, and it assumes the value at the beginning of the interval. In numerical analysis, this is called *left endpoint rectangle approximation* or *forward rectangle approximation*. In process control, it is also referred to as the *zero-order-hold (ZOH) approximation*. The other alternative would be the *right endpoint rectangle approximation* or *backward rectangle approximation* (see Figure 3.18b):

$$u(t) = \begin{cases} u(T_s), & \text{for } 0 < t \leq T_s \\ u(2T_s), & \text{for } T_s < t \leq 2T_s \\ u(3T_s), & \text{for } 2T_s < t \leq 3T_s \\ \vdots \\ u(jT_s), & \text{for } (j-1)T_s < t \leq jT_s \\ \vdots \end{cases} \tag{3.12.3}$$

Let's now calculate $y[j] = y(jT_S)$ using the solution formula (3.3.4) for the first-order differential equation $\tau(dy/dt) + y = ku$, and the forward rectangle approximation for the input function. Setting $t = jT_S$ and $t = (j-1)T_S$ in the solution formula, we get:

$$y(jT_S) = e^{-jT_S/\tau} y(0) + \int_0^{jT_S} \frac{k}{\tau} e^{-(jT_S - t')/\tau} u(t') \, dt' \tag{3.12.4}$$

$$y((j-1)T_S) = e^{-(j-1)T_S/\tau} y(0) + \int_0^{(j-1)T_S} \frac{k}{\tau} e^{-((j-1)T_S - t')/\tau} u(t') \, dt' \tag{3.12.5}$$

We then multiply Eq. (3.12.5) by $e^{-T_S/\tau}$ and subtract the result from Eq. (3.12.4) to obtain:

$$y(jT_S) = e^{-T_S/\tau} y((j-1)T_S) + \left[\frac{k}{\tau} \int_{(j-1)T_S}^{jT_S} e^{-(jT_S - t')/\tau} \, dt' \right] u((j-1)T_S) \tag{3.12.6}$$

We then change the integration variable in the integral ($t'' = jT_S - t'$) and perform the integration to finally obtain:

$$y(jT_S) = e^{-T_S/\tau} y((j-1)T_S) + (1 - e^{-T_S/\tau}) ku((j-1)T_S) \tag{3.12.7}$$

or

$$y[j] = e^{-T_S/\tau} y[j-1] + (1 - e^{-T_S/\tau}) ku[j-1] \tag{3.12.8}$$

The above relation allows the *recursive calculation* of the sequence of output values $y[j]$ from the sequence of input values $u[j]$. It is the time-discretization of the solution of the first-order differential equation $\tau(dy/dt) + y = ku$, with discretization time step T_S, under the *forward rectangle approximation*. If we want to use a *backward rectangle approximation*, the derivation is identical to the previous one, except that the input $u(t)$ is set equal to $u(jT_S)$

instead of $u((j-1)T_S)$ when performing the integration. This leads to the recursion (backward rectangle approximation):

$$y[j] = e^{-T_S/\tau} y[j-1] + (1 - e^{-T_S/\tau}) \, ku[j] \qquad (3.12.9)$$

Comparing the two versions of the rectangle approximation, we observe that they are virtually identical – the only difference is whether the input value appears in the current time step as $u[j]$ or in the previous time step as $u[j-1]$. A natural question to ask here is which version of the rectangle approximation would be better to use. With the sampling period T_S being very small ($T_S \ll \tau$), both discretization schemes give essentially the same answer, so it makes no difference from a practical point of view.

Example 3.3 Recursive calculation of the output response of a time-discretized first-order system

A first-order system with time constant $\tau = 2$ and static gain $k = 0.5$ is initially at steady state with the input $u = 2$ and the output $y = 1$. At $t = 0$, the input $u(t)$ starts changing, and it assumes the following values at times $t = 0, T_s, 2T_s, 3T_s, \ldots, 90T_s$, where $T_s = 0.2$:
$u = \{2.0, 2.2, 2.4, 2.6, 2.8, 3.0, 3.2, 3.4, 3.6, 3.8, 4.0, 4.2, 4.4, 4.6, 4.8, 5.0, 5.2, 5.4, 5.6, 5.8, 6.0, 6.2, 6.4, 6.6, 6.8, 7.0, 7.2, 7.4, 7.6, 7.8, 8.0, 7.9, 7.8, 7.7, 7.6, 7.5, 7.4, 7.3, 7.2, 7.1, 7.0, 6.9, 6.8, 6.7, 6.6, 6.5, 6.4, 6.3, 6.2, 6.1, 6.0, 5.9, 5.8, 5.7, 5.6, 5.5, 5.4, 5.3, 5.2, 5.1, 5.0, 5.0, 5.0, 5.0, 5.0, 5.0, 5.0, 5.0, 5.0, 5.0, 5.0, 5.0, 5.0, 5.0, 5.0, 5.0, 5.0, 5.0, 5.0, 5.0, 5.0, 5.0, 5.0, 5.0, 5.0, 5.0, 5.0, 5.0, 5.0, 5.0, 5.0\}$

Calculate the output at the same time instants via the forward rectangle approximation, applying the recursion (3.12.8) with the above input sequence $u[j], j = 0, \ldots, 90$ and initial output value $y[0] = 1$, and plot your results.

Solution

The given input sequence is plotted on the top graph of Figure 3.19. We see that it is a superposition of three ramps (a ramp applied at $t = 0$ and two delayed ramps applied at $t = 6$ and $t = 12$). Even though the response could be calculated analytically, the result would be rather complicated, and the formulas hardly useful. Also, since the input is given in the form of a data sequence, the most convenient calculation method is to apply the recursion (3.12.8) which, for the given parameter values, becomes:

$$y[j] = 0.90484 \, y[j-1] + 0.04758 \, u[j-1]$$

Applying the above recursion, with the given input values for $u[j], j = 0, \ldots, 90$ initialized at $y[0] = 1$, we obtain the values of the output $y[j], j = 1, \ldots, 90$. These are listed in the table below.

1.0000	1.0095	1.0276	1.0536	1.0865	1.1259	1.1710	1.2213	1.2764
1.3357	1.3990	1.4657	1.5356	1.6083	1.6836	1.7613	1.8411	1.9229

2.0063	2.0914	2.1778	2.2656	2.3545	2.4445	2.5354	2.6272	2.7198
2.8131	2.9070	3.0015	3.0965	3.1777	3.2465	3.3039	3.3511	3.3891
3.4187	3.4407	3.4558	3.4648	3.4681	3.4664	3.4601	3.4496	3.4354
3.4177	3.3970	3.3735	3.3475	3.3192	3.2888	3.2566	3.2226	3.1872
3.1503	3.1122	3.0730	3.0327	2.9916	2.9495	2.9068	2.8681	2.8330
2.8013	2.7727	2.7467	2.7232	2.7020	2.6828	2.6654	2.6496	2.6354
2.6225	2.6109	2.6003	2.5908	2.5821	2.5743	2.5672	2.5608	2.5550
2.5498	2.5451	2.5408	2.5369	2.5334	2.5302	2.5273	2.5247	2.5224

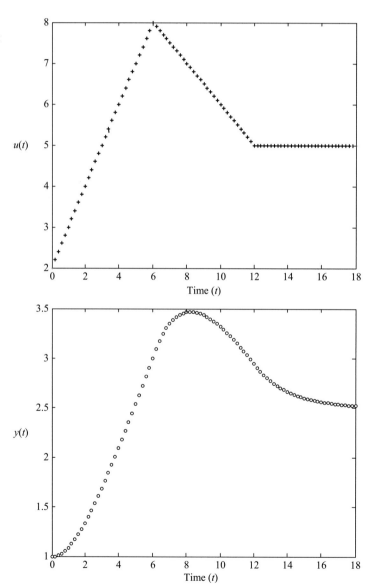

Figure 3.19 Input variation and output response for Example 3.3.

These output values are plotted on the bottom graph of Figure 3.19. The output response has the expected qualitative characteristics: it is a typical ramp response up to time $t = 6$, when the new ramp kicks in, and then the output turns around and follows the new direction of the input. Finally, at time $t = 12$, when the input levels off, the output goes towards steady state. The final output steady state is $\lim_{t \to \infty} y(t) = k \cdot \lim_{t \to \infty} u(t) = 2.5$, and this is clearly visible on the graph.

3.12.2 Discrete-Time First-Order Low-Pass Filter

Low-pass filters are extremely useful in industrial practice because very often, on-line measurements are corrupted with high-frequency noise. For a measurement signal to be useful (e.g. to be used by a controller), it may be necessary to remove its high-frequency component (noise) and keep the useful part of the signal that tells us how the physical variable is actually changing. For this purpose, filtering is a very important operation.

Even though sensors are sometimes connected to an electronic filter, like the ones illustrated in Figure 3.17, it is more common (and less expensive) to perform filtering in the control computer, given the measurement signal in digital form. This calls for a time-discretized form of the low-pass filter.

Applying the backward rectangle approximation (3.12.3) to the filter equation (3.11.7), we obtain the following discretized filter equation:

$$\sigma_f[j] = e^{-T_s/\tau_f} \sigma_f[j-1] + (1 - e^{-T_s/\tau_f}) \sigma[j] \tag{3.12.10}$$

Setting $f = 1 - e^{-T_s/\tau_f}$, the filter equation may be alternatively written in the form:

$$\sigma_f[j] = (1-f) \sigma_f[j-1] + f \sigma[j] \tag{3.12.11}$$

The parameter f is called the *filter factor*. Since $0 < \frac{T_s}{\tau_f} \ll 1$, this implies $0 < f \ll 1$.

At every point in time, the discrete-time filter calculates a weighted average of the previous value of the filtered signal $\sigma_f[j-1]$ and the current value of the signal, with weights $(1-f)$ and $0 < f \ll 1$, respectively. If a signal is too noisy, a very small weight is used for the current value of the signal and a heavy weight on the previous filtered value, i.e. a small value of f is to be used. In the case of a low-noise signal, we would have more faith in the current value of the signal, hence a higher value of f would be used.

As an illustration, consider the noisy signal depicted in Figure 3.20, which was sampled with sampling period $T_s = 0.01$. This signal has clearly visible high-frequency noise that needs to be cut out using a low-pass discrete-time filter. In Figure 3.20, the original noisy signal is compared to the filtered signal calculated from (3.12.10), for two different values of the filter time constant:

- $\tau_f = 0.1$ (one order of magnitude larger than the sampling period) and
- $\tau_f = 1$ (two orders of magnitude larger than the sampling period).

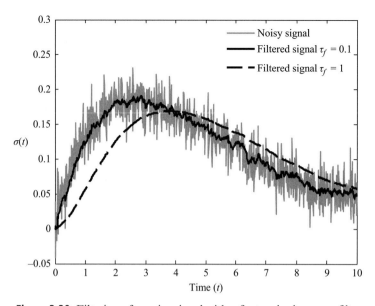

Figure 3.20 Filtering of a noisy signal with a first-order low-pass filter.

The filter with $\tau_f = 0.1$ has resulted in a major reduction of noise; however, it has not achieved complete elimination. The filter with $\tau_f = 1$ has completely eliminated the noise; however, the filtered signal lags behind the actual signal, and this is undesirable. Further search within the interval $0.1 < \tau_f < 1$ will be needed to optimize the selection of filter time constant τ_f. In general, for designing a low-pass filter, one should look for a best compromise between reduction of noise and speed of response of the filter.

3.13 Another Example of a First-Order System: Liquid Storage Tank

Liquid storage tanks, also known as surge tanks, are commonly used in the chemical and petrochemical industries in order to alleviate flow variations. Consider the cylindrical surge tank shown in Figure 3.21 which has a uniform cross-sectional area A and liquid level h. Under the assumption of constant density of the fluid, the following mass-balance equation can be written

$$A\frac{dh}{dt} = F_{in} - F_{out} \tag{3.13.1}$$

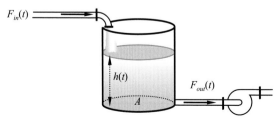

where F_{in} is the volumetric flowrate of the inlet stream and F_{out} is the volumetric flowrate of the

Figure 3.21 Liquid storage tank with a pump in the outflow stream.

outlet stream. The incoming flow is the input of the liquid storage tank, as its value is determined exogenously by an upstream process. In order to develop a complete mathematical description of the tank, an equation that determines the volumetric flowrate of the outlet stream is necessary. We will consider the following three cases.

Case 1: The volumetric flowrate of the outflow is proportional to the level of the liquid in the tank. In this case, the outlet volumetric flowrate is a linear function of the level:

$$F_{out} = \frac{1}{R}h \qquad (3.13.2)$$

where R is the flow resistance. Combining Eqs. (3.13.1) and (3.13.2) we obtain

$$A\frac{dh}{dt} = F_{in} - \frac{1}{R}h$$

or

$$AR\frac{dh}{dt} + h = RF_{in} \qquad (3.13.3)$$

If we set $y = h$, $u = F_{in}$, $\tau = AR$ and $k = R$, then Eq. (3.13.3) simplifies to Eq. (3.3.1). We then conclude that when the outlet volumetric flowrate depends linearly on the liquid level in the tank, the system is described by exactly the same form of first-order linear differential equation that we studied in the previous sections.

Case 2: There is a pump in the outflow. In this case the volumetric flowrate of the outlet stream is constant, i.e.

$$F_{out} = \text{constant} \qquad (3.13.4)$$

Then, the mathematical form of Eq. (3.13.1) does not match with Eq. (3.3.1). More specifically the system has the following characteristics.

(a) The system is at steady state only when $F_{in} = F_{out}$ = constant holds true. Under steady-state conditions, the liquid level h assumes a constant value which, however, cannot be determined by the steady-state form of Eq. (3.13.1). If the system is not at steady state, then

$$\text{if } F_{in} > F_{out} \Rightarrow \frac{dh}{dt} > 0 \Rightarrow \text{ the liquid level is increasing}$$

$$\text{if } F_{in} < F_{out} \Rightarrow \frac{dh}{dt} < 0 \Rightarrow \text{ the liquid level is decreasing.}$$

(b) The solution of the differential equation (3.13.1) is quite simple:

$$h(t) = h(0) + \frac{1}{A}\int_0^t \left(F_{in}(t') - F_{out}\right)dt' \qquad (3.13.5)$$

from which the response can be calculated by substituting the special form of the inlet flowrate variation.

(c) If we define the deviation variables $\bar{h} = h - h_s$ and $\bar{F}_{in} = F_{in} - F_{out}$, it follows that

$$A\frac{d\bar{h}}{dt} = \bar{F}_{in} \tag{3.13.6}$$

from which we can observe that the outlet flowrate does not appear explicitly in the final deviation form of the dynamical equation (3.13.6).

(d) By taking the Laplace transform of Eq. (3.13.6) under zero initial condition, we obtain

$$sA\bar{H}(s) = \bar{F}_{in}(s) \tag{3.13.7}$$

and the corresponding transfer function is the following

$$G(s) = \frac{\bar{H}(s)}{\bar{F}_{in}(s)} = \frac{1}{As} \tag{3.13.8}$$

From the form of Eq. (3.13.8) it follows again that the system is first order, but its transfer function does not match the one given by Eq. (3.6.3). This is because the original differential equations do not match. Equation (3.13.6) belongs to the class of linear first-order differential equations of the following general form

$$\frac{dy}{dt} = bu \tag{3.13.9}$$

where b is a constant. A system of the form (3.13.9) is called an integrating system.

Case 3: The outflow is proportional to the square root of the liquid level, i.e.

$$F_{out} = c\sqrt{h} \tag{3.13.10}$$

where c is a constant. Using Eqs. (3.13.1) and (3.13.10) we obtain

$$A\frac{dh}{dt} = F_{in} - c\sqrt{h} \tag{3.13.11}$$

The system is again a first-order system but the differential equation is nonlinear and an analytic solution can be obtained only in a limited number of cases. We can, however, approximate the nonlinear term in Eq. (3.13.11) by a linear term and effectively transform the nonlinear differential equation to an approximately linear one.

For a nonlinear function $f(x)$ and a point x_0, we can define a linear approximation

$$f(x) \approx f(x_0) + \left.\frac{df}{dx}\right|_{x_0}(x - x_0) \tag{3.13.12}$$

3.13 Another Example of a First-Order System: Liquid Storage Tank

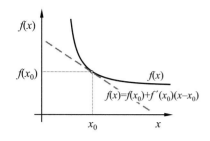

Figure 3.22 Tangent approximation of a nonlinear function around a point.

which is called tangent approximation and is shown in Figure 3.22. Consider what happens if we apply this idea to the surge tank example. First of all, we note that under steady-state conditions and using Eq. (3.13.10), it follows that $F_{out,s} = c\sqrt{h_s}$. We further consider an arbitrary state steady that is defined by the values of the inlet stream volumetric flowrate and liquid level ($F_{in,s}$, h_s). We apply the tangent approximation to the nonlinear term $\sqrt{h}$, to obtain

$$\sqrt{h} \approx \sqrt{h_s} + \left[\frac{d\sqrt{h}}{dh}\right]_{h=h_s} (h - h_s)$$

or

$$\sqrt{h} \approx \sqrt{h_s} + \left[\frac{1}{2\sqrt{h_s}}\right] (h - h_s) \tag{3.13.13}$$

Combining Eqs. (3.13.11) and (3.13.13) we obtain the approximate equation

$$A\frac{dh}{dt} = F_{in} - c\sqrt{h_s} - \left[\frac{c}{2\sqrt{h_s}}\right](h - h_s) \tag{3.13.14}$$

and taking into account that $F_{in,s} = c\sqrt{h_s}$, we have that

$$A\frac{d(h - h_s)}{dt} = (F_{in} - F_{in,s}) - \left[\frac{c}{2\sqrt{h_s}}\right](h - h_s) \tag{3.13.15}$$

We then define the deviation variables

$$\overline{h} = h - h_s, \quad \overline{F}_{in} = F_{in} - F_{in,s} \tag{3.13.16}$$

and Eq. (3.13.15) simplifies to

$$A\frac{d\overline{h}}{dt} = \overline{F}_i - \left[\frac{c}{2\sqrt{h_s}}\right]\overline{h} \tag{3.13.17}$$

which is a linear first-order differential equation with constant coefficients. Therefore, the use of the tangent approximation results in the same differential equation as in Case 1 with $R = 2\sqrt{h_s}/c$. The approach that has just been presented, the approximation of a nonlinear differential equation with a linear one, is called linearization.

Example 3.4 Comparison of the nonlinear and approximate linearized model of a liquid storage tank

A liquid stream of volumetric flowrate $F_{in} = 1$ m³/min is fed to a cylindrical liquid storage tank with cross-sectional area of 0.1 m². The liquid level at steady state is equal to $h = 1$ m. The volumetric flowrate of the outlet stream is proportional to the square root of the liquid level in the tank. Compare the predictions of the nonlinear model and its linear approximation under the following conditions:

(a) the system is initially at steady state and a step increase of 0.1 m³/min in the inlet volumetric flowrate takes place at $t = 0$,
(b) the system is initially at steady state and a step decrease of 0.9 m³/min in the inlet volumetric flowrate takes place at $t = 0$.

Solution

We assume that the dynamic behavior of the liquid storage tank is described by Eq. (3.13.11)

$$A\frac{dh}{dt} = F_{in} - c\sqrt{h}$$

From the given operating steady state, we can calculate the constant c

$$c = \frac{F_{in,s}}{\sqrt{h_s}} = \frac{1 \text{ m}^3/\text{min}}{1 \text{ m}^{1/2}} = 1\frac{\text{m}^{5/2}}{\text{min}}$$

The nonlinear differential equation can be written as

$$0.1\frac{dh}{dt} + \sqrt{h} = F_{in}$$

In case (a), we need to solve the following initial value problem

$$0.1\frac{dh}{dt} + \sqrt{h} = 1.1, \ h(0) = 1$$

whereas in case (b)

$$0.1\frac{dh}{dt} + \sqrt{h} = 0.1, \ h(0) = 1$$

The solution can be obtained numerically using e.g. MATLAB or Maple.
The linearized form of the differential equation (3.13.11) is the following (Eq. (3.13.17))

$$A\frac{d\bar{h}}{dt} = \bar{F}_{in} - \left[\frac{c}{2\sqrt{h_s}}\right]\bar{h}$$

or

$$0.1\frac{d\bar{h}}{dt} + \frac{1}{2}\bar{h} = \bar{F}_{in}$$

In case (a) we need to solve the following initial value problem

$$0.2\frac{d\bar{h}}{dt} + \bar{h} = 0.2, \quad \bar{h}(0) = 0$$

which has the following analytic solution (see Eq. (3.8.1) where $k = 2$, $M = 0.1$ and $\tau = 0.2$)

$$h(t) = 1 + 0.2\left(1 - e^{-t/0.2}\right)$$

In case (b) we obtain the following analytic solution ($k = 2$, $M = -0.9$ and $\tau = 0.2$)

$$h(t) = 1 - 1.8\left(1 - e^{-t/0.2}\right)$$

In Figures 3.23 and 3.24 the responses of the nonlinear and the approximate linear models are compared. We can observe that the closer the response is to the reference steady state, the better the approximation. When, however, the system moves away from the reference point, the error involved in the approximation becomes significant. It is important to observe that for case (b) the linear approximation predicts a new steady state where the liquid level becomes negative!

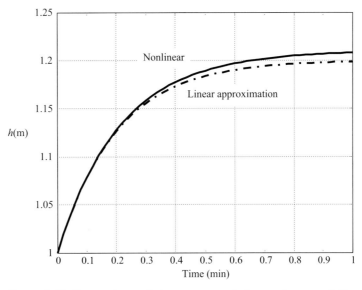

Figure 3.23 Comparison of the response of the nonlinear and linearized system to a small step input.

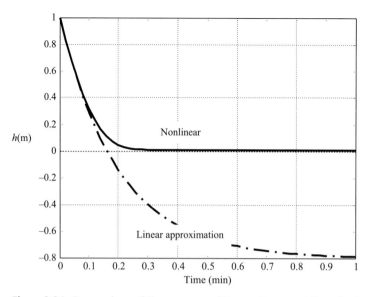

Figure 3.24 Comparison of the response of the nonlinear and linearized system to a large step input.

It is also important to compare the final values predicted from the nonlinear model:

$$h_s^{NL} = \left(\frac{F_{in} + M}{c}\right)^2 = \left(\sqrt{h(0)} + \frac{M}{c}\right)^2$$

and the linear approximation

$$h_s^L = h(0) + 2\sqrt{h(0)}\frac{M}{c}$$

We can observe that $h_s^{NL} > h_s^L$ and their difference is

$$e_s = h_s^{NL} - h_s^L = \left(\frac{M}{c}\right)^2$$

from which we conclude that the error becomes significant for large step changes in the inlet flowrate or small values of the model constant c.

3.14 Nonlinear First-Order Systems and their Linearization

Consider a nonlinear first-order system of the following general form:

$$\frac{dy}{dt} = f(y,u) \tag{3.14.1}$$

3.14 Nonlinear First-Order Systems and their Linearization

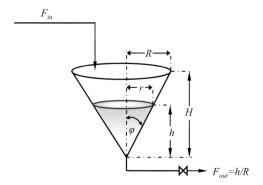

Figure 3.25 A cone-shaped liquid storage tank.

where u is the input, y is the output and $f(y,u)$ is a nonlinear function. Around a steady-state operating point denoted by (y_s, u_s), the f function can be approximated by the following linear function:

$$f(y,u) \approx f(y_s,u_s) + \left[\frac{\partial f}{\partial y}(y_s,u_s)\right](y-y_s) + \left[\frac{\partial f}{\partial u}(y_s,u_s)\right](u-u_s) \quad (3.14.2)$$

Since (y_s, u_s) is a steady state, it follows that

$$f(y_s,u_s) = 0 \quad (3.14.3)$$

Substituting Eq. (3.14.3) into Eq. (3.14.2), we obtain

$$f(y,u) \approx \left[\frac{\partial f}{\partial y}(y_s,u_s)\right](y-y_s) + \left[\frac{\partial f}{\partial u}(y_s,u_s)\right](u-u_s) \quad (3.14.4)$$

By defining the following deviation variables

$$\begin{aligned} \bar{y} &= y - y_s \\ \bar{u} &= u - u_s \end{aligned} \quad (3.14.5)$$

we finally obtain the linearized system

$$\frac{d\bar{y}}{dt} \approx \left[\frac{\partial f}{\partial y}(y_s,u_s)\right]\bar{y} + \left[\frac{\partial f}{\partial u}(y_s,u_s)\right]\bar{u} \quad (3.14.6)$$

Example 3.5 Model development and linearization for a cone-shaped liquid storage tank

Consider the cone-shaped liquid storage tank with height H (Figure 3.25). The outlet flowrate is given by $F_{out} = h/R$, where R is a constant. Develop a differential equation that describes the system dynamics and its linear approximation.

Solution

The volume of the liquid contained in a cone of height h is given by

$$V = \frac{1}{3}\pi r^2 h = \frac{1}{3}\pi h^3 \tan^2 \varphi$$

from which we obtain that

$$\frac{dV}{dh} = \pi h^2 \tan^2 \varphi$$

The mass balance is written as follows

$$\frac{dV}{dt} = \frac{dV}{dh}\frac{dh}{dt} = F_{in} - F_{out}$$

or

$$\pi h^2 \tan^2 \varphi \frac{dh}{dt} = F_{in} - \frac{h}{R}$$

Dividing by the coefficient that multiplies the derivative in the left-hand side, we can write

$$\frac{dh}{dt} = \frac{1}{\alpha h^2} F_{in} - \frac{1}{\alpha R h}$$

where $\alpha = \pi \tan^2 \varphi$. We finally define $u = F_{in}$ and $y = h$ and then

$$\frac{dy}{dt} = \frac{1}{\alpha y^2} u - \frac{1}{\alpha R y}$$

It should be observed that

$$f(y,u) = \frac{1}{\alpha y^2} u - \frac{1}{\alpha R y}$$

and therefore

$$\frac{\partial f(y,u)}{\partial y} = -\frac{2u}{\alpha y^3} + \frac{1}{\alpha R y^2}$$

$$\frac{\partial f(y,u)}{\partial u} = \frac{1}{\alpha y^2}$$

The linearized dynamics is given by Eq. (3.14.6), which in the present case is

$$\frac{d\bar{y}}{dt} = \left[-\frac{2u_s}{\alpha y_s^3} + \frac{1}{\alpha R y_s^2}\right]\bar{y} + \left[\frac{1}{\alpha y_s^2}\right]\bar{u}$$

But $y_s = h_s = RF_{in,s} = Ru_s$ hence the above equation simplifies to

$$\frac{d\bar{y}}{dt} = \left[-\frac{1}{\alpha y_s^2 R}\right]\bar{y} + \left[\frac{1}{\alpha y_s^2}\right]\bar{u}$$

3.15 Liquid Storage Tank with Input Bypass

The input variable to a system represents the "cause," while the output variable represents the "effect." It is not necessary for the output variable to be the dependent variable in the differential equation that describes the system dynamics. Consider the surge tank example and the case where the volumetric flowrate of the outlet stream is a linear function of the level of the liquid in the tank. The mathematical model of the tank is given by the following differential equation (see Eq. (3.13.3))

$$AR\frac{dh}{dt} + h = RF_{in} \tag{3.13.3}$$

where the input variable is the volumetric flowrate of the incoming stream ($u = F_{in}$). If the concern is to prevent possible overflow of the tank, then it is reasonable to select the height of the liquid in the tank as the output variable, i.e.

$$AR\frac{dh}{dt} + h = Ru \tag{3.15.1}$$

$$y = h \tag{3.15.2}$$

If the role of the storage tank is to dampen fluctuations in the volumetric flowrate and it is feeding a downstream unit, then it is reasonable to select $F_{out} = h/R$ as the output, i.e.

$$AR\frac{dh}{dt} + h = Ru \tag{3.15.3}$$

$$y = \frac{h}{R} \tag{3.15.4}$$

In this case, the output is proportional to the dependent variable h.

Consider now the liquid storage tank of Figure 3.26, where the inlet stream is split into two streams, one of which is fed to the tank while the other bypasses the tank and merges with its outlet stream. The mathematical description of this system is

$$A\frac{dh}{dt} + \frac{h}{R} = aF_{in} \tag{3.15.5}$$

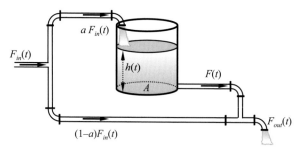

Figure 3.26 Liquid storage tank with bypass.

$$F_{out} = \frac{h}{R} + (1-a)F_{in} \qquad (3.15.6)$$

where a is the fraction of the feed stream that is actually fed to the tank. The volumetric flowrate of the incoming stream is again the input to the system ($u = F_{in}$). If we consider the case where the volumetric flowrate of the outgoing stream is the output of the system ($y = F_{out}$), then we have the following mathematical description

$$A\frac{dh}{dt} = -\frac{h}{R} + au \qquad (3.15.7)$$

$$y = \frac{h}{R} + (1-a)u \qquad (3.15.8)$$

Here the output of the system is a linear combination of the dependent variable h and the input of the system.

The height h of the liquid in the tank, which is the dependent variable in the differential equation, is called a state variable. The state variable is in essence an intermediate variable in the description of the relation between input and output. It is possible to eliminate the state variable h by taking the derivative of Eq. (3.15.8):

$$\frac{dy}{dt} = \frac{1}{R}\frac{dh}{dt} + (1-a)\frac{du}{dt} \qquad (3.15.9)$$

and then substituting Eq. (3.15.7), to obtain

$$AR\frac{dy}{dt} + y = u + (1-a)AR\frac{du}{dt} \qquad (3.15.10)$$

where the time derivative of the input signal appears in the right-hand side of Eq. (3.15.10). However, this elimination does not seem to offer any particular advantage.

Taking the Laplace transform of Eqs. (3.15.7) and (3.15.8), under zero initial condition, we obtain

$$AsH(s) = -\frac{H(s)}{R} + aU(s) \tag{3.15.11}$$

$$Y(s) = \frac{H(s)}{R} + (1-a)U(s) \tag{3.15.12}$$

where $H(s) = \mathcal{L}[h(t)]$. Equation (3.15.11) can be solved for $H(s)$:

$$H(s) = \frac{aR}{ARs+1} U(s) \tag{3.15.13}$$

and after substituting Eq. (3.15.13) into (3.15.12), we obtain

$$Y(s) = \frac{a}{ARs+1} U(s) + (1-a)U(s)$$

Hence

$$G(s) = \frac{Y(s)}{U(s)} = \frac{1 + (1-a)ARs}{ARs+1} \tag{3.15.14}$$

Equation (3.15.14) defines the transfer function of the liquid storage tank with bypass, in which the numerator is not constant, but depends on s (the presence of s comes from the derivative of the input signal – see Eq. (3.15.10)).

3.16 General Form of a First-Order System

The general form of a linear first-order system is the following

$$\frac{dx}{dt} = ax + bu \tag{3.16.1}$$

$$y = cx + du \tag{3.16.2}$$

where x is the state variable, u is the input variable and y is the output variable of the system. The differential equation

$$\tau \frac{dy}{dt} + y = ku \tag{3.16.3}$$

is a special case of the general description given by Eqs. (3.16.1) and (3.16.2), where

$$a = -\frac{1}{\tau}, \quad b = \frac{k}{\tau}, \quad c = 1, \quad d = 0 \tag{3.16.4}$$

The integrating system

$$\frac{dy}{dt} = bu \quad (3.16.5)$$

is also a special case with

$$a = 0,\ b = b,\ c = 1,\ d = 0 \quad (3.16.6)$$

In order to calculate the response of the general linear first-order system to an arbitrary input, we first multiply Eq. (3.16.1) by e^{-at} to obtain

$$e^{-at}\left(\frac{dx}{dt} - ax\right) = e^{-at}\,bu \Rightarrow \frac{d(e^{-at}x)}{dt} = e^{-at}\,bu$$

Integration from $t = 0$ ($x = x(0)$) to t gives

$$x(t) = e^{at}x(0) + \int_0^t e^{a(t-t')}bu(t')\,dt' \quad (3.16.7)$$

and finally

$$y(t) = ce^{at}x(0) + c\int_0^t e^{a(t-t')}bu(t')\,dt' + du(t) \quad (3.16.8)$$

The general form of a nonlinear first-order system is the following:

$$\frac{dx}{dt} = f(x,u) \quad (3.16.9)$$

$$y = h(x,u) \quad (3.16.10)$$

where f and h are nonlinear functions of x and/or u. When Eqs. (3.16.9) and (3.16.10) are linearized around a steady state, the resulting approximate linear system is expressed through Eqs. (3.16.1) and (3.16.2) where

$$a = \left.\frac{\partial f}{\partial x}\right|_{u=u_s,\,x=x_s},\ b = \left.\frac{\partial f}{\partial u}\right|_{u=u_s,\,x=x_s},\ c = \left.\frac{\partial h}{\partial x}\right|_{u=u_s,\,x=x_s},\ d = \left.\frac{\partial h}{\partial u}\right|_{u=u_s,\,x=x_s} \quad (3.16.11)$$

Transfer function calculation for the general linear first-order system: The Laplace transform of Eqs. (3.16.1) and (3.16.2) with zero initial condition is given by

$$sX(s) = aX(s) + bU(s) \quad (3.16.12a)$$

$$Y(s) = cX(s) + dU(s) \quad (3.16.12b)$$

from which we obtain

$$G(s) = \frac{Y(s)}{U(s)} = d + \frac{cb}{s-a} = \frac{ds + (cb - ad)}{s-a} \quad (3.16.13)$$

In most cases $d = 0$, and the transfer function is simplified to

$$G(s) = \frac{cb}{s-a} \quad (3.16.14)$$

It is important to note that under steady-state conditions ($dx/dt = 0$) from Eq. (3.16.1) it follows that

$$x_s = -a^{-1}bu_s \quad (3.16.15)$$

and substituting into Eq. (3.16.2) we obtain

$$y_s = ku_s \quad (3.16.16)$$

where

$$k = -ca^{-1}b + d \quad (3.16.17)$$

The last equation, which defines the static gain of the process, gives the relationship between the system input and output under steady-state conditions.

The time discretization of the general linear first-order system can be achieved by using the convolution integral in Eq. (3.16.7), which can be discretized in exactly the same way as it was done in Section 3.12 for the special form of first-order system. The discretized state equation can be combined with the algebraic output equation (3.16.2) in discrete form:

$$y[j] = c\, x[j] + d\, u[j] \quad (3.16.18)$$

The end results are the following.
Forward rectangle approximation:

$$\begin{aligned} x[j] &= e^{aT_s}\, x[j-1] + \left(\int_0^{T_s} e^{at'} b\, dt' \right) u[j-1] \\ y[j] &= c\, x[j] + d\, u[j] \end{aligned} \quad (3.16.19)$$

Backward rectangle approximation:

$$\begin{aligned} x[j] &= e^{aT_s}\, x[j-1] + \left(\int_0^{T_s} e^{at'} b\, dt' \right) u[j] \\ y[j] &= c\, x[j] + d\, u[j] \end{aligned} \quad (3.16.20)$$

3.17 Software Tools

3.17.1 Iterative Calculation of the Response of a Discretized First-Order System

The discretized form of the linear first-order system derived in Section 3.12 is particularly easy to apply in any software environment, and it is particularly convenient for real time simulation. The recursion (3.12.8) can be implemented as follows.

Given an array $u(j)$, $j = 1, \ldots, (N_{final} +1)$ that represents the values of the input at times $t = 0, \ldots, (j-1) \cdot T_S, \ldots, N_{final} \cdot T_S$, and an initial condition $y(1)$ that represents the given value of the output at time $t = 0$, the array of values of the output $y(j)$, $j = 2, \ldots, (N_{final} + 1)$, can be calculated by executing the following for loop in MATLAB:

```
y(1)=y0;
for    j = 2:Nfinal+1
       y(j) = exp(-Ts/tau)*y(j-1)+(1-exp(-Ts/tau))*k*u(j-1);
end
```

or the following do loop in Maple:

```
> y[1]:=y0:
> for j from 2 to Nfinal+1 do
> y[j]:=exp(-Ts/tau)*y[j-1]+(1-exp(-Ts/tau))*k*u[j-1];
> end do:
```

This looping procedure was used to calculate the solution of Example 3.3.

Of course, many other numerical methods are available, and many software packages are available that include ODE solvers. This will be seen in the next subsections.

3.17.2 Numerical Calculation of the Response of a First-Order System using the ODE Solvers in MATLAB

The linear first-order differential with constant coefficients given by Eq. (3.3.1) can also be written in the following form

$$\frac{dy(t)}{dt} = -\frac{1}{\tau}y(t) + \frac{k}{\tau}u(t) \tag{3.17.1}$$

In order to obtain the unit step response we create the following m-file in the current folder (or any folder in the MATLAB path)

```
function dydt=first_order(t,y)
% Prototype linear first order system
% Parameters: tau = time constant
%             k   = process gain
% define model constants
tau = 1;
```

```
k = 1;
% define input
u = 1; % unit step input
dydt = -(1/tau)*y+(k/tau)*u;
```

In the command window we only need to type

```
»ode45(@first_order,[0 10],0)
```

in order to obtain the step response shown in Figure 3.8. It should be noted that the structure of the ode commands available in MATLAB is

```
[t, y] = ode45(@odefilename,tspan,y0)
```

where

t is the independent variable, i.e. the time in most cases
y is the dependent variable(s)
tspan = [t0 tf] is the range of t (t0 is the initial value and tf is the final value of t) and
y0 is a initial value of the dependent variable.

In obtaining the step response we have used t0 = 0, tf = 10, y0 = 0.

To generate the unit pulse response with $\varepsilon = 0.001$ we modify the lines in the m-file first_order that define the input to the system as follows

```
% define input
% epsilon is the duration of the pulse
epsilon=0.001;
u = 1/epsilon;
if (t>epsilon)
        u=0;
end
```

and then use the ode45 ODE solver to obtain the solution in exactly the same way as we did in the unit step response. For the unforced response with initial condition $y(0) = 1$ we set the input equal to zero at all times in the m-file first_order and then in the command window we type

```
»ode45(@first_order,[0 10],1)
```

to obtain the response shown in Figure 3.5.

MATLAB features specific commands that can be used to obtain the response of linear systems to prototype inputs and these commands will be presented in the chapters that follow.

The same approach is followed when the system is a nonlinear first-order system. If we consider the case of the liquid storage tank studied in Section 3.13 where the volumetric flowrate of the outlet is proportional to the square root of the liquid level in the tank, the dynamics of the liquid level is described by Eq. (3.13.11) and the following m-file can be used to simulate the system dynamics:

```
function dhdt = tank(t,h)
% liquid level dynamics
% model constants
c=1;
A=0.1;
% input variation
Fin = 1.1;
dhdt = (Fin-c*sqrt(h))/A;
```

If we then type the following in the command window

```
» [t,x]=ode45(@tank,[0 1],1,[]);
» plot(t,x)
```

then we obtain the response shown in Figure 3.23.

3.17.3 Symbolic Calculation of the Response of a Linear First-Order System using Maple

All analytical calculations in the present chapter can be performed using symbolic computing. As an example, we will now show how to derive the ramp response formula (3.10.1) using Maple. The first method is through the convolution integral formula (3.6.1):

$$y(t) = \int_0^t g(t-t')u(t')\, dt'$$

where $g(t) = \dfrac{k}{\tau}e^{-t/\tau}$ and where, for the ramp input, $u(t) = Mt$. This calculation can be performed with Maple as follows. First, define the functions $u(t)$ and $g(t)$

```
> u(t):=M*t:
> g(t):=(k/tau)*exp(-t/tau):
```

Make the appropriate substitutions to form $u(t')$ and $g(t-t')$ and then integrate the product $g(t-t')u(t')$ from 0 to t using int command.

```
> u(tprime):=subs(t=tprime,u(t)):
> g(t-tprime):=subs(t=t-tprime,g(t)):
> y(t):=int(g(t-tprime)*u(tprime),tprime=0..t);
```

The result is

$$y(t) := k\, M\left(\tau - \tau e^{\frac{t}{\tau}} + t e^{\frac{t}{\tau}}\right) e^{-\frac{t}{\tau}}$$

which is equivalent to formula (3.10.1).

Symbolic calculation of the convolution integral for other standard inputs can be performed in exactly the same way, by just changing the line that defines the input function: u(t):=M: for a step input, u(t):=M*sin(omega*t): for a sine wave input, etc.

Alternatively, Maple can perform calculations in the Laplace domain. For a given input function $u(t)$, its Laplace transform $U(s)$ is calculated symbolically, this is multiplied by the transfer function $k/(\tau s + 1)$ to give the Laplace transform of the output $Y(s)$ and, finally, through inversion of the Laplace transform, the output response $y(t)$ is calculated. To perform Laplace transformation and inverse Laplace transformation symbolically, one must call the inttrans library at the beginning of the Maple program:

```
> with(inttrans):
```

Then, Laplace transformation is performed through the laplace command and inverse Laplace transformation through the invlaplace command. For example, the ramp response calculation goes as follows:

```
> u(t):=M*t:
> U(s):=laplace(u(t),t,s);
```

$$U(s) := \frac{M}{s^2}$$

```
> Y(s):=(k/(tau*s+1))*U(s);
```

$$Y(s) := \frac{k\,M}{(\tau s + 1)\,s^2}$$

```
> y(t):=invlaplace(Y(s),s,t);
```

$$y(t) := k\,M\left(t - \tau\left(1 - e^{-\frac{t}{\tau}}\right)\right)$$

Again Maple reproduces the result of formula (3.10.1).

3.17.4 Numerical Calculation of the Response of a Nonlinear First-Order System using an ODE Solver in Maple

The numerical solution of the nonlinear differential equation (3.13.11) that describes the dynamics of a liquid storage tank can be obtained with Maple as follows. First, define the ordinary differential equation to be solved (the derivative $dx(t)/dt$ can be written as diff(x(t),t), using the diff command):

```
> c:=1:A:=0.1:Fin:=1.1:
> ode:=diff(x(t),t)=(Fin-c*sqrt(x(t)))/A:
```

Solve the ordinary differential equation numerically, using the dsolve command, specifying the initial condition and the range of the independent variable over which the solution should be computed

```
> p:=dsolve({ode,x(0)=1},numeric,range=0..1):
```
Plot the result, using the `odeplot` command of the `plots` package

```
> with(plots): odeplot(p);
```

In this way, we obtain the result shown in Figure 3.23.

LEARNING SUMMARY

- Many chemical process systems can be modeled using a linear first-order differential equation of the following form:

$$\tau \frac{dy(t)}{dt} + y(t) = ku(t), \ y(0) = \text{known}$$

- τ is the time constant of the system, a measure of the speed of response, and k is the static gain of the system that relates the steady-state values of the input and output.
- The unforced response and the forced response of a first-order system to a number of standard input variations are summarized in Table 3.2.
- The transfer function $G(s)$ is the ratio of the Laplace transform of the output to the Laplace transform of the input for zero initial condition and for first-order systems is given by Eq. (3.6.3)

Table 3.2 Summary of output response formulas

	Input	Output
Unforced response	$u(t) = 0$	$y(t) = e^{-t/\tau} y(0), \ t \geq 0$
Forced response with zero initial condition: $y(0) = 0$	Step $u(t) = 0, \ t < 0$ $u(t) = M, \ t \geq 0$	$y(t) = kM(1 - e^{-t/\tau}), \ t \geq 0$
	Pulse of magnitude M and duration ε	$y(t) = \begin{cases} \dfrac{kM}{\varepsilon}\left(1 - e^{-t/\tau}\right), & 0 \leq t < \varepsilon \\ k\dfrac{M}{\varepsilon}\left(1 - e^{-\varepsilon/\tau}\right)e^{-(t-\varepsilon)/\tau}, & t \geq \varepsilon \end{cases}$
	Impulse of magnitude M $u(t) = M\,\delta(t)$	$y(t) = \dfrac{kM}{\tau} e^{-t/\tau}, \ t > 0$
	Ramp $u(t) = 0, \ t < 0$ $u(t) = Mt, \ t \geq 0$	$y(t) = kM(t - \tau + \tau e^{-t/\tau}), \ t \geq 0$
	Frequency response	$y(t) = \dfrac{kM}{\sqrt{1 + (\omega\tau)^2}} \sin(\omega t + \varphi), \quad \varphi = -\tan^{-1}(\omega\tau)$

- The impulse response function $g(t)$ of a linear system (output response to a unit impulse input) is the inverse Laplace transform of its transfer function $G(s)$.
- The response to any arbitrary input under zero initial condition can be calculated as the convolution of the impulse response function and the input, or equivalently as the inverse Laplace transform of the product of the transfer function times the Laplace transform of the input.
- The response of a linear first-order system to an arbitrary input can be calculated numerically by appropriately discretizing the convolution integral.
- The most general form of a linear first-order system is the following

$$\frac{dx}{dt} = ax + bu$$

$$y = cx + du$$

and its response to an arbitrary input is given by

$$x(t) = e^{\alpha t} x(0) + \int_0^t e^{\alpha(t-t')} bu(t') dt'$$

$$y(t) = ce^{\alpha t} x(0) + c\int_0^t e^{\alpha(t-t')} bu(t') dt' + du(t)$$

TERMS AND CONCEPTS

Amplitude ratio. The ratio of the amplitude of the output to the amplitude of the input in the long-time response of a linear system to sinusoidal input variation.

Bode diagrams. Plots of amplitude ratio versus frequency in log × log axes and phase angle versus frequency in lin × log axes, for the frequency response of a linear system.

Deviation variables. Variables defined as differences from a reference steady-state value of the variable.

First-order system. Any dynamical system described by a first-order differential equation (and possibly algebraic equations).

Forced response. The response of a system to a nonzero variation in the input variable.

Frequency response. The long-time response of a system to a sinusoidal input variation.

Impulse response. The limit of the pulse response as the duration of the pulse tends to zero, while preserving its magnitude.

Linear system. A dynamical system that is described by a linear differential equation or a set of linear differential (and possibly linear algebraic) equations. A linear system satisfies the principle of superposition.

Pulse response. The response of a system to a pulse input variation.

Sinusoidal response. The response of a system to sinusoidal input variation.

Static gain or steady-state gain. The multiplicative factor that relates the output value and the input value of a linear system at steady state. Equivalently it is ratio of Δ(output steady state) divided by Δ(input steady state).

Step response. The response of a system to a step change in the input.

Time constant. A measure of the speed of response of a linear system. For first-order linear systems, it is the time needed to reach 63.2% of the final value of the step response.

Unforced response or natural response or free response. The response of a dynamic system under zero input (or constant input if deviation variables are introduced in reference to the steady state corresponding to the value of the input).

FURTHER READING

A large number of additional systems that exhibit first-order dynamics can be found in the book

Ingham, J., Dunn, I. J., Heinzle, E. and Prenosil, J. E., *Chemical Engineering Dynamics, An Introduction to Modelling and Computer Simulation*, 2nd edn. Weinheim: Wiley-VCH, 2000.

Additional material on the solution of first-order differential equations and Laplace transform can be found in

Boyce W. E. and DiPrima, R. C., *Elementary Differential Equations and Boundary Value Problems*, 7th edn. New York: John Wiley and Sons, 2001.

Kreyszig, E., *Advanced Engineering Mathematics*, 10th edn. New Jersey: John Wiley & Sons, 2011.

Further analysis of first-order systems dynamics can be found in the books

Luyben, W. L. and Luyben, M. L., *Essentials of Process Control*. New York: McGraw Hill, 1997.

Ogata, K., *Modern Control Engineering, Pearson International Edition*, 5th edn, 2008.

PROBLEMS

3.1 Consider the constant-volume isothermal CSTR example of Section 3.1.2.
   (1) Calculate the transfer function between the concentration of the reactant in the feed stream and within the reactor.
   (2) If the reactor initially operates at steady state and suddenly there is a step change in the concentration of the reactant in the feed stream, calculate the response of the concentration in the reactor.
   (3) If the reactor operates at a different (constant) value of feed flow rate F, how would this affect the steady state and dynamic characteristics of the reactor?

3.2 Consider the constant-volume isothermal CSTR example of Section 3.1.2, but now with nth-order reaction kinetics:

$$V\frac{dC_R}{dt} = FC_{in,R} - FC_R - Vk_R C_R^n$$

(a) Linearize the system around a given reference steady state (C_{in,R_s}, C_{R_s}).
(b) Derive the transfer function of the CSTR.
(c) If the CSTR is initially at the reference steady state, but suddenly the feed concentration $C_{in,R}$ undergoes a small step change of size M, derive an approximate formula for the response of $C_R(t)$.

3.3 Calculate the response of a first-order system (3.3.1) with zero initial condition to the following variation in the input.

$$u(t) = \begin{cases} 0, & t < 0 \\ at, & 0 \le t < \dfrac{M}{a} \\ M, & t \ge \dfrac{M}{a} \end{cases}$$

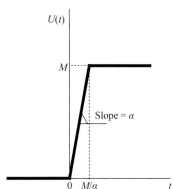

What happens in the limit as $a \to \infty$?

Hint: You can follow a similar approach to that in the derivation of the pulse response. Observe that $u(t) = at\mathcal{H}(t) - a\left(t - \dfrac{M}{a}\right)\mathcal{H}\left(t - \dfrac{M}{a}\right)$

3.4 A mixing tank initially operates with pure water. At time $t = 0$ a solution of a salt enters the tank. The feed concentration of the salt is given by $C_{in}(t) = C_0 e^{-\gamma t}$, where C_0 and γ are constants. Assuming that the tank operates under constant flowrate F and constant volume V, calculate the concentration of salt $C(t)$ in the tank.

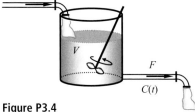

Figure P3.4

Hint: You need to distinguish two cases: (a) $\gamma \ne F/V$ and (b) $\gamma = F/V$.

3.5 Derive the transfer function for the hemispherical liquid storage tank shown in Figure P3.5 between F_{in} and h.

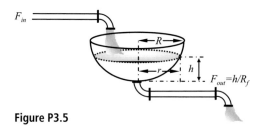

Figure P3.5

3.6 Consider the gas storage tank shown in Figure P3.6. The volumetric flowrate of the incoming gas (F_0) is determined by upstream processes. The volumetric flowrate of the outgoing stream through the control valve can be calculated by the following equation (which calculates the volumetric flowrate upstream of the valve)

$$F_1 = f(u) \, C_v \sqrt{P_1 - P_2}$$

where u is the fractional opening of the valve in the output pipe, $f(u)$ is a nonlinear function of u (for which $f(0) = 0$ and $f(1) = 1$ always hold true), C_v is a constant characteristic to the valve (with appropriate units), P_1 is the pressure in the tank and P_2 is the downstream pressure. The temperature is constant at T and the volume of the tank is V_1. Derive the transfer function between F_0 and P_1. Assume ideal behavior of the gas and that the temperature of the gas is approximately constant.

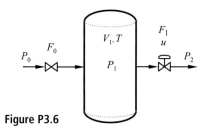

Figure P3.6

3.7 Consider a first-order system (3.3.1) that is initially at steady state. At $t = 0$ the input varies in the way shown in Figure P3.7a. The input shown in this figure is periodic and is known as square wave of amplitude M and half-period γ. The output of the system is also periodic for large times as shown in Figure P3.7b. Calculate an analytic expression for the output response for large times.

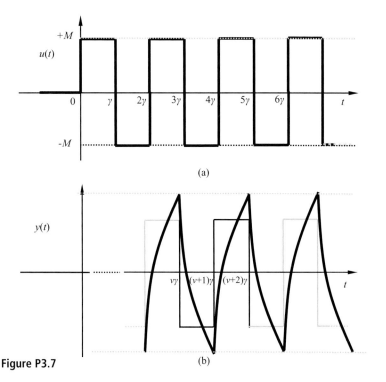

Figure P3.7

3.8 A first-order system (3.3.1) with zero initial condition is subject to the input

$$u(t) = \begin{cases} 0, & t \leq 0 \\ t, & 0 \leq t \leq 2 \\ -t+4, & 2 \leq t \leq 4 \\ 0, & t \geq 4 \end{cases}$$

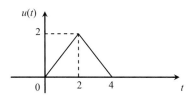

Calculate the response of the output. Plot your result for $k = 1$, $\tau = 0.5$.

3.9 A thermometer, initially at room temperature of 20 °C, is placed in a bath of temperature 50 °C for exactly 1 min, and then it is brought back to 20 °C. The thermometer reading is shown in Figure P3.9. Calculate the time constant of the thermometer.

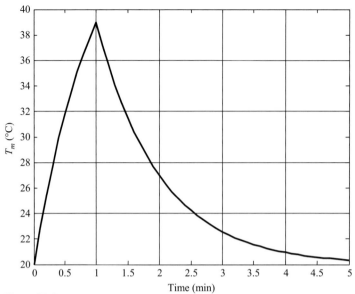

Figure P3.9

3.10 A thermometer, initially at room temperature of 20 °C, is placed in a liquid whose temperature varies sinusoidally, according to the equation $T_L(t) = 20+\sin(2t)$, in °C. The thermometer reading is recorded and shown in Figure P3.10. Calculate the time constant of the thermometer.

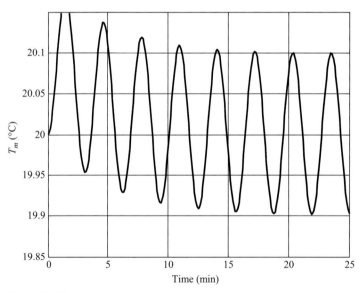

Figure P3.10

Table P3.11

Frequency (rad/s)	Amplitude ratio	Frequency (rad/s)	Amplitude ratio
0.01	10	10.0	7.04
0.10	9.99	15.0	5.55
1.0	9.95	20.0	4.47
5.0	8.94	40.0	2.43
7.0	8.19	50.0	1.96

3.11 In Table P3.11 experimental data of the amplitude ratio as a function of the frequency of the sinusoidal input variation are given. Using these data, determine the parameters k and τ of the system, assuming first-order dynamics of the form (3.3.1).

3.12 A thermometer, initially at room temperature of 20 °C, is placed in a liquid whose temperature varies according to the equation $T(t) = 20 + 5e^{-t}$, with T in °C, t in min. If the thermometer time constant is 10 s, calculate the thermometer reading as a function of time. Plot the thermometer reading and the actual temperature in a common plot.

3.13 A first-order system with time constant $\tau = 2$ and static gain $k = 0.5$ is initially at steady state with the input $u = 2$ and the output $y = 1$.

At $t = 0$, the input is stepped up to the value of $u = 8$, and remains constant up to $t = 5$, at which time it is stepped down to $u = 4$, and remains at that value afterwards.

Assuming that the input is given in discrete form, sampled every $T_s = 0.2$, write a simple MATLAB code to calculate the corresponding output values using the forward rectangle discretization. Plot your discrete input sequence and your calculated discrete output sequence over the range $0 \leq t \leq 12$.

3.14 Consider the liquid storage tank with constant outflow, described in deviation form by Eq. (3.13.6). If the tank is initially at steady state ($\bar{F}_{in} = 0$, $\bar{h} = 0$) and suddenly the inlet flow rate undergoes a pulse change, calculate the height in deviation form. Will it return to zero as $t \to \infty$?

3.15 Consider an integrating first-order system (3.13.9) under a sinusoidal input variation with frequency ω and amplitude M. Calculate the response of the output of the system. Does it oscillate with constant amplitude? At what frequency? Is it in phase with the input? What is the average value of the output? Can you draw Bode diagrams for the integrating system?

3.16 Discretize the integrating system (3.16.5), via forward rectangle and backward rectangle approximations.

3.17 Consider the circuit of Figure 3.17b. Using principles of electrical engineering, this can be modeled with the following equations:

$$RC\frac{dV_C}{dt} + V_C = V_{in}$$
$$V_{out} = V_{in} - V_C$$

where V_{in} is the inlet voltage (input variable), V_C is the voltage drop across the capacitor (state variable), V_{out} is the outlet voltage (output variable), R is the resistance and C is the capacitance (constant parameters).

(1) Calculate the transfer function of the circuit.
(2) If a step change in the inlet voltage V_{in} takes place under zero initial condition, calculate the response of the outlet voltage V_{out}.
(3) If a sinusoidal change in the inlet voltage V_{in} takes place, calculate response of the outlet voltage V_{out} for large t. Also, calculate amplitude ratio (AR) as a function of the frequency ω and plot AR versus $RC\omega$ in log×log scale. Is the name "high-pass filter" well justified?

Hint for question (3): First calculate $V_C(t)$ and then $V_{out}(t) = V_{in}(t) - V_C(t)$.

COMPUTATIONAL PROBLEMS

3.18 Consider the flash distillation unit studied in Chapter 2, Section 2.5.1 (Figure 2.10). The feed to the unit is $F = 100$ kmol/h and contains a binary mixture in which $z = 0.5$. The liquid holdup in the tank is constant and equal to $N_L = 10$ kmol. The vapor and liquid product streams have equal molar flowrates $V = L = 50$ kmol/h. Ideal vapor–liquid equilibrium of the mixture with constant relative volatility is valid with $a = 6$. At $t = 0$ there is a sudden drop in the composition of the volatile component in the feed from $z = 0.5$ to $z = 0.4$. Derive the nonlinear model of the unit and determine the steady-state solution. Use MATLAB to determine the response of the liquid composition to the step change in the feed composition. Derive the transfer function between z and x and calculate the response of the linearized system. Compare the response of the nonlinear and linearized systems.

3.19 Consider the SOFC model studied in Section 2.4.4 and Problem 2.13. Derive the transfer functions between the current density i and (a) the mole fraction of hydrogen in the fuel channel, (b) the mole fraction of oxygen in the air channel and (c) the output voltage. Calculate the unit step response of the system with $i(t)$ as the input and $y_{H_2}(t)$, $y_{O_2}(t)$ and $V(t)$ as the outputs.

4 Connections of First-Order Systems

A higher-order system can, in some cases, be composed of interconnected first-order systems. For example, two first-order systems connected in series form a second-order system. In this chapter we will examine

- first-order systems connected in series
- first-order systems connected in parallel
- interacting first-order systems

and we will study the properties of the overall system. Higher-order systems that are not composed of first-order systems connected with one another will be examined in the chapters that follow.

STUDY OBJECTIVES

After studying this chapter, you should be able to do the following.

- Identify series, parallel and interacting connections of dynamic systems.
- Derive the transfer function of the overall system from the transfer functions of its components, when they are connected in series or in parallel.
- Calculate the response of systems resulting from the series or parallel connection of first-order systems.
- Calculate and simulate the response of dynamic systems arising from the connection of first-order systems, using software tools like MATLAB or Maple.

4.1 First-Order Systems Connected in Series

In Figure 4.1 two liquid storage tanks connected in series are shown. The outlet stream from the first tank is fed to the second tank and therefore the first tank affects the second. The second tank, however, does not affect the operation of the first tank and therefore we say that the two tanks are connected in series. Our aim is to study the dynamics of the overall

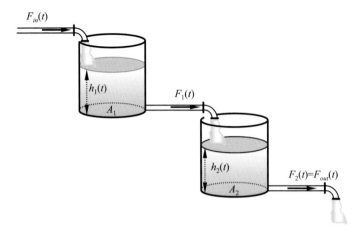

Figure 4.1 Two liquid storage tanks connected in series.

system including the effect of the volumetric flowrate of the feed to the volumetric flowrate of the exit stream. To this end we assume that the fluid has constant density and the two tanks constant (but not necessarily equal) cross-sectional areas. To simplify the presentation, we also assume that the volumetric flowrate of the outflow from each tank is proportional to the liquid level in the tank

$$F_1 = \frac{h_1}{R_1}, \quad F_2 = \frac{h_2}{R_2} \qquad (4.1.1)$$

where R_1, R_2 are the resistances. Application of the mass balance equation for each tank, assuming constant density, gives

$$A_1 \frac{dh_1}{dt} = F_{in} - F_1 \qquad (4.1.2)$$

$$A_2 \frac{dh_2}{dt} = F_1 - F_2 \qquad (4.1.3)$$

F_1 and F_2 are given by Eq. (4.1.1) and we then write

$$A_1 \frac{dh_1}{dt} = F_{in} - \frac{h_1}{R_1} \qquad (4.1.4)$$

$$A_2 \frac{dh_2}{dt} = \frac{h_1}{R_1} - \frac{h_2}{R_2} \qquad (4.1.5)$$

For the output of the system of tanks, we consider the volumetric flowrate of the stream exiting the system of two tanks $F_{out} = F_2$, i.e.

$$F_{out} = \frac{h_2}{R_2} \qquad (4.1.6)$$

Equations (4.1.4)–(4.1.6) provide a *state-space model* for the system of two tanks, with the input being the inlet flowrate F_{in}, the states being the liquid levels h_1 and h_2, and the output being the exit flowrate F_{out}. Alternatively, one can use the flowrates F_1 and F_2 as states, and express the state-space model as follows:

$$A_1 R_1 \frac{dF_1}{dt} = F_{in} - F_1 \tag{4.1.7}$$

$$A_2 R_2 \frac{dF_2}{dt} = F_1 - F_2 \tag{4.1.8}$$

$$F_{out} = F_2 \tag{4.1.9}$$

The state-space model (4.1.7)–(4.1.9) is completely equivalent to the one given by Eqs. (4.1.4)–(4.1.6), the only difference being that they use different state variables.

From Eq. (4.1.7) we observe that F_1 does not depend on F_2 and, furthermore, if F_{in} is known, the differential equation (4.1.7) can be solved independently without the knowledge of F_2. In order, however, to solve the differential equation (4.1.8) and determine F_2, we need to have already solved the differential equation for F_1. It is therefore clear that the physical connection of the two tanks and the fact that the first one affects the second, but not vice versa, is reflected in the structure of the differential equations that describe their dynamics. The serial structure of the tanks permits a sequential solution of the differential equations.

This serial structure will also become apparent in the Laplace domain. First of all, we observe that, at steady state, Eqs. (4.1.7)–(4.1.9) give

$$0 = F_{in,s} - F_{1,s} \tag{4.1.10}$$

$$0 = F_{1,s} - F_{2,s} \tag{4.1.11}$$

$$F_{out,s} = F_{2,s} \tag{4.1.12}$$

Subtracting Eqs. (4.1.10)–(4.1.12) from (4.1.7)–(4.1.9) and defining the following deviation variables

$$\bar{F}_{in} = F_{in} - F_{in,s} \tag{4.1.13}$$

$$\bar{F}_1 = F_1 - F_{1,s} \tag{4.1.14}$$

$$\bar{F}_2 = F_2 - F_{2,s} \tag{4.1.15}$$

$$\bar{F}_{out} = F_{out} - F_{out,s} \tag{4.1.16}$$

the system may be written in deviation form as follows:

$$A_1 R_1 \frac{d\bar{F}_1}{dt} = \bar{F}_{in} - \bar{F}_1 \tag{4.1.17}$$

$$A_2 R_2 \frac{d\bar{F}_2}{dt} = \bar{F}_1 - \bar{F}_2 \tag{4.1.18}$$

$$\bar{F}_{out} = \bar{F}_2 \tag{4.1.19}$$

Taking the Laplace transform of Eqs. (4.1.17)–(4.1.19) under zero initial conditions, we obtain the transfer-function description of each tank:

$$G_1(s) = \frac{\bar{F}_1(s)}{\bar{F}_{in}(s)} = \frac{1}{A_1 R_1 s + 1} \tag{4.1.20}$$

$$G_2(s) = \frac{\bar{F}_2(s)}{\bar{F}_1(s)} = \frac{1}{A_2 R_2 s + 1} \tag{4.1.21}$$

The overall transfer function between the input and the output of the system can easily be calculated by direct multiplication of G_1 and G_2

$$G(s) = G_1(s) G_2(s) = \frac{\bar{F}_{out}(s)}{\bar{F}_{in}(s)} = \frac{1}{(A_1 R_1 s + 1)} \cdot \frac{1}{(A_2 R_2 s + 1)} \tag{4.1.22}$$

We observe that when two systems are connected in series, the overall transfer function is equal to the product of the individual transfer functions. Figure 4.2 depicts the block diagram for the system of two tanks in series under consideration.

In general, in the serial connection of n systems with transfer functions $G_1(s)$, $G_2(s)$, … $G_n(s)$, the overall transfer function is:

$$G(s) = G_1(s) G_2(s) \cdots G_n(s) \tag{4.1.23}$$

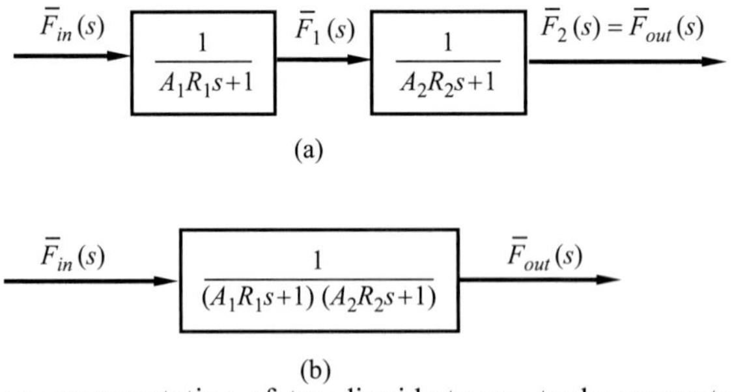

Figure 4.2 Block-diagram representation of two liquid storage tanks connected in series: (a) with the individual subsystems shown (b) overall system.

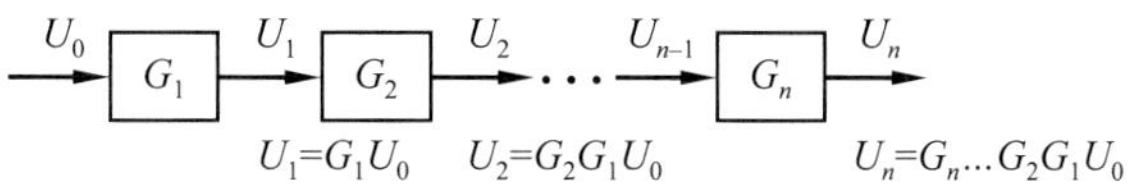

$$U_0 \to \boxed{G_1} \xrightarrow{U_1} \boxed{G_2} \xrightarrow{U_2} \cdots \xrightarrow{U_{n-1}} \boxed{G_n} \xrightarrow{U_n}$$

$$U_1 = G_1 U_0 \quad U_2 = G_2 G_1 U_0 \quad\quad U_n = G_n \cdots G_2 G_1 U_0$$

Figure 4.3 Block-diagram representation of n systems connected in series.

The corresponding block-diagram representation is shown in Figure 4.3. Note that in the block-diagram representation, the output of each block is obtained by multiplying the transfer function of the block with the input. Both the input and the output of the block (denoted by simple arrows) indicate a signal and not flow of mass (as is the case in the commonly used process flow diagrams in plant design).

4.2 First-Order Systems Connected in Parallel

We consider the physical system shown in Figure 4.4. The feed to the system has volumetric flowrate F_{in} and is split in two streams. The fraction of the feed that is fed to the first of the two tanks is a and the fraction that is fed to the second tank is $(1-a)$. The two tanks do not interact in any way. The streams that come out from the two tanks are merged to form the exit stream F_{out}. If we assume that the volumetric flowrate of each of the two product streams is given by Eq. (4.1.1), then we can write the mass balances for the two tanks as follows:

$$A_1 \frac{dh_1}{dt} = aF_{in} - \frac{h_1}{R_1} \tag{4.2.1}$$

$$A_2 \frac{dh_2}{dt} = (1-a)F_{in} - \frac{h_2}{R_2} \tag{4.2.2}$$

Alternatively, the mass balances may be written using the flow rates as dependent variables:

$$A_1 R_1 \frac{dF_1}{dt} = aF_{in} - F_1 \tag{4.2.3}$$

$$A_2 R_2 \frac{dF_2}{dt} = (1-a)F_{in} - F_2 \tag{4.2.4}$$

The volumetric flowrate F_{out} of the combined stream can be calculated as the sum of the individual flowrates, i.e.

$$F_{out} = F_1 + F_2 \tag{4.2.5}$$

Equations (4.2.3)–(4.2.5) provide a state-space model for the system of two tanks, with the input being the inlet flowrate F_{in}, the states being the flowrates F_1 and F_2, and the output being the exit flowrate F_{out}.

It is important to observe that each one of the differential equations (4.2.3) and (4.2.4) depend only on a and F_{in} and they do not involve any interaction terms. Therefore, they can

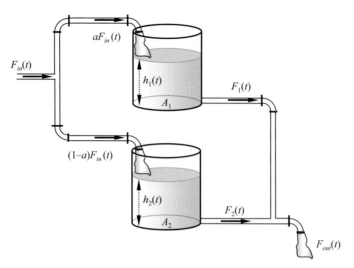

Figure 4.4 Two liquid storage tanks connected in parallel.

be solved independently and their solutions can be combined (added) to determine the exit stream's volumetric flowrate. The structure of the physical system is again reflected in the structure of the mathematical equations that describe its dynamics.

This feature is also directly visible in the Laplace domain. If we take the Laplace transformation of the differential equations (4.2.3) and (4.2.4) converted to deviation form, under zero initial conditions, we obtain:

$$G_1(s) = \frac{\overline{F}_1(s)}{\overline{F}_{in}(s)} = \frac{a}{A_1 R_1 s + 1} \tag{4.2.6}$$

$$G_2(s) = \frac{\overline{F}_2(s)}{\overline{F}_{in}(s)} = \frac{1-a}{A_2 R_2 s + 1} \tag{4.2.7}$$

The overall transfer function that connects the input F_{in} and the output F_{out} can be derived by observing that

$$G(s) = \frac{\overline{F}_{out}(s)}{\overline{F}_{in}(s)} = \frac{\overline{F}_1(s) + \overline{F}_2(s)}{\overline{F}_{in}(s)} = \frac{\overline{F}_1(s)}{\overline{F}_{in}(s)} + \frac{\overline{F}_2(s)}{\overline{F}_{in}(s)} = G_1(s) + G_2(s) \tag{4.2.8}$$

and therefore

$$G(s) = \frac{a}{A_1 R_1 s + 1} + \frac{1-a}{A_2 R_2 s + 1} \tag{4.2.9}$$

Figure 4.5 depicts the block diagram for the system of two tanks in parallel under consideration. In the block diagram shown in Figure 4.5 two types of signal manipulations have been introduced that will be used extensively in this book and thus deserve some discussion. The first point is a summing (or summation) point, which will be denoted by a circle. There

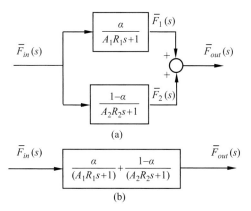

Figure 4.5 Block-diagram representation of two liquid storage tanks connected in parallel: (a) with the individual subsystems shown and (b) the overall system.

may be two or more incoming signals to a summation point but only one outgoing signal. The outgoing signal is obtained by performing a combination of algebraic addition or subtraction of all incoming signals depending on the sign indicated next to each arrow (addition for a positive sign and subtraction for a negative sign). A few examples of summing points are shown in Figure 4.6a.

The take-off point is a point from which the same input signal can be passed through more than one branch, as shown in two examples in Figure 4.6b. It is important to note that all signals emanating from the same take-off point are replicas of exactly the same signal.

Before closing this section, we must note the mathematical form of the overall transfer function of the two tanks system under consideration. When the two terms of the overall transfer function (4.2.9) are combined with a common denominator, it takes the form

$$G(s) = \frac{\left[(1-a)A_1 R_1 + a A_2 R_2\right] s + 1}{(A_1 R_1 s + 1)(A_2 R_2 s + 1)} \qquad (4.2.10)$$

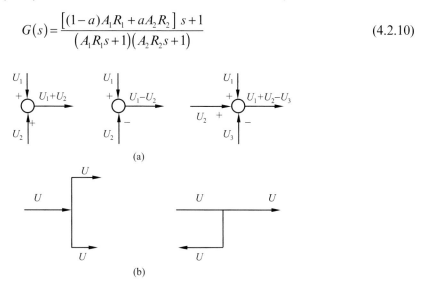

Figure 4.6 Examples of (a) summation points and (b) take-off points of a block diagram.

From Eq. (4.2.10) it becomes clear that, although each individual transfer function is first order, the overall transfer function is second order (the degree of the denominator polynomial is 2). The same holds true for first-order systems connected in series, as can be seen from Eq. (4.1.22). There is, however, an important difference in the overall transfer function for the parallel connection: the numerator of the transfer function depends on s, whereas for the series connection the numerator is a constant. It is only in the special case of identical tanks connected in parallel ($A_1 = A_2 = A$, $R_1 = R_2 = R$) that the overall transfer function simplifies to first order:

$$G(s) = \frac{[(1-\alpha)AR + \alpha AR]s + 1}{(ARs+1)(ARs+1)} = \frac{ARs+1}{(ARs+1)(ARs+1)} = \frac{1}{ARs+1} \qquad (4.2.11)$$

4.3 Interacting First-Order Systems

We now turn our attention to the system of two interacting tanks shown in Figure 4.7. The volumetric flowrate of the stream between the two tanks is a function of the difference between the liquid levels in the two tanks; for simplicity, we assume that it is proportional to the difference of the levels:

$$F_1 = \frac{h_1 - h_2}{R_1} \qquad (4.3.1)$$

The volumetric flowrate of the stream coming out of the second tank is assumed to be proportional to the level in the second tank:

$$F_2 = \frac{h_2}{R_2} \qquad (4.3.2)$$

The mass balances for the two tanks can then be written as

$$A_1 \frac{dh_1}{dt} = F_{in} - \frac{h_1 - h_2}{R_1} \qquad (4.3.3)$$

$$A_2 \frac{dh_2}{dt} = \frac{h_1 - h_2}{R_1} - \frac{h_2}{R_2} \qquad (4.3.4)$$

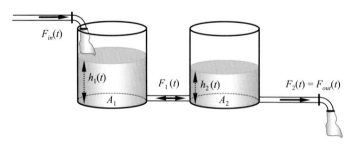

Figure 4.7 Two interacting tanks.

From Eqs. (4.3.3) and (4.3.4), we observe that the liquid level in each tank depends on the liquid level in the other tank and therefore we have interaction. Neither one of the two differential equations can be solved independently of the other.

The interaction is also visible in the Laplace domain. After introducing deviation variables and Laplace transforming under zero initial condition, we obtain

$$A_1 s \bar{H}_1(s) = \bar{F}_{in}(s) - \frac{\bar{H}_1(s) - \bar{H}_2(s)}{R_1} \quad (4.3.5)$$

$$A_2 s \bar{H}_2(s) = \frac{\bar{H}_1(s) - \bar{H}_2(s)}{R_1} - \frac{\bar{H}_2(s)}{R_2} \quad (4.3.6)$$

The above algebraic equations are coupled, and they need to be solved simultaneously.

Solving these equations, we find:

$$\frac{\bar{H}_1(s)}{\bar{F}_{in}(s)} = \frac{R_1(A_2 R_2 s + 1 + R_2/R_1)}{A_1 R_1 A_2 R_2 \, s^2 + (A_1 R_1 + A_1 R_2 + A_2 R_2)\, s + 1} \quad (4.3.7)$$

$$\frac{\bar{H}_2(s)}{\bar{F}_{in}(s)} = \frac{R_2}{A_1 R_1 A_2 R_2 \, s^2 + (A_1 R_1 + A_1 R_2 + A_2 R_2)\, s + 1} \quad (4.3.8)$$

Finally, if we consider $F_{out} = F_2 = h_2/R_2$ to be the system output, it follows that the transfer function of the system is:

$$G(s) = \frac{\bar{F}_{out}(s)}{\bar{F}_{in}(s)} = \frac{\bar{H}_2(s)/R_2}{\bar{F}_{in}(s)} = \frac{1}{A_1 R_1 A_2 R_2 \, s^2 + (A_1 R_1 + A_1 R_2 + A_2 R_2)\, s + 1} \quad (4.3.9)$$

This is again a second-order transfer function.

4.4 Response of First-Order Systems Connected in Series or in Parallel

Consider two first-order systems connected in series

$$\tau_1 \frac{dx_1}{dt} + x_1 = k_1 u \quad (4.4.1)$$

$$\tau_2 \frac{dx_2}{dt} + x_2 = k_2 x_1 \quad (4.4.2)$$

The first differential equation may be solved independently of the second and then, from the solution of the first equation, the second equation may be solved. Given an arbitrary input $u(t)$, the solution of Eq. (4.4.1) is given by (see Eq. (3.3.4)):

$$x_1(t) = e^{-\left(\frac{t}{\tau_1}\right)} x_1(0) + \int_0^t \frac{k_1}{\tau_1} e^{-\left(\frac{t-t'}{\tau_1}\right)} u(t')\, dt' \quad (4.4.3)$$

Given the above solution for $x_1(t)$, Eq. (4.4.2) may be solved in exactly the same way

$$x_2(t) = e^{-\left(\frac{t}{\tau_2}\right)} x_2(0) + \int_0^t \frac{k_2}{\tau_2} e^{-\left(\frac{t-t'}{\tau_2}\right)} x_1(t') dt' \qquad (4.4.4)$$

The calculations are generally simpler in the Laplace domain. Taking the Laplace transform of Eq. (4.4.1) we obtain

$$\tau_1 (sX_1(s) - x_1(0)) + X_1(s) = k_1 U(s)$$

or

$$X_1(s) = \frac{\tau_1}{\tau_1 s + 1} x_1(0) + \frac{k_1}{\tau_1 s + 1} U(s) \qquad (4.4.5)$$

Similarly, from Eq. (4.4.2)

$$X_2(s) = \frac{\tau_2}{\tau_2 s + 1} x_2(0) + \frac{k_2}{\tau_2 s + 1} X_1(s) \qquad (4.4.6)$$

Substituting the expression of $X_1(s)$ given by (4.4.5) into (4.4.6), we obtain

$$X_2(s) = \frac{\tau_2}{\tau_2 s + 1} x_2(0) + \frac{k_2 \tau_1}{(\tau_1 s + 1)(\tau_2 s + 1)} x_1(0) + \frac{k_1 k_2}{(\tau_1 s + 1)(\tau_2 s + 1)} U(s) \qquad (4.4.7)$$

For a specific input function and specific values of the initial conditions, the states $x_1(t)$ and $x_2(t)$ may be obtained by inverting the Laplace transforms of the expressions in Eqs. (4.4.5) and (4.4.7). Tables 4.1 and 4.2 give the responses of $x_2(t)$ to initial conditions, as well as step, impulse, ramp and sinusoidal inputs.

Considering the output of the overall system to be the second state x_2, we immediately obtain the overall transfer from (4.4.7), by setting the initial conditions equal to zero:

$$G(s) = \frac{X_2(s)}{U(s)} = \frac{k_1 k_2}{(\tau_1 s + 1)(\tau_2 s + 1)} \qquad (4.4.8)$$

For two first-order systems connected in parallel

$$\tau_1 \frac{dx_1}{dt} + x_1 = k_1 u \qquad (4.4.9)$$

$$\tau_2 \frac{dx_2}{dt} + x_2 = k_2 u \qquad (4.4.10)$$

$$y = x_1 + x_2 \qquad (4.4.11)$$

4.4 Response of First-Order Systems Connected in Series or in Parallel

the response of each state may be calculated independently

$$x_1(t) = e^{-\left(\frac{t}{\tau_1}\right)} x_1(0) + \int_0^t \frac{k_1}{\tau_1} e^{-\left(\frac{t-t'}{\tau_1}\right)} u(t') dt' \qquad (4.4.12)$$

$$x_2(t) = e^{-\left(\frac{t}{\tau_2}\right)} x_2(0) + \int_0^t \frac{k_2}{\tau_2} e^{-\left(\frac{t-t'}{\tau_2}\right)} u(t') dt' \qquad (4.4.13)$$

and the output is obtained as the sum of the two individual state responses. The same calculation may be alternatively preformed with the Laplace transform method, in which case the individual state responses are expressed as

$$x_1(t) = e^{-\frac{t}{\tau_1}} x_1(0) + \mathcal{L}^{-1}\left[\frac{k_1}{\tau_1 s + 1} U(s)\right] \qquad (4.4.14)$$

$$x_2(t) = e^{-\frac{t}{\tau_2}} x_2(0) + \mathcal{L}^{-1}\left[\frac{k_2}{\tau_2 s + 1} U(s)\right] \qquad (4.4.15)$$

Example 4.1 Response of two first-order systems connected in series to a step change of the input

Consider the system of two tanks in series shown in Figure 4.1, with the following parameter values: $A_1 = 1$ m², $A_2 = 2$ m² and $R_1 = R_2 = 1$ min·m⁻². The system is initially at steady state with $F_{in} = 1$ m³/min. At time $t = 0$, the feed flowrate is suddenly increased to $F_{in} = 1.1$ m³/min. Calculate the response of the liquid levels h_1 and h_2 in the two tanks.

Solution

Since we need to calculate liquid levels, we consider the state-space model with h_1 and h_2 as states. Converting Eqs. (4.1.4) and (4.1.5) to deviation-variable form relative to the steady state with $F_{in} = 1$ m³/min and corresponding steady-state levels $h_{1,s} = R_1 F_{in} = 1$ m and $h_{2,s} = R_2 F_{in} = 1$ m,

$$A_1 \frac{d\bar{h}_1}{dt} = \bar{F}_{in} - \frac{\bar{h}_1}{R_1}$$

$$A_2 \frac{d\bar{h}_2}{dt} = \frac{\bar{h}_1}{R_1} - \frac{\bar{h}_2}{R_2}$$

we can calculate the transfer functions of each individual tank, as well as the overall transfer function, which is their product. Under zero initial condition, the first differential equation gives the transfer function between F_{in} and h_1 for the first tank

$$G_1(s) = \frac{\bar{H}_1(s)}{\bar{F}_{in}(s)} = \frac{R_1}{A_1 R_1 s + 1}$$

and the second differential equation the one between h_1 and h_2 for the second tank

$$G_2(s) = \frac{\bar{H}_2(s)}{\bar{H}_1(s)} = \frac{R_2/R_1}{A_2 R_2 s + 1}$$

$$\bar{F}_{in}(s) \longrightarrow \boxed{\frac{R_1}{A_1 R_1 s + 1}} \xrightarrow{\bar{H}_1(s)} \boxed{\frac{R_2/R_1}{A_2 R_2 s + 1}} \xrightarrow{\bar{H}_2(s)}$$

For a step change in F_{in} of size $M = 0.1$ m³/min, $\bar{H}_1(s) = \frac{R_1}{A_1 R_1 s + 1} \cdot \frac{M}{s}$, hence

$$h_1(t) = h_{1,s} + \mathcal{L}^{-1}\left[\frac{R_1}{A_1 R_1 s + 1} \cdot \frac{M}{s}\right] = h_{1,s} + R_1 M\left(1 - e^{-t/A_1 R_1}\right) = 1 + 0.1\left(1 - e^{-t}\right)$$

and for the second tank,

$$\bar{H}_2(s) = \frac{R_2/R_1}{A_2 R_2 s + 1} \bar{H}_1(s) = \frac{R_2/R_1}{A_2 R_2 s + 1} \cdot \frac{R_1}{A_1 R_1 s + 1} \cdot \frac{M}{s} = \frac{R_2}{(A_2 R_2 s + 1)(A_1 R_1 s + 1)} \cdot \frac{M}{s}$$

hence

$$h_2(t) = h_{2,s} + \mathcal{L}^{-1}\left[\frac{R_2}{(A_2 R_2 s + 1)(A_1 R_1 s + 1)} \cdot \frac{M}{s}\right]$$

$$= h_{2,s} + R_2 M\left(1 - \frac{A_1 R_1 e^{-\frac{t}{A_1 R_1}} - A_2 R_2 e^{-\frac{t}{A_2 R_2}}}{A_1 R_1 - A_2 R_2}\right) = 1 + 0.1\left(1 + e^{-t} - 2e^{-(t/2)}\right)$$

The responses are shown in Figure 4.8 and we can see that the response of the first tank is faster than the response of the second tank. The response of h_2 is behind h_1 at all times. Moreover, we see that at $t = 0$, h_1 responds with a positive derivative whereas the response of h_2 is sluggish, with zero initial derivative.

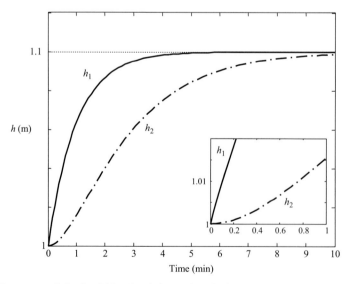

Figure 4.8 Response of the liquid levels of the tanks of Figure 4.1 to a step change in the volumetric flowrate of the feed.

4.4 Response of First-Order Systems Connected in Series or in Parallel

Example 4.2 Response of two identical first-order systems in series to various changes of the input

Consider the two mixing tanks in series shown in Figure 4.9. The tanks have constant volumes $V_1 = V_2 = V = 1$ m³ and constant volumetric flowrates $F = 0.1$ m³/min. The feed stream may contain a dye of concentration C_0 that can be adjusted by an upstream unit. Derive a mathematical model for the dynamics of the dye concentrations C_1 and C_2 in the two tanks, and calculate their responses for the following cases.

(1) The feed is pure water, $C_0(t) = 0$, $t > 0$, but there is leftover dye in the first tank from a previous experiment, with initial concentration of $C_1(0) = 10$ kg/m³. The second tank initially contains pure water $C_2(0) = 0$.
(2) The feed is initially pure water, the tanks initially contain pure water, but at time $t = 0$, the feed is suddenly switched to a solution of dye of concentration 10 kg/m³.
(3) The feed is initially pure water, the tanks initially contain pure water, but at time $t = 0$, the feed is suddenly switched to a solution of dye of concentration 10 kg/m³, up to time $t = \varepsilon = 2.5$ min, when it is switched back to pure water.
(4) The same amount of dye as in the previous question is delivered during an infinitesimally small time period ($\varepsilon \to 0$).
(5) The feed is initially pure water, the tanks initially contain pure water, but starting at time $t = 0$, the feed contains dye with linearly increasing concentration at a rate of 0.1 kg·m⁻³min⁻¹

Solution

In order to examine the dynamic behavior of the dye concentrations C_1 and C_2, we need to first develop the mathematical model. The model consists of a material balance for the dye in each tank, which can be written as follows:

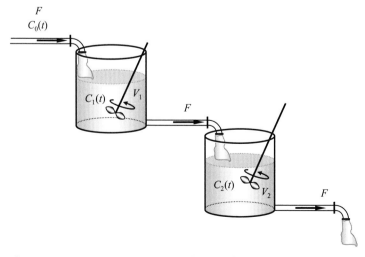

Figure 4.9 System of two mixing tanks in series.

$$V_1 \frac{dC_1}{dt} = FC_0 - FC_1$$

$$V_2 \frac{dC_2}{dt} = FC_1 - FC_2$$

Since the volumes $V_1 = V_2 = V$ and the volumetric flowrate F are constant, we may write

$$\tau \frac{dC_1}{dt} + C_1 = C_0$$

$$\tau \frac{dC_2}{dt} + C_2 = C_1$$

where $\tau = V/F = 10$ min. We have two first-order systems in series with the same time constant $\tau_1 = \tau_2 = \tau$ and unity steady-state gains $k_1 = k_2 = 1$. Applying (4.4.5) and (4.4.7) gives

$$C_1(s) = \frac{\tau}{\tau s + 1} C_1(0) + \frac{1}{\tau s + 1} C_0(s)$$

$$C_2(s) = \frac{\tau}{\tau s + 1} C_2(0) + \frac{\tau}{(\tau s + 1)^2} C_1(0) + \frac{1}{(\tau s + 1)^2} C_0(s)$$

For each of the scenarios to be examined in this problem, the above Laplace transforms will be evaluated and subsequently inverted to obtain $C_1(t)$ and $C_2(t)$.

(1) For zero input $C_0(t) = 0$ and initial conditions $C_1(0) \neq 0$ and $C_2(0) = 0$, we have:

$$C_1(s) = \frac{\tau}{\tau s + 1} C_1(0) \quad \Rightarrow \quad C_1(t) = e^{-t/\tau} C_1(0)$$

$$C_2(s) = \frac{\tau}{(\tau s + 1)^2} C_1(0) \quad \Rightarrow \quad C_2(t) = \frac{t}{\tau} e^{-t/\tau} C_1(0)$$

These two responses are shown in Figure 4.10 for $\tau = 10$ min and $C_1(0) = 10$ kg/m³.

(2) For a step input of size M and zero initial conditions,

$$C_1(s) = \frac{1}{\tau s + 1} \cdot \frac{M}{s} \quad \Rightarrow \quad C_1(t) = M(1 - e^{-t/\tau})$$

$$C_2(s) = \frac{1}{(\tau s + 1)^2} \cdot \frac{M}{s} \quad \Rightarrow \quad C_2(t) = M\left(1 - \left(1 + \frac{t}{\tau}\right) e^{-t/\tau}\right)$$

These two responses are shown in Figure 4.11 for $\tau = 10$ min and $M = 10$ kg/m³.

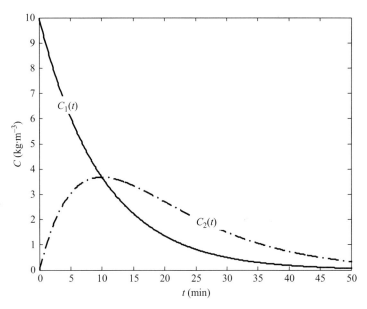

Figure 4.10 Response of the dye concentration in the two mixing tanks to an initial dye concentration in the first tank, under zero dye concentration in the feed.

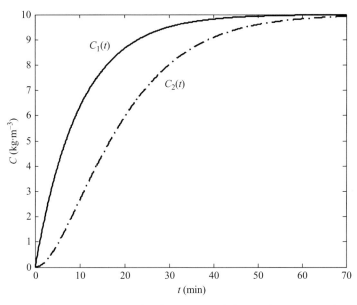

Figure 4.11 Response of the dye concentration in the two mixing tanks to a step change in the dye concentration in the feed stream.

(3) For a pulse input of magnitude M and duration ε, and zero initial conditions,

$$C_1(s) = \frac{1}{\tau s + 1} \cdot \left(\frac{M/\varepsilon}{s} - e^{-\varepsilon s} \frac{M/\varepsilon}{s} \right)$$

$$\Rightarrow C_1(t) = \begin{cases} \dfrac{M}{\varepsilon}\left(1 - e^{-t/\tau}\right), & 0 \le t < \varepsilon \\ \dfrac{M}{\varepsilon}\left(1 - e^{-t/\tau}\right) - \dfrac{M}{\varepsilon}\left(1 - e^{-(t-\varepsilon)/\tau}\right), & t \ge \varepsilon \end{cases}$$

$$C_2(s) = \frac{1}{(\tau s + 1)^2} \cdot \left(\frac{M/\varepsilon}{s} - e^{-\varepsilon s} \frac{M/\varepsilon}{s} \right)$$

$$\Rightarrow C_2(t) = \begin{cases} \dfrac{M}{\varepsilon}\left(1 - \left(1 + \dfrac{t}{\tau}\right) e^{-t/\tau}\right), & 0 \le t < \varepsilon \\ \dfrac{M}{\varepsilon}\left(1 - \left(1 + \dfrac{t}{\tau}\right) e^{-t/\tau}\right) - \dfrac{M}{\varepsilon}\left(1 - \left(1 + \dfrac{t-\varepsilon}{\tau}\right) e^{-(t-\varepsilon)/\tau}\right), & t \ge \varepsilon \end{cases}$$

These are shown in Figure 4.12, for $\tau = 10$ min, $M = 25$ min· kg/m³ and $\varepsilon = 2.5$ min.

(4) Using the answer of the previous question, we can take the limit as $\varepsilon \to 0$. To this end, we can first simplify the expressions of $C_1(t)$ and $C_2(t)$ for $t \ge \varepsilon$, and write

$$C_1(t) = \begin{cases} \dfrac{M}{\varepsilon}\left(1 - e^{-t/\tau}\right) & 0 \le t < \varepsilon \\ M\left(\dfrac{1 - e^{-\varepsilon/\tau}}{\varepsilon}\right) e^{-(t-\varepsilon)/\tau}, & t \ge \varepsilon \end{cases}$$

$$C_2(t) = \begin{cases} \dfrac{M}{\varepsilon}\left(1 - \left(1 + \dfrac{t}{\tau}\right) e^{-t/\tau}\right), & 0 \le t < \varepsilon \\ M\left[-\dfrac{1}{\tau} e^{-\varepsilon/\tau} + \left(\dfrac{1 - e^{-\varepsilon/\tau}}{\varepsilon}\right)\left(1 + \dfrac{t-\varepsilon}{\tau}\right)\right] e^{-(t-\varepsilon)/\tau} \end{cases}$$

Then, using the fact that $\displaystyle\lim_{\varepsilon \to 0} \frac{1 - e^{-\varepsilon/\tau}}{\varepsilon} = \frac{1}{\tau}$ (as a result of l'Hôpital's rule), it follows that, in the limit as $\varepsilon \to 0$,

$$C_1(t) \to \frac{M}{\tau} e^{-t/\tau}, \quad C_2(t) \to \frac{Mt}{\tau^2} e^{-t/\tau}.$$

The limiting responses are shown in Figure 4.13, for $\tau = 10$ min and $M = 25$ min· kg/m³.

The same result would have been obtained by considering the input to be $C_0(t) = M\delta(t)$, where $\delta(t)$ is the Dirac delta function:

$$C_1(s) = \frac{1}{\tau s + 1} \cdot M \quad \Rightarrow \quad C_1(t) = \frac{M}{\tau} e^{-t/\tau}$$

$$C_2(s) = \frac{1}{(\tau s + 1)^2} \cdot M \quad \Rightarrow \quad C_2(t) = \frac{Mt}{\tau^2} e^{-t/\tau}$$

4.4 Response of First-Order Systems Connected in Series or in Parallel

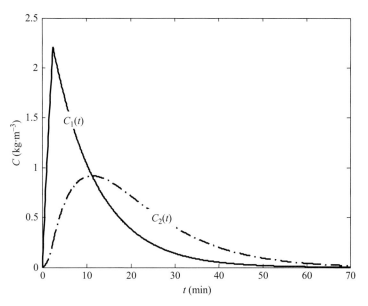

Figure 4.12 Response of the dye concentration in the two mixing tanks to a pulse change in the dye concentration in the feed stream.

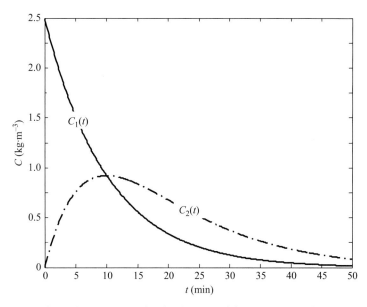

Figure 4.13 Response of the dye concentration in the two mixing tanks to an impulse change in the dye concentration in the feed stream.

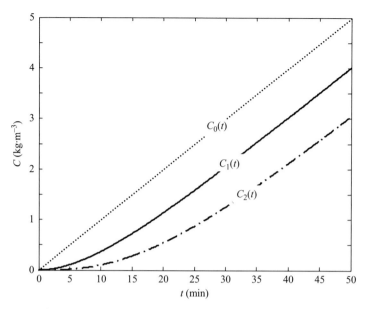

Figure 4.14 Response of the dye concentration in the two mixing tanks to a ramp change of the dye concentration in the feed stream.

(5) For a ramp input of slope M and zero initial conditions,

$$C_1(s) = \frac{1}{\tau s+1} \cdot \frac{M}{s^2} \quad \Rightarrow \quad C_1(t) = M\left(t - \tau + \tau e^{-t/\tau}\right)$$

$$C_2(s) = \frac{1}{(\tau s+1)^2} \cdot \frac{M}{s^2} \quad \Rightarrow \quad C_2(t) = M\left(t - 2\tau + (t+2\tau)e^{-t/\tau}\right)$$

The above responses, along with the feed concentration, are shown in Figure 4.14 for $\tau = 10$ min and $M = 0.1$ kg·m^{-3}min^{-1}.

4.5 Software Tools

4.5.1 Response Calculation for Two First-Order Systems Connected in Series using MATLAB

Equations (4.4.1) and (4.4.2) give the mathematical description of two first-order systems connected in series which can be written in the form

$$\frac{dx_1}{dt} = -\frac{1}{\tau_1}x_1 + \frac{k_1}{\tau_1}u \tag{4.5.1}$$

$$\frac{dx_2}{dt} = \frac{k_2}{\tau_2}x_1 - \frac{1}{\tau_2}x_2 \tag{4.5.2}$$

To obtain the response of this system to any input $u(t)$ numerically in MATLAB we need to first define the system in MATLAB by assuming specific numerical values for the constants τ_1, τ_2, k_1 and k_2. Consider the case that the constants assume the values given in Example 4.2: $\tau_1 = \tau_2 = \tau = 10$ and $k_1 = k_2 = 1$. The response to a step change in u is calculated in MATLAB as follows. We first create the following m-file:

```
function dxdt=SystemsInSeries(t,x)
% 2 noninteracting first order
% systems in series
% model constants
k1= 1;
k2= 1;
tau1 = 10;
tau2 = 10;
% step rsponse
u=1;
% equations
dxdt(1) = -(1/tau1)*x(1)+(k1/tau1)*u;
dxdt(2) = -(1/tau2)*x(2)+(k2/tau2)*x(1);
dxdt=dxdt(:);
```

In the command window we type

```
»ode45(@SystemsInSeries,[0 70],[0;0])
```

to obtain Figure 4.11. To obtain the response to other input variations we only need to change the lines in `SystemsInSeries.m` that define the input.

4.5.2 Response Calculation for Two First-Order Systems Connected in Series using Maple

The response of first-order systems in series can be calculated sequentially: the response of the first stage is calculated first, and the result is input for the calculation in the second stage. Under zero initial conditions, the response may be obtained by sequential evaluation of the convolution integrals (see Eqs. (4.4.3) and (4.4.4)):

$$x_1(t) = \int_0^t g_1(t-t')\, u(t')\, dt', \text{ where } g_1(t) = \frac{k_1}{\tau_1} e^{-t/\tau_1}$$

$$x_2(t) = \int_0^t g_2(t-t')\, x_1(t')\, dt', \text{ where } g_2(t) = \frac{k_2}{\tau_2} e^{-t/\tau_2}$$

In what follows, we give an example on how to perform this calculation symbolically for a ramp input, $u(t) = Mt$ and for equal time constants $\tau_1 = \tau_2 = \tau$, using Maple.

First, define the functions $u(t)$, $g_1(t)$ and $g_2(t)$:

```
> u(t):=M*t:
> g1(t):=(k1/tau)*exp(-t/tau):
> g2(t):=(k2/tau)*exp(-t/tau):
```

and then calculate the convolution integrals using the `int` command:

```
> g1(t-tprime):=subs(t=t-tprime,g1(t):
> x1(t):=int(g1(t-tprime)*subs(t=tprime,u(t)),tprime=0..t);
```

$$x1(t) := M\ kl\ \left(\tau - \tau e^{\frac{t}{\tau}} + t\ e^{\frac{t}{\tau}}\right) e^{-\frac{t}{\tau}}$$

```
> g2(t-tprime):=subs(t=t-tprime,g2(t)):
> x2(t):=int(g2(t-tprime)*subs(t=tprime,x1(t)),tprime=0..t);
```

$$x2(t) := M\ kl\ k2\ \left(2\tau + t - 2\tau e^{\frac{t}{\tau}} + t\ e^{\frac{t}{\tau}}\right) e^{-\frac{t}{\tau}}$$

Alternatively, the same calculation could be performed using the transfer functions of each stage and inversion of Laplace transforms, using the `inttrans` library.

LEARNING SUMMARY

- The connection of two first-order systems in series or in parallel results in an overall system that exhibits second-order dynamics.
- In a parallel connection, the response of each first-order subsystem is calculated separately using the results of Chapter 3 and the individual responses are added up.
- In the series connection, the response of the first subsystem is calculated independently, using the results of Chapter 3. Subsequently, the response of the second system is calculated, with the input response of the first subsystem. The response calculation under zero initial conditions can be performed either through sequential convolution integrals

$$x_1(t) = \int_0^t g_1(t-t')\ u(t')\ dt',\ x_2(t) = \int_0^t g_2(t-t')\ x_1(t')\ dt',$$

where $g_1(t) = \dfrac{k_1}{\tau_1} e^{-t/\tau_1}$, $g_2(t) = \dfrac{k_2}{\tau_2} e^{-t/\tau_2}$

or by using the inverse Laplace transform

$$x_2(t) = \mathcal{L}^{-1}\left[\frac{k_1 k_2}{(\tau_1 s + 1)(\tau_2 s + 1)} U(s)\right]$$

- The results of the above calculations for several standard inputs are summarized in Table 4.1 (unequal time constants) and Table 4.2 (equal time constants).

Table 4.1 Response of two first-order systems with unequal time constants $\tau_1 \neq \tau_2$ connected in series

	Input	Output of the second system
Unforced response	$u(t) = 0$	$x_2(t) = \dfrac{e^{-\left(\frac{t}{\tau_1}\right)} - e^{-\left(\frac{t}{\tau_2}\right)}}{\tau_1 - \tau_2} \tau_1 k_2 x_1(0) + e^{-\left(\frac{t}{\tau_2}\right)} x_2(0)$
Response to standard inputs under zero initial conditions $x_1(0) = x_2(0) = 0$	Step $u(t) = 0, t < 0$ $u(t) = M, t \geq 0$	$\dfrac{x_2(t)}{k_1 k_2 M} = \left(1 - \dfrac{\tau_1 e^{-\left(\frac{t}{\tau_1}\right)} - \tau_2 e^{-\left(\frac{t}{\tau_2}\right)}}{\tau_1 - \tau_2}\right)$
	Impulse $u(t) = M\delta(t)$	$\dfrac{x_2(t)}{k_1 k_2 M} = \dfrac{e^{-\left(\frac{t}{\tau_1}\right)} - e^{-\left(\frac{t}{\tau_2}\right)}}{\tau_1 - \tau_2}$
	Ramp $u(t) = 0, t < 0$ $u(t) = Mt, t \geq 0$	$\dfrac{x_2(t)}{k_1 k_2 M} = \left(t - (\tau_1 + \tau_2) + \dfrac{\tau_1^2 e^{-\left(\frac{t}{\tau_1}\right)} - \tau_2^2 e^{-\left(\frac{t}{\tau_2}\right)}}{\tau_1 - \tau_2}\right)$
	Sinusoidal $u(t) = 0, t < 0$ $u(t) = M\sin(\omega t), t \geq 0$	$\dfrac{x_2(t)}{k_1 k_2 M} = \left\{\dfrac{(1-\omega^2 \tau_1 \tau_2)\sin(\omega t)}{(1+(\omega\tau_1)^2)(1+(\omega\tau_2)^2)} - \dfrac{\omega(\tau_1+\tau_2)\cos(\omega t)}{(1+(\omega\tau_1)^2)(1+(\omega\tau_2)^2)} + \dfrac{1}{\tau_1 - \tau_2}\left(\dfrac{\omega\tau_1^2}{1+(\omega\tau_1)^2}e^{-\left(\frac{t}{\tau_1}\right)} - \dfrac{\omega\tau_2^2}{1+(\omega\tau_2)^2}e^{-\left(\frac{t}{\tau_2}\right)}\right)\right\}$

Table 4.2 Response of two first-order systems with equal time constants $\tau_1 = \tau_2 = \tau$ connected in series

		Input	Output of the second system
Unforced response		$u(t) = 0$	$x_2(t) = \dfrac{t}{\tau} e^{-\left(\frac{t}{\tau}\right)} k_2 x_1(0) + e^{-\left(\frac{t}{\tau}\right)} x_2(0)$
Response to standard inputs under zero initial conditions $x_1(0) = x_2(0) = 0$		Step $u(t) = 0,\ t < 0$ $u(t) = M,\ t \geq 0$	$\dfrac{x_2(t)}{k_1 k_2 M} = \left(1 - e^{-\left(\frac{t}{\tau}\right)} - \dfrac{t}{\tau} e^{-\left(\frac{t}{\tau}\right)}\right)$
		Impulse $u(t) = M\delta(t)$	$\dfrac{x_2(t)}{k_1 k_2 M} = \dfrac{t}{\tau^2} e^{-\left(\frac{t}{\tau}\right)}$
		Ramp $u(t) = 0,\ t < 0$ $u(t) = Mt,\ t \geq 0$	$\dfrac{x_2(t)}{k_1 k_2 M} = t - 2\tau + 2\tau e^{-\left(\frac{t}{\tau}\right)} + t e^{-\left(\frac{t}{\tau}\right)}$
		Sinusoidal $u(t) = 0,\ t < 0$ $u(t) = M\sin(\omega t),\ t \geq 0$	$\dfrac{x_2(t)}{k_1 k_2 M} = \left\{ \dfrac{(1-(\omega\tau)^2)\sin(\omega t) - 2\omega\tau\cos(\omega t)}{(1+(\omega\tau)^2)^2} + \dfrac{2\omega\tau}{(1+(\omega\tau)^2)^2} e^{-\left(\frac{t}{\tau}\right)} + \dfrac{\omega t}{1+(\omega\tau)^2} e^{-\left(\frac{t}{\tau}\right)} \right\}$

TERMS AND CONCEPTS

Interacting systems. Systems that affect each other both ways.

Systems connected in parallel. Systems with common input, whose dynamics evolve independently.

Systems connected in series. Systems where the first one affects the second but the second does not affect the first.

FURTHER READING

Additional material and further examples on first-order systems connected in series or with interaction can be found in the books

Ogata, K., *Modern Control Engineering*, Pearson International Edition, 5th edn, 2008.

Stephanopoulos, G., *Chemical Process Control, An Introduction to Theory and Practice*. New Jersey: Prentice Hall, 1984.

PROBLEMS

4.1 Consider the two liquid storage tanks shown in Figure P4.1. The volumetric flowrate of the stream that connects the two tanks is proportional to the difference between the two levels $F_1 = (h_2 - h_1)/R_1$, which is considered positive when liquid flows from the second to the first tank. The volumetric flow rate of the exit stream is $F_2 = h_2/R_2$.
  (a) Write down a state-space model for this system of tanks.
  (b) Derive the transfer function between F_{in} and h_2.
  (c) If $A_1 = 1$ m², $A_2 = 2$ m², $R_1 = 1$ min·m⁻² and $R_2 = 1/2$ min·m⁻², calculate the response of h_2 to a step increase in F_{in} by 1 m³/min.

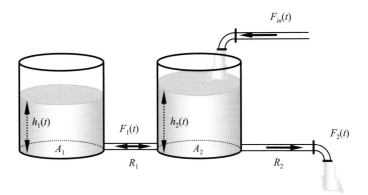

Figure P4.1 Interacting tanks, with feed and exit line in the same tank.

4.2 Consider a constant-volume continuous stirred tank reactor (CSTR), as shown in Figure P4.2, where the following irreversible consecutive reactions

$$R \xrightarrow{k_1} P \xrightarrow{k_2} B$$

take place isothermally, with reaction rates $r_1 = k_1 C_R$, $r_2 = k_2 C_P$. The intermediate product P is the desired product, whereas B is an undesirable byproduct. The feed is a solution of the reactant R of concentration $C_{in,R}$, that contains no P or B, and has a constant volumetric flowrate F.
The input of the system is the feed concentration $C_{in,R}$, the output is the concentration C_P of species P in the reactor.

  (a) Write down a state-space model for this system, consisting of mole balances for species R and P.
  (b) Derive the transfer function of the CSTR.
  (c) Suppose that the CSTR is initially at steady state. At $t = 0$, n_R moles of species R are quickly added to the reactor. Calculate the response of C_P.
  (d) Repeat stages (a) and (b) in case the first reaction is reversible:

$$R \underset{k_{-1}}{\overset{k_1}{\rightleftharpoons}} P \overset{k_2}{\longrightarrow} B$$

and the rate of the reverse reaction (from P to R) is $r_{-1} = k_{-1}C_P$.

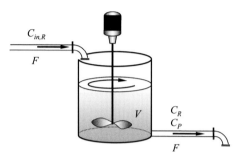

Figure P4.2 Isothermal CSTR.

4.3 In Figure P4.3 a heating/cooling tank is shown. Liquid with constant volumetric flowrate F and constant temperature T_0 is fed to the tank. The exit temperature is T, which is different from T_0, as an internal coil is used to increase/decrease the temperature of the liquid in the tank. A fluid is fed to the coil at temperature $T_{c,0}$ and exits the coil at temperature T_c. Derive the transfer function between the inlet temperature of the fluid in the coil $T_{c,0}$ (input) and the temperature T of the liquid content in the tank (output). Assume that all thermophysical properties are constant and that the liquid volume in the tank and in the coil are constant as well.

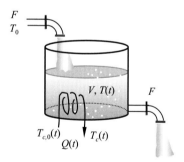

Figure P4.3

4.4 In the two continuous stirred tank reactors shown in Figure P4.4 the reaction R→P takes place under constant temperature, volume and inlet volumetric flowrates. The concentration of the reactant in the two feed streams is $C_{R,0}$ and is the main input to the system. Derive the transfer function between the fresh feed concentration $C_{R,0}$ and the concentrations of the reactant $C_{R,1}$ and $C_{R,2}$ in the two tanks. The reaction is first order with respect to the reactant concentration.

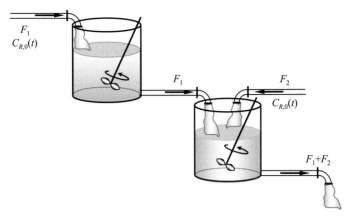

Figure P4.4

4.5 A system of two tanks in series as in Figure 4.1 has the following parameter values: $A_1 = 1$ m^2, $A_2 = 2$ m^2, $R_1 = 2$ min/m^2, $R_2 = 1$ min/m^2. The system is initially at steady state with $F_{in} = 1$ m^3/min. At time $t = 0$, the operator shuts a valve in the feed line by mistake, and this makes $F_{in} = 0$.
  (1) Calculate the response of the liquid levels h_1 and h_2 in the tanks.
  (2) After how much time will h_1 drop to 1 m? What will be the value of h_2 at that time? What will happen in the limit as $t \to \infty$ if the valve remains shut?
  (3) The operator realizes his mistake when h_1 reaches 1 m and immediately restores the feed flowrate to its normal value. What will happen afterwards? Compute h_1 and h_2. Will h_1 and h_2 ultimately reach their original steady state values?

4.6 In the system of two tanks in series shown in Figure P4.6, the outlet flow rate from the second tank F_{out} is set by a pump that has been installed, and is independent of the liquid level in the tank. The outflow from the first tank is proportional to the liquid level in the tank. The following parameter values are given for this system of tanks: $A_1 = A_2 = 0.1$ m^2, $R_1 = 10$ min/m^2.

The system is initially at steady state with $F_{out} = 0.05$ m^3/min. At time $t = 0$, a volume of 0.01 m^3 of liquid is added very quickly to the first tank. Calculate the response of the liquid levels in the tanks. What will happen as $t \to \infty$?

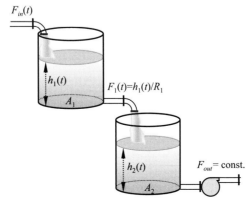

Figure P4.6

4.7 In the system of three tanks in series shown in Figure P4.7, the outlet flowrate from the third tank F_{out} is set by a pump that has been installed, and is independent of the liquid level in the tank. Derive the transfer function of the system with F_{in} as input and h_3 as output. If the system is initially at steady state but at time $t = 0$, F_{in} undergoes a step change of size M, calculate the response of h_3. What will happen as $t \to \infty$?

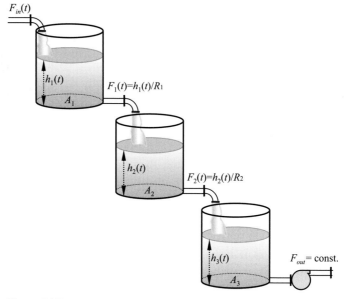

Figure P4.7

4.8 Consider a linear system that is represented in block-diagram form as shown in Figure P4.8, where $k_1 > k_2 > 0$, $\tau_1 > 0$, $\tau_2 > 0$.
  (a) If this system is initially at steady state, with all the signals equal to zero, and suddenly $u(t)$ undergoes a unit step increase, calculate the output response $y(t)$. Also, calculate the initial slope $(dy/dt)(0)$ as well as final value $y(\infty)$ of the response.
  (b) Plot your result for $k_1 = 2$, $k_2 = 1$, $\tau_1 = 4$, $\tau_2 = 1$. What do you observe? Please explain.
  (c) Same question as (b), for $k_1 = 2$, $k_2 = 1$, $\tau_1 = 1/4$, $\tau_2 = 1$.

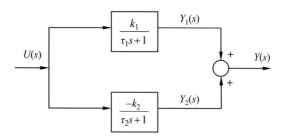

Figure P4.8

4.9 In the two continuous stirred tank reactors shown in Figure P4.9 the reaction A → B takes place under constant temperature, constant volumes (V and V_R) and constant volumetric flowrates (F and R). Part of the product stream of the first tank is feed for the second reactor and the product stream from the second reactor is mixed with the fresh feed and is fed back to the first reactor. The concentration of the reactant in the feed stream is C_0 and it is the main input to the system. Derive the transfer function between the fresh feed concentration and the concentration C of the reactant in the effluent. The reaction is first order with respect to the reactant concentration. Use the notation shown in Figure P4.9 where C_0, C and C_R refer to reactant concentrations.

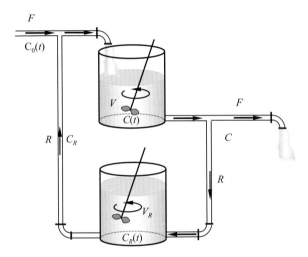

Figure P4.9

4.10 Consider a dynamic system arising from the serial connection of two first-order systems, and described by Eqs. (4.4.1) and (4.4.2). Show that when (4.4.3) is substituted into (4.4.4), the resulting expression for $x_2(t)$ may be simplified as follows:

$$x_2(t) = \begin{cases} e^{-\frac{t}{\tau_2}} x_2(0) + k_2\tau_1 \frac{e^{-\frac{t}{\tau_1}} - e^{-\frac{t}{\tau_2}}}{\tau_1 - \tau_2} x_1(0) + \int_0^t k_2 k_1 \frac{e^{-\frac{t-t'}{\tau_1}} - e^{-\frac{t-t'}{\tau_2}}}{\tau_1 - \tau_2} u(t')dt', & \text{if } \tau_2 \neq \tau_1 \\ e^{-\frac{t}{\tau_1}} x_2(0) + k_2 \frac{t}{\tau_1} e^{-\frac{t}{\tau_1}} x_1(0) + \int_0^t k_2 k_1 \frac{(t-t')}{\tau_1^2} e^{-\frac{t-t'}{\tau_1}} u(t')dt', & \text{if } \tau_2 = \tau_1 \end{cases}$$

4.11 Use Eq. (4.4.3) and the result of the previous problem to derive a forward rectangle discretization of the system described by (4.4.1) and (4.4.2).

4.12 Consider the heat exchanger tank examined in Problem 2.3. In an effort to improve the accuracy of our model, we split the tube in n sections and consider uniform temperature in each section. If the inlet temperature of the process stream, which has constant volumetric flowrate F and constant density ρ, is T_{in}, derive the transfer function between the inlet temperature and the temperature at the end of the tube (T_n). Use the notation shown in Figure P4.12 where n segments of the tube are shown; each one has volume V/n and temperature T_i, where i is the number of the segment. The temperature of the steam is constant at T_{st}, the overall heat transfer coefficient is U_H and the area available to heat transfer in each segment is A_H/n.

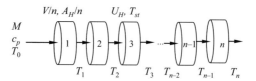

Figure P4.12 Approximation of a heat exchanger inner tube by n segments.

COMPUTATIONAL PROBLEMS

4.13 For the understanding of pharmacological phenomena in living organisms, a certain class of conceptual units has been used extensively: the so-called compartments. The organism to which a drug is administered is decomposed into a system of real or artificial interconnected pools, the compartments. The investigation of the properties of these compartments and of the material fluxes between them is termed compartment modeling. In most cases each compartment corresponds to a well-defined structure or organ interconnected by blood flow, lymph flow or other real-life material fluxes. The most important compartment in pharmacokinetic investigations is the systemic blood circulation, frequently termed the "central" compartment.

This is because it is easily accessible for continuous measurement. Into this central compartment the drug enters as an instantaneous influx burst (i.e. bolus administration), a constant rate influx (infusion), an exponentially decreasing first-order influx (oral route), etc. The most commonly used and simplest possible compartmental model is the two-compartment model with first-order linear fluxes shown in Figure P4.13a. A rectangle corresponds to a compartment and a circle to the source or sink of the drug administered to the organism. The arrows correspond to material fluxes that are normally linear in the concentration of the drug with proportionality constant k_{ij}, where i and j denote a flux from i to j (0 denotes the environment). It is important to note that the rate constants k_{ij} have units of inverse time and therefore they include the volume of the compartment (which is in most cases unknown). Derive the mathematical model of the compartmental model shown in Figure P4.13a and the transfer function between the dose and the concentration of the drug in the central compartment 1. Derive also the impulse response commonly used to simulate the intravenous (IV) bolus injection of a drug (the IV bolus is commonly used when rapid administration of a medication is needed, such as in an emergency).

In Figure P4.13b a more complicated compartmental model is shown where a third compartment has been added. The purpose of this third compartment is to simulate a metabolite of the parent drug that normally has a secondary activity. Derive the relative transfer functions and the concentration responses in an IV bolus injection.

Use Maple to validate your results.

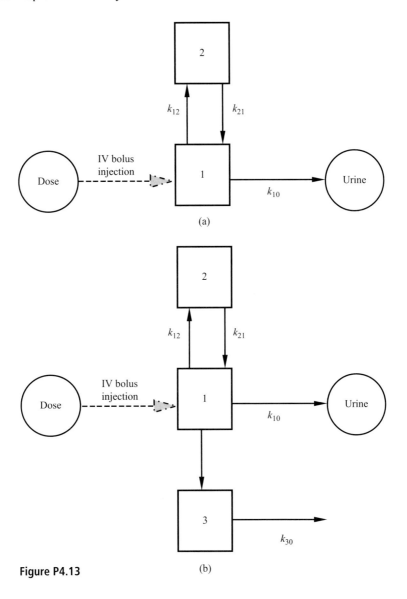

Figure P4.13

5 Second-Order Systems

In the previous chapter, we studied second-order systems that consist of

- two first-order systems connected in series
- two first-order systems connected in parallel
- two interacting first-order systems.

In all cases, the overall system is described by two first-order differential equations. There are, however, cases in which the system under study is inherently second order and cannot be analyzed or decomposed into simpler systems. Inherently second-order dynamic systems are obtained, in most cases, when the physical system follows Newton's second law of motion. In this chapter, we will study the dynamics of second-order systems. It is important to note that understanding first- and second-order systems dynamics is the basis of understanding the dynamics of higher-order systems.

First-order systems, in their simplest form, can be characterized by two parameters: the time constant (τ) and the static gain (k). We will see in this chapter that, in order to characterize simple second-order systems, an additional parameter is necessary: the damping factor ζ. This additional parameter is strongly related to the presence of oscillatory transients, which cannot be observed in first-order systems.

STUDY OBJECTIVES

After studying this chapter you should be able to do the following.

- Understand the qualitative and quantitative characteristics of the response of a second-order system, including the possibility of oscillatory behavior.
- Calculate the step response and the frequency response of a second-order system.
- Calculate and simulate the response of second-order systems using software tools, like MATLAB or Maple.

5.1 A Classical Example of a Second-Order System

In Figure 5.1 a mechanical system is shown in which an object of mass m is attached to one end of a spring with stiffness coefficient $\mathcal{K}$. The other end of the spring is attached to a wall. Friction forces are developed between the object and the surface over which it slides, with a friction coefficient that is equal to C. A time-varying external force $u(t)$ is applied to the object, that causes it to be displaced from its equilibrium position. We will use the following notation:

y is the displacement from the equilibrium position corresponding to $u = 0$
v is the velocity of the object
a is the acceleration of the object, and
p is the linear momentum of the object ($= m\,v$).

The velocity of the object is related to its displacement through the equation

$$v(t) = \frac{dy}{dt}(t) \qquad (5.1.1)$$

whereas the acceleration is given by

$$a(t) = \frac{dv}{dt}(t) = \frac{d^2 y}{dt^2}(t) \qquad (5.1.2)$$

From Newton's second law, the rate of change of the linear momentum must be equal to the sum of all forces acting on the system. The forces acting on the system are: the external force $u(t)$, the friction force $-Cv(t)$ and the spring compression force $-\mathcal{K}y(t)$. As the object has constant mass m, we can write the equations of motion as follows:

$$\frac{dy}{dt} = v \qquad (5.1.3)$$

$$m\frac{dv}{dt} = u - \mathcal{K}y - Cv \qquad (5.1.4)$$

The above equations completely characterize the motion, since their solution determines both the position and the velocity.

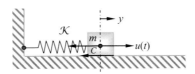

Figure 5.1 A classical example of a second-order system.

Alternatively, if only the displacement y is of interest (output of the system), it is possible to eliminate the velocity from (5.1.3) and (5.1.4) and obtain:

$$m\frac{d^2 y}{dt^2} + C\frac{dy}{dt} + \mathcal{K}y = u \tag{5.1.5}$$

Comparing Eqs. (5.1.3) and (5.1.4) to Eq. (5.1.5), we observe that we have two alternative ways of describing the dynamics:

- either through two first-order differential equations that represent the equations of motion (state space model with y, v as state variables)
- or through a single second-order differential equation that directly relates the external force u (input of the system) to the displacement y (output of the system).

Let's see now how we can describe the dynamics in the Laplace domain under zero initial conditions $y(0) = 0$ (initially at the equilibrium point corresponding to $u = 0$) and $v(0) = dy/dt(0)=0$ (initially still). Taking the Laplace transform of Eq. (5.1.5), we immediately see that

$$m s^2 Y(s) + Cs Y(s) + \mathcal{K} Y(s) = U(s) \tag{5.1.6}$$

therefore the transfer function of the system is

$$G(s) = \frac{Y(s)}{U(s)} = \frac{1}{ms^2 + Cs + \mathcal{K}} \tag{5.1.7}$$

In the study of second-order systems, we define the following parameters:

- the characteristic time: $\tau = \sqrt{m/\mathcal{K}}$, or alternatively the natural frequency $\omega_n = 1/\tau$
- the damping factor $\zeta = C/2\sqrt{m\mathcal{K}}$
- the static gain $k = 1/\mathcal{K}$

The differential equation (5.1.5) can then be written as

$$\tau^2 \frac{d^2 y}{dt^2} + 2\zeta\tau \frac{dy}{dt} + y = ku \tag{5.1.8}$$

while the transfer function (5.1.7) becomes

$$G(s) = \frac{Y(s)}{U(s)} = \frac{k}{\tau^2 s^2 + 2\zeta\tau s + 1} \tag{5.1.9}$$

It is important to note that $\tau > 0$ and $\zeta \geq 0$. Equations (5.1.8) and (5.1.9) are standard forms used to describe the effect of the input on the dynamic response of the output in a second-order system. When the natural frequency ω_n is used instead of the characteristic time τ, Eq. (5.1.8) is written as

$$\frac{d^2 y}{dt^2} + 2\zeta\omega_n \frac{dy}{dt} + \omega_n^2 y = k\omega_n^2 u \qquad (5.1.10)$$

while the transfer function given by Eq. (5.1.9) becomes

$$G(s) = \frac{Y(s)}{U(s)} = \frac{k\omega_n^2}{s^2 + 2\zeta\omega_n s + \omega_n^2} \qquad (5.1.11)$$

5.2 A Second-Order System can be Described by Either a Set of Two First-Order ODEs or a Single Second-Order ODE

In the mass-spring system of the previous section, we saw that the system dynamics can be described either by a single differential equation of second order (Eq. (5.1.8)) or by a set of two coupled differential equations of first order (Eqs. (5.1.3) and (5.1.4)). The conversion from one form to the other was very simple. Elimination of velocity from the equations of motion (5.1.3) and (5.1.4) immediately led to (5.1.8). Conversely, starting from (5.1.8), one can define the time derivative of the unknown function as an additional variable v, and this leads to (5.1.3) and (5.1.4).

Let's see now what happens in the case of two first-order systems in series, which, as we saw in the previous chapter, are described by:

$$\tau_1 \frac{dx_1}{dt} + x_1 = k_1 u \qquad (5.2.1)$$

$$\tau_2 \frac{dx_2}{dt} + x_2 = k_2 x_1 \qquad (5.2.2)$$

$$y = x_2 \qquad (5.2.3)$$

In order to eliminate the two state variables x_1 and x_2 we note that

$$\frac{dy}{dt} = \frac{dx_2}{dt} = \frac{k_2}{\tau_2} x_1 - \frac{1}{\tau_2} x_2 \qquad (5.2.4)$$

$$\frac{d^2 y}{dt^2} = \frac{d}{dt}\left(\frac{dy}{dt}\right) = \frac{k_2}{\tau_2} \frac{dx_1}{dt} - \frac{1}{\tau_2} \frac{dx_2}{dt} = -\frac{k_2}{\tau_2}\left(\frac{1}{\tau_1} + \frac{1}{\tau_2}\right) x_1 + \frac{1}{\tau_2^2} x_2 + \frac{k_1 k_2}{\tau_1 \tau_2} u \qquad (5.2.5)$$

Equations (5.2.3), (5.2.4) and (5.2.5) express y, dy/dt and d^2y/dt^2 as linear functions of x_1, x_2 and u. We can, therefore, eliminate x_1, x_2 and derive a linear expression that involves y, dy/dt,

d^2y/dt^2 and u only. More specifically, if we evaluate the following linear combination of y, dy/dt, and d^2y/dt^2, using Eqs. (5.2.3), (5.2.4) and (5.2.5),

$$\tau_1\tau_2 \frac{d^2y}{dt^2} + (\tau_1 + \tau_2)\frac{dy}{dt} + y$$

it turns out that it is independent of x_1, x_2 and is equal to k_1k_2u. Thus, we obtain the following second-order differential equation

$$\tau_1\tau_2 \frac{d^2y}{dt^2} + (\tau_1 + \tau_2)\frac{dy}{dt} + y = k_1k_2u \quad (5.2.6)$$

This differential equation is of the same form as the differential equation that describes the dynamics of the mechanical system of the previous section, and if we set

$$\tau = \sqrt{\tau_1\tau_2} \quad (5.2.7)$$

$$\zeta = \frac{\tau_1 + \tau_2}{2\sqrt{\tau_1\tau_2}} \quad (5.2.8)$$

$$k = k_1k_2 \quad (5.2.9)$$

then differential equation (5.2.6) becomes (5.1.8). It is interesting to note that the differential equation (5.1.8) can also be written in the form of Eq. (5.2.6) and therefore converted to the set (5.2.1)–(5.2.3) by setting

$$k_1k_2 = k \quad (5.2.10)$$

$$\tau_{1,2} = \left(\zeta \pm \sqrt{\zeta^2 - 1}\right)\tau \quad (5.2.11)$$

5.3 Calculating the Response of a Second-Order System – Step Response of a Second-Order System

In order to calculate the output response of a second-order system, we need to solve the corresponding second-order differential equation, or the two coupled first-order differential equations that describe the system. The solution can be calculated with a variety of methods, including time-domain and Laplace-transform techniques. In Chapter 6 we will show how the response of any linear system of any order can be calculated with time-domain techniques. In the present section, we will follow the Laplace-transform approach, the goal being to be able to derive specific formulas for the step response that will be useful in understanding and characterizing dynamic behavior of second-order systems.

5.3 Calculating the Response of a Second-Order System – Step Response of a Second-Order System

In order to calculate the solution of the differential equation (5.1.8), we first take the Laplace transform of both sides, and convert it into an algebraic equation

$$\tau^2\left(s^2 Y(s) - sy(0) - \frac{dy}{dt}(0)\right) + 2\zeta\tau(sY(s) - y(0)) + Y(s) = kU(s) \qquad (5.3.1)$$

which can be solved to obtain the Laplace transform of the unknown solution:

$$Y(s) = \frac{\tau^2 s + 2\zeta\tau}{\tau^2 s^2 + 2\zeta\tau s + 1} y(0) + \frac{\tau^2}{\tau^2 s^2 + 2\zeta\tau s + 1} \frac{dy}{dt}(0) + \frac{k}{\tau^2 s^2 + 2\zeta\tau s + 1} U(s) \qquad (5.3.2a)$$

In the case of zero initial conditions ($y(0) = dy/dt\,(0) = 0$), Eq. (5.3.2a) simplifies to

$$Y(s) = \frac{k}{\tau^2 s^2 + 2\zeta\tau s + 1} U(s) \qquad (5.3.2b)$$

The inverse Laplace transformation of Eq. (5.3.2b) gives the solution of Eq. (5.1.8).

But in order to invert the Laplace transform, we need to expand $Y(s)$ in partial fractions. To this end, we note that, for the polynomial $\tau^2 s^2 + 2\zeta\tau s + 1$,

- if $\zeta > 1$ there are two real roots: $-\dfrac{\zeta}{\tau} \pm \dfrac{\sqrt{\zeta^2 - 1}}{\tau}$
- if $\zeta = 1$ there is one double real root: $-\dfrac{1}{\tau}$
- if $0 \leq \zeta < 1$ there are two complex conjugate roots: $-\dfrac{\zeta}{\tau} \pm i\dfrac{\sqrt{1-\zeta^2}}{\tau}$.

The conclusion is that, depending on the value of ζ, the partial fraction will be different and, when inverted, the terms corresponding to the roots of $\tau^2 s^2 + 2\zeta\tau s + 1$ will give rise to different kinds of terms in the time domain: exponential terms when $\zeta > 1$, exponential times linear when $\zeta = 1$, exponential times sine and cosine when $0 < \zeta < 1$, and just sine and cosine when $\zeta = 0$. In what follows, we will calculate the step response of a second-order system, it will be unavoidable to distinguish cases, and we will end up with different formulas depending on the value of ζ.

For a second-order system initially at steady state ($y(0) = dy/dt(0) = 0$) and subject to a step input of size M, the Laplace transform of the output is given by

$$Y(s) = \frac{k}{\tau^2 s^2 + 2\zeta\tau s + 1} \cdot \frac{M}{s} \qquad (5.3.3)$$

For each case, we need to expand Eq. (5.3.3) in partial fractions and invert the Laplace transform. Table 5.1 gives an overview of the cases that will be considered.

Table 5.1 Classification of second-order systems on the basis of level of damping and corresponding qualitative characteristics of the step response

Case	Characterization of system	Damping factor	Roots of $\tau^2 s^2 + 2\zeta\tau s + 1$	Nature of roots	Step response
1	Overdamped	$\zeta > 1$	$-\dfrac{\zeta}{\tau} \pm \dfrac{\sqrt{\zeta^2-1}}{\tau}$	Simple, real and negative	No oscillations
2	Critically damped	$\zeta = 1$	$-\dfrac{1}{\tau}, -\dfrac{1}{\tau}$	Double, real and negative	No oscillations
3	Underdamped	$0 < \zeta < 1$	$-\dfrac{\zeta}{\tau} \pm i\dfrac{\sqrt{1-\zeta^2}}{\tau}$	Complex with negative real parts	Decaying oscillations
4	Undamped	$\zeta = 0$	$\pm i\dfrac{1}{\tau}$	Pure imaginary	Sustained oscillations

5.3.1 Case 1: $\zeta > 1$ (Overdamped System)

We define

$$p_1 = \frac{-\zeta - \sqrt{\zeta^2-1}}{\tau}, \quad p_2 = \frac{-\zeta + \sqrt{\zeta^2-1}}{\tau} \tag{5.3.4}$$

which are the roots of the polynomial $\tau^2 s^2 + 2\zeta\tau s + 1$, and they are both negative. We then expand in partial fractions:

$$\frac{1}{s(\tau^2 s^2 + 2\zeta\tau s + 1)} = \frac{1/\tau^2}{s(s-p_1)(s-p_2)} = \frac{C_1}{s} + \frac{C_2}{s-p_1} + \frac{C_3}{s-p_2} \tag{5.3.5}$$

where

$$C_1 = \left.\frac{1/\tau^2}{(s-p_1)(s-p_2)}\right|_{s=0} = \frac{1/\tau^2}{p_1 p_2} = 1 \tag{5.3.6}$$

$$C_2 = \left.\frac{1/\tau^2}{s(s-p_2)}\right|_{s=p_1} = \frac{1/\tau^2}{p_1(p_1-p_2)} = -\frac{1}{2}\left(1 - \frac{\zeta}{\sqrt{\zeta^2-1}}\right) \tag{5.3.7}$$

$$C_3 = \left.\frac{1/\tau^2}{s(s-p_1)}\right|_{s=p_2} = \frac{1/\tau^2}{p_2(p_2-p_1)} = -\frac{1}{2}\left(1 + \frac{\zeta}{\sqrt{\zeta^2-1}}\right) \tag{5.3.8}$$

and, therefore, inverting the Laplace transform

$$Y(s) = kM\left(\frac{C_1}{s} + \frac{C_2}{s-p_1} + \frac{C_3}{s-p_2}\right) \tag{5.3.9}$$

5.3 Calculating the Response of a Second-Order System – Step Response of a Second-Order System

we obtain

$$y(t) = kM\left(C_1 + C_2 e^{p_1 t} + C_3 e^{p_2 t}\right) \quad (5.3.10)$$

or

$$y(t) = kM\left[1 - \frac{1}{2}\left(1 - \frac{\zeta}{\sqrt{\zeta^2 - 1}}\right) e^{-\left(\zeta + \sqrt{\zeta^2 - 1}\right)\left(\frac{t}{\tau}\right)} - \frac{1}{2}\left(1 + \frac{\zeta}{\sqrt{\zeta^2 - 1}}\right) e^{-\left(\zeta - \sqrt{\zeta^2 - 1}\right)\left(\frac{t}{\tau}\right)}\right] \quad (5.3.11)$$

We see from Eq. (5.3.11) that the response involves two exponentially decaying terms coming from the roots of the polynomial $\tau^2 s^2 + 2\zeta\tau s + 1$. These will die out for large t, and the output will approach the final value $y(\infty) = kM$.

Figure 5.2 shows the step response of the second-order system for different values of $\zeta > 1$. We observe that the step response depends on both τ and ζ and that the step response of a second-order system is qualitatively similar to the step response of a first-order system. The smaller the value of ζ or τ, the faster is the response.

5.3.2 Case 2: $\zeta = 1$ (Critically Damped System)

For $\zeta = 1$, Eq. (5.3.3) can be written as

$$Y(s) = \frac{k}{(\tau s + 1)^2} \cdot \frac{M}{s} \quad (5.3.12)$$

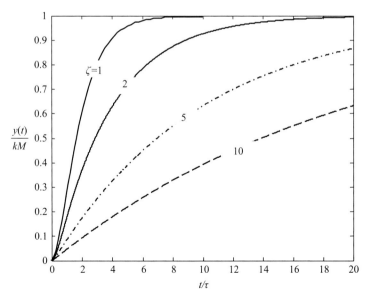

Figure 5.2 Step response of a second-order system for $\zeta \geq 1$.

Expanding in partial fractions,

$$Y(s) = kM\left[\frac{1}{s} - \frac{1}{s+\frac{1}{\tau}} - \frac{\frac{1}{\tau}}{\left(s+\frac{1}{\tau}\right)^2}\right] \tag{5.3.13}$$

and inverting the Laplace transform, we obtain

$$y(t) = kM\left[1 - \left(1 + \frac{t}{\tau}\right)e^{-\frac{t}{\tau}}\right] \tag{5.3.14}$$

We see from Eq. (5.3.14) that the response involves a linear-times-exponential term coming from the double root at $-1/\tau$. This will die out for large t, and the output will approach the final value $y(\infty) = kM$. Qualitatively, the response looks exactly like in the overdamped case (see Figure 5.2).

5.3.3 Case 3: $0 < \zeta < 1$ (Underdamped System)

Now the roots of the polynomial $\tau^2 s^2 + 2\zeta\tau s + 1$ are complex: $p_{1,2} = -\frac{\zeta}{\tau} \pm i\frac{\sqrt{1-\zeta^2}}{\tau}$.

The partial fraction expansion will be like Eq. (5.3.5), but because p_1 and p_2 are complex, so C_2 and C_3 will also be complex. After calculating the coefficients and inverting the Laplace transform, the terms $C_2 e^{p_1 t}$ and $C_3 e^{p_2 t}$ will involve exponential $e^{-\zeta t/\tau}$ (coming from the real part of the roots) multiplied by sine and cosine of $\frac{\sqrt{1-\zeta^2}}{\tau} t$ (coming from the imaginary part of the roots).

The procedure of calculating the partial fraction expansion and inverting involves lengthy algebra, which we will skip. The end result is:

$$y(t) = kM\left[1 - e^{-\frac{\zeta}{\tau}t}\left(\cos\left(\frac{\sqrt{1-\zeta^2}}{\tau}t\right) + \frac{\zeta}{\sqrt{1-\zeta^2}}\sin\left(\frac{\sqrt{1-\zeta^2}}{\tau}t\right)\right)\right] \tag{5.3.15}$$

From Eq. (5.3.15) we see that the response is oscillatory, with frequency $\omega_d = \sqrt{1-\zeta^2}/\tau$. Also, we see that because the sinusoidal functions are multiplied by the decaying exponential function $e^{-\zeta t/\tau}$, the oscillations will be decaying. They will die out for large t, and the output will approach the final value $y(\infty) = kM$. The frequency

$$\omega_d = \frac{\sqrt{1-\zeta^2}}{\tau} \tag{5.3.16}$$

is called the damped-oscillation frequency. In terms of ω_d, the step response formula takes the neater form:

$$y(t) = kM\left[1 - e^{-\frac{\zeta}{\tau}t}\left(\cos(\omega_d t) + \frac{\zeta}{\sqrt{1-\zeta^2}}\sin(\omega_d t)\right)\right] \tag{5.3.17}$$

5.3 Calculating the Response of a Second-Order System – Step Response of a Second-Order System

Also note that the sine and the cosine may be combined using trigonometry formulas, and Eq. (5.3.17) may be written equivalently as

$$y(t) = kM\left[1 - \frac{e^{-\frac{\zeta}{\tau}t}}{\sqrt{1-\zeta^2}}\sin\left(\omega_d t + \cos^{-1}\zeta\right)\right] \quad (5.3.18)$$

The response is shown in Figure 5.3 for different values of ζ between 0 and 1. We observe the oscillatory nature of the response and that the oscillations become more pronounced as $\zeta \to 0$. Furthermore, as $\zeta \to 0$ the response is initially faster, but takes more time to reach its final value. As $\zeta \to 1$, the response becomes similar to that of an overdamped system shown in Figure 5.2, which resembles that of the step response of first-order systems.

5.3.4 Case 4: $\zeta = 0$ (Undamped System)

For $\zeta = 0$, we obtain from Eq. (5.3.3) that

$$Y(s) = \frac{k}{\tau^2 s^2 + 1} \cdot \frac{M}{s} = kM\left(\frac{1}{s} - \frac{s}{s^2 + (1/\tau)^2}\right) \quad (5.3.19)$$

and inverting the Laplace transform, we obtain

$$y(t) = kM\left[1 - \cos\left(\frac{t}{\tau}\right)\right] = kM\left[1 - \cos(\omega_n t)\right] \quad (5.3.20)$$

i.e. the step response consists of sustained oscillations with frequency equal to the natural frequency ω_n of the system. Equation (5.3.20) could have been obtained from Eq. (5.3.15) by simply setting $\zeta = 0$. Also, from Eq. (5.3.16), $\omega_d = 1/\tau = \omega_n$ for $\zeta = 0$.

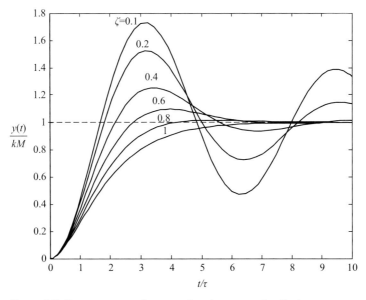

Figure 5.3 Step response of a second-order system for $\zeta \leq 1$.

5.4 Qualitative and Quantitative Characteristics of the Step Response of a Second-Order System

From the results of the previous section it becomes clear that the damping factor ζ determines the qualitative nature of the step response of a second-order system. More specifically

- when $\zeta = 0$ the step response involves sustained oscillations of frequency equal to the natural frequency ω_n,
- when $0 < \zeta < 1$ the step response is oscillatory with exponentially decaying amplitude and frequency equal to $\omega_d = \omega_n (1 - \zeta^2)^{1/2}$,
- when $\zeta \geq 1$ the step response is not oscillatory.

The damping factor is a measure of the level of damping in a second-order system. Oscillatory behavior is only observed when $0 < \zeta < 1$ and oscillations become more damped as ζ increases towards 1.

In Figure 5.4 the main characteristics of the step response of an underdamped second-order system are shown. These characteristics are

- the maximum overshoot (MOS)
- the decay ratio (DR)
- rise time (t_r)
- peak time (t_{p1})
- period of oscillation (T)
- the settling time (t_s).

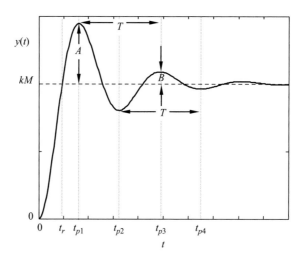

Figure 5.4 Step response characteristics of an underdamped second-order system ($0 < \zeta < 1$).

5.4 Qualitative and Quantitative Characteristics of the Step Response of a Second-Order System

The maximum overshoot (MOS) is defined as the maximum amount by which the output response exceeds the final value, divided by the final value (see Figure 5.4)

$$MOS = \frac{A}{kM} \tag{5.4.1}$$

The decay ratio is the ratio of the first two consecutive overshoots of the step response relative to its final value (see Figure 5.4)

$$DR = \frac{B}{A} \tag{5.4.2}$$

In order to derive the equations that relate MOS and DR with ζ and τ we take the time derivative of the step response $y(t)$ from Eq. (5.3.15) or (5.3.17):

$$\frac{dy}{dt}(t) = kM \frac{e^{-\frac{\zeta}{\tau}t}}{\tau\sqrt{1-\zeta^2}} \sin(\omega_d t) \tag{5.4.3}$$

At the extremal points of the response (either maxima or minima), the derivative of the response is equal to zero, in which case it follows from Eq. (5.4.3) that

$$\sin(\omega_d t) = 0 \tag{5.4.4}$$

or

$$t_{p_j} = \frac{j\pi}{\omega_d}, \; j = 0, 1, 2, \ldots \tag{5.4.5}$$

For $j = 0$ we have that $t_{p0} = 0$. For $k = 1$ we have that the maximum overshoot occurs at $t_{p1} = \pi/\omega_d$ and

$$\frac{y(t_{p_1})}{kM} = 1 + \frac{A}{kM} = 1 - e^{-\frac{\pi\zeta}{\omega_d\tau}}\left(\cos(\pi) + \frac{\zeta}{\sqrt{1-\zeta^2}}\sin(\pi)\right) = 1 + e^{-\frac{\pi\zeta}{\omega_d\tau}}$$

from which it follows that

$$MOS = \frac{A}{kM} = e^{-\frac{\pi\zeta}{\sqrt{1-\zeta^2}}} \tag{5.4.6}$$

As $j=2$ corresponds to a minimum, the second maximum in the response corresponds to $j=3$ or $t_{p3}=3\pi/\omega_d$ and

$$\frac{B}{kM} = e^{-3\frac{\pi\zeta}{\sqrt{1-\zeta^2}}} \tag{5.4.7}$$

From Eqs. (5.4.2), (5.4.6) and (5.4.7), we obtain the equation for the decay ratio

$$DR = \frac{B}{A} = e^{-2\frac{\pi\zeta}{\sqrt{1-\zeta^2}}} = MOS^2 \tag{5.4.8}$$

In Figure 5.5, the *MOS* and *DR* are shown as a function of the damping factor ζ. For small values of the damping factor, both *MOS* and *DR* approach one, while for values of damping factor greater than 0.7 *MOS* is less than 5% while *DR* is practically zero.

The period of oscillation T is the time between two consecutive maxima or minima of the step response and therefore can be calculated easily

$$T = t_{p_3} - t_{p_1} = \frac{2\pi}{\omega_d} = \frac{2\pi\tau}{\sqrt{1-\zeta^2}} \qquad (5.4.9)$$

For the rise time t_r, i.e. the time at which the output crosses its final value for the first time, $y(t_r) = kM$ holds true. From Eq. (5.3.18), we obtain the following condition:

$$\sin(\omega_d t + \cos^{-1}\zeta) = 0$$

The rise time t_r is the smallest positive t for which the above equation is satisfied. This corresponds to

$$\omega_d t_r + \cos^{-1}\zeta = \pi \qquad (5.4.10)$$

hence

$$t_r = \frac{\pi - \cos^{-1}\zeta}{\omega_d} = \frac{\tau(\pi - \cos^{-1}\zeta)}{\sqrt{1-\zeta^2}} \qquad (5.4.11)$$

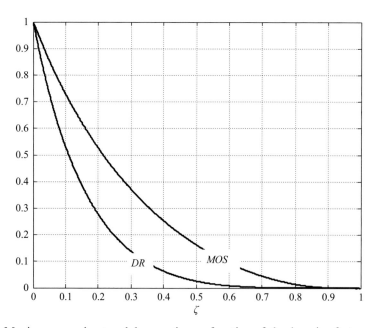

Figure 5.5 Maximum overshoot and decay ratio as a function of the damping factor.

5.4 Qualitative and Quantitative Characteristics of the Step Response of a Second-Order System

The settling time t_s is defined as the time necessary for the system output to reach and stay within a prespecified and narrow band about its final value (kM). The step response of an underdamped second-order system is given by Eq. (5.3.18) and taking the fact that $-1 \leq \sin(\omega_d t + \varphi) \leq 1$ into consideration, it follows that the response is always between the following two curves

$$1 \pm \frac{e^{-\frac{\zeta}{\tau}t}}{\sqrt{1-\zeta^2}}$$

i.e.

$$1 - \frac{e^{-\frac{\zeta}{\tau}t}}{\sqrt{1-\zeta^2}} \leq \frac{y(t)}{kM} \leq 1 + \frac{e^{-\frac{\zeta}{\tau}t}}{\sqrt{1-\zeta^2}} \quad (5.4.12)$$

These are called envelope curves. This is also shown in Figure 5.6. If we define the settling time as the time at which the step response reaches and remains within $\pm\gamma$ of its final value, then the following must hold true

$$1 - \gamma \leq \left.\frac{y(t)}{kM}\right|_{t \geq t_s} \leq 1 + \gamma \quad (5.4.13)$$

and using (5.4.12) we have that

$$e^{-\frac{\zeta}{\tau}t_s} = \gamma\sqrt{1-\zeta^2}$$

or

$$t_s = -\frac{\tau}{\zeta}\ln\left(\gamma\sqrt{1-\zeta^2}\right) \quad (5.4.14)$$

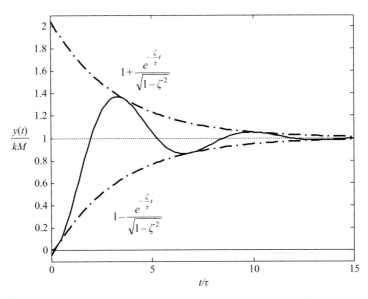

Figure 5.6 Step response of an underdamped second-order system and the pair of envelope curves.

Second-Order Systems

For small values of ζ we can write

$$t_s \approx -\frac{\tau}{\zeta}\ln(\gamma) \tag{5.4.15}$$

For a 5% strip we have that $\gamma = 0.05$ and from the last equation we obtain that

$$t_s \approx 3\frac{\tau}{\zeta} \tag{5.4.16}$$

while for a 2% strip the settling time is

$$t_s \approx 4\frac{\tau}{\zeta} \tag{5.4.17}$$

It is important to note that the settling time is inversely proportional to the damping factor.

Example 5.1 Estimating the parameters of a second-order underdamped system from step-response data

A unit step input is applied to a process system and the output shows a typical underdamped second-order behavior. The operator notes that the first peak in the output occurs 5 s after the step input was applied and that the peak value is 20% higher than the final value. What is the damping factor and the characteristic time of the system? Can you estimate the time necessary for the system to reach and stay within ±5% of the final value?

Solution

From the foregoing analysis, we have the following results for the MOS and the peak time

$$MOS = e^{-\frac{\pi\zeta}{\sqrt{1-\zeta^2}}}$$

$$t_{p1} = \pi/\omega_d = \pi\tau/\sqrt{1-\zeta^2}$$

Equation (5.4.6), which relates the MOS and ζ, can be written as follows

$$\zeta = \frac{-\ln(MOS)}{\sqrt{\pi^2 + \ln(MOS)^2}} = 0.456$$

From Eq. (5.4.5), which relates the extremal values with ζ and τ, we obtain

$$\tau = \frac{t_{p1}}{\pi}\sqrt{1-\zeta^2} = \frac{5}{\pi}\sqrt{1-0.456^2} = 1.416 \text{ s}$$

To calculate the settling time, we use Eq. (5.4.16):

$$t_s \approx \frac{3}{\zeta}\tau \approx 9.32 \text{ s}$$

The exact equation, (5.4.14), predicts a settling time of 9.66 s.

5.5 Frequency Response and Bode Diagrams of Second-Order Systems with $\zeta > 0$

Consider a second-order system initially at steady state ($y(0) = dy/dt(0) = 0$).
At time $t = 0$ the input starts varying sinusoidally, i.e. $u(t) = M\sin(\omega t)$. Using the fact that the Laplace transform of $u(t) = \sin(\omega t)$ is $U(s) = \omega/(s^2+\omega^2)$, Eq. (5.3.2) leads to

$$Y(s) = \frac{k}{\tau^2 s^2 + 2\zeta\tau s + 1} \cdot \frac{M\omega}{s^2 + \omega^2} \tag{5.5.1}$$

As in the case of the calculation of the step response, we need to expand to partial fractions before applying the inverse Laplace transformation. The partial fraction expansion will have the form

$$Y(s) = \frac{C_1}{s - i\omega} + \frac{C_2}{s + i\omega} + \left(\begin{array}{c}\text{Terms coming from the}\\ \text{roots of } \tau^2 s^2 + 2\zeta\tau s + 1\end{array}\right) \tag{5.5.2}$$

For large t, the contribution of the terms of the partial fraction expansion that come from roots of the polynomial $\tau^2 s^2 + 2\zeta\tau s + 1$ will be exponentially decaying. On the contrary, the first two terms will give rise to sustained oscillations, of the same frequency as the sinusoidal input function.

The long-time response of the output will be the inverse Laplace transformation of the first two terms. The calculation of the constants C_1 and C_2 is straightforward but requires some algebra, and the details are left as an exercise for the reader. The end result for the long-time response is given by the formula:

$$y(t) = kM \frac{\left(1 - \omega^2\tau^2\right)\sin(\omega t) - (2\zeta\omega\tau)\cos(\omega t)}{\left(1 - \omega^2\tau^2\right)^2 + (2\zeta\omega\tau)^2} \tag{5.5.3}$$

Using trigonometric identities, Eq. (5.5.3) can be written as

$$y(t) = kM \frac{\sin(\omega t + \varphi)}{\sqrt{\left(1 - \omega^2\tau^2\right)^2 + (2\zeta\omega\tau)^2}}, \text{ where } \varphi = -\cos^{-1}\left(\frac{1 - \omega^2\tau^2}{\sqrt{\left(1 - \omega^2\tau^2\right)^2 + (2\zeta\omega\tau)^2}}\right) \tag{5.5.4}$$

Equation (5.5.4) gives the frequency response of the second-order system.

We can now summarize the characteristics of frequency response. For simplicity, let's assume $k > 0$. We observe that, for large t,

- the output response is also sinusoidal with the same frequency as the input,
- the amplitude ratio is

$$AR = \frac{k}{\sqrt{\left(1 - \omega^2\tau^2\right)^2 + (2\zeta\omega\tau)^2}}, \tag{5.5.5}$$

- the output lags behind the input by an angle

$$\varphi = -\cos^{-1}\left(\frac{1-\omega^2\tau^2}{\sqrt{\left(1-\omega^2\tau^2\right)^2 + (2\zeta\tau\omega)^2}}\right) \tag{5.5.6}$$

These results can be used in order to construct the Bode diagrams of a second-order system. These are shown in Figure 5.7. From Eq. (5.5.5) it follows that

- at low frequencies where $\omega\tau \ll 1$, $AR \approx k$,
- at high frequencies where $\omega\tau \gg 1$, $AR \approx k/(\omega\tau)^2$.

Hence in logarithmic scales, the AR is constant and equal to k at low frequencies, while it is a straight line with slope -2 at high frequencies. The two asymptotes cross each other at the corner frequency $\omega_c = \omega_n = 1/\tau$, as shown in Figure 5.7. From the phase diagram, we have that at low frequencies $\varphi \approx 0°$, while $\varphi \approx -180°$ at high frequencies.

From the amplitude ratio diagram, we can observe that, for small ζ, the AR has a maximum that occurs close to $\omega = \omega_c = \omega_n = 1/\tau$, the natural frequency of the second-order system. This phenomenon is called resonance. In order to find the location and size of the resonance peak, we take the derivative of the expression for AR (Eq. (5.5.5)) with respect to $\omega\tau$.

$$\frac{d(AR)}{d(\omega\tau)} = \frac{d}{d(\omega\tau)}\left\{\frac{k}{\sqrt{\left[1-(\omega\tau)^2\right]^2 + \left[2\zeta(\omega\tau)\right]^2}}\right\} = k\frac{4\omega\tau\left[(\omega\tau)^2 + 2\zeta^2 - 1\right]}{\left[\left(1-\omega^2\tau^2\right)^2 + (2\zeta\omega\tau)^2\right]^{3/2}} = 0 \tag{5.5.7}$$

and we see that the derivative vanishes at the frequency

$$\omega_r = \frac{\sqrt{1-2\zeta^2}}{\tau} = \omega_n\sqrt{1-2\zeta^2} \tag{5.5.8}$$

as long as $1-2\zeta^2 > 0$ or $\zeta < 1/\sqrt{2}$. The frequency ω_r is called the resonant frequency. For $\zeta \geq 1/\sqrt{2}$, AR is a decreasing function of ω, and is maximized at $\omega = 0$. From Eq. (5.5.8) we can also note that the resonant frequency ω_r is always smaller than the natural frequency ω_n and $\omega_r \to \omega_n$ as $\zeta \to 0$. At $\omega = \omega_r$, the corresponding maximal value of AR is given by

$$AR_{max} = \frac{k}{2\zeta\sqrt{1-\zeta^2}} \tag{5.5.9}$$

from which it follows that $AR_{max} \to \infty$ as $\zeta \to 0$.

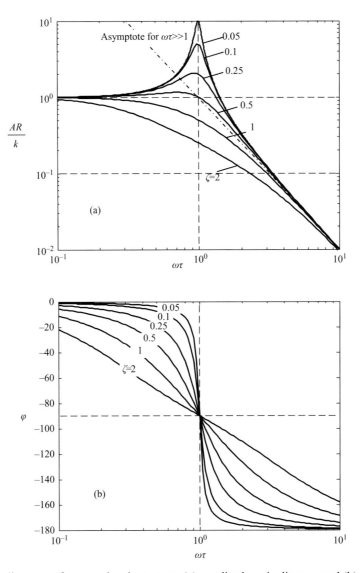

Figure 5.7 Bode diagrams of a second-order system: (a) amplitude ratio diagram and (b) phase diagram.

5.6 The General Form of a Linear Second-Order System

A linear second-order system with a single input and a single output has the following general form

$$\frac{dx_1}{dt} = a_{11}x_1 + a_{12}x_2 + b_1 u$$
$$\frac{dx_2}{dt} = a_{21}x_1 + a_{22}x_2 + b_2 u \quad (5.6.1a)$$

$$y = c_1 x_1 + c_2 x_2 + du \quad (5.6.1b)$$

or, using matrix notation,

$$\frac{d}{dt}\begin{bmatrix} x_1 \\ x_2 \end{bmatrix} = \begin{bmatrix} a_{11} & a_{12} \\ a_{21} & a_{22} \end{bmatrix}\begin{bmatrix} x_1 \\ x_2 \end{bmatrix} + \begin{bmatrix} b_1 \\ b_2 \end{bmatrix} u$$

$$y = \begin{bmatrix} c_1 & c_2 \end{bmatrix}\begin{bmatrix} x_1 \\ x_2 \end{bmatrix} + du$$

(5.6.2)

Equations (5.6.1) or (5.6.2) provide a state-space description of the second-order system, with input variable u, state variables x_1 and x_2, and output variable y. The system can be written in a more compact form as

$$\frac{dx}{dt} = Ax + bu$$

$$y = cx + du$$

(5.6.3)

where

$$x = \begin{bmatrix} x_1 \\ x_2 \end{bmatrix},\ A = \begin{bmatrix} a_{11} & a_{12} \\ a_{21} & a_{22} \end{bmatrix},\ b = \begin{bmatrix} b_1 \\ b_2 \end{bmatrix},\ c = \begin{bmatrix} c_1 & c_2 \end{bmatrix}$$

(5.6.4)

In the Laplace domain and under zero initial conditions, Eqs. (5.6.1) become

$$sX_1(s) = a_{11}X_1(s) + a_{12}X_2(s) + b_1U(s)$$
$$sX_2(s) = a_{21}X_1(s) + a_{22}X_2(s) + b_2U(s)$$
$$Y(s) = c_1X_1(s) + c_2X_2(s) + dU(s)$$

Solving the first two equations with respect to $X_1(s)$ and $X_2(s)$, we find

$$X_1(s) = \frac{b_1 s + (a_{12}b_2 - a_{22}b_1)}{s^2 - (a_{11} + a_{22})s + (a_{11}a_{22} - a_{12}a_{21})}U(s)$$

$$X_2(s) = \frac{b_2 s + (a_{21}b_1 - a_{11}b_2)}{s^2 - (a_{11} + a_{22})s + (a_{11}a_{22} - a_{12}a_{21})}U(s)$$

(5.6.5)

and substituting to the third, we obtain the system's transfer function description:

$$\frac{Y(s)}{U(s)} = \frac{(c_1 b_1 + c_2 b_2)s + (c_1 a_{12} b_2 + c_2 a_{21} b_1 - c_1 a_{22} b_1 - c_2 a_{11} b_2)}{s^2 - (a_{11} + a_{22})s + (a_{11}a_{22} - a_{12}a_{21})} + d$$

$$= \frac{ds^2 + \left[(c_1 b_1 + c_2 b_2) - d(a_{11} + a_{22})\right]s + (c_1 a_{12} b_2 + c_2 a_{21} b_1 - c_1 a_{22} b_1 - c_2 a_{11} b_2) + d(a_{11}a_{22} - a_{12}a_{21})}{s^2 - (a_{11} + a_{22})s + (a_{11}a_{22} - a_{12}a_{21})}$$

(5.6.6)

The denominator polynomial is a quadratic polynomial. If $d \neq 0$ then the numerator polynomial is quadratic as well. Usually, chemical process systems have $d = 0$ and the numerator polynomial is either of degree 1 or a constant (if $d = 0$ and $c_1 b_1 + c_2 b_2 = 0$).

5.7 Software Tools

5.7.1 Calculating the Response of Second-Order Systems Numerically in MATLAB

In order to be able to simulate the response of a prototype second-order system of the form of Eq. (5.1.8), we need to convert the second-order differential equation to a set of two first-order differential equations. This is achieved by defining the state variables

$$x_1 = y$$
$$x_2 = \frac{dy}{dt} \quad (5.7.1)$$

which leads to

$$\frac{dy}{dt} = \frac{dx_1}{dt} = x_2$$
$$\frac{d^2y}{dt^2} = \frac{dx_2}{dt} = -\frac{1}{\tau^2}x_1 - 2\zeta\frac{1}{\tau}x_2 + \frac{k}{\tau^2}u \quad (5.7.2)$$

The following m-file can be used to define the second-order system in state-space form, for the case where the input is a step of unit size:

```
function dxdt=SecondOrder(t,x)
% prototype second order system
% static gain
k=1;
% damping factor
zeta = 0.1;
% characteristic time
tau = 1;
% step input
if (t<0)
    u=0;
else
    u=1;
end
dxdt(1)= x(2);
dxdt(2)= -(1/tau^2)*x(1)-2*(zeta/tau)*x(2)+(k/tau^2)*u;
dxdt=dxdt(:);
```

We then type the following commands in the command window:

» [t,y]=ode45(@SecondOrder,[0 10],[0;0]);
» plot(t,y(:,1))

The result is the response shown in Figure 5.3 for $\zeta = 0.1$. By changing the value of ζ in the m-file we can obtain the other responses shown in Figures 5.2 and 5.3.

Second-Order Systems

The response to an impulse can be obtained by using a limiting pulse as shown in the following modified m-file of the second-order system:

```
function dxdt=SecondOrder(t,x)
% prototype second order system
% static gain
k=1;
% damping factor
zeta = 0.1;
% characteristic time
tau = 1;
% limiting pulse input
epsilon = 1e-6;
if (t<epsilon)
    u=1/epsilon;
else
    u=0;
end
dxdt(1)= x(2);
dxdt(2)= -(1/tau^2)*x(1)-2*(zeta/tau)*x(2)+(k/tau^2)*u;
dxdt=dxdt(:);
```

We then type the following in the command window to obtain the response shown in Figure 5.8 for $\zeta = 0.1$:

```
» [t,y]=ode45(@SecondOrder,[0 10],[0;0]);
» plot(t,y(:,1))
```

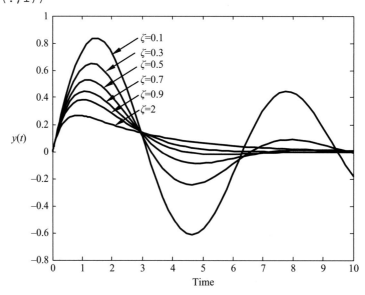

Figure 5.8 Impulse response of a second-order system.

By changing the value of ζ and repeating the last command we obtain Figure 5.8. Responses to other input variations, like ramp and sinusoidal, are obtained by changing the lines in SecondOrder.m that define the input.

5.7.2 Calculating the Response of Second-Order Systems Symbolically in Maple

The same calculation that was performed numerically with MATLAB will now be performed symbolically with Maple using the Laplace transform method.

We start by calling the inttrans library

```
> with(inttrans):
```

defining the system's transfer function

```
> G(s):= k/(s^2*tau^2+2*s*tau*zeta+1);
```

$$G(s) := \frac{k}{s^2\tau^2 + 2s\tau\zeta + 1}$$

and substituting the parameter values $k = 1$, $\tau = 1$, $\zeta = 0.1$:

```
> G(s):= subs([k=1,tau=1,zeta=1/10],G(s));
```

$$G(s) := \frac{1}{s^2 + \frac{1}{5}s + 1}$$

The response to a step input of unit size is then calculated through the laplace and invlaplace commands as follows:

```
> u(t):= 1:
> U(s):= laplace(u(t),t,s):
> Y(s):= G(s)*U(s);
```

$$Y(s) := \frac{1}{\left(s^2 + \frac{1}{5}s + 1\right)s}$$

```
> y(t):=invlaplace(Y(s),s,t);
```

leading to a formula for the unit step response,

$$y(t) := 1 - \frac{1}{33}e^{-\frac{1}{10}t}\left(33\cos\left(\frac{3}{10}\sqrt{11}\,t\right) + \sqrt{11}\sin\left(\frac{3}{10}\sqrt{11}\,t\right)\right)$$

which, when plotted

```
> plot(y(t),t=0..10);
```

gives the response curve shown in Figure 5.3 for $\zeta = 0.1$.

The response to a unit impulse input is calculated similarly:

```
> u(t):= Dirac(t):
> U(s):= laplace(u(t),t,s):
> Y(s):= G(s)*U(s);
```

$$Y(s) := \frac{1}{s^2 + \frac{1}{5}s + 1}$$

and, inverting the Laplace transform, gives a formula for the unit impulse response:

```
> y(t):=invlaplace(Y(s),s,t);
```

$$y(t) := \frac{10}{33}\sqrt{11}\, e^{-\frac{1}{10}t} \sin\left(\frac{3}{10}\sqrt{11}\, t\right)$$

When this is plotted, we obtain the response shown in Figure 5.8 for $\zeta = 0.1$.

LEARNING SUMMARY

- A second-order system may be described either as a system of two first-order differential equations or by a single second-order differential equation.
- Many second-order systems follow the differential equation

$$\tau^2 \frac{d^2 y}{dt^2} + 2\zeta\tau \frac{dy}{dt} + y = ku$$

where u is the input, y is the output, $\tau > 0$ is the characteristic time, $\zeta \geq 0$ is the damping factor and $k \neq 0$ is the static gain.
The response of these systems has been studied in the present chapter, including step response and sinusoidal response.
- Depending on the value ζ, the behavior of second-order systems is different. When $\zeta < 1$, oscillatory transients are observed in the step response of the system. However, when $\zeta \geq 1$, the step response is smooth (no oscillations).
- Under sinusoidal input, second-order systems exhibit the phenomenon of resonance when $\zeta < 1/\sqrt{2}$. Resonance occurs when the input frequency is close to the natural frequency $\omega_n = 1/\tau$.

TERMS AND CONCEPTS

Critically damped system. A system whose dynamic behavior is on the borderline between overdamped and underdamped. For second-order systems, this corresponds to the case of $\zeta = 1$.

Damping. The reduction of the amplitude of oscillations in an oscillatory system induced by external influences.

Damping factor. Also referred to as **damping ratio**, is the dimensionless parameter ζ that measures the amount of damping in a system.

Decay ratio. The ratio of two consecutive overshoots in the step response of an underdamped second-order system.

Natural frequency. The frequency ω_n at which a system oscillates in the absence of damping ($\zeta = 0$).

Overdamped system. A system that responds to external influences without exhibiting oscillations. When second-order systems are considered, this corresponds to cases where $\zeta > 1$.

Overshoot. The fractional deviation of the system output from its final value, at a peak of an oscillatory step response.

Resonant frequency. The frequency ω_r at which the amplitude ratio has a maximum.

Second-order system. Any system described by a single second-order differential equation or a set of two first-order differential equations.

Undamped system. A system where no damping mechanism is present and the system exhibits sustained oscillations when disturbed from equilibrium. When second-order systems are considered, this behavior is observed when $\zeta = 0$.

Underdamped system. A system with damped oscillatory response and, when second-order systems are considered, this corresponds to $0 < \zeta < 1$.

FURTHER READING

A large number of additional systems that exhibit second-order dynamics can be found in the book

Ingham, J., Dunn, I .J., Heinzle, E. and Prenosil, J. E., *Chemical Engineering Dynamics, An Introduction to Modelling and Computer Simulation*, 2nd edn. Weinheim: Wiley-VCH, 2000.

Additional material on the solution of second-order differential equations and Laplace transform can be found in

Boyce, W. E. and DiPrima, R. C., *Elementary Differential Equations and Boundary Value Problems*, 7th edn. New York: John Wiley and Sons, 2001.

Kreyszig, E., *Advanced Engineering Mathematics*, 10th edn. New Jersey: John Wiley & Sons, 2011.

Further analysis of second-order systems dynamics can be found in the books

Luyben, W. L. and Luyben, M. L., *Essentials of Process Control*. McGraw Hill, 1997.

Ogata, K., *Modern Control Engineering, Pearson International Edition*, 5th edn, 2008.

Stephanopoulos, G., *Chemical Process Control, An Introduction to Theory and Practice*. New Jersey: Prentice Hall, 1984.

PROBLEMS

5.1 Prove that the step response of an overdamped system ($\zeta > 1$) can be written as

$$\frac{y(t)}{kM} = 1 - e^{-\frac{\zeta}{\tau}t}\left[\frac{\zeta}{\sqrt{\zeta^2-1}}\sinh\left(\frac{\sqrt{\zeta^2-1}}{\tau}t\right) + \cosh\left(\frac{\sqrt{\zeta^2-1}}{\tau}t\right)\right] \tag{P5.1.1}$$

which resembles the step response (5.3.15) of an underdamped ($0 < \zeta < 1$) system.

5.2 Prove that the step response of an overdamped second-order system ($\zeta > 1$) has an inflection point. Calculate the time at the inflection point, as well as the corresponding value of the output and derivative of the output at the inflection point.

5.3 Prove that the response of an underdamped second-order system ($0 < \zeta < 1$) to an impulse input $u(t) = M\delta(t)$ is given by

$$y(t) = kM\frac{e^{-\frac{\zeta}{\tau}t}}{\tau\sqrt{1-\zeta^2}}\sin\left(\frac{\sqrt{1-\zeta^2}}{\tau}t\right) \tag{P5.3.1}$$

5.4 A second-order system with $\zeta > 0$, initially at steady state with $y(0) = dy/dt(0) = 0$, is subject to a ramp input $u(t) = Mt$. Prove that the long-time response (as $t \to \infty$) is given by

$$y(t) \approx kM(t - 2\zeta\tau) \tag{P5.4.1}$$

5.5 For the system of two interacting tanks that was studied in Chapter 4 (Section 4.3), derive a second-order differential equation that relates the feed flowrate F_{in} to the exit flowrate F_{out}. Calculate τ, ζ and k for this system. Is the system overdamped, critically damped or underdamped?

5.6 In a continuous stirred tank reactor the reversible reaction $R \rightleftarrows P$ takes place under constant temperature. The reaction rates are first order in both directions, with forward reaction rate constant k_f and backward reaction rate constant k_b. Assume that the reactor volume V and the volumetric flowrate F of the feed are both constant.
  (a) Derive the transfer function between the reactant concentration in the feed C_{R0} and the product concentration in the reactor C_P. Calculate τ, ζ and k for this system. Explain the similarities between the reactor under consideration and the system of two interacting tanks that was studied in Chapter 4 (Section 4.3).
  (b) Could the response of the product concentration C_P exhibit oscillations in response to a step change in the concentration of the reactant in the feed?

5.7 In Figure P5.7 a U-tube manometer that is open to the atmosphere is shown. The manometer is used to measure the pressure in a vessel where a compressed gas is stored.

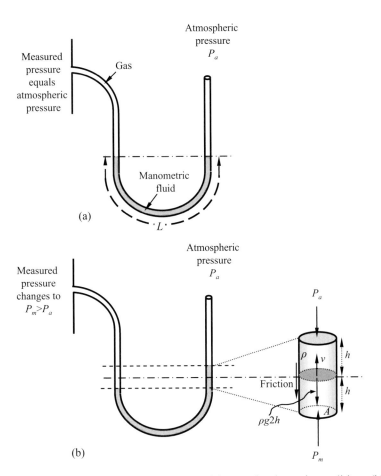

Figure P5.7 U-tube manometer at rest (a) or under dynamic conditions (b).

Initially the pressure inside the vessel is equal to the atmospheric pressure P_a and the manometric fluid is at rest as shown in Figure P5.7a. Suddenly the pressure inside the vessel is increased to $P_m > P_a$ and the manometric fluid is set in motion. Determine the differential equation for the motion of the manometric fluid. Assume laminar flow, Newtonian and incompressible manometric fluid and that Poiseuille's law can be used to interconnect friction forces and the velocity of the manometric fluid.

5.8 Figure P5.8 shows the response of a U-tube manometer reading to a change in the measured pressure from 1 atm to 1.5 atm. Based on the graph, calculate the characteristic time τ, the damping factor ζ and the static gain k of the manometer.

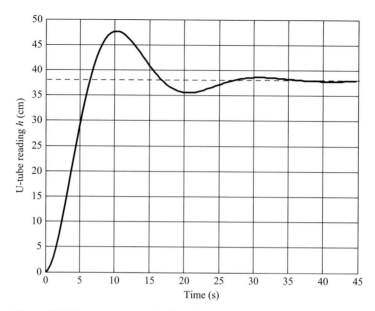

Figure P5.8 Step response of a U-tube manometer.

5.9 A U-tube manometer that is open to the atmosphere has a manometric fluid with the following characteristics:

- $L = 0.8$ m
- $\rho = 2000$ kg/m³
- $\mu = 0.002$ kg/(m s).

You are asked to design the tube diameter in the range 0.001 m to 0.010 m, so as to achieve

- rise time less than 0.5 s
- overshoot less than 50%.

Propose a final design based on the information that sinusoidal disturbances with amplitude of 10 000 Pa at various frequencies are common for the system under study.

Assume that $g \approx 10$ m/s² and that the design of an underdamped system has been selected (i.e. $0 < \zeta < 1$).

For all permissible designs, use MATLAB to generate
(a) the Bode diagram,
(b) the step response for a step change in the measured pressure of 10 000 Pa,
(c) the ramp response for a ramp input $u(t) = 10\,000t$ Pa that lasts for 2 s and
(d) the pulse response for a pulse of 1 000 000 Pa that lasts for 0.01 s.

6 Linear Higher-Order Systems

In the previous three chapters we studied the dynamics of simple first- and second-order systems, which are described by one or two first-order differential equations, or a single second-order differential equation. In the present chapter, we would like to generalize to systems of arbitrary order n. We will study systems described by n linear first-order differential equations, which constitute the state-space description of the system. We will use vectors and matrices to represent the system and to derive general formulas for the response, in a way that directly generalizes the results for first-order systems. The formulas are generally inconvenient for manual calculations, but they form the basis of computational methods that are available in modern software tools.

STUDY OBJECTIVES

After studying this chapter, you should be able to do the following.

- Describe a linear system in terms of vectors and matrices.
- Understand and make use of the properties of the matrix exponential function.
- Understand and be able to apply the general response formulas for a linear system.
- Calculate and simulate the response of general linear systems using software tools, such as MATLAB or Maple

6.1 Representative Examples of Higher-Order Systems – Using Vectors and Matrices to Describe a Linear System

In this section we will give representative examples of higher-order dynamics commonly encountered in process systems. Higher-order dynamics is the collective result of many underlying phenomena that can be expressed using first-order differential equations. The overall system is then modeled by a set of interconnected first-order differential equations. The description of dynamical systems using sets of first-order differential equations is known as the *state-space description*.

As a first example, consider the continuous stirred tank reactor shown in Figure 6.1 in which the following reactions are taking place

$$R \xrightarrow{k_1} I \underset{k_3}{\overset{k_2}{\rightleftharpoons}} P \tag{6.1.1}$$

where R is the reactant, I is an intermediate and P is the product. The reaction rates are first order with respect to the concentration of the corresponding reactant. The reactor operates isothermally and both the reactor volume and the feed and product volumetric flowrates are assumed constant. The feed contains reactant R only with concentration C_{R0}. The component mass balances can be written as follows

$$V \frac{dC_R}{dt} = FC_{R0} - FC_R - Vk_1 C_R \tag{6.1.2}$$

$$V \frac{dC_I}{dt} = -FC_I + Vk_1 C_R - Vk_2 C_I + Vk_3 C_P \tag{6.1.3}$$

$$V \frac{dC_P}{dt} = -FC_P + Vk_2 C_I - Vk_3 C_P \tag{6.1.4}$$

Equations (6.1.2), (6.1.3) and (6.1.4) may be written in matrix form as follows:

$$\frac{d}{dt}\begin{bmatrix} C_R \\ C_I \\ C_P \end{bmatrix} = \begin{bmatrix} -\left(\frac{F}{V}+k_1\right) & 0 & 0 \\ k_1 & -\left(\frac{F}{V}+k_2\right) & k_3 \\ 0 & k_2 & -\left(\frac{F}{V}+k_3\right) \end{bmatrix} \begin{bmatrix} C_R \\ C_I \\ C_P \end{bmatrix} + \begin{bmatrix} \frac{F}{V} \\ 0 \\ 0 \end{bmatrix} C_{R0} \tag{6.1.5}$$

Defining the following vectors and matrices

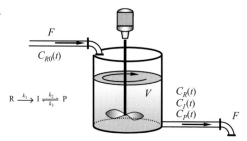

Figure 6.1 Example of a higher-order system.

$$x = \begin{bmatrix} C_R \\ C_I \\ C_P \end{bmatrix}, \quad A = \begin{bmatrix} -\left(\dfrac{F}{V}+k_1\right) & 0 & 0 \\ k_1 & -\left(\dfrac{F}{V}+k_2\right) & k_3 \\ 0 & k_2 & -\left(\dfrac{F}{V}+k_3\right) \end{bmatrix}, \quad b = \begin{bmatrix} \dfrac{F}{V} \\ 0 \\ 0 \end{bmatrix} \quad (6.1.6)$$

and $u = C_{R0}$, Eq. (6.1.5) takes the form

$$\frac{dx}{dt} = Ax + bu \quad (6.1.7)$$

This is the state-space description of the reacting system shown in Figure 6.1.

As a second example we consider the case of three liquid storage tanks connected in series (an extension of the system shown in Figure 4.1 and studied in Section 4.1) where the outflow from each tank depends linearly on the liquid level in the tank. The mathematical description of the system consists of a mass balance for each tank

$$A_1 R_1 \frac{dh_1}{dt} = R_1 F_{in} - h_1 \quad (6.1.8)$$

$$A_2 R_2 \frac{dh_2}{dt} = \frac{R_2}{R_1} h_1 - h_2 \quad (6.1.9)$$

$$A_3 R_3 \frac{dh_3}{dt} = \frac{R_3}{R_2} h_2 - h_3 \quad (6.1.10)$$

where A_1, A_2 and A_3 are the cross-sectional areas, R_1, R_2 and R_3 are the resistances to flow and h_1, h_2 and h_3 the liquid levels. By defining the constants

$$\tau_1 = A_1 R_1, \quad k_1 = R_1 \quad (6.1.11)$$

$$\tau_2 = A_2 R_2, \quad k_2 = R_2/R_1 \quad (6.1.12)$$

$$\tau_3 = A_3 R_3, \quad k_3 = R_3/R_2 \quad (6.1.13)$$

we can then write Eqs. (6.1.8)–(6.1.10) in the following matrix form

$$\frac{d}{dt}\begin{bmatrix} h_1 \\ h_2 \\ h_3 \end{bmatrix} = \begin{bmatrix} -1/\tau_1 & 0 & 0 \\ k_2/\tau_2 & -1/\tau_2 & 0 \\ 0 & k_3/\tau_3 & -1/\tau_3 \end{bmatrix} \begin{bmatrix} h_1 \\ h_2 \\ h_3 \end{bmatrix} + \begin{bmatrix} k_1/\tau_1 \\ 0 \\ 0 \end{bmatrix} F_{in} \quad (6.1.14)$$

Moreover, defining the following vectors and matrices

$$x = \begin{bmatrix} h_1 \\ h_2 \\ h_3 \end{bmatrix}, \quad A = \begin{bmatrix} -1/\tau_1 & 0 & 0 \\ k_2/\tau_2 & -1/\tau_2 & 0 \\ 0 & k_3/\tau_3 & -1/\tau_3 \end{bmatrix}, \quad b = \begin{bmatrix} k_1/\tau_1 \\ 0 \\ 0 \end{bmatrix} \quad (6.1.15)$$

as well as $u = F_{in}$, we see that, again, the system follows the general description of Eq. (6.1.7). If instead of three noninteracting tanks in series, we had an arbitrary number of n tanks in series then the mathematical description would be the following

$$\tau_1 \frac{dx_1}{dt} + x_1 = k_1 u$$

$$\tau_2 \frac{dx_2}{dt} + x_2 = k_2 x_1$$

$$\ldots$$

$$\tau_n \frac{dx_n}{dt} + x_n = k_n x_{n-1}$$

which can be written in matrix form as

$$\frac{d}{dt}\begin{bmatrix} x_1 \\ x_2 \\ \vdots \\ x_n \end{bmatrix} = \begin{bmatrix} -1/\tau_1 & 0 & \cdots & 0 \\ k_2/\tau_2 & -1/\tau_2 & \cdots & 0 \\ 0 & \ddots & \ddots & \vdots \\ 0 & 0 & k_n/\tau_n & -1/\tau_n \end{bmatrix} \begin{bmatrix} x_1 \\ x_2 \\ \vdots \\ x_n \end{bmatrix} + \begin{bmatrix} k_1/\tau_1 \\ 0 \\ \vdots \\ 0 \end{bmatrix} u \quad (6.1.16)$$

This is a state-space description of an nth order system consisting of n first-order systems in series.

In addition to the set of differential equations, one may wish to define an output variable, representing a variable of major significance that needs to be monitored during the operation of the process system. In the reactor example, an output could be the product concentration C_P, in the three tanks example an output could be the volumetric flow rate in the exit line $F_{out} = h_3/R_3$; more generally, an output can be any linear combination of the states and possibly the input.

In general, a linear system with one input variable u, n state variables $x_1, x_2, \ldots, x_n$ and one output y, is described by n linear first-order differential equations with dependent variables $x_1, x_2, \ldots, x_n$ and a linear algebraic equation that defines the output:

$$\frac{dx_1}{dt} = a_{11}x_1 + a_{12}x_2 + \ldots + a_{1n}x_n + b_1 u$$

$$\frac{dx_2}{dt} = a_{21}x_1 + a_{22}x_2 + \ldots + a_{2n}x_n + b_2 u$$

$$\vdots \qquad (6.1.17)$$

$$\frac{dx_n}{dt} = a_{n1}x_1 + a_{n2}x_2 + \ldots + a_{nn}x_n + b_n u$$

$$y = c_1 x_1 + c_2 x_2 + \ldots + c_n x_n + du$$

Defining the state vector x as the column vector that contains all the state variables $x_1, x_2, \ldots, x_n$, as well as appropriate vectors and matrices that contain all the constant parameters a_{ij}, b_i and c_j,

$$x = \begin{bmatrix} x_1 \\ x_2 \\ \vdots \\ x_n \end{bmatrix}, \quad A = \begin{bmatrix} a_{11} & a_{12} & \cdots & a_{1n} \\ a_{21} & a_{22} & \cdots & a_{2n} \\ \vdots & \vdots & \ddots & \vdots \\ a_{n1} & a_{n2} & \cdots & a_{nn} \end{bmatrix}, \quad b = \begin{bmatrix} b_1 \\ b_2 \\ \vdots \\ b_n \end{bmatrix}$$

$$(6.1.18)$$

$$c = \begin{bmatrix} c_1 & c_2 & \cdots & c_n \end{bmatrix}$$

the linear system may be represented in a more compact form as

$$\frac{dx}{dt} = Ax + bu \qquad (6.1.19)$$
$$y = cx + du$$

Equations (6.1.19) give the state-space description of a *single-input–single-output* (SISO) linear nth-order system, with input u, state vector x and output y. The term "linear" refers to the fact that the right-hand sides are linear combinations of state and input.

The term "*linear time-invariant*" (LTI) system is also used in reference to Eq. (6.1.19), to emphasize the fact that the parameters A, b, c, d are constant, independent of time t.

6.2 Steady State of a Linear System – Deviation Variables

Under steady-state conditions, $dx/dt = 0$ and Eqs. (6.1.19) become

$$0 = Ax_s + bu_s \qquad (6.2.1)$$

$$y_s = cx_s + du_s \qquad (6.2.2)$$

If the matrix A is nonsingular (i.e. invertible), Eq. (6.2.1) may be solved for the state vector

$$x_s = -A^{-1}bu_s \qquad (6.2.3)$$

Substituting (6.2.3) into (6.2.2), we obtain the relation between input and output at steady state:

$$y_s = \left(-cA^{-1}b + d\right)u_s \qquad (6.2.4)$$

This is a linear relation. The coefficient $k = -cA^{-1}b + d$ is called the static gain or steady-state gain of the system.

If we define deviation variables with respect to some reference steady state,

$$\begin{aligned} \bar{u} &= u - u_s \\ \bar{x} &= x - x_s \\ \bar{y} &= y - y_s \end{aligned} \qquad (6.2.5)$$

then, by subtracting Eqs. (6.2.1)–(6.2.2) from (6.1.19), we obtain

$$\begin{aligned} \frac{d(x - x_s)}{dt} &= A(x - x_s) + b(u - u_s) \\ y - y_s &= c(x - x_s) + d(u - u_s) \end{aligned} \qquad (6.2.6)$$

or

$$\frac{d\bar{x}}{dt} = A\bar{x} + b\bar{u} \qquad (6.2.7.\text{a})$$

$$\bar{y} = c\bar{x} + d\bar{u} \qquad (6.2.7.\text{b})$$

By direct comparison of Eq. (6.2.7) with (6.1.19), we see that exactly the same equations relate deviation variables $(\bar{x}, \bar{u}, \bar{y})$ and original variables (x, u, y).

The next step is to try to solve the equations in (6.1.19). The hard part is, of course, the solution of the vector differential equation $dx/dt = Ax + bu$. Once this is solved, the output is immediately obtained by substituting the solution into $y = cx + du$.

In what follows, two different solution approaches will be presented:

- solution based on the Laplace transform and
- solution based on the use of the matrix exponential function.

We will start with the Laplace transform method in the next section. We will derive a formula for the Laplace transform of the state vector and the output, which will need to be brought back to the time domain by inverting the Laplace transforms. At the inversion step, the matrix exponential function will emerge, and we will obtain a time-domain representation of the solution involving a convolution integral, in a way that directly generalizes the solution formula for first-order systems given in Chapter 3.

6.3 Using the Laplace-Transform Method to Solve the Linear Vector Differential Equation and Calculate the Response – Transfer Function of a Linear System

The Laplace-transform method will now be applied, using vectors and matrices. Laplace-transforming the vector differential equation (6.1.7), we obtain

$$sX(s) - x(0) = AX(s) + bU(s)$$

From the above equation, we can solve for the Laplace transform of the state vector:

$$X(s) = (sI - A)^{-1} x(0) + (sI - A)^{-1} bU(s) \tag{6.3.1}$$

The time response of the state vector can then be calculated by inverting the Laplace transform

$$x(t) = \mathcal{L}^{-1}\{(sI - A)^{-1}\} x(0) + \mathcal{L}^{-1}\{(sI - A)^{-1} bU(s)\} \tag{6.3.2}$$

Once the state vector has been calculated, one can immediately calculate the output from $y = cx + du$. The Laplace transform of the output is therefore given by

$$Y(s) = c(sI - A)^{-1} x(0) + \left[c(sI - A)^{-1} b + d \right] U(s) \tag{6.3.3}$$

Under zero initial conditions, Eq. (6.3.3) simplifies to

$$Y(s) = \left[c(sI - A)^{-1} b + d \right] U(s) \tag{6.3.4}$$

The function

$$G(s) = c(sI - A)^{-1} b + d \tag{6.3.5}$$

is the transfer function of the SISO linear system (6.1.19). It is important to observe that the matrix $(sI - A)^{-1}$ appears in the derived expressions for the Laplace transforms $X(s)$ and $Y(s)$, as well as in transfer function $G(s)$. This matrix is called the *resolvent matrix* and it will play an important role in characterizing the dynamic response of a linear system.

Example 6.1 Transfer function of a system of two interacting tanks

Consider the system of two interacting tanks that was studied in Chapter 4 (Section 4.3), whose dynamics is described by

$$A_1 \frac{dh_1}{dt} = F_{in} - \frac{h_1 - h_2}{R_1} \qquad (4.3.3)$$

$$A_2 \frac{dh_2}{dt} = \frac{h_1 - h_2}{R_1} - \frac{h_2}{R_2} \qquad (4.3.4)$$

and its output is the outlet flowrate $F_{out} = h_2/R_2$. Apply formula (6.3.5) to calculate the transfer function of the system.

Solution

The given model equations can be put in standard form (6.1.19) as follows:

$$\frac{d}{dt}\begin{bmatrix} h_1 \\ h_2 \end{bmatrix} = \begin{bmatrix} -\dfrac{1}{A_1 R_1} & \dfrac{1}{A_1 R_1} \\ \dfrac{1}{A_2 R_1} & -\left(\dfrac{1}{A_2 R_1} + \dfrac{1}{A_2 R_2}\right) \end{bmatrix} \begin{bmatrix} h_1 \\ h_2 \end{bmatrix} + \begin{bmatrix} \dfrac{1}{A_1} \\ 0 \end{bmatrix} F_{in}$$

$$F_{out} = \begin{bmatrix} 0 & \dfrac{1}{R_2} \end{bmatrix} \begin{bmatrix} h_1 \\ h_2 \end{bmatrix}$$

so we have

$$A = \begin{bmatrix} -\dfrac{1}{A_1 R_1} & \dfrac{1}{A_1 R_1} \\ \dfrac{1}{A_2 R_1} & -\left(\dfrac{1}{A_2 R_1} + \dfrac{1}{A_2 R_2}\right) \end{bmatrix}, \quad b = \begin{bmatrix} \dfrac{1}{A_1} \\ 0 \end{bmatrix}$$

$$c = \begin{bmatrix} 0 & \dfrac{1}{R_2} \end{bmatrix}, \quad d = 0$$

To apply formula (6.3.5), we must first calculate the matrix inverse $(sI - A)^{-1}$. We have:

$$A = \begin{bmatrix} -\dfrac{1}{A_1 R_1} & \dfrac{1}{A_1 R_1} \\ \dfrac{1}{A_2 R_1} & -\left(\dfrac{1}{A_2 R_1} + \dfrac{1}{A_2 R_2}\right) \end{bmatrix} \Rightarrow sI - A = \begin{bmatrix} s + \dfrac{1}{A_1 R_1} & -\dfrac{1}{A_1 R_1} \\ -\dfrac{1}{A_2 R_1} & s + \dfrac{1}{A_2 R_1} + \dfrac{1}{A_2 R_2} \end{bmatrix}$$

The determinant and the adjugate of $(sI - A)$ can now be calculated

$$\det(sI - A) = s^2 + \left(\frac{1}{A_1 R_1} + \frac{1}{A_2 R_1} + \frac{1}{A_2 R_2}\right)s + \frac{1}{A_1 R_1}\frac{1}{A_2 R_2}$$

$$\mathrm{Adj}(sI - A) = \begin{bmatrix} s + \dfrac{1}{A_2 R_1} + \dfrac{1}{A_2 R_2} & \dfrac{1}{A_1 R_1} \\ \dfrac{1}{A_2 R_1} & s + \dfrac{1}{A_1 R_1} \end{bmatrix}$$

from which we can calculate the matrix inverse

$$(sI - A)^{-1} = \frac{1}{\det(sI - A)}\mathrm{Adj}(sI - A)$$

$$= \frac{1}{s^2 + \left(\dfrac{1}{A_1 R_1} + \dfrac{1}{A_2 R_1} + \dfrac{1}{A_2 R_2}\right)s + \dfrac{1}{A_1 R_1}\dfrac{1}{A_2 R_2}} \begin{bmatrix} s + \dfrac{1}{A_2 R_1} + \dfrac{1}{A_2 R_2} & \dfrac{1}{A_1 R_1} \\ \dfrac{1}{A_2 R_1} & s + \dfrac{1}{A_1 R_1} \end{bmatrix}$$

We are now ready to apply formula (6.3.5):

$$G(s) = \frac{\begin{bmatrix} 0 & \dfrac{1}{R_2} \end{bmatrix} \begin{bmatrix} s + \dfrac{1}{A_2 R_1} + \dfrac{1}{A_2 R_2} & \dfrac{1}{A_1 R_1} \\ \dfrac{1}{A_2 R_1} & s + \dfrac{1}{A_1 R_1} \end{bmatrix} \begin{bmatrix} \dfrac{1}{A_1} \\ 0 \end{bmatrix}}{s^2 + \left(\dfrac{1}{A_1 R_1} + \dfrac{1}{A_2 R_1} + \dfrac{1}{A_2 R_2}\right)s + \dfrac{1}{A_1 R_1 A_2 R_2}} + 0$$

Performing the matrix multiplications in the numerator, we get the final answer:

$$G(s) = \frac{\dfrac{1}{A_1 R_1 A_2 R_2}}{s^2 + \left(\dfrac{1}{A_1 R_1} + \dfrac{1}{A_2 R_1} + \dfrac{1}{A_2 R_2}\right)s + \dfrac{1}{A_1 R_1 A_2 R_2}} = \frac{1}{A_1 R_1 A_2 R_2 s^2 + (A_1 R_1 + A_1 R_2 + A_2 R_2)s + 1}$$

The above is exactly the same with the result derived in Chapter 4 by Laplace-transforming each equation individually and solving the resulting linear algebraic equations (see Eq. (4.3.9)).

6.4 The Matrix Exponential Function

The formulas (6.3.1) and (6.3.3) derived in the previous section provide the Laplace transforms of the state vector and the output of the linear system (6.1.19). It would be nice to be able to write a formula for the corresponding inverse Laplace transforms. It turns out that this is indeed possible, but we need to make use of the matrix function that stands for the Laplace transform of $(sI - A)^{-1}$; this is the matrix exponential function.

The matrix exponential function is a function analogous to the usual scalar exponential function, but has a square matrix in the exponent. If A is a square matrix and t is a scalar variable, the matrix exponential function e^{At} is defined as the power series:

$$e^{At} = I + At + A^2 \frac{t^2}{2} + \cdots + A^k \frac{t^k}{k!} + \cdots = \sum_{k=0}^{\infty} A^k \frac{t^k}{k!} \tag{6.4.1}$$

where I is the identity matrix.

It can be shown that the above power series converges for every real number t and for every square matrix A. Also, it can be shown that convergence is uniform with respect to t, so the limit is a continuous function. The function e^{At} has the following properties:

(1) $e^{At}\big|_{t=0} = I$

(2) $\dfrac{d}{dt}\left(e^{At}\right) = Ae^{At} = e^{At}A$

(3) $A\left(\int_0^t e^{At'}\,dt'\right) = \left(\int_0^t e^{At'}\,dt'\right)A = e^{At} - I$

(4) $e^{A(t_1+t_2)} = e^{At_1}e^{At_2} = e^{At_2}e^{At_1}$ for every real numbers t_1, t_2

(5) $\left(e^{At}\right)^{-1} = e^{-At}$

(6) $\mathcal{L}\{e^{At}\} = (sI - A)^{-1}$, where $\mathcal{L}$ denotes the Laplace transform.

In what follows, the proofs of the foregoing properties will be outlined. Property (1) is an immediate consequence of the continuity of the function e^{At}; it is obtained by setting $t = 0$ in (6.4.1). Property (2) is obtained by differentiating the power series (6.4.1) term by term:

$$\frac{d}{dt}\left(e^{At}\right) = A + A^2 t + A^3 \frac{t^2}{2!} + \cdots + A^k \frac{t^{k-1}}{(k-1)!} + \cdots$$

$$= A\left(I + At + A^2 \frac{t^2}{2!} + \cdots + A^{k-1} \frac{t^{k-1}}{(k-1)!} + \cdots\right) = Ae^{At}$$

$$= \left(I + At + A^2 \frac{t^2}{2!} + \cdots + A^{k-1} \frac{t^{k-1}}{(k-1)!} + \cdots\right)A = e^{At}A$$

Property (3) is obtained by simply integrating the relation of property (2) from 0 to t and using property (1). Property (4) is proved as follows:

6.4 The Matrix Exponential Function

$$e^{A(t_1+t_2)} = \sum_{k=0}^{\infty} A^k \frac{(t_1+t_2)^k}{k!} = \sum_{k=0}^{\infty} A^k \left(\sum_{\ell=0}^{k} \frac{t_1^{\ell} t_2^{k-\ell}}{\ell!(k-\ell)!} \right)$$

$$= \left(\sum_{k=0}^{\infty} A^k \frac{t_1^k}{k!} \right) \left(\sum_{k=0}^{\infty} A^k \frac{t_2^k}{k!} \right) = e^{At_1} e^{At_2}$$

$$= \left(\sum_{k=0}^{\infty} A^k \frac{t_2^k}{k!} \right) \left(\sum_{k=0}^{\infty} A^k \frac{t_1^k}{k!} \right) = e^{At_2} e^{At_1}$$

Property (5) is obtained by applying property (4) for $t_1 = t$, $t_2 = -t$: $e^{At} e^{-At} = e^{-At} e^{At} = I$.
Property (6) is obtained by Laplace-transforming the power series (6.4.1) term by term:

$$\mathcal{L}\{e^{At}\} = \mathcal{L}\left\{ I + At + A^2 \frac{t^2}{2} + \ldots + A^k \frac{t^k}{k!} + \ldots \right\}$$

$$= I\frac{1}{s} + A\frac{1}{s^2} + A^2 \frac{1}{s^3} + \ldots + A^k \frac{1}{s^{k+1}} + \ldots$$

$$= \frac{1}{s}\left(I + A\frac{1}{s} + A^2 \frac{1}{s^2} + \ldots + A^k \frac{1}{s^k} + \ldots \right)$$

$$= \frac{1}{s}\left(I - A\frac{1}{s} \right)^{-1} = (sI - A)^{-1}$$

All the above properties are listed in Table 6.1, along with the corresponding properties of the usual scalar exponential function. The matrix exponential function inherits all the key properties of the usual scalar exponential function.

Table 6.1 Properties of the matrix exponential function

	Property of matrix exponential function	Property of usual scalar exponential function
Property (1)	$e^{At}\vert_{t=0} = I$	$e^0 = 1$
Property (2)	$\frac{d(e^{At})}{dt} = Ae^{At} = e^{At}A$	$\frac{d(e^{at})}{dt} = ae^{at} = e^{at}a$
Property (3)	$A\left(\int_0^t e^{At'} dt'\right) = \left(\int_0^t e^{At'} dt'\right) A = e^{At} - I$	$a\left(\int_0^t e^{at'} dt'\right) = \left(\int_0^t e^{at'} dt'\right) a = e^{at} - 1$
Property (4)	$e^{A(t_1+t_2)} = e^{At_1} e^{At_2} = e^{At_2} e^{At_1}$	$e^{a(t_1+t_2)} = e^{at_1} e^{at_2} = e^{at_2} e^{at_1}$
Property (5)	$(e^{At})^{-1} = e^{-At}$	$(e^{at})^{-1} = e^{-at}$
Property (6)	$\mathcal{L}[e^{At}] = (sI - A)^{-1}$	$\mathcal{L}[e^{at}] = \frac{1}{s-a}$

6.5 Solution of the Linear Vector Differential Equation using the Matrix Exponential Function

The solution of the vector differential equation

$$\frac{dx(t)}{dt} = Ax(t) + bu(t) \qquad (6.5.1)$$

where $u(t)$ is a given function, can be obtained using the method of integrating factor and properties (1)–(5) of the matrix exponential function. We first write Eq. (6.5.1) as follows

$$\frac{dx(t)}{dt} - Ax(t) = bu(t) \qquad (6.5.2)$$

and we multiply both sides by e^{-At} from the left to obtain

$$e^{-At}\frac{dx(t)}{dt} - e^{-At}Ax(t) = e^{-At}bu(t) \qquad (6.5.3)$$

Using property (2), we can write

$$\frac{d(e^{-At}x(t))}{dt} = e^{-At}bu(t) \qquad (6.5.4)$$

Integrating the preceding equation between 0 and t and using property (1), we get

$$e^{-At}x(t) - x(0) = \int_0^t e^{-At'}bu(t')\,dt'$$

The above can be solved for $x(t)$ if we multiply from the left by e^{At} and use property (5):

$$x(t) = e^{At}x(0) + e^{At}\int_0^t e^{-At'}bu(t')\,dt'$$

Finally, e^{At} can be inserted inside the integral and combined with the other matrix exponential using property (4):

$$x(t) = e^{At}x(0) + \int_0^t e^{A(t-t')}bu(t')\,dt' \qquad (6.5.5)$$

This is the solution formula for the vector differential equation. The first term accounts for the effect of initial condition $x(0)$, whereas the second term, which is a convolution integral, accounts for the effect of the input. In particular, the second term of (6.5.5) is exactly the convolution of the vector function $e^{At}b$ with the input $u(t)$.

Comparing the solution formula (6.5.5) to Eq. (6.3.1), which gives the Laplace transform of the state vector, we see that the second is the Laplace transform of the first, as a result of property (6):

6.5 Solution of the Linear Vector Differential Equation using the Matrix Exponential Function

$$\mathcal{L}\left[e^{At}x(0)\right] = (sI - A)^{-1}x(0)$$

$$\mathcal{L}\left[\int_0^t e^{A(t-t')}bu(t')dt'\right] = \mathcal{L}\left[e^{At}b\right] \cdot \mathcal{L}[u(t)] = (sI - A)^{-1}b \cdot U(s)$$

In fact, Eq. (6.5.5) could have been proved directly from Eq. (6.3.1) by inverting the Laplace transform.

Example 6.2 Solution of model equations for a system of two interacting tanks

Consider the system of two interacting tanks of Example 6.1, with the following parameter values:

$$A_1 = 1/2,\ R_1 = 1,\ A_2 = 1,\ R_2 = 1/2$$

Calculate the corresponding matrix exponential function and apply formula (6.5.5) to obtain the solution of the system equations.

Solution

For the given values of the parameters, the state space model takes the form:

$$\frac{d}{dt}\begin{bmatrix} h_1 \\ h_2 \end{bmatrix} = \begin{bmatrix} -2 & 2 \\ 1 & -3 \end{bmatrix}\begin{bmatrix} h_1 \\ h_2 \end{bmatrix} + \begin{bmatrix} 2 \\ 0 \end{bmatrix} F_{in}$$

$$F_{out} = \begin{bmatrix} 0 & 2 \end{bmatrix}\begin{bmatrix} h_1 \\ h_2 \end{bmatrix}$$

Here

$$A = \begin{bmatrix} -2 & 2 \\ 1 & -3 \end{bmatrix},\quad b = \begin{bmatrix} 2 \\ 0 \end{bmatrix}$$

$$c = \begin{bmatrix} 0 & 2 \end{bmatrix},\quad d = 0$$

Let's first calculate the matrix exponential function e^{At} for $A = \begin{bmatrix} -2 & 2 \\ 1 & -3 \end{bmatrix}$ by inverting the Laplace transform of $(sI - A)^{-1} = \dfrac{1}{\det(sI - A)}\text{Adj}(sI - A)$. We have:

$$(sI - A) = \begin{bmatrix} s+2 & -2 \\ -1 & s+3 \end{bmatrix}$$

$$\det(sI - A) = s^2 + 5s + 4 = (s+1)(s+4)$$

$$\mathrm{Adj}(sI - A) = \begin{bmatrix} s+3 & 2 \\ 1 & s+2 \end{bmatrix}$$

and so

$$(sI - A)^{-1} = \frac{1}{(s+1)(s+4)} \begin{bmatrix} s+3 & 2 \\ 1 & s+2 \end{bmatrix} = \begin{bmatrix} \dfrac{s+3}{(s+1)(s+4)} & \dfrac{2}{(s+1)(s+4)} \\ \dfrac{1}{(s+1)(s+4)} & \dfrac{s+2}{(s+1)(s+4)} \end{bmatrix}$$

Expanding all the elements of $(sI - A)^{-1}$ in partial fractions,

$$(sI - A)^{-1} = \begin{bmatrix} \dfrac{2/3}{s+1} + \dfrac{1/3}{s+4} & \dfrac{2/3}{s+1} - \dfrac{2/3}{s+4} \\ \dfrac{1/3}{s+1} - \dfrac{1/3}{s+4} & \dfrac{1/3}{s+1} + \dfrac{2/3}{s+4} \end{bmatrix}$$

We can invert all the Laplace transforms and obtain the exponential matrix function:

$$e^{At} = \begin{bmatrix} \dfrac{2}{3}e^{-t} + \dfrac{1}{3}e^{-4t} & \dfrac{2}{3}e^{-t} - \dfrac{2}{3}e^{-4t} \\ \dfrac{1}{3}e^{-t} - \dfrac{1}{3}e^{-4t} & \dfrac{1}{3}e^{-t} + \dfrac{2}{3}e^{-4t} \end{bmatrix}$$

Moreover, we have

$$e^{At}b = \begin{bmatrix} \dfrac{2}{3}e^{-t} + \dfrac{1}{3}e^{-4t} & \dfrac{2}{3}e^{-t} - \dfrac{2}{3}e^{-4t} \\ \dfrac{1}{3}e^{-t} - \dfrac{1}{3}e^{-4t} & \dfrac{1}{3}e^{-t} + \dfrac{2}{3}e^{-4t} \end{bmatrix} \begin{bmatrix} 2 \\ 0 \end{bmatrix} = \begin{bmatrix} \dfrac{4}{3}e^{-t} + \dfrac{2}{3}e^{-4t} \\ \dfrac{2}{3}e^{-t} - \dfrac{2}{3}e^{-4t} \end{bmatrix}$$

and therefore its convolution with the input function $F_{in}(t)$ is

6.5 Solution of the Linear Vector Differential Equation using the Matrix Exponential Function

$$\int_0^t e^{A(t-t')} b F_{in}(t') dt' = \begin{bmatrix} \int_0^t \left(\frac{4}{3} e^{-(t-t')} + \frac{2}{3} e^{-4(t-t')} \right) F_{in}(t') dt' \\ \int_0^t \left(\frac{2}{3} e^{-(t-t')} - \frac{2}{3} e^{-4(t-t')} \right) F_{in}(t') dt' \end{bmatrix}$$

Thus, the solution of the given system of differential equations is:

$$\begin{bmatrix} h_1(t) \\ h_2(t) \end{bmatrix} = e^{At} \begin{bmatrix} h_1(0) \\ h_2(0) \end{bmatrix} + \int_0^t e^{A(t-t')} b F_{in}(t') dt'$$

$$= \begin{bmatrix} \frac{2}{3}e^{-t} + \frac{1}{3}e^{-4t} & \frac{2}{3}e^{-t} - \frac{2}{3}e^{-4t} \\ \frac{1}{3}e^{-t} - \frac{1}{3}e^{-4t} & \frac{1}{3}e^{-t} + \frac{2}{3}e^{-4t} \end{bmatrix} \begin{bmatrix} h_1(0) \\ h_2(0) \end{bmatrix} + \begin{bmatrix} \int_0^t \left(\frac{4}{3} e^{-(t-t')} + \frac{2}{3} e^{-4(t-t')} \right) F_{in}(t') dt' \\ \int_0^t \left(\frac{2}{3} e^{-(t-t')} - \frac{2}{3} e^{-4(t-t')} \right) F_{in}(t') dt' \end{bmatrix}$$

or

$$\begin{bmatrix} h_1(t) \\ h_2(t) \end{bmatrix} = \begin{bmatrix} \left(\frac{2}{3}e^{-t} + \frac{1}{3}e^{-4t} \right) h_1(0) + \left(\frac{2}{3}e^{-t} - \frac{2}{3}e^{-4t} \right) h_2(0) + \int_0^t \left(\frac{4}{3} e^{-(t-t')} + \frac{2}{3} e^{-4(t-t')} \right) F_{in}(t') dt' \\ \left(\frac{1}{3}e^{-t} - \frac{1}{3}e^{-4t} \right) h_1(0) + \left(\frac{1}{3}e^{-t} + \frac{2}{3}e^{-4t} \right) h_2(0) + \int_0^t \left(\frac{2}{3} e^{-(t-t')} - \frac{2}{3} e^{-4(t-t')} \right) F_{in}(t') dt' \end{bmatrix}$$

The above is the solution for the state vector.
The output $F_{out} = 2h_2$ can be immediately obtained:

$$F_{out}(t) = \left(\frac{2}{3}e^{-t} - \frac{2}{3}e^{-4t} \right) h_1(0) + \left(\frac{2}{3}e^{-t} + \frac{4}{3}e^{-4t} \right) h_2(0) + \int_0^t \left(\frac{4}{3} e^{-(t-t')} - \frac{4}{3} e^{-4(t-t')} \right) F_{in}(t') dt'$$

Example 6.3 Solution of model equations for a system consisting of two first-order systems in series with unequal time constants

Consider two first-order systems series, with the following mathematical description:

$$\tau_1 \frac{dx_1}{dt} + x_1 = k_1 u$$

$$\tau_2 \frac{dx_2}{dt} + x_2 = k_2 x_1$$

where the time constants are unequal, $\tau_1 \neq \tau_2$. Calculate the corresponding matrix exponential function and apply formula (6.5.5) to obtain the solution of the system equations.

Solution

We write the mathematical model in the form of a vector differential equation:

$$\frac{d}{dt}\begin{bmatrix} x_1 \\ x_2 \end{bmatrix} = \begin{bmatrix} -1/\tau_1 & 0 \\ k_2/\tau_2 & -1/\tau_2 \end{bmatrix}\begin{bmatrix} x_1 \\ x_2 \end{bmatrix} + \begin{bmatrix} k_1/\tau_1 \\ 0 \end{bmatrix} u$$

Here

$$A = \begin{bmatrix} -1/\tau_1 & 0 \\ k_2/\tau_2 & -1/\tau_2 \end{bmatrix}, \quad b = \begin{bmatrix} k_1/\tau_1 \\ 0 \end{bmatrix}$$

We first calculate the matrix exponential function e^{At}. We have

$$(sI - A)^{-1} = \left(\begin{bmatrix} s & 0 \\ 0 & s \end{bmatrix} - \begin{bmatrix} -1/\tau_1 & 0 \\ k_2/\tau_2 & -1/\tau_2 \end{bmatrix}\right)^{-1} = \begin{bmatrix} s+1/\tau_1 & 0 \\ -k_2/\tau_2 & s+1/\tau_2 \end{bmatrix}^{-1}$$

$$= \frac{1}{\left(s+\frac{1}{\tau_1}\right)\left(s+\frac{1}{\tau_2}\right)} \begin{bmatrix} \left(s+\frac{1}{\tau_2}\right) & 0 \\ k_2/\tau_2 & \left(s+\frac{1}{\tau_1}\right) \end{bmatrix} = \begin{bmatrix} \dfrac{1}{\left(s+\frac{1}{\tau_1}\right)} & 0 \\ \dfrac{k_2/\tau_2}{\left(s+\frac{1}{\tau_1}\right)\left(s+\frac{1}{\tau_2}\right)} & \dfrac{1}{\left(s+\frac{1}{\tau_2}\right)} \end{bmatrix}$$

from which, calculating the inverse Laplace transform of each element of the matrix, we obtain

$$e^{At} = \mathcal{L}^{-1}\begin{bmatrix} \dfrac{1}{\left(s+\frac{1}{\tau_1}\right)} & 0 \\ \dfrac{k_2/\tau_2}{\left(s+\frac{1}{\tau_1}\right)\left(s+\frac{1}{\tau_2}\right)} & \dfrac{1}{\left(s+\frac{1}{\tau_2}\right)} \end{bmatrix} = \begin{bmatrix} e^{-t/\tau_1} & 0 \\ \dfrac{k_2\tau_1}{\tau_1-\tau_2}\left(e^{-t/\tau_1}-e^{-t/\tau_2}\right) & e^{-t/\tau_2} \end{bmatrix}$$

Moreover, we have:

$$e^{At}b = \begin{bmatrix} e^{-t/\tau_1} & 0 \\ \dfrac{k_2\tau_1}{\tau_1-\tau_2}\left(e^{-t/\tau_1}-e^{-t/\tau_2}\right) & e^{-t/\tau_2} \end{bmatrix} \begin{bmatrix} \dfrac{k_1}{\tau_1} \\ 0 \end{bmatrix} = \begin{bmatrix} \dfrac{k_1}{\tau_1}e^{-t/\tau_1} \\ \dfrac{k_1 k_2}{\tau_1-\tau_2}\left(e^{-t/\tau_1}-e^{-t/\tau_2}\right) \end{bmatrix}$$

Substituting the above to the solution formula (6.5.5), we obtain:

$$\begin{bmatrix} x_1(t) \\ x_2(t) \end{bmatrix} = e^{At}\begin{bmatrix} x_1(0) \\ x_2(0) \end{bmatrix} + \int_0^t e^{A(t-t')}bu(t')\,dt'$$

$$= \begin{bmatrix} e^{-t/\tau_1} & 0 \\ \dfrac{k_2\tau_1}{\tau_1-\tau_2}\left(e^{-t/\tau_1}-e^{-t/\tau_2}\right) & e^{-t/\tau_2} \end{bmatrix}\begin{bmatrix} x_1(0) \\ x_2(0) \end{bmatrix} + \begin{bmatrix} \displaystyle\int_0^t \dfrac{k_1}{\tau_1}e^{-(t-t')/\tau_1}u(t')\,dt' \\ \displaystyle\int_0^t \dfrac{k_1 k_2}{\tau_1-\tau_2}\left(e^{-(t-t')/\tau_1}-e^{-(t-t')/\tau_2}\right)u(t')\,dt' \end{bmatrix}$$

$$= \begin{bmatrix} e^{-t/\tau_1}x_1(0) + \displaystyle\int_0^t \dfrac{k_1}{\tau_1}e^{-(t-t')/\tau_1}u(t')\,dt' \\ \dfrac{k_2\tau_1}{\tau_1-\tau_2}\left(e^{-t/\tau_1}-e^{-t/\tau_2}\right)x_1(0) + e^{-t/\tau_2}x_2(0) + \displaystyle\int_0^t \dfrac{k_1 k_2}{\tau_1-\tau_2}\left(e^{-(t-t')/\tau_1}-e^{-(t-t')/\tau_2}\right)u(t')\,dt' \end{bmatrix}$$

6.6 Dynamic Response of a Linear System

Equation (6.5.5) provides the response of the state vector of a linear *n*-th-order system to any combination of initial conditions and/or input variation. The effects of initial condition and input are additive (superposition principle), which allows us to study them separately. In particular, we have the following.

(1) The *unforced response*, i.e. the response of the system to zero input and any nonzero initial conditions, is given by

$$x(t) = e^{At}x(0) \qquad (6.6.1)$$

We observe that, in the unforced response, the matrix exponential function is exactly the matrix that relates the state at time *t* to the state at time 0.

(2) The *forced response*, i.e. the response to any nonzero input, is obtained through the convolution integral. Under zero initial condition, the response is given by

$$x(t) = \int_0^t e^{A(t-t')}bu(t')\,dt' \qquad (6.6.2)$$

Because the dependence of the response on the input is rather complicated, this has created a strong motivation to study representative scenarios, as captured by the idealized input functions, like the step, impulse, ramp and sinusoidal functions.

Table 6.2 summarizes the results for the unforced response and the response to standard idealized inputs. These are a direct generalization of the results derived in Chapter 3 for first-order systems. Indeed, setting $A = -1/\tau$, $b = k/\tau$, one obtains the response formulas for a first-order system given in Table 3.2.

Some of the formulas of Table 6.2, especially the ramp and sinusoidal response formulas, are quite inconvenient for manual calculations because of the heavy matrix algebra. However, because software packages, including MATLAB and Maple, have a built-in function for the exponential matrix as well as one for matrix inversion, these formulas can be directly typed in, and the function $x(t)$ can be subsequently plotted.

It is not hard to derive the results of the table from formula (6.5.5). One will need to perform the necessary integrations for every input function, using properties (1)–(5) of the matrix exponential function. A good starting point for all derivations is to put Eq. (6.6.2) in an equivalent form by changing the variable of integration to $t'' = t - t'$. This gives

$$x(t) = -\int_t^0 e^{At''} bu(t-t'') dt'' = \int_0^t e^{At''} bu(t-t'') dt'' \tag{6.6.3}$$

In the case of a step input of size M, the response is calculated by setting $u(t) = M$ in Eq. (6.6.3)

Table 6.2 Response formulas for the state vector of *n*-th-order linear systems

		Solution of the vector differential equation: $\dfrac{dx}{dt} = Ax + bu$, $x(0)$ = given
Unforced response	$u(t) = 0$	$x(t) = e^{At} x(0)$
Impulse response	$u(t) = M\,\delta(t)$ $x(0) = 0$	$x(t) = e^{At} bM$
Step response	$u(t) = M$ $x(0) = 0$	$x(t) = \left(e^{At} - I\right) A^{-1} bM$
Ramp response	$u(t) = M\,t$ $x(0) = 0$	$x(t) = \left[\left(e^{At} - I\right) A^{-1} - tI\right] A^{-1} bM$
Sinusoidal response	$u(t) = M \sin \omega t$ $x(0) = 0$	$x(t) = \left(\omega e^{At} - A \sin \omega t - \omega I \cos \omega t\right)\left(A^2 + \omega^2 I\right)^{-1} bM$

$$x(t) = \int_0^t e^{At''} bM dt'' = \left(\int_0^t e^{At''} dt''\right) bM$$

If the matrix A is invertible, property (3) of the matrix exponential function leads to

$$x(t) = (e^{At} - I) A^{-1} bM$$

To derive the impulse response formula, one can start from an orthogonal pulse

$$u(t) = \begin{cases} \dfrac{M}{\varepsilon}, & \text{when } 0 \le t < \varepsilon \\ 0, & \text{when } t \ge \varepsilon \end{cases}$$

which, when substituted into Eq. (6.6.3), gives

$$x(t) = \begin{cases} \displaystyle\int_0^t e^{At''} b \frac{M}{\varepsilon} dt'' , & \text{for } 0 \le t < \varepsilon \\ \displaystyle\int_{t-\varepsilon}^t e^{At''} b \frac{M}{\varepsilon} dt'' , & \text{for } t \ge \varepsilon \end{cases}$$

$$= \begin{cases} \dfrac{1}{\varepsilon}\displaystyle\int_0^t e^{At''} bM dt'' , & \text{for } 0 \le t < \varepsilon \\[2mm] \dfrac{\displaystyle\int_0^t e^{At''} bM dt'' - \displaystyle\int_0^{t-\varepsilon} e^{At''} bM dt''}{\varepsilon} , & \text{for } t \ge \varepsilon \end{cases}$$

Taking the limit as $\varepsilon \to 0$, the interval $[0, \varepsilon]$ shrinks to the point 0, whereas the response for $t \ge \varepsilon$ tends to the derivative of $\int_0^t e^{At''} bM dt''$, i.e. to $e^{At} bM$. Thus, we conclude that the impulse response is given by

$$x(t) = e^{At} bM$$

For a ramp input of slope M i.e. for $u(t) = Mt$, Eq. (6.6.3) gives

$$x(t) = \int_0^t e^{At''} bM(t-t'') dt'' = \left(t\int_0^t e^{At''} dt'' - \int_0^t e^{At''} t'' dt''\right) bM$$

If the matrix A is invertible, the integrals can be evaluated using property (3):

$$\int_0^t e^{At''} dt'' = (e^{At} - I) A^{-1} , \quad \int_0^t e^{At''} t'' dt'' = \left(te^{At} - (e^{At} - I) A^{-1}\right) A^{-1}$$

and these lead to the result of Table 6.2. The derivation of the sinusoidal response formula is left as an exercise for the reader.

Once the response of the state vector has been calculated, the output is immediately obtained from $y = cx + du$. The solution formula (6.5.5) leads to

$$y(t) = ce^{At}x(0) + \int_0^t ce^{A(t-t')}bu(t')dt' + du(t) \tag{6.6.4}$$

Equation (6.6.4) involves a convolution integral, in particular the convolution of the function

$$g(t) = ce^{At}b \tag{6.6.5}$$

with the input function $u(t)$. The function $g(t)$ equals the unit impulse response if $d = 0$.

Table 6.3 gives the output response formulas that are immediately obtained when the results of Table 6.2 are substituted into the output equation $y = cx + du$.

Table 6.3 Output response formulas for *n*th-order linear systems

	Output response of the linear system: $\frac{dx}{dt} = Ax + bu, \ x(0) = \text{given}$ $y = cx + du$
Unforced response $u(t) = 0$	$y(t) = ce^{At}x(0)$
Impulse response $u(t) = M\,\delta(t),\ x(0) = 0$	$\dfrac{y(t)}{M} = ce^{At}b + d\delta(t)$
Step response $u(t) = M,\ x(0) = 0$	$\dfrac{y(t)}{M} = c\left(e^{At} - I\right)A^{-1}b + d$
Ramp response $u(t) = M\,t,\ x(0) = 0$	$\dfrac{y(t)}{M} = c\left[\left(e^{At} - I\right)A^{-1} - tI\right]A^{-1}b + dt$
Sinusoidal response $u(t) = M \sin \omega t,\ x(0) = 0$	$\dfrac{y(t)}{M} = c\left(\omega e^{At} - A\sin\omega t - \omega I \cos\omega t\right)\left(A^2 + \omega^2 I\right)^{-1}b + d\sin\omega t$

Example 6.4 Calculation of impulse response and step response of two first-order systems in series with unequal time constants

Consider the two first-order systems connected in series with unequal time constants studied in Example 6.3. Considering $y = x_2$ to be the system output, calculate the unit impulse response and the unit step response.

Solution

From Example 6.3 we have that

$$e^{At} = \begin{bmatrix} e^{-t/\tau_1} & 0 \\ \dfrac{k_2\tau_1}{\tau_1-\tau_2}\left(e^{-t/\tau_1}-e^{-t/\tau_2}\right) & e^{-t/\tau_2} \end{bmatrix}$$

Moreover, we have

$$A^{-1} = \begin{bmatrix} -1/\tau_1 & 0 \\ k_2/\tau_2 & -1/\tau_2 \end{bmatrix}^{-1} = \begin{bmatrix} -\tau_1 & 0 \\ -k_2\tau_1 & -\tau_2 \end{bmatrix}, \quad b = \begin{bmatrix} k_1/\tau_1 \\ 0 \end{bmatrix}, \quad c = \begin{bmatrix} 0 & 1 \end{bmatrix}, \quad d = 0$$

From the impulse response formula of Table 6.3, we obtain

$$y(t) = ce^{At}b = \begin{bmatrix} 0 & 1 \end{bmatrix} \begin{bmatrix} e^{-t/\tau_1} & 0 \\ \dfrac{k_2\tau_1}{\tau_1-\tau_2}\left(e^{-t/\tau_1}-e^{-t/\tau_2}\right) & e^{-t/\tau_2} \end{bmatrix} \begin{bmatrix} \dfrac{k_1}{\tau_1} \\ 0 \end{bmatrix} = \dfrac{k_1 k_2}{\tau_1-\tau_2}\left(e^{-t/\tau_1}-e^{-t/\tau_2}\right)$$

From the step response formula of Table 6.3, we obtain

$$y(t) = c\left(e^{At}-I\right)A^{-1}bM$$

$$= \begin{bmatrix} 0 & 1 \end{bmatrix} \left(\begin{bmatrix} e^{-t/\tau_1} & 0 \\ \dfrac{k_2\tau_1}{\tau_1-\tau_2}\left(e^{-t/\tau_1}-e^{-t/\tau_2}\right) & e^{-t/\tau_2} \end{bmatrix} - \begin{bmatrix} 1 & 0 \\ 0 & 1 \end{bmatrix} \right) \begin{bmatrix} -\tau_1 & 0 \\ -k_2\tau_1 & -\tau_2 \end{bmatrix} \begin{bmatrix} \dfrac{k_1}{\tau_1} \\ 0 \end{bmatrix} M$$

$$= \begin{bmatrix} \dfrac{k_2\tau_1}{\tau_1-\tau_2}\left(e^{-t/\tau_1}-e^{-t/\tau_2}\right) & e^{-t/\tau_2}-1 \end{bmatrix} \begin{bmatrix} -k_1 \\ -k_1 k_2 \end{bmatrix} M$$

or

$$y(t) = k_1 k_2 M \left[1 - \dfrac{\tau_1 e^{-t/\tau_1} - \tau_2 e^{-t/\tau_2}}{\tau_1 - \tau_2} \right]$$

6.7 Response to an Arbitrary Input – Time Discretization of a Linear System

The formulas for the idealized standard inputs derived in the previous section are useful in connection with software packages like MATLAB or Maple, which have a built-in function for the exponential matrix. One can simply enter the appropriate system matrices and plot the formula over a given time interval. In this way, one can gain valuable insights on the dynamic behavior of the system, since these idealized inputs capture (at least approximately) common situations encountered in practice. However, there are also situations where the input variation does not resemble an idealized input, and the output response needs to be accurately calculated. One will then need to apply a numerical algorithm to calculate the output response. In what follows, we will derive a numerical algorithm in exactly the same way as was done in Chapter 3 for first-order systems. Suppose that we know the input values as a set of numbers (say the values at every second) that we read from a data file. Specifically, suppose that the values of $u(t)$ are given in the form of a sequence of numbers, at times $t = 0, T_s, 2T_s, \ldots, jT_s, \ldots$, where T_s is the *sampling period*:

$$u[0] = u(0)$$
$$u[1] = u(T_s)$$
$$u[2] = u(2T_s)$$
$$\vdots$$
$$u[j] = u(jT_s)$$
$$\vdots$$

and we would like to calculate the corresponding values of the state $x[j]$ and the output $y[j]$ at times $t = T_s, 2T_s, \ldots, jT_s, \ldots$, again in the form of a sequence of numbers.

The understanding is, of course, that the sampling period T_s is very small, so that the input data provide sufficiently detailed information on the input variation.

Piecewise constant approximation of the input function – rectangle approximation of the convolution integral: As in first-order systems, we can do a rectangle approximation of the convolution integral, holding the input constant over each time interval and equal to the value at the beginning of the interval:

$$u(t) = \begin{cases} u(0) & , \text{ for } 0 \leq t < T_s \\ u(T_s) & , \text{ for } T_s \leq t < 2T_s \\ u(2T_s) & , \text{ for } 2T_s \leq t < 3T_s \\ \vdots \\ u(jT_s) & , \text{ for } jT_s \leq t < (j+1)T_s \\ \vdots \end{cases}$$

This is the so-called left endpoint rectangle approximation or forward rectangle approximation or zero-order-hold (ZOH) approximation. The other alternative would be a backward rectangle approximation (see Chapter 3), which can be treated similarly.

Let's now calculate $x[j] = x(jT_s)$ using the solution formula (6.5.5) for the vector differential equation $\dfrac{dx}{dt} = Ax + bu$, and the forward rectangle approximation for the input function. Setting $t = jT_s$ and $t = (j-1)T_s$ in the solution formula, we get:

6.7 Response to an Arbitrary Input – Time Discretization of a Linear System

$$x(jT_s) = e^{A(jT_s)}x(0) + \int_0^{jT_s} e^{A(jT_s-t')}bu(t')dt' \tag{6.7.1}$$

$$x((j-1)T_s) = e^{A(j-1)T_s}x(0) + \int_0^{(j-1)T_s} e^{A((j-1)T_s-t')}bu(t')dt' \tag{6.7.2}$$

We then multiply Eq. (6.7.2) by e^{AT_s} and subtract the result from (6.7.1) to obtain:

$$x(jT_s) = e^{AT_s} x((j-1)T_s) + \left[\int_{(j-1)T_s}^{jT_s} e^{A(jT_s-t')}b\,dt'\right]u((j-1)T_s)$$

We then change the integration variable to $t'' = jT_s - t'$ and we conclude:

$$x(jT_s) = e^{AT_s} x((j-1)T_s) + \left(\int_0^{T_s} e^{At''}b\,dt''\right)u((j-1)T_s)$$

or

$$x[j] = e^{AT_s} x[j-1] + \left(\int_0^{T_s} e^{At''}b\,dt''\right)u[j-1] \tag{6.7.3}$$

The above relation allows the recursive calculation of the sequence of state vectors $x[j]$ from the sequence of input values $u[j]$. It is the time-discretization of the solution of the vector differential equation $dx/dt = Ax + bu$, with discretization time step T_s, under the forward rectangle approximation. To complete the discretization of the linear system, we just need to observe that the output equation $y = cx + du$, being a linear algebraic equation, can be immediately discretized as

$$y[j] = cx[j] + du[j] \tag{6.7.4}$$

In summary, the following is the zero-order hold discretization of the linear system of Eg. (6.5.1), with time step T_s:

$$\begin{cases} x[j] = A_d\, x[j-1] + b_d u[j-1] \\ y[j] = cx[j] + du[j] \end{cases} \tag{6.7.5}$$

where

$$A_d = e^{AT_s} \quad , \quad b_d = \int_0^{T_s} e^{At}b\,dt \tag{6.7.6}$$

Example 6.5 Time discretization of a system of two interacting tanks

Use the zero-order hold (forward rectangle) approximation to time-discretize the system of interacting tanks studied in Example 6.2.

Solution

Given the expressions for e^{At} and $e^{At}b$ derived in Example 6.2, we have:

$$A_d = e^{AT_s} = \begin{bmatrix} \frac{2}{3}e^{-T_s} + \frac{1}{3}e^{-4T_s} & \frac{2}{3}e^{-T_s} - \frac{2}{3}e^{-4T_s} \\ \frac{1}{3}e^{-T_s} - \frac{1}{3}e^{-4T_s} & \frac{1}{3}e^{-T_s} + \frac{2}{3}e^{-4T_s} \end{bmatrix}$$

$$b_d = \int_0^{T_s} e^{At} b \, dt = \begin{bmatrix} \int_0^{T_s} \left(\frac{4}{3}e^{-t} + \frac{2}{3}e^{-4t}\right) dt \\ \int_0^{T_s} \left(\frac{2}{3}e^{-t} - \frac{2}{3}e^{-4t}\right) dt \end{bmatrix} = \begin{bmatrix} \frac{3}{2} - \frac{4}{3}e^{-T_s} - \frac{1}{6}e^{-4T_s} \\ \frac{1}{2} - \frac{2}{3}e^{-T_s} + \frac{1}{6}e^{-4T_s} \end{bmatrix}$$

Therefore, the time discretization of the system with time step T_s is:

$$\begin{bmatrix} h_1[j] \\ h_2[j] \end{bmatrix} = \begin{bmatrix} \frac{2}{3}e^{-T_s} + \frac{1}{3}e^{-4T_s} & \frac{2}{3}e^{-T_s} - \frac{2}{3}e^{-4T_s} \\ \frac{1}{3}e^{-T_s} - \frac{1}{3}e^{-4T_s} & \frac{1}{3}e^{-T_s} + \frac{2}{3}e^{-4T_s} \end{bmatrix} \begin{bmatrix} h_1[j-1] \\ h_2[j-1] \end{bmatrix} + \begin{bmatrix} \frac{3}{2} - \frac{4}{3}e^{-T_s} - \frac{1}{6}e^{-4T_s} \\ \frac{1}{2} - \frac{2}{3}e^{-T_s} + \frac{1}{6}e^{-4T_s} \end{bmatrix} F_{in}[j-1]$$

$$F_{out}[j] = \begin{bmatrix} 0 & 2 \end{bmatrix} \begin{bmatrix} h_1[j] \\ h_2[j] \end{bmatrix}$$

Example 6.6 Time discretization of a system consisting of two first-order systems in series with unequal time constants

Use the zero-order-hold (forward rectangle) approximation to time-discretize the system consisting of two first-order systems in series with unequal time constants that was studied in Example 6.3.

Solution

Given the expressions for e^{At} and $e^{At}b$ derived in Example 6.3, we have:

$$A_d = e^{AT_s} = \begin{bmatrix} e^{-T_s/\tau_1} & 0 \\ \dfrac{k_2 \tau_1}{\tau_1 - \tau_2}\left(e^{-T_s/\tau_1} - e^{-T_s/\tau_2}\right) & e^{-T_s/\tau_2} \end{bmatrix}$$

$$b_d = \int_0^{T_s} e^{At} b\, dt = \begin{bmatrix} \int_0^{T_s} \dfrac{k_1}{\tau_1} e^{-t/\tau_1}\, dt \\ \int_0^{T_s} \dfrac{k_1 k_2}{\tau_1 - \tau_2}\left(e^{-t/\tau_1} - e^{-t/\tau_2}\right) dt \end{bmatrix} = \begin{bmatrix} k_1(1 - e^{-T_s/\tau_1}) \\ k_1 k_2 \left(1 - \dfrac{\tau_1 e^{-T_s/\tau_1} - \tau_2 e^{-T_s/\tau_2}}{\tau_1 - \tau_2}\right) \end{bmatrix}$$

6.8 Calculating the Response of a Second-Order System via the Matrix Exponential Function

In this section we will apply the results of this chapter to the case of a prototype second-order system. Consider the second-order system

$$\tau^2 \frac{d^2 y}{dt^2} + 2\zeta\tau \frac{dy}{dt} + y = ku \quad (6.8.1)$$

with $\tau > 0$ and $\zeta \geq 0$.

Defining the following state variables

$$\begin{aligned} x_1 &= y \\ x_2 &= \frac{dy}{dt} \end{aligned} \quad (6.8.2)$$

the second-order differential equation (6.8.1) can be written as a set of first-order differential equations

$$\begin{aligned} \frac{dx_1}{dt} &= x_2 \\ \tau^2 \frac{dx_2}{dt} &= -x_1 - 2\zeta\tau x_2 + ku \end{aligned} \quad (6.8.3)$$

Then, in standard state-space form, a second-order system can be written as follows:

$$\frac{d}{dt}\begin{bmatrix} x_1 \\ x_2 \end{bmatrix} = \begin{bmatrix} 0 & 1 \\ -\dfrac{1}{\tau^2} & -\dfrac{2\zeta}{\tau} \end{bmatrix} \begin{bmatrix} x_1 \\ x_2 \end{bmatrix} + \begin{bmatrix} 0 \\ \dfrac{k}{\tau^2} \end{bmatrix} u$$

$$y = \begin{bmatrix} 1 & 0 \end{bmatrix} \begin{bmatrix} x_1 \\ x_2 \end{bmatrix} \quad (6.8.4)$$

which is of the general form of Eq. (6.1.19) with

$$x = \begin{bmatrix} x_1 \\ x_2 \end{bmatrix}, \quad A = \begin{bmatrix} 0 & 1 \\ -\dfrac{1}{\tau^2} & -\dfrac{2\zeta}{\tau} \end{bmatrix}, \quad b = \begin{bmatrix} 0 \\ \dfrac{k}{\tau^2} \end{bmatrix}, \quad c = \begin{bmatrix} 1 & 0 \end{bmatrix}, \quad d = 0 \qquad (6.8.5)$$

To derive a time-domain formula for the response of a second-order system, we need to first calculate the matrix exponential function $e^{At} = \mathcal{L}^{-1}\left[(sI - A)^{-1}\right]$, where

$$(sI - A)^{-1} = \begin{bmatrix} s & -1 \\ \dfrac{1}{\tau^2} & s + \dfrac{2\zeta}{\tau} \end{bmatrix}^{-1} = \begin{bmatrix} \dfrac{\tau^2 s + 2\zeta\tau}{\tau^2 s^2 + 2\zeta\tau s + 1} & \dfrac{\tau^2}{\tau^2 s^2 + 2\zeta\tau s + 1} \\ -\dfrac{1}{\tau^2 s^2 + 2\zeta\tau s + 1} & \dfrac{\tau^2 s}{\tau^2 s^2 + 2\zeta\tau s + 1} \end{bmatrix}$$

To invert the Laplace transform, all four elements of the matrix must be expanded in partial fractions, depending of course on the value of ζ. The derivations are straightforward but rather tedious, so they will not be given here. The end result is as follows.

- Case $\zeta > 1$ (overdamped system):

$$e^{At} = \begin{bmatrix} \dfrac{1}{2}\left(1 - \dfrac{\zeta}{\sqrt{\zeta^2 - 1}}\right)e^{p_1 t} + \dfrac{1}{2}\left(1 + \dfrac{\zeta}{\sqrt{\zeta^2 - 1}}\right)e^{p_2 t} & \dfrac{\tau}{2\sqrt{\zeta^2 - 1}}\left(e^{p_2 t} - e^{p_1 t}\right) \\ \dfrac{1}{2\tau\sqrt{\zeta^2 - 1}}\left(e^{p_1 t} - e^{p_2 t}\right) & \dfrac{1}{2}\left(1 + \dfrac{\zeta}{\sqrt{\zeta^2 - 1}}\right)e^{p_1 t} + \dfrac{1}{2}\left(1 - \dfrac{\zeta}{\sqrt{\zeta^2 - 1}}\right)e^{p_2 t} \end{bmatrix}$$

where $p_1 = \dfrac{-\zeta - \sqrt{\zeta^2 - 1}}{\tau}, \quad p_2 = \dfrac{-\zeta + \sqrt{\zeta^2 - 1}}{\tau}$

- Case $\zeta = 1$ (critically damped system):

$$e^{At} = \begin{bmatrix} \left(1 + \dfrac{t}{\tau}\right)e^{-\frac{t}{\tau}} & t e^{-\frac{t}{\tau}} \\ -\dfrac{t}{\tau^2}e^{-\frac{t}{\tau}} & \left(1 - \dfrac{t}{\tau}\right)e^{-\frac{t}{\tau}} \end{bmatrix} = e^{-\frac{t}{\tau}}\begin{bmatrix} 1 + \dfrac{t}{\tau} & t \\ -\dfrac{t}{\tau^2} & 1 - \dfrac{t}{\tau} \end{bmatrix}$$

- Case $0 < \zeta < 1$ (underdamped system):

$$e^{At} = e^{-\frac{\zeta}{\tau}t}\begin{bmatrix} \cos\left(\dfrac{\sqrt{1-\zeta^2}}{\tau}t\right) + \dfrac{\zeta}{\sqrt{1-\zeta^2}}\sin\left(\dfrac{\sqrt{1-\zeta^2}}{\tau}t\right) & \dfrac{\tau}{\sqrt{1-\zeta^2}}\sin\left(\dfrac{\sqrt{1-\zeta^2}}{\tau}t\right) \\ -\dfrac{1}{\tau\sqrt{\zeta^2-1}}\sin\left(\dfrac{\sqrt{1-\zeta^2}}{\tau}t\right) & \cos\left(\dfrac{\sqrt{1-\zeta^2}}{\tau}t\right) - \dfrac{\zeta}{\sqrt{1-\zeta^2}}\sin\left(\dfrac{\sqrt{1-\zeta^2}}{\tau}t\right) \end{bmatrix}$$

- Case $\zeta = 0$ (undamped system):

$$e^{At} = \begin{bmatrix} \cos\left(\dfrac{t}{\tau}\right) & \tau \sin\left(\dfrac{t}{\tau}\right) \\ -\dfrac{1}{\tau}\sin\left(\dfrac{t}{\tau}\right) & \cos\left(\dfrac{t}{\tau}\right) \end{bmatrix}$$

From the above results, one can directly apply the response formulas of Table 6.3 to calculate the response to initial conditions and to standard inputs. Also, one can derive time-discretization formulas. All these calculations are straightforward and will be left as exercises.

6.9 Multi-Input–Multi-Output Linear Systems

So far, we have restricted our attention to single-input–single-output (SISO) linear systems. But dynamic systems could have more than one input and/or output.

Consider for example the system of two interacting tanks with two feeds, as shown in Figure 6.2. Under standard assumptions, mass balances for the two tanks can then be written as follows

$$A_1 \frac{dh_1}{dt} = F_{in,1} - \frac{h_1 - h_2}{R_1} \tag{6.9.1}$$

$$A_2 \frac{dh_2}{dt} = F_{in,2} + \frac{h_1 - h_2}{R_1} - \frac{h_2}{R_2} \tag{6.9.2}$$

The inputs of the system are $F_{in,1}$ and $F_{in,2}$, and suppose that we consider the exit flowrate F_{out} as the system's output:

$$F_{out} = \frac{h_2}{R_2} \tag{6.9.3}$$

The system equations can be represented using vectors and matrices as

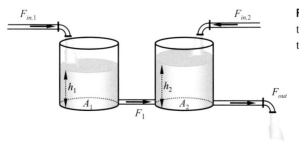

Figure 6.2 Two interacting tanks with feeds in both tanks.

$$\frac{d}{dt}\begin{bmatrix} h_1 \\ h_2 \end{bmatrix} = \begin{bmatrix} -\frac{1}{A_1 R_1} & \frac{1}{A_1 R_1} \\ \frac{1}{A_2 R_1} & -\left(\frac{1}{A_2 R_1} + \frac{1}{A_2 R_2}\right) \end{bmatrix}\begin{bmatrix} h_1 \\ h_2 \end{bmatrix} + \begin{bmatrix} \frac{1}{A_1} \\ 0 \end{bmatrix}F_{in,1} + \begin{bmatrix} 0 \\ \frac{1}{A_2} \end{bmatrix}F_{in,2}$$

$$F_{out} = \begin{bmatrix} 0 & \frac{1}{R_2} \end{bmatrix}\begin{bmatrix} h_1 \\ h_2 \end{bmatrix}$$

(6.9.4)

This is a two-input–one-output system, of the following form:

$$\frac{dx}{dt} = Ax + b_1 u_1 + b_2 u_2$$
$$y = cx + d_1 u_1 + d_2 u_2$$

(6.9.5)

and its response can be calculated in exactly the same way as for SISO systems but, of course, more calculations are involved. In the Laplace domain,

$$X(s) = (sI - A)^{-1} x(0) + (sI - A)^{-1} b_1 U_1(s) + (sI - A)^{-1} b_2 U_2(s)$$
$$Y(s) = c(sI - A)^{-1} x(0) + \left[c(sI - A)^{-1} b_1 + d_1\right] U_1(s) + \left[c(sI - A)^{-1} b_2 + d_2\right] U_2(s)$$

and we now have two transfer functions:

$$G_1(s) = c(sI - A)^{-1} b_1 + d_1$$
$$G_2(s) = c(sI - A)^{-1} b_2 + d_2$$

(6.9.6)

and the system may be represented in block-diagram form as shown in Figure 6.3.

In the time domain, the response of the states and outputs is expressed in exactly the same way as in the SISO case:

$$x(t) = e^{At} x(0) + \int_0^t e^{A(t-t')} b_1 u_1(t') dt' + \int_0^t e^{A(t-t')} b_2 u_2(t') dt'$$

(6.9.7)

$$y(t) = c e^{At} x(0) + \left(\int_0^t c e^{A(t-t')} b_1 u_1(t') dt' + d_1 u_1(t)\right) + \left(\int_0^t c e^{A(t-t')} b_2 u_2(t') dt' + d_2 u_2(t)\right)$$

(6.9.8)

The effect of initial conditions is again captured in the exponential matrix, whereas the effects of the two inputs are additive. This means we can separately calculate the response

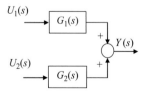

Figure 6.3 Block-diagram representation of a two-input–one-output linear system.

of each input and add them up. The bottom line is that the presence of more than one input does not make the problem more difficult; it just involves more calculations.

Likewise, if we have two outputs instead one (e.g. in the example under consideration, if we view h_1 and h_2 as the outputs), we calculate them separately. We just have more work to do. A linear system with two inputs and two outputs is of the following form:

$$\begin{aligned}\frac{dx}{dt} &= Ax + b_1 u_1 + b_2 u_2 \\ y_1 &= c_1 x + d_{11} u_1 + d_{12} u_2 \\ y_2 &= c_2 x + d_{21} u_1 + d_{22} u_2\end{aligned} \quad (6.9.9)$$

and it has four transfer functions, two for each output:

$$\begin{aligned} G_{11}(s) &= c_1 (sI - A)^{-1} b_1 + d_{11}, \quad G_{12}(s) = c_1 (sI - A)^{-1} b_2 + d_{12} \\ G_{21}(s) &= c_2 (sI - A)^{-1} b_1 + d_{21}, \quad G_{22}(s) = c_2 (sI - A)^{-1} b_2 + d_{22} \end{aligned} \quad (6.9.10)$$

and the system may be represented in block-diagram form as shown in Figure 6.4.

A linear system with m input variables $u_1, u_2, \ldots, u_m$, n state variables $x_1, x_2, \ldots, x_n$, and p output variables $y_1, y_2, \ldots, y_p$, is described by n linear first-order differential equations with dependent variables $x_1, x_2, \ldots, x_n$ and p linear algebraic equations that define the outputs:

$$\begin{aligned}\frac{dx_1}{dt} &= a_{11} x_1 + a_{12} x_2 + \ldots + a_{1n} x_n + b_{11} u_1 + b_{12} u_2 + \ldots + b_{1m} u_m \\ \frac{dx_2}{dt} &= a_{21} x_1 + a_{22} x_2 + \ldots + a_{2n} x_n + b_{21} u_1 + b_{22} u_2 + \ldots + b_{2m} u_m \\ &\vdots \\ \frac{dx_n}{dt} &= a_{n1} x_1 + a_{n2} x_2 + \ldots + a_{nn} x_n + b_{n1} u_1 + b_{n2} u_2 + \ldots + b_{nm} u_m \\ y_1 &= c_{11} x_1 + c_{12} x_2 + \ldots + c_{1n} x_n + d_{11} u_1 + d_{12} u_2 + \ldots + d_{1m} u_m \\ y_2 &= c_{21} x_1 + c_{22} x_2 + \ldots + c_{2n} x_n + d_{21} u_1 + d_{22} u_2 + \ldots + d_{2m} u_m \\ &\vdots \\ y_p &= c_{p1} x_1 + c_{p2} x_2 + \ldots + c_{pn} x_n + d_{p1} u_1 + d_{p2} u_2 + \ldots + d_{pm} u_m \end{aligned} \quad (6.9.11)$$

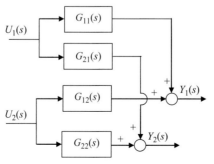

Figure 6.4 Block-diagram representation of a two-input–two-output linear system.

Defining the state vector x as the column vector that contains all the state variables $x_1, x_2, \ldots, x_n$, as well as appropriate vectors and matrices that contain all the constant parameters a_{ij}, b_{ij} and c_{ij},

$$x = \begin{bmatrix} x_1 \\ x_2 \\ \vdots \\ x_n \end{bmatrix}, \quad A = \begin{bmatrix} a_{11} & a_{12} & \cdots & a_{1n} \\ a_{21} & a_{22} & \cdots & a_{2n} \\ \vdots & \vdots & \ddots & \vdots \\ a_{n1} & a_{n2} & \cdots & a_{nn} \end{bmatrix}, \quad b_j = \begin{bmatrix} b_{1j} \\ b_{2j} \\ \vdots \\ b_{nj} \end{bmatrix}, \quad j = 1, \ldots, m \qquad (6.9.12)$$

$$c_i = \begin{bmatrix} c_{i1} & c_{i2} & \cdots & c_{in} \end{bmatrix}, \quad i = 1, \ldots, p$$

the linear system may be represented in a more compact form as

$$\frac{dx}{dt} = Ax + \sum_{j=1}^{m} b_j u_j$$

$$y_i = c_i x + \sum_{j=1}^{m} d_{ij} u_j, \quad i = 1, \ldots, p \qquad (6.9.13)$$

Equations (6.9.13) are the state space description of a *multi-input–multi-output* (MIMO) *linear* system, with inputs u_j, $j = 1, \ldots, m$, state vector x and outputs y_i, $i = 1, \ldots, p$.

The response of MIMO linear systems is calculated in exactly the same way as for SISO but, of course, more calculations are needed. In the Laplace domain,

$$X(s) = (sI - A)^{-1} x(0) + \sum_{j=1}^{m} (sI - A)^{-1} b_j U_j(s) \qquad (6.9.14)$$

$$Y_i(s) = c_i (sI - A)^{-1} x(0) + \sum_{j=1}^{m} \left(c_i (sI - A)^{-1} b_j + d_{ij} \right) U_j(s), \quad i = 1, \ldots, p \qquad (6.9.15)$$

and there are $p \cdot m$ transfer functions, one for each input–output pair:

$$G_{ij}(s) = c_i (sI - A)^{-1} b_j + d_{ij} \qquad (6.9.16)$$

$G_{ij}(s)$ is the transfer function between the input u_j and the output y_i.

In the time domain, the response of the states and outputs is expressed in exactly the same way as in the SISO case:

$$x(t) = e^{At} x(0) + \sum_{j=1}^{m} \int_0^t e^{A(t-t')} b_j u_j(t') dt' \qquad (6.9.17)$$

$$y_i(t) = c_i e^{At} x(0) + \sum_{j=1}^{m} \left(\int_0^t c_i e^{A(t-t')} b_j u_j(t') dt' + d_{ij} u_j(t) \right), \quad i = 1, \ldots, p \qquad (6.9.18)$$

Again, we see that the effect of initial conditions is captured in the exponential matrix, whereas the effects of the m inputs are additive. The p outputs can be calculated separately of each other.

Finally, it should be noted that the state-space description of the linear MIMO system (6.9.13) may be written in a more compact form by collecting all the inputs into one vector and all the outputs into one vector,

$$u = \begin{bmatrix} u_1 \\ u_2 \\ \vdots \\ u_m \end{bmatrix}, \quad y = \begin{bmatrix} y_1 \\ y_2 \\ \vdots \\ y_p \end{bmatrix} \tag{6.9.19}$$

and collecting all the column vectors b_j to form a B matrix, all the row vectors c_i to form a C matrix, and all the coefficients d_{ij} to form a D matrix:

$$B = \begin{bmatrix} b_1 & b_2 & \cdots & b_m \end{bmatrix}$$

$$C = \begin{bmatrix} c_1 \\ c_2 \\ \vdots \\ c_p \end{bmatrix}, \quad D = \begin{bmatrix} d_{11} & d_{12} & \cdots & d_{1m} \\ d_{21} & d_{22} & \cdots & d_{2m} \\ \vdots & \vdots & \ddots & \vdots \\ d_{p1} & d_{p2} & \cdots & d_{pm} \end{bmatrix} \tag{6.9.20}$$

Then (6.9.13) takes the form:

$$\frac{dx}{dt} = Ax + Bu$$
$$y = Cx + Du \tag{6.9.21}$$

In the new notation, the state and output vectors are expressed in the Laplace domain as

$$X(s) = (sI - A)^{-1} x(0) + (sI - A)^{-1} BU(s) \tag{6.9.22}$$

$$Y(s) = C(sI - A)^{-1} x(0) + \left(C(sI - A)^{-1} B + D\right) U(s) \tag{6.9.23}$$

and the transfer function that relates the input and output vectors under zero initial condition

$$G(s) = C(sI - A)^{-1} B + D \tag{6.9.24}$$

is now a $p \times m$ matrix. Its ij-element is the function $G_{ij}(s)$ given by Eq. (6.9.16): the transfer function between the input u_j and the output y_i.

6.10 Software Tools

6.10.1 Calculating the Step, Impulse and Unforced Response of SISO LTI Systems in MATLAB using the State-Space Description

In order to demonstrate how we can calculate the response of a system to any input or initial conditions using a state-space description in MATLAB, we will consider the special case of the second-order system given by Eq. (6.8.4) where $k = \tau = 1$ and $\zeta = 0.6$. We first define the system in MATLAB:

```
>> k=1;
>> tau=1;
>> zeta=0.6;
>> A=[0 1; -1/tau^2 -2*zeta/tau];
>> b=[0; k/tau^2];
>> c=[1 0];
>> d=0;
```

We will first calculate the unit step response calculation using directly the results derived in this chapter and more specifically the result of Table 6.3. To this end, we note that MATLAB has the expm(X) command to calculate the matrix exponential of a square matrix X. Therefore, direct application of step response formula of Table 6.3 is possible in MATLAB, as shown in the following script file:

```
t=linspace(0,10,100);
for i=1:length(t)
    y(i)=-c*(eye(size(A))-expm(A*t(i)))*inv(A)*b+d;
end
plot(t,y,'o')
```

The Control Systems Toolbox of MATLAB contains a number of special commands for calculating the response of systems to some commonly used inputs. To calculate the step response, for instance, we first generate a state-space (SS) object using the command ss available in MATLAB

```
>> sys=ss(A,b,c,d)
a =
         x1    x2
   x1     0     1
   x2    -1  -1.2

b =
         u1
   x1     0
   x2     1

c =
```

```
           x1    x2
     y1     1    0
d =
           u1
     y1     0
Continuous-time model.
```

Then the step response can be immediately obtained using the command `step`

```
>> step(sys)
```

In the same way we can obtain the impulse response (`>> impulse(sys)`) and the unforced response (`>> initial(sys,x0)`). We can also determine the transfer function numerator and denominator polynomials using the `ss2tf` command (**s**tate **s**pace **to t**ransfer **f**unction)

```
>> [num,den] =ss2tf(A,b,c,d)
num =
           0    0.0000    1.0000
den =
       1.0000    1.2000    1.0000
```

`num` and `den` contain the coefficients of the numerator and denominator polynomials in descending powers of s, i.e. the corresponding transfer function is the following:

$$G(s) = \frac{1}{s^2 + 1.2s + 1}$$

6.10.2 Calculating the Zero-Order Hold Discretization of SISO LTI Systems in MATLAB using the State-Space Description

The MATLB command `c2d(sys,T_s)` discretizes the system sys using the zero-order-hold approximation on the input and sampling period T_s.

Consider the interacting tanks of Example 6.5, where we derived the zero-order-hold discretization analytically. We first define the continuous time system:

```
>> a=[-2 2;1 -3]; b=[2;0]; c=[0 2]; d=0;
>> sys=ss(a,b,c,d)
sys =
a =
           x1    x2
     x1    -2    2
     x2     1    -3
b =
           u1
     x1     2
```

```
                x2   0
   c =
                x1   x2
          y1    0    2
   d =
                u1
          y1    0
   Continuous-time state-space model.
```

and then we can apply the c2d command for a given sampling period, say $T_s = 0.1$, to obtain the discrete time system:

```
>> Ts=0.1;
>> sysd=c2d(sys,Ts)
>> sysd =
   a =
                x1          x2
          x1    0.8267      0.1563
          x2    0.07817     0.7485
   b =
                u1
          x1    0.1818
          x2    0.008495
   c =
                x1   x2
          y1    0    2
   d =
                u1
          y1    0
   Sample time: 0.1 seconds
   Discrete-time state-space model.
```

These are exactly the same numbers that the analytical result gives for the sampling period of $T_s = 0.1$.

6.10.3 Calculating the Step, Impulse and Unforced Response of SISO LTI Systems in Maple using the State-Space Description

Consider again the second-order system of Eq. (6.8.4) with $k = \tau = 1$ and $\zeta = 0.6$ that we used for the numerical calculation of responses with MATLAB. To perform the response calculations symbolically with Maple, we start by calling the LinearAlgebra library, which is needed for the matrix manipulations,

```
> with(LinearAlgebra):
```

and defining the system's matrices A, b, c, d

```
> k:=1:tau:=1:zeta:=3/5:
> A:=Matrix(2,[[0,1],[-1/tau^2,-2*zeta/tau]]);
```

$$A := \begin{bmatrix} 0 & 1 \\ -1 & -\dfrac{6}{5} \end{bmatrix}$$

```
> b:=Vector([0,k/tau^2]);
```

$$b := \begin{bmatrix} 0 \\ 1 \end{bmatrix}$$

```
> c:=Vector[row](2,[1,0]);
```

$$c := \begin{bmatrix} 1 & 0 \end{bmatrix}$$

```
> d:=0;
```

The exponential matrix is calculated symbolically using the special command

```
> expAt:=MatrixExponential(A,t);
```

$$expAt := \begin{bmatrix} e^{-\frac{3}{5}t}\cos\left(\frac{4}{5}t\right) + \frac{3}{4}e^{-\frac{3}{5}t}\sin\left(\frac{4}{5}t\right) & \frac{5}{4}e^{-\frac{3}{5}t}\sin\left(\frac{4}{5}t\right) \\ -\frac{5}{4}e^{-\frac{3}{5}t}\sin\left(\frac{4}{5}t\right) & e^{-\frac{3}{5}t}\cos\left(\frac{4}{5}t\right) - \frac{3}{4}e^{-\frac{3}{5}t}\sin\left(\frac{4}{5}t\right) \end{bmatrix}$$

If we want to calculate the unit impulse response from formula (6.6.5), we have to use the function `Multiply(X,Y)` to perform matrix multiplications $X \cdot Y$:

```
> unit_impulse_response:=Multiply(c,Multiply(expAt,b));
```

$$unit\_impulse\_response := \frac{5}{4}e^{-\frac{3}{5}t}\sin\left(\frac{4}{5}t\right)$$

If we want to apply the unit step response formula of Table 6.3, in addition to the `Multiply` function, we need to use the function `MatrixInverse(A)` to perform inversion of the matrix A, as well as the symbol `IdentityMatrix(2)` for the 2×2 identity matrix:

```
> unit_step_response:=Multiply(c,Multiply(Multiply((expAt-
  IdentityMatrix(2)),MatrixInverse(A)),b))+d;
```

$$unit\_step\_response := -e^{-\frac{3}{5}t}\cos\left(\frac{4}{5}t\right) - \frac{3}{4}e^{-\frac{3}{5}t}\sin\left(\frac{4}{5}t\right) + 1$$

Calculation of the system's transfer function form formula (6.3.5) is performed similarly:

```
> G:=Multiply(c,Multiply(MatrixInverse(s*IdentityMatrix
(2)- A),b))+d;
```

$$G := \frac{5}{5s^2 + 6s + 5}$$

LEARNING SUMMARY

In this chapter, methodologies for calculating the response of higher-order systems have been presented, based on either the matrix exponential-function approach or the Laplace-transform approach. Tables 6.2 and 6.3 give the pertinent response formulas. Software packages such as MATLAB and Maple provide useful tools for performing calculations for higher-order systems.

TERMS AND CONCEPTS

Matrix exponential function. A matrix function that directly generalizes the well-known scalar exponential function. It involves an $n \times n$ (square) matrix in the exponent.

MIMO (multi-input–multi-output) systems. Systems with more than one input and/or output.

SISO (single-input–single-output) systems. Systems with one input and one output.

Linear system. A system where the right-hand sides of its state-space representation are linear combinations of state and input variables.

LTI (linear time-invariant) system. A linear system with time-independent (constant) parameters.

State-space representation. A mathematical description of a physical system where input, output and state variables are related through first-order differential equations and algebraic equations.

FURTHER READING

Additional material on the state-space description of SISO LTI systems and the matrix exponential function can be found in the following books

Chen, C. T., *Linear Systems Theory and Applications*, 3rd edn. New York: Oxford University Press, 1999.

DeCarlo, R. A., *Linear Systems, A State Variable Approach with Numerical Implementation*. New Jersey: Prentice Hall International Editions, 1989.

Ogata, K., *Modern Control Engineering*, Pearson International Edition, 5th edn, 2008.

PROBLEMS

6.1 Prove the sinusoidal response formula of Table 6.2.

6.2 Prove the results on the exponential matrix of a second-order system given in Section 6.8. Based on these results, derive impulse response formulas for a second-order system for each case of ζ.

6.3 Use the `impulse` command of MATLAB to calculate the impulse response of a prototype second-order system with $k = \tau = 1$ and $\zeta = 0.3$. Use also the `initial` command to calculate the unforced response to the initial conditions $x_0 = [0\ 1]^T$. Compare the responses obtained and comment on the reason why they are identical.

6.4 Propose a method that can be used to calculate the impulse, step and ramp response of any SISO LTI system with $d = 0$ in MATLAB using the `impulse` command only.

6.5 Consider the following dynamical system

$$\frac{d}{dt}\begin{bmatrix} x_1 \\ x_2 \end{bmatrix} = \begin{bmatrix} 0 & -\alpha \\ 1 & -2\sqrt{\alpha} \end{bmatrix}\begin{bmatrix} x_1 \\ x_2 \end{bmatrix} + \begin{bmatrix} 0 \\ 1 \end{bmatrix} u, \ \alpha > 0 \quad\quad (P6.5.1)$$

Calculate the matrix exponential function and the response to a unit step input (a) analytically for $\alpha > 0$ and (b) numerically for $\alpha = 1$.

6.6 Consider a constant-volume continuous stirred tank reactor (CSTR) shown in Figure P6.6.

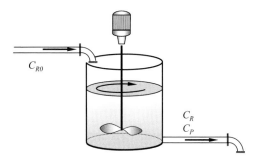

Figure P6.6

The following reaction takes place isothermally in the CSTR:

$$R \underset{k_{-1}}{\overset{k_1}{\rightleftharpoons}} P$$

with the reaction rate being first order in both directions: $r_1 = k_1 C_R$, $r_{-1} = k_{-1} C_P$. The inlet concentration of the reactant C_{R0} is the input of the system, the concentration of species P in the reactor C_P is the output.

(a) Write down a state-space model for this system, consisting of mole balances for species R and P. Put your model in standard matrix form

$$\frac{dx}{dt} = Ax + bu$$
$$y = cx + du$$

(b) Derive the transfer function of the CSTR using the formula $G(s) = c(sI - A)^{-1}b + d$.

6.7 A system of two interacting tanks, with cross-sectional areas $A_1 = 1$ m² and $A_2 = 2$ m², initially operates at steady state with feed flow rate $F_{in} = 1$ m³/min and levels $h_1 = 1.5$ m, $h_2 = 0.5$ m. At time $t = 0$, the operator shuts off the feed by mistake, and this makes $F_{in} = 0$.

(a) Calculate the response of the liquid levels h_1 and h_2 in the tanks.
(b) After how much time will h_2 drop to 0.2 m? What will be the value of h_1 at that time?
(c) An alarm rings when h_2 reaches 0.2 m, the operator realizes his mistake, and immediately restores the feed flow rate to its normal value. What will happen afterwards? Compute h_1 and h_2. Will h_1 and h_2 ultimately reach their original steady-state values?

6.8 The dynamics of two identical first-order processes in series is described by

$$\tau \frac{dx_1}{dt} + x_1 = ku$$

$$\tau \frac{dx_2}{dt} + x_2 = kx_1$$

(a) Put the above differential equations in standard vector form $\frac{dx}{dt} = Ax + bu$ and calculate the corresponding matrix exponential function e^{At}.
(b) If the input $u(t) = 0$ but the initial conditions for the states $x_1(0)$ and $x_2(0)$ are nonzero, calculate the solution of the differential equations.
(c) If the input is given in discrete form with sampling period T_s, use the zero-order-hold approximation to time-discretize the differential equations.

6.9 A two-input–one-output linear system is described by the following equations:

$$\frac{dx_1}{dt} = -83x_1 + 21u_1 + 4u_2$$

$$\frac{dx_2}{dt} = 50x_1 - 75x_2 - u_2$$

$$y = x_2$$

(a) Calculate the transfer functions of the system and draw its block diagram.
(b) Calculate the output response in the following cases:

- $u_1(t) = 1$, $u_2(t) = 0$, $x_1(0) = 0$, $x_2(0) = 0$
- $u_1(t) = 0$, $u_2(t) = 1$, $x_1(0) = 0$, $x_2(0) = 0$

and plot your results.

6.10 Consider a system of three interacting tanks as shown in Figure P6.10.

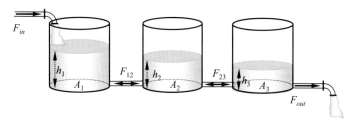

Figure P6.10

The dynamics of the tanks are described by the following mass balances:

$$\begin{cases} A_1 \dfrac{dh_1}{dt} = -\dfrac{h_1 - h_2}{R_1} + F_{in} \\ A_2 \dfrac{dh_2}{dt} = \dfrac{h_1 - h_2}{R_1} - \dfrac{h_2 - h_3}{R_2} \\ A_3 \dfrac{dh_3}{dt} = \dfrac{h_2 - h_3}{R_2} - \dfrac{h_3}{R_3} \end{cases}$$

with $A_1 = A_2 = A_3 = 1$ m², $R_1 = R_2 = 1$ min·m⁻² and $R_3 = 1/2$ min·m⁻². The input of the system is the volumetric feed flowrate F_{in}.

(a) Suppose that the tanks are initially at steady state with $F_{in} = 1$ m³/min. Suddenly, at $t = 0$, the feed is shut off. Use MATLAB to calculate and plot, in a common graph, the response of the heights $h_1(t)$, $h_2(t)$, $h_3(t)$, by applying the general formula $x(t) = e^{At}x(0)$.

(b) Suppose that the tanks are initially at steady state with $F_{in} = 1$ m³/min. Suddenly, at $t = 0$, the feed flowrate is increased to 1.1 m³/min. Plot the response of the heights h_1, h_2, h_3 using the `step` command of MATLAB. Do separate graphs for each height, in nondeviation form, over the range $0 \le t \le 10$ min.

(c) If $F_{out} = h_3/R_3$ is the output, use MATLAB to obtain: (i) the transfer function of the system; (ii) a formula for the unit impulse response of the system.

(d) Suppose that the tanks are initially at steady state with $F_{in} = 1$ m³/min. Suddenly, at $t = 0$, the feed flow rate starts varying sinusoidally around the value of 1 m³/min, with amplitude $M = 0.1$ L/min and frequency $\omega = 1$ rad/min. Use MATLAB to calculate the response of F_{out} by applying formula

$$y(t) = \left[c[\omega e^{At} - A\sin\omega t - \omega I \cos\omega t](A^2 + \omega^2 I)^{-1}b + d\sin\omega t \right]M$$

and plot your result, in nondeviation form, over the range $0 \le t \le 20$ min.

(e) Use the `c2d` command of MATLAB to obtain a zero-order-hold discretization of the system, with sampling period $T_s = 1/6$ min.

6.11 Consider the system of interacting tanks shown in Figure P6.11.

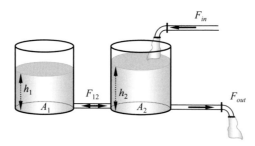

Figure P6.11

Assume that the outlet flowrates are proportional to the pressure drop:

$$F_{12} = \frac{h_1 - h_2}{R_1}, \quad F_{out} = \frac{h_2}{R_2}$$

The input of the system is the inlet flowrate F_{in} and the output of the system is the height h_2.
(a) Write down a state-space model for this system.
(b) Derive the transfer function of the system
(c) For the following values of the parameters: $A_1 = 1$, $R_1 = 1$, $A_2 = 2$, $R_2 = 1/2$, calculate the response of the output to a unit step increase in the inlet flowrate.

6.12 Consider a system of interacting tanks as shown in Figure P6.12.

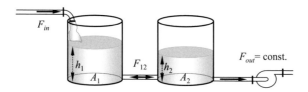

Figure P6.12

The pump in the outflow of the second tank removes fluid at a constant flowrate F_{out}. The flowrate between the two tanks is proportional to the difference of levels: $F_{12} = (h_1 - h_2)/R$. The input of the system is the inlet flowrate F_{in} and the output is the height h_1.
(a) Write down a state-space model for this system.
(b) Put your model in deviation variable form.
(c) Calculate the transfer function of the system.
(d) For the parameter values $A_1 = A_2 = 1/2$ and $R = 1$, calculate the response of the output h_1 to a unit impulse change in the inlet flowrate F_{in} and sketch your result.

6.13 Consider a constant-volume CSTR shown in Figure P6.13.

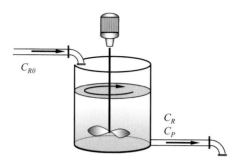

Figure P6.13

The following reactions are taking place in a CSTR under constant temperature:

$$R \underset{k_2}{\overset{k_1}{\rightleftharpoons}} P$$
$$R \overset{k_3}{\longrightarrow} B$$

Under the assumptions of first-order kinetics, constant reactor volume V and constant inlet and outlet volumetric flowrates F, the mole balances of species R and P take the form:

$$V\frac{dC_R}{dt} = F(C_{R0} - C_R) - Vk_1 C_R + Vk_2 C_P - Vk_3 C_R$$

$$V\frac{dC_P}{dt} = -FC_P + Vk_1 C_R - Vk_2 C_P$$

The inlet concentration of the reactant C_{R0} is the input variable, the concentrations C_R and C_P in the reactor are the states, the reactant concentration C_R is the output variable (V, F, k_1, k_2, k_3 are constant parameters). Calculate the transfer function of this system.

6.14 The dynamics of a process is described by the following model:

$$\frac{d}{dt}\begin{bmatrix} x_1 \\ x_2 \end{bmatrix} = \begin{bmatrix} -1 & 0 \\ 2 & -2 \end{bmatrix}\begin{bmatrix} x_1 \\ x_2 \end{bmatrix} + \begin{bmatrix} 1 \\ 0 \end{bmatrix} u$$

$$y = \begin{bmatrix} 0 & 1 \end{bmatrix}\begin{bmatrix} x_1 \\ x_2 \end{bmatrix}$$

and we have calculated the exponential matrix

$$e^{\begin{bmatrix} -1 & 0 \\ 2 & -2 \end{bmatrix} t} = \begin{bmatrix} e^{-t} & 0 \\ 2(e^{-t} - e^{-2t}) & e^{-2t} \end{bmatrix}$$

(a) If $x_1(0) = x_2(0) = 0$, calculate the output response $y(t)$ to an arbitrary input change $u(t)$ in the form of an integral.

(b) If the input is given in discrete form with sampling period T_s, use the zero-order-hold approximation to discretize the differential equations.

6.15 A two-tank mixing process is shown in the Figure P6.15.

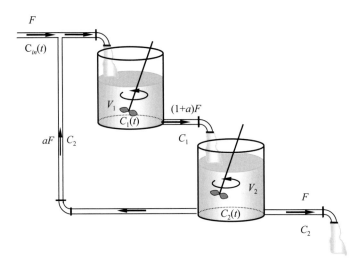

Figure P6.15

The dynamics of this process is described by the following differential equations:

$$V_1 \frac{dC_1}{dt} = FC_{in} + aFC_2 - (1+a)FC_1$$

$$V_2 \frac{dC_2}{dt} = (1+a)FC_1 - (1+a)FC_2$$

The feed concentration C_{in} is the input variable, whereas the concentration C_2 in the second tank is the output variable (V_1, V_2, F and a are constant parameters). Calculate the transfer function of this system.

6.16 Consider a system of two continuous stirred tank reactors with recycle, shown in Figure P6.16, where the irreversible reaction R → P takes place isothermally. The reaction has first-order kinetics with rate constant k. It is assumed that the inlet and outlet volumetric flowrates F, the recycle volumetric flowrate F_R and the reactor volumes

V_1, V_2 are maintained constant during the operation of the reactors. Under the above assumptions, mole balances of the reactant R for each reactor take the form:

$$V_1 \frac{dC_1}{dt} = FC_{in} + F_R C_2 - (F + F_R)C_1 - V_1 k C_1$$

$$V_2 \frac{dC_2}{dt} = (F + F_R)C_1 - (F + F_R)C_2 - V_2 k C_2$$

The inlet concentration of the reactant C_{in} is the input of the system, the concentrations C_1 and C_2 of the reactant in the reactors are the states, and the outlet reactant concentration C_2 is the output (V_1, V_2, F, F_R and k are constant parameters). Calculate the transfer function of this system.

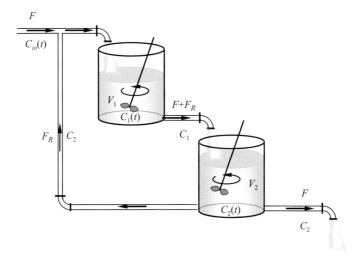

Figure P6.16

6.17 In the absence of friction, the dynamics of a mass–spring system is described by following equations of motion:

$$\frac{d}{dt}\begin{bmatrix} y \\ v \end{bmatrix} = \begin{bmatrix} 0 & 1 \\ -\omega_n^2 & 0 \end{bmatrix}\begin{bmatrix} y \\ v \end{bmatrix} + \begin{bmatrix} 0 \\ \frac{1}{m} \end{bmatrix} f$$

The external force f is the input, the displacement y and the velocity v of the center of mass are the states, and m and ω_n are constant parameters.

(a) Calculate the matrix exponential function $e^{\begin{bmatrix} 0 & 1 \\ -\omega_n^2 & 0 \end{bmatrix} t}$.

(b) If the external force $f(t) = 0$ and the initial velocity $v(0) = 0$, but the initial displacement $y(0)$ is nonzero, what will be the response of the states $y(t)$ and $v(t)$?

(c) Discretize the system equations with sampling period T_s. Use the zero-order-hold approximation.

6.18 Using the power-series expansion (6.4.1) of the matrix exponential function, derive power series expansions for the matrices A_d and b_d defined by (6.7.6), in powers of T_s. If T_s is very small, it is possible to truncate the power series, keeping up to linear in T_s terms. Evaluate the accuracy of this approximation for the matrix

$$A = \begin{bmatrix} -2 & 2 \\ 1 & -3 \end{bmatrix}$$

and for $T_s = 0.001, 0.01, 0.1, 1$.

6.19 Consider the system of two heaters in series shown in Figure P6.19. The dynamics of the heaters can be described by the following energy balances:

$$V_1 \rho c_p \frac{dT_1}{dt} = F \rho c_p (T_0 - T_1) + Q_1$$

$$V_2 \rho c_p \frac{dT_2}{dt} = F \rho c_p (T_1 - T_2) + Q_2$$

The inlet temperature T_0 and the heating rates Q_1, Q_2 are the system inputs, whereas the temperatures T_1 and T_2 are the outputs. V_1, V_2, F, ρ and c_p are constant parameters.

(a) Calculate the transfer functions and draw the block diagram of the system.

(b) If the system is initially at steady state, and suddenly the heating rate in the first tank Q_1 undergoes a step change of size M_1, calculate the response of the outputs $T_1(t)$ and $T_2(t)$. Assume that $V_1 \neq V_2$.

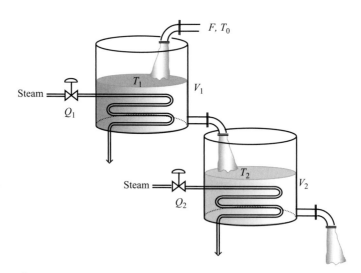

Figure P6.19

7   Eigenvalue Analysis – Asymptotic Stability

The aim of this chapter is to study qualitative aspects of dynamic response and, in particular, to introduce the notion of stability, which is of paramount importance in the analysis of process dynamics and the design of controllers. We begin by introducing the connection between the eigenvalues of the system matrix and the unforced response of a linear system. We then introduce the physical meaning of stability and the mathematical conditions for asymptotic stability of linear systems and local asymptotic stability of nonlinear systems. For nonlinear systems, local asymptotic stability is determined through the eigenvalues of the linearization of the system around a reference steady state.

STUDY OBJECTIVES

After studying this chapter you should be able to do the following.

- Explain the notion of stability and its importance when studying system dynamics.
- Use simple criteria to determine stability of linear systems.
- Perform local stability analysis of nonlinear systems using linearization.

7.1   Introduction

In the previous chapter, we focused on how to calculate the response of a linear higher-order system, analytically or numerically. In the present chapter, we would like to focus on the qualitative characteristics of the response, including the property of stability. Stability is an absolute prerequisite for a process to operate properly and safely. When designing controllers, the number one requirement is stability of the feedback control system. Stability will first be studied for a linear system, and a simple criterion will be derived in terms of the system eigenvalues. For nonlinear systems, stability will be defined in a local sense, in the vicinity of a steady state, and will be determined through the linearization of the nonlinear system around that steady state.

7.2 The Role of System Eigenvalues on the Characteristics of the Response of a Linear System

In the previous chapter, we saw that a linear system may be represented in state-space form as

$$\frac{dx}{dt} = Ax + Bu$$
$$y = Cx + Du \qquad (7.2.1)$$

where u is the input vector, x is the state vector, y is the output vector and A, B, C, D are constant matrices of appropriate dimensions. The state vector may be calculated in the Laplace domain as

$$X(s) = (sI - A)^{-1} x(0) + (sI - A)^{-1} BU(s) \qquad (7.2.2)$$

and, since $(sI - A)^{-1} = \mathcal{L}[e^{At}]$, the response of the state in the time domain is:

$$x(t) = e^{At} x(0) + \int_0^t e^{A(t-t')} Bu(t') dt' \qquad (7.2.3)$$

The qualitative and quantitative characteristics of the response (7.2.3) depend on the properties of the exponential matrix function e^{At}, which multiplies the initial conditions and also enters in the convolution integral. Does e^{At} grow or decay as time progresses? And how fast? Does it oscillate or not? These are questions of key significance in studying dynamic behavior.

Remember that the matrix $(sI - A)^{-1}$, which is the Laplace transform of e^{At}, is called the resolvent matrix and can be calculated as

$$(sI - A)^{-1} = \frac{\text{Adj}(sI - A)}{\det(sI - A)} \qquad (7.2.4)$$

From (7.2.4), we see that there is a common denominator of all elements of $(sI - A)^{-1}$: it is $\det(sI - A)$, i.e. the characteristic polynomial of the matrix A. The ij-th element of $(sI - A)^{-1}$ is therefore given by:

$$\left[(sI - A)^{-1}\right]_{ij} = \frac{[\text{Adj}(sI - A)]_{ij}}{\det(sI - A)} \qquad (7.2.5)$$

hence the ij-th element of e^{At} is:

$$\left[e^{At}\right]_{ij} = \mathcal{L}^{-1}\left\{\frac{[\text{Adj}(sI - A)]_{ij}}{\det(sI - A)}\right\} \qquad (7.2.6)$$

In order to find the inverse Laplace transform, we need to expand in partial fractions:

$$\frac{[\text{Adj}(sI - A)]_{ij}}{\det(sI - A)} = \sum_\kappa \frac{c_\kappa}{s - \lambda_\kappa} \qquad (7.2.7)$$

7.2 The Role of System Eigenvalues on the Characteristics of the Response of a Linear System

where λ_κ are the roots of $\det(sI - A)$, i.e. the eigenvalues of the matrix A (assuming that they are simple roots). Consequently, every element of e^{At} is a linear combination of scalar exponential functions:

$$\left[e^{At}\right]_{ij} = \sum_\kappa c_\kappa e^{\lambda_\kappa t} \tag{7.2.8}$$

with the eigenvalues $\lambda_1, \lambda_2, \ldots, \lambda_n$ of the matrix A being the coefficients in the exponents.

In the event that some of the roots of $\det(sI - A)$ are multiple roots (double, triple, …, etc.), the partial fraction expansion (7.2.7) will also have powers of $1/(s-\lambda_\kappa)$ up to the multiplicity and $[e^{At}]_{ij}$ will be a linear combination of $e^{\lambda_\kappa t}, te^{\lambda_\kappa t}, t^2 e^{\lambda_\kappa t}, \ldots, t^{m_\kappa - 1} e^{\lambda_\kappa t}$, where m_κ is the multiplicity of eigenvalue λ_κ.

The eigenvalues of the matrix A of a linear system will be called system eigenvalues. Their presence in the exponents of the expression of e^{At} indicates their key role in assessing the characteristics of dynamic response.

At this point, it will be instructive to revisit Examples 6.2 and 6.3 in light of the foregoing observation on the eigenvalues and the exponential matrix function.

- For the system of interacting tanks of Example 6.2, the system matrix was

$$A = \begin{bmatrix} -2 & 2 \\ 1 & -3 \end{bmatrix}$$

and it was found that $\det(sI - A) = (s+1)(s+4)$, hence the system eigenvalues are $\lambda_1 = -1$ and $\lambda_2 = -4$. On the other hand, the exponential matrix function was found to be

$$e^{At} = \begin{bmatrix} \frac{2}{3}e^{-t} + \frac{1}{3}e^{-4t} & \frac{2}{3}e^{-t} - \frac{2}{3}e^{-4t} \\ \frac{1}{3}e^{-t} - \frac{1}{3}e^{-4t} & \frac{1}{3}e^{-t} + \frac{2}{3}e^{-4t} \end{bmatrix}$$

so indeed, all entries of e^{At} are linear combinations of $e^{\lambda_1 t} = e^{-t}$ and $e^{\lambda_2 t} = e^{-4t}$.

- For the system consisting of two first-order systems in series with unequal time constants that was considered in Example 6.3, the system matrix was

$$A = \begin{bmatrix} -1/\tau_1 & 0 \\ k_2/\tau_2 & -1/\tau_2 \end{bmatrix}$$

and we found that $\det(sI - A) = (s + 1/\tau_1)(s + 1/\tau_2)$, hence the system eigenvalues are $\lambda_1 = -1/\tau_1$ and $\lambda_2 = -1/\tau_2$. On the other hand, the exponential matrix function is

$$e^{At} = \begin{bmatrix} e^{-t/\tau_1} & 0 \\ \dfrac{k_2 \tau_1}{\tau_1 - \tau_2}\left(e^{-t/\tau_1} - e^{-t/\tau_2}\right) & e^{-t/\tau_2} \end{bmatrix}$$

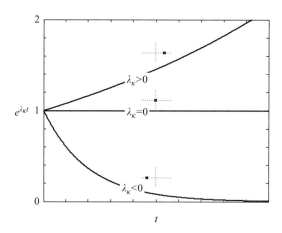

Figure 7.1 Qualitative behavior of the function $e^{\lambda_\kappa t}$, when λ_κ is a real number.

so all entries of e^{At} are linear combinations of $e^{\lambda_1 t} = e^{-t/\tau_1}$ and $e^{\lambda_2 t} = e^{-t/\tau_2}$.

In both examples, the eigenvalues are negative, which implies that the corresponding e^{At} is composed of exponentially decaying functions. This is not always the case.

Figure 7.1 depicts a plot of the exponential function $e^{\lambda_\kappa t}$. We see that when $\lambda_\kappa < 0$, then $e^{\lambda_\kappa t} \to 0$ as time increases, while when $\lambda_\kappa > 0$, then $e^{\lambda_\kappa t} \to \infty$ as time increases. When $\lambda_\kappa = 0$, then $e^{\lambda_\kappa t}$ is constant.

If it so happens that all the system eigenvalues λ_κ are negative real numbers, then every entry of e^{At} will be a decaying function of time and therefore the unforced response of the system's state vector $x(t) = e^{At}x(0)$ will be a decaying function of time, approaching zero for large t, for every initial condition $x(0)$. This will not be the case if there is even one nonnegative system eigenvalue.

Eigenvalues are not always real numbers. It could happen that some eigenvalue is complex, say $\lambda_\kappa = \rho_\kappa + i\omega_\kappa$. In this case, its complex conjugate $\lambda_\kappa^* = \rho_\kappa - i\omega_\kappa$ will also be an eigenvalue. Moreover, their corresponding coefficients c_κ and c_κ^* in the partial fraction expansion (7.2.7) will also be complex conjugate of each other. Thus, the sum of $c_\kappa e^{\lambda_\kappa t} + c_\kappa^* e^{\lambda_\kappa^* t}$, after applying Euler's formula on the exponentials, will end up being a real function, and it will be a linear combination of $\sin(\omega_\kappa t)e^{\rho_\kappa t}$ and $\cos(\omega_\kappa t)e^{\rho_\kappa t}$.

Figure 7.2 depicts the function $\sin(\omega_\kappa t)e^{\rho_\kappa t}$. If the real part ρ_κ of the complex conjugate pair of eigenvalues $\lambda_\kappa = \rho_\kappa \pm i\omega_\kappa$ is negative, then $\sin(\omega_\kappa t)e^{\rho_\kappa t} \to 0$ as time increases, and it is oscillatory with decaying amplitude. If the real part ρ_κ of the complex conjugate eigenvalues is positive, then we have oscillations of increasing amplitude. If the complex conjugate eigenvalues are purely imaginary (zero real part, $\lambda_\kappa = \pm i\omega_\kappa$), then the oscillations have constant amplitude.

The same type of behavior can be observed for the function $\cos(\omega_\kappa t)e^{\rho_\kappa t}$. Thus we conclude that the presence of complex eigenvalues gives rise to oscillations that may be decaying, growing or steady, depending on the real part of the eigenvalues, whether it is negative,

7.2 The Role of System Eigenvalues on the Characteristics of the Response of a Linear System

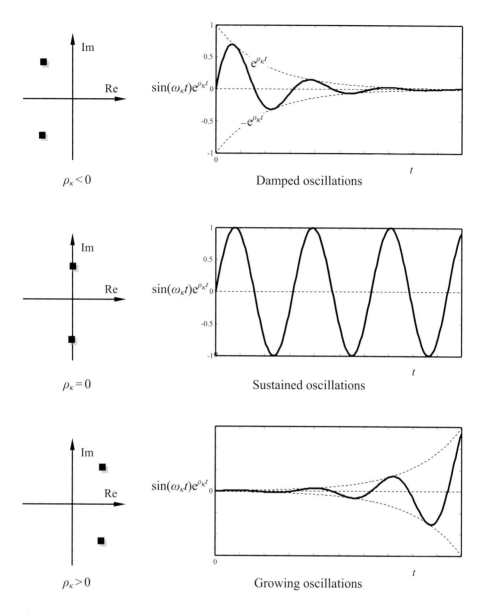

Figure 7.2 Qualitative behavior of the function $\sin(\omega_\kappa t)e^{\rho_\kappa t}$ depending on the sign of ρ_κ.

positive or zero. Even one pair of complex eigenvalues will make the exponential matrix e^{At} oscillatory. If all the system eigenvalues are real, then the exponential matrix function e^{At} and therefore the unforced response will be nonoscillatory.

The same analysis may be performed in the presence of eigenvalue multiplicity, leading to the same conclusions: the functions $e^{\lambda_\kappa t}, te^{\lambda_\kappa t}, t^2 e^{\lambda_\kappa t}, \ldots, t^{m_\kappa - 1} e^{\lambda_\kappa t}$ will be decaying to zero when λ_κ is either real and negative or complex with negative real part; in all other cases they will not be decaying to zero. These functions will be oscillatory when λ_κ is complex, nonoscillatory when λ_κ is real.

The bottom line is that the location of the eigenvalues of the system matrix A in the complex plane determines the qualitative characteristics of a system. Figures 7.1 and 7.2 also depict the location of the eigenvalue(s) in the complex plane for each of the cases considered.

7.3 Asymptotic Stability of Linear Systems

When we study the dynamics of linear systems, we can pose the following questions relative to the qualitative and quantitative characteristics of their unforced response.

- If the system starts away from its steady state, will it return to the steady state or not?
- If the system returns to its initial steady state, what will be the qualitative shape of the response? Will it be smooth or oscillatory?
- If the system returns to its steady state, how fast will this happen?

The notion of stability and the pertinent mathematical analysis tools will provide answers to the aforementioned questions. Two example systems that can be used to explain the notion of stability are depicted in Figure 7.3.

In the system shown in Figure 7.3a, a ball rests at the lowest point of a valley. If we move the ball from its initial point and then let it go free, it will finally, after some oscillations, return to its initial position. This is the result of the well-known fact that the initial potential energy of the ball is transformed to kinetic energy and at the same time to thermal energy due to friction. As time progresses, the total (kinetic and potential) energy of the ball decreases and the amplitude of the oscillations decreases. The rate at which energy is lost as thermal energy determines the rate at which the ball returns to its initial position. The same observations can be made for the pendulum system shown in Figure 7.3c.

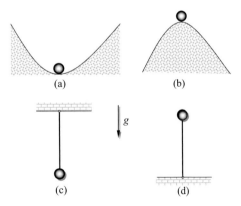

Figure 7.3 Physical examples of stable ((a) and (c)) and unstable ((b) and (d)) systems.

For the systems shown in either Figure 7.3b or 7.3d, any deviation from their initial steady state will have undesirable results. The ball will keep moving away continuously from its initial position and will never return. The inverted pendulum will eventually hit the ground.

There are, therefore, fundamental differences between the systems of Figures 7.3a and 7.3c and those of Figures 7.3b or 7.3d. In the case of Figure 7.3a, the ball is somehow trapped in the valley and no external finite force will be ever possible to prevent the ball from returning at its equilibrium point. The ball of Figure 7.3b, on the other hand, needs only a tiny force to make it move away from its initial position and never return.

Imagine now that the ball corresponds to the state of a chemical, biochemical or nuclear reactor. It is clear that we would like the reactor to behave in a manner similar to the ball shown in Figure 7.3a. This means that any disturbance to the reactor will only cause a temporal deviation of the reactor from its normal point of operation and the reactor will always return to it. It will be very uncomfortable to have to deal with a reactor that runs away from its normal operating point and behaves similarly to the ball shown in Figure 7.3b.

It is therefore important to develop methodologies that can be used to classify the behavior of a system similar to the ball shown in either Figure 7.3a or 7.3b. Systems that behave similarly to the ball of Figure 7.3a are called stable systems while systems that behave similarly to the ball of Figure 7.3b are called unstable systems.

In the literature there are many different definitions of stability; some are stricter, some are looser. In the present chapter we define and characterize the notion of asymptotic stability of a dynamic system. In the next chapter, we will discuss the notion of input–output stability, which refers to the input–output behavior of the system. Loosely speaking, a system is asymptotically stable if it behaves similarly to the ball shown in Figure 7.3a and a physical mechanism exists that dissipates any additional energy that is stored to the system due to external disturbances. The effect of this inherent dissipation mechanism is that the system will always return to its steady state when it starts away from it. We will first state the definition of asymptotic stability and then we will connect it to the results of the previous section.

A linear system with state-space description (7.2.1) is asymptotically stable if its unforced response $x(t) = e^{At}x(0)$ has the property that $\lim_{t \to \infty} x(t) = 0$, for any initial condition $x(0)$.

From the definition, a linear system will be asymptotically stable if and only if

$$\lim_{t \to \infty} e^{At} = 0 \qquad (7.3.1)$$

Based on the analysis of the previous section, this is equivalent to

$$\text{Re}\{\lambda_\kappa\} < 0, \text{ for all eigenvalues } \lambda_\kappa \text{ of the matrix } A \qquad (7.3.2)$$

This is a very important and strong result, and it can be used to determine whether a linear system is stable or not, using information on the system matrix only.

Thus, so far, we have given the answer to the first question posed at the beginning of this section: the system will always return if and only if it is asymptotically stable, which in turn is equivalent to *all* the system eigenvalues having negative real parts. The system eigenvalues also provide the answer to the second question: the unforced response will be nonoscillatory if and only if all the system eigenvalues are real. It will be oscillatory if the system has at least one pair of complex eigenvalues.

We now turn to the third question: how fast will an asymptotically stable system return to its steady state? This is also very important, since we would like to be able to get some indication of the time needed for the system to go to rest at its steady state. Again, the answer will come from the system eigenvalues and specifically from the rate of decay of each of $e^{\lambda_\kappa t}$.

Figure 7.4 provides plots of the function $e^{\lambda_\kappa t}$, for different negative values of λ_κ. We observe that the larger the absolute value of the eigenvalue, the faster is the response of the system. Furthermore, we observe that the initial slope of the function $e^{\lambda_\kappa t}$ is

$$\left.\frac{de^{\lambda_\kappa t}}{dt}\right|_{t=0} = \lambda_\kappa \qquad (7.3.3)$$

If the system were moving with constant speed equal to its initial speed, then it would return to zero at time t equal to

$$\tau_\kappa = 1/(-\lambda_k) \qquad (7.3.4)$$

Equation (7.3.4) provides an indication of the time needed for the function $e^{\lambda_\kappa t}$ to decay to 0. The inverse of the absolute value of the eigenvalue acts like a time constant.

For a pair of complex conjugate eigenvalues $\lambda_\kappa = \rho_\kappa \pm i\omega_\kappa$, the corresponding linear combination of the complex exponential functions $e^{\lambda_\kappa t}$ will be a linear combination of the real functions $\sin(\omega_\kappa t)e^{\rho_\kappa t}$ and $\cos(\omega_\kappa t)e^{\rho_\kappa t}$, and these are absolutely bounded by $e^{\rho_\kappa t}$: $|\sin(\omega_\kappa t)e^{\rho_\kappa t}| \leq e^{\rho_\kappa t}$ and $|\cos(\omega_\kappa t)e^{\rho_\kappa t}| \leq e^{\rho_\kappa t}$ for every t. Thus, in the complex eigenvalue case, it is its real part $\rho_\kappa = \text{Re}\{\lambda_\kappa\}$ that determines the speed of response, instead of λ_κ itself that we saw in the real eigenvalue case.

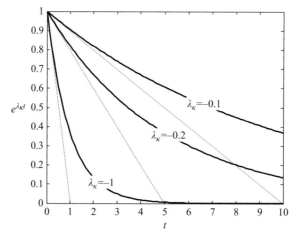

Figure 7.4 The function $e^{\lambda_\kappa t}$ for different negative values of λ_κ.

In an asymptotically stable higher-order system, each eigenvalue has its own rate of decay. So, the overall speed of unforced response will be determined by the slowest of the eigenvalues λ_K. This is like the rate-limiting step in a sequence of chemical reactions: it is the slowest reaction step that determines the overall speed at which the reaction progresses.

Example 7.1 Dynamic characteristics of the unforced response from the system eigenvalues

For a system of the form (7.2.1), check for asymptotic stability and discuss the general characteristics of its unforced response, for the following cases:

(i) $A = \begin{bmatrix} 0 & 1 \\ -0.5 & -1.5 \end{bmatrix}$

(ii) $A = \begin{bmatrix} 0 & 1 \\ -1.25 & -1 \end{bmatrix}$

(iii) $A = \begin{bmatrix} 0 & 1 \\ -1.25 & 1 \end{bmatrix}$

Solution

(i) The characteristic polynomial of the matrix A is

$$\det(\lambda I - A) = \lambda^2 + 1.5\lambda + 0.5 = (\lambda + 1)(\lambda + 0.5)$$

hence the system eigenvalues are $\lambda_1 = -1$ and $\lambda_2 = -0.5$. As they are both negative, the system is asymptotically stable. Because they are both real, the unforced response will be nonoscillatory; it will involve linear combinations of the exponentially decaying functions e^{-t} and $e^{-0.5t}$. The slowest decaying exponential is $e^{-0.5t}$, so the effective time constant is $\tau = 2$.

(ii) The characteristic polynomial of the matrix A is

$$\det(\lambda I - A) = \lambda^2 + \lambda + 1.25 = (\lambda + 0.5 - i)(\lambda + 0.5 + i)$$

hence the system eigenvalues are $\lambda_{1,2} = -0.5 \pm i$. As they have negative real part, the system is asymptotically stable. Because they are complex, the unforced response will be oscillatory; it will involve linear combinations of the exponentially decaying oscillatory functions $e^{-0.5t}\sin(t)$ and $e^{-0.5t}\cos(t)$. The effective time constant is $\tau = 2$.

(iii) The characteristic polynomial of the matrix A is

$$\det(\lambda I - A) = \lambda^2 - \lambda + 1.25 = (\lambda - 0.5 - i)(\lambda - 0.5 + i)$$

hence the system eigenvalues are $\lambda_{1,2} = 0.5 \pm i$. Because they have positive real part, the system is unstable. Because they are complex, the unforced response will be oscillatory; it will involve linear combinations of the exponentially growing oscillatory functions $e^{0.5t}\sin(t)$ and $e^{0.5t}\cos(t)$.

7.4 Properties of the Forced Response of Asymptotically Stable Linear Systems

Asymptotic stability was defined in terms of the unforced response of a linear system: the system is asymptotically stable if and only if it returns to equilibrium whenever it starts away from it. But what does this property mean to forced response? Will the forced response of an asymptotically stable system be well behaved in some sense? The answer is that it is indeed well behaved, and in two different ways.

Forced response of asymptotically stable systems has two key properties. The first is the so-called bounded-input–bounded-state (BIBS) property: whenever the input variation is bounded, the response of the states is also bounded. The second property is that the state of an asymptotically stable linear system will vary like the input for long times: if the input varies linearly with constant slope, so will the states for large t; if the input varies sinusoidally, so will the states for large t, etc. In this section, we will discuss these two important properties.

An input $u(t)$ is said to be bounded if there is a finite positive constant M such that $|u(t)| \leq M$ for every t (see Figure 7.5).

Considering for simplicity a single-input asymptotically stable linear system under zero initial condition ($x(0) = 0$), then from (7.2.3)

$$x(t) = \int_0^t e^{A(t-t')} b u(t') dt' = \int_0^t e^{At''} b u(t-t'') dt'' \qquad (7.4.1)$$

From Eq. (7.4.1) it follows that if $|u(t)| \leq M$ for all t, then the i-th component of $x(t)$ satisfies the following inequalities:

$$|x_i(t)| = \left| \int_0^t [e^{At''} b]_i u(t-t'') dt'' \right| \leq \int_0^t \left| [e^{At''} b]_i u(t-t'') \right| dt'' \leq \left(\int_0^t \left| [e^{At''} b]_i \right| dt'' \right) M$$
$$\leq \left(\int_0^\infty \left| [e^{At''} b]_i \right| dt'' \right) M \qquad (7.4.2)$$

Because $[e^{At''} b]_i$ is a linear combination of $e^{\lambda_\kappa t}$ with $\text{Re}\{\lambda_\kappa\} < 0$, the function $[e^{At''} b]_i$ will be exponentially decaying and therefore the last integral is finite. Thus we see that the response of state vector for an asymptotically stable system remains bounded when the input is bounded.

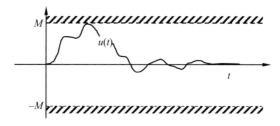

Figure 7.5 Graphical representation of a bounded function $u(t)$ satisfying $|u(t)| \leq M$.

7.5 The Role of Eigenvalues in Time Discretization of Linear Systems

Table 7.1 Ultimate response of asymptotically stable linear systems

		Response for $t \to \infty$ (ultimate response)	Comments
Impulse response	$u(t) = M\delta(t)$ $x(0) = 0$	$x(t) \approx 0$	The effect of the impulse ultimately vanishes
Step response	$u(t) = M$ $x(0) = 0$	$x(t) \approx -A^{-1}bM$	System is led to a new steady state that corresponds to $u_s = M$
Ramp response	$u(t) = Mt$ $x(0) = 0$	$x(t) = (A^{-1} + tI)(-A^{-1}bM)$	The state varies linearly
Sinusoidal response	$u(t) = M \sin \omega t$ $x(0) = 0$	$x(t) = -\sin\omega t \cdot A(A^2 + \omega^2 I)^{-1} bM$ $\quad - \cos\omega t \cdot \omega(A^2 + \omega^2 I)^{-1} bM$	The state varies sinusoidally

The ultimate response of a linear system is the response for large t. For an asymptotically stable system, e^{At} tends to zero as $t \to \infty$ (see (7.3.1)), therefore one can use the results of Table 6.2 to derive formulas for the ultimate response to impulse, step, ramp and sinusoidal inputs. These are summarized in Table 7.1. We see that the state of asymptotically stable systems varies like the input at large times.

7.5 The Role of Eigenvalues in Time Discretization of Linear Systems – Stability Test on a Discretized Linear System

We saw in Chapter 6 how to derive a time discretization of a linear system. Under a zero-order-hold (forward rectangle) approximation of the input, the time discretization of the general linear system (7.2.1) is

$$x[j] = A_d\, x[j-1] + B_d u[j-1]$$
$$y[j] = Cx[j] + Du[j] \qquad (7.5.1)$$

where

$$A_d = e^{AT_s}, \quad B_d = \int_0^{T_s} e^{At} B\, dt \qquad (7.5.2)$$

where T_s is the discretization time step (sampling period). One question that was left unanswered is how small T_s should be, so that the discretized system can capture the transient behavior of the system.

In Chapter 3, in the study of first-order systems, we made the assumption that T_s is much smaller than the system time constant τ, so that we can get enough data points to describe the exponential transient $e^{-t/\tau}$. For a higher-order system, the transients are in e^{At}, but e^{At} is a linear combination of scalar exponential functions $e^{\lambda_\kappa t}$. To be able to capture all the transients, we must sample at a high enough rate, relative to the size of every eigenvalue. For a real eigenvalue λ_κ, choosing $1/T_s$ to be much larger than $|\lambda_\kappa|$ is sufficient to capture the variation of the real exponential function $e^{\lambda_\kappa t}$. For a pair of complex eigenvalues $\lambda_\kappa = \rho_\kappa \pm i\omega_\kappa$, the contribution of the complex exponential functions $e^{(\rho_\kappa + i\omega_\kappa)t}$ and $e^{(\rho_\kappa - i\omega_\kappa)t}$ will be a linear combination of $\sin(\omega_\kappa t)e^{\rho_\kappa t}$ and $\cos(\omega_\kappa t)e^{\rho_\kappa t}$, hence the sampling rate $1/T_s$ should be fast enough not only relative to $|\rho_\kappa|$, but also relative to the frequency ω_κ of $\sin(\omega_\kappa t)$ and $\cos(\omega_\kappa t)$. Therefore by choosing $1/T_s$ much larger than the absolute value $|\lambda_\kappa| = \sqrt{\rho_\kappa^2 + \omega_\kappa^2} \geq \max(|\rho_\kappa|, \omega_\kappa)$ will accommodate both. The conclusion is that choosing

$$\frac{1}{T_s} \gg |\lambda_\kappa|, \text{ for all eigenvalues } \lambda_\kappa \text{ of the matrix } A \tag{7.5.3}$$

or equivalently,

$$T_s \ll \frac{1}{\max_\kappa |\lambda_\kappa|} \tag{7.5.4}$$

will guarantee that the discretized system provides sufficiently detailed information that can capture all transients.

Once a linear system has been discretized, it is possible to do the eigenvalue analysis and the stability test directly on the eigenvalues of the matrix A_d.

In fact, as a result of a property of the exponential matrix (see Appendix B, Section B.4), the eigenvalues μ_κ of the matrix $A_d = e^{AT_s}$ are related to the eigenvalues λ_κ of the matrix A through the following relation

$$\mu_\kappa = e^{\lambda_\kappa T_s} \tag{7.5.5}$$

From the above equation, we can see how the eigenvalues of A are mapped to eigenvalues of A_d:

- An eigenvalue at the origin, $\lambda_\kappa = 0$, is mapped to $\mu_\kappa = e^0 = 1$.
- Imaginary eigenvalues, $\lambda_\kappa = \pm i\omega_\kappa$, are mapped to $\mu_\kappa = e^{\pm i\omega_\kappa T_s} = \cos(\omega_\kappa T_s) \pm i\sin(\omega_\kappa T_s)$, which have unity magnitude, $|\mu_\kappa| = 1$, i.e. lie on the circumference of the unit circle.
- Real eigenvalues λ_κ are mapped to $\mu_\kappa = e^{\lambda_\kappa T_s}$, which are positive real numbers. They are less than 1 if $\lambda_\kappa < 0$, and greater than 1 if $\lambda_\kappa > 0$.
- Complex eigenvalues, $\lambda_\kappa = \rho_\kappa \pm i\omega_\kappa$, are mapped to $\mu_\kappa = e^{\rho_\kappa T_s}(\cos(\omega_\kappa T_s) \pm i\sin(\omega_\kappa T_s))$. They have magnitude $|\mu_\kappa| < 1$ if $\rho_\kappa < 0$, but $|\mu_\kappa| > 1$ if $\rho_\kappa > 0$.

Summarizing the above observations, we can say that those eigenvalues λ_κ of the continuous-time system that have negative real part (lying in the shaded region of Figure 7.6a), are mapped to eigenvalues μ_κ of the discrete-time system that lie in the interior of the unit circle (shaded region in Figure 7.6b), those λ_κ that have positive real part are mapped to μ_κ outside

7.5 The Role of Eigenvalues in Time Discretization of Linear Systems

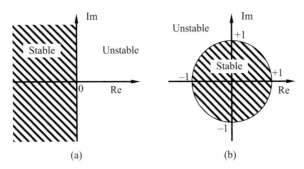

Figure 7.6 Graphical representation of the asymptotic stability condition on the eigenvalues of (a) the original continuous-time system and (b) the time-discretized system.

the unit circle, and those λ_κ with zero real part (on imaginary axis) are mapped to μ_κ on the circumference of the unit circle.

Also notice that the unforced response of the state of the discretized system (7.5.1)

$$x[j] = (A_d)^j\, x(0) = \left(e^{AT_s}\right)^j x(0) = e^{A(jT_s)}x(0) \qquad (7.5.6)$$

exactly matches the response of the continuous-time system $x(t) = e^{At}x(0)$ at every $t = jT_s$, hence $x[j] \to 0$ as $j \to \infty$ if and only if the continuous-time state $x(t) \to 0$ as $t \to \infty$.

Thus we conclude that the asymptotic stability criterion (7.3.2) on the eigenvalues of the matrix A is equivalent to

$$|\mu_\kappa| < 1, \text{ for all eigenvalues } \mu_\kappa \text{ of the matrix } A_d \qquad (7.5.7)$$

where $|\cdot|$ denotes the magnitude of the (in general) complex number μ_κ.

Example 7.2 Check for asymptotic stability from the discretized system equations

For a discretized system of the form (7.5.1), check for asymptotic stability for the following cases:

(i) $A_d = \begin{bmatrix} 0.99 & 0.18 \\ -0.09 & 0.72 \end{bmatrix}$

(ii) $A_d = \begin{bmatrix} 0.979 & 0.198 \\ -0.2475 & 0.781 \end{bmatrix}$

(iii) $A_d = \begin{bmatrix} 0.9656 & 0.2448 \\ -0.306 & 1.2104 \end{bmatrix}$

Solution

(i) The characteristic polynomial of the matrix A_d is

$$\det(\mu I - A_d) = \mu^2 - 1.71\mu + 0.729 = (\mu - 0.9)(\mu - 0.81)$$

hence the discrete system eigenvalues are $\mu_1 = 0.9$ and $\mu_2 = 0.81$, real and positive, and smaller than 1. Because both eigenvalues lie in the interior of the unit disk, the system is asymptotically stable.

(ii) The characteristic polynomial of the matrix A_d is

$$\det(\mu I - A_d) = \mu^2 - 1.76\mu + 0.813604 = (\mu - 0.88 - 0.198i)(\mu - 0.88 + 0.198i)$$

hence the discrete system eigenvalues are $\mu_{1,2} = 0.88 \pm 0.198i$, complex conjugate with magnitude $|\mu_{1,2}| = \sqrt{0.88^2 + 0.198^2} = 0.902$, which is less than 1. Because both eigenvalues lie in the interior of the unit disk, the system is asymptotically stable.

(iii) The characteristic polynomial of the matrix A_d is

$$\det(\mu I - A_d) = \mu^2 - 2.176\mu + 1.24367104$$
$$= (\mu - 1.088 - 0.2448i)(\mu - 1.088 + 0.2448i)$$

hence the discrete system eigenvalues are $\mu_{1,2} = 1.088 \pm 0.2448i$, complex conjugate with magnitude $|\mu_{1,2}| = \sqrt{1.088^2 + 0.2448^2} = 1.1152$, which is larger than 1. Because both eigenvalues lie outside the unit disk, the system is unstable.

7.6 Nonlinear Systems and their Linearization

7.6.1 Linearization of Nonlinear Systems

In Chapter 3, we studied single-input, first-order nonlinear systems of the form

$$\frac{dx}{dt} = f(x, u) \tag{7.6.1}$$

and we saw liquid storage tank examples where the differential equation had a nonlinear function in the right-hand side (see Sections 3.13 and 3.14). We also saw that such a system can be approximated around a steady-state point of operation (x_s, u_s) (at which $f(x_s, u_s) = 0$) by the linear system

$$\frac{d\bar{x}}{dt} = a\bar{x} + b\bar{u} \tag{7.6.2}$$

where the overbar denotes deviation variables $(\bar{x} = x - x_s, \ \bar{u} = u - u_s)$ and the constants a and b are the partial derivatives of the right-hand side

$$a = \frac{\partial f}{\partial x}(x_s, u_s), \quad b = \frac{\partial f}{\partial u}(x_s, u_s) \tag{7.6.3}$$

evaluated, as indicated, at the steady-state point (x_s, u_s).

If we consider two nonlinear liquid storage tanks, in series or interconnected, then we have a nonlinear second-order system, and the idea of linearization approximation can be applied in the same way. In particular, considering a single-input, second-order nonlinear system of the form

7.6 Nonlinear Systems and their Linearization

$$\frac{dx_1}{dt} = f_1(x_1, x_2, u)$$
$$\frac{dx_2}{dt} = f_2(x_1, x_2, u)$$
(7.6.4)

one can follow exactly the same steps to obtain a linear approximation of the nonlinear system around a given steady state-point. The right-hand sides can be approximated by linear functions according to

$$f_1(x_1, x_2, u) \approx f_1(x_{1s}, x_{2s}, u_s) + \left[\frac{\partial f_1}{\partial x_1}(x_{1s}, x_{2s}, u_s)\right](x_1 - x_{1s}) + \left[\frac{\partial f_1}{\partial x_2}(x_{1s}, x_{2s}, u_s)\right](x_2 - x_{2s})$$

$$+ \left[\frac{\partial f_1}{\partial u}(x_{1s}, x_{2s}, u_s)\right](u - u_s)$$

$$f_2(x_1, x_2, u) \approx f_2(x_{1s}, x_{2s}, u_s) + \left[\frac{\partial f_2}{\partial x_1}(x_{1s}, x_{2s}, u_s)\right](x_1 - x_{1s}) + \left[\frac{\partial f_2}{\partial x_2}(x_{1s}, x_{2s}, u_s)\right](x_2 - x_{2s})$$

$$+ \left[\frac{\partial f_2}{\partial u}(x_{1s}, x_{2s}, u_s)\right](u - u_s) \quad (7.6.5)$$

By substituting these approximations into the differential equations, and also noting that $f_1(x_{1s}, x_{2s}, u_s) = f_2(x_{1s}, x_{2s}, u_s) = 0$, we obtain

$$\begin{bmatrix} \frac{dx_1}{dt} \\ \frac{dx_2}{dt} \end{bmatrix} = \begin{bmatrix} \frac{\partial f_1}{\partial x_1}(x_{1s}, x_{2s}, u_s) & \frac{\partial f_1}{\partial x_2}(x_{1s}, x_{2s}, u_s) \\ \frac{\partial f_2}{\partial x_1}(x_{1s}, x_{2s}, u_s) & \frac{\partial f_2}{\partial x_2}(x_{1s}, x_{2s}, u_s) \end{bmatrix} \begin{bmatrix} x_1 - x_{1s} \\ x_2 - x_{2s} \end{bmatrix} + \begin{bmatrix} \frac{\partial f_1}{\partial u}(x_{1s}, x_{2s}, u_s) \\ \frac{\partial f_2}{\partial u}(x_{1s}, x_{2s}, u_s) \end{bmatrix}(u - u_s) \quad (7.6.6)$$

We then define the vector of state variables and the input in deviation form

$$\bar{x} = \begin{bmatrix} \bar{x}_1 \\ \bar{x}_2 \end{bmatrix} = \begin{bmatrix} x_1 - x_{1s} \\ x_2 - x_{2s} \end{bmatrix}, \quad \bar{u} = u - u_s \quad (7.6.7)$$

as well as the matrix A and the vector b according to

$$A = \begin{bmatrix} \frac{\partial f_1}{\partial x_1}(x_{1s}, x_{2s}, u_s) & \frac{\partial f_1}{\partial x_2}(x_{1s}, x_{2s}, u_s) \\ \frac{\partial f_2}{\partial x_1}(x_{1s}, x_{2s}, u_s) & \frac{\partial f_2}{\partial x_2}(x_{1s}, x_{2s}, u_s) \end{bmatrix}, \quad b = \begin{bmatrix} \frac{\partial f_1}{\partial u}(x_{1s}, x_{2s}, u_s) \\ \frac{\partial f_2}{\partial u}(x_{1s}, x_{2s}, u_s) \end{bmatrix} \quad (7.6.8)$$

Using these definitions, Eq. (7.6.6) can be written as

$$\frac{d\bar{x}}{dt} = A\bar{x} + b\bar{u} \quad (7.6.9)$$

which is the linearized form of the nonlinear system given by Eq. (7.6.4). Equation (7.6.9) may be used to approximately calculate the response of (7.6.4) in the vicinity of the steady-state point (x_{1s}, x_{2s}, u_s).

The most general case is an n-th-order nonlinear system having n state variables $(x_1, x_2, \ldots, x_n)$ and m inputs $(u_1, u_2, \ldots, u_m)$ described by

$$\frac{dx_1}{dt} = f_1(x_1, x_2, \ldots, x_n, u_1, u_2, \ldots, u_m)$$

$$\frac{dx_2}{dt} = f_2(x_1, x_2, \ldots, x_n, u_1, u_2, \ldots, u_m)$$

$$\vdots$$

$$\frac{dx_n}{dt} = f_n(x_1, x_2, \ldots, x_n, u_1, u_2, \ldots, u_m)$$

(7.6.10)

which can also be written in compact form as

$$\frac{dx}{dt} = f(x, u) \tag{7.6.11}$$

where x is the vector of state variables, u is the vector of the input variables and f is the vector function of the right-hand sides of the differential equations. Following the steps of the linearization of a first- and a second-order nonlinear system, we obtain the following linearized system

$$\frac{d\bar{x}}{dt} = A\bar{x} + B\bar{u} \tag{7.6.12}$$

where the constant matrices A and B are the Jacobian matrices of the vector function f with respect to the state vector x and the input vector u, respectively, evaluated at the steady-state point:

$$A = \frac{\partial f}{\partial x}(x_s, u_s) = \begin{bmatrix} \frac{\partial f_1}{\partial x_1}(x_s, u_s) & \cdots & \frac{\partial f_1}{\partial x_n}(x_s, u_s) \\ \vdots & & \vdots \\ \frac{\partial f_n}{\partial x_1}(x_s, u_s) & \cdots & \frac{\partial f_n}{\partial x_n}(x_s, u_s) \end{bmatrix}$$

$$B = \frac{\partial f}{\partial u}(x_s, u_s) = \begin{bmatrix} \frac{\partial f_1}{\partial u_1}(x_s, u_s) & \cdots & \frac{\partial f_1}{\partial u_m}(x_s, u_s) \\ \vdots & & \vdots \\ \frac{\partial f_n}{\partial u_1}(x_s, u_s) & \cdots & \frac{\partial f_n}{\partial u_m}(x_s, u_s) \end{bmatrix}$$

(7.6.13)

The outputs $(y_1, y_2, \ldots, y_p)$ of a nonlinear system are, in general, possibly nonlinear algebraic functions of the system states and inputs, i.e.

$$y_1 = h_1(x_1, x_2, \ldots, x_n, u_1, u_2, \ldots, u_m)$$
$$y_2 = h_2(x_1, x_2, \ldots, x_n, u_1, u_2, \ldots, u_m)$$
$$\vdots$$
$$y_p = h_p(x_1, x_2, \ldots, x_n, u_1, u_2, \ldots, u_m)$$

or, in vector form, in terms of the input, state and output vectors,

$$y = h(x, u) \quad (7.6.14)$$

These may be approximated by linear functions as

$$\bar{y} = C\bar{x} + D\bar{u} \quad (7.6.15)$$

where $\bar{y} = y - y_s = y - h(x_s, u_s)$ is the output vector in deviation form and

$$C = \frac{\partial h}{\partial x}(x_s, u_s) \;,\; D = \frac{\partial h}{\partial u}(x_s, u_s) \quad (7.6.16)$$

are the $p \times n$ and $p \times m$ Jacobian matrices with respect to the state vector and the input vector, respectively, evaluated at the steady state (defined similarly to (7.6.13)).

Equations (7.6.12) and (7.6.15), with the A, B, C, D matrices given by Eqs. (7.6.13) and (7.6.16), constitute the linearization approximation of the nonlinear system of Eqs. (7.6.11) and (7.6.14). The linearization is used to calculate the approximate dynamic response of the nonlinear system in the vicinity of a steady-state point.

Also, we can use the linearization to calculate an approximate transfer-function matrix

$$G(s) = C(sI - A)^{-1} B + D \quad (7.6.17)$$

and use it to calculate approximate responses to standard inputs.

7.6.2 Asymptotic Stability of Nonlinear Systems

In the previous section, we studied the asymptotic stability of linear systems of the form given by Eq. (7.2.1). Asymptotic stability of nonlinear systems can be defined in the same vein, based on the idea that a stable system, in the absence of input variation, will return to steady state when starting away from it. However, as nonlinear systems can have more than one steady state or equilibrium point, it is meaningful to define stability of a nonlinear system as a local property, in the vicinity of a specific steady state: When starting close enough to the steady state, a stable system's response will stay close enough at all times and eventually will approach the steady state. Moreover, in the study of stability of nonlinear systems it is more appropriate to refer to stability of a steady state (rather than stability of the entire system). Keeping these remarks in mind, we can define the asymptotic stability of the steady state (x_s, u_s) (which, in terms of the deviation variables, is the origin of the state space).

Given a nonlinear system of the form (7.6.11), a steady state (x_s, u_s) is locally asymptotically stable when the response of the constant-input system

$$\frac{dx}{dt} = f(x, u_s) \qquad (7.6.18)$$

has the following properties.
(a) For every $\varepsilon > 0$, there is a $\delta > 0$ such that

$$\|x(t) - x_s\| < \varepsilon, \quad \forall t > 0 \qquad (7.6.19)$$

for every initial state $x(0)$ satisfying $\|x(0) - x_s\| < \delta$.

(b) There exists a $\delta > 0$ such that

$$\lim_{t \to \infty} x(t) = x_s \qquad (7.6.20)$$

for every initial state $x(0)$ satisfying $\|x(0) - x_s\| < \delta$.

In the above conditions, $\|\cdot\|$ denotes a norm of a vector, e.g. the Euclidian norm of a vector $v = (v_1, v_2, \ldots, v_n)$ given by

$$\|v\| = \sqrt{v_1^2 + v_2^2 + \ldots + v_n^2} \qquad (7.6.21)$$

Condition (a) requires that the entire response $x(t)$, $t > 0$ can get arbitrarily close to the steady state x_s when the initial state $x(0)$ is sufficiently close to x_s (see the left-hand drawing of Figure 7.7). Condition (b) requires that $x(t)$ tends to x_s in the limit as $t \to \infty$ when the initial state $x(0)$ is sufficiently close to x_s (see the right-hand drawing of Figure 7.7).

Note that condition (a) ensures that the entire response is also bounded (i.e. does not become infinite before returning close to the steady state). It was redundant in the definition of asymptotic stability of linear systems, where satisfaction of condition (b) implies condition (a). However, in nonlinear systems, (a) is an essential condition.

Once local asymptotic stability has been defined, we need a test to check stability of a steady state of a nonlinear system. For this purpose, the linearization of the nonlinear system is a valuable tool, since it approximately describes dynamic behavior in the vicinity of a

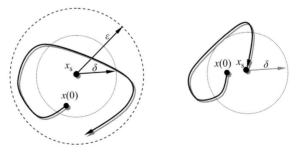

Figure 7.7 Graphical representation of the conditions of local asymptotic stability.

steady state. The following theorem due to Lyapunov gives simple sufficient conditions for stability or instability based on the linearization. Consider a nonlinear system of the form (7.6.11) and its steady state (x_s, u_s):

- if all the eigenvalues of the matrix $A = \partial f/\partial x(x_s, u_s)$ have negative real parts, the steady state (x_s, u_s) is locally asymptotically stable
- if at least one of the eigenvalues of the matrix $A = \partial f/\partial x(x_s, u_s)$ has positive real part, the steady state (x_s, u_s) is unstable.

In case the matrix $A = \partial f/\partial x(x_s, u_s)$ has one eigenvalue with zero real part and all the other eigenvalues have negative real parts, then linearization does not provide enough information to conclude stability or instability.

The results presented above are known as the First or Indirect Method of Lyapunov. A more advanced technique for studying the stability of nonlinear systems is the Second or Direct Method of Lyapunov. This is outside the scope of this book.

As a final remark, it should be noted that a number of nonequivalent concepts of stability have been developed in the literature. Some are stricter, some are looser; some are local, some are global; some refer to the effect of initial conditions, some refer to the effect of the input. In the context of process dynamics and control, the concept of asymptotic stability discussed in this chapter and the concept of input–output stability that will be discussed in the next chapter, have proved to be the most meaningful.

Example 7.3 Linearization and stability analysis of a continuous bioreactor

Biomass grows in a continuous bioreactor, like the one shown in Figure 2.7. Assuming that the cell growth rate follows Monod kinetics (2.4.21) and that the substrate consumption rate is proportional to the cell growth rate with yield coefficient $Y_{X/S}$, cell mass and substrate balances take the form (see Eqs. (2.4.18) and (2.4.20)):

$$\frac{dX}{dt} = -DX + \mu(S)X \qquad (7.6.22)$$

$$\frac{dS}{dt} = D(S_{in} - S) - \frac{\mu(S)}{Y_{X/S}}X \qquad (7.6.23)$$

where X and S are the cell mass and substrate concentrations, respectively (state variables), S_{in} is the feed substrate concentration and $D = F_{in}/V$ is the dilution rate (input variables). The specific growth rate is given by the Monod equation

$$\mu(S) = \mu_{max}\left(\frac{S}{S + K_S}\right)$$

(a) For a given fixed set of values for S_{in} and D, calculate the corresponding bioreactor steady state(s) and check local asymptotic stability of the steady state(s).

(b) Assuming that the dilution rate D is constant but the feed substrate concentration S_{in} is varying (input variable), and considering the biomass concentration X as the system output, calculate the transfer function of the linearized system.

(c) Repeat question (b) in case the dilution rate D is the input, whereas S_{in} is constant.

Solution

(a) Apart from the trivial steady state $X_s = 0$, $S_s = S_{in}$, which is undesirable since no biomass is produced, the bioreactor's steady states are determined from the solution of the algebraic equations

$$0 = -D + \mu(S_s)$$

$$0 = D(S_{in} - S_s) - \frac{\mu(S_s)}{Y_{X/S}} X_s$$

or, equivalently,

$$\mu(S_s) = D$$
$$X_s = Y_{X/S}(S_{in} - S_s)$$

The first equation is solvable only when $D < \mu_{max}$, in which case $S_s = \frac{D}{\mu_{max} - D} K_S$. When $D \geq \mu_{max}$, the bioreactor does not have a nontrivial steady state.

To linearize the bioreactor under fixed S_{in} and D, note that the state vector is

$$x = \begin{bmatrix} x_1 \\ x_2 \end{bmatrix} = \begin{bmatrix} X \\ S \end{bmatrix}$$

and that the right-hand sides of the differential equations are

$$f_1(x_1, x_2) = -DX + \mu(S)X$$

$$f_2(x_1, x_2) = D(S_{in} - S) - \frac{\mu(S)}{Y_{X/S}} X$$

The A matrix of the linearized process dynamics around a bioreactor steady state is

$$A = \begin{bmatrix} \frac{\partial f_1}{\partial x_1}(x_{1,s}, x_{2,s}) & \frac{\partial f_1}{\partial x_2}(x_{1,s}, x_{2,s}) \\ \frac{\partial f_2}{\partial x_1}(x_{1,s}, x_{2,s}) & \frac{\partial f_2}{\partial x_2}(x_{1,s}, x_{2,s}) \end{bmatrix} = \begin{bmatrix} -D + \mu(S_s) & X_s \frac{d\mu}{dS}(S_s) \\ -\frac{\mu(S_s)}{Y_{X/S}} & -D - \frac{X_s}{Y_{X/S}} \frac{d\mu}{dS}(S_s) \end{bmatrix}$$

and because $\mu(S_s) = D$, it follows that

$$A = \begin{bmatrix} 0 & X_s \frac{d\mu}{dS}(S_s) \\ -\frac{D}{Y_{X/S}} & -D - \frac{X_s}{Y_{X/S}} \frac{d\mu}{dS}(S_s) \end{bmatrix}$$

7.6 Nonlinear Systems and their Linearization

The system eigenvalues are calculated as the solution of the characteristic equation

$$\det(\lambda I - A) = \lambda^2 + \left(D + \frac{X_s}{Y_{X/S}}\frac{d\mu}{dS}(S_s)\right)\lambda + D\frac{X_s}{Y_{X/S}}\frac{d\mu}{dS}(S_s) = 0$$

from which we find:

$$\lambda_1 = -D, \quad \lambda_2 = -\frac{X_s}{Y_{X/S}}\frac{d\mu}{dS}(S_s)$$

λ_1 is always negative, while the sign of λ_2 depends on the derivative of the specific growth rate with respect to the substrate concentration. But from the Monod equation

$$\frac{d\mu}{dS}(S) = \frac{d}{dS}\left(\mu_{max}\frac{S}{S+K_S}\right) = \mu_{max}\frac{K_S}{(S+K_S)^2} > 0$$

Hence λ_2 is always negative and the bioreactor's nontrivial steady state is locally asymptotically stable.

(b) When the system has input $u = S_{in}$, the right-hand sides of the differential equations f_1 and f_2 are functions of the states x_1, x_2 and the input u. Note, however, that all partial derivatives with respect to the states $\partial f_i / \partial x_j$ have already been calculated in the previous question and they are actually independent of u. Therefore, the A matrix of the linearized system is exactly the same.

We also need to calculate the b vector. Observe that S_{in} enters linearly in the state equations, so

$$b = \begin{bmatrix} \dfrac{\partial f_1}{\partial u} \\ \dfrac{\partial f_2}{\partial u} \end{bmatrix}_s = \begin{bmatrix} 0 \\ D \end{bmatrix}$$

Thus, the linearization of the bioreactor dynamics around a steady state $(X_s, S_s, S_{in,s})$ is

$$\frac{d}{dt}\begin{bmatrix} \bar{X} \\ \bar{S} \end{bmatrix} = \begin{bmatrix} 0 & X_s\dfrac{d\mu}{dS}(S_s) \\ -\dfrac{D}{Y_{X/S}} & -D - \dfrac{X_s}{Y_{X/S}}\dfrac{d\mu}{dS}(S_s) \end{bmatrix}\begin{bmatrix} \bar{X} \\ \bar{S} \end{bmatrix} + \begin{bmatrix} 0 \\ D \end{bmatrix}\bar{S}_{in}$$

The output is

$$y = \bar{X} = \begin{bmatrix} 1 & 0 \end{bmatrix}\begin{bmatrix} \bar{X} \\ \bar{S} \end{bmatrix}$$

and the transfer function is given by

$$G(s) = \begin{bmatrix} 1 & 0 \end{bmatrix} \begin{bmatrix} s & -X_s \dfrac{d\mu}{dS}(S_s) \\ \dfrac{D}{Y_{X/S}} & s + D + \dfrac{X_s}{Y_{X/S}} \dfrac{d\mu}{dS}(S_s) \end{bmatrix}^{-1} \begin{bmatrix} 0 \\ D \end{bmatrix}$$

$$= \dfrac{\begin{bmatrix} 1 & 0 \end{bmatrix} \begin{bmatrix} s + D + \dfrac{X_s}{Y_{X/S}} \dfrac{d\mu}{dS}(S_s) & X_s \dfrac{d\mu}{dS}(S_s) \\ -\dfrac{D}{Y_{X/S}} & s \end{bmatrix} \begin{bmatrix} 0 \\ D \end{bmatrix}}{s^2 + \left(D + \dfrac{X_s}{Y_{X/S}} \dfrac{d\mu}{dS}(S_s)\right)s + D\dfrac{X_s}{Y_{X/S}} \dfrac{d\mu}{dS}(S_s)} = \dfrac{DX_s \dfrac{d\mu}{dS}(S_s)}{(s+D)\left(s + \dfrac{X_s}{Y_{X/S}} \dfrac{d\mu}{dS}(S_s)\right)}$$

(c) When $u = D$ is the input, repeating the derivation of the linearization, we find

$$A = \begin{bmatrix} \dfrac{\partial f_1}{\partial x_1} & \dfrac{\partial f_1}{\partial x_2} \\ \dfrac{\partial f_2}{\partial x_1} & \dfrac{\partial f_2}{\partial x_2} \end{bmatrix}_s = \begin{bmatrix} 0 & X_s \dfrac{d\mu}{dS}(S_s) \\ -\dfrac{D_s}{Y_{X/S}} & -D_s - \dfrac{X_s}{Y_{X/S}} \dfrac{d\mu}{dS}(S_s) \end{bmatrix}, \quad b = \begin{bmatrix} \dfrac{\partial f_1}{\partial u} \\ \dfrac{\partial f_2}{\partial u} \end{bmatrix}_s = \begin{bmatrix} -X_s \\ S_{in} - S_s \end{bmatrix}$$

and, repeating the calculation of the transfer function, we now have

$$G(s) = \dfrac{-X_s s - X_s D_s - X_s \left(\dfrac{X_s}{Y_{X/S}} - (S_{in} - S_s)\right) \dfrac{d\mu}{dS}(S_s)}{(s + D_s)\left(s + \dfrac{X_s}{Y_{X/S}} \dfrac{d\mu}{dS}(S_s)\right)} = \dfrac{-X_s(s + D_s)}{(s + D_s)\left(s + \dfrac{X_s}{Y_{X/S}} \dfrac{d\mu}{dS}(S_s)\right)}$$

In the above, the steady-state condition $X_s = Y_{X/S}(S_{in} - S_s)$ has been used to simplify the expression in the numerator.

It is noteworthy that $G(s)$ has a common factor $(s + D_s)$ in the numerator and the denominator, therefore it may be further simplified by canceling the common factor.

Table 7.2 Parameters of the CSTR of Example 7.4

Parameter	Value	Units	Parameter	Value	Units
F	1	m³·h⁻¹	k_0	7.2·10¹⁰	h⁻¹
V	1	m³	E/R	10 000	K
$C_{in,R}$	1	kmol·m⁻³	ρc_p	1000	kJ·m⁻³ K⁻¹
T_{in}	350	K	$A_H U_H$	1000	kJ·h⁻¹ K⁻¹
T_J	350	K	$-\Delta h_{rxn}$	200 000	kJ·kmol⁻¹

Example 7.4 Stability analysis of a nonisothermal continuous chemical reactor

Consider a constant-volume CSTR shown in Figure 7.8 where the irreversible exothermic reaction R → P takes place, with first-order kinetics and Arrhenius-type dependence of the rate constant on temperature. A cooling jacket around the reactor, with jacket fluid temperature T_J, heat transfer coefficient U_H and heat transfer area A_H, is used to remove the heat released by the chemical reaction. The mathematical model of the reactor consists of a mole balance for the reactant R and an energy balance for the reactor (see Chapter 2, Sections 2.4.1 and 2.4.2):

$$V\frac{dC_R}{dt} = F\left(C_{R,in} - C_R\right) - Vk_0 e^{-\frac{E}{RT}} C_R \qquad (7.6.24)$$

$$\rho c_p V \frac{dT}{dt} = \rho c_p F\left(T_{in} - T\right) + (-\Delta h_{rxn}) V k_0 e^{-\frac{E}{RT}} C_R + A_H U_H \left(T_J - T\right) \qquad (7.6.25)$$

The reactor operates under constant jacket temperature T_J, constant feed temperature T_{in} and reactant concentration $C_{R,in}$, and constant flowrate F. From the data given in Table 7.2,

(a) calculate the steady state(s) of the CSTR,
(b) check local asymptotic stability of the reactor steady state(s).

Solution

(a) To determine the steady state(s) of the CSTR, we need to solve the nonlinear algebraic equations arising from (7.6.24) and (7.6.25) at steady state:

$$F\left(C_{R,in} - C_{R,s}\right) - Vk_0 e^{-\frac{E}{RT_s}} C_{R,s} = 0 \qquad (7.6.26)$$

$$\rho c_p F\left(T_{in} - T_s\right) + (-\Delta h_{rxn}) V k_0 e^{-\frac{E}{RT_s}} C_{R,s} + A_H U_H \left(T_J - T_s\right) = 0 \qquad (7.6.27)$$

Equation (7.6.26) can be solved for the steady-state reactant concentration:

$$C_{R,s} = \frac{FC_{R,in}}{F + Vk_0 e^{-\frac{E}{RT_s}}} \qquad (7.6.28)$$

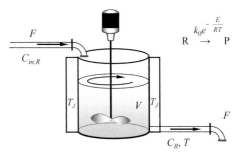

Figure 7.8 Exothermic CSTR.

Equation (7.6.27) can be rearranged as follows:

$$\underbrace{(-\Delta \hat{h}_{rxn})Vk_0 e^{-\frac{E}{RT_s}} C_{R,s}}_{Q_g=\text{heat released by the chemical reaction}} = \underbrace{\rho c_p F(T_s - T_{in}) + A_H U_H (T_s - T_J)}_{Q_r=\text{heat removed from the CSTR}} \qquad (7.6.29)$$

The term on the left-hand side is the heat released by the chemical reaction. The first term on the right-hand side corresponds to the energy change between the incoming and outgoing streams, while the second term corresponds to the heat removed by the cooling medium. At steady state, the left-hand side must be equal to the right-hand side. When $C_{R,s}$ from (7.6.28) is substituted into (7.6.29), we obtain a single equation with one unknown, T_s, which can be solved graphically.

Figure 7.9 depicts the heat released Q_g and the heat removed Q_r as a function of the temperature T_s. We see that there are three intersections, which means that the CSTR has three steady states. The values of the temperature at the points of intersection are:

$$T_s = 353.64 \text{ (point A)}, \; T_s = 400.00 \text{ (point B)}, \; T_s = 441.15 \text{ (point C)}$$

and the corresponding reactant concentrations are obtained from Eq. (7.6.28):

$$C_{R,s} = 0.964, \; C_{R,s} = 0.5, \; C_{R,s} = 0.089$$

(b) To linearize the CSTR equations, note that the state vector is $x = \begin{bmatrix} C_R \\ T \end{bmatrix}$ and that the vector of the right-hand sides of differential equations (7.6.24) and (7.6.25) is

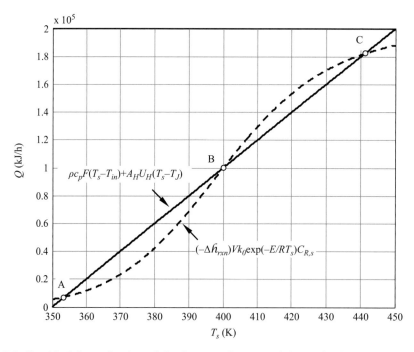

Figure 7.9 Graphical determination of the three steady states of the nonisothermal CSTR model.

$$f(x) = \begin{bmatrix} \frac{F}{V}(C_{R,in} - C_R) - k_0 e^{-\frac{E}{RT}} C_R \\ \frac{F}{V}(T_{in} - T) + \frac{(-\Delta h_{rxn})}{\rho c_p} k_0 e^{-\frac{E}{RT}} C_R + \frac{A_H U_H}{\rho c_p V}(T_J - T) \end{bmatrix}$$

The A matrix of the linearized process dynamics around a CSTR steady state is

$$A = \frac{\partial f}{\partial x}(x_s) = \begin{bmatrix} -\left(\frac{F}{V} + k_0 e^{-\frac{E}{RT_s}}\right) & -\frac{E}{RT_s^2} k_0 e^{-\frac{E}{RT_s}} C_{R,s} \\ \frac{(-\Delta h_{rxn})}{\rho c_p} k_0 e^{-\frac{E}{RT_s}} & -\left(\frac{F}{V} - \frac{(-\Delta h_{rxn})}{\rho c_p} \frac{E}{RT_s^2} k_0 e^{-\frac{E}{RT_s}} C_{R,s} + \frac{A_H U_H}{\rho c_p V}\right) \end{bmatrix}$$

Table 7.3 Summary of the steady states for the CSTR case study

Steady state point	T_s (K)	C_{Rs} (kmol/m³)	λ_1	λ_2
A (stable)	353.64	0.964	−1.1084	−1.3481
B (unstable)	400.00	0.500	−0.7500	+3.0000
C (stable)	441.15	0.089	−1.9639 ± 3.0604i	

The eigenvalues of the matrix A are the roots of its characteristic equation:

$$\lambda^2 + a_1 \lambda + a_2 = 0$$

where

$$a_1 = \left(\frac{F}{V} + k_0 e^{-\frac{E}{RT_s}}\right) + \left(\frac{F}{V} + \frac{A_H U_H}{\rho c_p V} - \frac{(-\Delta h_{rxn})}{\rho c_p} \frac{E}{RT_s^2} k_0 e^{-\frac{E}{RT_s}} C_{R,s}\right)$$

$$a_2 = \left(\frac{F}{V} + k_0 e^{-\frac{E}{RT_s}}\right)\left(\frac{F}{V} + \frac{A_H U_H}{\rho c_p V}\right) - \frac{F}{V}\frac{(-\Delta h_{rxn})}{\rho c_p}\frac{E}{RT_s^2} k_0 e^{-\frac{E}{RT_s}} C_{R,s}$$

This is a quadratic equation that is easily solvable once numerical values are substituted. The roots of the quadratic equation will both have negative real parts if and only if $a_1 > 0$ and $a_2 > 0$.

Substituting the numerical values of each of the steady states calculated in the previous question, we calculate the corresponding eigenvalues. These are given in Table 7.3. We see that the steady-state point B is unstable since it has a positive eigenvalue, whereas points A and C are stable since both their eigenvalues have negative real parts (real and negative for A, complex conjugate with negative real parts for C).

It is interesting to point out here that in the given problem, stability and instability can be interpreted physically. Consider, for example, the point B. We see from Figure 7.9 that, when the system deviates from B and has higher temperature, the heat generated by the chemical reaction is larger than the heat removed, resulting in further temperature increase, and the system moves away from point B towards point C. When the system deviates from B and has lower temperature, then the heat generated by the chemical reaction is lower than the heat removed, resulting in further temperature decrease, and the system moves away from point B towards point A. In summary, any deviation from point B results in the system moving away from point B towards either points C or A, which are stable points.

Similar physical arguments may be used to justify stability of points A and C. We see from Figure 7.9 that, when the system deviates from either A or C and has higher (lower) temperature, the heat generated by chemical reaction is smaller (larger) than the heat removed, and the system returns to the steady state.

It is possible to elaborate further on these arguments and note that at the stable points the heat-removed curve has higher slope than the heat-generated curve (see Figure 7.9), i.e.

$$\frac{d}{dT_s}\left[\rho c_p F(T_s - T_{in}) + A_H U_H(T_s - T_J)\right] > \frac{d}{dT_s}\left[V k_0 e^{-\frac{E}{RT_s}} C_{R,s}(-\Delta h_{rxn})\right]$$

Substituting Eq. (7.6.28), calculating the derivatives and doing some algebraic manipulations, this leads to

$$\left(\frac{F}{V} + k_0 e^{-\frac{E}{RT_s}}\right)\left(\frac{F}{V} + \frac{A_H U_H}{\rho c_p V}\right) - \frac{F}{V}\frac{(-\Delta h_{rxn})}{\rho c_p}\frac{E}{RT_s^2} k_0 e^{-\frac{E}{RT_s}} C_{R,s} > 0$$

which is the same with condition $a_2 > 0$ that was obtained previously in the calculation of the eigenvalues.

7.7 Software Tools

7.7.1 Calculating the Steady States of the CSTR Example and their Stability Characteristics in MATLAB

The calculations of Example 7.4 can be easily performed with MATLAB.
First, we define the process parameters:

» F=1;
» V=1;
» CRin=1;
» Tin=350;
» Tj=350;
» k0=7.2e10;

```
» R=8.314;
» E=10000*R;
» rhoCp=1000;
» AUH=1000;
» Dhrxn=-200000;
```

Over the temperature range $250 < T < 450$ we calculate the heat Q_g released by the chemical reaction and the heat Q_r removed from the CSTR as a function of T:

```
T=linspace(350,450,101);
for i=1:101
    k(i)  = k0*exp(-E/(R*T(i)));
    CR(i) = F*CRin/(F+V*k(i));
    Qr(i) = rhoCp*F*(T(i)-Tin) + AUH*(T(i)-Tj);
    Qg(i) = (-Dhrxn)*k(i)*CR(i)*V;
end
```

By plotting the two terms (» `plot(T,Qr)` and » `plot(T,Qg)`), Figure 7.9 is obtained. For each point of intersection, we calculate the eigenvalues of the A matrix. For example, for point B,

```
» Ts=400.003;
» ks   = k0*exp(-E/(R*Ts));
» CRs  = F*CRin/(F+V*ks);
» A(1,1) = -(F/V+ks);
» A(1,2) = -ks*CRs*E/(R*Ts^2);
» A(2,1) = (-Dhrxn/(rhoCp))*ks;
» A(2,2) = -(F/V+AUH/(rhoCp*V) -
           (-Dhrxn/(rhoCp))*ks*CRs*E/(R*Ts^2));
» eig(A)
ans =
   -0.7500
    3.0000
```

Thus, the results of Table 7.3 are obtained. It is also interesting to demonstrate the use of the command `fsolve` to solve the nonlinear equation (7.6.29). We first define the nonlinear equation through the use of an anonymous function in MATLAB:

```
» f=@(T) (-Dhrxn)*V*k0*exp(-E/(R*T))*F*CRin/
  (F+V*k0*exp(-E/(R*T)))-rhoCp*F*(T-Tin)-AUH*(T-Tj);
```

We then take T0 = 350 K as an initial guess and use `fsolve` to obtain a solution:

```
» T0=350;
» Ts=fsolve(f,T0)
Ts = 353.6336
```

By using different initial guesses for the temperature we can determine all potential steady-state points:

- ```
  T0=400;
  ```
- ```
  Ts=fsolve(f,T0)
  Ts  =  400.0030
  ```
- ```
  T0=450;
  ```
- ```
  Ts=fsolve(f,T0)
  Ts  =  441.1475
  ```

7.7.2 Linearization, Steady-State Calculation and Local Stability Analysis in the CSTR Example using Maple

The Jacobian matrix in Example 7.3 can be computed symbolically using Maple as follows.

We first define the nonlinear functions of the right-hand sides of the differential equations (7.6.24) and (7.6.25), as functions of x1 = C_R and x2 = T:

```
> f1:=(F/V)*(CRin-x1)-k0*exp(-E/(R*x2))*x1:
> f2:=(F/V)*(Tin-x2)+(-Dhrxn/rhoCp)*k0*exp(-E/(R*x2))*x1-(AUH/
(rhoCp*V))*(x2-Tj):
```

Then, the Jacobian matrix of f1 and f2 with respect to x1 and x2 can be obtained by computing all pertinent partial derivatives using the diff command:

```
> fx:=Matrix(2,[[diff(f1,x1),diff(f1,x2)],[diff(f2,x1),
diff(f2,x2)]]);
```

$$fx := \begin{bmatrix} -\dfrac{F}{V} - k0\,e^{-\frac{E}{Rx2}} & -\dfrac{k0\,E\,e^{-\frac{E}{Rx2}}\,x1}{Rx2^2} \\ -\dfrac{Dhrxn\,k0\,e^{-\frac{E}{Rx2}}}{rhoCp} & -\dfrac{F}{V} - \dfrac{Dhrxn\,k0\,E\,e^{-\frac{E}{Rx2}}\,x1}{rhoCp\,Rx2^2} - \dfrac{AUH}{rhoCp\,V} \end{bmatrix}$$

Alternatively, the Jacobian matrix can be computed using the Jacobian command of the VectorCalculus library of Maple, as follows:

```
> with(VectorCalculus):
> fx:=Jacobian([f1,f2],[x1,x2]);
```

leading to exactly the same result as above. When specific parameter values are given, like the ones of Table 7.2,

```
> F:=1:
> V:=1:
```

```
> CRin:=1:
> Tin:=350:
> Tj:=350:
> k0:=7.2e10:
> R:=8.314:
> E:=10000*R:
> rhoCp:=1000:
> AUH:=1000:
> Dhrxn:=-200000:
```

the reactor steady states can be found graphically by plotting the heat Q_g released by the chemical reaction and the heat Q_r removed from the CSTR versus temperature, as defined in Eq. (7.6.29), with C_R given by Eq. (7.6.28):

```
> CR:=F*CRin/(F+V*k0*exp(-E/(R*T))):
> Qr:=rhoCp*F*(T-Tin)+AUH*(T-Tj):
> Qg:=(-Dhrxn)*k0*exp(-E/(R*T))*CR*V:
> plot([Qr,Qg],T=350..450,linestyle=[solid,dash]);
```

In this way, the plot of Figure 7.9 is obtained, in which the three points of intersection give the three steady-state values of temperature. More accurate values of the steady states can be obtained by numerical solution of the equation $Q_r - Q_g = 0$ using the fsolve command:

```
> Ts_1:=fsolve(Qr-Qg,T,350..370);
  CRs_1:=evalf(subs(T=Ts_1,CR));
```

$$Ts\_1 := 353.6336472$$
$$CRs\_1 := 0.9636635277$$

```
> Ts_2:=fsolve(Qr-Qg,T,390..410);
  CRs_2:=evalf(subs(T=Ts_2,CR));
```

$$Ts\_2 := 400.0030242$$
$$CRs\_2 := 0.4999697586$$

```
> Ts_3:=fsolve(Qr-Qg,T,430..450);
  CRs_3:=evalf(subs(T=Ts_3,CR));
```

$$Ts\_3 := 441.1474666$$
$$CRs\_3 := 0.08852533423$$

Finally, for a specific steady state of interest, direct substitution into the Jacobian gives the A matrix of the linearized system, whose eigenvalues determine the local stability or instability of the steady state. For example, if the middle steady state is of interest, we find:

```
> A:=subs(x1=CRs_2,x2=Ts_2,fx);
```

$$A := \begin{bmatrix} -2.000120973 & -0.03125141754 \\ 200.0241946 & 4.250283508 \end{bmatrix}$$

```
> Eigenvalues(A);
```

$$\begin{bmatrix} 3.000141112 \\ -0.7499785770 \end{bmatrix}$$

LEARNING SUMMARY
- When the eigenvalues of the system matrix A are simple, the matrix exponential function is a linear combination of exponential functions $e^{\lambda_\kappa t}$, each one corresponding to a distinct eigenvalue λ_κ.
- When the system matrix A has repeated eigenvalues λ_κ of multiplicity m_κ, the matrix exponential function is a linear combination of $e^{\lambda_\kappa t}$, $te^{\lambda_\kappa t}$, …, $t^{m_\kappa - 1} e^{\lambda_\kappa t}$.
- It is the real part of the eigenvalues that determines the stability characteristics of a linear system. Linear systems are asymptotically stable if and only if all the eigenvalues λ_κ of the system matrix A satisfy $\text{Re}\{\lambda_\kappa\} < 0$ (see Figure 7.10).

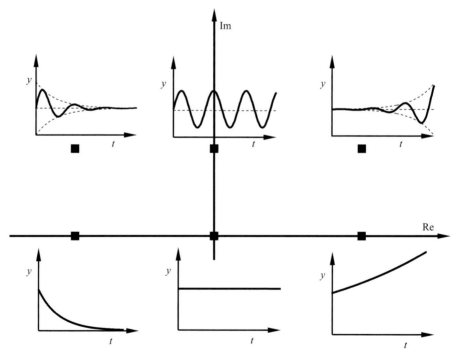

Figure 7.10 Contributions to the unforced response from eigenvalues at various locations in the complex plane. (The conjugate root is not shown.)

- The inverse of the magnitude of an eigenvalue of an asymptotically stable system is like a time constant. The overall speed of response of the system is determined by the slowest eigenvalue.
- Asymptotically stable systems have the bounded-input–bounded-state property, which ensures that the system states will exhibit bounded response to a bounded input.
- Asymptotic stability of time-discretized linear systems may be checked through the eigenvalues of the matrix A_d of the discrete system.
- A steady state of a nonlinear system is locally asymptotically stable when the linearized system around that steady state is asymptotically stable.

TERMS AND CONCEPTS

Asymptotically stable systems. Systems that, after a perturbation of their initial conditions, finally return to their steady-state point while remaining in a region close to the steady state.

Asymptotically stable systems have the **bounded-input–bounded-state property**: the system states remain bounded in response to a bounded change in the input.

FURTHER READING

Further discussion of the alternative concepts of stability is presented in the books
Khalil, H. K., *Nonlinear Systems*, Prentice-Hall International Editions, 3rd edn, 2000.
Vidyasagar, M., *Nonlinear Systems Analysis*, Prentice-Hall International Editions, 2nd edn, 1993.

PROBLEMS

7.1 The variable volume, variable temperature tank shown in Figure P7.1 is used to heat a process stream from temperature T_{in} to temperature T. Assuming constant thermophysical properties of the liquid and square-root dependence of the outlet flowrate on the height, mass and energy balances on the tank will give the following process model:

$$A\frac{dh}{dt} = F_{in} - c\sqrt{h}$$

$$Ah\frac{dT}{dt} = F_{in}(T_{in} - T) + \frac{Q}{\rho c_p}$$

where h is the height of the liquid in the tank, F_{in} is the feed flowrate, T_{in} is the feed temperature, A is the cross-sectional area of the tank, T is the temperature in the tank, ρ is the density, c_p is the heat capacity and Q is the heat input to the tank.

(a) Considering F_{in}, T_{in} and Q to be the inputs of the system, linearize the system equations around a given steady state and check for local asymptotic stability.

(b) If the temperature T is the output, derive the transfer functions of the system.

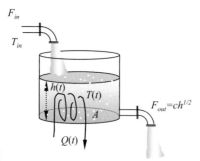

Figure P7.1

7.2 In a variable volume but isothermal CSTR, the first-order irreversible reaction $R \to P$ takes place, with reaction rate $r = k_R C_R$, where k_R is the rate constant and C_R the concentration of the reactant.

Assuming that the outflow of the CSTR is proportional to the square root of the liquid level in the tank, material balances on the CSTR give the following process model:

$$A\frac{dh}{dt} = F_{in} - c\sqrt{h}$$

$$Ah\frac{dC_R}{dt} = F_{in}\left(C_{R,in} - C_R\right) - Ah \cdot k_R C_R$$

where h is the height of the liquid in the tank, F_{in} is the feed flowrate, $C_{R,in}$ is the reactant concentration in the feed and A is the cross-sectional area of the tank

(a) Considering F_{in} and $C_{R,in}$ to be the inputs of the system, linearize the system equations around a given steady state and check for local asymptotic stability.
(b) If the reactant concentration C_R is the output, derive the transfer functions of the system.

7.3 Consider the bioreactor of Example 7.3, with the following numerical values:

$\mu_{max} = 0.05$ d^{-1}, $K_S = 100$ mg/L, $Y_{X/S} = 0.5$ g/g, $D = 0.04$ d^{-1} and $S_{in} = 12400$ mg/L.

(a) Calculate the nontrivial steady state of the bioreactor.
(b) If $X(0)=6100$ mg/L and $S(0)=350$ mg/L, calculate the approximate (linearized) response of the system. Will the system return to steady state?

7.4 Consider the bioreactor of Example 7.3, but with the specific growth rate of the biomass $\mu(S, X)$ following the Contois equation

$$\mu(S, X) = \mu_{max}\left(\frac{S}{S + K_S X}\right)$$

where μ_{max} is the maximum specific growth rate and K_S is the saturation constant. For a specific microorganism, the following numerical values of the parameters are given $\mu_{max} = 1$ h^{-1}, $K_S = 1$ g/g and $Y_{X/S} = 1$ g/g.

(a) If the substrate concentrate in the feed stream is $S_{in}=0.1$ g/L and the dilution rate is $D=0.08$ h^{-1}, calculate the nontrivial bioreactor steady state.
(b) Check local asymptotic stability of the steady state of the previous question.
(c) If the dilution rate D is the process input, whereas S_{in} remains constant, derive the linearized bioreactor equations around the steady state. Calculate the bioreactor's transfer function, considering the biomass X as the output.

7.5 Consider an isothermal, constant-volume CSTR in which the following autocatalytic reaction is taking place

$$R + P \to P + P$$

where R is the reactant and P is the product. The rate of consumption of the reactant is given by the following equation

$$r = kC_R C_P$$

where k is the reaction rate constant and C_R and C_P are the reactant and product concentrations in the reactor. Component mass balances for species R and P give the following state-space model for the reactor:

$$V\frac{dC_R}{dt} = F(C_{R,in} - C_R) - VkC_R C_P$$

$$V\frac{dC_P}{dt} = F(C_{P,in} - C_P) + VkC_R C_P$$

where V is the reactor volume, F is the feed flowrate, and $C_{R,in}$ and $C_{P,in}$ are the reactant and product concentrations in the feed. The following parameter values are given:

$V = 1$ m³, $F = 1$ m³/h, $k = 1.8$ m³/(kmol·h), $C_{R,in} = 9.5$ kmol/m³, $C_{P,in} = 1$ kmol/m³

(a) For the given parameter values, verify that the reactor steady state is:

$$C_{R,s} = 0.5 \text{ kmol/m}^3, \ C_{P,s} = 10 \text{ kmol/m}^3.$$

(b) Linearize the system equations around the steady state. Is the steady state locally asymptotically stable?
(c) If the initial concentrations are $C_R(0) = 0.4$ kmol/m³ and $C_P(0) = 10.5$ kmol/m³, calculate the response of the system based on the linearized equations.

7.6 Consider an isothermal, constant-volume CSTR in which the following consecutive irreversible reactions take place

$$R \to P \to B$$

where R is the reactant, P is the product and B is an undesirable byproduct. The reaction R → P is not elementary and its rate is given by

$$r_P = \frac{k_1 C_R}{1 + k_2 C_R}$$

whereas the reaction P → B has first-order kinetics

$$r_B = k_3 C_P$$

where k_1, k_2 and k_3 are constants. Component mass balances for species R and P give the following state space model for the reactor:

$$V\frac{dC_R}{dt} = F(C_{R,in} - C_R) - V\frac{k_1 C_R}{1 + k_2 C_R}$$

$$V\frac{dC_P}{dt} = -FC_P + V\frac{k_1 C_R}{1 + k_2 C_R} - Vk_3 C_P$$

where V is the reactor volume, F is the feed flowrate and $C_{R,in}$ is the reactant concentration in the feed.

(a) Considering $C_{R,in}$ as the input, linearize the system equations around a given steady state.
(b) Determine the steady state for $C_{R,in} = 1$ kmol/m³, $F = 1$ m³/h, $V = 1$ m³, $k_1 = 9$ h⁻¹, $k_2 = 8$ m³/kmol, $k_3 = 2$ h⁻¹ and check if it is locally asymptotically stable.
(c) If the system is initially at the steady state but an impulsive disturbance in the feed system shifts the reactant concentration to $C_R(0) = 1/2$ kmol/m³, calculate the approximate response of the system. What will be the system state as $t \to \infty$?

7.7 Consider an isothermal, constant-volume CSTR in which the following series–parallel reactions take place

$$R \to P \to B_1$$

$$2R \to B_2$$

where R is the reactant, P is the product and B_1, B_2 are undesirable byproducts. The reactions R → P and P → B_1 have first-order kinetics, whereas 2R → B_2 has second-order kinetics. Component mass balances for species R and P give the following state-space model for the reactor:

$$V\frac{dC_R}{dt} = F(C_{R,in} - C_R) - Vk_1 C_R - Vk_3 C_R^2$$

$$V\frac{dC_P}{dt} = -FC_P + Vk_1 C_R - Vk_2 C_P$$

The reactor has two input variables:

(1) the reactant concentration in the feed $u_1 = C_{R,in}$
(2) the feed flowrate $u_2 = F$

and one output variable: the concentration of the product P in the reactor $y = C_P$.
The following parameter values are given for the system:

$$V = 1 \text{ L}, k_1 = 50 \text{ min}^{-1}, k_2 = 54 \text{ min}^{-1}, k_3 = 4 \text{ min}^{-1} \text{ L/mol}$$

(a) Calculate the reactor steady state when $C_{R,in} = 5.5$ mol/L and $F = 21$ L/min.
(b) Linearize the system equations around the steady state of part (a). Check if it is locally asymptotically stable.
(c) Suppose that the reactor is initially at the steady state of part (a). Suddenly, at $t = 0$, the flowrate F increases to 28 L/min, while the feed concentration $C_{R,in}$ remains unchanged. Use MATLAB to simulate the output response for both the nonlinear and linearized systems, and plot the responses on a common graph.

COMPUTATIONAL PROBLEMS

7.8 A three interacting tank system is shown in Figure P7.8. A control pump is used to deliver water to the first tank at volumetric rates (F_{in}) between 0 and 110 cm³/s and the tanks were all of the same size, with cross-sectional area $A = 154$ cm².

The mathematical model of the system is given by the following nonlinear equations:

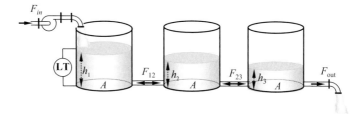

Figure P7.8

$$A\frac{dh_1}{dt} = F_{in} - k_{12} \operatorname{sgn}(h_1 - h_2)\sqrt{|h_1 - h_2|}$$

$$A\frac{dh_2}{dt} = k_{12} \operatorname{sgn}(h_1 - h_2)\sqrt{|h_1 - h_2|} - k_{23} \operatorname{sgn}(h_2 - h_3)\sqrt{|h_2 - h_3|}$$

$$A\frac{dh_3}{dt} = k_{23} \operatorname{sgn}(h_2 - h_3)\sqrt{|h_2 - h_3|} - k_3\sqrt{h_3}$$

where sgn $(\cdot)$ denotes the sign of a real number:

$$\operatorname{sgn}(\chi) = \begin{cases} 1, & \text{if } \chi > 0 \\ 0, & \text{if } \chi = 0 \\ -1, & \text{if } \chi < 0 \end{cases}$$

h_1, h_2 and h_3 are the liquid levels in the tanks in cm, and k_{12}, k_{23} and k_3 are constant parameters with the following values

$$k_{12} = 10.1 \, \frac{cm^{5/2}}{s}, \quad k_{23} = 11 \, \frac{cm^{5/2}}{s}, \quad k_3 = 19.7 \, \frac{cm^{5/2}}{s}.$$

(a) Show that steady-state liquid levels are determined by the value of the inlet flowrate F_{in} according to

$$h_{1,s} = \left(\frac{F_{in}}{k_{12}}\right)^2 + \left(\frac{F_{in}}{k_{23}}\right)^2 + \left(\frac{F_{in}}{k_3}\right)^2, \quad h_{2,s} = \left(\frac{F_{in}}{k_{23}}\right)^2 + \left(\frac{F_{in}}{k_3}\right)^2, \quad h_{3,s} = \left(\frac{F_{in}}{k_3}\right)^2.$$

(b) Linearize the differential equations around a given steady state.
(c) Calculate the steady state that corresponds to F_{in} = 46.166 cm³/s and use the linearized equations to calculate the transfer function $G(s) = \bar{H}_1(s) / \bar{F}_{in}(s)$.
(d) Evaluate the accuracy of the linear approximation for the steady state of the previous question and for a step increase in F_{in} by 5 cm³/s. Use MATLAB or Maple for the numerical simulation.

8 Transfer-Function Analysis of the Input–Output Behavior

The aim of this chapter is to present methods of analysis of the dynamic behavior of a linear system using the transfer-function description. The notion of stability, which was introduced in the previous chapter for the state-space description, will be reformulated in an input–output sense to the transfer-function description. The asymptotic response of a stable dynamical system to the most common input signals, using the transfer-function description, will be calculated.

STUDY OBJECTIVES

After studying this chapter you should be able to do the following.

- Derive the response of a system, whose transfer function is known, to the commonly used input variations.
- Study the input–output stability of systems from their transfer-function description.
- Calculate the asymptotic response of a linear system to the commonly used idealized inputs.

8.1 Introduction

We consider a linear system with one input u and one output y. The state-space description of such a system has already been given in Chapter 6:

$$\frac{dx}{dt} = Ax + bu$$
$$y = cx + du \qquad (6.1.19)$$

Moreover, we saw in Chapters 6 and 7 how to analyze the dynamics of such systems. We saw that the eigenvalues of the system matrix A can be used to test asymptotic stability as well as other characteristics of the response, such as presence or absence of oscillations and estimates of the speed of response. We also saw that the solution of Eqs. (6.1.19) under zero initial condition can be represented as

$$y(t) = \int_0^t ce^{A(t-t')}bu(t')dt' + du(t) \tag{6.6.4}$$

and this can form the basis for deriving formulas for step, ramp, sinusoidal, etc., variations in $u(t)$.

In the Laplace domain under zero initial condition, we saw that input and output are related through the relation

$$Y(s) = \left[c(sI - A)^{-1}b + d \right] U(s) \tag{6.3.4}$$

where the function

$$G(s) = c(sI - A)^{-1}b + d \tag{6.3.5}$$

is the transfer function. The transfer-function description (6.3.4) emerges directly from the Laplace transformation of (6.1.19) and elimination of the state variables. The inverse Laplace transform of (6.3.4) gives the solution formula (6.6.4).

In the present chapter, we will study the transfer function as a tool for analyzing input–output behavior of linear dynamical systems. We will see that, if the transfer function is calculated upfront, it is possible to analyze the input–output behavior of any linear system in terms of stability and other dynamic response characteristics. As we will see, transfer-function analysis, in many cases, is simpler than the analysis using the state-space description. Transfer functions are generally more "friendly" for manual calculations, but, of course, when the order of the system gets too high, one needs to resort to software packages, symbolic or numeric.

8.2 A Transfer Function is a Higher-Order Differential Equation in Disguise

In the state-space description (6.1.19) of a linear system, the state variables x are intermediate variables in the system equations, which can be eliminated if we want to. By eliminating the state variables, we can obtain a higher-order differential equation that directly relates the input and the output. The elimination can be done in the same manner as it was done in Chapter 5, Section 5.2, for two first-order systems in series. We take the output equation and we differentiate it n times:

$$\begin{aligned}
y &= cx + du \\
\frac{dy}{dt} &= c\frac{dx}{dt} + d\frac{du}{dt} = cAx + cbu + d\frac{du}{dt} \\
\frac{d^2y}{dt^2} &= cA\frac{dx}{dt} + cb\frac{du}{dt} + d\frac{d^2u}{dt^2} = cA^2x + cAbu + cb\frac{du}{dt} + d\frac{d^2u}{dt^2} \\
&\vdots \\
\frac{d^ny}{dt^n} &= cA^nx + cA^{n-1}bu + cA^{n-2}b\frac{du}{dt} + \cdots + cb\frac{d^{n-1}u}{dt^{n-1}} + d\frac{d^nu}{dt^n}
\end{aligned} \tag{8.2.1}$$

Let $a_1, a_2, \ldots, a_n$ be the coefficients of the characteristic polynomial of the matrix A, i.e. $s^n + a_1 s^{n-1} + \cdots + a_{n-1} s + a_n = \det(sI - A)$. Multiplying the first equation of (8.2.1) by a_n, the second equation by $a_{n-1}, \ldots$, the nth equation by a_1 and the last equation by 1, and adding them together, the states are eliminated as a result of the Cayley–Hamilton theorem (see Appendix B, Eq. (B.5.2)), and we obtain:

$$\frac{d^n y}{dt^n} + a_1 \frac{d^{n-1} y}{dt^{n-1}} + \cdots + a_n y = \beta_0 \frac{d^n u}{dt^n} + \beta_1 \frac{d^{n-1} u}{dt^{n-1}} + \beta_2 \frac{d^{n-2} u}{dt^{n-2}} + \cdots + \beta_n u \tag{8.2.2}$$

where

$$\beta_0 = d$$
$$\beta_1 = cb + a_1 d$$
$$\beta_2 = cAb + a_1 cb + a_2 d$$
$$\vdots$$
$$\beta_n = cA^{n-1}b + a_1 cA^{n-2}b + \cdots + a_{n-1} cb + a_n d \tag{8.2.3}$$

If the elimination of the state variables is to be performed in the Laplace domain under zero initial condition, we start from the Laplace transform of (6.1.19)

$$sX(s) = AX(s) + bU(s)$$
$$Y(s) = cX(s) + dU(s)$$

and then, eliminating the state $X(s)$, we immediately obtain Eq. (6.3.4). Using the resolvent identity (see Appendix B, Eq. (B.5.5))

$$(sI - A)^{-1} = \frac{Is^{n-1} + (A + a_1 I)s^{n-2} + \cdots + (A^{n-1} + a_1 A^{n-2} + \cdots + a_{n-1} I)}{s^n + a_1 s^{n-1} + \cdots + a_{n-1} s + a_n}$$

where, again, $s^n + a_1 s^{n-1} + \cdots + a_{n-1} s + a_n = \det(sI - A)$ is the characteristic polynomial of matrix A, Eq. (6.3.4) is written equivalently as

$$Y(s) = \frac{\beta_0 s^n + \beta_1 s^{n-1} + \cdots + \beta_{n-1} s + \beta_n}{s^n + a_1 s^{n-1} + \cdots + a_{n-1} s + a_n} U(s) \tag{8.2.4}$$

where the coefficients $\beta_0, \beta_1, \ldots, \beta_{n-1}, \beta_n$ are the same as those given by (8.2.3). The transfer function is therefore

$$G(s) = \frac{\beta_0 s^n + \beta_1 s^{n-1} + \cdots + \beta_{n-1} s + \beta_n}{s^n + a_1 s^{n-1} + \cdots + a_{n-1} s + a_n} \tag{8.2.5}$$

which is a rational function (fraction of two polynomials). Observe that the same coefficients of the differential equation (8.2.2) are the coefficients of the polynomials in the expression for the transfer function. In fact, if we Laplace transform the differential equation (8.2.2) under zero initial conditions for y and its derivatives up to order $n - 1$, we immediately obtain the transfer-function representation (8.2.4).

The main point to be made here is that a transfer function and a higher-order differential equation are, in essence, the same mathematical model. The term *input–output model* is often used to refer to either one of them, no matter if it is described in the time domain or in the Laplace domain. In manual calculations of output response to idealized inputs, the Laplace transform is generally the most convenient method and, for this reason, the usual starting point of the calculations is the transfer function.

Example 8.1 Input–output model of two interacting tanks

Consider the system of two interacting tanks studied in Chapter 4 (Section 4.3) and also in Chapter 6, Example 6.1:

$$\frac{d}{dt}\begin{bmatrix} h_1 \\ h_2 \end{bmatrix} = \begin{bmatrix} -\dfrac{1}{A_1 R_1} & \dfrac{1}{A_1 R_1} \\ \dfrac{1}{A_2 R_1} & -\left(\dfrac{1}{A_2 R_1} + \dfrac{1}{A_2 R_2}\right) \end{bmatrix} \begin{bmatrix} h_1 \\ h_2 \end{bmatrix} + \begin{bmatrix} \dfrac{1}{A_1} \\ 0 \end{bmatrix} F_{in}$$

$$F_{out} = \begin{bmatrix} 0 & \dfrac{1}{R_2} \end{bmatrix} \begin{bmatrix} h_1 \\ h_2 \end{bmatrix}$$

Apply the formulas derived in this section to calculate the coefficients of the input–output model for this process.

Solution

Here
$$A = \begin{bmatrix} -\dfrac{1}{A_1 R_1} & \dfrac{1}{A_1 R_1} \\ \dfrac{1}{A_2 R_1} & -\left(\dfrac{1}{A_2 R_1} + \dfrac{1}{A_2 R_2}\right) \end{bmatrix}, \quad b = \begin{bmatrix} \dfrac{1}{A_1} \\ 0 \end{bmatrix}$$

$$c = \begin{bmatrix} 0 & \dfrac{1}{R_2} \end{bmatrix}, \quad d = 0$$

We first calculate the characteristic polynomial of the matrix A:

$$\det(sI - A) = \det\begin{bmatrix} s + \dfrac{1}{A_1 R_1} & -\dfrac{1}{A_1 R_1} \\ -\dfrac{1}{A_2 R_1} & s + \dfrac{1}{A_2 R_1} + \dfrac{1}{A_2 R_2} \end{bmatrix} = s^2 + \underbrace{\left(\dfrac{1}{A_1 R_1} + \dfrac{1}{A_2 R_1} + \dfrac{1}{A_2 R_2}\right)}_{=a_1} s + \underbrace{\dfrac{1}{A_1 R_1} \cdot \dfrac{1}{A_2 R_2}}_{=a_2}$$

and then, applying formulas (8.2.3),

$$\beta_0 = d = 0$$

$$\beta_1 = cb + a_1 d = \begin{bmatrix} 0 & \dfrac{1}{R_2} \end{bmatrix} \begin{bmatrix} \dfrac{1}{A_1} \\ 0 \end{bmatrix} + a_1 \cdot 0 = 0$$

$$\beta_2 = cAb + a_1 cb + a_2 d$$

$$= \begin{bmatrix} 0 & \dfrac{1}{R_2} \end{bmatrix} \begin{bmatrix} -\dfrac{1}{A_1 R_1} & \dfrac{1}{A_1 R_1} \\ \dfrac{1}{A_2 R_1} & -\left(\dfrac{1}{A_2 R_1} + \dfrac{1}{A_2 R_2}\right) \end{bmatrix} \begin{bmatrix} \dfrac{1}{A_1} \\ 0 \end{bmatrix} + a_1 \cdot 0 + a_2 \cdot 0 = \dfrac{1}{A_1 A_2 R_1 R_2}$$

Therefore the input–output model in transfer function form will be

$$\dfrac{F_{out}(s)}{F_{in}(s)} = G(s) = \dfrac{\beta_0 s^2 + \beta_1 s + \beta_2}{s^2 + a_1 s + a_2} = \dfrac{\dfrac{1}{A_1 A_2 R_1 R_2}}{s^2 + \left(\dfrac{1}{A_1 R_1} + \dfrac{1}{A_2 R_1} + \dfrac{1}{A_2 R_2}\right)s + \dfrac{1}{A_1 A_2 R_1 R_2}}$$

and in differential equation form

$$\dfrac{d^2 y}{dt^2} + a_1 \dfrac{dy}{dt} + a_2 y = \beta_0 \dfrac{d^2 u}{dt^2} + \beta_1 \dfrac{du}{dt} + \beta_2 u$$

i.e. $\dfrac{d^2 F_{out}}{dt^2} + \left(\dfrac{1}{A_1 R_1} + \dfrac{1}{A_2 R_1} + \dfrac{1}{A_2 R_2}\right) \dfrac{dF_{out}}{dt} + \dfrac{1}{A_1 A_2 R_1 R_2} F_{out} = \dfrac{1}{A_1 A_2 R_1 R_2} F_{in}$

Example 8.2 Input–output model of an isothermal chemical reactor where consecutive reactions take place

Consider the third-order system studied in Chapter 6, Section 6.1, whose state-space description is given by Eq. (6.1.5), with output $y = C_P$. Derive the input–output model for this process.

Solution

The characteristic polynomial of the matrix A is calculated first:

$$\det(sI - A) = \det \begin{bmatrix} s + \dfrac{F}{V} + k_1 & 0 & 0 \\ -k_1 & s + \dfrac{F}{V} + k_2 & -k_3 \\ 0 & -k_2 & s + \dfrac{F}{V} + k_3 \end{bmatrix}$$

$$= \left(s+\frac{F}{V}+k_1\right)\det\begin{bmatrix} s+\frac{F}{V}+k_2 & -k_3 \\ -k_2 & s+\frac{F}{V}+k_3 \end{bmatrix}$$

$$= \left(s+\frac{F}{V}+k_1\right)\left[\left(s+\frac{F}{V}+k_2\right)\left(s+\frac{F}{V}+k_3\right)-k_2 k_3\right]$$

$$= \left(s+\frac{F}{V}+k_1\right)\left(s+\frac{F}{V}\right)\left(s+\frac{F}{V}+k_2+k_3\right)$$

This can be expanded as a third-degree polynomial in s, i.e. as

$$\det(sI-A) = s^3 + a_1 s^2 + a_2 s + a_3,$$

where

$$a_1 = 3\left(\frac{F}{V}\right) + k_1 + k_2 + k_3$$

$$a_2 = 3\left(\frac{F}{V}\right)^2 + 2(k_1 + k_2 + k_3)\left(\frac{F}{V}\right) + k_1(k_2 + k_3)$$

$$a_3 = \left(\frac{F}{V}\right)^3 + (k_1 + k_2 + k_3)\left(\frac{F}{V}\right)^2 + k_1(k_2 + k_3)\left(\frac{F}{V}\right) \qquad (8.2.6)$$

For the coefficients of the numerator polynomial, we note that $c = [0\ 0\ 1]$ and $d = 0$ from which it follows that $cb = 0$ and $cAb = 0$, and thus, applying Eq. (8.2.3),

$$\beta_0 = d = 0$$
$$\beta_1 = cb + a_1 d = 0$$
$$\beta_2 = cAb + a_1 cb + a_2 d = 0$$
$$\beta_3 = cA^2 b + a_1 cAb + a_2 cb + a_3 d = cA^2 b$$

and we calculate

$$\beta_3 = cA^2 b = \begin{bmatrix} 0 & 0 & 1 \end{bmatrix} \begin{bmatrix} -\left(\frac{F}{V}+k_1\right) & 0 & 0 \\ k_1 & -\left(\frac{F}{V}+k_2\right) & k_3 \\ 0 & k_2 & -\left(\frac{F}{V}+k_3\right) \end{bmatrix}^2 \begin{bmatrix} F/V \\ 0 \\ 0 \end{bmatrix} = k_1 k_2 \frac{F}{V}$$

The conclusion is that the input–output model of the CSTR in transfer function form is

$$\frac{C_P(s)}{C_{R0}(s)} = G(s) = \frac{k_1 k_2 \dfrac{F}{V}}{\left(s + \dfrac{F}{V} + k_1\right)\left(s + \dfrac{F}{V}\right)\left(s + \dfrac{F}{V} + k_2 + k_3\right)} = \frac{k_1 k_2 \dfrac{F}{V}}{s^3 + a_1 s^2 + a_2 s + a_3}$$

with $a_1, a_2, a_3, \ldots$ given by (8.2.6), and in differential equation form

$$\frac{d^3 C_P}{dt^3} + a_1 \frac{d^2 C_P}{dt^2} + a_2 \frac{dC_P}{dt} + a_3 C_P = k_1 k_2 \frac{F}{V} C_{R0}$$

8.3 Proper and Improper Transfer Functions – Relative Order

The formula for the transfer function derived in the previous section leads to some important conclusions. The first is that (degree of numerator) ≤ (degree of denominator) with the equality holding only when $d \neq 0$.

A rational function

$$G(s) = \frac{\text{Polynomial of degree } m}{\text{Polynomial of degree } n}$$

is called improper if $m > n$, proper if $m \leq n$ and strictly proper if $m < n$.

An improper transfer function cannot arise from any state-space model; it is in this sense that it is "improper." A strictly proper transfer function arises from state-space models with $d = 0$.

An example of improper transfer function is $G(s) = k(1 + \tau s)$, which is rational as it can be written in the form $G(s) = k(1 + \tau s)/1$, and it has degree of numerator $m = 1$ and degree of denominator $n = 0$. When the relation $Y(s) = k(1 + \tau s)U(s)$ is converted to the time domain, it translates to $y(t) = k\left(u(t) + \tau \dfrac{du}{dt}(t)\right)$ and it involves the time derivative of the input.

Improper transfer functions cannot be realized exactly: it is not possible to build a physical device (electrical, mechanical, …), whose input–output behavior exactly follows an improper transfer function, because the physical device will follow a state-space model.

The issue of properness will arise in subsequent chapters when studying transfer functions of controllers and in particular the so-called proportional-integral-derivative controllers (PID). At that point, we will make the distinction between "ideal" PID controllers that have an improper transfer function that cannot be realized, and "real" PID controllers that are proper approximations of the ideal ones.

The second important conclusion from the analysis of the previous section has to do with the difference of degrees between denominator and numerator polynomials. This is called the relative order or relative degree of the system and is denoted by r.

When $d \neq 0$, the coefficient $\beta_0 \neq 0$, the numerator and denominator degrees are equal, so the relative order $r = 0$.

When $d = 0$, the transfer function will be strictly proper, but the degree of the numerator will not necessarily be $n - 1$, as some of the coefficients $\beta_1, \beta_2, \ldots$ in (8.2.3) may be zero. The numerator might even be a constant, as happens in Examples 8.1 and 8.2.

- If $d = 0$ and $cb \neq 0$, the coefficients $\beta_0 = 0$ but $\beta_1 \neq 0$. Thus (degree of numerator) = $n - 1$, i.e. the difference in degrees is $r = 1$.
- If $d = 0$, $cb = 0$ and $cAb \neq 0$, the coefficients $\beta_0 = \beta_1 = 0$ but $\beta_2 \neq 0$. This means that (degree of numerator) = $n - 2$, i.e. the difference in degrees is $r=2$.
- If $d = 0$, $cb = cAb = 0$ and $cA^2b \neq 0$, the coefficients $\beta_0 = \beta_1 = \beta_2 = 0$ but $\beta_3 \neq 0$. This means that (degree of numerator) = $n - 3$, i.e. the difference in degrees is $r = 3$.
- In general, if $d = 0$ and r is the smallest positive integer for which $cA^{r-1}b \neq 0$, the coefficients $\beta_0 = \beta_1 = \ldots = \beta_{r-1} = 0$ but $\beta_r \neq 0$. Therefore (degree of numerator) = $n - r$, i.e. the difference in degrees is r.

In summary, the relative order is $r = $ (degree of denominator) – (degree of numerator) of the transfer function. For a system described by a state space model, $r = 0$ only when $d \neq 0$. When $d = 0$, the relative order r is the smallest positive integer for which $cA^{r-1}b \neq 0$.

Example 8.3 Relative order of a system of two interacting tanks

Consider the system of Example 8.1. For this system,

$$d = 0, \quad cb = \begin{bmatrix} 0 & \dfrac{1}{R_2} \end{bmatrix} \begin{bmatrix} \dfrac{1}{A_1} \\ 0 \end{bmatrix} = 0$$

but

$$cAb = \begin{bmatrix} 0 & \dfrac{1}{R_2} \end{bmatrix} \begin{bmatrix} -\dfrac{1}{A_1 R_1} & \dfrac{1}{A_1 R_1} \\ \dfrac{1}{A_2 R_1} & -\left(\dfrac{1}{A_2 R_1} + \dfrac{1}{A_2 R_2}\right) \end{bmatrix} \begin{bmatrix} \dfrac{1}{A_1} \\ 0 \end{bmatrix} = \dfrac{1}{A_1 A_2 R_1 R_2} \neq 0$$

Therefore, the relative order of the system is $r = 2$.

This is also evident from the transfer function of the system that was calculated in Example 8.1. The transfer function has a constant numerator (degree = 0) and a quadratic denominator (degree = 2).

8.4 Poles, Zeros and Static Gain of a Transfer Function

Consider a rational transfer function

$$G(s) = \frac{N(s)}{D(s)} \tag{8.4.1}$$

where $N(s)$ and $D(s)$ are polynomials in s. When the transfer function is derived from a state-space model, the denominator polynomial $D(s)$ is exactly the characteristic polynomial of the matrix A

$$D(s) = \det(sI - A) = s^n + a_1 s^{n-1} + \cdots + a_{n-1} s + a_n \tag{8.4.2}$$

and the numerator polynomial

$$N(s) = \beta_0 s^n + \beta_1 s^{n-1} + \cdots + \beta_{n-1} s + \beta_n \tag{8.4.3}$$

where $\beta_0, \beta_1, \ldots, \beta_n$ are given by Eq. (8.2.3).

The degree n of the denominator polynomial $D(s)$ is equal to the order of the process (number of state variables), but the degree m of the numerator polynomial can be anything between 0 and n since some of its leading coefficients may be 0.

The roots of the polynomials $N(s)$ and $D(s)$ determine the poles and the zeros of the transfer function. Specifically,

- a number p is called a *pole* of the transfer function $G(s)$ if $G(p)$ is infinite;
- a number z is called a *zero* of the transfer function $G(s)$ if $G(z)$ is zero.

If the polynomials $N(s)$ and $D(s)$ do not have any common factors, the poles $p_1, \ldots, p_n$ of the transfer function are exactly the roots of $D(s)$ and the zeros $z_1, \ldots, z_m$ of the transfer function are exactly the roots of $N(s)$.

If $N(s)$ and $D(s)$ have common factors, the roots of the common factors are neither poles nor zeros. We can cancel the common factors if we want to, and simplify the transfer function as an irreducible fraction of polynomials.

It is often convenient to factorize the transfer function in the following form

$$G(s) = k_{zp} \frac{(s-z_1)(s-z_2)\cdots(s-z_m)}{(s-p_1)(s-p_2)\cdots(s-p_n)} \tag{8.4.4}$$

where $p_1, \ldots, p_n$ are the poles of the transfer function, $z_1, \ldots, z_m$ are the zeros of the transfer function and k_{zp} is the ratio of the leading coefficients of the polynomials $N(s)$ and $D(s)$.

The poles and the zeros encode very important information for characterizing the transient response of the output of a dynamic system. The poles are eigenvalues of the A matrix of the state-space description, therefore they are expected to play a key role in the qualitative and quantitative characteristics of the output response, including stability, presence or absence of oscillations and speed of response. The role of the zeros is more subtle, and it has to do with the initial shape of the output response. In the next two sections, the significance of the poles and the zeros will be discussed.

Finally, a very important parameter is the static gain or steady-state gain of the system:

$$k = G(0) \tag{8.4.5}$$

The static gain is therefore the ratio of the constant terms of numerator and denominator of the transfer function, or the ratio of the zero-derivative coefficients in the higher-order differential equation (8.2.2):

$$k = \frac{\beta_n}{a_n} \tag{8.4.6}$$

or, in terms of the parameters of the state-space model, $k = -cA^{-1}b + d$. Either way, the static gain relates input and output at steady state:

$$y_s = ku_s \tag{8.4.7}$$

The role of the static gain in the output response to standard inputs will also be discussed in subsequent sections.

Example 8.4 Poles and zeros of a system of two tanks in parallel

In Chapter 4 (Section 4.2) we derived the transfer function of two mixing tanks connected in parallel (see Figures 4.4 and 4.5). The transfer function that relates the feed volumetric flowrate $F_{in}(t)$ and the outlet volumetric flowrate $F_{out}(t)$ is the following:

$$G(s) = \frac{\left[(1-a)A_1 R_1 + aA_2 R_2\right]s + 1}{(A_1 R_1 s + 1)(A_2 R_2 s + 1)} \tag{4.2.10}$$

What are the poles and the zeros of the system? Are there any common factors? Under what conditions?

Solution

If we define the time constants of the two tanks $\tau_1 = A_1 R_1$ and $\tau_2 = A_2 R_2$, then Eq. (4.2.10) can be written as

$$G(s) = \frac{\left[(1-a)\tau_1 + a\tau_2\right]s + 1}{(\tau_1 s + 1)(\tau_2 s + 1)}$$

The denominator has two roots: $p_1 = -\dfrac{1}{\tau_1}$, $p_2 = -\dfrac{1}{\tau_2}$.

The numerator has one root: $z = -\dfrac{1}{\tau_{12}} = -\dfrac{1}{(1-a)\tau_1 + a\tau_2}$, where $\tau_{12} = (1-a)\tau_1 + a\tau_2$.

The transfer function can be factored according to (8.4.4) as

$$G(s) = \dfrac{\tau_{12}}{\tau_1 \tau_2} \dfrac{s + \dfrac{1}{\tau_{12}}}{\left(s + \dfrac{1}{\tau_1}\right)\left(s + \dfrac{1}{\tau_2}\right)}$$

When $\tau_1 \neq \tau_2$ and $0 < a < 1$, $\min\{\tau_1, \tau_2\} < \tau_{12} < \max\{\tau_1, \tau_2\}$ and there are no common factors. The system has two poles p_1 and p_2 and one zero z, as calculated previously.

However, when $\tau_1 = \tau_2 = \tau$, we will have that $\tau_{12} = \tau$ and there will be cancellations in the transfer function:

$$G(s) = \dfrac{\left[(1-a)\tau_1 + a\tau_2\right]s + 1}{(\tau_1 s + 1)(\tau_2 s + 1)} = \dfrac{\left[(1-a)\tau + a\tau\right]s + 1}{(\tau s + 1)(\tau s + 1)} = \dfrac{1}{\tau s + 1}$$

The system then has one pole at $-1/\tau$ and no zeros. We see that cancellation can occur when there is symmetry (identical tanks) in the system.

8.5 Calculating the Output Response to Common Inputs from the Transfer Function – the Role of Poles in the Response

Because the poles of a transfer function are eigenvalues of the state-space model from which it originates, we expect to see exponential functions $e^{p_i t}$ in the output response to impulse, step, ramp, sinusoidal, etc., inputs. In this section, we will use a third-order example to examine how the poles affect the output response characteristics.

Let us consider the reactor studied in Section 6.1 (Figure 6.1) whose state-space description is given by Eq. (6.1.5). In Example 8.2 the transfer function of the reactor has been derived when the concentration of the reactant in the feed is the input and the concentration of the product in the reactor is the output

$$G(s) = \dfrac{\bar{C}_P(s)}{\bar{C}_{R0}(s)} = \dfrac{k_1 k_2 F / V}{\left(s + \dfrac{F}{V} + k_1\right)\left(s + \dfrac{F}{V}\right)\left(s + \dfrac{F}{V} + k_2 + k_3\right)} \quad (8.5.1)$$

where F is the volumetric flowrate of the feed stream, V is the volume of the reacting mixture and k_1, k_2 and k_3 are the reaction rate constants. Here the input and output variables are understood to be in deviation form

$$u(t) = \bar{C}_{R0}(t) = C_{R0}(t) - C_{R0,s}$$
$$y(t) = \bar{C}_P(t) = C_P(t) - C_{P,s}$$

and the assumption is that the system is initially at the reference steady state.

The transfer function (8.5.1) has three real poles:

$$p_1 = -\left(\frac{F}{V} + k_1\right), \quad p_2 = -\left(\frac{F}{V}\right), \quad p_3 = -\left(\frac{F}{V} + k_2 + k_3\right) \tag{8.5.2}$$

and no zeros. Furthermore, the degree of the denominator polynomial is $n = 3$, the degree of the numerator polynomial is 0, hence the relative order is $r = 3$. The transfer function is strictly proper and there are no common factors in the numerator and denominator.

8.5.1 Response to Impulse Input

We consider the case where the reactor is at steady state and at time $t = 0$ there is an impulse change in the incoming reactant concentration, $u(t) = M\delta(t)$, whose Laplace transform is $U(s) = M$. In order to calculate the response of the product concentration in the reactor, we multiply the transfer function by the Laplace transform of the input to obtain

$$Y(s) = G(s)U(s) = M \frac{k_1 k_2 F / V}{(s - p_1)(s - p_2)(s - p_3)} \tag{8.5.3}$$

To calculate the inverse Laplace transform, we perform a partial fraction expansion of the transfer function $G(s)$:

$$G(s) = \frac{k_1 k_2 F / V}{(s - p_1)(s - p_2)(s - p_3)} = \frac{\gamma_1}{s - p_1} + \frac{\gamma_2}{s - p_2} + \frac{\gamma_3}{s - p_3} \tag{8.5.4}$$

where

$$\gamma_1 = \frac{F}{V} \cdot \frac{k_2}{k_1 - k_2 - k_3}, \quad \gamma_2 = \frac{F}{V} \cdot \frac{k_2}{k_2 + k_3}, \quad \gamma_3 = -\frac{F}{V} \cdot \frac{k_1 k_2}{(k_2 + k_3)(k_1 - k_2 - k_3)} \tag{8.5.5}$$

For an impulsive input of size M, using the expansion (8.5.4), we obtain the output response

$$y(t) = M \mathcal{L}^{-1}(G(s)) = M\left(\gamma_1 e^{p_1 t} + \gamma_2 e^{p_2 t} + \gamma_3 e^{p_3 t}\right) \tag{8.5.6}$$

Note that the impulse response is a weighted sum of the exponential functions $e^{p_i t}$. In the system under consideration, the poles p_i are all real and negative therefore the exponential functions e^{p_it} decay to 0 as time increases, and the impulse response tends to 0 as time tends to ∞. The effect of the impulse vanishes as $t \to \infty$, and the system returns to its original steady state.

8.5.2 Response to Step Input

Let us now consider the case of a step change in the reactant concentration in the feed of the reactor of Figure 6.1. The change occurs at time $t = 0$ and has magnitude M: $u(t) = M$, $t > 0$. The Laplace transform of the product concentration in the reactor in deviation form is then given by

$$Y(s) = G(s)U(s) = M\frac{G(s)}{s} = M\frac{k_1 k_2 F/V}{s(s-p_1)(s-p_2)(s-p_3)} \qquad (8.5.7)$$

To calculate the inverse Laplace transform, we perform a partial fraction expansion of $G(s)/s$. Using the expansion of $G(s)$ given by Eq. (8.5.4), we obtain

$$\frac{G(s)}{s} = \frac{1}{s}\left(\frac{\gamma_1}{s-p_1} + \frac{\gamma_2}{s-p_2} + \frac{\gamma_3}{s-p_3}\right) = \frac{G(0)}{s} + \frac{\gamma_1/p_1}{s-p_1} + \frac{\gamma_2/p_2}{s-p_2} + \frac{\gamma_3/p_3}{s-p_3} \qquad (8.5.8)$$

where

$$G(0) = G(s)\big|_{s=0} = \frac{k_1 k_2 F/V}{-p_1 p_2 p_3} = -\left(\frac{\gamma_1}{p_1} + \frac{\gamma_2}{p_2} + \frac{\gamma_3}{p_3}\right) \qquad (8.5.9)$$

and where p_i and γ_i are defined by (8.5.2) and (8.5.5), respectively.

Inverting the Laplace transform in (8.5.8) and multiplying by the step size M, we obtain the step response

$$y(t) = M\mathcal{L}^{-1}\left(\frac{G(s)}{s}\right) = M\left(G(0) + \frac{\gamma_1}{p_1}e^{p_1 t} + \frac{\gamma_2}{p_2}e^{p_2 t} + \frac{\gamma_3}{p_3}e^{p_3 t}\right) \qquad (8.5.10)$$

In the system under consideration, the poles p_i are all real and negative, therefore the exponential functions e^{p_it} decay to 0 as time increases, and the step response tends to $MG(0)$ as time tends to ∞, i.e. the output reaches a new steady state.

8.5.3 Response to Ramp Input

Similar arguments can be made to show that for the case of a linear input variation of the reactant concentration in the feed of the reactor of Figure 6.1, $u(t) = Mt$, $t > 0$. The Laplace transform of the product concentration in the reactor in deviation form is then given by

$$Y(s) = G(s)U(s) = M\frac{G(s)}{s^2} = M\frac{k_1 k_2 F/V}{s^2(s-p_1)(s-p_2)(s-p_3)} \qquad (8.5.11)$$

To calculate the inverse Laplace transform, we perform a partial fraction expansion of $G(s)/s^2$. Using the expansion of $G(s)$ given by (8.5.4), we obtain

$$\frac{G(s)}{s^2} = \frac{1}{s^2}\left(\frac{\gamma_1}{s-p_1} + \frac{\gamma_2}{s-p_2} + \frac{\gamma_3}{s-p_3}\right) = \frac{G'(0)}{s} + \frac{G(0)}{s^2} + \frac{\gamma_1/p_1^2}{s-p_1} + \frac{\gamma_2/p_2^2}{s-p_2} + \frac{\gamma_3/p_3^2}{s-p_3} \qquad (8.5.12)$$

where $G(0)$ is given by Eq. (8.5.9) and $G'(0)$ is given by

$$G'(0) = \left.\frac{dG(s)}{ds}\right|_{s=0} = \left.\frac{d}{ds}\left(\frac{\gamma_1}{s-p_1} + \frac{\gamma_2}{s-p_2} + \frac{\gamma_3}{s-p_3}\right)\right|_{s=0} = -\left(\frac{\gamma_1}{p_1^2} + \frac{\gamma_2}{p_2^2} + \frac{\gamma_3}{p_3^2}\right) \qquad (8.5.13)$$

Inverting the Laplace transform and multiplying by the slope M, we obtain the ramp response

$$y(t) = M\mathcal{L}^{-1}\left(\frac{G(s)}{s^2}\right) = M\left(G'(0) + G(0)t + \frac{\gamma_1}{p_1^2}e^{p_1 t} + \frac{\gamma_2}{p_2^2}e^{p_2 t} + \frac{\gamma_3}{p_3^2}e^{p_3 t}\right) \qquad (8.5.14)$$

In the system under consideration, the poles p_i are all real and negative, therefore the exponential functions $e^{p_i t}$ decay to 0 as time increases, and the ramp response approaches $M(G'(0) + G(0)t)$ for large t, i.e. the output changes linearly for large t.

8.5.4 Response to Sinusoidal Input

We now consider the case of a sinusoidal variation $u(t) = M\sin(\omega t)$ of the reactant concentration in the feed of the reactor of Figure 6.1. The Laplace transform of the product concentration in the reactor in deviation form is then given by

$$Y(s) = G(s)U(s) = M\frac{\omega G(s)}{s^2+\omega^2} = M\frac{\omega k_1 k_2 F/V}{(s-i\omega)(s+i\omega)(s-p_1)(s-p_2)(s-p_3)} \qquad (8.5.15)$$

To calculate the inverse Laplace transform, we perform a partial fraction expansion of $\frac{\omega G(s)}{s^2+\omega^2}$. Using the expansion of $G(s)$ given by (8.5.4), we obtain

$$\frac{\omega G(s)}{(s-i\omega)(s+i\omega)} = \frac{\omega}{s^2+\omega^2}\left(\frac{\gamma_1}{s-p_1} + \frac{\gamma_2}{s-p_2} + \frac{\gamma_3}{s-p_3}\right)$$

$$= \left(\frac{G(i\omega)}{2i}\right)\frac{1}{s-i\omega} + \left(-\frac{G(-i\omega)}{2i}\right)\frac{1}{s+i\omega} + \frac{\gamma_1'}{s-p_1} + \frac{\gamma_2'}{s-p_2} + \frac{\gamma_3'}{s-p_3} \qquad (8.5.16)$$

where

$$\gamma_i' = \left.\frac{\omega G(s)(s-p_i)}{(s^2+\omega^2)}\right|_{s=p_i} = \frac{\omega \gamma_i}{p_i^2+\omega^2} \qquad (8.5.17)$$

8.5 Calculating the Output Response to Common Inputs from the Transfer Function

Inverting the Laplace transform and multiplying by the amplitude M, we obtain

$$y(t) = M\left(\frac{G(i\omega)}{2i}e^{i\omega t} - \frac{G(-i\omega)}{2i}e^{-i\omega t} + \gamma'_1 e^{p_1 t} + \gamma'_2 e^{p_2 t} + \gamma'_3 e^{p_3 t}\right) \quad (8.5.18)$$

Remembering from Euler's formula that $e^{\pm i\theta} = \cos\theta \pm i\sin\theta$, the first two terms may be rearranged as follows:

$$\begin{aligned}\frac{G(i\omega)}{2i}e^{i\omega t} - \frac{G(-i\omega)}{2i}e^{-i\omega t} &= \frac{G(i\omega)}{2i}(\cos\omega t + i\sin\omega t) - \frac{G(-i\omega)}{2i}(\cos\omega t - i\sin\omega t)\\ &= \frac{G(i\omega)+G(-i\omega)}{2}\sin\omega t + \frac{G(i\omega)-G(-i\omega)}{2i}\cos\omega t\\ &= \operatorname{Re}\{G(i\omega)\}\sin\omega t + \operatorname{Im}\{G(i\omega)\}\cos\omega t\end{aligned}$$

where $\operatorname{Re}\{G(i\omega)\}$ and $\operatorname{Im}\{G(i\omega)\}$ are the real and imaginary part of $G(i\omega)$. Thus Eq. (8.5.18) takes the form

$$y(t) = M\left(\operatorname{Re}\{G(i\omega)\}\sin\omega t + \operatorname{Im}\{G(i\omega)\}\cos\omega t + \gamma'_1 e^{p_1 t} + \gamma'_2 e^{p_2 t} + \gamma'_3 e^{p_3 t}\right) \quad (8.5.19)$$

The method and the results of the example directly generalize to any strictly proper transfer function $G(s)$ with simple poles p_i:

$$G(s) = \sum_i \frac{\gamma_i}{s - p_i} \quad (8.5.20)$$

The results for the output response to standard idealized inputs are summarized in Table 8.1. Further generalizations are possible, to include the possibility of repeated poles and also proper, but not strictly proper, transfer functions.

Table 8.1 Summary of the results for the output response to standard idealized inputs

Input	Output	Output for a strictly proper $G(s)$ with simple poles p_i
Impulse $u(t) = M\delta(t)$	$y(t) = M\mathcal{L}^{-1}\{G(s)\}$	$y(t) = M \sum_i \gamma_i e^{p_i t}$
Step $u(t) = M$	$y(t) = M\mathcal{L}^{-1}\left\{\dfrac{G(s)}{s}\right\}$	$y(t) = M\left(G(0) + \sum_i \dfrac{\gamma_i}{p_i} e^{p_i t}\right)$ (assuming $p_i \neq 0$)
Ramp $u(t) = Mt$	$y(t) = M\mathcal{L}^{-1}\left\{\dfrac{G(s)}{s^2}\right\}$	$y(t) = M\left(G'(0) + G(0)t + \sum_i \dfrac{\gamma_i}{p_i^2} e^{p_i t}\right)$ (assuming $p_i \neq 0$)
Sinusoidal $u(t) = M\sin\omega t$	$y(t) = M\mathcal{L}^{-1}\left\{\dfrac{\omega G(s)}{s^2+\omega^2}\right\}$	$y(t) = M(\operatorname{Re}\{G(i\omega)\}\sin\omega t + \operatorname{Im}\{G(i\omega)\}\cos\omega t + \sum_i \dfrac{\omega\gamma_i}{p_i^2+\omega^2} e^{p_i t})$ (assuming $p_i \neq \pm i\omega$)

Example 8.5 Calculation of the product concentration response to variations in the reactant feed concentration

For the reactor studied in this section and the following numerical values for the parameters: $V/F = 10$ s, $k_1 = 1/10$ s^{-1}, $k_2 = 5/100$ s^{-1} and $k_3 = 2/100$ s^{-1}, calculate the transfer function and the unit impulse response, unit step response, unit ramp response and sinusoidal response with unit amplitude and frequency $\omega = 0.1$ s^{-1}.

Solution

From Example 8.2 and Eq. (8.5.1) we have that

$$G(s) = \frac{5/10000}{\left(s+\dfrac{1}{5}\right)\left(s+\dfrac{1}{10}\right)\left(s+\dfrac{17}{100}\right)}$$

and the poles are $p_1 = -1/5$, $p_2 = -1/10$ and $p_3 = -17/100$. The transfer function $G(s)$ is expanded in partial fractions:

$$G(s) = \frac{1}{6} \cdot \frac{1}{s+\dfrac{1}{5}} + \frac{1}{14} \cdot \frac{1}{s+\dfrac{1}{10}} - \frac{5}{21} \cdot \frac{1}{s+\dfrac{17}{100}}$$

so $\gamma_1 = 1/6$, $\gamma_2 = 1/14$, $\gamma_3 = -5/21$.

The response of the product concentration to a unit impulse change in the reactant concentration in the feed is obtained from (8.5.6):

$$y(t) = \gamma_1 e^{p_1 t} + \gamma_2 e^{p_2 t} + \gamma_3 e^{p_3 t} = \frac{1}{6} e^{-\frac{t}{5}} + \frac{1}{14} e^{-\frac{t}{10}} - \frac{5}{21} e^{-\frac{17t}{100}}$$

A plot of the unit impulse response is shown in Figure 8.1.

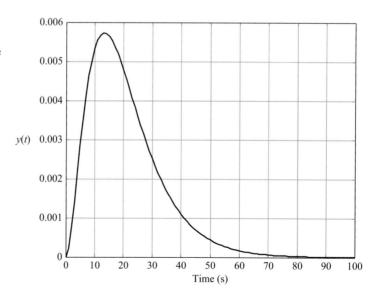

Figure 8.1 Impulse response of the product concentration – Example 8.5.

8.5 Calculating the Output Response to Common Inputs from the Transfer Function

The response of the product concentration to a unit step change in feed concentration is calculated from (8.5.10):

$$y(t) = G(0) + \frac{\gamma_1}{p_1} e^{p_1 t} + \frac{\gamma_2}{p_2} e^{p_2 t} + \frac{\gamma_3}{p_3} e^{p_3 t} = \frac{5}{34} - \frac{5}{6} e^{-\frac{t}{5}} - \frac{5}{7} e^{-\frac{t}{10}} + \frac{500}{357} e^{-\frac{17t}{100}}$$

A plot of the unit step response of the product concentration is shown in Figure 8.2.

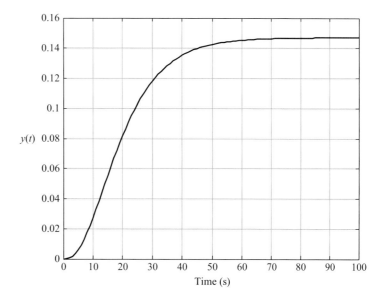

Figure 8.2 Unit step response of the product concentration – Example 8.5.

To calculate the response of the product concentration to a unit ramp change in the feed concentration we use Eq. (8.5.14), with $G'(0)$ from (8.5.13), to obtain

$$y(t) = G'(0) + G(0)t + \frac{\gamma_1}{p_1^2} e^{p_1 t} + \frac{\gamma_2}{p_2^2} e^{p_2 t} + \frac{\gamma_3}{p_3^2} e^{p_3 t}$$

$$= -\frac{1775}{578} + \frac{5}{34} t + \frac{25}{6} e^{-\frac{t}{5}} + \frac{50}{7} e^{-\frac{t}{10}} - \frac{50000}{6069} e^{-\frac{17t}{100}}$$

The unit ramp response of the product concentration is shown in Figure 8.3.

To calculate the response of the product concentration to the sinusoidal input $u(t) = \sin(0.1t)$ we use Eq. (8.5.19). We first calculate $G(i\omega)$ for $\omega = 1/10$,

$$G\left(\frac{i}{10}\right) = \frac{5/10000}{\left(\frac{i}{10} + \frac{1}{5}\right)\left(\frac{i}{10} + \frac{1}{10}\right)\left(\frac{i}{10} + \frac{17}{100}\right)} = -\frac{13 + 61i}{778}$$

and then applying (8.5.19) with γ_i' given by (8.5.17), we obtain

$$y_\omega(t) = M\left[\text{Re}\{G(i\omega)\}\sin\omega t + \text{Im}\{G(i\omega)\}\cos\omega t + \sum_{i=1}^{n} \gamma_i' e^{p_i s}\right]$$

$$= -\left(\frac{13}{778}\sin\left(\frac{t}{10}\right) + \frac{61}{778}\cos\left(\frac{t}{10}\right)\right) + \frac{1}{3} e^{-\frac{t}{5}} + \frac{5}{14} e^{-\frac{t}{10}} - \frac{5000}{8169} e^{-\frac{17t}{100}}$$

Figure 8.3 Unit ramp response of the product concentration – Example 8.5.

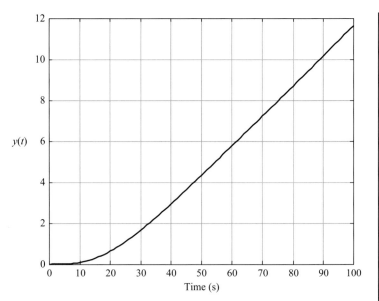

The result is plotted in Figure 8.4. In this figure, the sinusoidal response is also analyzed into the periodic part and the part contributed by the summation of the exponential terms. It should be noted that the exponential terms decay rapidly and for time greater than 60 s the response coincides with the periodic solution. In addition, all responses resemble the form of the imposed input (compare Figures 8.1–8.4 with the impulse, step, ramp and sinusoidal input).

Figure 8.4 Sinusoidal response of the product concentration – Example 8.5.

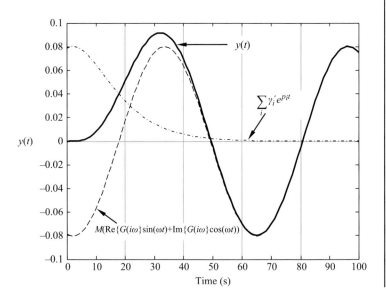

8.6 Effect of Zeros on the Step Response

In this chapter we have introduced the term pole to characterize any root of the denominator of the transfer function. We have seen that the location of poles plays an important role in the dynamic response. We have also introduced the term zero to characterize the roots of the numerator polynomial and in this section we will discuss briefly their effects on the step response of a system. To this end, we consider the case study of the constant volume CSTR studied in Example 7.4. In this example we have derived the following state-space description for the reactor

$$\frac{dC_R}{dt} = \frac{F}{V}(C_{R,in} - C_R) - k_0 e^{-\frac{E}{RT}} C_R$$

$$\frac{dT}{dt} = \frac{F}{V}(T_{in} - T) + \frac{(-\Delta h_{rxn})}{\rho c_p} k_0 e^{-\frac{E}{RT}} C_R + \frac{A_H U_H}{\rho c_p V}(T_J - T)$$

(8.6.1)

where the state variables are the reactant concentration and the temperature: $x = [C_R \; T]^T$. The A matrix of the linearized state-space model has also been derived in Example 7.4:

$$A = \begin{bmatrix} a_{11} & a_{12} \\ a_{21} & a_{22} \end{bmatrix} = \begin{bmatrix} -\left(\frac{F_s}{V} + k_0 e^{-\frac{E}{RT_s}}\right) & -\frac{E}{RT_s^2} k_0 e^{-\frac{E}{RT_s}} C_{R,s} \\ \beta k_0 e^{-\frac{E}{RT_s}} & -\left(\frac{F_s}{V} - \beta \frac{E}{RT_s^2} k_0 e^{-\frac{E}{RT_s}} C_{R,s} + \alpha\right) \end{bmatrix}$$

(8.6.2)

where $\alpha = U_H A_H/(\rho c_p V)$ and $\beta = (-\Delta h_{rxn})/(\rho c_p)$. Considering the volumetric flowrate as the input $u = F$ and the temperature of the reactor as the output $y = T$, the corresponding b, c and d matrices are

$$b = \begin{bmatrix} b_1 \\ b_2 \end{bmatrix} = \begin{bmatrix} (C_{R,in} - C_{R,s})/V \\ (T_{in} - T_s)/V \end{bmatrix}, \; c = \begin{bmatrix} c_1 & c_2 \end{bmatrix} = \begin{bmatrix} 0 & 1 \end{bmatrix}, \; d = 0$$

(8.6.3)

The transfer function of a second-order system has been derived in Chapter 5 (see Eq. (5.6.6)) and is given by

$$G(s) = \frac{(c_1 b_1 + c_2 b_2)s + (c_1 a_{12} b_2 + c_2 a_{21} b_1 - c_1 a_{22} b_1 - c_2 a_{11} b_2)}{s^2 - (a_{11} + a_{22})s + (a_{11} a_{22} - a_{12} a_{21})} + d$$

When $d = c_1 = 0$, $c_2 = 1$, this simplifies to

$$G(s) = \frac{b_2 s + (a_{21} b_1 - a_{11} b_2)}{s^2 - (a_{11} + a_{22})s + (a_{11} a_{22} - a_{12} a_{21})}$$

(8.6.4)

We observe that the transfer function has two poles and one zero, which we will calculate in what follows. We will use the numerical values given in Table 7.2 except for a different feed volumetric flowrate $F_s = 0.15$ m³/h. Under these conditions, the reactor has only one steady state $C_{R,s} = 0.765643$ kmol/m³, $T_s = 356.114$ K, and we can calculate the corresponding matrices A and b:

$$A = \begin{bmatrix} a_{11} & a_{12} \\ a_{21} & a_{22} \end{bmatrix} = \begin{bmatrix} -0.195915 & -0.002772 \\ 9.183 & -0.5956 \end{bmatrix}, \quad b = \begin{bmatrix} b_1 \\ b_2 \end{bmatrix} = \begin{bmatrix} 0.234357 \\ -6.114 \end{bmatrix} \quad (8.6.5)$$

With these numerical values, the denominator polynomial of the transfer function is

$$D(s) = s^2 - (a_{11} + a_{22})s + (a_{11}a_{22} - a_{12}a_{21}) = s^2 + 0.791515s + 0.1421425 \quad (8.6.6)$$

with roots (poles) $p_1 = -0.51610$, $p_2 = -0.27541$ and the numerator polynomial

$$N(s) = b_2 s + (a_{21}b_1 - a_{11}b_2) = -6.11366s + 0.95429 \quad (8.6.7)$$

with root (zero) $z = a_{11} - a_{21}\dfrac{b_1}{b_2} = 0.15609$.

The transfer function can be factorized as

$$G(s) = k_{zp}\frac{s-z}{(s-p_1)(s-p_2)} = k\frac{(\tau_z s + 1)}{(\tau_1 s + 1)(\tau_2 s + 1)} \quad (8.6.8)$$

where

$$\tau_1 = -\frac{1}{p_1} = 1.9376 \text{ h}, \quad \tau_2 = -\frac{1}{p_2} = 3.6309 \text{ h}, \quad \tau_z = -\frac{1}{z} = -6.4065 \text{ h}$$

$$k_{zp} = b_2 = -6.11366 \frac{K}{m^3}, \quad k = G(0) = \frac{a_{21}b_1 - a_{11}b_2}{a_{11}a_{22} - a_{12}a_{21}} = 6.7137 \frac{K}{m^3/h} \quad (8.6.9)$$

The next step is to calculate the unit step response of the transfer function given by (8.6.8). Following the standard procedure, we find that the unit step response is

$$y(t) = k\left[1 - \left(\frac{\tau_1 - \tau_z}{\tau_1 - \tau_2}\right)e^{-t/\tau_1} - \left(\frac{\tau_2 - \tau_z}{\tau_2 - \tau_1}\right)e^{-t/\tau_2}\right] \quad (8.6.10)$$

The response, for the particular values of the reactor problem, is shown in Figure 8.5. We note that at $t = 0$, the output is $y(0) = 0$, while for large times, the exponential terms in (8.6.10) decay to zero and the response asymptotically approaches a new steady-state value $y(\infty) = k$. At the early part of the response, the unit step response is significantly different from that of a classical second-order overdamped system (or two first-order systems in series) studied in Chapters 4 and 5. More specifically, we see from Figure 8.5 that the initial response is in the opposite direction compared to the asymptotic output value. As we will see in what follows, this feature, which is not possible for a classical second-order system, is due to the presence of a zero in the right half plane.

To this end, we consider a transfer function $G(s)$ without zeros, and also the same transfer function but with an added factor that adds a zero at $s = z = -1/\tau_z$, i.e.

$$G_z(s) = (\tau_z s + 1)G(s) \quad (8.6.11)$$

8.6 Effect of Zeros on the Step Response

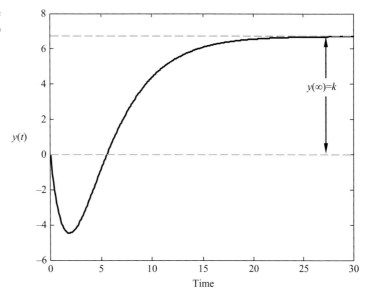

Figure 8.5 Response of the temperature of an exothermic reactor to a unit step change in the feed flowrate.

In order to calculate the unit step response $y_z(t)$ of $G_z(s)$, we observe that

$$y_z(t) = \mathcal{L}^{-1}\left\{\frac{G_z(s)}{s}\right\} = \mathcal{L}^{-1}\left\{\frac{(\tau_z s + 1)G(s)}{s}\right\} = \mathcal{L}^{-1}\left\{\frac{G(s)}{s}\right\} + \tau_z \mathcal{L}^{-1}\{G(s)\}$$

$$= (\text{unit step response of } G(s)) - \frac{1}{z}(\text{unit impulse response of } G(s)) \quad (8.6.12)$$

For the particular case where $G(s) = 1/(\tau_1 s + 1)(\tau_2 s + 1)$, we have:

- (unit step response of $G(s)$) = $1 - (\tau_1 e^{-t/\tau_1} - \tau_2 e^{-t/\tau_2})/(\tau_1 - \tau_2)$
- (unit impulse response of $G(s)$) = $1 - (e^{-t/\tau_1} - e^{-t/\tau_2})/(\tau_1 - \tau_2)$

and so

$$y_z(t) = \left(1 - \frac{\tau_1 e^{-t/\tau_1} - \tau_2 e^{-t/\tau_2}}{\tau_1 - \tau_2}\right) - \frac{1}{z}\left(\frac{e^{-t/\tau_1} - e^{-t/\tau_2}}{\tau_1 - \tau_2}\right) \quad (8.6.13)$$

which is the result found in Eq. (8.6.10) but in a rearranged form. Notice that for the system under consideration, the unit impulse response function is a bell-shaped function that vanishes at both $t = 0$ and $t = \infty$, hence the initial and final values of $y_z(t)$ are unaffected by the zero: $y_z(0) = 0$ and $y_z(\infty) = 1$. The presence of the zero has an effect in intermediate times.

If we calculate the initial slope of the response by differentiating Eq. (8.6.13), we find

$$\frac{dy_z}{dt}(0) = -\frac{1}{z} \cdot \frac{1}{\tau_1 \tau_2} \quad (8.6.14)$$

Figure 8.6 Step response of Eq. (8.6.13) with $\tau_1 = 1$, $\tau_2 = 1/5$, for different values of z.

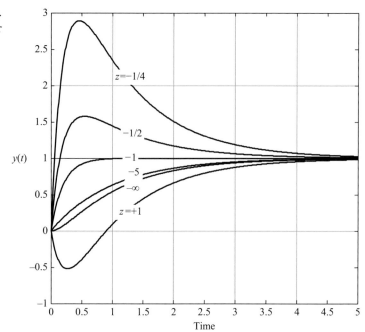

and we observe that the initial slope can be positive or negative, depending on the sign of z. If $z < 0$, the initial slope is positive, but if $z > 0$, the initial slope is negative. This leads us to the conclusion that the shape of the response at short times $t > 0$ will be strongly affected by the value of the zero. Figure 8.6 depicts the unit step response (8.6.13) for $\tau_1 = 1$, $\tau_2 = 1/5$ and several different values of z.

We observe that for z tending to $-\infty$, i.e. for $-1/z \to 0$, the response of $y_z(t)$ tends to the response of a classic second-order system without zeros, which is monotonically increasing. However, for small negative z, the response overshoots, whereas for positive z, the response undershoots. Nonmonotonicity of the response is observed for $z > \max\{p_1, p_2\}$. As $z \to 0^-$, $y_z(t)$ resembles the impulse response of the system without the zero, multiplied by $-1/z > 0$ (note that this holds true for the early part of the response as the final response is unaffected by the zero). On the other hand, for $z > 0$, the short-time response resembles an upside-down impulse response (now $-1/z < 0$).

The effect of the zero is more important when the zero is located to the right of the poles in the complex plane. If the zero is negative and small in magnitude, there is an overshoot, whereas if the zero is positive, there is an undershoot. The undershoot is also referred to as an *inverse response* to indicate the fact that the response is initially in the opposite direction compared to the long-term response.

A system consisting of two opposing first-order systems connected in parallel, shown in Figure 8.7, is a classical example of a second-order system with a zero.

The transfer function of the overall system of Figure 8.7 is

$$Y(s) = \left(\frac{k_1}{\tau_1 s + 1} - \frac{k_2}{\tau_2 s + 1} \right) U(s) = \frac{k_1 \tau_2 s + k_1 - k_2 \tau_1 s - k_2}{(\tau_1 s + 1)(\tau_2 s + 1)} U(s)$$

Figure 8.7 Two opposing first-order systems connected in parallel ($k_1, k_2 > 0$).

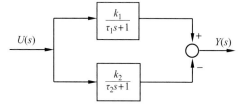

or

$$Y(s) = k \frac{(\tau_z s + 1)}{(\tau_1 s + 1)(\tau_2 s + 1)} U(s) \tag{8.6.15}$$

where

$$\tau_z = \frac{(k_1 \tau_2 - k_2 \tau_1)}{(k_1 - k_2)}, \quad k = k_1 - k_2 \tag{8.6.16}$$

We observe that the overall transfer function is the same as the one of Eq. (8.6.8) and an inverse response is observed whenever $z = -1/\tau_z > 0$, i.e. when

$$\frac{k_1 - k_2}{k_1 \tau_2 - k_2 \tau_1} < 0 \Rightarrow (k_1 - k_2)\left(\frac{k_1}{\tau_1} - \frac{k_2}{\tau_2}\right) < 0 \tag{8.6.17}$$

8.7 Bounded-Input–Bounded-Output (BIBO) Stability

In Chapter 7 we introduced the asymptotic stability in the context of studying the response of a linear system to nonzero initial conditions. We concluded that, if all eigenvalues of the A matrix lie in the left half of the complex plane, then the system will return to the origin. We have also noted that when a bounded input is imposed on an asymptotically stable system then the state vector is also bounded (bounded-input–bounded-state property), therefore the output is also bounded. When an input–output model such as the transfer function is used, we need to examine the response of the output of the system (and not the state that is not modeled) to any bounded input. As the initial state is set to zero when the transfer function is derived, the idea of asymptotic stability is not directly applicable. We therefore need to define stability in a slightly different way when an input–output description is used. In particular, we will say that the input–output description of a system is stable if for every bounded input the output is also bounded.

For a system described by the transfer function $G(s)$, its response to an arbitrary input $u(t)$ is calculated using the inverse Laplace transform of the transfer function times the Laplace transform of the input $U(s)$. Using the convolution theorem of the Laplace transform (see Appendix A), the response $y(t)$ can be calculated in terms of a convolution integral. In particular, denoting

$$g(t) = \mathcal{L}^{-1}\{G(s)\} \tag{8.7.1}$$

the inverse Laplace transform of the transfer function, the transfer function description $Y(s) = G(s)U(s)$ translates to

$$y(t) = g(t) * u(t) = \int_0^t g(t-t')u(t')dt' = \int_0^t g(t'')u(t-t'')\, dt'' \tag{8.7.2}$$

In order to simplify the exposition of the arguments, we will restrict our attention to strictly proper transfer functions $G(s)$ with distinct poles represented by Eq. (8.5.20), whose inverse Laplace transform is

$$g(t) = \sum_i \gamma_i e^{p_i t} \tag{8.7.3}$$

The output response to any arbitrary input is then given by

$$y(t) = \sum_{i=1}^n \gamma_i e^{p_i t} * u(t) = \int_0^t \sum_i \gamma_i e^{p_i t''} u(t-t'')dt'' = \sum_i \gamma_i \int_0^t e^{p_i t''} u(t-t'')dt'' \tag{8.7.4}$$

We then note that when the input is bounded, i.e. $|u(t)| < M$, then each term satisfies the following inequalities:

$$\left| \gamma_i \int_0^t e^{p_i t''} u(t-t'')dt'' \right| \leq |\gamma_i| \int_0^t \left| e^{p_i t''} u(t-t'') \right| dt'' \leq |\gamma_i| \left(\int_0^t \left| e^{p_i t''} \right| dt'' \right) M$$

$$\leq \left(\int_0^\infty \left| e^{p_i t''} \right| dt'' \right) |\gamma_i| M$$

If $\operatorname{Re}\{p_i\} < 0$, the function $e^{p_i t}$ is exponentially decaying and therefore the last integral is finite. If all the poles p_i of the transfer function satisfy $\operatorname{Re}\{p_i\} < 0$, then all the terms in Eq. (8.7.4) are absolutely bounded, hence $y(t)$ is bounded.

Thus we proved that if all the poles are in the open left half plane (i.e. $\operatorname{Re}\{p_i\} < 0, \forall i$), the output response to any bounded input is bounded. This is precisely the defining property of bounded-input–bounded-output (BIBO) stability.

One can easily see that the condition $\operatorname{Re}\{p_i\} < 0, \forall i$ is also necessary for BIBO stability.

– If there is a pole p_i with positive real part, then the corresponding term in Eq. (8.7.4) under a unit step input will be

$$\gamma_i \int_0^t e^{p_i t''} dt'' = \gamma_i \frac{e^{p_i t} - 1}{p_i}$$

which grows exponentially without a bound. Hence the output response will be unbounded.

– If there is a pole $p_i = 0$, then the corresponding term in Eq. (8.7.4) under a unit step input will be

$$\gamma_i \int_0^t e^{p_i t''} dt'' = \gamma_i \int_0^t dt'' = \gamma_i t$$

which grows linearly with time. Hence the output response will be unbounded.
- If there is a pole $p_i = i\omega$, then the corresponding term in Eq. (8.7.4) under a sinusoidal input of frequency ω (frequency equal to the imaginary part of p_i) will be

$$\gamma_i \int_0^t e^{p_i t''} \sin[\omega(t-t'')]dt'' = \gamma_i \int_0^t e^{i\omega t''} \sin[\omega(t-t'')]dt'' = \frac{\gamma_i}{2i}\left(te^{i\omega t} - \frac{1}{\omega}\sin\omega t\right)$$

which oscillates with linearly growing amplitude. Hence the output response will be unbounded.

The conclusion is that for a system to be BIBO stable the condition $\text{Re}\{p_i\} < 0$, $\forall i$ is both necessary and sufficient. Since all poles are eigenvalues of the A matrix of the state-space description, an asymptotically stable linear system is also BIBO stable for any output.

8.8 Asymptotic Response of BIBO-Stable Linear Systems

The response of a linear system to standard idealized inputs, such as the step or the sinusoidal input, is a powerful tool in analyzing the dynamic behavior of linear systems. In Section 8.5, the calculation of the output response from the transfer function description was discussed. We now restrict our attention to BIBO-stable systems, the aim being to calculate the asymptotic or ultimate response, i.e. the response for large t. Under the assumption that all poles of the system have negative real parts, all exponential terms $e^{p_i t}$ tend to zero for large t. Based on this observation, we obtain the following results, as an immediate consequence of the results of Table 8.1:

- asymptotic impulse response

$$y(t) = 0, \text{ for large } t \qquad (8.8.1)$$

- asymptotic step response

$$y(t) = MG(0), \text{ for large } t \qquad (8.8.2)$$

- asymptotic ramp response

$$y(t) = MG'(0) + MG(0)\,t, \text{ for large } t \qquad (8.8.3)$$

- asymptotic sinusoidal response

$$y(t) = M\left(\text{Re}\{G(i\omega)\}\sin\omega t + \text{Im}\{G(i\omega)\}\cos\omega t\right), \text{ for large } t \qquad (8.8.4)$$

From the asymptotic step and ramp responses we observe that $G(0)$, which is equal to the static gain k of the system, relates the change in the input to the change in the output. When the input is a step input of size M, for instance, then the output steady state changes by $G(0)M$, i.e. $k = G(0)$ is the proportionality constant that relates a change in the input with the resulting change of the output at steady state. More specifically, for the step response we observe that

$$k = G(0) = \frac{\text{steady-state change of the output}}{\text{steady-state change of the input}} \qquad (8.8.5)$$

On the other hand, for the ramp response

$$k = G(0) = \frac{\text{slope of the output}}{\text{slope of the input}} \qquad (8.8.6)$$

which means that the static gain also relates the steady rates of change of input and output. Following the same reasoning, one might expect that the static gain relates the amplitudes of the input and output in the sinusoidal response. However, this is not the case. If we write $G(i\omega)$ in polar form as $G(i\omega) = |G(i\omega)|e^{i\varphi} = |G(i\omega)|(\cos\varphi + i\sin\varphi)$, so that $\text{Re}\{G(i\omega)\} = |G(i\omega)|\cos\varphi$ and $\text{Im}\{G(i\omega)\} = |G(i\omega)|\sin\varphi$, where $\varphi = \arg\{G(i\omega)\}$ (see Figure 8.8), then the sinusoidal response given by Eq. (8.8.4) is expressed in the form $y(t) = M|G(i\omega)|(\cos\varphi\sin\omega t + \sin\varphi\cos\omega t))$, or

$$y(t) = M|G(i\omega)|\sin(\omega t + \varphi) \qquad (8.8.7)$$

We note that when the input is sinusoidal, the output is also sinusoidal with a different amplitude that depends on the frequency of the input signal and, more specifically, the amplitude ratio (AR) is a function of the frequency ω and is given by

$$AR = \frac{\text{amplitude of the output}}{\text{amplitude of the input}} = |G(i\omega)| \qquad (8.8.8)$$

We also note that the output signal has the same frequency as the input (ω), but is not in phase with the input. The phase shift φ is equal to the argument of the transfer function at $i\omega$, and also depends on the frequency

$$\varphi = \arg\{G(i\omega)\} \qquad (8.8.9)$$

The results presented in this section are summarized in Table 8.2. It is interesting to note from this table that the asymptotic response shares common characteristics with the corresponding input signal. For the impulse response, the input quickly recovers from the impulse and returns to zero asymptotically. For the step, ramp or sinusoidal inputs the asymptotic response is ultimately a step, ramp or sinusoid, correspondingly. It is also important to remember that all these asymptotic properties hold true under the assumption that the system is BIBO stable (all poles have negative real parts). The key inherent characteristic of the system that affects the asymptotic response is the static gain for the case of step or ramp inputs and the magnitude and argument of the system at the frequency of the input for the case of the sinusoidal input.

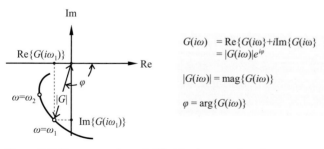

Figure 8.8 Representation of $G(i\omega)$ in the complex plane.

8.8 Asymptotic Response of BIBO-Stable Linear Systems

Table 8.2 Summary of the results for the asymptotic (long-time) output response of BIBO-stable linear systems to standard idealized inputs

Input	Asymptotic output response (large t)
Impulse, $u(t) = M\delta(t)$	$y(t) = 0$
Step, $u(t) = M$	$y(t) = M\, G(0)$
Ramp, $u(t) = Mt$	$y(t) = M(G'(0) + G(0)t)$
Sinusoidal, $u(t) = M\sin\omega t$	$y(t) = M\left(\text{Re}\{G(i\omega)\}\sin\omega t + \text{Im}\{G(i\omega)\}\cos\omega t\right)$
	$= M\lvert G(i\omega)\rvert \sin(\omega t + \varphi)$, where $\varphi = \arg\{G(i\omega)\}$

A final remark must be made before closing this section. For a system that has a single pole at $s = 0$, and all other poles have negative real parts, when a sinusoidal signal $u(t) = M\sin\omega t$ is applied as an input then the asymptotic response is given by (see Problem 8.8):

$$y(t) = \frac{M}{\omega}\lim_{s\to 0}[sG(s)] + M\left(\text{Re}\{G(i\omega)\}\sin\omega t + \text{Im}\{G(i\omega)\}\cos\omega t\right)$$

hence the amplitude and the phase shift are still determined by the magnitude and the argument of $G(i\omega)$ according to Eqs. (8.8.8) and (8.8.9). The difference is that the system oscillates around $(M/\omega)\lim_{s\to 0}[sG(s)]$ and not around zero. This result allows computing AR and φ for systems with a single pole at zero in the same way as BIBO-stable systems.

Example 8.6 Calculation of the product concentration asymptotic response to a sinusoidal input

For the reactor studied in Example 8.5 calculate the asymptotic response to sinusoidal input with unit amplitude and frequencies $\omega_1 = 1/10$ and $\omega_2 = 2/10$.

Solution

For the transfer function of Example 8.5,

$$G(i\omega) = \frac{\frac{5}{10000}}{\left(i\omega + \frac{1}{5}\right)\left(i\omega + \frac{1}{10}\right)\left(i\omega + \frac{17}{100}\right)}$$

from which

$$\text{Re}\{G(i\omega)\} = \frac{5}{2}\cdot\frac{17 - 2350\omega^2}{(1 + 5^2\omega^2)(1 + 10^2\omega^2)(17^2 + 100^2\omega^2)}$$

$$\text{Im}\{G(i\omega)\} = \frac{25}{2} \frac{(1000\omega^2 - 71)\omega}{(1+5^2\omega^2)(1+10^2\omega^2)(17^2+100^2\omega^2)}$$

At $\omega_1 = 1/10$, $\text{Re}\{G(\omega_1)\} = -13/778 = -0.0167$ and $\text{Im}\{G(\omega_1)\} = -61/778 = -0.0784$. The asymptotic response can be calculated using the result of Table 8.2:

$$y(t) = M\left(\text{Re}\{G(i\omega)\}\sin\omega t + \text{Im}\{G(i\omega)\}\cos\omega t\right)$$
$$= -0.0167\sin(0.1t) - 0.0784\cos(0.1t)$$

At $\omega_2 = 2/10$, $\text{Re}\{G(i\omega_2)\} = -77/2756 = -0.0279$, $\text{Im}\{G(i\omega_2)\} = -31/2756 = -0.0112$ and the asymptotic response is

$$y(t) = -0.0279\sin(0.2t) - 0.0112\cos(0.2t)$$

In Figures 8.9 and 8.10 the sinusoidal input variation and the asymptotic response of the system studied in this example are shown for the two different frequencies $\omega_1 = 1/10$ and $\omega_2 = 2/10$. It should be clear that both the magnitude and the frequency of the input signal are important in determining the asymptotic response of a system. We can observe that when the frequency of the input signal varies, the magnitude of the output signal as well as the phase lag are affected and can vary significantly. An efficient and compact way of representing this dependence on the frequency of the input signal is the use of the polar plot that is shown in Figure 8.8. Other equivalent representations will be presented in a later chapter.

As far as the computation of $G(i\omega)$ is concerned, it became evident that when the substitution $s = i\omega$ is made, obtaining the real and imaginary parts is rather laborious and error prone. A methodology that is easier to apply will be presented in the next chapter.

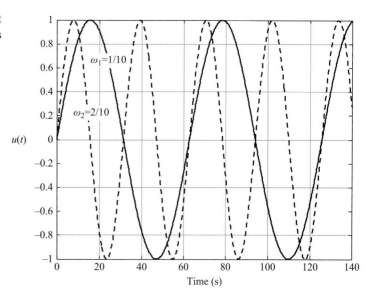

Figure 8.9 Sinusoidal input variations for two frequencies considered in Example 8.6.

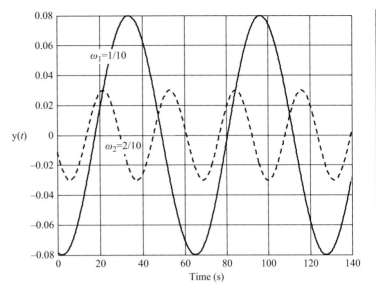

Figure 8.10 Asymptotic sinusoidal response for the two input variations shown in Figure 8.9.

8.9 Software Tools

8.9.1 MATLAB Commands Relative to Transfer Function Description

The first MATLAB command that is relevant to the subject of this chapter is the tf (transfer function) command, which uses the coefficients of the numerator and denominator polynomials in order to build the transfer function description of a system. We consider the following transfer function:

$$G(s) = \frac{(s+3)}{(s+1)(s+2)} = \frac{s+3}{s^2+3s+2} = \frac{\underbrace{\begin{bmatrix} 1 & 3 \end{bmatrix}}_{\text{num}} \begin{bmatrix} s^1 \\ s^0 \end{bmatrix}}{\underbrace{\begin{bmatrix} 1 & 3 & 2 \end{bmatrix}}_{\text{den}} \begin{bmatrix} s^2 \\ s^1 \\ s^0 \end{bmatrix}}$$

```
>> num=[1 3];
>> den=[1 3 2];
>> G=tf(num,den)
Transfer function:
     s + 3
  -------------
  s^2 + 3 s + 2
```

When the transfer function is available we can calculate its zeros, poles and the static gain as follows:

```
>> z=zero(G)
z =
    -3
>> p=pole(G)
p =
    -2
    -1
>> dcgain(G)
ans =
    1.5000
```

The `tf2pz` (transfer function to pole–zero) command can also be used to obtain the factorization (8.4.4) of the transfer function in terms of zeros and poles:

```
>> [z,p,kzp]=tf2zp(num,den)
z =
    -3
p =
    -2
    -1
kzp =
    1
```

It should be emphasized that the static gain $k = G(0)$ of the transfer function is different from the coefficient k_{zp} in the zero–pole factorization.

Another useful command is the `residue` command, which calculates the partial fraction expansion of a transfer function. The syntax of the command is

$$[\text{gamma},\text{p},\text{q}] = \text{residue}(\text{num},\text{den})$$

where p is the vector with the poles of the transfer function, gamma is the vector with the corresponding coefficients in the partial fraction expansion and q is the quotient of the division of the numerator and denominator polynomial (which is zero for proper transfer functions), i.e.

$$G(s) = \frac{\text{num}(s)}{\text{den}(s)} = q(s) + \sum_k \frac{\text{gamma}(k)}{s - p_k}$$

For the transfer function used above we have that

```
>> [gamma,p,q]=residue(num,den)
gamma =
    -1
     2
p =
    -2
    -1
q =
    []
```

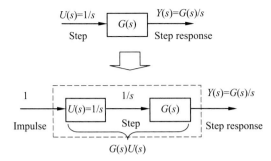

Figure 8.11 Alternative representation of the step response through the use of the impulse response of the system multiplied by the Laplace transform of the step input.

i.e. $G(s) = \dfrac{(s+3)}{(s+1)(s+2)} = \dfrac{-1}{s+2} + \dfrac{2}{s+1}$

To obtain the step and impulse response of a system whose transfer function is known we may use the step, impulse commands in MATLAB that we introduced in Chapter 6. We can use impulse to compute the response of any system and input $u(t)$ (with Laplace transform $U(s)$) based on the property (see also Figure 8.11)

$$y(t) = \mathcal{L}^{-1}\{G(s)U(s)\} = \text{impulse}\{G(s)U(s)\}$$

If we consider the transfer function considered in Example 8.2

$$G(s) = \dfrac{\dfrac{5}{10000}}{\left(s + \dfrac{2}{10}\right)\left(s + \dfrac{1}{10}\right)\left(s + \dfrac{17}{100}\right)}$$

then the step response is obtained using the following commands:

```
>> num=5/10000;
>> den=conv([1 2/10],conv([1 1/10],[1 17/100]));
>> G=tf(num,den)

Transfer function:
         0.0005
---------------------------------
s^3 + 0.47 s^2 + 0.071 s + 0.0034
>> U=tf(1,[1 0])
Transfer function:
1
-
s
>> Y=series(G,U)
Transfer function:
            0.0005
-------------------------------------
s^4 + 0.47 s^3 + 0.071 s^2 + 0.0034 s
>> impulse(Y)
```

The result is shown in Figure 8.2. The impulse response shown in Figure 8.1 can be obtained by using

```
>> impulse(G)
```

To obtain the ramp response shown in Figure 8.3 we follow the steps

```
>> U=tf(1,[1 0 0])
Transfer function:
   1
  ---
  s^2
>> Y=series(G,U)
Transfer function:
                    0.0005
   -------------------------------------------
   s^5 + 0.47 s^4 + 0.071 s^3 + 0.0034 s^2
>> impulse(Y)
>> axis([0 100 0 12])
```

For the sinusoidal response, shown in Figure 8.4, we have to use the following commands (note that $\mathcal{L}\{M\sin\omega t\} = M\omega/(s^2+\omega^2)$)

```
>> M=1;
>> wmega=0.1;
>> U=tf(M*wmega,[1 0 wmega^2])
Transfer function:
    0.1
  ----------
  s^2 + 0.01
>> Y=series(G,U)
Transfer function:
                            5e-005
   ----------------------------------------------------------------
   s^5 + 0.47 s^4 + 0.081 s^3 + 0.0081 s^2 + 0.00071 s + 3.4e-005
>> impulse(Y)
>> axis([0 300 -0.1 0.1])
```

8.9.2 Transfer Function Analysis using Maple

Consider again the transfer function

$$G(s) = \frac{s+3}{s^2 + 3s + 2}$$

The poles and the zeros can be calculated by simply finding the roots of the numerator and denominator polynomials

```
> G:=(s+3)/(s^2+3*s+2);
```

$$G := \frac{s+3}{s^2 + 3s + 2}$$

```
> zeros:=solve(numer(G)=0,s);
```

$$zeros := -3$$

```
> poles:=solve(denom(G)=0,s);
```

$$poles := -1, \ -2$$

whereas the steady-state gain is calculated by evaluating the transfer function at $s = 0$

```
> Static_Gain:=subs(s=0,G);
```

$$Static\_Gain := \frac{3}{2}$$

The zeros, the poles and the coeffient k_{zp} in the factorization of Eq. (8.4.4) can be computed using the ZeroPoleGain command of the DynamicSystems library of Maple, as follows:

```
> with(DynamicSystems):
> zeros:=ZeroPoleGain(G):-z; poles:=ZeroPoleGain(G):-p;
  kzp:=ZeroPoleGain(G):-k;
```

$$zeros := [[-3]]$$

$$poles := [[-2, \ -1]]$$

$$kzp := [1]$$

which means that the zero–pole representation of the system is

$$G(s) = 1\frac{(s-(-3))}{(s-(-1))(s-(-2))} = 1\frac{(s+3)}{(s+1)(s+2)}$$

Partial fraction expansions, as well as response calculations, are also straightforward to obtain. Considering now the transfer function of Example 8.5,

$$G(s) = \frac{\dfrac{5}{10000}}{\left(s+\dfrac{2}{10}\right)\left(s+\dfrac{1}{10}\right)\left(s+\dfrac{17}{100}\right)}$$

the partial fraction expansion can be computed using the convert/parfrac command as follows:

```
> F:=10:V:=100:k1:=1/10:k2:=5/100:k3:=2/100:
> G(s):=k1*k2*(F/V)/((s+F/V+k1)*(s+F/V)*(s+F/V+k2+k3));
```

$$G(s) := \frac{1}{2000\left(s + \frac{1}{5}\right)\left(s + \frac{1}{10}\right)\left(s + \frac{17}{100}\right)}$$

```
> convert(G(s),parfrac,s);
```

$$-\frac{500}{21(100s+17)} + \frac{5}{6(5s+1)} + \frac{5}{7(10s+1)}$$

Alternatively, one could first compute the poles p_i, and then the coefficients γ_i using the `residue` command:

```
> p:=solve(denom(G(s))=0,s);
```

$$p := -\frac{1}{5}, -\frac{1}{10}, -\frac{17}{100}$$

```
> gamma1:=residue(G(s),s=p[1]);gamma2:=
residue(G(s),s=p[2]);gamma3:=residue(G(s),s=p[3]);
```

$$\gamma 1 := \frac{1}{6}$$

$$\gamma 2 := \frac{1}{14}$$

$$\gamma 3 := -\frac{5}{21}$$

With the above calculation, we immediately have the impulse response function from Eq. (8.5.4):

```
> g(t):=gamma1*exp(p[1]*t)+gamma2*exp(p[2]*t)+
gamma3*exp(p[3]*t);
```

$$g(t) := \frac{1}{6} e^{-\frac{1}{5}t} + \frac{1}{14} e^{-\frac{1}{10}t} - \frac{5}{21} e^{-\frac{17}{100}t}$$

In general, responses to any standard input can be obtained through Laplace transform inversion, using the `invlaplace` command of the `inttrans` library of Maple. For example, the unit step response can be obtained as follows:

```
> with(inttrans):
> U(s):=1/s;
```

$$U(s) := \frac{1}{s}$$

```
> y_step:=invlaplace(G(s)*U(s),s,t);
```

$$y\_step := \frac{5}{34} + \frac{500}{357}e^{-\frac{17}{100}t} - \frac{5}{7}e^{-\frac{1}{10}t} - \frac{5}{6}e^{-\frac{1}{5}t}$$

and similarly the unit ramp response as

> U(s):=1/s^2;

$$U(s) := \frac{1}{s^2}$$

> y_ramp:=invlaplace(G(s)*U(s),s,t);

$$y\_ramp := -\frac{50000}{6069}e^{-\frac{17}{100}t} + \frac{50}{7}e^{-\frac{1}{10}t} + \frac{5}{34}t + \frac{25}{6}e^{-\frac{1}{5}t} - \frac{1775}{578}$$

When the above results for unit impulse, step and ramp responses are plotted, we obtain the graphs of Figures 8.1, 8.2 and 8.3. Alternatively, the above results for unit impulse, step and ramp responses can be obtained using the ImpulseResponse command of the DynamicSystems library of Maple, as follows:

> y_impulse:=ImpulseResponse(TransferFunction(G(s)));

$$y\_impulse := \left[\frac{1}{6}e^{-\frac{1}{5}t} + \frac{1}{14}e^{-\frac{1}{10}t} - \frac{5}{21}e^{-\frac{17}{100}t}\right]$$

> y_step:=ImpulseResponse(TransferFunction(G(s)/s));

$$y\_step := \left[\frac{500}{357}e^{-\frac{17}{100}t} - \frac{5}{7}e^{-\frac{1}{10}t} - \frac{5}{6}e^{-\frac{1}{5}t} + \frac{5}{34}\right]$$

> y_ramp:=ImpulseResponse(TransferFunction(G(s)/s^2));

$$y\_ramp := \left[-\frac{50000}{6069}e^{-\frac{17}{100}t} + \frac{50}{7}e^{-\frac{1}{10}t} + \frac{5}{34}t + \frac{25}{6}e^{-\frac{1}{5}t} - \frac{1775}{578}\right]$$

Finally, it should be noted that Maple can assist in the application of the asymptotic response formulas outlined in Table 8.2. For example, in the asymptotic response formula to a sinusoidal input, the real and imaginary parts of $G(i\omega)$ are involved, whose manual calculation could be rather tedious.

For the same transfer function under consideration, substitution of $s = i\omega$ leads to

> G(I*omega):=subs(s=I*omega,G(s));

$$G(I\omega) := \frac{1}{2000\left(I\omega + \frac{1}{5}\right)\left(I\omega + \frac{1}{10}\right)\left(I\omega + \frac{17}{1000}\right)}$$

However, further algebraic manipulation is needed, in order to identify real and imaginary parts: multiply and divide by the complex conjugate of the denominator, and collect terms. This can be done through the application of the evalc and simplify commands on the result of the substitution:

> simplify(evalc(G(I*omega)));

$$\frac{5}{2} \frac{17 - 2350\omega^2 - 355I\omega + 5000I\omega^3}{(1+25\omega^2)(1+100\omega^2)(289+10000\omega^2)}$$

With the above rearrangement, the real and imaginary parts of $G(i\omega)$ become clearly visible. In order to isolate the real and imaginary parts of $G(i\omega)$ and use them in subsequent calculations, we can apply the formulas $\text{Re}\{G(i\omega)\} = \dfrac{G(i\omega)+G(-i\omega)}{2}$ and $\text{Im}\{G(i\omega)\} = \dfrac{G(i\omega)-G(-i\omega)}{2i}$:

> ReG(I*omega):=simplify(evalc((subs(s=I*omega,G(s))+ subs(s=-I*omega,G(s)))/2));

$$\text{Re}\,G(I\omega) := -\frac{5}{2} \frac{-17 + 2350\omega^2}{(1+25\omega^2)(1+100\omega^2)(289+10000\omega^2)}$$

> ImG(I*omega):=simplify(evalc((subs(s=I*omega,G(s))- subs(s=-I*omega,G(s)))/(2*I)));

$$\text{Im}\,G(I\omega) := \frac{25}{2} \frac{\omega(-71+1000\omega^2)}{(1+25\omega^2)(1+100\omega^2)(289+10000\omega^2)}$$

With the above results, considering a sinusoidal input of unit amplitude, we can immediately apply Eq. (8.8.4):

> y_sine_asymptotic:=ReG(I*omega)*sin(omega*t)+ ImG(I*omega)*cos(omega*t);

$$y\_sine\_asymptotic := -\frac{5}{2} \frac{(-71+2350\omega^2)\sin(\omega t)}{(1+25\omega^2)(1+100\omega^2)(289+25\omega^2)}$$
$$+ \frac{25}{2} \frac{\omega(-71+1000\omega^2)\cos(\omega t)}{(1+25\omega^2)(1+100\omega^2)(289+10000\omega^2)}$$

The above expression can be plotted for specific values of the frequency, as was done in Example 8.6 (see Figure 8.10).

LEARNING SUMMARY
- In Tables 8.1 and 8.2, the calculation of complete and asymptotic output responses for systems described by transfer functions are summarized.
- A system is bounded-input–bounded-output (BIBO) stable if for every bounded input, the output is bounded. The necessary and sufficient condition for BIBO stability is that all the poles of the transfer function have negative real parts: $\text{Re}\{p_i\} < 0$, $\forall i$. Because all poles of the transfer function are eigenvalues of the state-space description, every asymptotically stable system is also BIBO stable.
- Zeros close to the origin have a significant effect on shape of the output response. Right-half-plane (RHP) zeros may result in inverse response.
- When the input to a system with transfer function $G(s)$ is a sinusoidal signal $u(t) = M\sin(\omega t)$, the output is also sinusoidal with a different amplitude that depends on the frequency of the input signal. Specifically, the amplitude ratio (AR) is a function of the frequency ω and is given by

$$AR = \frac{\text{amplitude of the output}}{\text{amplitude of the input}} = |G(i\omega)|$$

The output signal has the same frequency as the input (ω) but is not in phase with the input. The phase shift φ is equal to the argument of the transfer function at $i\omega$, and also depends on the frequency

$$\varphi = \arg\{G(i\omega)\}$$

TERMS AND CONCEPTS

Bounded-input–bounded-output (BIBO) stable systems. Systems for which the output is bounded for any bounded input. A system is BIBO stable if and only if all its poles p_i have negative real parts, $\text{Re}\{p_i\} < 0$.

Inverse response. This may be observed in systems with a right-half-plane (RHP) zero. Inverse response refers to the situation where the small-time response is in the opposite direction relative to the long-time response.

Pole. A pole of a transfer function is a number p such that the transfer function becomes infinite at $s = p$.

Proper systems. Systems for which the degree of the numerator polynomial of the transfer function is less than or equal to the degree of the denominator polynomial.

Relative order. The difference in the degrees of the denominator and numerator polynomials of a transfer function.

Right-half-plane (RHP) zero. A zero of a transfer function that lies in the right half of the complex plane, i.e. it has positive real part.

Strictly proper systems. Systems for which the degree of the numerator polynomial is less than the degree of the denominator polynomial.

Transfer function. The transfer function of a SISO linear system is the Laplace transform of the output divided by the Laplace transform of the input under zero initial conditions.
Zero. A zero of a transfer function is a number z such that the transfer function vanishes at $s = z$.

FURTHER READING

Additional material on the transfer function description of SISO systems can be found in

Chen, C. T., *Linear Systems Theory and Applications*, 3rd edn. New York: Oxford University Press, 1999.

Ogata, K., *Modern Control Engineering*, Pearson International Edition, 5th edn, 2008.

Ogunnaike, B. and Ray, H., *Process Dynamics, Modelling and Control*. New York: Oxford University Press, 1994.

PROBLEMS

8.1 Consider the following transfer function

$$G(s) = \frac{s(s+1)}{s^2 + 4s + 4}$$

(a) Calculate the poles, zeros, relative order and static gain of the system. Is the system BIBO stable?
(b) Write down a differential equation that describes the input–output behavior of the system in the time domain.
(c) Calculate and plot the unit step response. Determine the initial value and the final value of the step response. Can you explain the qualitative shape of the step response?
(d) If the input is a ramp with unit slope, what is the asymptotic response of the output?
(e) If the input is sinusoidal with unit amplitude and frequency ω, what is the asymptotic response of the output?
(f) Check your results with MATLAB or Maple.

8.2 Consider the following transfer function

$$G(s) = \frac{\tau_1 s}{(\tau_1 s + 1)(\tau_2 s + 1)}, \quad \tau_1, \tau_2 > 0$$

(a) Calculate the poles, zeros, relative order and static gain of the system. Is the system BIBO stable?
(b) Write down a differential equation that describes the input/output behavior of the system in the time domain.
(c) Calculate and plot the unit step response. Determine the initial value and the final value of the step response. Can you explain the qualitative shape of the step response?

(d) If the input is a ramp with unit slope, what is the asymptotic response of the output?
(e) If the input is sinusoidal with unit amplitude and frequency ω, what is the asymptotic response of the output?
(f) Check your results with MATLAB or Maple. Use $\tau_1 = 10$ and $\tau_2 = 0.01$.

8.3 Consider the following transfer function

$$G(s) = k \frac{\dfrac{\tau^2}{3}s^2 - \tau s + 1}{\dfrac{\tau^2}{3}s^2 + \tau s + 1}$$

(a) Calculate the poles, zeros, relative order and static gain of the system. Is the system BIBO stable?
(b) Write down a differential equation that describes the input/output behavior of the system in the time domain.
(c) Calculate and plot the unit step response. Determine the initial value and the final value of the step response. Can you explain the qualitative shape of the step response?
(d) If the input is a ramp with unit slope, what is the asymptotic response of the output?
(e) If the input is sinusoidal with unit amplitude and frequency ω, what is the asymptotic response of the output?
(f) Check your results with MATLAB or Maple.

8.4 Consider the following transfer function

$$G(s) = k \frac{1 - \tau' s}{s(\tau s + 1)}, \quad \tau, \tau' > 0$$

(a) Calculate the poles, zeros, relative order and static gain of the system. Is the system BIBO stable?
(b) Write down a differential equation that describes the input/output behavior of the system in the time domain.
(c) Calculate and plot the unit step response. Determine the initial value and the final value of the step response. Can you explain the qualitative shape of the step response?
(d) If the input is sinusoidal with unit amplitude and frequency ω, what is the asymptotic response of the output?
(e) Check your results with MATLAB or Maple. Use $k = 1/2$, $\tau = 1$ and $\tau' = 4$.

8.5 Consider the following transfer function

$$G(s) = \frac{101}{(s+2)(s^2 + 2s + 101)}$$

(a) Calculate the poles, zeros, relative order and static gain of the system. Is the system BIBO stable?
(b) Write down a differential equation that describes the input/output behavior of the system in the time domain.

(c) Calculate and plot the unit step response. Determine the initial value and the final value of the step response. Can you explain the qualitative shape of the step response?
(d) If the input is a ramp with unit slope, what is the asymptotic response of the output?
(e) If the input is sinusoidal with unit amplitude and frequency ω, what is the asymptotic response of the output?
(f) Check your results with MATLAB or Maple.

8.6 In Figure P8.6 the experimental unit step response of a process system is shown.
   (a) What is the transfer function that can be used to best describe the system dynamics among the following.

   (1) $G_1(s) = k \dfrac{1}{(\tau s + 1)^2}$, $\tau > 0$

   (2) $G_2(s) = k \dfrac{1}{s(\tau s + 1)}$, $\tau > 0$

   (3) $G_3(s) = k \dfrac{Ts + 1}{(\tau s + 1)^2}$, $\tau > 0$, $T > 0$

   (4) $G_4(s) = k \dfrac{-Ts + 1}{(\tau s + 1)^2}$, $\tau > 0$, $T > 0$

   (5) $G_5(s) = k \dfrac{Ts + 1}{s(\tau s + 1)}$, $\tau > 0$, $T > 0$

   (6) $G_6(s) = k \dfrac{-Ts + 1}{s(\tau s + 1)}$, $\tau > 0$, $T > 0$

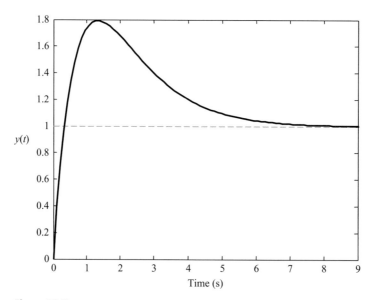

Figure P8.6

Discuss the reason(s) for which you exclude some of the transfer function and the reason(s) why you propose a specific one.
  (b) Can you estimate the parameters of the transfer function you have selected?
8.7 In Figure P8.7 the experimental unit step response of a process system is shown. The following transfer function can be used to best describe the system dynamics:

$$G_1(s) = k\frac{-Ts+1}{s(\tau s+1)}, \quad k, \tau, T > 0$$

  (a) Explain the reason(s) why this is a potential transfer function for the system that exhibits the given unit step response.
  (b) What is the initial and final slope of the unit step response?
  (c) Determine the analytic form of the unit step response.
  (d) Determine the time at which the minimum of the response takes place.
  (e) Based on the above determine the three parameters (k, T and τ) of the transfer function.

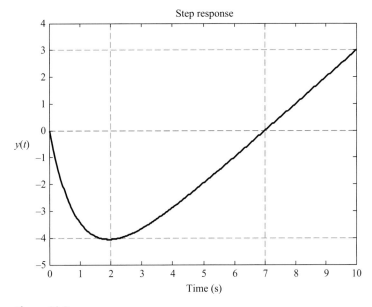

Figure P8.7

8.8 Consider a system with transfer function $G(s)$ that has a pole at 0 and all the other poles with negative real parts. When a sinusoidal input $u(t) = M\sin\omega t$ is applied to the system, prove that the asymptotic response of the output is given by

$$y(t) = \frac{M}{\omega}\lim_{s \to 0}[sG(s)] + M\left(\text{Re}\{G(i\omega)\}\sin\omega t + \text{Im}\{G(i\omega)\}\cos\omega t\right)$$

8.9 Consider the system of N mixing tanks in series shown in Figure P8.9.
(a) Show that the transfer function of each tank is

$$G_1(s) = \frac{H_1(s)}{F_0(s)} = \frac{k_1}{\tau_1 s + 1}, \quad \tau_1 = A_1 R_1, \, k_1 = R_1$$

$$G_i(s) = \frac{H_i(s)}{H_{i-1}(s)} = \frac{k_i}{\tau_i s + 1}, \quad \tau_i = A_i R_i, \, k_i = \frac{R_i}{R_{i-1}}, \, 2 \leq i \leq N$$

(b) Consider the overall transfer function $H_N(s)/F_0(s)$ given by

$$G(s) = \frac{H_N(s)}{F_0(s)} = \prod_{i=1}^{N} \frac{k_i}{\tau_i s + 1} = \frac{k}{\prod_{i=1}^{N}(\tau_i s + 1)}, \quad \text{where } k = \prod_{i=1}^{N} k_i,$$

Show that the unit step response is given by the following expression

$$y(t) = k\left(1 + \sum_{i=1}^{N} \psi_i e^{-t/\tau_i}\right), \text{ where } \psi_i = \lim_{s \to -\frac{1}{\tau_i}} \left[(\tau_i s + 1)\frac{G(s)}{ks}\right]$$

(c) Show that the step response has the following property:

$$\int_0^\infty \left(1 - \frac{y(t)}{k}\right) dt = \sum_{i=1}^{N} \tau_i$$

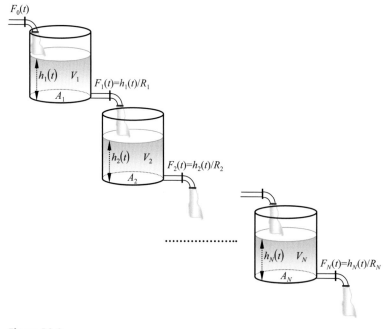

Figure P8.9

(d) For the special case

$$G(s) = \frac{1}{\prod_{i=1}^{N}\left(\dfrac{T}{N}s+1\right)} = \frac{1}{\left(\dfrac{T}{N}s+1\right)^{N}}$$

calculate the impulse and step responses and determine the maximum of the impulse response (which is also the infection point of the step response). Use MATLAB or Maple to validate your results and plot the response for $N = 1$, 10 and 100. What is the limit of the response as $N \to \infty$?

(e) For the system studied in (d) find the asymptotic response to a sinusoidal input with unit amplitude and frequency ω.

8.10 Consider a linear system with state-space description

$$\frac{d}{dt}\begin{bmatrix} x_1 \\ x_2 \end{bmatrix} = \begin{bmatrix} 1 & 0 \\ 1 & -1 \end{bmatrix}\begin{bmatrix} x_1 \\ x_2 \end{bmatrix} + \begin{bmatrix} 0 \\ 1 \end{bmatrix}u$$

$$y = \begin{bmatrix} 1 & 1 \end{bmatrix}\begin{bmatrix} x_1 \\ x_2 \end{bmatrix}$$

(a) Calculate the relative order of the system.
(b) Is the system asymptotically stable?
(c) Calculate the transfer function of the system, its poles and zeros.
(d) Is the system input–output stable?

8.11 Consider a system with transfer function

$$G(s) = k\frac{Ts+1}{\tau s+1}, \quad \tau > 0$$

Derive the step response of the system and comment on the initial and final value. What is the shape of the step response for (a) $T < 0$, (b) $0 < T < \tau$ and (c) $0 < \tau < T$? Please explain.

8.12 Use MATLAB or Maple to generate the step response of the following systems with right-half-plane (RHP) zeros:

(1) $G_1(s) = \dfrac{(-10s+1)}{(s+1)(2s+1)(3s+1)(4s+1)(5s+1)}$

(2) $G_2(s) = \dfrac{(-10s+1)(-20s+1)}{(s+1)(2s+1)(3s+1)(4s+1)(5s+1)}$

(3) $G_3(s) = \dfrac{(-10s+1)(-20s+1)(-30s+1)}{(s+1)(2s+1)(3s+1)(4s+1)(5s+1)}$

Can you draw any conclusions about the effect of RHP zeros on the shape of the step response?

8.13 Consider the process shown in Figure P8.13 in which steam with mass flowrate $\dot{M}_{st}$ is condensing in the coil in order to heat liquid which is fed with mass flowrate $\dot{M}$ and temperature T_0 in the well-stirred tank. The temperature is measured using a thermocouple that is placed in a thermowell. Both the thermocouple and the thermowell exhibit first-order dynamics and the mathematical model of the temperature dynamics and its measurement is

$$MC_p \frac{dT}{dt} = \dot{M}C_p T_0 - \dot{M}C_p T + \dot{M}_{st} \Delta h_{vap,st}$$

$$\tau_w \frac{dT_w}{dt} = T - T_w$$

$$\tau_m \frac{dT_m}{dt} = T_w - T_m$$

Data: M = mass of the liquid in the tank = 100 kg
C_p = specific heat capacity of the liquid = 1 kJ·kg^{-1} °C^{-1}
T_0 = temperature of the feed stream = 20 °C
T = temperature of the liquid in the tank = 60 °C
$\dot{M}$ = mass flowrate of the liquid = 10 kg·min^{-1}
$\Delta h_{vap,st}$ = latent heat of vaporization of steam = 2000 kJ·kg^{-1}
τ_w = time constant of the thermowell = 1 min
τ_m = time constant of the thermocouple = 0.1 min

If the mass flowrate of the steam $\dot{M}_{st}$ is the input and the measured temperature T_m is the output, express the model in deviation variable form, after determining the steady state, and calculate

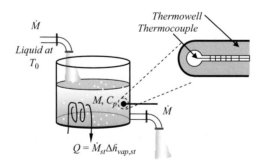

Figure P8.13

(a) the transfer function
(b) the unit impulse response
(c) the response to a step change of $\dot{M}_{st}$ by +10% of its steady-state value
(d) the asymptotic response to a sinusoidal input.

8.14 Air is used to strip ammonia from a wastewater stream using the gas stripping column with three equilibrium stages shown in Figure P8.14. The air stream has a molar flowrate of $V = 1$ kmol air/min and is free of ammonia. The liquid wastewater stream has a molar flowrate $L = 1$ kmol/min and is 0.1% in ammonia. The mole fraction of ammonia in the liquid (x) and in the air stream (y) exhibit linear equilibrium $y = mx$, where $m = 1.5$ is a constant. The liquid holdup on each equilibrium tray is $M=1$ kmol.
   (a) Use the notation given in Figure P8.14 to derive the following material balance for ammonia

$$M\frac{dx_i}{dt} = Lx_{i-1} - (L+mV)x_i + Vmx_{i+1}$$

   (b) Derive the state-space description of the system using $y = x_3$ and $u = x_0$. What are the steady-state values of the mole fractions of ammonia on the trays?
   (c) Derive the transfer function $X_3(s)/X_0(s)$.
   (d) Calculate the impulse and the step response of the system.
   Note: the state-space description is given by:

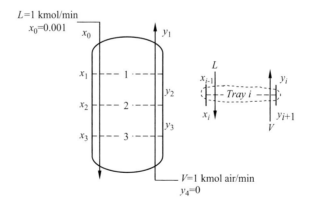

Figure P8.14

$$\begin{bmatrix} \dfrac{dx_1}{dt} \\ \dfrac{dx_2}{dt} \\ \dfrac{dx_3}{dt} \end{bmatrix} = \begin{bmatrix} -\dfrac{(L+mV)}{M} & \dfrac{mV}{M} & 0 \\ \dfrac{L}{M} & -\dfrac{(L+mV)}{M} & \dfrac{mV}{M} \\ 0 & \dfrac{L}{M} & -\dfrac{(L+mV)}{M} \end{bmatrix} \begin{bmatrix} x_1 \\ x_2 \\ x_3 \end{bmatrix} + \begin{bmatrix} \dfrac{L}{M} \\ 0 \\ 0 \end{bmatrix} x_0$$

$$y = \begin{bmatrix} 0 & 0 & 1 \end{bmatrix} \begin{bmatrix} x_1 \\ x_2 \\ x_3 \end{bmatrix}$$

8.15 Repeat Problem 8.14 using ten trays and an air molar flowrate of $V = 0.5$ kmol air/min (free of ammonia). Use MATLAB or Maple to perform your calculations. *Note*: use the `diag` command in MATLAB to build the A matrix.

9 Frequency Response

The aim of this chapter is to present frequency response analysis methods for studying the dynamic behavior of a linear system. Two are the most popular methods of representing the frequency response: the Bode diagram, which is a plot of the amplitude ratio and the phase of the response as a function of the frequency of the sinusoidal input, and the Nyquist diagram, which is the plot of the amplitude ratio and phase in polar coordinates. An approximate technique will be presented that can be used to sketch the Bode and Nyquist plots, which is useful in verifying computer generated plots.

STUDY OBJECTIVES

After studying this chapter you should be able to do the following.

- Understand the concept of frequency response.
- Be able sketch the Bode plots of a linear system.
- Be able to sketch the Nyquist diagrams of a linear system.

9.1 Introduction

Consider a linear system with transfer function $G(s)$ that is subject to a sinusoidal input $u(t) = M\sin(\omega t)$ (see Figure 9.1). It was shown in Chapter 8 that, under the assumption of BIBO stability, the output response for large t is given by

$$y(t) = M|G(i\omega)|\sin(\omega t + \varphi) \tag{8.8.7}$$

where

$$\varphi = \arg\{G(i\omega)\} \tag{8.8.9}$$

The output varies sinusoidally with the same frequency ω. The ratio of output and input amplitudes is called the amplitude ratio (AR) and is equal to the magnitude of the complex

Figure 9.1 Response of a stable linear system to a sinusoidal input variation.

number $G(s)|_{s=i\omega}$. The phase angle φ is equal to the argument of the complex number $G(s)|_{s=i\omega}$. Both AR and φ are functions of the frequency ω of the input signal. The long-time output response of a stable linear system to a sinusoidal input is called *frequency response*.

The frequency response depends on the frequency ω of the input signal, and can be determined by making the substitution $s = i\omega$ in $G(s)$ and then evaluating the magnitude and argument of the complex quantity $G(i\omega)$:

$$G(i\omega) = X(\omega) + iZ(\omega) = |G(i\omega)| e^{i\varphi} \qquad (9.1.1)$$

where

$$\begin{aligned} X(\omega) &= \mathrm{Re}\{G(i\omega)\} \\ Z(\omega) &= \mathrm{Im}\{G(i\omega)\} \\ AR &= |G(i\omega)| = \sqrt{[X(\omega)]^2 + [Z(\omega)]^2} \\ \varphi &= \arg\{G(i\omega)\} = \tan^{-1}\frac{Z(\omega)}{X(\omega)} = \cos^{-1}\frac{X(\omega)}{\sqrt{[X(\omega)]^2 + [Z(\omega)]^2}} \end{aligned} \qquad (9.1.2)$$

There are several alternatives for plotting graphs of the frequency response of a system. We can use a polar plot to represent the frequency response of a system using the real and imaginary parts of $G(i\omega)$ according to Eq. (9.1.1). These are called Nyquist plots and will be presented in the latter part of this chapter. The plots of $AR = |G(i\omega)|$ and $\varphi = \arg\{G(i\omega)\}$ versus ω, called Bode plots, are the most convenient method for portraying the frequency response and will be presented first.

9.2 Frequency Response and Bode Diagrams

Based on the summary of results given in the previous section, we may conclude that the asymptotic response of linear systems to a sinusoidal input variation can be calculated in terms of the magnitude and the phase angle of the transfer function $G(s)$ evaluated at $s = i\omega$. Bode was the first to propose the use of diagrams of the magnitude and the phase as a function of the frequency. These diagrams, called Bode diagrams, consist of

- a log–log plot of the magnitude of $G(i\omega)$ as a function of frequency ω (magnitude plot)
- a semilog plot of the argument of $G(i\omega)$ as a function of the frequency ω (phase plot).

9.2.1 Bode Diagrams of a First-Order System

In what follows we will start with the Bode diagrams of simple systems and then we will introduce a systematic technique for constructing the Bode diagrams of more complex systems.

We consider a first-order system

$$G(s) = \frac{1}{\tau s + 1}, \quad \tau > 0 \qquad (9.2.1)$$

We make the substitution $s = i\omega$ to obtain

$$G(i\omega) = \frac{1}{1+i\tau\omega} = \frac{1}{1+i\tau\omega} \cdot \frac{1-i\tau\omega}{1-i\tau\omega} = \frac{1}{1+(\tau\omega)^2} + i\frac{-\tau\omega}{1+(\tau\omega)^2}$$

hence here $X(\omega) = 1/(1+(\tau\omega)^2)$ and $Z(\omega) = -\omega\tau/(1+(\tau\omega)^2)$. Using Eqs. (9.1.2) we obtain the following expressions for the amplitude ratio and phase of a first-order transfer function

$$AR = \frac{1}{\sqrt{1+(\tau\omega)^2}} \qquad (9.2.2a)$$

$$\varphi = -\tan^{-1}(\tau\omega) \qquad (9.2.2b)$$

This result agrees with the results derived in Chapter 3 (Eqs. (3.11.4) and (3.11.5)). The Bode diagrams of a first-order system are shown in Figure 3.15. From this figure we have observed that the magnitude plot can be approximated by the following asymptotes:

- at low frequencies ($\tau\omega \ll 1$) $AR \approx 1$
- at high frequencies ($\tau\omega \gg 1$) $AR \approx 1/(\tau\omega)$

i.e. at low frequencies the magnitude diagram (in log–log scales) is a straight line, parallel to the frequency axis, while at high frequencies it is also a straight line but with slope equal to -1. These two straight lines cross at the corner frequency $\omega = 1/\tau$. From the phase diagram of Figure 3.15 we see that $\varphi \approx 0°$ at low frequencies and $\varphi \approx -90°$ at high frequencies. The phase plot may also be approximated by straight lines: there is a straight line starting from $\varphi = 0°$ at $\tau\omega = 1/10$, passing through the point ($\tau\omega = 1$, $\varphi = -45°$) and reaching $-90°$ at $\tau\omega = 10$, which approximates the phase in the intermediate frequency range.

9.2.2 Bode Diagrams of a Second-Order System

We now consider a second-order system with transfer function

$$G(s) = \frac{1}{\tau^2 s^2 + 2\zeta\tau s + 1}, \quad \tau > 0, \, \zeta > 0 \qquad (9.2.3a)$$

We make the substitution $s = i\omega$ to obtain

$$G(i\omega) = \frac{\left(1-\tau^2\omega^2\right)}{\left(1-\tau^2\omega^2\right)^2 + \left(2\zeta\tau\omega\right)^2} + i\frac{\left(-2\zeta\tau\omega\right)}{\left(1-\tau^2\omega^2\right)^2 + \left(2\zeta\tau\omega\right)^2} \quad (9.2.3b)$$

Using Eqs. (9.1.2) we obtain the following expressions, which agree with the results derived in Chapter 5 (Eqs. (5.5.5) and (5.5.6)):

$$AR = \frac{1}{\sqrt{\left(1-\tau^2\omega^2\right)^2 + \left(2\zeta\tau\omega\right)^2}} \quad (9.2.4a)$$

$$\varphi = -\cos^{-1}\left(\frac{1-\tau^2\omega^2}{\sqrt{(1-\tau^2\omega^2)^2 + (2\zeta\tau\omega)^2}}\right) \quad (9.2.4b)$$

The Bode diagram of a second-order system is shown in Figure 5.7. From this figure we have observed that the magnitude plot can be approximated by the following asymptotes

- at low frequencies ($\omega\tau \ll 1$) $AR \approx 1$
- at high frequencies ($\omega\tau \gg 1$) $AR \approx 1/(\tau\omega)^2$

i.e. at low frequencies the magnitude diagram (in log–log scales) is a straight line, parallel to the frequency axis, while at high frequencies it is also a straight line but with slope equal to −2. These two straight lines cross at the corner frequency $\omega = 1/\tau$. From the phase diagram of Figure 5.7 we see that $\varphi \approx 0°$ at low frequencies and $\varphi \approx -180°$ at high frequencies. The Bode phase plot can be approximated by straight lines as in the case of a first-order system.

9.2.3 Bode Diagrams of an Integrator

Consider a first-order system with integrating behavior, with transfer function

$$G(s) = \frac{1}{s} \quad (9.2.5)$$

We saw in Chapter 8 that for systems with a simple pole at 0 and any other poles with negative real parts, the formulas $AR = |G(i\omega)|$ and $\varphi = \arg\{G(i\omega)\}$, still hold. Thus, for the given system we can calculate

$$G(i\omega) = \frac{1}{i\omega} = -i\frac{1}{\omega} \quad (9.2.6)$$

The magnitude and phase of $G(i\omega)$ are therefore given by

$$AR = 1/\omega, \quad \varphi = -90° \quad (9.2.7)$$

We observe that the phase is constant, independent of the frequency, while the amplitude ratio is inversely proportional to the frequency ω.

In Table 9.1, the analytic results for AR and φ and the corresponding Bode diagrams of several simple transfer functions are summarized.

9.2 Frequency Response and Bode Diagrams

Table 9.1 Magnitude and phase of simple transfer functions

$G(s)$	$AR = |G(i\omega)|$	$\varphi = \arg G(i\omega)$
k Gain	$|k|$	$\begin{cases} 0, & \text{if } k > 0 \\ -180°, & \text{if } k < 0 \end{cases}$
$\dfrac{1}{s}$ Integrator	$\dfrac{1}{\omega}$	$-90°$
s Differentiator	ω	$+90°$
$\dfrac{1}{\tau s + 1}$, $\tau > 0$ First-order lag	$\dfrac{1}{\sqrt{\tau^2 \omega^2 + 1}}$	$-\tan^{-1}(\tau\omega)$
$\tau s + 1$, $\tau > 0$ First-order lead	$\sqrt{\tau^2 \omega^2 + 1}$	$\tan^{-1}(\tau\omega)$
$\dfrac{1}{\tau^2 s^2 + 2\zeta\tau s + 1}$, $\tau, \zeta > 0$ Second-order lag	$\dfrac{1}{\sqrt{(1-\tau^2\omega^2)^2 + (2\zeta\tau\omega)^2}}$	$-\cos^{-1}\left(\dfrac{1-\tau^2\omega^2}{\sqrt{(1-\tau^2\omega^2)^2 + (2\zeta\tau\omega)^2}}\right)$

Table 9.1 *(Cont.)*

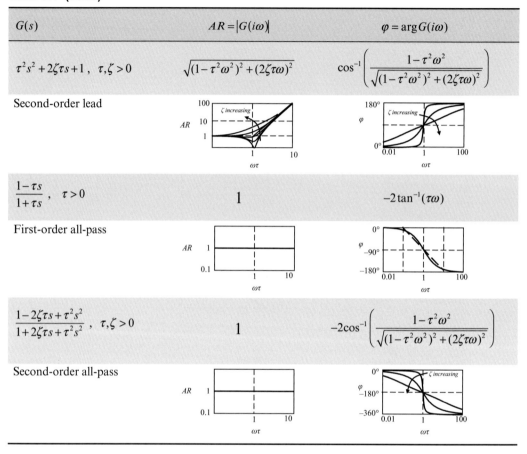

9.2.4 Bode Diagrams of Higher-Order Systems

Consider the case where the transfer function of the system under study has been factored as the product of simple transfer functions like the ones in Table 9.1, i.e.

$$G(s) = G_1(s) \cdot G_2(s) \cdot \ldots G_m(s) \qquad (9.2.8)$$

Substituting $s = i\omega$ in Eq. (9.2.8) and expressing each of the factors in polar form as $G(i\omega) = |G(i\omega)|e^{i\varphi}$, we can rewrite it as follows

$$\begin{aligned}G(i\omega) &= |G(i\omega)|e^{i\varphi} = |G_1(i\omega)|e^{i\varphi_1} \cdot |G_2(i\omega)|e^{i\varphi_2} \cdot \ldots |G_m(i\omega)|e^{i\varphi_m} \\ &= |G_1(i\omega)| \cdot |G_2(i\omega)| \cdot \ldots |G_m(i\omega)| \cdot e^{i(\varphi_1+\varphi_2+\cdots+\varphi_m)}\end{aligned} \qquad (9.2.9)$$

from which we conclude that

$$|G(i\omega)| = |G_1(i\omega)| \cdot |G_2(i\omega)| \cdot \ldots |G_m(i\omega)|$$
$$\varphi = \varphi_1 + \varphi_2 + \cdots + \varphi_m$$
(9.2.10)

i.e. the magnitude of the overall transfer function is the product of the magnitudes of the individual factors and the phase of the overall transfer function is the sum of the phases of the individual transfer functions.

Also note that taking logarithms of the magnitude equation in (9.2.10) leads to

$$\log_{10} |G(i\omega)| = \log_{10} |G_1(i\omega)| + \log_{10} |G_2(i\omega)| + \cdots + \log_{10} |G_m(i\omega)| \qquad (9.2.11)$$

which shows that in log scale, the individual magnitudes are added to form the magnitude of the overall transfer function.

These fundamental results are useful for drawing the Bode diagrams of complex transfer functions, expressed in factored form. Bode plots can be easily obtained by using computer software, but it is also possible to do reasonably accurate sketches of the magnitude and the phase versus frequency without relying on a computer. Using simple straight-line approximations, one can sketch the Bode diagrams of each individual factor and add them up, to obtain a sketch of the overall. This will be illustrated in the next section.

Two remarks must be made before closing this section. First, it should be noted that some mechanical and electrical engineers like to express the amplitude ratio in decibel (dB) units, defined as

$$20 \log_{10} |G(i\omega)| \qquad (9.2.12)$$

Then, the magnitude plot has a linear scale in decibels and, from Eq. (9.2.11),

$$20 \log_{10} |G(i\omega)| = 20 \log_{10} |G_1(i\omega)| + 20 \log_{10} |G_2(i\omega)| + \cdots + 20 \log_{10} |G_m(i\omega)| \qquad (9.2.13)$$

Second, it should be noted that the argument of a complex number, which is used to determine the phase angle, is not uniquely defined. Any multiple of 2π can be added to the argument of a complex number, and it is also an argument. When plotting $\varphi(\omega)$, the main convention is that it must be a continuous function of ω. Factorizing the transfer function into simple factors and using Table 9.1 for each factor, will always give a continuous overall plot, and will be consistent with the other usual conventions.

9.3 Straight-Line Approximation Method for Sketching Bode Diagrams

The straight-line approximation method is best introduced through examples. Let's consider the transfer function

$$G(s) = \frac{10s+1}{s(s+1)} \qquad (9.3.1)$$

This transfer function consists of three simple factors, the Bode diagrams of which are known. These factors are $1/s$, $(10s+1)$ and $1/(s+1)$. Each factor will be analyzed first.

The amplitude ratio of the factor $1/s$ is a straight line with a slope of -1 in a plot of AR versus $\log\omega$ (see Table 9.1) that passes through the point ($\omega = 1$, $AR = 1$). The phase angle is constant and equal to $-90°$ at all frequencies.

The straight-line approximation of AR for the factor $(10s+1)$ is (see also Table 9.1):

for $\omega \ll 1/10$: $\sqrt{(10\omega)^2 + 1} \approx 1$, slope $= 0$,

for $\omega \gg 1/10$: $\sqrt{(10\omega)^2 + 1} \approx 10\omega$, slope $= +1$.

The corner frequency is $\omega = 1/10 = 0.1$.

The straight-line approximation of AR for the factor $1/(s+1)$ is:

for $\omega \ll 1$: $1/\sqrt{\omega^2 + 1} \approx 1$, slope $= 0$,

for $\omega \gg 1$: $1/\sqrt{\omega^2 + 1} \approx 1/\omega$, slope $= -1$.

The corner frequency is $\omega = 1$.

The slopes of the straight-line approximation of the overall Bode magnitude plot can be found as shown in Table 9.2. We first note that there are three distinct intervals in the Bode diagram. One is the interval between $\omega = 0$ and the smallest corner frequency, which is the corner frequency of the factor $10s+1$ ($\omega = 0.1$). In this interval, the magnitude is determined by the magnitude of the factor $1/s$ as the magnitude of the other two terms is equal to 1. Starting from $\omega = 0.1$ and up to $\omega = 1$ the slope is zero and the approximation is a straight line parallel to the frequency axis. For frequencies $\omega > 1$ the slope is -1 and the final part of the approximation can easily be generated. The result is shown in Figure 9.2.

To construct the approximate Bode phase plot we can follow a similar approach. We start by noting that the approximate phase plot of the factor $10s+1$, as shown in Table 9.1, consists of a phase of $0°$ for frequencies up to about $(1/10)(1/\tau) = 0.01$ as well as of a phase of $+90°$ for frequencies greater than $10(1/\tau) = 1$. The phase varies approximately linearly from $0°$ to $+90°$ in the intermediate frequencies. Similar arguments apply for the $1/(s+1)$ factor. The

Table 9.2 Slopes of the straight-line approximation of the magnitude plot of the transfer function given by Eq. (9.3.1)

	Frequency range (rad/s)		
	0 to 0.1	0.1 to 1	1 to $+\infty$
$1/s$	-1	-1	-1
$10s+1$	0	$+1$	$+1$
$1/(s+1)$	0	0	-1
TOTAL	**-1**	**0**	**-1**

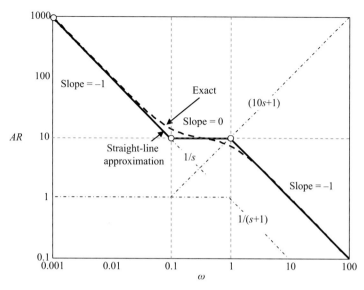

Figure 9.2 Magnitude plot: exact (dashed) and straight-line approximation (continuous line) for the transfer function of Eq. (9.3.1) and the individual factors (dash-dotted).

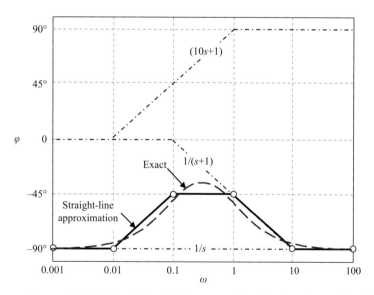

Figure 9.3 Phase plot: exact (dashed) and straight-line approximation (continuous line) for the transfer function of Eq. (9.3.1) and the individual factors (dash-dotted).

factor $1/s$ contributes $-90°$ at all frequencies. We add the phase angles of $10s + 1$ and $1/(s + 1)$ and then subtract $-90°$ at all frequencies to obtain the straight line approximation shown in Figure 9.3.

It is also important to note that, using the results of Table 9.1, we can easily derive analytic expressions of the magnitude and the phase. For the transfer function given by Eq. (9.3.1),

$$|G(i\omega)| = |i10\omega + 1| \cdot \left|\frac{1}{i\omega}\right| \cdot \left|\frac{1}{i\omega + 1}\right| = \sqrt{10^2\omega^2 + 1} \cdot \frac{1}{\omega} \cdot \frac{1}{\sqrt{\omega^2 + 1}} = \frac{\sqrt{10^2\omega^2 + 1}}{\omega\sqrt{\omega^2 + 1}} \quad (9.3.2)$$

$$\varphi = -90° + \tan^{-1} 10\omega - \tan^{-1} \omega \quad (9.3.3)$$

The method followed in this example is very general. After the transfer function is factorized in terms of simple factors, the results of Table 9.1 are applied for each factor and the overall frequency response is obtained from Eq. (9.2.10).

As a more general example, consider a system with M negative zeros, N negative poles, Q pairs of complex conjugate poles with negative real parts, and a pole at 0:

$$G(s) = \frac{\prod_{m=1}^{M}(\tau_m s + 1)}{s\prod_{n=1}^{N}(\tau_n s + 1)\prod_{q=1}^{Q}(\tau_q^2 s^2 + 2\zeta_q \tau_q s + 1)} \quad (9.3.4)$$

where $\tau_m > 0$, $\tau_p > 0$, $\tau_q > 0$, $\zeta_q > 0$. Then, applying the results from Table 9.1,

$$|G(i\omega)| = \frac{1}{\omega} \cdot \prod_{m=1}^{M}\sqrt{\tau_m^2\omega^2 + 1} \cdot \prod_{n=1}^{N}\frac{1}{\sqrt{\tau_n^2\omega^2 + 1}} \cdot \prod_{q=1}^{Q}\frac{1}{\sqrt{(1 - \tau_q^2\omega^2)^2 + (2\zeta_q\tau_q\omega)^2}} \quad (9.3.5)$$

$$\varphi = -90° + \sum_{m=1}^{M}\tan^{-1}\tau_m\omega - \sum_{n=1}^{N}\tan^{-1}\tau_n\omega - \sum_{q=1}^{Q}\cos^{-1}\frac{1 - \tau_q^2\omega^2}{\sqrt{(1 - \tau_q^2\omega^2)^2 + (2\zeta_q\tau_q\omega)^2}} \quad (9.3.6)$$

Example 9.1 Bode plots of a lead–lag element

The transfer function

$$G(s) = \frac{\tau' s + 1}{\tau s + 1}, \quad \tau > 0, \quad \tau' > 0$$

is called a lead–lag element as it consists of a lead element and a lag element in series. The dynamic behavior of a lead–lag element depends on the ratio $\alpha = \tau'/\tau$, whether it is smaller or larger than 1. When $\alpha > 1$, the numerator is dominant and the system is also called a phase-lead element, whereas when $0 < \alpha < 1$, the denominator is dominant and the system is also called a phase-lag element. Lead–lag elements are often encountered in the transfer functions of controllers, as will be seen in subsequent chapters.

9.3 Straight-Line Approximation Method for Sketching Bode Diagrams

Derive analytic expressions for $AR(\omega)$ and $\varphi(\omega)$ for the transfer function

$$G(s) = \frac{\alpha\tau s + 1}{\tau s + 1}, \quad \tau > 0, \quad \alpha > 0 \qquad (9.3.7)$$

and construct the corresponding Bode plots. Distinguish two cases: $\alpha > 1$ (phase-lead) and $0 < \alpha < 1$ (phase-lag).

Solution

Using the results of Table 9.1,

$$AR(\omega) = |G(i\omega)| = |i\alpha\tau\omega + 1| \cdot \left|\frac{1}{i\tau\omega + 1}\right| = \sqrt{\frac{1 + \alpha^2(\tau\omega)^2}{1 + (\tau\omega)^2}}$$

$$\varphi(\omega) = \arg\{G(i\omega)\} = \tan^{-1}\alpha\tau\omega - \tan^{-1}\tau\omega$$

We observe that

- for low frequencies $\omega \to 0$, $AR \approx 1$ and $\varphi \approx 0$
- for high frequencies $\omega \to \infty$, $AR \approx \alpha$ and $\varphi \approx 0$.

Case 1: $\alpha > 1$

Then, for all ω, $AR(\omega) \geq 1$ and $\varphi(\omega) \geq 0$ (which explains the term phase-lead element).
The function $AR(\omega)$ is monotonically increasing, going from 1 to α.
The function $\varphi(\omega)$ increases from 0 to a maximum value and then decreases back to 0. The maximum value of the phase can be calculated through the derivative of $\varphi(\omega)$.

We find: $(\tau\omega)_{max} = \dfrac{1}{\sqrt{\alpha}} < 1$, $\varphi_{max} = \sin^{-1}\left(\dfrac{\alpha - 1}{\alpha + 1}\right)$

and the corresponding $AR((\tau\omega)_{max}) = \sqrt{\alpha}$.

Case 2: $0 < \alpha < 1$

Then, for all ω, $AR(\omega) \leq 1$ and $\varphi(\omega) \leq 0$ (which explains the term phase-lag element).
The function $AR(\omega)$ is monotonically decreasing, going from 1 to α.
The function $\varphi(\omega)$ decreases from 0 to a minimum value and then increases back to 0. The minimum value of the phase can be calculated through the derivative of $\varphi(\omega)$.

We find: $(\tau\omega)_{min} = \dfrac{1}{\sqrt{\alpha}} > 1$, $\varphi_{min} = \sin^{-1}\left(\dfrac{\alpha - 1}{\alpha + 1}\right) < 0$

and the corresponding $AR((\tau\omega)_{min}) = \sqrt{\alpha}$.

Figures 9.4–9.7 depict the exact Bode diagrams of the lead–lag element for various representative values of the parameter α.

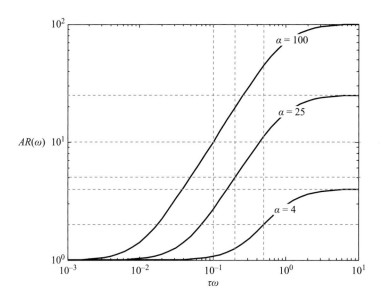

Figure 9.4 Bode magnitude plot of phase-lead element.

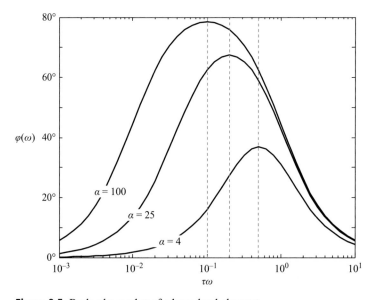

Figure 9.5 Bode phase plot of phase-lead element.

9.3 Straight-Line Approximation Method for Sketching Bode Diagrams

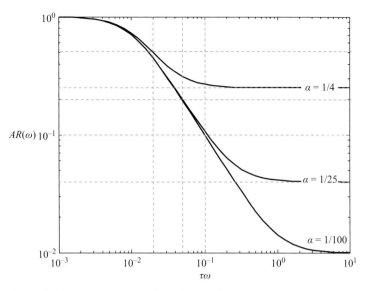

Figure 9.6 Bode magnitude plot of phase-lag element.

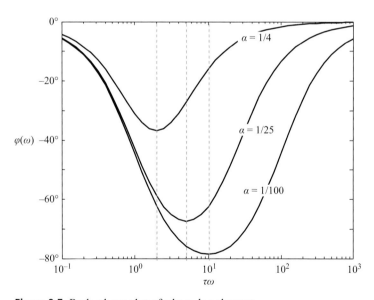

Figure 9.7 Bode phase plot of phase-lag element.

Example 9.2 Bode Plots of a Nonminimum Phase Element

The transfer function

$$G(s) = \frac{-\alpha\tau s + 1}{\tau s + 1}, \quad \tau > 0, \ \alpha > 0 \qquad (9.3.8)$$

is often called a nonminimum phase element. Derive analytic expressions for $AR(\omega)$ and $\varphi(\omega)$ and construct the corresponding Bode plots.

Solution

We first note that the given transfer function has a positive zero at $z = 1/\alpha\tau$ and a negative pole at $p = -1/\tau < 0$. The transfer function is the product of a first-order all-pass and a lead–lag element (see Example 9.1):

$$G(s) = \left(\frac{-\alpha\tau s + 1}{\alpha\tau s + 1}\right) \cdot \left(\frac{\alpha\tau s + 1}{\tau s + 1}\right)$$

Thus from the results of Table 9.1 and Example 9.1, we immediately obtain:

$$AR(\omega) = |G(i\omega)| = 1 \cdot \sqrt{\frac{1 + \alpha^2 (\tau\omega)^2}{1 + (\tau\omega)^2}}$$

$$\varphi(\omega) = \arg\{G(i\omega)\} = (-2\tan^{-1}\alpha\tau\omega) + (\tan^{-1}\alpha\tau\omega - \tan^{-1}\tau\omega)$$

$$= -\tan^{-1}\alpha\tau\omega - \tan^{-1}\tau\omega$$

We observe that, because $\alpha > 0$, it follows that $\varphi \le 0$ at all frequencies. In addition, $\varphi \to 0$ when $\omega \to 0$, and $\varphi \to -180°$ when $\omega \to \infty$.

Systems with right-half-plane (RHP) zeros are called nonminimum phase (NMP) systems. To justify the term "nonminimum phase," let's compare the system under consideration that

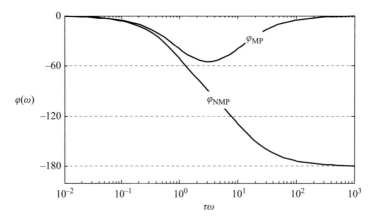

Figure 9.8 Phase plot of a nonminimum phase (NMP) element and the corresponding lead–lag element (MP) for $\alpha = 0.1$.

has a positive zero at $1/\alpha\tau$, with the corresponding lead–lag element of Example 9.1 that has a negative zero at $-1/\alpha\tau$. The two systems have exactly the same $AR(\omega)$, but the present system has an additional phase lag: $-2\tan^{-1}\alpha\tau\omega$. This additional phase lag is due to the positivity of the system's zero. Systems without RHP zeros are called minimum phase (MP) since they have the smallest phase for their $AR(\omega)$. Figure 9.8 depicts the phase plots of the two systems for the case $\alpha = 0.1$. Nonminimum phase characteristics are also exhibited by systems with dead time; this will be seen in Chapter 14.

9.4 Low-Frequency and High-Frequency Response

The factorization method outlined in the previous section also indicates what happens at the low end and at the high end of the frequency range. At the low end, when $\omega \to 0$, lead, lag and all-pass elements have $AR \approx 1$ and $\varphi \approx 0°$ and therefore do not contribute; it is only gain and differentiator or integrator (if any) that contribute. On the other hand, at the high end, when $\omega \to \infty$, each of the factors gives a slope of 0, ±1 or ±2 in the AR, depending on the powers of s and whether s is in the numerator or the denominator, and correspondingly multiples of ±90° in the phase φ. More precisely, the following conclusions can be drawn.

Low-frequency asymptotic behavior ($\omega \to 0$)

Consider a transfer function $G(s)$ factorized as a product of elementary factors from Table 9.1, with positive gain factor k.

- If $G(s)$ has no integrators or differentiators (no pole or zero at 0), then for $\omega \to 0$, $AR \approx k = G(0)$ ($\Rightarrow$ slope = 0 in the AR plot) and $\varphi \approx 0°$.
- If $G(s)$ has one integrator factor (one pole at 0), then for $\omega \to 0$, $AR \approx k/\omega$ ($\Rightarrow$ slope = -1 in the AR plot) and $\varphi \approx -90°$.
- If $G(s)$ has one differentiator factor (one zero at 0), then for $\omega \to 0$, $AR \approx k\omega$ ($\Rightarrow$ slope = $+1$ in the AR plot) and $\varphi \approx +90°$.

In the presence of a negative sign in the gain factor k, a phase of ($-180°$) must be added to the above φ, and k must be replaced by $|k|$ in AR.

High-frequency asymptotic behavior ($\omega \to \infty$)

Consider a transfer function $G(s)$ factorized as a product of elementary factors from Table 9.1 and denote by r the relative order of $G(s)$.

Superimposing the contributions of each of the elementary factors that are present, i.e. multiplying their ARs (adding up the slopes of AR plots), and summing up their φ's, the end result is as follows, for $\omega \to \infty$.

- If $G(s)$ has no all-pass factors (no RHP zeros),
  then $AR \propto \omega^{-r}$ ($\Rightarrow$ slope $= -r$ in the AR plot) and $\varphi \approx -r \cdot 90°$.
- If $G(s)$ has one RHP zero (one first-order all-pass factor),
  then $AR \propto \omega^{-r}$ ($\Rightarrow$ slope $= -r$ in the AR plot) and $\varphi \approx -r \cdot 90° - 180°$.
- If $G(s)$ has N_{RHP} right-half-plane zeros (all pass factors of total order N_{RHP}),
  then $AR \propto \omega^{-r}$ ($\Rightarrow$ slope $= -r$ in the AR plot) and $\varphi \approx -r \cdot 90° - N_{RHP} \cdot 180°$.

As an example, consider the system (9.3.4). This has gain factor $k = 1$, one integrator factor, relative order $r = 1 + N + 2Q - M$ and no RHP zeros. Applying the foregoing results, we conclude the following.

- For $\omega \to 0$, $AR \approx 1/\omega$ ($\Rightarrow$ slope $= -1$ in the AR plot) and $\varphi \approx -90°$.
- For $\omega \to \infty$, $AR \propto \omega^{-(1+N+2Q-M)}$ ($\Rightarrow$ slope $= -(1 + N + 2Q - M)$ in the AR plot) and $\varphi \approx -(1 + N + 2Q - M) \cdot 90°$.

If the M zeros were positive instead of being negative (i.e. if $\tau_m < 0$), then the transfer function (9.3.4) would be factorized as follows:

$$G(s) = \frac{\prod_{m=1}^{M}(-\tau_m s + 1)}{s \prod_{n=1}^{N}(\tau_n s + 1) \prod_{q=1}^{Q}(\tau_q^2 s^2 + 2\zeta_q \tau_q s + 1)} \prod_{m=1}^{M}\left(\frac{\tau_m s + 1}{-\tau_m s + 1}\right)$$

involving M first-order all-pass elements in addition to the integrator, the M first-order leads, the N first-order lags and the Q second-order lags. The low-frequency response characteristics would remain the same, as well as the high-frequency amplitude ratio. However, the high-frequency phase would have additional $M(-180°)$, coming from the M all-pass factors: $\varphi \approx -(1 + N + 2Q - M) \cdot 90° - M \cdot 180° = -(1 + N + 2Q + M) \cdot 90°$.

9.5 Nyquist Plots

The polar plot $G(i\omega)$, which is a complex function, is a graph of $\text{Im}\{G(i\omega)\}$ versus $\text{Re}\{G(i\omega)\}$ for $0 \leq \omega \leq +\infty$. The polar plot of $G(i\omega)$ can also be obtained by plotting the magnitude of $G(i\omega)$ versus the phase angle of $G(i\omega)$. The polar plots obtained are identical; only the coordinates used are different, as shown in Figure 9.9. It should be noted that in a polar plot positive phase angles are measured counterclockwise while negative phase angles are measured clockwise (with respect to the positive real axis). Nyquist plots are polar plots generated for $-\infty \leq \omega \leq +\infty$. However, owing to the fact that the polar plot exhibits conjugate symmetry (i.e. the graph for $-\infty \leq \omega \leq 0$ is the mirror image about the real axis of the graph for $0 \leq \omega \leq +\infty$), only the graph for $0 \leq \omega \leq +\infty$ needs to be constructed. This can be achieved by using the data necessary to construct the corresponding Bode plot. The frequency graduation is normally shown on the plot (note that constant frequency graduations are not separated by equal intervals along the plot). In what follows, some examples of Nyquist plots will be presented.

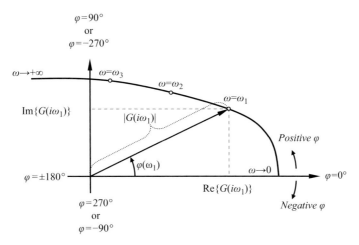

Figure 9.9 Polar plots of $G(i\omega)$ in rectangular or polar coordinates are identical.

9.5.1 Nyquist Plot of a First-Order System

For a first-order system

$$G(s) = \frac{1}{\tau s + 1}, \quad \tau > 0 \tag{9.5.1}$$

we have shown that

$$\text{Re}\{G(i\omega)\} = \frac{1}{1+(\tau\omega)^2}, \quad \text{Im}\{G(i\omega)\} = -\frac{\tau\omega}{1+(\tau\omega)^2}$$

and also that

$$|G(i\omega)| = \frac{1}{\sqrt{1+(\tau\omega)^2}}, \quad \varphi = -\tan^{-1}(\tau\omega)$$

The Nyquist plot is shown in Figure 9.10. Note the two parts of the plot: the one corresponding to $0 \leq \omega \leq +\infty$ (lower part – continuous line) and the one corresponding to $-\infty \leq \omega \leq 0$ (upper part – dashed line) and the fact that they exhibit conjugate symmetry. For $\omega = 0$ we have that $\text{Re}\{G(i\omega)\} = 1$, $\text{Im}\{G(i\omega)\} = 0$ or alternatively $|G(i\omega)| = 1$, $\varphi = 0$. For $\omega = 1/\tau$ we have that $\text{Re}\{G(i\omega)\} = 0.5$, $\text{Im}\{G(i\omega)\} = -0.5$ or alternatively $|G(i\omega)| = 1/\sqrt{2}$, $\varphi = -45°$. Finally, for $\omega \to +\infty$ we have that $\text{Re}\{G(i\omega)\} = 0$, $\text{Im}\{G(i\omega)\} = 0$ or alternatively $|G(i\omega)| = 0$, $\varphi = -90°$. To prove that the Nyquist plot is actually a cycle with center at $(1/2, 0)$ and radius $1/2$, as is evident from Figure 9.10, we note that

$$\left(\text{Re}\{G(i\omega)\} - \frac{1}{2}\right)^2 + \text{Im}\{G(i\omega)\}^2 = \left(\frac{1}{1+(\tau\omega)^2} - \frac{1}{2}\right)^2 + \left(\frac{-\tau\omega}{1+(\tau\omega)^2}\right)^2$$

$$= \left(\frac{1}{2} \cdot \frac{1-(\tau\omega)^2}{1+(\tau\omega)^2}\right)^2 + \left(\frac{2}{2} \cdot \frac{-\tau\omega}{1+(\tau\omega)^2}\right)^2 = \left(\frac{1}{2}\right)^2$$

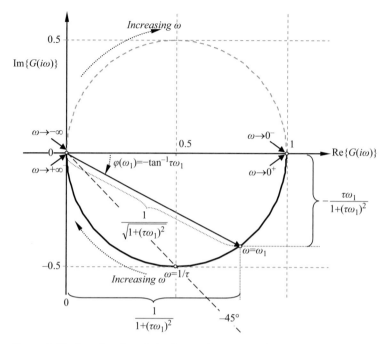

Figure 9.10 Nyquist plot of the first-order system $1/(\tau s + 1)$.

9.5.2 Nyquist Plot of a Second-Order System

For a second-order system

$$G(s) = \frac{1}{\tau^2 s^2 + 2\zeta\tau s + 1}, \quad \tau > 0, \ \zeta > 0 \qquad (9.5.2)$$

we have shown that

$$\text{Re}\{G(i\omega)\} = \frac{1-\tau^2\omega^2}{\left(1-\tau^2\omega^2\right)^2 + (2\zeta\tau\omega)^2}, \quad \text{Im}\{G(i\omega)\} = -\frac{2\zeta\tau\omega}{\left(1-\tau^2\omega^2\right)^2 + (2\zeta\tau\omega)^2}$$

and also

$$|G(i\omega)| = \frac{1}{\sqrt{\left(1-\tau^2\omega^2\right)^2 + (2\zeta\tau\omega)^2}}, \quad \varphi = -\cos^{-1}\left(\frac{1-\tau^2\omega^2}{\sqrt{(1-\tau^2\omega^2)^2 + (2\zeta\tau\omega)^2}}\right)$$

The Nyquist plot is shown in Figure 9.11 for several values of the damping factor ζ. For large ζ the Nyquist plot of the second-order system resembles that of a first-order system. Note again the two conjugate symmetric parts of the plot. For $\omega = 0$ we have that $\text{Re}\{G(i\omega)\} = 1$, $\text{Im}\{G(i\omega)\} = 0$ or alternatively $|G(i\omega)| = 1$, $\varphi = 0$. For $\omega = 1/\tau$ we have that $\text{Re}\{G(i\omega)\} = 0$, $\text{Im}\{G(i\omega)\} = -1/(2\zeta)$ or alternatively $|G(i\omega)| = 1/(2\zeta)$, $\varphi = -90°$. Finally, for $\omega \to +\infty$ we have that $\text{Re}\{G(i\omega)\} = 0$, $\text{Im}\{G(i\omega)\} = 0$ or alternatively $|G(i\omega)| = 0$, $\varphi = -180°$.

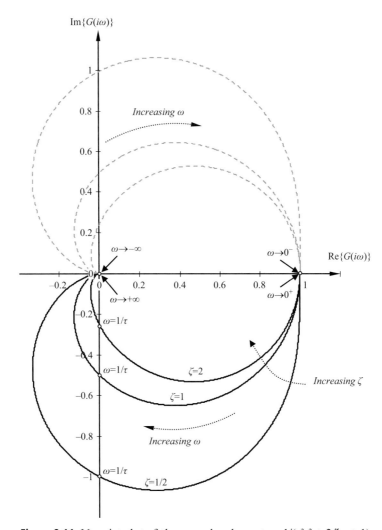

Figure 9.11 Nyquist plot of the second-order system $1/(\tau^2 s^2 + 2\zeta\tau s + 1)$.

9.5.3 Nyquist Plot of an Integrator

For the integrator system

$$G(s) = \frac{1}{s} \qquad (9.5.3)$$

we have shown that

$$\text{Re}\{G(i\omega)\} = 0, \quad \text{Im}\{G(i\omega)\} = -\frac{1}{\omega}$$

$$|G(i\omega)| = \frac{1}{\omega}, \quad \varphi = -90°$$

i.e. the polar plot is the negative imaginary axis.

In the same way it can be shown that the polar plot of the differentiator system $G(s)=s$ is the positive imaginary axis as

$$\text{Re}\{G(i\omega)\} = 0,\ \text{Im}\{G(i\omega)\} = \omega$$

$$|G(i\omega)| = \omega,\ \varphi = +90°$$

9.5.4 Nyquist Plot of a First-Order Lead System

For the first-order lead system

$$G(s) = 1 + \tau s,\ \tau > 0 \qquad (9.5.4)$$

we have that

$$G(i\omega) = 1 + i\tau\omega \qquad (9.5.5)$$

from which it follows that

$$\text{Re}\{G(i\omega)\} = 1,\ \text{Im}\{G(i\omega)\} = \tau\omega$$

and also that

$$|G(i\omega)| = \sqrt{1+(\tau\omega)^2},\ \varphi = \tan^{-1}(\tau\omega)$$

The polar plot is shown in Figure 9.12. By comparing Figures 9.10 and 9.12 we observe that there is no resemblance between the polar plots of a first-order lag system and a first-order lead system. The same holds true for the polar plot of a second-order lag system (shown in Figure 9.11) and a second-order lead system

$$G(s) = \tau^2 s^2 + 2\zeta\tau s + 1 \qquad (9.5.6)$$

the construction of which is left as an exercise for the reader.

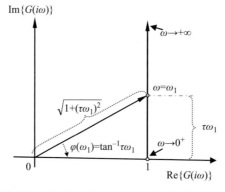

Figure 9.12 Nyquist plot of the first-order lead system $(1+\tau s)$.

9.5.5 Polar Plots of Higher-Order Systems

When the transfer function of the system under study has been factored as the product of simple transfer functions like those in Table 9.1, the polar plot can be constructed using Eqs. (9.2.10). A simple demonstration is presented in Figure 9.13 where the overall transfer function $G(i\omega) = G_1(i\omega)G_2(i\omega)$ is determined in polar coordinates using complex-algebra multiplication.

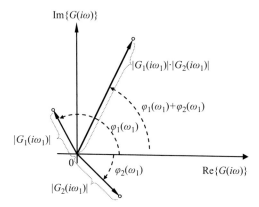

Figure 9.13 Multiplication of two transfer functions in a polar plot.

If the polar plot of $G(i\omega) = G_1(i\omega)G_2(i\omega)\ldots G_m(i\omega)$ is to be constructed, it is advisable to determine the logarithmic plot of $G_1(i\omega)G_2(i\omega)\ldots G_m(i\omega)$ first (as is described in the previous sections) and then convert the result into a polar plot.

Example 9.3 Construction of the polar plot of a system consisting of an integrator in series with a first-order lag

The transfer function

$$G(s) = \frac{1}{s(\tau s + 1)}, \quad \tau > 0$$

is a second-order system with integrating behavior. Construct the polar plot for $\tau = 1$.

Solution

The transfer function given can be factored as $G(i\omega) = G_1(i\omega)G_2(i\omega)$, where $G_1(i\omega) = 1/(i\omega)$ and $G_2(i\omega) = 1/(i\tau\omega+1)$. Using Table 9.1 we obtain

$$|G(i\omega)| = \left|\frac{1}{i\omega}\right| \cdot \left|\frac{1}{i\tau\omega+1}\right| = \frac{1}{\omega\sqrt{\tau^2\omega^2+1}}$$

$$\varphi = \varphi_1 + \varphi_2 = -90° - \tan^{-1}\tau\omega$$

Alternatively, we can calculate directly the real and imaginary part of $G(i\omega)$

$$G(i\omega) = \frac{1}{i\omega(i\tau\omega+1)} = \frac{1}{-\tau\omega^2 + i\omega} \cdot \frac{-\tau\omega^2 - i\omega}{-\tau\omega^2 - i\omega}$$

$$= -\frac{\tau}{1+(\tau\omega)^2} - i\frac{1}{\omega[1+(\tau\omega)^2]}$$

i.e. $\operatorname{Re}\{G(i\omega)\} = -\dfrac{\tau}{1+(\tau\omega)^2}$, $\operatorname{Im}\{G(i\omega)\} = -\dfrac{1}{\omega[1+(\tau\omega)^2]}$

The polar plot of the system under study is shown in Figure 9.14 for $\tau = 1$.

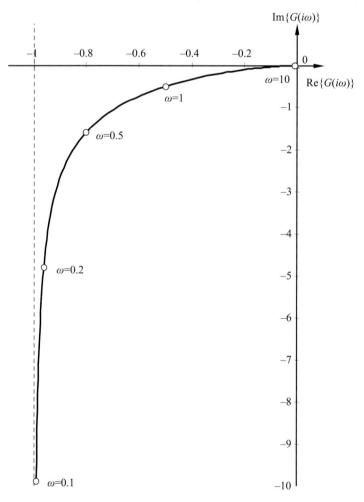

Figure 9.14 Polar plot of the second-order system with integrating behavior studied in Example 9.3.

9.6 Software Tools

9.6.1 MATLAB Commands Relative to Bode and Nyquist Plots

Bode plots can be generated in MATLAB using the `bode` command. Let us consider the transfer function given by Eq. (9.3.1). We can generate its Bode plot as follows:

```
>> num=[10 1];
>> den=[1 1 0];
>> bode(num,den)
```

When bode is invoked without using left-hand-side arguments then the Bode plot is automatically generated. However, the magnitude plot is given in terms of decibel (dB). If the use of dB is to be avoided then we must use left-hand-side arguments and the command

```
>> [magnitude,phase,frequency]=bode(num,den);
>> loglog(frequency,magnitude)
```

and the magnitude plot shown in Figure 9.2 is obtained. To obtain the phase plot of Figure 9.3 we then type

```
>> semilogx(frequency,phase)
```

The two plots can be constructed in the same figure using the `subplot` command

```
>> subplot(2,1,1)
>> loglog(frequency,magnitude)
>> subplot(2,1,2)
>> semilogx(frequency,phase)
```

To generate the Nyquist plot of a system the command `nyquist` can be used

```
>> num=1;
>> den=[1 1];
>> nyquist(num,den)
```

and Figure 9.10 is generated. When the `nyquist` command is invoked with left-hand arguments then the real and imaginary parts of the sinusoidal are obtained for positive frequencies only

```
>> [re,im]= nyquist(G); ReG = squeeze (re); ImG = squeeze (im);
>> plot(ReG,ImG)
```

To obtain the complete Nyquist plot (including the plot for negative frequencies) we use the commands

```
>> hold on
>> plot(ReG,-ImG,'-')
```

9.6.2 Bode Plots and Nyquist Plots using Maple

Bode plots can be generated using the `BodePlot` command of the `DynamicSystems` package of Maple, as follows: Consider, for example, the transfer function given by Eq. (9.3.1), we first define the dynamic system:

```
> with(DynamicSystems): G:=(10*s+1)/(s*(s+1));
```

$$G := \frac{10s + 1}{s(s+1)}$$

```
> sys:=TransferFunction(G):
```

Then, the command

```
> BodePlot(sys);
```

will generate the Bode plots of the system, with the magnitude expressed in decibel (dB). If the use of dB is undesirable, and the absolute magnitude must be plotted, one must use the `BodePlot` command, with the corresponding option:

```
> BodePlot(sys,decibels=false);
```

This will generate the plots shown in Figures 9.2 and 9.3.

Nyquist plots can be generated using the `NyquistPlot` command of the `DynamicSystems` package of Maple. After the system is defined in terms of its transfer function (as above), one can use the command

```
> NyquistPlot(sys);
```

to generate the Nyquist plot.

Note, however, that, since Nyquist plots are simple parametric plots of $\text{Im}\{G(i\omega)\}$ versus $\text{Re}\{G(i\omega)\}$, the use of the `DynamicSystems` package of Maple is not necessary; a Nyquist plot can be generated through a simple plot command. For the transfer function of Example 9.3,

```
> G:=1/(s*(s+1)):
```

after calculating $\text{Re}\{G(i\omega)\} = \dfrac{G(i\omega) + G(-i\omega)}{2}$ and $\text{Im}\{G(i\omega)\} = \dfrac{G(i\omega) - G(-i\omega)}{2i}$,

```
> ReG:=(subs(s=I*omega,G)+subs(s=-I*omega,G))/2:
> ImG:=(subs(s=I*omega,G)-subs(s=-I*omega,G))/(2*I):
```

we can simply use the plot command:

```
> plot([ReG,ImG,omega=0..infinity]);
```

to generate the Nyquist plot over the range for $0 \leq \omega \leq +\infty$, as shown in Figure 9.14. If we wish to generate the plot both for $0 \leq \omega \leq +\infty$ and for $-\infty \leq \omega \leq 0$, and with different line styles:

```
> plot([[ReG,ImG,omega=0..infinity],
  [ReG,ImG,omega=-infinity..0]],linestyle=[solid,dash]);
```

LEARNING SUMMARY

- When a system with transfer function $G(s)$ is subject to a sinusoidal input $u(t) = M\sin(\omega t)$, the output is also sinusoidal with the same frequency ω but a different amplitude. The ratio of the amplitudes (AR) is a function of the frequency ω and is given by

$$AR = \frac{\text{amplitude of the output}}{\text{amplitude of the input}} = |G(i\omega)|$$

The output signal is not in phase with the input signal. The phase difference is a function of the frequency ω and is given by

$$\varphi = \arg\{G(i\omega)\}$$

- The Bode diagrams consist of

  ◊ a log–log diagram of $AR = |G(i\omega)|$ as a function of frequency ω (magnitude plot)
  ◊ a lin–log diagram of the phase $\varphi = \arg\{G(i\omega)\}$ as a function of frequency ω (phase plot).

- When the transfer function is factored as the product of simple transfer functions as

$$G(s) = G_1(s) \cdot G_2(s) \cdot \ldots G_m(s)$$

then

$$|G(i\omega)| = |G_1(i\omega)| \cdot |G_2(i\omega)| \cdot \ldots |G_m(i\omega)|$$
$$\arg\{G(i\omega)\} = \arg\{G_1(i\omega)\} + \arg\{G_2(i\omega)\} + \ldots + \arg\{G_m(i\omega)\}$$

- These results, combined with the Bode plots of simple transfer functions given in Table 9.1, can be used to construct the Bode plot of any stable higher-order system in factored form.
- Nyquist plots can be generated easily if the logarithmic plots are available by using either rectangular or polar coordinates, to represent the real and imaginary part of $G(i\omega)$.

TERMS AND CONCEPTS

All-pass element. A system with amplitude ratio $AR = 1$ at all frequencies.
Bode diagrams. These consist of (i) a plot of the magnitude $|G(i\omega)|$ versus the frequency ω, in log–log scales, and (ii) a plot of the phase $\arg\{G(i\omega)\}$ versus the frequency ω, in lin–log scales.

Break or corner frequency. A frequency at which the linear approximation of the AR plot changes slope.
Lag element. A system with phase $\varphi \leq 0$ for all $\omega > 0$.
Lead element. A system with phase $\varphi \geq 0$ for all $\omega > 0$.
Minimum-phase transfer function. A transfer function whose poles and zeros all have negative real parts.
Nonminimum-phase transfer function. A transfer function that has at least one zero with positive real part.
Nyquist diagram. A polar plot of $G(i\omega)$ for $-\infty \leq \omega \leq \infty$.
Polar plot. A plot of the real part of $G(i\omega)$ versus the imaginary part of $G(i\omega)$.

FURTHER READING

Additional material on frequency response analysis of linear systems can be found in
Ogata, K., *Modern Control Engineering*, Pearson International Edition, 5th edn, 2008.
Ogunnaike, B. and Ray, H., *Process Dynamics, Modelling and Control*. New York: Oxford University Press, 1994.

PROBLEMS

9.1 For a system with transfer function

$$G(s) = \frac{10s}{(10s+1)\left(\frac{1}{10}s+1\right)}$$

calculate the amplitude ratio and the phase angle as a function of frequency. Follow the method of Section 9.3 to sketch the corresponding the Bode diagrams. Verify your results using MATLAB or Maple.

9.2 Use MATLAB or Maple to draw the Nyquist diagram for the system of Problem 9.1. Compare the Nyquist diagram with the corresponding Bode diagrams, in terms of how they depict the low-frequency, high-frequency and medium-frequency response characteristics.

9.3 Calculate the amplitude ratio and the phase angle as a function of frequency, and sketch the corresponding the Bode diagrams, for the following transfer functions:

(i) $G(s) = \dfrac{1}{s(0.1s+1)(10s+1)}$

(ii) $G(s) = \dfrac{10s+1}{(s+1)^2}$

(iii) $G(s) = \dfrac{-10s+1}{(s+1)^2}$

(iv) $G(s) = \dfrac{1}{(s^2+0.1s+1)(9s^2+0.3s+1)}$

Verify your results using MATLAB or Maple.

9.4 Use MATLAB or Maple to draw the Nyquist diagrams of the systems of Problem 9.3. For each case, compare the corresponding Bode and Nyquist diagrams, in terms of how they depict low-frequency, high-frequency and medium-frequency response characteristics.

9.5 The following are the transfer functions of first-, second- and third-order Butterworth filters, respectively:

(a) $G_1(s) = \dfrac{1}{\tau s + 1}$

(b) $G_2(s) = \dfrac{1}{\tau^2 s^2 + \sqrt{2}\tau s + 1}$

(c) $G_3(s) = \dfrac{1}{(\tau s + 1)(\tau^2 s^2 + \tau s + 1)}$

where $\tau > 0$.

Calculate the amplitude ratio and the phase angle as a function of frequency and sketch the corresponding the Bode diagrams. Validate your results using MATLAB or Maple. Are all of the above transfer functions low-pass filters? What is the cut-off frequency in each case?

9.6 Calculate the amplitude ratio and the phase angle as a function of frequency for a system with transfer function

$$G(s) = \dfrac{1}{\left(\dfrac{T}{N}s + 1\right)^N}$$

with $T > 0$, and N positive integer. Discuss the low-frequency and the high-frequency behavior of the amplitude ratio. Is the above transfer function a low-pass filter? What is the cut-off frequency? Also, derive approximate expressions for the phase angle for low and high frequency.

9.7 Use MATLAB or Maple to draw the Nyquist diagram for the system of Problem 9.6 for $N = 1, 2, 5, 10, 20, 50$. Compare with the corresponding Bode diagrams. What happens as $N \to \infty$?

9.8 Calculate the amplitude ratio and the phase angle as a function of frequency for

(a) a proportional-integral controller with transfer function $G_{PI}(s) = k_c\left(1 + \dfrac{1}{\tau_I s}\right)$

(b) an ideal proportional-derivative controller with transfer function $G_{PD}(s) = k_c(1 + \tau_D s)$

(c) an ideal proportional-integral-derivative controller with transfer function

$$G_{PID}(s) = k_c\left(1 + \dfrac{1}{\tau_I s} + \tau_D s\right)$$

and sketch the corresponding Bode diagrams. Validate your results using MATLAB or Maple with specific values for the controller parameters.

9.9 Use MATLAB or Maple to draw the Nyquist diagrams for the systems of Problem 9.8. Compare with the corresponding Bode diagrams.

9.10 In Figure P9.10 the Bode plots of an unknown system are given. Select a transfer function that can be used to best describe the given Bode diagrams among the following (explain qualitatively and quantitatively your selection):

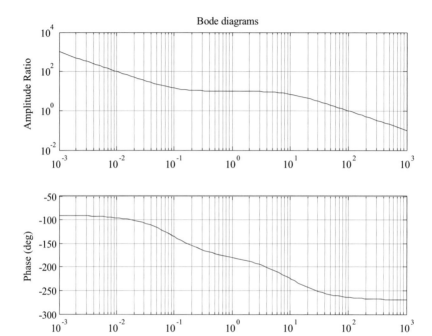

Figure P9.10

(a) $G_1(s) = k \dfrac{(Ts+1)}{(\tau s+1)^2}$

(b) $G_2(s) = k \dfrac{(-Ts+1)}{(\tau s+1)^2}$

(c) $G_3(s) = k \dfrac{(Ts+1)}{s(\tau s+1)}$

(d) $G_4(s) = k \dfrac{(-Ts+1)}{s(\tau s+1)}$

where $k > 0$, $T > 0$ and $\tau > 0$ in all cases. For the transfer function selected, estimate the values of the parameters k, T and τ.

9.11 In Figure P9.11 the Bode plots of an unknown system are given. Select a transfer function that can be used to best describe the given Bode diagrams among the following (explain qualitatively and quantitatively your selection):

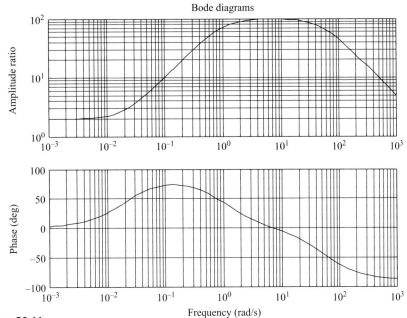

Figure P9.11

(a) $G_1(s) = \dfrac{k}{(\tau_1 s + 1)(\tau_2 s + 1)}$

(b) $G_2(s) = \dfrac{k(Ts+1)}{(\tau_1 s + 1)(\tau_2 s + 1)}$

(c) $G_3(s) = k\dfrac{(Ts+1)}{s(\tau_1 s + 1)}$

(d) $G_4(s) = \dfrac{k}{s(\tau_1 s + 1)(\tau_2 s + 1)}$

(e) $G_5(s) = \dfrac{ks}{(\tau_1 s + 1)(\tau_2 s + 1)}$

where $k > 0$, $T > 0$, $\tau_1 > 0$ and $\tau_2 > 0$ in all cases. For the transfer function selected, estimate the values of the parameters k, T, τ_1, τ_2.

9.12 In Figure P9.12 the Bode plots of an unknown system are given. Select a transfer function that can be used to best describe the given Bode diagrams among the following (explain qualitatively and quantitatively your selection):

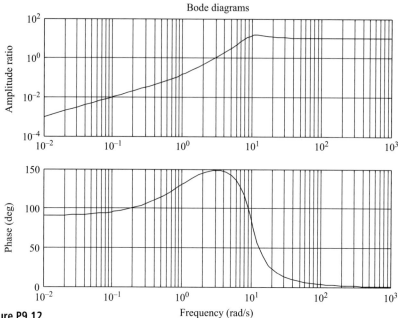

Figure P9.12

(a) $G_1(s) = \dfrac{k}{(Ts+1)(\tau^2 s^2 + 2\zeta\tau s + 1)}$

(b) $G_2(s) = \dfrac{k}{s(Ts+1)(\tau^2 s^2 + 2\zeta\tau s + 1)}$

(c) $G_3(s) = \dfrac{ks}{(Ts+1)(\tau^2 s^2 + 2\zeta\tau s + 1)}$

(d) $G_4(s) = \dfrac{k(Ts+1)}{s(\tau^2 s^2 + 2\zeta\tau s + 1)}$

(e) $G_5(s) = \dfrac{k(Ts+1)s}{(\tau^2 s^2 + 2\zeta\tau s + 1)}$

where $k > 0$, $T > 0$, $\tau > 0$, $\zeta > 0$ in all cases. For the transfer function selected, estimate the values of the parameters k, T and τ.

10 The Feedback Control System

The aim of this chapter is to introduce the hardware elements that constitute a feedback control system. These elements are related to measuring the main process variables, implementing the control law and finally applying the corrective action to the physical system. More specifically the temperature, pressure and level sensors are briefly discussed, followed by the most common final control element, which is the control valve. Standard control laws, involving proportional, integral and derivative actions are also introduced.

STUDY OBJECTIVES

After studying this chapter, you should be able to do the following.

- Identify the main building blocks of a feedback control system.
- Develop basic skills in building block-diagram representations of feedback control systems.
- Use the terminology related to standard control laws such as proportional, proportional-integral and proportional-integral-derivative control.

10.1 Heating Tank Process Example

In Figure 10.1 a simple heating tank is shown. The feed stream to the tank has flowrate F and temperature T_0. The tank is equipped with an internal coil where saturated steam is condensed and removed as saturated liquid at the same temperature. The heat released by steam condensation is taken up by the fluid in the tank and its temperature is increased from T_0 to T. The temperature of the feed stream $T_0(t)$ varies significantly and, as a result, the exit temperature will vary unless it is controlled by some external means (by varying the steam flowrate). For this reason a temperature-measuring device (thermocouple) has been installed; this monitors continuously the temperature, producing a small electrical voltage (mV) that is proportional to the temperature of the fluid in the tank. A device called a temperature transmitter (TT) is used to transform the electrical voltage into an appropriate electrical current (4–20 mA) that can be sent to a computer system that can be located near or very far away from the physical system. This "computer system" is the controller. The controller uses the input signal and an internally stored algorithm (calibration curve) to calculate the measured temperature of the tank ($T_m(t)$). The difference between the measured temperature and the desired temperature or temperature "set point" T_{sp} (which is supplied by external means) is the error $e(t)$ signal

$$e(t) = T_{sp} - T_m(t) \tag{10.1.1}$$

The controller then uses a mathematical algorithm or "control law" in order to determine the flowrate of the steam that is necessary in order to achieve the desired temperature in the liquid content of the tank. The results of this calculation are finally used to set the valve opening, which determines the flowrate of the steam fed to the coil and therefore the heat input to the liquid in the tank.

This heating tank example, despite its simplicity, contains all these elements that are common to all process control systems. First of all, a specification is set to the operation of the process under study. In the heating tank the objective is to achieve a predetermined temperature in the liquid product stream. This local objective is set by the more general operating strategy of the plant. In our case study the heating tank may be used in order to prepare the liquid stream before entering a chemical reactor or a mixing operation. The desired or setpoint temperature is then determined by the optimal conditions of these, more important, downstream processes. Secondly, disturbances cause variations in the specification that are unacceptable for the downstream processes as they tend to worsen their performance. To alleviate the problems caused by the potential variability in the important variable (outlet temperature, in our case) a controller is installed that redirects the variability to a less-important variable (steam flowrate, in our case). Needless to say, perfect elimination of the variability in the important variable cannot be achieved but it can, hopefully, be minimized to a level that is insignificant for the successful operation of the plant.

To implement a control strategy, we need hardware that is related to measuring the controlled variable (sensor), transforming the measurement to a proper form to be transmitted to a computer system (transmitter), a computer system that implements the error-correction algorithm (controller) and a final control element (a valve in most cases) that implements physically the corrective action to the process. All these constitute a closed loop around the process that is shown in the block diagram in Figure 10.2. An in-depth analysis of the types and characteristics of either sensors/transmitters or final control elements is not among the objectives of this book. A short introduction to temperature, pressure and level sensors and

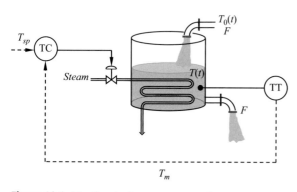

Figure 10.1 Heating tank process example.

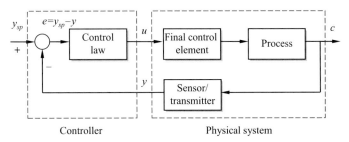

Figure 10.2 Block-diagram representation of a controlled system.

valves follows. The reader may consult the references at the end of the chapter for a comprehensive introduction to the hardware involved in process control systems.

10.2 Common Sensors and Final Control Elements

Temperature, pressure and level sensors are among the most common, reliable and essential measuring devices in a plant, and control valves are by far the most common final control elements. They come in many different types, but the most important will be reviewed next.

10.2.1 Temperature Sensors

Thermocouples (TC) are the most commonly used temperature sensors as they can be used over a wide range of temperatures, they are reliable, and they are easy to manufacture and install in process equipment. Their operation is based on the "thermoelectric circuit" discovered by Thomas Seebeck in 1821. Seebeck found that a circuit made from two dissimilar metals would deflect a compass magnet when the two ends are at different temperatures. It was later realized that an electric potential (voltage) is produced by the temperature difference, and this drives an electric current in the closed circuit (see Figure 10.3). The material of the two wires used in a TC determines its type: type J (iron–constantan, range: −40 °C to +750 °C, sensitivity of about ~55 μV/°C) and type K (chromel–alumel, the most common general-purpose thermocouple, sensitivity ~40 μV/°C, range −200 °C to +1350 °C) are among the more commonly used.

Resistance thermometer detectors (RTDs) are made of materials that exhibit almost linear variation of their resistance with temperature, a phenomenon known as thermoresistivity. Temperature can, therefore, be inferred by simply measuring the resistance of the RTD element. The most accurate temperature sensor belongs to this class and is made of platinum. Apart from the fact that RTDs can be used at lower temperatures (generally below 850 °C) and have slower response when compared to TCs, they are more

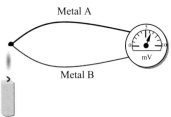

Figure 10.3 A simplistic representation of the Seebeck effect.

accurate, stable and less susceptible to noise. Arguably, the main reason for using a TC is that the temperature range is beyond what is reasonable for an RTD.

Temperature sensors are modeled as first-order systems with unity static gain and a time constant that varies depending on the application. The sensor input is the actual value of the measured temperature and the output is the measurement.

10.2.2 Pressure and Level Sensors

Next to temperature, pressure is the second most common measurement in process plants. In the old days manometers, Bourdon tubes and bellows were used for measuring pressure. Modern pressure transmitters are extremely high-tech, accurate and include smart electronics that can measure pressures from 0.1 Pa to 1 GPa. The physical principle behind the pressure-measurement devices is the conversion of pressure to either displacement or force, and the use of a signal conditioner to finally convert the force or displacement into a voltage or current. There are three methods for measuring pressure: absolute, gauge and differential. Absolute pressure is referenced to the pressure in a vacuum, whereas gauge and differential pressures are referenced to another pressure such as the ambient atmospheric pressure or pressure at a different point of a vessel.

Strain-gauge pressure sensors are based on the fact that if a metal wire is stretched then its electrical resistance changes in an almost linear way. Crystals of the type known as "piezo-electric crystals" produce an electrical signal when mechanically deformed. The voltage of the signal is proportional to the deformation. Such crystals are mechanically attached to a metal diaphragm with one side of the diaphragm connected to the process fluid (to sense pressure). Deflection-type pressure sensors consist of two parts: the first part converts the measured pressure into a (proportional) displacement and the second part converts the displacement into a change in the measurable characteristics of an electrical element. A representative example is the potentiometric-type sensor, one of the oldest types of pressure sensors, shown in Figure 10.4. It converts pressure through a mechanical device (a diaphragm that is used to move the wiper arm of a potentiometer) to a variable voltage drop. The voltage drop is sent to an electronic unit, whose output is normally a 4–20 mA DC current.

Level-measurement devices share many common characteristics with pressure-measuring devices. A characteristic example is shown in Figure 10.5, where a differential pressure (DP cell)

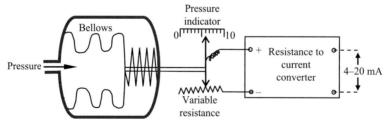

Figure 10.4 A simplistic representation of potentiometric-type pressure sensor.

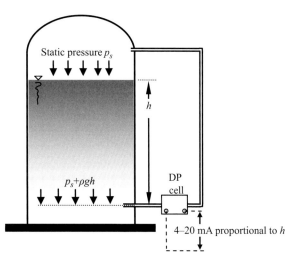

Figure 10.5 A simplistic representation of liquid-level measurement that is based on differential pressure measurement.

sensor is used to infer the level of the liquid in the tank from the pressure variation between the bottom of the tank and the free space above the liquid. In this case the DP cell senses only changes to the liquid level ($\rho g h$) and not changes in the static pressure (p_s) on the liquid surface. The electrical measurement can be transformed to an appropriate electrical signal.

10.2.3 Control Valves

A control valve is used to regulate the flow of a process stream. Figure 10.6 illustrates a representative globe-type control valve of air-to-close type. In a globe valve the plug is attached to the valve stem which, when moved, increases or decreases the resistance to flow. A pneumatic actuator is used to move the stem as shown in Figure 10.6. A pneumatic control signal (3–15 psi) is used to regulate the pressure in the space above the diaphragm and therefore to move the stem by applying a force against the spring.

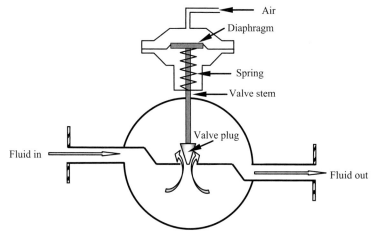

Figure 10.6 A simplistic representation of a control valve.

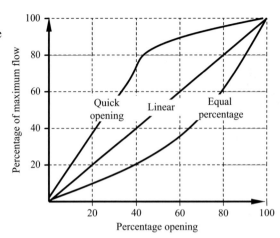

Figure 10.7 Control-valve performance curves.

The equation that is most commonly used to relate the volumetric flowrate of the process fluid through the control valve (Q) and the pressure drop across the valve (ΔP) is the following:

$$Q = C_v f(x) \sqrt{\frac{\Delta P}{G}} \qquad (10.2.1)$$

where G is the specific gravity of the fluid, C_v is the valve coefficient and $f(x)$ is the fraction of the total area of the valve, which is a function of the stem position (x). A curve that shows how $f(x)$ varies as a function of x is called the "inherent characteristics" of the valve and representative examples are shown in Figure 10.7. In the case of a linear valve the flow through the valve depends linearly on the valve stem position, i.e. the volumetric flowrate of the process fluid through the control valve is

$$Q = Q_{max}\, x \qquad (10.2.2)$$

where x is the fractional opening of the valve. Linear valves are used when the pressure drop across the valve remains fairly constant. When the pressure drop is significant equal-percentage valves are most commonly used. The term "equal-percentage" stems from the fact that the slope of the curve is a constant fraction of $f(x)$. The meaning of this is that a given percentage change in the stem position produces the same percentage change in the flow through the valve. A quick-opening valve is predominantly used when an aggressive, almost on/off, response of the control system is desirable.

10.3 Block-Diagram Representation of the Heating Tank Process Example

Returning to the example of the heating tank, we note that, in order to be able to analyze the behavior of the closed-loop system, we need to develop the overall process dynamics.

This can be achieved by using the block-diagram representation shown in Figure 10.2. Each block in this representation denotes a mathematical model that describes the input–output behavior of the underlying system.

We will begin developing these mathematical models by starting from the dynamics of the heating tank. The controlled (output) variable is the temperature of the liquid in the tank while there are two inputs: the manipulated variable (the valve stem position or the steam flowrate) and the disturbance (temperature of the feed stream). The energy balance of the tank, under the assumption of constant volume and constant thermophysical properties, can be written as follows (see Chapter 2, Eq. (2.3.14))

$$V\rho c_p \frac{dT}{dt} = F\rho c_p (T_0 - T) + Q \tag{10.3.1}$$

The amount of heat Q transferred from the heating medium to the liquid content in the tank can be related to the mass flowrate of the steam $\dot{M}_{st}$ through the following equation

$$Q = \dot{M}_{st} \Delta \hat{h}_{vap,st} \tag{10.3.2}$$

where $\Delta \hat{h}_{vap,st}$ is the latent heat of the steam that can be assumed constant. After substituting Eq. (10.3.2) into Eq. (10.3.1), we obtain

$$\left(\frac{V}{F}\right)\frac{dT}{dt} + T = T_0 + \left(\frac{\Delta \hat{h}_{vap,st}}{F\rho c_p}\right)\dot{M}_{st} \tag{10.3.3}$$

The terms that appear in parenthesis in the last equation are constants and therefore the differential equation is linear. After introducing deviation variables around a reference steady state and taking the Laplace transform we obtain

$$\overline{T}(s) = \frac{1}{\left(\frac{V}{F}s+1\right)}\overline{T}_0(s) + \frac{\frac{\Delta \hat{h}_{vap,st}}{F\rho c_p}}{\left(\frac{V}{F}s+1\right)}\overline{\dot{M}}_{st}(s) \tag{10.3.4}$$

or

$$\overline{T}(s) = \frac{1}{(\tau_p s + 1)}\overline{T}_0(s) + \frac{k_p}{(\tau_p s + 1)}\overline{\dot{M}}_{st}(s) \tag{10.3.5}$$

where $\tau_p = V/F$ is the process time constant and $k_p = \Delta \hat{h}_{vap,st}/(F\rho c_p)$ is the process gain.

It has already been mentioned that the dynamics of the temperature sensor can be described by a first-order differential equation with unity static gain

$$\tau_m \frac{dT_m}{dt} + T_m = T \tag{10.3.6a}$$

or, in the Laplace domain

$$\bar{T}_m(s) = \frac{1}{(\tau_m s + 1)} \bar{T}(s) \qquad (10.3.6b)$$

where τ_m is the sensor time constant.

Finally, assuming that a linear control valve with very fast dynamics is used, this can be described by the following algebraic equation

$$\overline{\dot{M}}_{st} = k_v u \qquad (10.3.7a)$$

or, in the Laplace domain

$$\overline{\dot{M}}_{st}(s) = k_v U(s) \qquad (10.3.7b)$$

where k_v is a constant (with appropriate units) and u is the control signal (in deviation form) sent to the valve from the controller.

The next step is to substitute the mathematical description of each element of the closed-loop system into the block-diagram representation shown in Figure 10.2 to obtain the more informative block diagram shown in Figure 10.8. The reader must carefully observe the correspondence between each block and the underlying mathematical model. It is also important to note that the input signals to the closed-loop system are

- the set point temperature T_{sp}
- the feed temperature T_0

and the output is the actual temperature of the liquid content of the heating tank T. It should be noted that up to this point no particular form of the controller has been assumed. The simplest possible controller is a proportional-only controller and will be presented next.

Before closing this paragraph it is important to note that, regarding the block-diagram representation of Figure 10.8, it appears that the output of the system is the actual temperature of the process fluid (T). However, only the measured temperature (T_m) is accessible to the operator or the computer control system. The temperature T can, in the best case, be inferred using the mathematical model of the process and is never known exactly. It is, therefore, common to use the equivalent block-diagram representation shown in Figure 10.9 where the output of the system is the measured temperature. Although the two representations are equivalent, the one shown in Figure 10.9, referred to as "unity feedback" system, is the one that will be mainly used in this book as the sensor dynamics (as well as the final control element dynamics) is considered as part of the process under study. Furthermore, sensor dynamics can usually be neglected as process systems tend to have time constants that are several orders of magnitude larger than the time constants of the commonly used sensors. Most classical chemical process control textbooks use the representation shown in Figure 10.8 but they neglect sensor dynamics and in effect implicitly use the representation shown in Figure 10.9. To avoid confusion we will be using the symbol $c(t)$ ($C(s)$) to denote the actual value of the controlled variable and the symbol $y(t)$ ($Y(s)$) to denote its measurement.

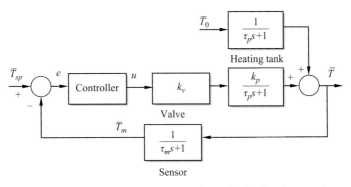

Figure 10.8 Block-diagram representation of a feedback control system.

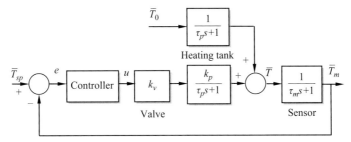

Figure 10.9 Alternative block-diagram representation of a feedback control system.

10.4 Further Examples of Process Control Loops

10.4.1 Liquid-Level Control Loop

In Figure 10.10 a liquid-level control loop is shown. The liquid level in the tank is measured by using a differential pressure sensor/transmitter (LT) and the measurement (h_m) is fed to a level controller (LC), which is responsible for taking the necessary actions to restore the liquid level in the tank to its desired (h_{sp}) value. In this particular example the tank has two feeds. A controlled feed stream ($F_{in}(t)$) and a disturbance stream ($F_d(t)$). When there is a change in the disturbance stream then the liquid level will deviate from its set point value. This deviation is calculated by the controller and a corrective action is taken by changing the flowrate of the manipulated stream $F_{in}(t)$. Using the results presented in Chapter 3 we can develop the following model for the liquid level in the tank

$$AR\frac{dh}{dt} + h = RF_{in} + RF_d \tag{10.4.1a}$$

or by introducing deviation variables and taking the Laplace transform

Figure 10.10 A liquid-level control example.

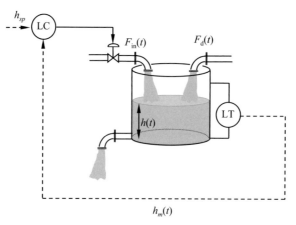

$$\bar{h}(s) = \frac{k_p}{(\tau_p s + 1)} \bar{F}_{in}(s) + \frac{k_p}{(\tau_p s + 1)} \bar{F}_d(s) \tag{10.4.1b}$$

where $k_p = R$ and $\tau_p = AR$. If we further assume first-order dynamics of the liquid level sensor and a linear valve we obtain (see Eqs. (10.3.6) and (10.3.7))

$$\bar{h}_m(s) = \frac{1}{(\tau_m s + 1)} \bar{h}(s) \tag{10.4.2}$$

$$\bar{F}_{in}(s) = k_v U(s) \tag{10.4.3}$$

where τ_m is the sensor time constant and k_v is the valve constant. By using Eqs. (10.4.1), (10.4.2) and (10.4.3) we obtain the block-diagram representation shown in Figure 10.11, which has exactly the same structure as the block diagram representation of the temperature-control system shown in Figures 10.1 and 10.8.

Figure 10.11 Block-diagram representation of the liquid-level control system.

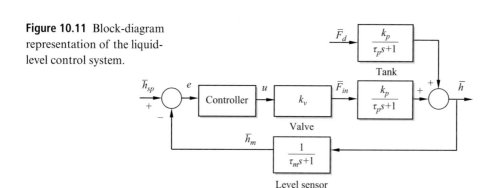

10.4.2 An Example of a Pressure Control Loop

A spherical gas vessel of volume V is shown in Figure 10.12. A gas with variable volumetric flowrate $F_{in}(t)$ is fed to the vessel which has a desired pressure P_{sp}. The pressure of the gas in the vessel is $P(t)$. A pressure sensor/transmitter (PT) monitors continuously the pressure in the vessel and transmits the measured pressure ($P_m(t)$) to a pressure controller (PC), which regulates the opening of the valve installed in the exit line. In this manner, by adjusting the volumetric flowrate of the outgoing stream ($F_{out}(t)$) a constant pressure inside the vessel can be achieved. Under the assumption of constant temperature it can be shown that the pressure dynamics is first order and a block diagram similar to the ones shown in Figures 10.8 and 10.11 can be developed.

10.4.3 Flow Control Loop

In Figure 10.13 a flow control loop is shown. A differential-pressure flowmeter, such as an orifice plate, is installed that infers the flowrate from the pressure drop across an obstruction in the process pipe. The measured flowrate is sent to a flow controller (FC), which adjusts the opening of a valve located downstream of the flowrate sensor/transmitter (FT). By changing the valve stem position the pressure drop, and therefore the flowrate, are adjusted so as to obtain a desired value (F_{sp}). The flow control loop shown in Figure 10.13 is almost universally applied to process streams to achieve fast regulation of the flowrate. The set point of these control loops is supplied externally by another control loop that controls the temperature, pressure or liquid level in a process vessel. This nested structure, called cascade control, has a number of important advantages that will be analyzed later in this book.

A simple cascade control system is shown in Figure 10.14. This is the example considered in Section 10.4.1 and shown in Figure 10.10. In an inner loop the flowrate of the incoming stream is measured and transmitted to a flow controller that regulates the flowrate. The set point for this inner controller is supplied by an external controller that is used to control the liquid level in the tank. In this outer loop the measured level is transmitted to a level controller whose output is not the stem valve position but the desired flowrate.

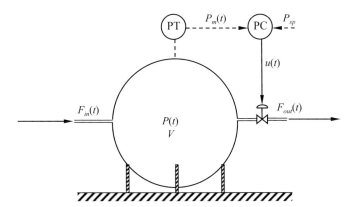

Figure 10.12 A pressure control system in a gas vessel.

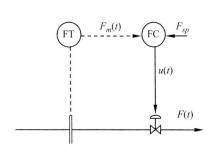

Figure 10.13 A typical flow control system.

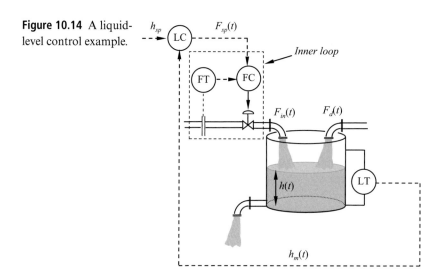

Figure 10.14 A liquid-level control example.

10.5 Commonly Used Control Laws

The control law or control algorithm incorporates in a (usually) simple mathematical formula the relationship between the deviation of the controlled variable ($y(t)$) from the desired or set-point value (y_{sp}) and the controller action. Imagine the situation where you have been requested to control the temperature of a heating tank by adjusting manually the opening of the valve that regulates the flowrate of steam in an internal coil. You then monitor carefully the temperature reading of the installed sensor and you observe an increase in the temperature of the liquid in the tank. Action is clearly needed by you to correct the observed deviation from the desired temperature. Based on common sense, you start decreasing slowly the valve opening and, as a result of reducing the steam flowrate, you observe that the temperature starts decreasing. A significant effort is then needed in order to adjust the valve opening until the temperature returns to the desired value. After achieving that, the temperature remains fairly constant for a short period of time and then a sudden decrease in the temperature is observed. You react to this by quickly increasing the valve opening to send more steam to the system and restore the temperature to its desired value. This is clearly something that you would not like to be doing for the rest of your professional life and something that computers can perform routinely in a much better and sophisticated way. The logic behind the reaction of the operator to any change in the controlled variable needs to be codified in mathematical expressions that can easily be programmed in a computer.

Up to the late 1970s most controllers were mechanical and pneumatic, and were the result of ingenious engineering design. They demanded tubing for instrument air to drive them; this tubing had to be run back and forth between the equipment and the control room. Computer electronics vendors that were looking to expand their potential customers beyond military application decided to market their early computers (1950s) as process control computers. However, only when microprocessors were incorporated in the

Programmable Logic Controllers (PLCs) in the mid 1970s did the execution of complicated math and data-manipulation operations become possible. Improvements in communications and circuit design also made it possible to place PLCs away from harsh industrial conditions. Communication cards, PLCs, microprocessors and memory modules are now designed to be small in size, cheap and extremely reliable, and easy to communicate with through high-speed data highways. They are mounted in metal enclosures in the proximity of process equipment and are part of modern distributed control systems or DCS (a distributed structure that was first introduced by Honeywell in the late 1960s). High-level software facilitates monitoring, controller configuration tuning and reporting through high-resolution graphics units.

The simplest possible control law is the proportional control law, where the control signal is directly proportional to the error signal

$$u(t) = u_b + k_c e(t) \tag{10.5.1}$$

Here, u_b is the controller output when the error signal is zero ($e(t) = 0$) and is called bias. Observe that when the error signal is zero the controller output cannot be zero as this case would correspond to completely closing (or completely opening) the valve. When $e(t) = 0$ the controller output (u_b) must have the value that corresponds to the steady-state solution of the corresponding mathematical model with the controlled variable equal to its desired value. The controller gain, k_c, is the measure of how fast the controller reacts to a change in the error signal. The controller signal as a function of the error signal for a proportional controller is shown in Figure 10.15. Several cases are depicted in Figure 10.15 depending on the value of k_c. In the figure, u_{max} and u_{min} are the values of the control signal that cause saturation of the final control element (fully closed or fully opened control valve, for instance). The interval $[u_{max}, u_{min}]$ is referred to as the proportional band. It can be easily observed from the same figure that, in order to achieve equal variation of the manipulated variable in both negative error and positive error directions, u_b must be selected to be as close to the middle of the proportional band as possible. We can furthermore observe that as $k_c \to 0$ the controller becomes very loose, whereas when $k_c \to \infty$ the controller behaves as an on–off controller where either a small positive or negative error causes saturation of the final control element (complete opening or complete closing of a control valve). This is the simplest possible form of control law as it has no design parameters and is extremely easy to implement in practice.

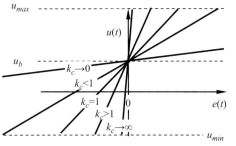

Figure 10.15 Proportional control as a function of controller gain.

However, on–off control tends to produce an oscillatory response and its use is very limited in process control applications. In what follows we will be using the symbol $u(t)$ to denote the deviation of the controller output from the nominal variable or bias (u_b).

Apart from the proportional controller, two other control laws are commonly used in process systems. The proportional-integral (PI) control is used when it is desirable to eliminate completely the steady-state deviation of the controlled variable from its set point. As will be shown in the chapter that follows, one of the main drawbacks of proportional control is its inability to eliminate steady-state error. PI control manages to eliminate steady-state error by acting not only against the error but also against the time integral of the error variable

$$u(t) = k_c \left[e(t) + \frac{1}{\tau_I} \int_0^t e(t')dt' \right] \tag{10.5.2}$$

The first term is the proportional part of the PI controller and the second term is the integral part of the PI controller. The constant $\tau_I > 0$ that appears in the integral part is called the integral (or reset) time and has units of time. To understand the fundamental idea of PI control, consider a step change in the error. The integral of the step will be a ramp signal, which when divided by the integral time and multiplied by the controller gain, gives a linearly growing contribution to the control signal (see Figure 10.16). This linear growth of the control action can only stop when the error becomes zero. Alternatively, one can calculate the time derivative of the control signal in (10.5.2):

$$\frac{du(t)}{dt} = k_c \left[\frac{de(t)}{dt} + \frac{1}{\tau_I} e(t) \right] \tag{10.5.3}$$

From the last equation it is clear that the PI controller keeps acting until both the derivative of the error and the error itself become zero, and thus it manages to bring the error to

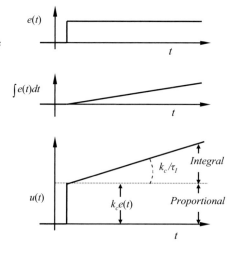

Figure 10.16 Proportional-integral control action for a step change in the error.

zero without any offset. A proportional controller will rest when the derivative of the error becomes zero irrespective of the value of the error, and this results in a permanent offset.

The transfer function of a proportional (P) controller follows directly from Eq. (10.5.1):

$$G_P(s) = \frac{U(s)}{E(s)} = k_c \tag{10.5.4}$$

The transfer function of a PI controller also follows directly from Eq. (10.5.2) by using the formula for the Laplace transform of the integral and keeping in mind that the error signal is initially zero

$$G_{PI}(s) = \frac{U(s)}{E(s)} = k_c \left(1 + \frac{1}{\tau_I s}\right) \tag{10.5.5}$$

To derive an appropriate state-space description of the P and PI controllers we first need to observe that, in the most general case, the controller has the set point $y_{sp}(t)$ and the measurement $y(t)$ as inputs and the control signal $u(t)$ as output. In the case of P or PI control the input to the controller is the error signal $e(t)$ and, in effect, the state-space description of a P and PI controller must have the following general structure

$$\frac{dx_c}{dt} = A_c x_c + b_c e$$
$$u = c_c x_c + d_c e \tag{10.5.6}$$

where x_c is the state vector of the controller. A P controller does not involve dynamics and has no state vector. Only d_c is nonzero and is equal to the controller gain k_c. For the case of the PI controller one needs to note that the integral of the error signal can be calculated through the following equation

$$e_I = \int_0^t e(t')dt' \quad \Rightarrow \quad \frac{de_I}{dt} = e \tag{10.5.7}$$

Using the integral of the error as the only state variable we can write the state-space description of a PI controller as follows:

$$\frac{de_I}{dt} = e$$
$$u = \frac{k_c}{\tau_I} e_I + k_c e \tag{10.5.8}$$

i.e. $A_c = 0$, $b_c = 1$, $c_c = k_c/\tau_I$ and $d_c = k_c$.

A PI controller uses information about the past dynamics of the system under control through its integral part and also information about the present state of the system through the proportional part. In order to forecast the future dynamic behavior of the system, we may use the following approximation of the value of the error at τ_D time units ahead

$$e(t+\tau_D) \approx e(t) + \tau_D \frac{de(t)}{dt} \tag{10.5.9}$$

By incorporating predictive capability into the PI controller described by Eq. (10.5.2) we obtain the mathematical description of the proportional-integral-derivative or PID controller

$$u(t) = k_c \left(e(t) + \frac{1}{\tau_I} \int_0^t e(t') dt' + \tau_D \frac{de(t)}{dt} \right) \tag{10.5.10}$$

where the constant $\tau_D \geq 0$ is called derivative time and is a measure of the size of the derivative part of the control action. Taking the Laplace transform we obtain the following transfer function description of the PID controller

$$G_{PID}(s) = \frac{U(s)}{E(s)} = k_c \left(1 + \frac{1}{\tau_I s} + \tau_D s \right) \tag{10.5.11}$$

The PID form given by Eqs. (10.5.10) and (10.5.11), usually called the ideal PID controller form, has theoretical value only, as step changes in the set point, for instance, will create serious problems when implementing the derivative part (the derivative of the step is the impulse function with infinite magnitude and infinitesimal duration). To derive a realistic form of the PID controller we first consider the case of a PD controller with the following transfer function

$$G_{PD}(s) = \frac{U(s)}{E(s)} = k_c (1 + \tau_D s) \tag{10.5.12}$$

and the corresponding time-domain description

$$u(t) = k_c \left(e(t) + \tau_D \frac{de(t)}{dt} \right) \tag{10.5.13}$$

A realistic PD controller approximates the time dynamics of the ideal PD controller through the following differential equation

$$\alpha \tau_D \frac{du(t)}{dt} + u(t) = k_c \left(e(t) + \tau_D \frac{de(t)}{dt} \right) \tag{10.5.14}$$

where α is a small fixed parameter (e.g. $\alpha = 0.05$ or 0.1 are commonly used values in electronic controllers). When $\alpha = 0$, Eq. (10.5.14) simplifies to Eq. (10.5.13). The transfer function of the real PD controller is

$$G_{PD}(s) = \frac{U(s)}{E(s)} = k_c \frac{(1 + \tau_D s)}{\alpha \tau_D s + 1} \tag{10.5.15}$$

An alternative way of looking into Eq. (10.5.15) is to retain the ideal form of a PD controller (Eq. (10.5.12)) but instead of feeding the error to the controller, one can use a filtered error signal e_F from a first-order low-pass filter (see Chapter 3):

$$\alpha \tau_D \frac{de_F(t)}{dt} + e_F(t) = e(t) \quad (10.5.16)$$

The effect of filtering the error signal is that significant variations of the error are smoothed and in effect the filtered error signal can now be fed to the ideal PD controller. The error signal is then approximated by $e_F(t)$, and the derivative of the error signal is approximated by $de_F(t)/dt$, which because of Eq. (10.5.16) is given by

$$\frac{de_F(t)}{dt} = \frac{e(t) - e_F(t)}{\alpha \tau_D} \quad (10.5.17)$$

Using this approximation, which can easily be implemented, the time-domain description of the PD controller is as follows:

$$\frac{de_F(t)}{dt} = \frac{1}{\alpha \tau_D}\left(e(t) - e_F(t)\right)$$
$$u(t) = k_c\left(e_F(t) + \frac{e(t) - e_F(t)}{\alpha}\right) \quad (10.5.18)$$

The situation is depicted in Figure 10.17, which clearly shows the real PD controller being generated from a first-order low-pass filter followed by an ideal PD controller. The time-domain description of the real PD controller follows the standard state-space template (10.5.6). Indeed, under a slight rearrangement of (10.5.18), we obtain

$$\frac{de_F}{dt} = -\frac{1}{\alpha \tau_D}e_F + \frac{1}{\alpha \tau_D}e$$
$$u = k_c\left(\frac{\alpha - 1}{\alpha}\right)e_F + \frac{k_c}{\alpha}e \quad (10.5.19)$$

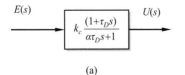

Figure 10.17 Alternative representations of a real PD controller.

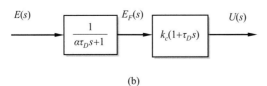

To obtain the state-space description of a real PID controller we only need to add the integral of the error and its contribution to the controller output:

$$\frac{de_I}{dt} = e$$
$$\frac{de_F}{dt} = -\frac{1}{\alpha\tau_D}e_F + \frac{1}{\alpha\tau_D}e \qquad (10.5.20)$$
$$u = \frac{k_c}{\tau_I}e_I + k_c\left(\frac{\alpha-1}{\alpha}\right)e_F + \frac{k_c}{\alpha}e$$

The corresponding transfer function is the following

$$G_{PID}(s) = \frac{U(s)}{E(s)} = k_c\left(\frac{1+\tau_D s}{\alpha\tau_D s+1} + \frac{1}{\tau_I s}\right) \qquad (10.5.21)$$

Table 10.1 summarizes the mathematical descriptions of the P, PI, PD and PID controllers. From the state-space descriptions, one can derive appropriate time-discretized forms of the

Table 10.1 Mathematical description of classical control laws

Controller	Transfer function	State space description
P	$G_P(s) = k_c$	$u = k_c e$
PI	$G_{PI}(s) = k_c\left(1 + \dfrac{1}{\tau_I s}\right)$	$\dfrac{de_I}{dt} = e$ $u = \dfrac{k_c}{\tau_I}e_I + k_c e$
Ideal PD	$G_{PD}(s) = k_c(1+\tau_D s)$	does not exist
Ideal PID	$G_{PID}(s) = k_c\left(1 + \dfrac{1}{\tau_I s} + \tau_D s\right)$	does not exist
Real PD	$G_{PD}(s) = k_c\left(\dfrac{1+\tau_D s}{\alpha\tau_D s+1}\right)$	$\dfrac{de_F}{dt} = -\dfrac{1}{\alpha\tau_D}e_F + \dfrac{1}{\alpha\tau_D}e$ $u = k_c\left(\dfrac{\alpha-1}{\alpha}\right)e_F + \dfrac{k_c}{\alpha}e$
Real PID	$G_{PID}(s) = k_c\left(\dfrac{1+\tau_D s}{\alpha\tau_D s+1} + \dfrac{1}{\tau_I s}\right)$	$\dfrac{de_I}{dt} = e$ $\dfrac{de_F}{dt} = -\dfrac{1}{\alpha\tau_D}e_F + \dfrac{1}{\alpha\tau_D}e$ $u = \dfrac{k_c}{\tau_I}e_I + k_c\left(\dfrac{\alpha-1}{\alpha}\right)e_F + \dfrac{k_c}{\alpha}e$

controllers, which are necessary for computer implementation. The time-discretized controller will receive a sampled error signal $e[j]$ and generate a discrete control signal $u[j]$. In particular, applying a backward rectangle discretization with sampling period T_s (see Chapter 6)

- the PI controller (10.5.8) is discretized as

$$\begin{cases} e_I[j] = e_I[j-1] + T_s\, e\,[j] & [Integration] \\ u\,[j] = k_c e\,[j] + \dfrac{k_c}{\tau_I} e_I[j] & [Controller\ output] \end{cases} \quad (10.5.22)$$

- the real PD controller (10.5.19) is discretized as

$$\begin{cases} e_F[j] = e^{-T_s/\alpha\tau_D} e_F[j-1] + \left(1 - e^{-T_s/\alpha\tau_D}\right) e[j] & [Filter] \\ u[j] = k_c \left(\dfrac{\alpha - 1}{\alpha}\right) e_F[j] + \dfrac{k_c}{\alpha} e[j] & [Controller\ output] \end{cases} \quad (10.5.23)$$

- the real PID controller (10.5.20) is discretized as

$$\begin{cases} e_I[j] = e_I[j-1] + T_s e[j] & [Integration] \\ e_f[j] = e^{-T_s/\alpha\tau_D} e_f[j-1] + \left(1 - e^{-T_s/\alpha\tau_D}\right) e[j] & [Filter] \\ u[j] = \dfrac{k_c}{\tau_I} e_I[j] + k_c \left(\dfrac{\alpha - 1}{\alpha}\right) e_f[j] + \dfrac{k_c}{\alpha} e[j] & [Controller\ output] \end{cases} \quad (10.5.24)$$

LEARNING SUMMARY
- The typical block-diagram representation of a feedback control system is shown in Figure 10.2 and involves the controller, the final control element, the process and the sensor.
- The dynamics of the most common sensor/transmitter elements can be described as first-order systems.
- The simplest form of control law is the PID control law, which has the mathematical description summarized in Table 10.1.

TERMS AND CONCEPTS

Bias. The output of the controller when the error signal is zero.
Derivative time. The constant that multiplies the derivative of the error signal in the ideal form of the PID controller.

Final control element. This is a control valve in most chemical process control systems, and is the device used to implement the controller action.

Integral time. The constant that divides the integral of the error signal in the PID controller.

Sensor. A device used to measure a variable by producing an electrical signal that is related to the value of the measured variable.

Transmitter. A device used to transmit an electrical signal from the sensor to the controller.

FURTHER READING

A comprehensive coverage of the block-diagram representation methodology is presented in the book

Ogata, K., *Modern Control Engineering, Pearson International Edition*, 5th edn, 2008.

Additional examples of block-diagram representation of single process control loops can be found in

Smith, C. A. and Corripio, A. B., *Principles and Practice of Automatic Process Control*, 2nd edn. New Jersey: John Wiley & Sons, Inc., 1997.

Stephanopoulos, G., *Chemical Process Control, An Introduction to Theory and Practice*. New Jersey: Prentice Hall, 1984.

A comprehensive discussion of PID controllers is presented in

Åström, K. J. and Hägglund, T., *PID-Controllers: Theory, Design and Tuning*, 2nd edn. Research Triangle Park, NC: International Society of Automation (ISA), 1995.

PROBLEMS

10.1 Consider the two tanks in series shown in Figure P10.1 that have been studied in Chapter 4. A control system has been installed where a level sensor/transmitter (LT) measures and transmits the liquid level in the second tank and the controller manipulates the volumetric flowrate of a stream fed to the first tank ($F_0(t)$). A second stream is also fed to the first tank and varies uncontrollably ($F_w(t)$). Given a controller transfer function $G_c(s)$, derive the transfer functions of all other hardware elements and sketch the block diagram of the feedback control system.

10.2 Repeat Problem 10.1 for the case of two tanks in series with interaction (see Figure 4.5).

10.3 Repeat Problem 10.1 for the case where the outflow of the second tank is determined by a pump (see Figure P4.6).

10.4 Repeat Problem 10.1 for the case of three tanks in series where the sensor measures the liquid level in the third tank and the controller manipulates the incoming flowrate into the first tank.

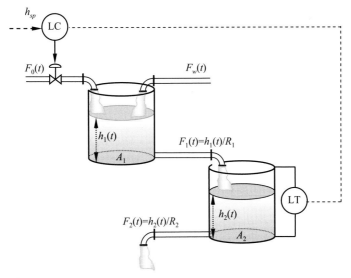

Figure P10.1

10.5 Derive the block-diagram representation of the three heating tanks system shown in Figure P10.5. The controlled variable is the temperature of the third tank and the manipulated variable the steam flowrate in the first tank. The steam flowrates in the second and third tank are fixed. T_0 is a disturbance while F and V are constant.

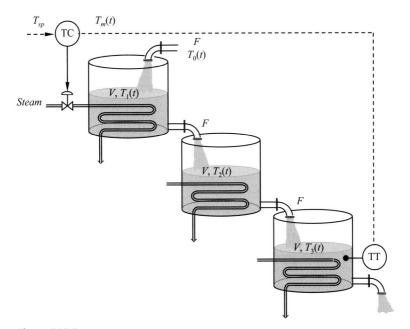

Figure P10.5

10.6 Derive the block-diagram representation for the system shown in Figure 10.12.

10.7 Derive the block-diagram representation for the system shown in Figure 10.14

10.8 The real PD controller:

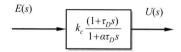

has been developed as an approximation of the ideal PD controller

$$E(s) \longrightarrow \boxed{k_c(1+\tau_D s)} \longrightarrow U(s)$$

in the limit as $\alpha \to 0$.

(a) Calculate the responses of the real PD and the ideal PD controller to a unit step change in the error. Plot the response of the real PD for various small values of α (e.g. $\alpha = 0.1, 0.02, 0.01, 0.005, \ldots$) in a $u(t)/k_c$ versus t/τ_D diagram. Does it approximate the response of the ideal PD as $\alpha \to 0$?

(b) Calculate and compare the frequency response of the real PD and the ideal PD controller. Use MATLAB to plot the Bode diagrams AR/k_c and φ versus $\omega \tau_D$ for both the ideal PD and the real PD for various small values of α (e.g. $\alpha = 0.1, 0.05, 0.02, 0.01, 0.005, \ldots$). In what sense does the real PD approximate the ideal PD?

10.9 A PID controller is sometimes implemented as series connection of a first-order low-pass filter with the ideal PID transfer function

$$E(s) \longrightarrow \boxed{k_c \frac{1+\dfrac{1}{\tau_I s}+\tau_D s}{\tau_F s+1}} \longrightarrow U(s)$$

The above is referred to as "PID with filter."

(a) Derive a state-space description for the "PID with filter" controller.
(b) Derive a zero-order-hold discretization of the controller equations of the previous question.

10.10 Consider the following version of "real PID controller":

(a) Derive a state-space description for this controller
(b) Derive a zero-order-hold discretization of the controller equations of the previous question.

10.11 An I controller involves using integral action only. It has a transfer function of the form

$$\frac{U(s)}{E(s)} = \frac{k_I}{s}$$

(a) Derive a state-space description for the I controller.
(b) Derive a zero-order-hold discretization of the controller equations of the previous question.

10.12 Derive a state-space description for a controller with proportional and double integral action, whose transfer function is of the form:

$$\frac{U(s)}{E(s)} = k_c \left(1 + \frac{1}{\tau_I^2 s^2}\right)$$

11 Block-Diagram Reduction and Transient-Response Calculation in a Feedback Control System

The aim of this chapter is to introduce block-diagram simplification or reduction, to calculate the overall transfer functions of a feedback control loop, and to use them to calculate transient responses to set point or disturbance changes. The block-diagram reduction is performed by simple algebraic manipulations of the transfer functions of the individual components. A simple temperature-control example for a heater is used to illustrate the procedure. In the same example, transient responses to step changes in the set point and in the disturbance are studied both under P and under PI control, with the primary emphasis on testing the ability of the control system to bring the output to set point.

STUDY OBJECTIVES

After studying this chapter you should be able to do the following.

- Simplify complex block diagrams that describe feedback control systems.
- Use the overall closed-loop transfer functions to derive the transient response of a feedback control system under P or PI control, for step changes in set point or disturbance.
- Use software tools, like MATLAB and Maple, to simulate the closed-loop response under P or PI control.

11.1 Calculation of the Overall Closed-Loop Transfer Functions in a Standard Feedback Control Loop

In the previous chapter, we saw examples of feedback control loops represented in block-diagram form. We will now see how to calculate the overall transfer functions of the entire feedback control loop; these relate the output of the loop to its inputs. The procedure that we will follow is referred to as block-diagram simplification or reduction. It involves simple algebraic manipulations of the transfer-function description of the subsystems comprising the feedback loop. We will first consider the case where the measurement of the controlled variable is the output of the feedback loop and then the case where the actual physical variable is the output.

11.1 Calculation of the Overall Closed-Loop Transfer Functions in a Standard Feedback

11.1.1 Block-Diagram Reduction when the Measured Variable is the Output of the Control System

We consider the block-diagram representation of the feedback control loop shown in Figure 11.1, where the blocks (subsystems) are the following:

- the controller with transfer function $G_c(s)$ relating the error ($E(s)$) and the controller output ($U(s)$)
- the final control element (usually a valve) with transfer function $G_v(s)$ relating the controller output ($U(s)$) and the manipulated input of the process ($M(s)$)
- the process with transfer functions $G_p(s)$ relating the manipulated input ($M(s)$) and the process output ($C(s)$), and $G_w(s)$ relating the disturbance input ($W(s)$) and the process output
- the sensor with transfer function $G_m(s)$ relating the process output ($C(s)$) and its measurement ($Y(s)$).

The inputs to the feedback control loop are the disturbance $W(s)$ and the set point $Y_{sp}(s)$. The output is the measurement of the controlled variable $Y(s)$. Our goal is to eliminate algebraically all other signals appearing in the block diagram of Figure 11.1 and derive the overall transfer functions between the inputs and the output.

We begin by first simplifying the relationship between $U(s)$, $W(s)$ and $Y(s)$, which is the part of the block diagram enclosed by the dotted line in Figure 11.1, and represents the overall instrumented process. This is achieved as follows:

$$\begin{aligned} Y(s) &= G_m(s)\, C(s) \\ &= G_m(s)\left[G_w(s)\, W(s) + G_p(s)\, M(s)\right] \\ &= G_m(s)G_w(s)\, W(s) + G_m(s)G_p(s)G_v(s)\, U(s) \end{aligned}$$

from which it follows that

$$Y(s) = G(s)\, U(s) + G'(s)\, W(s) \qquad (11.1.1)$$

where

$$G(s) = G_m(s)G_p(s)G_v(s) \qquad (11.1.2)$$

$$G'(s) = G_m(s)G_w(s) \qquad (11.1.3)$$

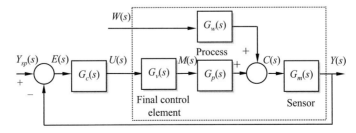

Figure 11.1 Block-diagram representation of a feedback control system.

After this simplification, the block diagram shown in Figure 11.1 can be reduced to that of Figure 11.2. It is important at this point to observe that the overall transfer functions of the instrumented process involve products of individual transfer functions, which are connected in series. We continue the algebraic simplification as follows

$$Y(s) = G(s)\,U(s) + G'(s)\,W(s)$$
$$= G(s)G_c(s)\,E(s) + G'(s)\,W(s)$$
$$= G(s)G_c(s)\left[Y_{sp}(s) - Y(s)\right] + G'(s)\,W(s)$$

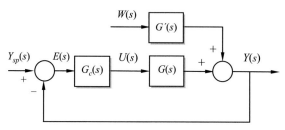

Figure 11.2 Block-diagram representation of a feedback control system after the initial simplification.

With the foregoing algebraic manipulations, all intermediate signals of the loop have been eliminated. We only need to solve the last equation for $Y(s)$, to obtain

$$Y(s) = \frac{G(s)G_c(s)}{1 + G(s)G_c(s)} Y_{sp}(s) + \frac{G'(s)}{1 + G(s)G_c(s)} W(s) \tag{11.1.4}$$

This is a very important result: it tells us how the output $Y(s)$ depends on the set point $Y_{sp}(s)$ and the disturbance $W(s)$. The two transfer functions: $\dfrac{G(s)G_c(s)}{1 + G(s)G_c(s)}$ and $\dfrac{G'(s)}{1 + G(s)G_c(s)}$ are called closed-loop transfer functions. They share the same denominator, which involves the product of the controller and instrumented process transfer functions $G_c(s)G(s)$.

With the result of Eq. (11.1.4), the block diagram of Figure 11.2 is simplified (reduced) to that of Figure 11.3. The transfer functions $G(s)$ and $G'(s)$ are given by Eqs. (11.1.2) and (11.1.3), so in terms of the original block diagram of Figure 11.1, the result can be expressed as

$$Y(s) = \frac{G_m(s)G_p(s)G_v(s)G_c(s)}{1 + G_m(s)G_p(s)G_v(s)G_c(s)} Y_{sp}(s) + \frac{G_m(s)G_w(s)}{1 + G_m(s)G_p(s)G_v(s)G_c(s)} W(s) \tag{11.1.5}$$

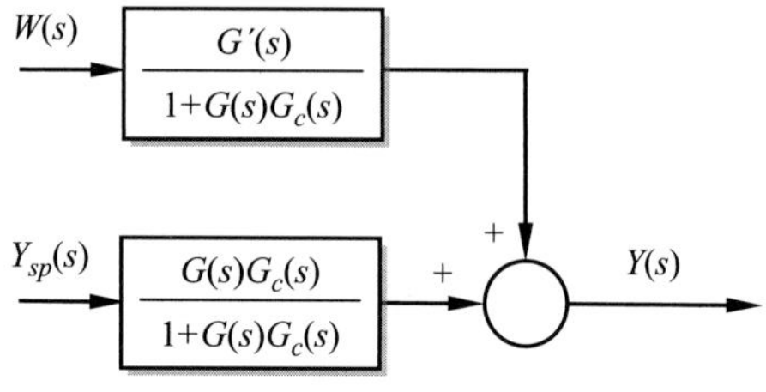

Figure 11.3 Block-diagram representation of a feedback control system after final reduction.

11.1 Calculation of the Overall Closed-Loop Transfer Functions in a Standard Feedback

It is interesting to note that the numerator of each closed-loop transfer function is the product of all transfer functions of the path connecting the input and output, while the denominator equals 1 plus the product of all transfer functions appearing in the loop shown in Figure 11.1. This mnemonic rule, which applies only for the standard feedback control loop, is worth noting as it will be useful throughout this chapter.

Figure 11.4 summarizes useful simplification rules, some of which have already been derived while proving Eq. (11.1.4). This is not a complete set, but can be used to simplify most of the common block diagrams in process control.

Using the rules given in Figure 11.4, we can derive the closed-loop transfer functions more easily, as follows. We first observe from Figure 11.1 that the transfer functions $G_c(s)$, $G_v(s)$, $G_p(s)$ and $G_m(s)$, as well as $G_w(s)$ and $G_m(s)$, are connected in series and by applying the corresponding rules (rule (a) and then rule (c) of Figure 11.4), we obtain the block diagram shown

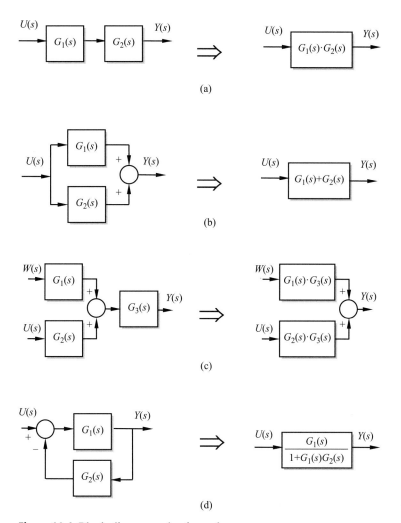

Figure 11.4 Block-diagram reduction rules.

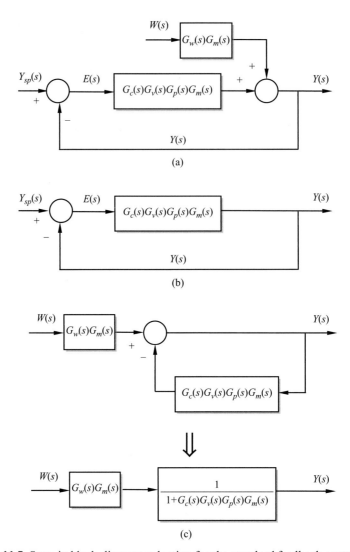

Figure 11.5 Steps in block-diagram reduction for the standard feedback control loop.

in Figure 11.5a. We can then proceed with a separate calculation of the transfer function between Y_{sp} and Y and the one between W and Y.

For the transfer function between Y_{sp} and Y, we set the disturbance signal to zero ($W(s) = 0$), in which case the block diagram of Figure 11.5a simplifies to that of Figure 11.5b. From rule (d) we can directly obtain the transfer function between Y_{sp} and Y. For the transfer function between W and Y, we set the set-point signal to zero ($Y_{sp}(s) = 0$), in which case the block diagram of Figure 11.5a simplifies to that of Figure 11.5c. From this representation, the transfer function between W and Y is obtained.

As an application of our result for the closed-loop transfer functions, we will consider the heating tank process example studied in Section 10.3, which has the block-diagram

11.1 Calculation of the Overall Closed-Loop Transfer Functions in a Standard Feedback

representation shown in Figure 10.9. By comparing Figures 10.9 and 11.1 we can identify the individual transfer functions:

$$G_p(s) = \frac{k_p}{\tau_p s + 1} \tag{11.1.6}$$

$$G_w(s) = \frac{1}{\tau_p s + 1} \tag{11.1.7}$$

$$G_m(s) = \frac{1}{\tau_m s + 1} \tag{11.1.8}$$

$$G_v(s) = k_v \tag{11.1.9}$$

From the above, we calculate the transfer functions of the instrumented process

$$G(s) = \frac{k_v k_p}{(\tau_p s + 1)(\tau_m s + 1)} \tag{11.1.10}$$

$$G'(s) = \frac{1}{(\tau_p s + 1)(\tau_m s + 1)} \tag{11.1.11}$$

and substituting into Eq. (11.1.4), we obtain

$$\frac{G(s)G_c(s)}{1+G(s)G_c(s)} = \frac{\dfrac{k_v k_p}{(\tau_p s + 1)(\tau_m s + 1)} G_c(s)}{1 + \dfrac{k_v k_p}{(\tau_p s + 1)(\tau_m s + 1)} G_c(s)} = \frac{k_v k_p G_c(s)}{(\tau_p s + 1)(\tau_m s + 1) + k_v k_p G_c(s)}$$

$$\frac{G'(s)}{1+G(s)G_c(s)} = \frac{\dfrac{1}{(\tau_p s + 1)(\tau_m s + 1)}}{1 + \dfrac{k_v k_p}{(\tau_p s + 1)(\tau_m s + 1)} G_c(s)} = \frac{1}{(\tau_p s + 1)(\tau_m s + 1) + k_v k_p G_c(s)}$$

and finally (see also Figure 11.6)

$$\bar{T}_m(s) = \frac{k_v k_p G_c(s)}{(\tau_p s + 1)(\tau_m s + 1) + k_v k_p G_c(s)} \bar{T}_{sp}(s) + \frac{1}{(\tau_p s + 1)(\tau_m s + 1) + k_v k_p G_c(s)} \bar{T}_0(s) \tag{11.1.12}$$

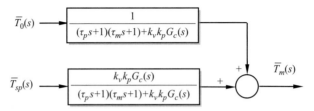

Figure 11.6 Simplified block diagram of the feedback loop of the heating tank.

11.1.2 Block-Diagram Reduction when the Controlled Process Output is the Output of the Feedback Loop

When the controlled process output – the actual physical variable – is considered to be the output of the feedback control loop, the block diagram is as shown in Figure 11.7 (see also Figure 10.8 for the heating tank example).

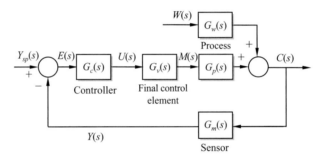

Figure 11.7 Alternative block-diagram representation of a feedback control system.

Using the block-diagram reduction rules of Figure 11.4, we can obtain the overall transfer function description of the feedback loop, which relates the output C to the inputs Y_{sp} and W:

$$C(s) = \frac{G_p(s)G_v(s)G_c(s)}{1+G_m(s)G_p(s)G_v(s)G_c(s)} Y_{sp}(s) + \frac{G_w(s)}{1+G_m(s)G_p(s)G_v(s)G_c(s)} W(s) \qquad (11.1.13)$$

Note that (11.1.13) and (11.1.5) are really the same equation, since $Y(s) = G_m(s)\,C(s)$.

11.2 Calculation of Overall Transfer Functions in a Multi-Loop Feedback Control System

In this section we will show how the block-diagram simplification or reduction techniques can be applied to a multi-loop system such as the cascade control structure shown in Figure 10.14 of the previous chapter. In the inner loop a flowmeter is used to measure the flow and

11.2 Calculation of Overall Transfer Functions in a Multi-Loop Feedback Control System

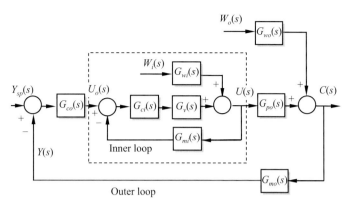

Figure 11.8 Block-diagram representation of a cascade control system.

transmit the measurement to a flow controller that manipulates the valve opening. The set point for this inner loop is supplied by an outer loop where the liquid level is measured and transmitted to a level controller. A block-diagram representation of this cascade control system is shown in Figure 11.8. In this figure, subscript i denotes the inner loop and subscript o denotes the outer loop.

Using Eq. (11.1.13) we can simplify the inner loop

$$U(s) = G_i(s) U_o(s) + G'_i(s) W_i(s) \tag{11.2.1}$$

where

$$G_i(s) = \frac{G_v(s)G_{ci}(s)}{1+G_{mi}(s)G_v(s)G_{ci}(s)}, \quad G'_i(s) = \frac{G_{wi}(s)}{1+G_{mi}(s)G_v(s)G_{ci}(s)} \tag{11.2.2}$$

and obtain the simplified block diagram shown in Figure 11.9. Recall that in a feedback loop, the closed-loop transfer function between any input and the output is the product of

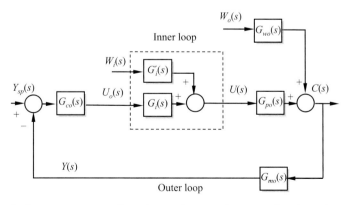

Figure 11.9 Block-diagram representation of cascade control system after simplifying the inner loop.

all transfer functions appearing in the path that connects them directly, divided by 1 plus the product of all transfer functions appearing in the loop. Thus, we obtain the closed-loop transfer functions shown in Figure 11.10.

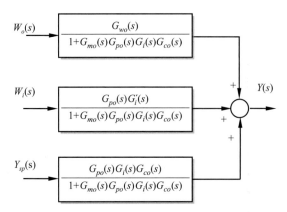

Figure 11.10 Simplified block-diagram representation of a cascade control system.

Example 11.1 Block-diagram reduction of a multi-loop system

Simplify the multi-loop system shown in Figure 11.11.

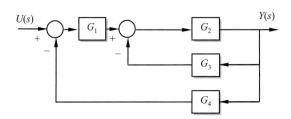

Figure 11.11 Block-diagram representation of Example 11.1.

Solution

To simplify the block diagram, we start from the inner loop, where transfer functions G_2 and G_3 are involved. The resulting block diagram is shown in Figure 11.12a from which Figure 11.12b is obtained by incorporating transfer function G_1 (as it is connected in series with the feedback connection of G_2 and G_3). Finally, from Figure 11.12b we obtain the simplified overall transfer function shown in Figure 11.12c.

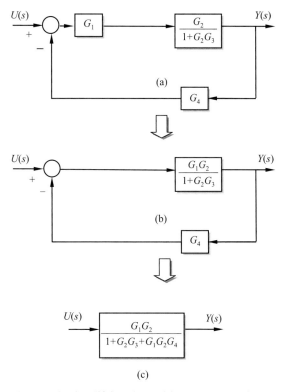

Figure 11.12 Steps in simplifying the multi-loop system of Example 11.1.

11.3 Stirred Tank Heater under Negligible Sensor Dynamics: Closed-Loop Response Calculation under P or PI Control

In this section we will consider the heating tank example with the assumption that the sensor dynamics can be neglected, i.e. the time constant of the process τ_p is much larger than τ_m. More specifically, we will consider the extreme case where $\tau_m = 0$. Two types of controllers will be examined. P control will be examined first, followed by PI control. At this point we need to remind ourselves that the transfer function of a P controller is constant, equal to the controller gain:

$$G_P(s) = \frac{U(s)}{E(s)} = k_c \tag{11.3.1}$$

whereas that of a PI controller involves an extra term proportional to $1/s$ that accounts for the integral of the error:

$$G_{PI}(s) = \frac{U(s)}{E(s)} = k_c \left(1 + \frac{1}{\tau_I s}\right) \tag{11.3.2}$$

The transfer function description of the temperature control loop in the heating tank example is given by Eq. (11.1.12), from which by setting $\tau_m = 0$ we obtain

$$\bar{T}_m(s) = \frac{k_v k_p G_c(s)}{\tau_p s + 1 + k_v k_p G_c(s)} \bar{T}_{sp}(s) + \frac{1}{\tau_p s + 1 + k_v k_p G_c(s)} \bar{T}_0(s) \qquad (11.3.3)$$

11.3.1 P Control of the Heating Tank

Substituting the P controller transfer function $G_c(s) = k_c$ into Eq. (11.3.3), we obtain

$$\bar{T}_m(s) = \frac{k_v k_p k_c}{\tau_p s + 1 + k_v k_p k_c} \bar{T}_{sp}(s) + \frac{1}{\tau_p s + 1 + k_v k_p k_c} \bar{T}_0(s) \qquad (11.3.4)$$

Note that the transfer functions of the instrumented process (see Eqs. (11.1.10) and (11.1.11)) are given by

$$G(s) = \frac{k_v k_p}{\tau_p s + 1} \qquad (11.3.5)$$

$$G'(s) = \frac{1}{\tau_p s + 1} \qquad (11.3.6)$$

and they are first order, both having time constant τ_p, but different static gains.

It will be interesting to compare the above open-loop transfer functions to the closed-loop transfer functions in Eq. (11.3.4). We observe that the closed-loop transfer functions are also first order, but with different static gains and time constant. Indeed, the expressions of the closed-loop transfer functions in Eq. (11.3.4) may be rearranged as

$$\bar{T}_m(s) = \frac{k_{CL}}{\tau_{CL} s + 1} \bar{T}_{sp}(s) + \frac{k'_{CL}}{\tau_{CL} s + 1} \bar{T}_0(s) \qquad (11.3.7)$$

where the closed-loop gains and time constant are given by

$$k_{CL} = \frac{k_v k_p k_c}{1 + k_v k_p k_c} \qquad (11.3.8)$$

$$k'_{CL} = \frac{1}{1 + k_v k_p k_c} \qquad (11.3.9)$$

$$\tau_{CL} = \frac{\tau_p}{1 + k_v k_p k_c} \qquad (11.3.10)$$

The following observations can be made here.

(1) The order of the closed-loop system is the same as the order of the open-loop system, i.e. P control does not change the order of the system.

(2) The pole of the closed-loop system is $-(1+k_v k_p k_c)/\tau_p$, therefore it will be BIBO stable if and only if $k_v k_p k_c > -1$.

(3) When $k_v k_p k_c > 0$, the closed-loop time constant τ_{CL} is smaller than the process time constant τ_p (the closed-loop system becomes faster), otherwise the closed-loop time constant τ_{CL} is larger than the process time constant τ_p (the controller slows down the process).

(4) Assuming $k_v k_p k_c > 0$, as controller gain k_c increases in magnitude, the closed-loop time constant gets smaller and the disturbance gain k'_{CL} gets smaller.

These facts are significant since they provide guidance on the selection of the controller gain k_c. It is desirable to choose k_c so that $k_v k_p k_c$ is positive and fairly large.

We can now use this simple heater example to calculate the response of the closed-loop system, and try to understand what happens when a P controller is applied to the process. Before we proceed, we must define what the control system is expected to do. The control system is expected to perform two functions.

(a) When the set point is changed, the control system should bring the process to the new set point. This is referred to as the set-point tracking or servo problem.
(b) When an external disturbance acts on the process, the control system should apply corrective action and bring the process back to the set point. This is referred to as the disturbance rejection or regulator problem.

Let's see what happens in our example. We will assume that $k_v k_p k_c > 0$.

Set-point tracking or servo problem. Consider a step change in the set point of size ΔT_{sp}, in the absence of disturbance. This means $\bar{T}_{sp}(s) = \Delta T_{sp}/s$ and $\bar{T}_0(s) = 0$, hence

$$\bar{T}_m(s) = \frac{k_{CL}}{\tau_{CL} s + 1} \frac{\Delta T_{sp}}{s} \qquad (11.3.11)$$

Inverting the Laplace transform, we find

$$\bar{T}_m(t) = k_{CL} \Delta T_{sp} \left(1 - e^{-t/\tau_{CL}}\right) \qquad (11.3.12)$$

i.e. that

$$\bar{T}_m(t) = \left(\frac{k_v k_p k_c}{1 + k_v k_p k_c}\right) \Delta T_{sp} \left[1 - e^{-t / \left(\frac{\tau_p}{1 + k_v k_p k_c}\right)}\right] \qquad (11.3.13)$$

A qualitative representation of the response is shown in Figure 11.13. From the last equation, taking the limit as $t \to \infty$, it follows that

$$\bar{T}_m \to \frac{k_v k_p k_c}{1 + k_v k_p k_c} \Delta T_{sp} \qquad (11.3.14)$$

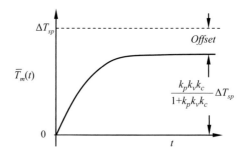

Figure 11.13 Response to a step change in set point under P control.

The offset or steady-state error is defined as the difference between the new set point and the final value of the output, i.e. here

$$Offset = \left(1 - \frac{k_v k_p k_c}{1 + k_v k_p k_c}\right) \Delta T_{sp} = \frac{1}{1 + k_v k_p k_c} \Delta T_{sp} \neq 0 \tag{11.3.15}$$

We observe that the offset is nonzero but approaches zero as the magnitude of $k_c \to \infty$. On the other hand, when $k_c \to 0$ (i.e. open-loop operation) the offset becomes equal to ΔT_{sp}.

Disturbance rejection or regulator problem. Consider a step change in the feed temperature (disturbance) of size ΔT_0, while the set point remains unchanged. This means $\bar{T}_0(s) = \Delta T_0 / s$ and $\bar{T}_{sp}(s) = 0$, therefore

$$\bar{T}_m(s) = \frac{k'_{CL}}{\tau_{CL} s + 1} \frac{\Delta T_0}{s} \tag{11.3.16}$$

Inverting the Laplace transform, we find

$$\bar{T}_m(t) = k'_{CL} \Delta T_0 \left(1 - e^{-t/\tau_{CL}}\right) \tag{11.3.17}$$

i.e. that

$$\bar{T}_m(t) = \left(\frac{1}{1 + k_v k_p k_c}\right) \Delta T_0 \left[1 - e^{-t/\left(\frac{\tau_p}{1 + k_v k_p k_c}\right)}\right] \tag{11.3.18}$$

A qualitative representation of the response is shown in Figure 11.14. The figure also includes the open-loop response, i.e. the response to the same step change in T_0 in the absence of a controller, as prescribed by the transfer function (11.3.6).

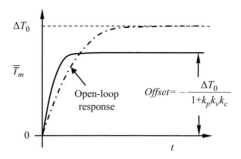

Figure 11.14 Response to a step disturbance under P control.

From the last equation, taking the limit as $t \to \infty$, it follows that

$$\bar{T}_m \to \frac{1}{1+k_p k_v k_c}\Delta T_0 \tag{11.3.19}$$

The offset or steady-state error is the difference between the set point, which is zero, and the final value of the output, i.e.

$$\text{Offset} = 0 - \frac{1}{1+k_p k_v k_c}\Delta T_0 = -\frac{\Delta T_0}{1+k_p k_v k_c} \neq 0 \tag{11.3.20}$$

We observe that the offset is nonzero but approaches zero as the magnitude of $k_c \to \infty$. Note that, when the controller is "off" (i.e. the controller output $U(s)=0$), the effect of the disturbance is described by $\bar{T}_m(s) = G'(s)\bar{T}_0(s) = \bar{T}_0(s)/(\tau_p s+1)$, hence for a step change in T_0 of size ΔT_0, the new steady state of the output will be ΔT_0, which corresponds to a steady-state error of $(-\Delta T_0)$. Thus Eq. (11.3.20) can be written as

$$\text{Offset} = \frac{1}{1+k_p k_v k_c}(\text{Offset in open loop}) \tag{11.3.21}$$

Because of the assumption that $k_v k_p k_c > 0$, it follows that the offset in closed loop is always smaller than the offset in open loop.

11.3.2 PI Control of the Heating Tank

Substituting the PI controller transfer function $G_c(s) = k_c(1+1/\tau_I s)$ to Eq. (11.3.3), we obtain the closed-loop transfer functions. In particular, we have the following.

Set-point tracking or servo problem ($\bar{T}_0 = 0$)

$$\bar{T}_m(s) = \frac{k_v k_p k_c\left(1+\dfrac{1}{\tau_I s}\right)}{\tau_p s + 1 + k_v k_p k_c\left(1+\dfrac{1}{\tau_I s}\right)}\bar{T}_{sp}(s) = \frac{k_v k_p k_c s + \dfrac{k_v k_p k_c}{\tau_I}}{\tau_p s^2 + (1+k_v k_p k_c)s + \dfrac{k_v k_p k_c}{\tau_I}}\bar{T}_{sp}(s) \tag{11.3.22}$$

Disturbance rejection or regulator problem ($\bar{T}_{sp} = 0$)

$$\bar{T}_m(s) = \frac{1}{\tau_p s + 1 + k_v k_p k_c\left(1+\dfrac{1}{\tau_I s}\right)}\bar{T}_0(s) = \frac{s}{\tau_p s^2 + (1+k_v k_p k_c)s + \dfrac{k_v k_p k_c}{\tau_I}}\bar{T}_0(s) \tag{11.3.23}$$

Based on Eqs. (11.3.22) and (11.3.23) it follows that:

(1) a first-order instrumented process controlled by a PI controller makes a second-order system, i.e. the PI controller increases the order of the system by 1;
(2) the common denominator of the closed-loop transfer functions is a quadratic polynomial. Assuming that $k_v k_p k_c > 0$ and $\tau_I > 0$, its roots (closed-loop poles) will either

be both real and negative, or complex conjugate with negative real parts, therefore the closed-loop system will be BIBO stable;

(3) it is possible to define characteristic time and damping factor for the closed-loop transfer functions in the same spirit as in Chapter 5. These will depend on the specific values of integral time τ_I and controller gain k_c, therefore the qualitative nature of the response (overdamped, critically damped or underdamped) will depend on the controller parameters.

The derivation of analytic expressions for the closed-loop response can be tedious, leading to complicated expressions; however, it is possible to derive the behavior of the closed-loop system as $t \to \infty$ without much pain. To this end, the final-value theorem can be used to show that for a step change in the set point

$$\lim_{t \to \infty} \overline{T}_m(t) = \lim_{s \to 0}\left[s\overline{T}_m(s) \right] = \lim_{s \to 0} \frac{k_v k_p k_c s + \dfrac{k_v k_p k_c}{\tau_I}}{\tau_p s^2 + (1 + k_v k_p k_c)s + \dfrac{k_v k_p k_c}{\tau_I}} \Delta T_{sp} = \Delta T_{sp} \quad (11.3.24)$$

while for a step change in the disturbance

$$\lim_{t \to \infty} \overline{T}_m(t) = \lim_{s \to 0}\left[s\overline{T}_m(s) \right] = \lim_{s \to 0} \frac{s}{\tau_p s^2 + (1 + k_v k_p k_c)s + \dfrac{k_v k_p k_c}{\tau_I}} \Delta T_0 = 0 \quad (11.3.25)$$

and, in effect, the offset is zero in both cases. The PI controller manages, therefore, to bring the system to set point exactly, without any offset. This was not the case under proportional control.

Example 11.2 P control of an integrating process

Consider the liquid storage tank shown in Figure 11.15. In the outlet stream, there is a pump that holds the flowrate F_{out} constant. There are two incoming streams: a manipulated stream with flowrate $F_u(t)$ and a disturbance stream with flowrate $F_w(t)$. If the dynamics of the sensor and the control valve are very fast, calculate the offset under P control for a unit step change in the set point or the disturbance.

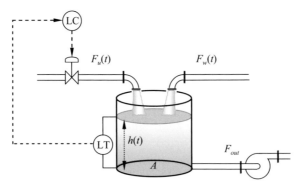

Figure 11.15 Liquid storage tank with a pump in the outflow stream.

Solution

The mass balance for the tank has been derived in Chapter 3:

$$A\frac{dh}{dt} = F_u + F_w - F_{out}$$

Using deviation variables and noticing that F_{out} is constant, we obtain

$$A\frac{d\bar{h}}{dt} = \bar{F}_u + \bar{F}_w$$

Laplace transforming the differential equation,

$$\bar{h}(s) = \frac{1}{As}\bar{F}_u(s) + \frac{1}{As}\bar{F}_w(s)$$

hence the process transfer functions are

$$G_p(s) = G_w(s) = \frac{1}{As}$$

Since both sensor and valve dynamics are considered to be very fast, we also have that $G_v(s) = G_m(s) = 1$. From Eqs. (11.1.2)–(11.1.4) we conclude that

$$G(s) = G_m(s)G_p(s)G_v(s) = G_p(s) = \frac{1}{As}$$

$$G'(s) = G_m(s)G_w(s) = G_w(s) = \frac{1}{As}$$

$$\bar{H}(s) = \frac{G_c(s)}{As + G_c(s)}\bar{H}_{sp}(s) + \frac{1}{As + G_c(s)}\bar{F}_w(s)$$

For a P controller,

$$\bar{H}(s) = \frac{1}{(A/k_c)s + 1}\bar{H}_{sp}(s) + \frac{(1/k_c)}{(A/k_c)s + 1}\bar{F}_w(s)$$

Both closed-loop transfer functions are BIBO stable as long as $k_c > 0$.
  For a unit step change in the set point, the offset can be calculated as follows:

$$\text{Offset} = 1 - \lim_{s \to 0}\left[s\bar{H}(s)\right] = 1 - \lim_{s \to 0}\left[s\frac{1}{(A/k_c)s + 1}\cdot\frac{1}{s}\right] = 0$$

while for a unit step change in the disturbance

$$\text{Offset} = 0 - \lim_{s \to 0}\left[s\bar{H}(s)\right] = 0 - \lim_{s \to 0}\left[s\frac{(1/k_c)}{(A/k_c)s + 1}\cdot\frac{1}{s}\right] = -1/k_c \neq 0$$

So we observe that the offset for a step change in the set point is zero, while for a step change in the disturbance the offset is nonzero and it is inversely proportional to the controller gain.

Liquid storage systems with integrating behavior are quite common in industry and this is one of the reasons for the popularity of P controllers for level control (also holds true for pressure control for controlling gas holdup).

11.4 Software Tools

11.4.1 Calculating the Closed-Loop Transfer Function and the Closed-Loop Response in MATLAB

There are two commands available for calculating the closed-loop transfer function in MATLAB. The first one is the `series` command that calculates the transfer function of systems connected in series, as shown in Figure 11.16a. The `series` function is equivalent to the direct multiplication. When the individual systems are described by transfer functions, with known numerator and denominator polynomials, we first define the individual transfer functions from the known polynomials and then we connect them in series using the `series` function. Consider the heating tank example with negligible sensor time constant ($\tau_m = 0$) where $k_p = 1$, $\tau_p = 1$ and $k_v = 0.2$ under P control with $k_c = 10$. From Eqs. (11.1.10) and (11.1.11) we have that

$$G(s) = \frac{k_v k_p}{(\tau_p s + 1)(\tau_m s + 1)} = \frac{0.2}{s+1} \tag{11.4.1}$$

$$G'(s) = \frac{1}{(\tau_p s + 1)(\tau_m s + 1)} = \frac{1}{s+1} \tag{11.4.2}$$

and

$$G_c(s) = k_c = 10$$

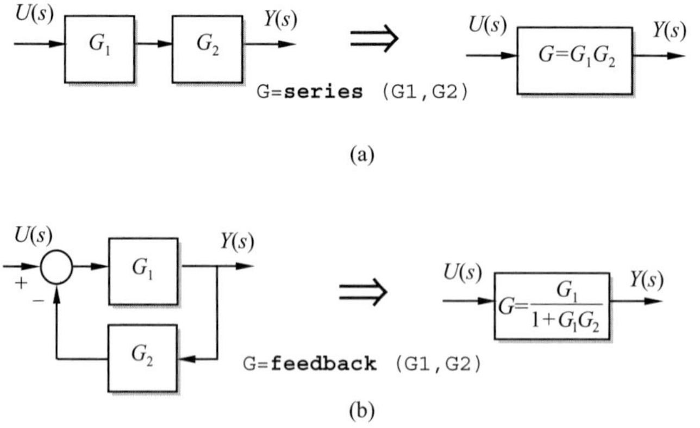

Figure 11.16 The (a) `series` and (b) `feedback` commands available in MATLAB.

To calculate the transfer function between the set point and the measured temperature in MATLAB we first define the individual transfer functions

```
» G=tf(0.2,[1 1])
Transfer function:
 0.2
-----
s + 1
» Gc=tf(10,1)
Transfer function:
10
```

and we then connect them in series to obtain the open-loop transfer function

```
» GOL=series(G,Gc)
Transfer function:
  2
-----
s + 1
```

Finally, we can obtain the closed-loop transfer function by noting that the feedback loop transfer function G_2 shown in Figure 11.16b is equal to unity. This is achieved by using the `feedback` function whose functionality is shown in Figure 11.16b (note that `feedback` assumes a negative feedback, but a third argument can be used to denote explicitly the sign of the feedback loop)

```
» GCL=feedback(GOL,1)
Transfer function:
  2
-----
s + 3
```

The validity of the result can be checked by direct substitution of the constants into Eq. (11.3.4). The closed-loop response can now be obtained using the `step` command

```
» [y,t]=step(GCL,5);
» plot(t,y)
```

By repeating the same process for different values of the controller gain, Figure 11.17 is obtained. From this figure it can be observed that as the controller gain increases, the closed-loop system becomes faster and the offset becomes smaller.

When a PI controller is used the procedure is the same apart from the definition of the controller transfer function. Let's consider a PI controller with $k_c = 10$ and $\tau_I = 0.1$:

```
» G=tf(0.2,[1 1])
Transfer function:
```

Block-Diagram Reduction and Transient-Response Calculation in a Feedback Control System

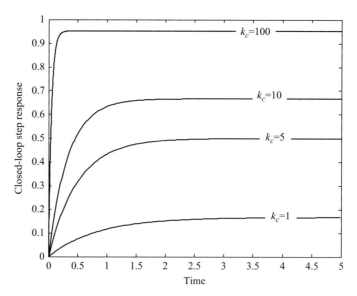

Figure 11.17 Response of the heater temperature to a unit step change in set point under P control, for different values of the controller gain.

```
    0.2
   -----
   s + 1
» kc=10; tau_I=0.1;
» Gc=tf(kc*[tau_I 1],[tau_I 0])

Transfer function:
  s + 10
  ------
  0.1 s

» GOL=series(Gc,G)

Transfer function:
       0.2 s + 2
   -------------------
   0.1 s^2 + 0.1 s

» GCL=feedback(GOL,1)

Transfer function:
       0.2 s + 2
   -------------------
   0.1 s^2 + 0.3 s + 2

» [y,t]=step(GCL,4);
» plot(t,y)
```

By repeating the same procedure for constant controller gain ($k_c = 10$) and increasing values of integral time we obtain Figure 11.18.

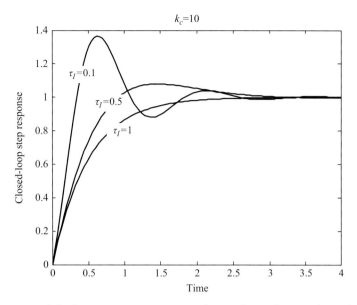

Figure 11.18 Response of the heater temperature to a unit step change in set point under PI control, for different values of integral time.

We turn our attention now to the calculation of the regulatory response. We will follow exactly the same approach as for the servo response by using the closed-loop system block-diagram representation shown in Figure 11.19. Observe that this structure has also been used in Figure 11.5c and is obtained from Figure 11.2 by setting $Y_{sp}(s) = 0$ and rearranging the position of the blocks. We start by defining the three transfer functions involved in the determination of the closed-loop transfer function between the disturbance $T_0(s)$ and the output $T_m(s)$ for the case of P control with $k_c = 10$

```
» G=tf(0.2,[1 1]);
» Gprime=tf(1,[1 1]);
» Gc=tf(10,1);
```

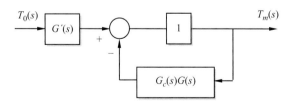

Figure 11.19 Steps in block-diagram reduction for the heating tank example.

Then, the closed-loop transfer function is calculated and connected in series with the G' transfer function to obtain in effect the overall closed-loop transfer function

```
» G1=feedback(1,series(Gc,G))
Transfer function:
 s + 1
 -----
 s + 3
» GCL=series(Gprime,G1)
Transfer function:
      s + 1
 -------------
 s^2 + 4 s + 3
```

Observe that the result obtained might look different than the result obtained analytically and given by Eq. (11.3.4). However, -1 is a root of the denominator polynomial obtained in MATLAB and the two results are identical. Finally the step response can be obtained by using the step command.

We will now turn our attention to the regulator problem with the use of PI control with $k_c = 10$ and $\tau_I = 0.1$. Following the same steps as for the case of P control we obtain

```
» G=tf(0.2,[1 1]);
» Gprime=tf(1,[1 1]);
» kc=10; tau_I=0.1;
» Gc=tf(kc*[tau_I 1],[tau_I 0]);
» G1=feedback(1,series(Gc,G));
» GCL=series(Gprime,G1)
Transfer function:
       0.1 s^2 + 0.1 s
 -----------------------------
 0.1 s^3 + 0.4 s^2 + 2.3 s + 2
» [y,t]=step(GCL,5);
» plot(t,y)
```

By repeating the same approach for different values of the integral time τ_I we obtain Figure 11.20. It can be observed that the offset is zero in all cases independent of the value of the integral time. Furthermore, as the integral is increased the response becomes progressively slower but at the same time the oscillations disappear.

11.4.2 Calculating the Closed-Loop Transfer Functions and the Closed-Loop Response in Maple

Calculation of the closed-loop transfer functions $\dfrac{G(s)G_c(s)}{1+G(s)G_c(s)}$ and $\dfrac{G'(s)}{1+G(s)G_c(s)}$ can be performed symbolically in Maple. Referring to the same example as in the previous

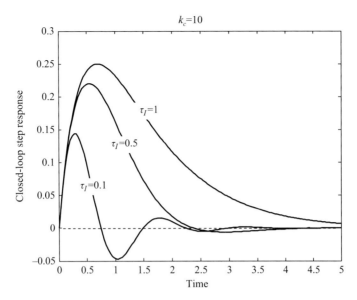

Figure 11.20 Response of the heater temperature to a unit step change in the disturbance under PI control, for different values of integral time.

subsection, with the process transfer functions given by (11.4.1) and (11.4.2), calculation for P control ($G_c(s)=k_c$) goes as follows:

```
> G:=(1/5)/(s+1):
> GPrime:=1/(s+1):
> Gc:=kc:
> GCL_sp:=simplify(G*Gc/(1+G*Gc));
```

$$GCL\_sp := \frac{kc}{5s+5+kc}$$

```
> GCL_w:=simplify(GPrime/(1+G*Gc));
```

$$GCL\_w := \frac{5}{kc+5s+5}$$

The closed-loop response can now be obtained analytically via Laplace-transform inversion using the inttrans package, and subsequently plotted for a specific value of the controller gain ($k_c = 10$). For a unit step change in the set point,

```
> with(inttrans):
> yStep_sp:=invlaplace(eval(GCL_sp/s,[kc=10]),s,t);
```

$$yStep\_sp := \frac{2}{3} - \frac{2}{3}e^{-3t}$$

```
> plot(yStep_sp,t=0..5);
```

and by repeating the same procedure for different values of the controller gain, Figure 11.17 can be obtained. Similarly, for a unit step disturbance,

```
> yStep_w:=invlaplace(eval(GCL_w/s,[kc=10]),s,t);
```

$$yStep\_w := \frac{1}{3} - \frac{1}{3}e^{-3t}$$

and the result can be subsequently plotted. When a PI controller is used, the calculation procedure is exactly the same, apart from the definition of the controller transfer function:

```
> Gc:=kc*(1+1/(tauI*s));
```

$$Gc := kc\left(1 + \frac{1}{tauIs}\right)$$

In this way, the closed-loop transfer functions under PI control can be calculated

$$GCL\_sp := \frac{kc(tauIs + 1)}{5\,tauI\,s^2 + 5\,tauI\,s + kc\,tauI\,s + kc}$$

$$GCL\_w := \frac{5\,tauI\,s}{5\,tauI\,s^2 + 5\,tauI\,s + kc\,tauI\,s + kc}$$

and, subsequently, the responses to unit step changes in the set point or the disturbance for specific values of the controller parameters ($k_c = 10$ and $\tau_I = 0.1$) are calculated:

```
> yStep_sp:=invlaplace(eval(GCL_sp/s,[kc=10,tauI=1/10]),s,t);
```

$$yStep\_sp := 1 + \frac{1}{71}e^{-\frac{3}{2}t}\left(-71\cos\left(\frac{1}{2}\sqrt{71}t\right) + \sqrt{71}\sin\left(\frac{1}{2}\sqrt{71}t\right)\right)$$

```
> yStep_w:=invlaplace(eval(GCL_w/s,[kc=10,tauI=1/10]),s,t);
```

$$yStep\_w := \frac{2}{71}\sqrt{71}e^{-\frac{3}{2}t}\sin\left(\frac{1}{2}\sqrt{71}\,t\right)$$

By repeating the same approach for different values of τ_I, Figures 11.18 and 11.20 can be obtained.

LEARNING SUMMARY
- Rules for simplifying block diagrams are summarized in Figure 11.4.
- The closed-loop transfer function between an input and an output of a single-loop feedback control system equals the product of all transfer functions appearing in the path that connects them directly, divided by 1 plus the product of all transfer functions appearing in the loop.

- When a P controller is used, the order of the closed-loop system is the same as the order of the open-loop system. This is not the case when a PI controller is used, which increases the order of the system by 1.
- In general, P control cannot drive the system exactly to the set point, resulting in steady-state error or offset. On the other hand, the PI controller can bring the system exactly at set point, without any offset.
- P control can track set-point changes without offset when the process system exhibits integrating behavior.

TERMS AND CONCEPTS

Closed-loop transfer function. The overall transfer function relating an input to an output of a closed-loop system.

Multi-loop systems. Systems involving more than one feedback loop in their block-diagram representation.

Offset or steady-state error. The steady-state (sustained) deviation of the measured output from the set point.

Reduction of a block diagram. The systematic simplification of a block diagram that leads to transfer functions directly relating each input to each output of the overall system.

Regulatory control or disturbance rejection. This refers to the control of the system output at a constant set point in the face of disturbance variations.

Servo control or set-point tracking. This refers to the control of the system output when the set point changes, in the absence of disturbances.

Simplification rules. Rules, some presented in Figure 11.4, that can be used to systematically simplify or reduce complex block diagrams.

FURTHER READING

A comprehensive coverage of the block-diagram reduction methodology is presented in the book

Ogata, K., *Modern Control Engineering, Pearson International Edition*, 5th edn, 2008.
   Additional examples of multi-loop systems in process systems are presented in

Smith, C. A. and Corripio, A. B., *Principles and Practice of Automatic Process Control*, 2nd edn. New Jersey: John Wiley & Sons, Inc., 1997.

Stephanopoulos, G., *Chemical Process Control, An Introduction to Theory and Practice*. New Jersey: Prentice Hall, 1984.

PROBLEMS

11.1 Simplify the block diagram shown in Figure P11.1.

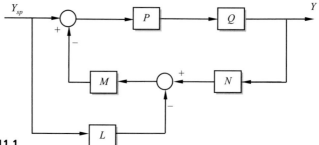

Figure P11.1

11.2 Simplify the block diagrams shown in Figure P11.2.

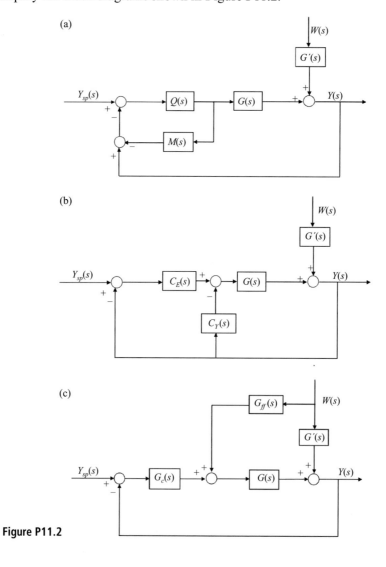

Figure P11.2

11.3 Simplify the block diagram shown in Figure P11.3.

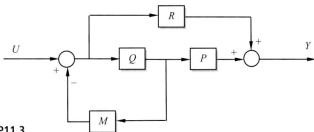

Figure P11.3

11.4 Simplify the block diagram shown in Figure P11.4.

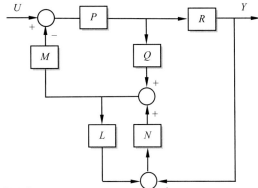

Figure P11.4

11.5 Simplify the block diagram shown in Figure P11.5.

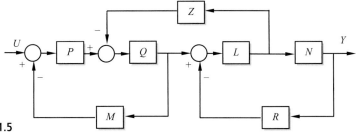

Figure P11.5

11.6 Consider the closed-loop system shown in Figure P11.6, where the transfer function of the process is that of a second-order system, i.e.

$$G_p(s) = \frac{k_p}{\tau^2 s^2 + 2\zeta\tau s + 1}$$

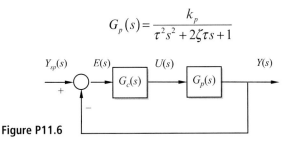

Figure P11.6

If the controller is a P controller ($G_c(s)=k_c$)

(a) determine the closed-loop transfer function
(b) determine the offset to a unit step change of the set point.

11.7 Repeat Problem 11.6 for the case of an integral-only controller with transfer function $G_c(s)=k_I/s$.

11.8 Repeat Problem 11.6 for the case of a real proportional-derivative controller with transfer function $G_c(s)=k_c(1+\tau_D s)/(1+\alpha\tau_D s)$.

11.9 Consider the closed-loop system shown in Figure P11.6 where the process is a first-order system, i.e.

$$G_p(s) = \frac{k_p}{\tau_p s + 1}$$

If the controller is an integral-only controller ($G_c(s)= k_I/s$)

(a) determine the closed-loop transfer function
(b) determine the offset to a unit step change of the set point.

11.10 Repeat Problem 11.9 for the case of a real PD controller with transfer function $G_c(s)=k_c(1+\tau_D s)/(1+\alpha\tau_D s)$.

11.11 Consider the system of two tanks in series shown in Figure P11.11, where the volume of the liquid in each tank (V_1 and V_2) is constant. The volumetric flowrate of the feed stream is F (constant) and its temperature is $T_0(t)$ and varies significantly (disturbance). A heating coil has been installed in the first tank to control the temperature of its liquid content by adjusting the amount of heat exchanged between the heating medium and the liquid in the tank is $Q(t)$ (manipulated variable).

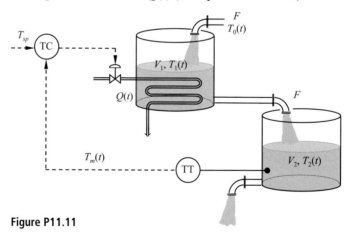

Figure P11.11

(a) Derive the transfer function between T_{sp} and T_2.
(b) Derive the transfer function between T_0 and T_2.
(c) If the controller is a P controller, calculate the offset for a unit step change in either T_{sp} or T_0.

12 Steady-State and Stability Analysis of the Closed-Loop System

The aim of this chapter is to introduce steady-state analysis of closed-loop systems as well as basic tools for stability analysis of closed-loop systems. The sensitivity function and its importance in determining the offset in a closed-loop system will be introduced and discussed. The Routh stability criterion is also introduced. This is a powerful criterion that can be used to determine conditions on the controller parameters for which closed-loop stability can be guaranteed. This chapter concludes the ideas and methodologies introduced in Chapters 10 and 11, with emphasis given in transfer-function representations. The chapter that follows will concentrate on methodologies that can be used in the state-space approach.

STUDY OBJECTIVES

After studying this chapter you should be able to do the following.

- Calculate the sensitivity function and the sensitivity coefficient of a closed-loop system.
- Find the controller parameter combinations for which closed-loop stability is guaranteed.
- Use software tools such as MATLAB or Maple to calculate the sensitivity coefficient.
- Use Maple to derive closed-loop stability conditions via the Routh criterion.

12.1 Steady-State Analysis of a Feedback Control System

Consider the general block-diagram representation of a feedback control loop shown in Figure 12.1. In the previous chapter it was shown that the closed-loop system has the following transfer-function description:

$$Y(s) = \frac{G(s)G_c(s)}{1+G(s)G_c(s)} Y_{sp}(s) + \frac{G'(s)}{1+G(s)G_c(s)} W(s) \qquad (12.1.1)$$

Figure 12.1 Block-diagram representation of a feedback control loop.

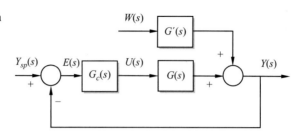

The error signal $E(s)$ can be calculated as follows

$$E(s) = Y_{sp}(s) - Y(s) = \frac{1}{1+G(s)G_c(s)} Y_{sp}(s) - \frac{G'(s)}{1+G(s)G_c(s)} W(s) \qquad (12.1.2)$$

This section is dedicated to steady-state analysis of the closed-loop system, under the assumption that the closed-loop transfer functions in Eqs. (12.1.1) and (12.1.2) are BIBO stable. This will be a standing assumption throughout the present section; in the next sections, we will see how to find the conditions that guarantee BIBO stability for the closed-loop system.

Suppose now that the set point y_{sp} and/or the disturbance w are changing, moving to new steady states. Because of the stability assumption, the output y and therefore the error e will reach new steady-state values. The final value of the error is called steady-state error or offset, and it is the main goal of this section to show how to calculate it.

12.1.1 Steady-State Error under P Control

Consider a feedback control system as shown in Figure 12.1, where all the variables are in deviation form, and suppose that a P controller, $G_c(s) = k_c$, is used. In what follows, we will calculate the offset or steady-state error for step changes in the set point y_{sp} and/or the disturbance w.

Set-point tracking or servo problem. Suppose the closed-loop system under consideration is initially at the reference steady state, with the output equal to the set point and all variables (in deviation from) equal to zero. If the set point undergoes a step change of size Δy_{sp}, i.e. $Y_{sp}(s) = \Delta y_{sp}/s$, and in the absence of disturbances, i.e. $W(s) = 0$, Eqs. (12.1.1) and (12.1.2) give

$$Y(s) = \frac{G(s)k_c}{1+G(s)k_c} \frac{\Delta y_{sp}}{s} \text{ and } E(s) = \frac{1}{1+G(s)k_c} \frac{\Delta y_{sp}}{s} \qquad (12.1.3)$$

Applying the final-value theorem, we conclude that

$$\lim_{t \to \infty}[y(t)] = \lim_{s \to 0}[sY(s)] = \lim_{s \to 0}\left[\frac{G(s)k_c}{1+G(s)k_c}\right]\Delta y_{sp} \qquad (12.1.4a)$$

and

$$\lim_{t \to \infty}[e(t)] = \lim_{s \to 0}[sE(s)] = \lim_{s \to 0}\left[\frac{1}{1+G(s)k_c}\right]\Delta y_{sp} \qquad (12.1.4b)$$

If the process transfer function $G(s)$ has finite static gain $k = G(0)$, then

$$\lim_{t\to\infty}[y(t)] = \frac{kk_c}{1+kk_c}\Delta y_{sp} \text{ and } \lim_{t\to\infty}[e(t)] = \frac{1}{1+kk_c}\Delta y_{sp} \quad (12.1.5)$$

So the conclusion is that, in the servo problem,

$$\text{Offset} = \frac{1}{1+kk_c} \cdot \left[\begin{array}{c}\text{magnitude of the step}\\ \text{change in the set point}\end{array}\right] \neq 0 \quad (12.1.6)$$

Disturbance rejection or regulator problem. Suppose the closed-loop system under consideration is initially at the reference steady state, with the output equal to the set point, and all variables (in deviation from) equal to zero. If the disturbance undergoes a step change of size Δw, i.e. $W(s) = \Delta w/s$, while the set point remains unchanged, i.e. $Y_{sp}(s) = 0$, Eqs. (12.1.1) and (12.1.2) give

$$Y(s) = \frac{G'(s)}{1+G(s)k_c}\frac{\Delta w}{s} \text{ and } E(s) = -\frac{G'(s)}{1+G(s)k_c}\frac{\Delta w}{s} \quad (12.1.7)$$

Applying the final-value theorem, we conclude that

$$\lim_{t\to\infty}[y(t)] = \lim_{s\to 0}[sY(s)] = \lim_{s\to 0}\left[\frac{G'(s)}{1+G(s)k_c}\right]\Delta w \quad (12.1.8a)$$

and

$$\lim_{t\to\infty}[e(t)] = \lim_{s\to 0}[sE(s)] = \lim_{s\to 0}\left[\frac{-G'(s)}{1+G(s)k_c}\right]\Delta w \quad (12.1.8b)$$

If the process transfer functions have finite static gains, $k = G(0)$ and $k' = G'(0)$, then

$$\lim_{t\to\infty}[y(t)] = \frac{1}{1+kk_c}\cdot k'\Delta w \text{ and } \lim_{t\to\infty}[e(t)] = \frac{1}{1+kk_c}\cdot(-k'\Delta w) \quad (12.1.9)$$

Note that, when the controller is "off" (i.e. when $G_c(s) = 0$, which implies $U(s) = 0$), the effect of the disturbance on the output is described by $Y(s) = G'(s)W(s)$, hence for a step change in w of size Δw, the new steady state of the output y will be $G'(0)\Delta w = k'\Delta w$, which corresponds to a steady-state error of $(-k'\Delta w)$.

So the conclusion is that, in the regulator problem,

$$\text{Offset} = \frac{1}{1+kk_c}\cdot\left[\begin{array}{c}\text{Offset in the}\\ \text{absence of control}\end{array}\right] \neq 0 \quad (12.1.10)$$

Combining the results of the servo and regulator problems, we found that the offset or steady-state error under P control is nonzero. It is a fraction of the set-point change (in the servo problem) or the offset in open loop (in the regulator problem), the coefficient being in both cases equal to $1/(1 + kk_c)$. To be able to have a decent performance of the P controller, we need $kk_c > 0$ and large.

It must be emphasized that, to be able to reach these general conclusions, at some point we have assumed the process has finite static gains. In the event that the process has a pole at 0, as in the tank system discussed in Example 11.2, then it will not have finite steady-state gains, therefore formulas (12.1.5) and (12.1.9) are not applicable; we will need to apply Eqs. (12.1.4) or (12.1.8) to the specific transfer functions, and calculate the limit.

12.1.2 Steady-State Error under PI Control

Suppose now that a PI controller, $G_c(s) = k_c(1+1/\tau_I s)$, is used. Then, substituting into (12.1.2), we can calculate the Laplace transform of the error

$$E(s) = \frac{1}{1+G(s)k_c\left(1+\frac{1}{\tau_I s}\right)} Y_{sp}(s) - \frac{G'(s)}{1+G(s)k_c\left(1+\frac{1}{\tau_I s}\right)} W(s)$$

$$= \frac{s}{s+G(s)k_c\left(s+\frac{1}{\tau_I}\right)} Y_{sp}(s) - \frac{sG'(s)}{s+G(s)k_c\left(s+\frac{1}{\tau_I}\right)} W(s) \qquad (12.1.11)$$

$$= \frac{s}{s+G(s)k_c\left(s+\frac{1}{\tau_I}\right)} \cdot \left(Y_{sp}(s) - G'(s)W(s)\right)$$

When the set point and/or the disturbance undergo a step change i.e. for $Y_{sp}(s) = \Delta y_{sp}/s$, $W(s) = \Delta w/s$, applying the final-value theorem, we can calculate the steady-state error:

$$\lim_{t\to\infty}[e(t)] = \lim_{s\to 0}[sE(s)] = \lim_{s\to 0}\left(\frac{s}{s+G(s)k_c\left(s+\frac{1}{\tau_I}\right)} \cdot \left(sY_{sp}(s) - G'(s)sW(s)\right)\right)$$

$$= \lim_{s\to 0}\left(\frac{s}{s+G(s)k_c\left(s+\frac{1}{\tau_I}\right)}\right) \cdot \left(\Delta y_{sp} - \lim_{s\to 0}[G'(s)]\cdot \Delta w\right)$$

As long as $\lim_{s\to 0}[G(s)] \neq 0$ and $\lim_{s\to 0}[G'(s)]$ is finite, the limit of the first factor is zero and the limit of the second factor is finite, hence the steady-state error is zero:

$$\text{Offset} = \lim_{t\to\infty}[e(t)] = 0 \qquad (12.1.12)$$

This is a very important conclusion. The PI controller is generally capable of controlling the system without permanent error, whereas with P action alone, the permanent error was inevitable.

12.1.3 General Steady-State Analysis of a Feedback Control System – Sensitivity Function and Sensitivity Coefficient

Consider again the feedback loop of Figure 12.1. The sensitivity function is defined as

$$\Sigma(s) = \frac{1}{1+G(s)G_c(s)} \qquad (12.1.13)$$

In terms of the sensitivity function, Eq. (12.1.2) can be expressed as

$$E(s) = \Sigma(s)\left[Y_{sp}(s) - G'(s)W(s)\right] \qquad (12.1.14)$$

The static gain of the sensitivity function

$$\Sigma_0 = \Sigma(0) = \lim_{s \to 0}\left[\frac{1}{1+G(s)G_c(s)}\right] \qquad (12.1.15)$$

will be called the sensitivity coefficient.

The sensitivity function is a very important function for predicting how a disturbance or set-point change affects the response of the error of the closed-loop system. The sensitivity coefficient directly relates to the steady-state error.

Suppose now that the closed-loop system is initially at the reference steady state, with the output equal to the set point, and all variables (in deviation from) equal to zero. When the set point and/or the disturbance undergo a step change i.e. $Y_{sp}(s) = \Delta y_{sp}/s$ and $W(s) = \Delta w/s$, applying the final-value theorem, we can calculate the steady-state error:

$$\begin{aligned}\lim_{t\to\infty}[e(t)] &= \lim_{s\to 0}[sE(s)] = \lim_{s\to 0}\left[\Sigma(s)\left(sY_{sp}(s) - G'(s)sW(s)\right)\right] \\ &= \lim_{s\to 0}[\Sigma(s)] \cdot \left(\lim_{s\to 0}[sY_{sp}(s)] - \lim_{s\to 0}[G'(s)] \cdot \lim_{s\to 0}[sW(s)]\right) \\ &= \Sigma_0\left(\Delta y_{sp} - \lim_{s\to 0}[G'(s)] \cdot \Delta w\right)\end{aligned} \qquad (12.1.16)$$

From Eq. (12.1.15), we see that in the absence of disturbances ($\Delta w = 0$),

$$\textit{Offset} = \Sigma_0 \cdot \begin{bmatrix}\text{magnitude of the step} \\ \text{change in the set point}\end{bmatrix} \qquad (12.1.17)$$

whereas for unchanged set point ($\Delta y_{sp} = 0$),

$$\textit{Offset} = \Sigma_0 \cdot \left(-\lim_{s\to 0}[G'(s)]\Delta w\right) = \Sigma_0 \cdot \begin{bmatrix}\textit{Offset}\ \text{in the} \\ \text{absence of control}\end{bmatrix} \qquad (12.1.18)$$

As long as $\lim_{s\to 0}[G'(s)]$ is finite, Eq. (12.1.16) implies that

$$\text{Offset} = 0 \quad \text{if and only if} \quad \Sigma_0 = 0 \qquad (12.1.19)$$

In P control of a process $G(s)$ with finite static gain $G(0) = k$, the sensitivity coefficient is $\Sigma_0 = 1/(1+kk_c)$ and we see that Eqs. (12.1.17) and (12.1.18) reproduce (12.1.6) and (12.1.10), which were derived previously specifically for P control.

In PI control of a process $G(s)$ with nonzero static gain, the sensitivity coefficient is

$$\Sigma_0 = \lim_{s\to 0}\frac{1}{1+G(s)k_c\left(1+\dfrac{1}{\tau_I s}\right)} = \lim_{s\to 0}\frac{s}{s+G(s)k_c\left(s+\dfrac{1}{\tau_I}\right)} = 0 \qquad (12.1.20)$$

and this is what leads to the vanishing steady-state error in Eq. (12.1.12).

The general steady-state error analysis performed in this subsection in terms of the sensitivity coefficient, enables quick conclusions for any controller $G_c(s)$, without having to repeat the procedure with the final-value theorem.

Example 12.1 Sensitivity function of a feedback-controlled first-order process

Consider the feedback control loop of Figure 12.2, where a first-order process

$$G(s) = \frac{k}{\tau s + 1}, \tau > 0$$

is controlled by a controller with transfer function $G_c(s)$.

(a) Calculate the sensitivity function and the sensitivity coefficient under P control. Also calculate the response of the error to a unit step change of the set point.
(b) Calculate the sensitivity function and the sensitivity coefficient under PI control.

Solution

(a) For P control and a first-order system we have that $G_c(s) = k_c$, therefore the sensitivity function is

$$\Sigma(s) = \frac{1}{1+G(s)G_c(s)} = \frac{1}{1+\dfrac{kk_c}{\tau s+1}} = \frac{\tau s+1}{\tau s+1+kk_c}$$

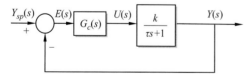

Figure 12.2 A closed-loop system consisting of a first-order system and a controller with transfer function $G_c(s)$.

and the sensitivity coefficient

$$\Sigma_0 = \lim_{s \to 0}[\Sigma(s)] = \frac{1}{1+kk_c}$$

To calculate the response of the error to a unit step change in the set point we must note that the sensitivity function is the transfer function between the set point and the error. Therefore the calculation proceeds in a standard manner, through partial fraction expansion and inversion of the Laplace transform:

$$E(s) = \Sigma(s)\frac{1}{s} = \frac{\tau s + 1}{\tau s + 1 + kk_c} \cdot \frac{1}{s} = \frac{1}{1+kk_c}\left(\frac{1}{s} + \frac{kk_c \tau}{\tau s + 1 + kk_c}\right)$$

leading to

$$e(t) = \frac{1}{1+kk_c}\left(1 + kk_c e^{-\left(\frac{1+kk_c}{\tau}\right)t}\right)$$

We observe that for $t = 0$, $e(0) = 1$, whereas as $t \to \infty$, $e(t) \to 1/(1 + kk_c) = \Sigma_0$. The qualitative shape of the response of the error response is shown in Figure 12.3. The error becomes equal to 1 at the moment the unit step change in the set point is applied, and approaches Σ_0 as time tends to infinity.

(b) For the case of PI control, $G_c(s) = k_c(1+1/\tau_I s)$, therefore the sensitivity function is

$$\Sigma(s) = \frac{1}{1+G(s)G_c(s)} = \frac{1}{1+\frac{k}{\tau s + 1}k_c\left(1+\frac{1}{\tau_I s}\right)} = \frac{s(\tau s + 1)}{\tau s^2 + (1+kk_c)s + \frac{kk_c}{\tau_I}}$$

and the sensitivity coefficient is

$$\Sigma_0 = \lim_{s \to 0}[\Sigma(s)] = \lim_{s \to 0}\left[\frac{s(\tau s + 1)}{\tau s^2 + (1+kk_c)s + \frac{kk_c}{\tau_I}}\right] = 0$$

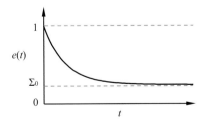

Figure 12.3 Qualitative shape of the response of the error to a unit step change in the set point, under P control.

It is the presence of integral action in the controller that has resulted in zero sensitivity coefficient and therefore zero offset, as expected.

12.1.4 Integral Action

We have seen in the previous subsections that the PI controller, which has integral action, achieves zero offset, as opposed to the P controller, which does not have integral action and results in nonzero offset. In what follows, we will see that this is indeed a general property of integral action.

We say that a controller with transfer function $G_c(s)$ has integral action if $G_c(s)$ has a pole at $s = 0$, i.e. $s = 0$ is a root of the denominator polynomial of the transfer function of the controller.

To investigate the effect of integral action on the steady-state error of the closed-loop system, we consider a process with a rational transfer function with nonzero static gain:

$$G(s) = \frac{N(s)}{D(s)}, \quad \text{with} \quad N(0) \neq 0 \tag{12.1.21}$$

where $N(s)$ is the numerator polynomial and $D(s)$ is the denominator polynomial. Any controller with integral action has a pole at $s = 0$ and it can, in a similar way, be written as a ratio of two polynomials as

$$G_c(s) = \frac{N_c(s)}{D_c(s)} = \frac{N_c(s)}{s \cdot D_c'(s)}, \quad \text{with} \quad N_c(0) \neq 0 \tag{12.1.22}$$

The sensitivity function, using Eqs. (12.1.21) and (12.1.22), can be written as

$$\Sigma(s) = \frac{1}{1 + G(s)G_c(s)} = \frac{s \cdot D(s)D_c'(s)}{s \cdot D(s)D_c'(s) + N(s)N_c(s)} \tag{12.1.23}$$

Since

$$\lim_{s \to 0}[N(s)N_c(s)] \neq 0 \text{ and } \lim_{s \to 0}[s \cdot D(s)D_c'(s)] = 0$$

it follows that the sensitivity coefficient

$$\Sigma_0 = \lim_{s \to 0}[\Sigma(s)] = \lim_{s \to 0}\left[\frac{s \cdot D(s)D_c'(s)}{s \cdot D(s)D_c'(s) + N(s)N_c(s)}\right] = 0 \tag{12.1.24}$$

Thus we have proved that, when the controller has integral action and the process has a nonzero static gain, then it always follows that the sensitivity coefficient $\Sigma_0 = 0$ and therefore from Eqs. (12.1.17) and (12.1.18), it also follows that the offset is zero.

12.2 Closed-Loop Stability, Characteristic Polynomial and Characteristic Equation

Consider again the general feedback control system shown in Figure 12.1. To be able to say that the closed-loop system is stable, *all* closed-loop transfer functions, for the output $Y(s)$ as well as the internal signals $E(s)$ and $U(s)$, must be BIBO stable. From the block diagram, we have that

$$E(s) = \frac{1}{1+G(s)G_c(s)} Y_{sp}(s) - \frac{G'(s)}{1+G(s)G_c(s)} W(s) \tag{12.2.1}$$

$$U(s) = \frac{G_c(s)}{1+G(s)G_c(s)} Y_{sp}(s) - \frac{G_c(s)G'(s)}{1+G(s)G_c(s)} W(s) \tag{12.2.2}$$

$$Y(s) = \frac{G(s)G_c(s)}{1+G(s)G_c(s)} Y_{sp}(s) + \frac{G'(s)}{1+G(s)G_c(s)} W(s) \tag{12.2.3}$$

To be able to see under what conditions they are all stable, it is convenient to represent all functions involved, $G(s)$, $G'(s)$ and $G_c(s)$ as ratios of polynomials:

$$G(s) = \frac{N(s)}{D(s)}, \quad G'(s) = \frac{N'(s)}{D(s)} \quad \text{and} \quad G_c(s) = \frac{N_c(s)}{D_c(s)} \tag{12.2.4}$$

where $G(s)$ and $G'(s)$ have been represented with the same denominator polynomial, with the understanding that they originate from the same state-space model that describes the process dynamics (with $D(s)$ being the characteristic polynomial of the A matrix of the state-space description of the process). Then Eqs. (12.2.1)–(12.2.3) can then be rewritten as

$$E(s) = \frac{D(s)D_c(s)}{D(s)D_c(s)+N(s)N_c(s)} Y_{sp}(s) - \frac{N'(s)D_c(s)}{D(s)D_c(s)+N(s)N_c(s)} W(s) \tag{12.2.5}$$

$$U(s) = \frac{D(s)N_c(s)}{D(s)D_c(s)+N(s)N_c(s)} Y_{sp}(s) - \frac{N'(s)N_c(s)}{D(s)D_c(s)+N(s)N_c(s)} W(s) \tag{12.2.6}$$

$$Y(s) = \frac{N(s)N_c(s)}{D(s)D_c(s)+N(s)N_c(s)} Y_{sp}(s) + \frac{N'(s)D_c(s)}{D(s)D_c(s)+N(s)N_c(s)} W(s) \tag{12.2.7}$$

We immediately observe that all transfer functions share a common denominator:

$$D(s)D_c(s) + N(s)N_c(s) \tag{12.2.8}$$

This is called the closed-loop characteristic polynomial of the feedback control loop. In order for the feedback control loop to be stable, all transfer functions appearing in Eqs. (12.2.5)–(12.2.7) must be stable, hence all roots of the characteristic polynomial must have negative real parts. The roots of the characteristic polynomial, which are called closed-loop poles, are the solutions of the following equation

$$D(s)D_c(s) + N(s)N_c(s) = 0 \qquad (12.2.9)$$

Equation (12.2.9) is called the closed-loop characteristic equation. The characteristic equation may also be written in the form

$$1 + G(s)G_c(s) = 0 \qquad (12.2.10)$$

as long as no cancellations of common factors have been made between $G(s)$ and $G_c(s)$.

It is important to stress at this point that the fact that a system is open-loop stable does not ensure that any closed-loop system formed using a controller is also stable. Inappropriate selection of the controller parameters can result in unstable closed-loop dynamics, as demonstrated in the example that follows.

Example 12.2 Closed-loop stability of the heating tank under PI temperature control

Consider the heating tank temperature-control loop shown in Figure 10.9. In Section 11.1, the closed-loop transfer functions were derived for a general controller with transfer function $G_c(s)$; they are given by Eq. (11.1.12) and in Figure 11.6 in block-diagram form. In Section 11.3, this example was studied under the assumption of negligible sensor dynamics ($\tau_m = 0$) and under P and PI control. Now consider the same example with $\tau_m \neq 0$, and a PI controller, so here

$$G(s) = \frac{k_p k_v}{(\tau_p s + 1)(\tau_m s + 1)}, \quad G'(s) = \frac{1}{(\tau_p s + 1)(\tau_m s + 1)}, \quad G_c(s) = k_c\left(1 + \frac{1}{\tau_I s}\right)$$

where $\tau_p > 0$, $\tau_m > 0$, $\tau_I > 0$. For the following numerical values of the process parameters $\tau_p = 1$, $\tau_m = 0.1$, $k_v = 0.2$ and $k_p = 1$, study the response of the closed-loop system to a unit step change in T_{sp} for the following cases of controller parameters: $k_c = 50$ and (a) $\tau_I = 1$, (b) $\tau_I = 0.1$ and (c) $\tau_I = 0.01$.

Solution

Applying Eq. (12.1.1), we obtain the transfer-function description of the closed-loop system:

$$\overline{T}_m(s) = \frac{\dfrac{k_p k_v k_c\left(1 + \dfrac{1}{\tau_I s}\right)}{(\tau_p s + 1)(\tau_m s + 1)}}{1 + \dfrac{k_p k_v k_c\left(1 + \dfrac{1}{\tau_I s}\right)}{(\tau_p s + 1)(\tau_m s + 1)}} \overline{T}_{sp}(s) + \frac{\dfrac{1}{(\tau_p s + 1)(\tau_m s + 1)}}{1 + \dfrac{k_p k_v k_c\left(1 + \dfrac{1}{\tau_I s}\right)}{(\tau_p s + 1)(\tau_m s + 1)}} \overline{T}_0(s)$$

Simplifying the expressions in the closed-loop transfer functions,

$$\bar{T}_m(s) = \frac{k_p k_v k_c s + k_p k_v \dfrac{k_c}{\tau_I}}{\tau_m \tau_p s^3 + (\tau_m + \tau_p)s^2 + (1 + k_p k_v k_c)s + \dfrac{k_p k_v k_c}{\tau_I}} \bar{T}_{sp}(s)$$

$$+ \frac{s}{\tau_m \tau_p s^3 + (\tau_m + \tau_p)s^2 + (1 + k_p k_v k_c)s + \dfrac{k_p k_v k_c}{\tau_I}} \bar{T}_0(s)$$

(12.2.11)

For change in the set point T_{sp} in the absence of disturbance ($\bar{T}_0(s) = 0$), the second term is zero.

(a) Substituting the given numerical values of the process parameters and the controller parameters $k_c = 50$ and $\tau_I = 1$ in the closed-loop transfer function, we obtain

$$\bar{T}_m(s) = \frac{s+1}{0.01s^3 + 0.11s^2 + 1.1s + 1} \bar{T}_{sp}(s)$$

The roots of the denominator polynomial are calculated numerically; they are equal to $-5 \pm 8.6603i$ and -1. All poles of the closed-loop system have negative real parts, therefore the closed-loop system is stable.

(b) Substituting the given numerical values of the process parameters and the controller parameters $k_c = 50$ and $\tau_I = 0.1$ in the closed-loop transfer function, we obtain

$$\bar{T}_m(s) = \frac{0.1s+1}{0.001s^3 + 0.011s^2 + 0.11s + 1} \bar{T}_{sp}(s)$$

The roots of the denominator polynomial are equal to $-0.5 \pm 9.9875i$ and -10. All poles of the closed-loop system have negative real parts, therefore the closed-loop system is again stable.

(c) Substitution of the controller parameters $k_c = 50$ and $\tau_I = 0.01$ in this case gives the following closed-loop transfer function

$$\bar{T}_m(s) = \frac{0.01s+1}{0.0001s^3 + 0.0011s^2 + 0.011s + 1} \bar{T}_{sp}(s)$$

The roots of the denominator polynomial are equal to $+6.4509 \pm 19.4104i$ and -23.9018. It follows that the system is unstable as the real part of the complex conjugate roots is positive.

The unit step response for the three cases is shown in Figure 12.4 from which it can be seen that the gradual decrease of the integral time drives the closed-loop system to instability.

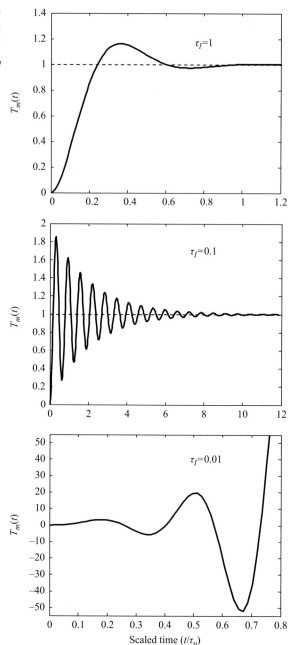

Figure 12.4 Unit step response to set point changes of the tank heating example under PI control for different values of τ_I ($k_c = 50$ in all cases).

An important conclusion follows directly from Example 12.2: even for an open-loop stable process, there can be combinations of controller parameters for which the closed-loop system is unstable. In the cases considered in Example 12.2, it is the decrease of the integral time that drives the system to instability. It is therefore important to be able to identify the combinations of the controller parameters that result in a stable closed-loop system. A simple method that can be used for this purpose is presented in the next section. This method is called the Routh–Hurwitz criterion, and the particular form that will be presented is the Routh criterion.

12.3 The Routh Criterion

The Routh criterion can be used to determine whether all roots of a given polynomial have negative real parts. The Routh criterion can achieve this without having to calculate the roots, and this is what makes it very powerful.

Suppose that we are given an nth-degree polynomial of the following general form:

$$a_0 s^n + a_1 s^{n-1} + a_2 s^{n-2} + \cdots + a_{n-1} s^1 + a_n \tag{12.3.1}$$

where $a_0 > 0$ (if $a_0 < 0$ we then multiply the polynomial by -1).

(i) *A necessary condition for all roots of the polynomial (12.3.1) to have negative real parts is that all the coefficients a_i are positive.*

With this criterion, we can draw the conclusion, for example, that the polynomials $(s^2 - s + 1)$ and $(s^2 + 1)$ have roots with nonnegative real parts, as the coefficient of the s^1 term is negative in the first case and zero in the second. If, however, the coefficients of the polynomial are all positive, we cannot conclude that all its roots have negative real parts.

(ii) *Necessary and sufficient conditions* for all the roots to have negative real parts can be derived using the Routh array.

	column 1	column 2	column 3	column 4	...
row 1	a_0	a_2	a_4	a_6	...
row 2	a_1	a_3	a_5	a_7	...
row 3	B_1	B_2	B_3	B_4	...
row 4	C_1	C_2	C_3	:	...
row 5	D_1	D_2	:	:	...
:	:	:	:	:	...
row n	:	:	:	:	...
row $n+1$	:	:	:	:	...

The first two rows of the Routh array contain the coefficients of the polynomial (12.3.1) with the order and at the places shown in the table. The rest of the entries of the first two rows are understood to be 0.

The elements of the third, fourth, ... rows are calculated from the following formulas

$$B_1 = \frac{a_1 a_2 - a_0 a_3}{a_1}, \quad B_2 = \frac{a_1 a_4 - a_0 a_5}{a_1}, \quad B_3 = \frac{a_1 a_6 - a_0 a_7}{a_1}, \ldots \tag{12.3.2}$$

$$C_1 = \frac{B_1 a_3 - a_1 B_2}{B_1}, \quad C_2 = \frac{B_1 a_5 - a_1 B_3}{B_1}, \ldots \tag{12.3.3}$$

$$D_1 = \frac{C_1 B_2 - B_1 C_2}{C_1}, \ldots \tag{12.3.4}$$

The reader must observe the repeated structure in forming the elements of the Routh array. The Routh array comprises $n+1$ rows, where n is the degree of the polynomial. (If the procedure were to continue, all subsequent rows would be all 0s.)

The necessary and sufficient condition for all roots to have negative real parts is that all elements of the first column of the Routh array are positive.

For $n = 3$, where the given polynomial is $a_0 s^3 + a_1 s^2 + a_2 s + a_3$, with $a_0 > 0$, the Routh array is as follows.

	column 1	column 2
row 1	a_0	a_2
row 2	a_1	a_3
row 3	$B_1 = \dfrac{a_1 a_2 - a_0 a_3}{a_1}$	0
row 4	$C_1 = \dfrac{B_1 a_3 - a_1 B_2}{B_1} = a_3$	0

Thus, for the case of a third-degree polynomial, the necessary and sufficient condition for all its roots to have negative real parts is the following set of inequalities:

$$a_1 > 0, \quad a_1 a_2 - a_0 a_3 > 0, \quad a_3 > 0 \tag{12.3.5}$$

For $n = 4$, where the given polynomial is $a_0 s^4 + a_1 s^3 + a_2 s^2 + a_3 s + a_4$, with $a_0 > 0$, the Routh array is as follows.

	column 1	column 2	column 3
row 1	a_0	a_2	a_4
row 2	a_1	a_3	0
row 3	$B_1 = \dfrac{a_1 a_2 - a_0 a_3}{a_1}$	$B_2 = a_4$	0
row 4	$C_1 = \dfrac{B_1 a_3 - a_1 B_2}{B_1} = \dfrac{a_1 a_2 a_3 - a_0 a_3^2 - a_4 a_1^2}{a_1 a_2 - a_0 a_3}$	0	0
row 5	$D_1 = \dfrac{C_1 B_2 - B_1 C_2}{C_1} = a_4$	0	0

Thus, for the case of a fourth-degree polynomial, the necessary and sufficient condition for all its roots to have negative real parts is the following set of inequalities:

$$a_1 > 0, \quad a_1 a_2 - a_0 a_3 > 0, \quad a_1 a_2 a_3 - a_0 a_3^2 - a_4 a_1^2 > 0, \quad a_4 > 0 \qquad (12.3.6)$$

Example 12.3 Closed-loop stability of the heating tank under PI control

Consider the heating tank temperature-control problem studied in Example 12.2. Use the Routh criterion to find the combinations of controller gain k_c and integral time τ_I for which the closed-loop system is stable. The values of the process parameters are $k_p = 1$, $\tau_p = 1$, $\tau_m = 0.1$ and $k_v = 0.2$.

Solution

From the closed-loop transfer functions (12.2.11) derived in Example 12.2, we see that their common denominator is the following polynomial:

$$\tau_m \tau_p s^3 + (\tau_m + \tau_p)s^2 + (1 + k_p k_v k_c)s + \frac{k_p k_v k_c}{\tau_I} \qquad (12.3.7)$$

For the specific numerical values of the process parameters,

$$a_0 = \tau_m \tau_p = 0.1, \; a_1 = \tau_m + \tau_p = 1.1, \; a_2 = 1 + k_p k_v k_c = 1 + 0.2 k_c, \; a_3 = \frac{k_p k_v k_c}{\tau_I} = 0.2 \frac{k_c}{\tau_I}$$

The Routh array is the following.

	column 1	column 2
row 1	$a_0 = 0.1$	$a_2 = 1 + 0.2 k_c$
row 2	$a_1 = 1.1$	$a_3 = 0.2 k_c / \tau_I$
row 3	$B_1 = \dfrac{a_1 a_2 - a_0 a_3}{a_1} = \dfrac{1.1(1 + 0.2 k_c) - 0.02 k_c / \tau_I}{1.1}$	0
row 4	$C_1 = \dfrac{B_1 a_3 - a_1 B_2}{B_1} = a_3 = 0.2 k_c / \tau_I$	0

For stability, the elements of the first column need to be positive, hence we must have $1.1 + 0.22 k_c > 0.02 k_c / \tau_I$ and $0.2 k_c / \tau_I > 0$, or equivalently,

$$55 + 11 k_c > \frac{k_c}{\tau_I} > 0 \qquad (12.3.8)$$

When $0 < 1/\tau_I < 11$, the above inequalities are satisfied for all $k_c > 0$, while when $1/\tau_I > 11$, the inequalities are equivalent to

$$0 < k_c < \frac{55}{\dfrac{1}{\tau_I} - 11} \tag{12.3.9}$$

These results are shown in Figure 12.5 for a wide range of the controller gain and the inverse integral time. For the case studied in Example 12.2 where $k_c = 50$, it follows that the system is at the verge of instability when $(1/\tau_I) = 12.1$ or $\tau_I = 0.0826$. For smaller values of the integral time, the system is unstable; for larger values it is stable.

Let's now solve this problem more generally, with symbols. The Routh array for the polynomial (12.3.7) is the following.

	column 1	column 2
row 1	$a_0 = \tau_m \tau_p$	$a_2 = 1 + k_p k_v k_c$
row 2	$a_1 = \tau_m + \tau_p$	$a_3 = \dfrac{k_p k_v k_c}{\tau_I}$
row 3	$B_1 = \dfrac{a_1 a_2 - a_0 a_3}{a_1} = \dfrac{(\tau_m + \tau_p)(1 + k_p k_v k_c) - \tau_m \tau_p \dfrac{k_p k_v k_c}{\tau_I}}{\tau_m + \tau_p}$	0
row 4	$C_1 = \dfrac{B_1 a_3 - a_1 B_2}{B_1} = a_3 = \dfrac{k_p k_v k_c}{\tau_I}$	0

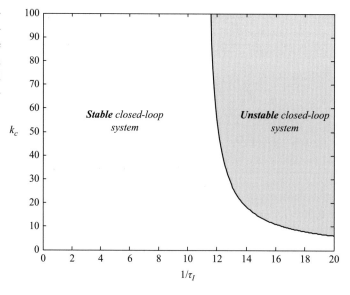

Figure 12.5 Pictorial representation of the conditions resulting from the application of the Routh criterion to the PI control of the heating tank example.

Because the time constants τ_p and τ_m are positive, the first two elements of the first column are already positive. From the third and the fourth, we obtain the conditions

$$(\tau_m + \tau_p)(1 + k_p k_v k_c) - \tau_m \tau_p \frac{k_p k_v k_c}{\tau_I} > 0$$

$$\frac{k_p k_v k_c}{\tau_I} > 0$$

which can be combined as

$$\left(\frac{1}{\tau_p} + \frac{1}{\tau_m}\right)(1 + k_p k_v k_c) > \frac{k_p k_v k_c}{\tau_I} > 0 \qquad (12.3.10)$$

When $0 < \dfrac{1}{\tau_I} < \dfrac{1}{\tau_p} + \dfrac{1}{\tau_m}$, the closed-loop system is stable for all k_c such that $k_p k_v k_c > 0$.

When $\dfrac{1}{\tau_I} > \dfrac{1}{\tau_p} + \dfrac{1}{\tau_m}$, the closed-loop system is stable if $0 < k_p k_v k_c < \dfrac{\dfrac{1}{\tau_p} + \dfrac{1}{\tau_m}}{\dfrac{1}{\tau_I} - \left(\dfrac{1}{\tau_p} + \dfrac{1}{\tau_m}\right)}$.

Example 12.4 Closed-loop stability of the heating tank with important sensor and valve dynamics under P control

Consider again the heating tank example, now with first-order dynamics in both the sensor and the valve, with time constants τ_m and τ_v, respectively. Suppose that a P controller is used. For what range of values of the controller gain k_c is the closed-loop system stable?

Solution

The transfer functions of the process, valve and sensor are the following

$$G_p(s) = \frac{k_p}{\tau_p s + 1}, \; G_v(s) = \frac{k_v}{\tau_v s + 1}, \; G_m(s) = \frac{1}{\tau_m s + 1}$$

hence

$$G(s) = \frac{N(s)}{D(s)} = \frac{k_v k_p}{(\tau_v s + 1)(\tau_p s + 1)(\tau_m s + 1)}$$

For a P controller we have that

$$G_c(s) = \frac{N_c(s)}{D_c(s)} = \frac{k_c}{1}$$

We substitute into Eq. (12.2.9) to obtain the characteristic equation of the closed-loop system

$$(\tau_v s+1)(\tau_p s+1)(\tau_m s+1) + k_v k_p k_c = 0$$

or

$$\tau_v \tau_p \tau_m s^3 + (\tau_v \tau_p + \tau_p \tau_m + \tau_m \tau_v)s^2 + (\tau_v + \tau_p + \tau_m)s + 1 + k_v k_p k_c = 0$$

This is a third-degree polynomial equation. Application of the Routh criterion results in the following necessary and sufficient conditions for stability:

$$a_0 = \tau_v \tau_m \tau_p > 0$$

$$a_1 = \tau_v \tau_p + \tau_p \tau_m + \tau_m \tau_v > 0$$

$$a_1 a_2 - a_0 a_3 = (\tau_v \tau_p + \tau_p \tau_m + \tau_m \tau_v) \cdot (\tau_v + \tau_p + \tau_m) - \tau_v \tau_m \tau_p (1 + k_v k_p k_c) > 0$$

$$a_3 = 1 + k_v k_p k_c > 0$$

The first two inequalities are satisfied because of the positivity of the time constants. Solving the last two for ($k_v k_p k_c$) leads to

$$-1 < k_v k_p k_c < \left(\frac{1}{\tau_v} + \frac{1}{\tau_p} + \frac{1}{\tau_m}\right) \cdot (\tau_v + \tau_p + \tau_m) - 1$$

Observe that when all time constants are equal, then the above condition simplifies to $-1 < k_v k_p k_c < 8$. For any other combination of time constants, the upper limit on $k_v k_p k_c$ is larger.

12.4 Calculating Stability Limits via the Substitution $s = i\omega$

The Routh array gives us a simple and useful criterion for performing stability analysis on a closed-loop system, to be able to evaluate the effect of controller parameters on closed-loop stability. The result of the Routh criterion is a set of inequalities on the controller parameters that are necessary and sufficient for stability of the closed-loop system. These are strict inequalities (they are > or <, *not* ≥ or ≤). One question that immediately arises is what happens at the limits of these inequalities, where the system is actually unstable, but it is at the verge of instability.

The answer is quite simple: the verge of instability corresponds to the presence of some root(s) with zero real part, with all other roots having negative real parts.

In fact, it is possible to use this property to actually calculate the stability limits, as an alternative to the Routh criterion.

Let's consider again Example 12.3, where the denominator polynomial is (12.3.7). At the limit of stability, there will be a root $s = \omega i$: a pure imaginary root or a root equal to zero. Therefore, at the limit of stability, it will hold that

$$-\tau_m \tau_p \omega^3 i - (\tau_m + \tau_p)\omega^2 + (1 + k_p k_v k_c)\omega i + \frac{k_p k_v k_c}{\tau_I} = 0, \text{ for some real number } \omega.$$

The above equation implies two equations, since both the real part and the imaginary part of the left-hand side must vanish:

- real part $= 0 \Rightarrow -(\tau_m + \tau_p)\omega^2 + \dfrac{k_p k_v k_c}{\tau_I} = 0$

- imaginary part $= 0 \Rightarrow -\tau_m \tau_p \omega^3 + (1 + k_p k_v k_c)\omega = 0$

From the imaginary-part equation, we conclude that either $\omega = 0$, or $\omega^2 = \dfrac{1 + k_p k_v k_c}{\tau_m \tau_p}$.

In the first case, the real-part equation gives $k_p k_v k_c = 0$, in the second case it gives

$$-\left(\frac{1}{\tau_p} + \frac{1}{\tau_m}\right)(1 + k_p k_v k_c) + \frac{k_p k_v k_c}{\tau_I} = 0.$$

For the specific values of parameters, we get $k_c = 0$ and $-11(1 + 0.2k_c) + \dfrac{0.2k_c}{\tau_I} = 0$, which are exactly the boundaries of the stability region of Figure 12.5. (The other boundary of $(1/\tau_I) = 0$ is due to the assumed positivity of τ_I.)

Generally speaking, the method of substitution $s = i\omega$ is not the best method for finding the stability limits. It can get tricky for high-order systems, as some of the calculated controller parameter relations may involve other roots with positive real parts, and therefore they may not represent stability limits. The Routh criterion is simple and rigorous, and it gives us complete information.

The main lesson to be learned in this section is that when the inequalities derived from the Routh array are replaced with equalities, this corresponds to the verge of instability, i.e. to the presence of a root with zero real part. In this case, there will be either sustained oscillations (in case of imaginary roots) or a linearly growing step response (in case of a simple root at 0) of the closed-loop system.

12.5 Some Remarks about the Role of Proportional, Integral and Derivative Actions

With what we have seen so far in the analysis of the present and the previous chapter, including the examples of P and PI control, we are now in a position to summarize our understanding on the role of proportional and integral actions.

We saw in the examples and in the general steady-state analysis, that proportional action alone cannot bring the output exactly to the set point: it gives permanent offset. The magnitude of the offset is strongly dependent on the sign and magnitude of the controller gain. The controller gain should be chosen to be of the same sign as the process gain, and it should be of large magnitude in order to have a small percent offset. However, using too-large controller gains, which would make the controller too aggressive, is not permitted by hardware limitations and possibly by stability limitations (as in Example 12.4). In some applications, e.g. in level control of storage tanks, the presence of offset may not be a significant issue, as long as it is not too large. In other applications, however, e.g. in temperature control or composition control in chemical reactors, tight control may be necessary, and the presence of offset might create product quality issues.

Integral action can eliminate the offset. However, we saw in the heater example that integral action has a definite effect on closed-loop stability. In Example 12.2 and Figure 12.4, we saw that increasing $1/\tau_I$ makes the response more oscillatory, and if it is increased too much, it can make the closed-loop system unstable. In Example 12.3, we derived the exact closed-loop stability conditions on k_c and τ_I, which were depicted in the parametric diagram of Figure 12.5. It is clear from this figure that large $1/\tau_I$ limits the range of controller gains that make the closed-loop system stable.

In most cases, one would expect to see a destabilizing effect from integral action, inducing oscillatory behavior, or even complete loss of stability. To be able to precisely assess the effect of integral action in a specific problem, a rigorous stability analysis will be needed.

So far, we have not considered derivative action. In what follows, we will return to the heater example, and redo the stability analysis when ideal derivative action is used, in addition to proportional and integral actions.

Example 12.5 Closed-loop stability of the heating tank under PID control

Consider again the heating tank temperature control problem studied in Examples 12.2 and 12.3, but now with an ideal PID controller:

$$G_c(s) = k_c \left(1 + \frac{1}{\tau_I s} + \tau_D s\right), \quad \tau_I > 0, \quad \tau_D \geq 0$$

For the numerical values of the process parameters $\tau_p = 1$, $\tau_m = 0.1$, $k_v = 0.2$ and $k_p = 1$, derive necessary and sufficient conditions for closed-loop stability. Draw the stability boundary in a k_c versus $1/\tau_I$ diagram, for the following values of the derivative time: $\tau_D = 0, 0.1, 0.25, 0.5, 1$. Compare the closed-loop responses to a unit step change in T_{sp} for $\tau_D = 0$ and 0.1.

Solution

Substituting the PID controller transfer function in (11.1.12), we obtain the closed-loop transfer functions:

12.5 Some Remarks about the Role of Proportional, Integral and Derivative Actions

$$\bar{T}_m(s) = \frac{k_p k_v k_c \left(s + \frac{1}{\tau_I} + \tau_D s^2\right)}{s(\tau_m s + 1)(\tau_p s + 1) + k_p k_v k_c \left(s + \frac{1}{\tau_I} + \tau_D s^2\right)} \bar{T}_{sp}(s)$$

$$+ \frac{s}{s(\tau_m s + 1)(\tau_p s + 1) + k_p k_v k_c \left(s + \frac{1}{\tau_I} + \tau_D s^2\right)} \bar{T}_0(s)$$

(12.5.1)

The common denominator is the polynomial

$$s(\tau_m s + 1)(\tau_p s + 1) + k_p k_v k_c \left(s + \frac{1}{\tau_I} + \tau_D s^2\right)$$

$$= \tau_m \tau_p s^3 + (\tau_m + \tau_p + k_p k_v k_c \tau_D) s^2 + (1 + k_p k_v k_c) s + \frac{k_p k_v k_c}{\tau_I}$$

(12.5.2)

For the specific numerical values of the process parameters, it becomes

$$0.1 s^3 + (1.1 + 0.2 k_c \tau_D) s^2 + (1 + 0.2 k_c) s + \frac{0.2 k_c}{\tau_I}$$

Its Routh array is the following:

	column 1	column 2
row 1	$a_0 = 0.1$	$a_2 = 1 + 0.2 k_c$
row 2	$a_1 = 1.1 + 0.2 k_c \tau_D$	$a_3 = 0.2 k_c / \tau_I$
row 3	$B_1 = \frac{a_1 a_2 - a_0 a_3}{a_1} = \frac{(1.1 + 0.2 k_c \tau_D)(1 + 0.2 k_c) - 0.02 k_c / \tau_I}{1.1 + 0.2 k_c \tau_D}$	0
row 4	$C_1 = \frac{B_1 a_3 - a_1 B_2}{B_1} = a_3 = 0.2 k_c / \tau_I$	0

and from the Routh criterion all elements of the first column have to be positive. Combining the resulting inequalities leads to

$$(11 + 2 k_c \tau_D)(5 + k_c) > \frac{k_c}{\tau_I} > 0$$

(12.5.3)

In case $\tau_D = 0$, where the controller becomes PI, inequalities (12.5.3) become (12.3.8), which were the result of the stability analysis under PI control. When $\tau_D > 0$, conditions (12.5.3) are weaker than (12.3.8): they will be satisfied by more pairs (k_c, τ_I).

Figure 12.6 depicts the stability boundary in a parametric k_c versus $1/\tau_I$ diagram, for various values of τ_D. The stability boundary for $\tau_D = 0$ is exactly the same as in Figure 12.5. As τ_D increases, the stability boundary shifts to the right, making the stability region larger and the instability region smaller.

Figure 12.7 compares the closed-loop response for $\tau_D = 0$ (PI control) and $\tau_D = 0.1$. We see that the presence of derivative action has made the response less oscillatory, with smaller overshoot.

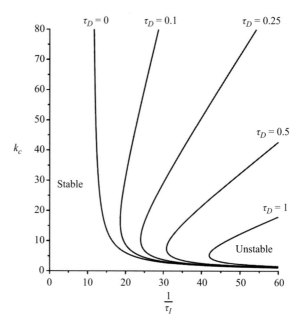

Figure 12.6 Stability boundaries for ideal PID control of the heating tank for different values of the derivative time τ_D.

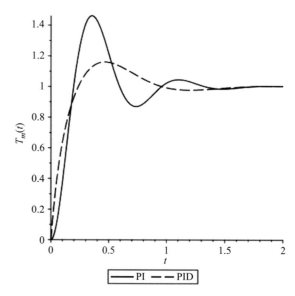

Figure 12.7 Heating tank response to a unit step change in the set point for PI control and PID control with $\tau_D = 0.1$. In both cases, $k_c = 50$ and $\tau_I = 0.25$.

From the previous example, we can make a very important observation: that derivative action has enlarged the stability region, and also its presence makes the response better damped. In most cases, one would expect to see a stabilizing effect from derivative action, inducing a higher level of damping for the same k_c and τ_I, and permitting us to use larger gains k_c and larger integral action $1/\tau_I$ without loss of stability. A rigorous stability analysis will be needed to be able to assess precisely the effect of derivative action in a specific problem.

To summarize our observations, we have seen (i) that P action alone is associated with the presence of offset, (ii) that I action eliminates offset at the expense of reducing damping and limiting stability and (iii) that D action increases damping and enlarges stability. These features are quite typical in practice and, of course, any conclusions regarding stability must be based on a rigorous stability analysis. In general, the question of finding the best "amounts" of P, I and D actions, to achieve the best possible controller performance for a given process, does not have an easy answer. In subsequent chapters, we will explore methods for optimizing controller performance.

12.6 Software Tools

12.6.1 Calculating the Sensitivity Function and the Sensitivity Coefficient in MATLAB

The sensitivity function is defined by Eq. (12.1.13) and refers to the closed-loop system shown in Figure 12.1. It can be seen by inspecting the definition that the calculation of the sensitivity function in MATLAB can be performed by using the `feedback` function introduced in Figure 11.16b with $G_1 = 1$ and $G_2 = G_c G$. This can also be seen by calculating the closed-loop transfer function of the configuration shown in Figure 12.8. The steps are therefore the following: we first form $G_c(s)G(s)$, which is the product of all transfer functions appearing between $E(s)$ and $Y(s)$ (including that of the controller), by using the `series` command and then we use the `feedback(1, `$G_c(s)G(s)$`)` to calculate the sensitivity transfer function.

To demonstrate the idea, consider the first-order system shown in Figure 12.2 that is controlled by a P controller, with process parameter values $k_p = 1$, $k_c = 1$ and $\tau_p = 2$ (see also Example 12.1). First we form the individual transfer functions

```
>> kp=1;
>> taup=2;
>> kc=1;
>> Gp=tf(kp,[taup 1])
Transfer function:
```

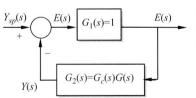

Figure 12.8 Block-diagram representation of the sensitivity transfer function.

```
      1
   -------
   2 s + 1
>> Gc=tf(kc,1)
Transfer function:
1
```

The overall transfer function of the elements of the feedback path that are connected in series can be calculated as follows

```
>> Gey=series(Gp,Gc)
Transfer function:
      1
   -------
   2 s + 1
```

The sensitivity function is then calculated using the `feedback` command

```
>> S=feedback(1,Gey)
Transfer function:
   2 s + 1
   ---------
   2 s + 2
```

To calculate the sensitivity coefficient we can use the `dcgain` command

```
>> dcgain(S)
ans =
     0.5000
```

We can also calculate the response of the error to a unit step change in the set point shown in Figure 12.9

```
>> step(S)
>> axis([0 6 0.3 1])
```

Figure 12.9 agrees well with the qualitative response of the error to a step change in the set point shown in Figure 12.3. We can follow the same approach to reproduce the results of Example 12.2. We first define the constants

```
>> kp=1;
>> kv=0.2;
>> taup=1;
>> taum=0.1;
>> kc=50;
>> tauI=1;
```

and then the transfer functions of the individual processes

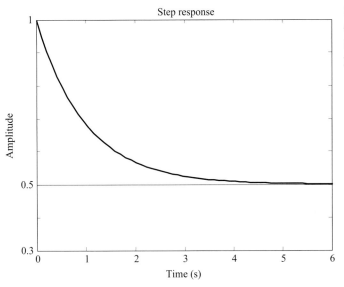

Figure 12.9 Response of the error to a unit step change in the set point under P control.

```
>> Gp=tf(kp,[taup 1])
Transfer function:
   1
-----
s + 1
>> Gv=tf(kv,1)
Transfer function:
0.2
>> Gm=tf(1,[taum 1])
Transfer function:
     1
---------
0.1 s + 1
>> Gc=tf(kc*[tauI 1],[tauI 0])
Transfer function:
 50 s + 50
-----------
     s
```

We now form the open-loop and finally the closed-loop transfer functions

```
>> GOL=series(Gm,series(Gp,Gv))
Transfer function:
         0.2
---------------------
0.1 s^2 + 1.1 s + 1
```

```
>> G=series(Gc,GOL)
Transfer function:
      10 s + 10
---------------------
0.1 s^3 + 1.1 s^2 + s
>> GCL=feedback(G,1)
Transfer function:
         10 s + 10
-------------------------------
0.1 s^3 + 1.1 s^2 + 11 s + 10
```

Figure 12.4 can be reproduced using the step command

```
>> step(GCL)
```

To calculate the sensitivity function we can use the `feedback` function

```
>> S=feedback(1,G)
Transfer function:
    0.1 s^3 + 1.1 s^2 + s
-------------------------------
0.1 s^3 + 1.1 s^2 + 11 s + 10
>> dcgain(S)
ans =
     0
>> step(S)
```

The unit step response of the error is shown in Figure 12.10.

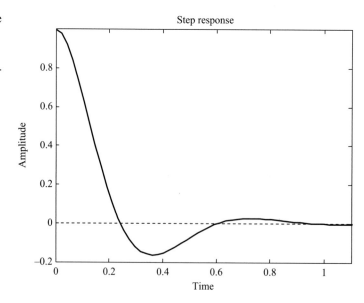

Figure 12.10 Response of the error to a unit step change in the set point under PI control.

12.6.2 Sensitivity Function, Sensitivity Coefficient and Stability Analysis with Maple

The sensitivity function, defined by Eq. (12.1.13), can be computed symbolically in Maple, as can the sensitivity coefficient by simply substituting $s = 0$ to the sensitivity function. Referring again to Example 12.2 as in the previous subsection, calculation for P control ($G_c(s) = k_c$) goes as follows.

```
> G:=(1/5)/((s+1)*((s/10)+1)):
> Gc:=kc:
> Sigma:=simplify(1/(1+G*Gc));
```

$$\Sigma := \frac{(s+1)(s+10)}{s^2 + 11s + 10 + 2kc}$$

```
> Sigma0:=subs(s=0,Sigma);
```

$$\Sigma 0 := \frac{5}{5 + kc}$$

Repeating the same calculation for a PI controller `Gc:=kc*(1+1/(tauI*s))`, gives the sensitivity function

$$\Sigma := \frac{(s+1)(s+10)s}{s^3 + 11s^2 + 10s + 2kcs + \frac{2kc}{taul}}$$

and, as expected,

$$\Sigma 0 := 0$$

The sensitivity function can be used to calculate the response of the error to changes in the set point. For example, a unit step change in the set point and controller parameter values $k_c = 50$ and $\tau_I = 1$,

```
> with(inttrans):>
  Error_StResp:=invlaplace(eval(Sigma/s,[kc=50,tauI=1]),s,t);
```

$$Error\_StResp := \frac{1}{3}e^{-5t}\left(3\cos\left(5\sqrt{3}\,t\right) + \sqrt{3}\sin\left(5\sqrt{3}\,t\right)\right)$$

and, plotting the above result,

```
> plot(Error_StResp,t=0..1.1);
```

the plot of Figure 12.10 is obtained.

To perform stability analysis, Maple has the special command `RouthTable` in the `DynamicSystems` package. In the example under consideration, under PI control, where the characteristic polynomial is

```
> Charact_Poly:=s^3/10+11*s^2/10+(1+kc/5)*s+(1/5)*(kc/ tauI);
```

$$Charact\_Poly := \frac{1}{10}s^3 + \frac{11}{10}s^2 + \left(1 + \frac{1}{5}kc\right)s + \frac{1}{5}\frac{kc}{taul}$$

the Routh table and the stability conditions are obtained as follows:

```
> with(DynamicSystems):
> RouthTable(Charact_Poly,s);
```

$$\begin{vmatrix} \dfrac{1}{10} & 1+\dfrac{1}{5}kc & s^3 \\ \dfrac{11}{10} & \dfrac{1}{5}\dfrac{kc}{taul} & s^2 \\ 1+\dfrac{1}{5}kc-\dfrac{1}{55}\dfrac{kc}{taul} & 0 & s \\ \dfrac{1}{5}\dfrac{kc}{taul} & 0 & 1 \end{vmatrix}$$

```
> RouthTable(Charact_Poly,s,'stablecondition'=true);
```

$$0 < \frac{kc}{taul} \quad \text{and} \quad 0 < 55 + 11 - \frac{kc}{taul}$$

which are equivalent to the results of Example 12.3.

LEARNING SUMMARY

- The sensitivity function of the closed-loop system shown in Figure 12.1 is defined as

$$\Sigma(s) = \frac{1}{1+G(s)G_c(s)}$$

  The sensitivity coefficient $\Sigma_0 = \Sigma(0)$ is the static gain of the sensitivity function.
- The offset for step changes in the set point or the disturbance is proportional to the sensitivity coefficient.
- Integral action results in zero sensitivity coefficient, therefore zero offset.
- Routh's stability criterion can be applied directly to the characteristic polynomial of the closed-loop system (common denominator of the closed-loop transfer functions) to investigate stability. The necessary and sufficient condition for stability is that all elements of the first column of the Routh array are positive.
- For third-order systems with denominator polynomial $a_0 s^3 + a_1 s^2 + a_2 s + a_3$, where $a_0 > 0$, the stability conditions are: $a_1 > 0$, $a_1 a_2 - a_0 a_3 > 0$, $a_3 > 0$.
- For fourth-order systems with denominator polynomial $a_0 s^4 + a_1 s^3 + a_2 s^2 + a_3 s + a_4$,

where $a_0 > 0$, the stability conditions are:

$$a_1 > 0, \; a_1 a_2 - a_0 a_3 > 0, \; a_1 a_2 a_3 - a_0 a_3^2 - a_4 a_1^2 > 0, \; a_4 > 0.$$

TERMS AND CONCEPTS

Characteristic polynomial. The characteristic polynomial of a closed-loop system is the polynomial in s that appears as common denominator in all closed-loop transfer functions.
Offset. The steady-state error under a step change of the set point and/or the disturbance.
Routh's criterion. A simple algorithm to determine whether all the roots of a polynomial have negative real parts.
Sensitivity coefficient. The value of the sensitivity function at $s = 0$.
Sensitivity function. The transfer function relating the set point and the error signal in a feedback control loop.

FURTHER READING

A comprehensive discussion of the sensitivity function and further examples of its calculation can be found in
Ogata, K., *Modern Control Engineering, Pearson International Edition*, 5th edn, 2008.
Rohrs, C. E., Melsa, J. L. and Schultz, D. G., *Linear Control Systems*, McGraw Hill, 1993.
  Further examples of the Routh criterion can be found in
Smith, C. A. and Corripio, A. B., *Principles and Practice of Automatic Process Control*, 2nd edn. New Jersey: John Wiley & Sons, Inc., 1997.
Stephanopoulos, G., *Chemical Process Control, An Introduction to Theory and Practice*. New Jersey: Prentice Hall, 1984.

PROBLEMS

12.1 Consider the two-tank level-control system of Problem 10.1, where the tanks are identical ($A_1 = A_2 = 1$, $R_1 = R_2 = 1$), the level sensor exhibits first-order dynamics with unity static gain and time constant $\tau_m = 0.1$ and the valve has negligible time constant and static gain $k_v = 1$. If the controller is a P controller, determine the offset as a function of the controller gain for a step change in the set point and the value of the controller gain for which the system will exhibit sustained oscillations.

12.2 Consider again Problem 12.1. If a PI controller is used, derive necessary and sufficient conditions on the controller gain and the integral time under which the closed-loop system is stable.

12.3 Consider a two-tank level control system like the one studied in Problem 10.1, but with interacting tanks, with $A_1 = 1$, $A_2 = 2$, $R_1 = 2$, $R_2 = 1$. The level sensor exhibits first-order dynamics with unity static gain and time constant $\tau_m = 0.1$ and the valve has negligible time constant and static gain $k_v = 1$. Determine the offset as a function of the gain of a P controller, for a step change in the set point. What is the minimum possible offset?

12.4 Consider the two interacting tanks control system of Problem 12.3 and a PID controller. Determine the conditions under which the closed-loop system is stable.

12.5 Consider the level-control system of Problem 12.1, but with a pump in the outflow of the second tank, delivering constant flowrate, and a P controller. Calculate the offset as a function of the controller gain (i) for a step change in the set point and (ii) for a step disturbance. Determine the range of values of the controller gain for which the closed-loop system is stable.

12.6 Repeat Problem 12.1 for the case of a PD controller with $\tau_D = 0.2$. Does the addition of derivative action have any effect on the offset?

12.7 Consider the closed-loop system shown in Figure P12.7, where the controller is a P controller and $k_p = 1$, $\tau_p = 1$. What is the range of values of the controller gain for which the closed-loop system is stable?

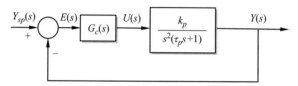

Figure P12.7

12.8 Repeat Problem 12.7 for the case of an ideal PD controller with transfer function

$$G_{PD} = k_c\left(1 + \tau_D s\right)$$

12.9 Consider the closed-loop system shown in Figure P12.9 where $\tau_1 = 3$, $\tau_2 = 2$ and $k_p = 1$. Derive necessary and sufficient conditions for closed-loop stability under the following types of controllers: (a) P controller, (b) ideal PD controller and (c) ideal PID controller. What is the effect of the derivative action? For those controllers that can guarantee stability, calculate the offset for a unit step change in the set point.

12.10 For the closed-loop system shown in Figure P12.10, find the range of values of the gain of a P controller for closed-loop stability. What is the offset as a function of the controller gain? Repeat the stability analysis for the case of a PI controller.

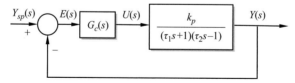

Figure P12.9

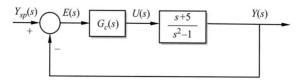

Figure P12.10

12.11 For the system shown in Figure P12.11, find the range of values of a P controller gain for which the closed-loop system is stable. Repeat the stability analysis for an ideal PD controller with $\tau_D = 1/6$.

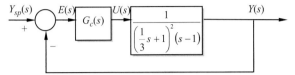

Figure P12.11

12.12 Process fluid is heated in a series of three stirred tank heaters of the same size.

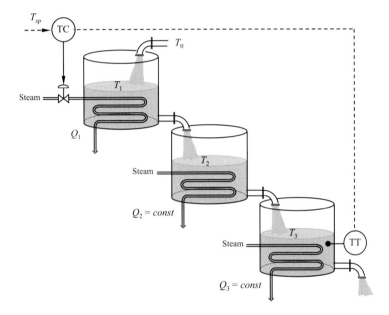

The outlet temperature is measured and controlled to set point by manipulating the heating rate of the first heater. The heating rates of the second and third heaters are constant.

Assuming very fast sensor and actuator dynamics, the control system is represented in block-diagram form as follows.

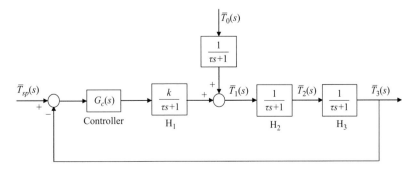

where:

- $\tau > 0$ are the time constants of the heaters (all equal)
- $k > 0$ is the steady-state gain between heating rate and temperature in each heater.

  (a) For the following types of controllers:
  - P controller
  - ideal PD controller, with derivative time $\tau_D = \tau/3$

    calculate the range of values of the controller gain that guarantee closed-loop stability.

    What is the effect of derivative action on closed-loop stability?

  (b) If the controller is P only, the set point is the same at all times, and the feed temperature T_0 undergoes a step change of size M, calculate the offset as a function of the controller gain. What is the smallest feasible offset?

12.13 In an isothermal CSTR, the concentration of the product is measured and controlled to set point by adjusting the flowrate of the feed stream, in the presence of disturbances. The control system is represented in block-diagram form as follows.

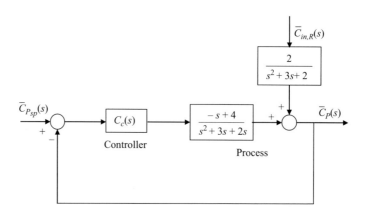

(a) For the following types of controllers:
- P controller
- PI controller, with integral time $\tau_I = 1/4$

calculate the range of values of the controller gain k_c that guarantee closed-loop stability.

What is the effect of integral action on closed-loop stability?

(b) If the controller is P only, and the set point undergoes a step change of size M, in the absence of disturbances, calculate the offset as a function of k_c.

What is the smallest feasible offset?

13 State-Space Description and Analysis of the Closed-Loop System

The aim of this chapter is to introduce the description of closed-loop systems in the time domain in terms of differential equations (state-space description). In addition, steady-state analysis and stability analysis of the closed-loop system using the state-space description will be presented. One of the advantages of the state-space description is that it retains the physical variables of the system as state variables enhancing clarity of presentation and physical understanding. The present chapter is parallel to the previous two chapters as it involves the same study tasks, but in the time domain instead of the Laplace domain.

STUDY OBJECTIVES

After studying this chapter you should be able to do the following.

- Describe the closed-loop system in state space, in terms of a system of first-order differential equations.
- Perform steady-state analysis of the closed-loop system using the state-space description and, in particular, calculate the offset under P control.
- Perform stability analysis of the closed-loop system using the state-space description.
- Use software tools, like MATLAB or Maple, to simulate and analyze the closed-loop system in state space.

13.1 State-Space Description and Analysis of the Heating Tank

13.1.1 Heating Tank under Proportional Control

Consider the steam-heated tank example studied in Chapter 10 that is shown in Figure 10.1, and suppose that we are using a P controller to vary the steam flowrate in order to control the temperature of the liquid content of the tank

$$u = u_b + k_c \left(T_{sp} - T_m \right) \tag{13.1.1}$$

The dynamics of the physical components of the control system have been presented in Chapter 10 (see Eqs. (10.3.1), (10.3.2), (10.3.6a) and (10.3.7a)):

control valve

$$\dot{M}_{st} = k_v u \tag{13.1.2}$$

heated tank

$$\frac{dT}{dt} = \frac{F}{V}(T_0 - T) + \frac{\Delta h_{vap,st}}{V \rho c_p} \dot{M}_{st} \tag{13.1.3}$$

temperature sensor

$$\tau_m \frac{dT_m}{dt} + T_m = T \tag{13.1.4}$$

The above set of equations (13.1.1)–(13.1.4) describe each component of the control system separately. Equation (13.1.1) is the controller, whereas (13.1.2)–(13.1.4) describe the instrumented process (process plus sensor and valve) or open-loop system. After substituting (13.1.2) into (13.1.3) and rearranging Eq. (13.1.4), we obtain the open-loop system dynamics, which is governed by

$$\frac{dT}{dt} = \frac{F}{V}(T_0 - T) + \frac{\Delta h_{vap,st}}{V \rho c_p} k_v u \tag{13.1.5a}$$

$$\frac{dT_m}{dt} = \frac{1}{\tau_m}(T - T_m) \tag{13.1.5b}$$

The inputs are the control signal u and the disturbance T_0. The output of the open-loop system is the measured temperature T_m of the tank:

$$y = T_m \tag{13.1.5c}$$

In matrix form, the open-loop system can be written as

$$\frac{d}{dt}\begin{bmatrix} T \\ T_m \end{bmatrix} = \begin{bmatrix} -\frac{F}{V} & 0 \\ \frac{1}{\tau_m} & -\frac{1}{\tau_m} \end{bmatrix} \begin{bmatrix} T \\ T_m \end{bmatrix} + \begin{bmatrix} \frac{\Delta h_{vap,st}}{V \rho c_p} k_v \\ 0 \end{bmatrix} u + \begin{bmatrix} \frac{F}{V} \\ 0 \end{bmatrix} T_0 \tag{13.1.6a}$$

$$y = \begin{bmatrix} 0 & 1 \end{bmatrix} \begin{bmatrix} T \\ T_m \end{bmatrix} \tag{13.1.6b}$$

To obtain the state-space description of the closed-loop (CL) system, we need to include the controller equation(s) into the mathematical description. Because the P controller has no dynamic states and is described by a simple algebraic equation (13.1.1), a simple substitution is only needed. The dynamics of the closed-loop system is governed by two differential equations: Eq. (13.1.5a) (with (13.1.1) substituted in) and Eq. (13.1.5b):

$$\frac{dT}{dt} = \frac{F}{V}(T_0 - T) + \frac{\Delta \hat{h}_{vap,st}}{V \rho c_p} k_v u_b + \frac{\Delta \hat{h}_{vap,st}}{V \rho c_p} k_v k_c (T_{sp} - T_m) \quad (13.1.7a)$$

$$\frac{dT_m}{dt} = \frac{1}{\tau_m}(T - T_m) \quad (13.1.7b)$$

The inputs to the closed-loop system are the set-point temperature T_{sp} and the disturbance T_0. The output of the closed-loop system could be the temperature measurement T_m or the actual temperature T or the error $e = T_{sp} - T_m$, etc., depending on what we wish to calculate.

The closed-loop system is expressed in terms of the physical variables of the instrumented process, with each of the differential equations having its physical significance. The simulation of the closed-loop system can be performed numerically with a differential equation solver.

For analysis purposes, it is convenient to express the physical variables in deviation form, relative to a reference steady state:

$$\frac{d\bar{T}}{dt} = \frac{F}{V}(\bar{T}_0 - \bar{T}) + \frac{\Delta \hat{h}_{vap,st}}{V \rho c_p} k_v k_c (\bar{T}_{sp} - \bar{T}_m) \quad (13.1.8a)$$

$$\frac{d\bar{T}_m}{dt} = \frac{1}{\tau_m}(\bar{T} - \bar{T}_m) \quad (13.1.8b)$$

where the overbar is used to denote deviation variable. Note that, after the introduction of deviation variables, the constant term arising from the controller bias drops out. It is often useful to express the closed-loop dynamics in matrix form. Assuming that T_m is considered to be the output of the closed-loop system, the matrix form of the closed-loop system is

$$\frac{d}{dt}\begin{bmatrix} \bar{T} \\ \bar{T}_m \end{bmatrix} = \begin{bmatrix} -\dfrac{F}{V} & -\dfrac{\Delta \hat{h}_{vap,st}}{V \rho c_p} k_v k_c \\ \dfrac{1}{\tau_m} & -\dfrac{1}{\tau_m} \end{bmatrix} \begin{bmatrix} \bar{T} \\ \bar{T}_m \end{bmatrix} + \begin{bmatrix} \dfrac{\Delta \hat{h}_{vap,st}}{V \rho c_p} k_v k_c \\ 0 \end{bmatrix} \bar{T}_{sp} + \begin{bmatrix} \dfrac{F}{V} \\ 0 \end{bmatrix} \bar{T}_0 \quad (13.1.9a)$$

$$y = \begin{bmatrix} 0 & 1 \end{bmatrix} \begin{bmatrix} T \\ T_m \end{bmatrix} \quad (13.1.9b)$$

In the previous two chapters we studied how to calculate the offset and how to derive closed-loop stability conditions when the control system is defined in the Laplace domain in terms of transfer functions. In what follows we will demonstrate how to do the same analysis using the state-space description.

Steady-state analysis. When the closed-loop system is at steady state, Eqs. (13.1.8a) and (13.1.8b) give:

$$0 = \frac{F}{V}\left(\bar{T}_{0,s} - \bar{T}_s\right) + \frac{\Delta h_{vap,st}}{V\rho c_p} k_v k_c \left(\bar{T}_{sp,s} - \bar{T}_{m,s}\right) \qquad (13.1.10a)$$

$$0 = \bar{T}_s - \bar{T}_{m,s} \qquad (13.1.10b)$$

from which, solving for $\bar{T}_s$ and $\bar{T}_{m,s}$, we obtain

$$\bar{T}_s = \bar{T}_{m,s} = \frac{\bar{T}_{0,s} + \dfrac{\Delta h_{vap,st}}{F\rho c_p} k_v k_c \bar{T}_{sp,s}}{1 + \dfrac{\Delta h_{vap,st}}{F\rho c_p} k_v k_c} \qquad (13.1.11)$$

and therefore the steady-state error is given by

$$e_s = \bar{T}_{sp,s} - \bar{T}_{m,s} = \frac{1}{1 + \dfrac{\Delta h_{vap,st}}{F\rho c_p} k_v k_c}\left(\bar{T}_{sp,s} - \bar{T}_{0,s}\right) \qquad (13.1.12)$$

Assume that the system is initially at the reference steady state, with all deviation variables being equal to zero. Then, the set point undergoes a step change of magnitude ΔT_{sp} while the disturbance input remains unchanged. Their new steady-state values in deviation form are $\bar{T}_{sp,s} = \Delta T_{sp}$, $\bar{T}_{0,s} = 0$, and Eq. (13.1.12) gives:

$$\text{Offset} = e_s = \bar{T}_{sp,s} - \bar{T}_{m,s} = \frac{1}{1 + \dfrac{\Delta h_{vap,st}}{F\rho c_p} k_v k_c} \Delta T_{sp} \qquad (13.1.13)$$

On the other hand, when the disturbance input undergoes a step change of magnitude ΔT_0, while the set point remains unchanged, which means that their new steady-state values in deviation form are $\bar{T}_{sp,s} = 0$, $\bar{T}_{0,s} = \Delta T_0$, Eq. (13.1.2) gives:

$$\text{Offset} = e_s = \bar{T}_{sp,s} - \bar{T}_{m,s} = -\frac{1}{1 + \dfrac{\Delta h_{vap,st}}{F\rho c_p} k_v k_c} \Delta T_0 \qquad (13.1.14)$$

The above results for the offset are the same as the results of Chapter 11 given by Eqs. (11.3.15) and (11.3.20).

Eigenvalue analysis – Asymptotic stability analysis. Calculating the offset is meaningful only if the closed-loop system is asymptotically stable. Stability of a closed-loop system, and in general the qualitative characteristics of the response, depend upon the eigenvalues of the system matrix. In the problem under consideration, we see from Eqs. (13.1.9a, b) that the system matrix of the closed-loop system is

$$A_{CL} = \begin{bmatrix} -\dfrac{F}{V} & -\dfrac{\Delta \hat{h}_{vap,st}}{V\rho c_p}k_v k_c \\ \dfrac{1}{\tau_m} & -\dfrac{1}{\tau_m} \end{bmatrix} \quad (13.1.15)$$

Notice that A_{CL} depends upon the controller gain k_c and therefore the eigenvalues of the closed-loop system depend on the controller gain k_c. Calculating the characteristic polynomial of A_{CL}, we find:

$$\det(\lambda I - A_{CL}) = \lambda^2 + \left(\dfrac{F}{V} + \dfrac{1}{\tau_m}\right)\lambda + \dfrac{1}{\tau_m}\cdot\left(\dfrac{F\rho c_p + \Delta \hat{h}_{vap,st} k_v k_c}{V\rho c_p}\right) = 0 \quad (13.1.16)$$

Because $V/F > 0$ and $\tau_m > 0$, the roots of (13.1.16) will both have negative real parts, hence the closed-loop system will be stable if and only if the following is satisfied:

$$\dfrac{F\rho c_p + \Delta \hat{h}_{vap,st} k_v k_c}{V\rho c_p} > 0 \quad \Leftrightarrow \quad 1 + \dfrac{\Delta \hat{h}_{vap,st}}{F\rho c_p}k_v k_c > 0 \quad (13.1.17)$$

13.1.2 Heating Tank under PI Control

We consider again the steam-heated tank where the temperature of the liquid content is controlled using a PI controller. The state-space description of the instrumented process or open-loop system is again given by Eqs. (13.1.5a–c). The state-space description of a PI controller has been presented in Chapter 10 and is given by the following equations (see Table 10.1):

$$\dfrac{de_I}{dt} = T_{sp} - T_m \quad (13.1.18a)$$

$$u = u_b + \dfrac{k_c}{\tau_I}e_I + k_c(T_{sp} - T_m) \quad (13.1.18b)$$

To obtain the state-space description of the closed-loop system, we need to combine the equations of the PI controller with the open-loop system equations, (13.1.5a–c). Notice that the control signal u can be eliminated by substituting (13.1.18b) into (13.1.5a) to obtain the following three differential equations:

$$\dfrac{de_I}{dt} = T_{sp} - T_m \quad (13.1.19a)$$

$$\dfrac{dT}{dt} = \dfrac{F}{V}(T_0 - T) + \dfrac{\Delta \hat{h}_{vap,st}}{V\rho c_p}k_v\left[u_b + \dfrac{k_c}{\tau_I}e_I + k_c(T_{sp} - T_m)\right] \quad (13.1.19b)$$

$$\dfrac{dT_m}{dt} = \dfrac{1}{\tau_m}(T - T_m) \quad (13.1.19c)$$

The first differential equation comes from the controller, whereas the other two come from the process. The closed-loop system is of order 3, which is the sum of the orders of controller and process. The above equations can be written in deviation variable form

$$\frac{d\bar{e}_I}{dt} = \bar{T}_{sp} - \bar{T}_m \tag{13.1.20a}$$

$$\frac{d\bar{T}}{dt} = \frac{F}{V}(\bar{T}_0 - \bar{T}) + \frac{\Delta h_{vap,st}}{V\rho c_p} k_v \left[\frac{k_c}{\tau_I} \bar{e}_I + k_c(\bar{T}_{sp} - \bar{T}_m) \right] \tag{13.1.20b}$$

$$\frac{d\bar{T}_m}{dt} = \frac{1}{\tau_m}(\bar{T} - \bar{T}_m) \tag{13.1.20c}$$

and can also be put in matrix form

$$\frac{d}{dt}\begin{bmatrix} \bar{e}_I \\ \bar{T} \\ \bar{T}_m \end{bmatrix} = \begin{bmatrix} 0 & 0 & -1 \\ \frac{\Delta h_{vap,st}}{V\rho c_p} k_v \frac{k_c}{\tau_I} & -\frac{F}{V} & -\frac{\Delta h_{vap,st}}{V\rho c_p} k_v k_c \\ 0 & \frac{1}{\tau_m} & -\frac{1}{\tau_m} \end{bmatrix} \begin{bmatrix} \bar{e}_I \\ \bar{T} \\ \bar{T}_m \end{bmatrix} + \begin{bmatrix} 1 \\ \frac{\Delta h_{vap,st}}{V\rho c_p} k_v k_c \\ 0 \end{bmatrix} \bar{T}_{sp} + \begin{bmatrix} 0 \\ \frac{F}{V} \\ 0 \end{bmatrix} \bar{T}_0 \tag{13.1.21}$$

Steady-state analysis. When the closed-loop system is at steady state, Eqs. (13.1.20a–c) give:

$$0 = \bar{T}_{sp,s} - \bar{T}_{m,s} \tag{13.1.22a}$$

$$0 = \frac{F}{V}(\bar{T}_{0,s} - \bar{T}_s) + \frac{\Delta h_{vap,st}}{V\rho c_p} k_v \left[\frac{k_c}{\tau_I} \bar{e}_{I,s} + k_c(\bar{T}_{sp,s} - \bar{T}_{m,s}) \right] \tag{13.1.22b}$$

$$0 = \bar{T}_s - \bar{T}_{m,s} \tag{13.1.22c}$$

From (13.1.22a) and (13.1.22c), we immediately see that

$$\bar{T}_s = \bar{T}_{sp,s} = \bar{T}_{m,s} \tag{13.1.23}$$

which implies the steady-state error

$$e_s = \bar{T}_{sp,s} - \bar{T}_{m,s} = 0 \tag{13.1.24}$$

The offset is always zero as expected due to the presence of integral action.

Eigenvalue analysis – Asymptotic stability analysis. The foregoing steady-state analysis is meaningful and the nice property of zero offset is achievable only when the closed-loop system is asymptotically stable. We see from Eq. (13.1.21) that the A matrix of the closed-loop system is

$$A_{CL} = \begin{bmatrix} 0 & 0 & -1 \\ \dfrac{\Delta h_{vap,st}}{V\rho c_p} k_v \dfrac{k_c}{\tau_I} & -\dfrac{F}{V} & -\dfrac{\Delta h_{vap,st}}{V\rho c_p} k_v k_c \\ 0 & \dfrac{1}{\tau_m} & -\dfrac{1}{\tau_m} \end{bmatrix} \qquad (13.1.25)$$

which depends upon the controller gain k_c and the integral time τ_I. The characteristic polynomial of A_{CL} can be calculated and the result is:

$$\det(\lambda I - A_{CL}) = \lambda^3 + \left(\dfrac{F}{V} + \dfrac{1}{\tau_m}\right)\lambda^2 + \left(\dfrac{F\rho c_p + \Delta h_{vap,st} k_v k_c}{V\rho c_p \tau_m}\right)\lambda + \dfrac{\Delta h_{vap,st}}{V\rho c_p \tau_m} k_v \dfrac{k_c}{\tau_I} \qquad (13.1.26)$$

The eigenvalues of the closed-loop system are the roots of the above polynomial. Closed-loop stability conditions can be derived via the Routh array. The result is

$$\left(\dfrac{F}{V} + \dfrac{1}{\tau_m}\right)(F\rho c_p + \Delta h_{vap,st} k_v k_c) > \Delta h_{vap,st} k_v \dfrac{k_c}{\tau_I} > 0$$

or

$$\left(\dfrac{F}{V} + \dfrac{1}{\tau_m}\right)\left(1 + \dfrac{\Delta h_{vap,st}}{F\rho c_p} k_v k_c\right) > \dfrac{\Delta h_{vap,st}}{F\rho c_p} k_v \dfrac{k_c}{\tau_I} > 0 \qquad (13.1.27)$$

This is equivalent to the result found through the closed-loop transfer functions in the previous chapter (see Eq. (12.3.10)).

13.2 State-Space Analysis of Closed-Loop Systems

In the previous section we studied the particular case of the steam-heated tank under P or PI control. In this section we will consider the general open-loop system with the following state-space description

$$\dfrac{dx}{dt} = Ax + bu + b_w w \qquad (13.2.1a)$$

$$y = cx \qquad (13.2.1b)$$

where x is the state vector of the process including the sensor/transmitter and the final control element, u is the controller output (scalar), w is the disturbance to the process and y is the measurement of the process output (scalar), all in deviation variable form. When a P controller is used, also in deviation variable form, then the control law is described by the following simple algebraic formula

$$u = k_c\left(y_{sp} - y\right) \tag{13.2.2}$$

Substituting (13.2.2) into (13.2.1a) gives the matrix differential equation of the closed-loop system

$$\frac{dx}{dt} = (A - bk_c c)x + bk_c y_{sp} + b_w w \tag{13.2.3}$$

This, along with the output equation (13.2.1b), forms a state-space description of the closed-loop system. The result is summarized in Figure 13.1. Notice that the action of the controller changes the system matrix; for the closed-loop system it is given by

$$A_{CL} = A - bk_c c \tag{13.2.4}$$

and it depends on the value of the controller gain k_c.

When the system is at steady state, we obtain from Eqs. (13.2.1a, b) and (13.2.2) that

$$0 = Ax_s + bu_s + b_w w_s \tag{13.2.5a}$$

$$y_s = cx_s \tag{13.2.5b}$$

$$u_s = k_c\left(y_{sp,s} - y_s\right) \tag{13.2.5c}$$

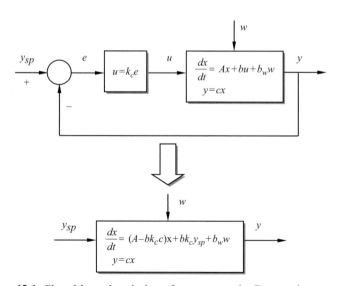

Figure 13.1 Closed-loop description of a system under P control.

Elimination of x_s and u_s from Eqs. (13.2.5a–c) leads to

$$y_s = \frac{(-cA^{-1}b)k_c}{1+(-cA^{-1}b)k_c} y_{sp,s} + \frac{(-cA^{-1}b_w)}{1+(-cA^{-1}b)k_c} w_s \qquad (13.2.6)$$

which relates the set point, the disturbance and the controlled output in closed loop. Moreover, we can calculate the steady-state error:

$$e_s = y_{sp,s} - y_s = \frac{y_{sp,s} - (-cA^{-1}b_w)w_s}{1+(-cA^{-1}b)k_c} \qquad (13.2.7)$$

Some important conclusions can be drawn from Eq. (13.2.7). Suppose that the closed-loop system is initially at the reference steady state (in which all deviation variables are zero) and a step change in either the set point (Δy_{sp}) or the disturbance (Δw) is imposed, so that their new steady-state values in deviation form are $y_{sp,s} = \Delta y_{sp}$, $w_s = \Delta w$. Then, provided that the closed-loop system is stable, the error will finally rest at a new steady state, and

$$\textit{Offset} = e_s = \frac{\Delta y_{sp} - (-cA^{-1}b_w)\Delta w}{1+(-cA^{-1}b)k_c} \qquad (13.2.8)$$

Notice that

$$k = -cA^{-1}b, \ k' = -cA^{-1}b_w \qquad (13.2.9)$$

are exactly the static gains of the open-loop system (13.2.1a, b) with respect to the manipulated input and the disturbance, respectively (see Chapter 6, Section 6.2), so Eq. (13.2.8) may be written equivalently as

$$\textit{Offset} = e_s = \frac{\Delta y_{sp} - k'\Delta w}{1+ kk_c} \qquad (13.2.10)$$

This is exactly what we found in the previous chapter (see Eqs. (12.1.5) and (12.1.9)). Another important conclusion from (13.2.7) is that the static gain between set point and the error is

$$\Sigma_0 = \frac{1}{1+(-cA^{-1}b)k_c} = \frac{1}{1+kk_c} \qquad (13.2.11)$$

As far as the asymptotic stability of the closed-loop system is concerned, it follows from Eq. (13.2.4) that it is determined by the eigenvalues of the closed-loop system matrix $A_{CL} = A - bk_c c$: they must all have negative real parts for stability.

The case of PI control is slightly more complicated as the controller incorporates dynamics increasing the size of the state vector needed to describe the closed-loop system. The open-loop dynamics is again described by Eqs. (13.2.1a, b) while the PI controller in deviation form is described by (see Table 10.1)

$$\frac{de_I}{dt} = y_{sp} - y \qquad (13.2.12a)$$

$$u = \frac{k_c}{\tau_I} e_I + k_c \left(y_{sp} - y \right) \qquad (13.2.12b)$$

Combining Eqs. (13.2.12a, b) with (13.2.1a, b) and doing the appropriate substitutions to eliminate u and y, we obtain the differential equations of the closed-loop system

$$\frac{de_I}{dt} = y_{sp} - cx \qquad (13.2.13a)$$

$$\frac{dx}{dt} = b\frac{k_c}{\tau_I}e_I + (A - bk_c c)x + bk_c y_{sp} + b_w w \qquad (13.2.13b)$$

These, along with the output equation (13.2.1b), form a state-space description of the closed-loop system under PI control that can be written in matrix form

$$\frac{d}{dt}\begin{bmatrix} e_I \\ x \end{bmatrix} = \begin{bmatrix} 0 & -c \\ b\dfrac{k_c}{\tau_I} & A - bk_c c \end{bmatrix}\begin{bmatrix} e_I \\ x \end{bmatrix} + \begin{bmatrix} 1 \\ bk_c \end{bmatrix} y_{sp} + \begin{bmatrix} 0 \\ b_w \end{bmatrix} w \qquad (13.2.14a)$$

$$y = \begin{bmatrix} 0 & c \end{bmatrix}\begin{bmatrix} e_I \\ x \end{bmatrix} \qquad (13.2.14b)$$

This result is summarized in Figure 13.2. Notice that the system matrix of the closed-loop system, A_{CL}, is given by

$$A_{CL} = \begin{bmatrix} 0 & -c \\ b\dfrac{k_c}{\tau_I} & A - bk_c c \end{bmatrix} \qquad (13.2.15)$$

and it depends on the values of the controller parameters k_c and τ_I.

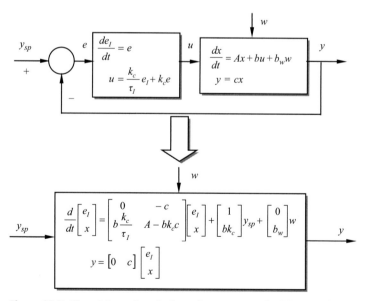

Figure 13.2 Closed-loop description of a system under PI control.

If the closed-loop system is asymptotically stable, then a step change in the set point or in the disturbance will cause the system to move to a different steady state. At any steady state, however, the steady-state equations from (13.2.13a, b) imply that $y = y_{sp}$ at steady state. This means that the offset is always zero under PI control. Asymptotic stability of the closed-loop system is again determined by the eigenvalues of the matrix A_{CL}, which can easily be calculated numerically in MATLAB or Maple.

Before closing this section we will consider the more general case where the process state-space description is again given by Eqs. (13.2.1a, b) and the controller is described by the following general state-space description

$$\frac{dx_c}{dt} = A_c x_c + b_c (y_{sp} - y) \qquad (13.2.16a)$$

$$u = c_c x_c + d_c (y_{sp} - y) \qquad (13.2.16b)$$

where x_c is the state vector of the controller. To determine the state-space description of the closed-loop system, we first use (13.2.16b) to eliminate u from the state equations of the open-loop process and to obtain the following

$$\frac{dx}{dt} = (A - bd_c c) x + bc_c x_c + bd_c y_{sp} + b_w w \qquad (13.2.17)$$

Putting Eq. (13.2.17) together with Eq. (13.2.16a) and writing them in matrix form we obtain

$$\frac{d}{dt} \begin{bmatrix} x_c \\ x \end{bmatrix} = \begin{bmatrix} A_c & -b_c c \\ bc_c & A - bd_c c \end{bmatrix} \begin{bmatrix} x_c \\ x \end{bmatrix} + \begin{bmatrix} b_c \\ bd_c \end{bmatrix} y_{sp} + \begin{bmatrix} 0 \\ b_w \end{bmatrix} w \qquad (13.2.18a)$$

$$y = \begin{bmatrix} 0 & c \end{bmatrix} \begin{bmatrix} x_c \\ x \end{bmatrix} \qquad (13.2.18b)$$

Equations (13.2.18a, b) constitute the state-space description of the closed-loop system.

Example 13.1 P and PI control of a system of two liquid storage tanks

Consider the system of two liquid storage tanks shown in Figure 13.3. The volumetric flowrate out of each tank is a linear function of the liquid level in the tank. The liquid level in the second tank is measured by a differential pressure sensor that exhibits first-order dynamics with unity static gain and time constant τ_m. The controller used is a P controller that is driven by the liquid-level measurement and manipulates the flowrate of one of the two streams fed in the first tank (F_0). F_w is a disturbance for the system. For the case where $A_1 = R_1 = A_2 = R_2 = k_v = 1$ and $\tau_m = 0.1$, calculate the sensitivity coefficient and the range of values of the controller gain k_c for which asymptotic stability of the closed-loop system is guaranteed. Repeat for the case of PI control.

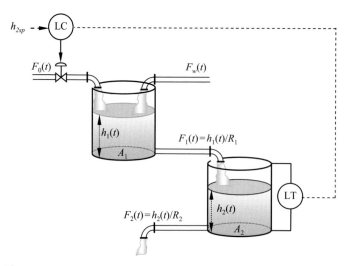

Figure 13.3 Two liquid storage tanks under P or PI control.

Solution

Application of the material balance to each tank results in the following differential equations that describe the liquid level dynamics and measurement dynamics

$$A_1 \frac{dh_1}{dt} = -\frac{h_1}{R_1} + k_v u + F_w$$

$$A_2 \frac{dh_2}{dt} = \frac{h_1}{R_1} - \frac{h_2}{R_2}$$

$$\tau_m \frac{dh_{2,m}}{dt} + h_{2,m} = h_2$$

These equations can be written in the following matrix form

$$\frac{d}{dt}\begin{bmatrix} h_1 \\ h_2 \\ h_{2,m} \end{bmatrix} = \begin{bmatrix} -\frac{1}{A_1 R_1} & 0 & 0 \\ \frac{1}{A_2 R_1} & -\frac{1}{A_2 R_2} & 0 \\ 0 & \frac{1}{\tau_m} & -\frac{1}{\tau_m} \end{bmatrix} \begin{bmatrix} h_1 \\ h_2 \\ h_{2,m} \end{bmatrix} + \begin{bmatrix} \frac{k_v}{A_1} \\ 0 \\ 0 \end{bmatrix} u + \begin{bmatrix} \frac{1}{A_1} \\ 0 \\ 0 \end{bmatrix} F_w$$

13.2 State-Space Analysis of Closed-Loop Systems

and by substituting the numerical values we obtain

$$\frac{d}{dt}\begin{bmatrix} h_1 \\ h_2 \\ h_{2,m} \end{bmatrix} = \begin{bmatrix} -1 & 0 & 0 \\ 1 & -1 & 0 \\ 0 & 10 & -10 \end{bmatrix} \begin{bmatrix} h_1 \\ h_2 \\ h_{2,m} \end{bmatrix} + \begin{bmatrix} 1 \\ 0 \\ 0 \end{bmatrix} u + \begin{bmatrix} 1 \\ 0 \\ 0 \end{bmatrix} F_w$$

The output is given by

$$y = \begin{bmatrix} 0 & 0 & 1 \end{bmatrix} \begin{bmatrix} h_1 \\ h_2 \\ h_{2,m} \end{bmatrix}$$

We first calculate the sensitivity coefficient under P control using Eq. (13.2.11)

$$\Sigma_0 = \frac{1}{1+\left(-cA^{-1}b\right)k_c} = \frac{1}{1+\begin{bmatrix} -\begin{bmatrix} 0 & 0 & 1 \end{bmatrix}\begin{bmatrix} -1 & 0 & 0 \\ 1 & -1 & 0 \\ 0 & 10 & -10 \end{bmatrix}^{-1}\begin{bmatrix} 1 \\ 0 \\ 0 \end{bmatrix} \end{bmatrix}k_c} = \frac{1}{1+k_c}$$

To determine the stability conditions we first calculate the A_{CL} as follows

$$A_{CL} = A - bk_c c = \begin{bmatrix} -1 & 0 & 0 \\ 1 & -1 & 0 \\ 0 & 10 & -10 \end{bmatrix} - \begin{bmatrix} 1 \\ 0 \\ 0 \end{bmatrix}k_c\begin{bmatrix} 0 & 0 & 1 \end{bmatrix} = \begin{bmatrix} -1 & 0 & -k_c \\ 1 & -1 & 0 \\ 0 & 10 & -10 \end{bmatrix}$$

Its eigenvalues are roots of

$$\det(\lambda I - A_{CL}) = \det\begin{bmatrix} \lambda+1 & 0 & k_c \\ -1 & \lambda+1 & 0 \\ 0 & -10 & \lambda+10 \end{bmatrix} = (\lambda+1)(\lambda+1)(\lambda+10) + 10k_c = 0$$

or

$$\lambda^3 + 12\lambda^2 + 21\lambda + 10(1+k_c) = 0$$

Using the Routh criterion for a third-order polynomial we derive the following condition for stability

$$\left.\begin{aligned} 12\cdot 21 - 10(1+k_c) &> 0 \\ 1+k_c &> 0 \end{aligned}\right\} \Rightarrow -1 < k_c < 24.2$$

When a PI controller is used, A_{CL} is slightly more complicated

$$A_{CL} = \begin{bmatrix} 0 & -c \\ b\dfrac{k_c}{\tau_I} & A - bk_c c \end{bmatrix} = \begin{bmatrix} 0 & 0 & 0 & -1 \\ \dfrac{k_c}{\tau_I} & -1 & 0 & -k_c \\ 0 & 1 & -1 & 0 \\ 0 & 0 & 10 & -10 \end{bmatrix}$$

and following the same approach as before we have that the eigenvalues of the A_{CL} matrix are the roots of the following polynomial equation

$$\lambda^4 + 12\lambda^3 + 21\lambda^2 + 10(1+k_c)\lambda + 10\dfrac{k_c}{\tau_I} = 0$$

Using the Routh criterion again, we determine that the stability conditions are:

$$-1 < k_c < 24.2$$

$$0 < \dfrac{k_c}{\tau_I} < \dfrac{(242 - 10k_c)(1+k_c)}{144}$$

13.3 Time Discretization of the Closed-Loop System

Once the closed-loop system is derived in state-space form (given by Eq. (13.2.3) for P control, Eq. (13.2.14) for PI control and Eq. (13.2.18) for the general case), it can be simulated for different scenarios of set point or disturbance changes. For standard inputs like step, impulse or sinusoidal changes, it is possible to use special commands of software packages. For arbitrary inputs, the closed-loop system is appropriately discretized under a small enough time step (sampling period), and then simulated by running the corresponding recursion, as we have already seen in dynamics (Chapters 3 and 6). In particular, if

$$\dfrac{dx_{CL}}{dt} = A_{CL} x_{CL} + b_{CL} y_{sp} + b_{wCL} w \qquad (13.3.1)$$

$$y = c_{CL} x_{CL}$$

is the state-space description of the closed-loop system, applying the zero-order-hold discretization (see Chapter 6, Section 6.7), we obtain the discretized closed-loop system:

$$x_{CL}[j] = A_{CL\_d} x_{CL}[j-1] + b_{CL\_d} y_{sp}[j-1] + b_{wCL\_d} w[j-1]$$
$$y[j] = c_{CL} x_{CL}[j] \qquad (13.3.2)$$

where

$$A_{CL\_d} = e^{A_{CL} T_s}, \quad b_{CL\_d} = \int_0^{T_s} e^{A_{CL} t} b_{CL} \, dt, \quad b_{wCL\_d} = \int_0^{T_s} e^{A_{CL} t} b_{wCL} \, dt \qquad (13.3.3)$$

and where T_s is the sampling period.

13.3 Time Discretization of the Closed-Loop System

Note that, in a computer control system, the controller is implemented in discrete form, with the controller output held constant in between sampling instants (zero-order-hold implementation). In the chemical industry, the sampling period is very small, and thus the discretized controller equations (see Chapter 10) accurately represent the continuous P, PI, PID, etc., functions. This means that a continuous-time description of the closed-loop system of the form (13.3.1) is already an approximation of the actual closed-loop system, and it is accurate because of the small sampling period.

But since the controller is actually discrete, an alternative approach is to combine the discrete controller equations to the discretized process equations and obtain a discretized closed-loop system.

Suppose for example that a process of the form (13.2.1a, b) is controlled by a PI controller. The continuous-time description of the closed-loop system is given by Eqs. (13.2.14a, b), which are of the form of Eq. (13.3.1). Instead of using the discretization of Eqs. (13.3.2) and (13.3.3), an alternative approach would be to combine a zero-order-hold discretized PI controller

$$e_I[j] = e_I[j-1] + T_s\left(y_{sp}[j-1] - y[j-1]\right)$$
$$u[j] = \frac{k_c}{\tau_I} e_I[j] + k_c\left(y_{sp}[j] - y[j]\right) \quad (13.3.4)$$

with the zero-order-hold discretized process equations

$$x[j] = A_d x[j-1] + b_d u[j-1] + b_{wd} w[j-1]$$
$$y[j] = cx[j] \quad (13.3.5)$$

where

$$A_d = e^{AT_s}, \quad b_d = \int_0^{T_s} e^{At} b \, dt, \quad b_{wd} = \int_0^{T_s} e^{At} b_w \, dt \quad (13.3.6)$$

This leads to the following discrete-time closed-loop system

$$\begin{bmatrix} e_I[j] \\ x[j] \end{bmatrix} = \begin{bmatrix} 1 & -T_s c \\ b_d \frac{k_c}{\tau_I} & A_d - b_d k_c c \end{bmatrix} \begin{bmatrix} e_I[j-1] \\ x[j-1] \end{bmatrix} + \begin{bmatrix} T_s \\ b_d k_c \end{bmatrix} y_{sp}[j-1] + \begin{bmatrix} 0 \\ b_{wd} \end{bmatrix} w[j-1]$$

$$y[j] = \begin{bmatrix} 0 & c \end{bmatrix} \begin{bmatrix} e_I[j] \\ x[j] \end{bmatrix} \quad (13.3.7)$$

The two alternative discretization schemes will agree as long as the sampling period T_s is very small. In fact, it is possible to derive small-T_s approximations for the matrices A_{CL_d}, b_{CL_d} and b_{wCL_d} of Eq. (13.3.3) for the A_{CL}, b_{CL} and b_{wCL} defined via Eqs. (13.2.14a, b), and compare them to small-T_s approximations of the corresponding matrices of (13.3.7), and finally conclude that they match up to first-order terms in T_s (see Problem 13.8).

Example 13.2 Discretized closed-loop system for a heating tank under PI temperature control

Consider again the heating tank example discussed in Section 13.1, where the instrumented process is described by Eqs. (13.1.6a, b), and for the following values of the process parameters:

$$\frac{F}{V} = 1, \quad \frac{\Delta h_{vap,s}}{V \rho c_p} k_v = 0.2, \quad \tau_m = 0.1$$

Suppose that the temperature of the heating tank is controlled by a PI controller with $k_c = 50$ and $\tau_I = 1$. Discretize the closed-loop system with sampling period $T_s = 0.01$, in two different ways: discretization at the beginning (discrete PI combined with discretized process) and discretization at the end (of the continuous-time closed-loop system (13.1.21)). Use both discretization schemes to simulate the output response to a unit step change in the set point. Do they agree?

Solution

For the given parameter values, the process is described by Eq. (13.2.1) with

$$A = \begin{bmatrix} -1 & 0 \\ 10 & -10 \end{bmatrix}, \quad b = \begin{bmatrix} 0.2 \\ 0 \end{bmatrix}, \quad b_w = \begin{bmatrix} 1 \\ 0 \end{bmatrix}, \quad c = \begin{bmatrix} 0 & 1 \end{bmatrix}$$

When the process is discretized with sampling period $T_s = 0.01$,

$$A_d = e^{AT_s} = \begin{bmatrix} 0.9900 & 0 \\ 0.0947 & 0.9048 \end{bmatrix}, \quad b_d = \int_0^{T_s} e^{At} b \, dt = \begin{bmatrix} 0.0020 \\ 0.0001 \end{bmatrix}, \quad b_{wd} = \int_0^{T_s} e^{At} b_w \, dt = \begin{bmatrix} 0.0100 \\ 0.0005 \end{bmatrix}$$

Applying Eq. (13.3.7) for the given controller parameters leads to

$$\begin{bmatrix} e_I[j] \\ \bar{T}[j] \\ \bar{T}_m[j] \end{bmatrix} = \begin{bmatrix} 1 & 0 & -0.01 \\ 0.0995 & 0.9900 & -0.0995 \\ 0.0048 & 0.0947 & 0.9000 \end{bmatrix} \begin{bmatrix} e_I[j-1] \\ \bar{T}[j-1] \\ \bar{T}_m[j-1] \end{bmatrix} + \begin{bmatrix} 0.01 \\ 0.0995 \\ 0.0048 \end{bmatrix} \bar{T}_{sp}[j-1] + \begin{bmatrix} 0 \\ 0.0100 \\ 0.0005 \end{bmatrix} \bar{T}_0[j-1]$$

$$y[j] = \begin{bmatrix} 0 & 0 & 1 \end{bmatrix} \begin{bmatrix} e_I[j] \\ \bar{T}[j] \\ \bar{T}_m[j] \end{bmatrix} \qquad (13.3.8)$$

The other approach is to discretize the continuous-time system (13.1.21), which for the given process and controller parameter values, becomes

13.3 Time Discretization of the Closed-Loop System

$$\frac{d}{dt}\begin{bmatrix} \bar{e}_I \\ \bar{T} \\ \bar{T}_m \end{bmatrix} = \begin{bmatrix} 0 & 0 & -1 \\ 10 & -1 & -10 \\ 0 & 10 & -10 \end{bmatrix}\begin{bmatrix} \bar{e}_I \\ \bar{T} \\ \bar{T}_m \end{bmatrix} + \begin{bmatrix} 1 \\ 10 \\ 0 \end{bmatrix}\bar{T}_{sp} + \begin{bmatrix} 0 \\ 1 \\ 0 \end{bmatrix}\bar{T}_0$$

$$y = \begin{bmatrix} 0 & 0 & 1 \end{bmatrix}\begin{bmatrix} \bar{e}_I \\ \bar{T} \\ \bar{T}_m \end{bmatrix}$$

(13.3.9)

Its zero-order-hold discretization for the given sampling period $T_s = 0.01$ is

$$\begin{bmatrix} e_I[j] \\ \bar{T}[j] \\ \bar{T}_m[j] \end{bmatrix} = \begin{bmatrix} 1.0000 & -0.0005 & -0.0095 \\ 0.0993 & 0.9852 & -0.0950 \\ 0.0048 & 0.0945 & 0.9002 \end{bmatrix}\begin{bmatrix} e_I[j-1] \\ \bar{T}[j-1] \\ \bar{T}_m[j-1] \end{bmatrix} + \begin{bmatrix} 0.0100 \\ 0.0998 \\ 0.0048 \end{bmatrix}\bar{T}_{sp}[j-1] + \begin{bmatrix} 0.0000 \\ 0.0100 \\ 0.0005 \end{bmatrix}\bar{T}_0[j-1]$$

$$y[j] = \begin{bmatrix} 0 & 0 & 1 \end{bmatrix}\begin{bmatrix} e_I[j] \\ \bar{T}[j] \\ \bar{T}_m[j] \end{bmatrix}$$

(13.3.10)

Comparing the two discretized systems (13.3.8) and (13.3.10), we see that the numbers are close but not identical. Some entries differ in the third or fourth decimal places. Figure 13.4

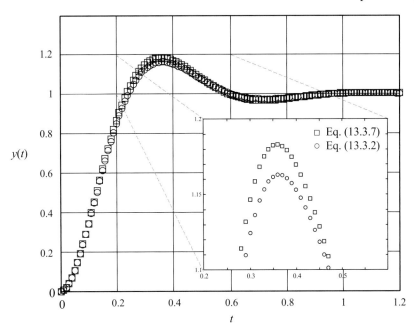

Figure 13.4 Calculation of the output response to a unit step change in set point with the two alternative discretization schemes.

compares the predictions of the two discretized systems for the output response to a unit step change in the set point. Again, they are very close (with mismatch up to 2% of the set-point change), but not identical. Because discretization (13.3.10) has been derived from (13.3.9) under the assumption of piecewise constant input and this is satisfied for the applied step change, the response of (13.3.10) exactly matches the corresponding response of the continuous-time system (13.3.9).

The simulation results indicate that the value of $T_s = 0.01$ is small enough for the discretized system to capture the transient behavior of the system. This can also be checked through the eigenvalues of the continuous-time closed-loop system (13.3.9):

$\lambda_1 = -1$, $\lambda_{2,3} = -5 \pm i5\sqrt{3}$, with absolute values $|\lambda_1|=1$, $|\lambda_{2,3}|=10$, hence $1/(\max|\lambda_\kappa|) = 0.1$.

The value of $T_s = 0.01$ is exactly one tenth of $1/(\max|\lambda_\kappa|)$, hence it is small enough for discretization (see Chapter 7). A smaller T_s would give higher accuracy and better agreement between the two discretizations; a larger T_s would be less accurate.

13.4 State-Space Description of Nonlinear Closed-Loop Systems

The state-space description of closed-loop dynamics is the only option when it comes to nonlinear processes. Even though nonlinear process models are usually linearized around normal operating steady-state conditions and the control system is analyzed by using the linearized process model, at the end we always want to test the performance of the controller against the actual nonlinear system. For this purpose, we need to write down the nonlinear closed-loop system in state-space form, and then simulate it with a differential equation solver.

Consider a nonlinear process described by the following dynamic model, not necessarily in deviation form:

$$\frac{dx}{dt} = f(x,u,w) \qquad (13.4.1)$$

where x is the state vector, u is the (scalar) manipulated input and w is a disturbance.

Also, consider for simplicity that the process output is one of the states or a linear combination of the states, i.e. of the form

$$y = cx \qquad (13.4.2)$$

Suppose, for example, that this process is controlled with a PI controller

$$\frac{de_I}{dt} = y_{sp} - y$$
$$u = u_b + \frac{k_c}{\tau_I}e_I + k_c(y_{sp} - y) \qquad (13.4.3)$$

again, not necessarily in deviation form, possibly involving a constant bias term u_b in its output. Combining Eqs. (13.4.3) with (13.4.1) and (13.4.2), and doing the appropriate substitutions to eliminate u and y, we obtain the differential equations of the closed-loop system

$$\frac{de_I}{dt} = y_{sp} - cx$$
$$\frac{dx}{dt} = f\left(x, \left(u_b + \frac{k_c}{\tau_I}e_I + k_c(y_{sp} - cx)\right), w\right) \quad (13.4.4)$$
$$y = cx$$

The nonlinear closed-loop system (13.4.4) can be simulated using a differential equation solver to test the performance of the controller on the nonlinear process. Other types of controllers (P, PD, PID, …) can be handled similarly.

Example 13.3 Nonlinear closed-loop system for PI control of an exothermic CSTR

Consider a constant-volume exothermic CSTR shown in Figure 7.8 whose linearized dynamics was studied in Chapter 7 (Example 7.4) and Chapter 8 (Section 8.6). The original nonlinear state-space model consists of a mole balance for the reactant and an energy balance for the reactor:

$$\frac{dC_R}{dt} = \frac{F}{V}(C_{R,in} - C_R) - k_0 e^{-\frac{E}{RT}} C_R$$
$$\frac{dT}{dt} = \frac{F}{V}(T_{in} - T) + \frac{(-\Delta h_{rxn})}{\rho c_p} k_0 e^{-\frac{E}{RT}} C_R + \frac{A_H U_H}{\rho c_p V}(T_J - T) \quad (8.6.1)$$

The volumetric flowrate $u = F$ is an input variable that can be manipulated and the reactor temperature $y = T$ is the output that must be controlled. The design steady-state conditions of the reactor are $(F = F_s, C_R = C_{R,s}, T = T_s)$ and these are the normal operating conditions, and we need a temperature controller to correct for disturbances in the feed conditions and/or handle possible changes in the reactor temperature set point.

If we use a PI controller for this purpose, write down a state-space model for the resulting closed-loop system.

Solution

Combining the PI controller
$$\frac{de_I}{dt} = T_{sp} - T$$
$$F = F_s + \frac{k_c}{\tau_I} e_I + k_c(T_{sp} - T)$$

with the process equations, we obtain the closed-loop system equations

$$\frac{de_I}{dt} = T_{sp} - T$$

$$\frac{dC_R}{dt} = \frac{F_s + \frac{k_c}{\tau_I} e_I + k_c(T_{sp} - T)}{V}(C_{R,in} - C_R) - k_0 e^{-\frac{E}{RT}} C_R$$

$$\frac{dT}{dt} = \frac{F_s + \frac{k_c}{\tau_I} e_I + k_c(T_{sp} - T)}{V}(T_{in} - T) + \frac{(-\Delta h_{rxn})}{\rho c_p} k_0 e^{-\frac{E}{RT}} C_R + \frac{A_H U_H}{\rho c_p V}(T_J - T)$$

The output of the closed-loop system is $y = T$.

13.5 Software Tools

In this section, we consider again the heating tank example, where the open-loop system is described by Eqs. (13.1.6a, b), and for the following values of process parameters:

$$\frac{F}{V} = 1, \quad \frac{\Delta h_{vap,s}}{V \rho c_p} k_v = 0.2, \quad \tau_m = 0.1$$

Thus, the open-loop system is described by Eqs. (13.2.1a, b) with

$$A = \begin{bmatrix} -1 & 0 \\ 10 & -10 \end{bmatrix}, \quad b = \begin{bmatrix} 0.2 \\ 0 \end{bmatrix}, \quad b_w = \begin{bmatrix} 1 \\ 0 \end{bmatrix}, \quad c = \begin{bmatrix} 0 & 1 \end{bmatrix} \quad (13.5.1)$$

All necessary matrix manipulations and calculations presented in this chapter can be performed using software:

(a) the construction of the appropriate matrices in the state-space description of the closed-loop system given by Eqs. (13.2.3) and (13.2.16) for P control, Eqs. (13.2.14a, b) for PI control and Eq. (13.2.18) for the general case;
(b) calculation of the sensitivity coefficient under P control via Eq. (13.2.11);
(c) testing for stability of the closed-loop system;
(d) simulation of the closed-loop system's response to changes in the set point or the disturbance.

All the above can be performed numerically or symbolically, as will be shown below.

13.5.1 Closed-Loop System Analysis using in MATLAB

We first define the matrices appearing in the state-space description of the process

```
» A=[-1 0; 10-10]
A =
    -1    0
    10  -10
```

```
» b=[0.2; 0]
b =
    0.2000
         0
» bw=[1; 0]
bw =
    1
    0
» c=[0 1]
c =
    0    1
```

Assuming that a P controller with $k_c = 50$ is used then the closed-loop description is given by Eq. (13.2.3), which can be constructed in MATLAB:

```
» kc=50;
» ACL=A-b*kc*c
ACL =
    -1   -10
    10   -10
» bCL=b*kc
bCL =
    10
     0
» bwCL=bw;
```

Having the open-loop and closed-loop matrices available we can calculate the sensitivity coefficient through Eq. (13.2.11)

```
» SIGMA0=1/(1+(-c*inv(A)*b)*kc)
SIGMA0=
    0.0909
```

and we can check stability by calculating the eigenvalues of the closed-loop system matrix

```
» eig(ACL)
ans =
    -5.5000 + 8.9303i
    -5.5000 - 8.9303i
```

and calculate the unit step response shown in Figure 13.5.

```
» step(ACL,bCL,c,0)
```

When a PI controller with $k_c = 50$ and $\tau_I = 1$ is used then we have the following:

```
» kc=50;
» tauI=1;
» ACL=[0 -c; b*kc/tauI A-b*kc*c]
```

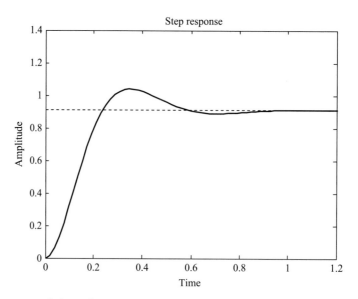

Figure 13.5 Response of the tank temperature under P control to a unit step change in the set point.

```
ACL =
      0    0   -1
     10   -1  -10
      0   10  -10
» bCL = [1;b*kc]
bCL =
      1
     10
      0
» bwCL = [0; bw]
bwCL =
      0
      1
      0
» cCL = [0 c]
cCL =
      0  0  1
```

After defining the closed-loop system description we calculate the eigenvalues of the closed-loop system matrix

```
» eig(ACL)
ans =
   -1.0000
   -5.0000 + 8.6603i
   -5.0000 - 8.6603i
```

and finally the unit step response in the set-point change shown in Figure 13.6

» `step(ACL,bCL,cCL,0)`

or the unit step response to disturbance change shown in Figure 13.7.

» `step(ACL,bwCL,cCL,0)`

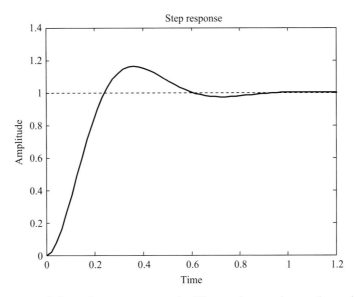

Figure 13.6 Response of the tank temperature under PI control to a unit step change in the set point.

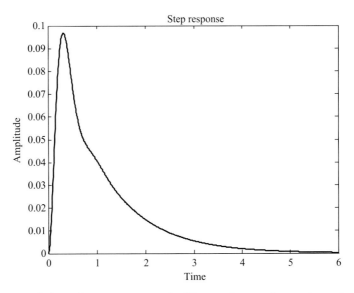

Figure 13.7 Response of the tank temperature under PI control to a unit step change in the disturbance.

13.5.2 Closed-Loop System Analysis using Maple

To perform linear algebra calculations with Maple, we start by calling the LinearAlgebra library and defining the open-loop system's matrices A, b, b_w, c given in Eq. (13.5.1)

```
> with(LinearAlgebra):
> A:=Matrix(2,[[-1,0],[10,-10]]):
> b:=Vector(2,[1/5,0]): bw:=Vector(2,[1,0]):
> c:=Vector[row](2,[0,1]):
```

When the system is controlled with a P controller, we compute the closed-loop system's matrices as given in (13.2.3), with c being the same as in the open-loop system

```
> ACL:=A-kc*Multiply(b,c);
```

$$ACL := \begin{bmatrix} -1 & -\frac{1}{5}kc \\ 10 & -10 \end{bmatrix}$$

```
> bCL:=kc*b;bwCL:=bw;
```

$$bCL := \begin{bmatrix} \frac{1}{5}kc \\ 0 \end{bmatrix}$$

$$bwCL := \begin{bmatrix} 1 \\ 0 \end{bmatrix}$$

```
> cCL:=c;
```

$$cCl := \begin{bmatrix} 0 & 1 \end{bmatrix}$$

The sensitivity coefficient is calculated by direct application of formula (13.2.9)

```
> Sigma0:=1/(1-Multiply(c,Multiply(MatrixInverse(A),b))*kc);
```

$$\Sigma 0 := \frac{1}{1+\frac{1}{5}kc}$$

To perform stability analysis, we first calculate the characteristic polynomial of the A matrix of the closed-loop system

```
> Char_Poly_ACL:=CharacteristicPolynomial(ACL,lambda);
```

$$Char\_Poly\_ACL := \lambda^2 + 11\lambda + 10 + 2kc$$

and then obtain the stability condition using the special command `RouthTable` of the `DynamicSystems` package of Maple:

```
> with(DynamicSystems):
> RouthTable(Char_Poly_ACL,lambda,'stablecondition'=true);
```

$$0 < 5 + kc$$

When the system is controlled with a PI controller, the closed-loop system is given by Eqs. (13.2.14a, b). The closed-loop system's matrices are calculated according to

```
> ACL:=Matrix([[0,-c],[(kc/tauI)*b,A-kc*Multiply(b,c)]]);
```

$$ACL := \begin{bmatrix} 0 & 0 & -1 \\ \frac{1}{5}\frac{kc}{taul} & -1 & -\frac{1}{5}kc \\ 0 & 10 & -10 \end{bmatrix}$$

```
> bCL:=Vector([1,kc*b]);bwCL:=Vector([0,bw]);
```

$$bCL := \begin{bmatrix} 1 \\ \frac{1}{5}kc \\ 0 \end{bmatrix}$$

$$bwCL := \begin{bmatrix} 0 \\ 1 \\ 0 \end{bmatrix}$$

```
> cCL:=Vector[row]([0,c]);
```

$$cCL := \begin{bmatrix} 0 & 0 & 1 \end{bmatrix}$$

The sensitivity coefficient is zero. Closed-loop stability conditions are obtained by using exactly the same commands as in the case of P control, the result being

$$0 < \frac{kc}{taul} \text{ and } 0 < 55 + 11kc - \frac{kc}{taul}$$

For given values of controller parameters, the closed-loop system's matrices A_{CL}, b_{CL}, b_{wCL}, c_{CL} can be used to calculate the response to changes in the set point or the disturbance through direct application of the methods and results of Chapter 6. For example, in the case of PI control and for

```
> kc:=50: tauI:=1:
```

we can use the step response formula (see Table 6.3) to calculate the response to a unit step change in the set point

```
> yStep_sp:=Multiply(cCL,Multiply(Multiply((MatrixExponential
(ACL,t)-IdentityMatrix(3)),MatrixInverse(ACL)),bCL));
```

$$yStep\_sp := -\frac{1}{3}e^{-5t}\sin(5t\sqrt{3})\sqrt{3} - e^{-5t}\cos(5t\sqrt{3}) + 1$$

or in the disturbance

```
> yStep_w:=Multiply(cCL,Multiply(Multiply((MatrixExponential
(ACL,t)-IdentityMatrix(3)), MatrixInverse(ACL)),bwCL));
```

$$yStep\_w := -\frac{8}{273}e^{-5t}\sin(5t\sqrt{3})\sqrt{3} - \frac{10}{91}e^{-5t}\cos(5t\sqrt{3}) + \frac{10}{91}e^{-t}$$

and, plotting the results, we obtain Figures 13.6 and 13.7.

LEARNING SUMMARY

For systems having the following open-loop description in state space

$$\frac{dx}{dt} = Ax + bu + b_w w$$

$$y = cx$$

- the closed-loop, state-space description under P-only control is

$$\frac{dx}{dt} = A_{CL}x + bk_c y_{sp} + b_w w$$

where $A_{CL} = A - bk_c c$
- the closed-loop, state-space description under PI control is

$$\frac{d}{dt}\begin{bmatrix} e_I \\ x \end{bmatrix} = A_{CL}\begin{bmatrix} e_I \\ x \end{bmatrix} + \begin{bmatrix} 1 \\ bk_c \end{bmatrix} y_{sp} + \begin{bmatrix} 0 \\ b_w \end{bmatrix} w$$

where

$$A_{CL} = \begin{bmatrix} 0 & -c \\ b\dfrac{k_c}{\tau_I} & A - bk_c c \end{bmatrix}$$

- under PI control the offset is zero, whereas under P control the fractional offset or sensitivity coefficient is given by

$$\Sigma_0 = \frac{1}{1+(-cA^{-1}b)k_c}$$

- closed-loop stability is determined by the eigenvalues of A_{CL}.

FURTHER READING

A comprehensive discussion of the state-space methodologies can be found in the following textbooks

Chen, C. T., *Linear Systems Theory and Applications*, 3rd edn. New York: Oxford University Press, 1999.

DeCarlo, R. A., *Linear Systems, A State Variable Approach with Numerical Implementation*. New Jersey: Prentice Hall International Editions, 1989.

Ogata, K., *Modern Control Engineering, Pearson International Edition*, 5th edn, 2008.

PROBLEMS

13.1 Consider the general state-space description of an open-loop system where the input directly affects the output ($d \neq 0$):

$$\frac{dx}{dt} = Ax + bu + b_w w$$

$$y = cx + du$$

Derive the description of the closed-loop system when a P-only controller is used. Calculate the offset for step change in the set point and/or in the disturbance.

13.2 Repeat Problem 13.1 for the case of PI control.

13.3 Consider the system of two liquid storage tanks studied in Example 13.1. Use MATLAB to simulate the closed-loop response under a unit step change in the set point or the disturbance
  (i) under P control with k_c = 10, 24.2 and 30,
  (ii) under PI control with k_c = 5 and τ_I = 2 and 0.2.

13.4 Consider again the system of tanks of Example 13.1, but with a (real) PID controller. Derive the closed-loop system in state-space form.

13.5 Consider the system of two interacting tanks of Problem 12.3, with $A_1 = 1$, $A_2 = 2$, $R_1 = 2$, $R_2 = 1$, where the sensor exhibits first-order dynamics with unity static gain and time constant $\tau_m = 0.1$ and the valve has negligible time constant and static gain $k_v = 1$. When this system is controlled with a P controller, derive the state-space description and determine the offset as a function of the controller gain. What is the minimum possible offset?

13.6 Consider the system of tanks of Example 13.1, under PI control with $k_c = 5$ and $\tau_I = 2$. Derive two alternative time-discretization schemes, following the steps outlined in Section 13.3, with $T_s = 0.02$. Simulate the closed-loop response for a unit step change in the set point and a unit step change in the disturbance.

13.7 Derive the linearization of the nonlinear closed-loop system (13.4.4) around the steady state $x = x_s$ corresponding to $u = u_s$, $w = w_s$.

13.8 Consider a system of the form (13.3.1) that represents a PI-controlled process, where A_{CL}, b_{CL} and b_{wCL} are defined via (13.2.14a, b). Also, consider its discretization (13.3.2) with A_{CL_d}, b_{CL_d} and b_{wCL_d} given by (13.3.3). Use the approximation $e^{At} \approx I + At$ to derive small-T_s approximations for A_{CL_d}, b_{CL_d} and b_{wCL_d}, keeping up to first-order terms in T_s. Do a similar small-T_s approximation for the matrices of the discretization (13.3.7), again keeping up to first-order terms in T_s. Compare the two approximately discretized systems. Do they agree?

13.9 Redo Example 13.2 with $T_s = 0.002$ and $T_s = 0.05$ (one fifth and five times the sampling period used in that example). Compare the responses of the two alternative discretization schemes and comment on the effect of the sampling period.

13.10 In the example of the heating tank discussed in Section 13.1, suppose the operation of the valve is subject to random disturbances, so that the steam flowrate is given by

$$\dot{M}_{st} = k_v u + v$$

where v is random noise.

(a) Rederive the state-space description of the closed-loop system under PI control, with v being an additional disturbance input.

(b) For the following values of process and controller parameters

$$\frac{F}{V} = 1, \; \frac{\Delta h_{vap,s}}{V\rho c_p} = 1, \; k_v = 0.2, \; \tau_m = 0.1, \; k_c = 50, \; \tau_I = 1$$

discretize the closed-loop system with sampling period $T_s = 0.01$.

(c) Use the discretized closed-loop system equations to simulate the closed-loop response of $T_m(t)$ to a unit step change in the set point in the presence of valve noise, where the discrete values of v are normally distributed random numbers with zero mean and standard deviation 0.8 (use MATLAB's `normrnd` or `randn` function). Comment on the ability of the control system to handle the given level of valve noise.

14 Systems with Dead Time

The aim of this chapter is to introduce an element of paramount importance to the dynamics of chemical process systems, which is the dead time or time delay. The presence of dead time is usually attributed to the distributed nature of some process systems. In many cases, dead time is introduced as an approximation, to represent collectively higher-order dynamics. In the present chapter, several systems with inherent dead time are presented. Owing to the fact that the presence of dead time results in transfer functions that are not rational functions, several approximation schemes are introduced to facilitate dynamic analysis. Closed-loop dynamics in the presence of dead time in the feedback loop is also studied, including closed-loop stability.

STUDY OBJECTIVES

After studying this chapter you should be able to do the following.

- Identify the underlying reasons for the presence of dead time in process dynamics.
- Select appropriate rational approximations of dead time to perform dynamic analysis of systems with dead time.
- Perform closed-loop stability analysis when dead time is present.
- Use software tools, such as MATLAB or Maple, to simulate and analyze closed-loop systems involving dead time.

14.1 Introduction

14.1.1 A Heating Tank with Dead Time

For many systems in the process industries there is a significant delay between the time a change in a manipulated variable is implemented and the time that a noticeable change in the output is observed. In other words, no response is observed for a definite period of time even though a change in the input of the system has taken place. We refer to such systems as time-delay or dead-time systems. The underlying reason for the observed delay in the response can be attributed to the process itself or to the sensor used. Some simple examples are presented first to familiarize ourselves with the notion of dead time.

In the process shown in Figure 14.1, cold water is fed to a vessel at mass flowrate $\dot{M}$ and temperature T_{in}. Saturated steam at temperature T_{st} and mass flowrate $\dot{M}_{st}$ is fed to an internal coil in order to increase the water temperature up to temperature $T(t)$. There is no temperature sensor installed inside the tank, but there is a temperature sensor located downstream at distance L from the tank, as shown in Figure 14.1. The velocity of the hot water in the exit pipe is v. The energy balance in the tank results in the following first-order differential equation

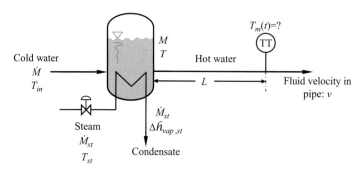

Figure 14.1 Heating tank equipped with a temperature sensor located at distance L in the exit pipe.

$$\frac{d(Mc_p T)}{dt} = \dot{M}c_p T_{in} + \dot{M}_{st} \Delta h_{vap,st} - \dot{M}c_p T \tag{14.1.1}$$

where M is the water holdup in the tank, c_p is the heat capacity of water and $\Delta h_{vap,st}$ is the heat of condensation of steam at T_{st}. If the heat capacity and heat of condensation of steam are assumed constant, then Eq. (14.1.1) simplifies to

$$\tau \frac{dT}{dt} + T = T_{in} + k\dot{M}_{st} \tag{14.1.2}$$

where $\tau = M/\dot{M}$ and $k = \Delta h_{vap,st}/\dot{M}c_p$. If we denote by y the deviation of the tank temperature from its steady-state value and denote by u the deviation of the steam mass flowrate from its steady value then Eq. (14.1.2) becomes

$$\tau \frac{dy}{dt} + y = ku \tag{14.1.3}$$

i.e. the classic form of a first-order system dynamics is obtained.

We then consider the case where a step change is applied to the flowrate of the heating steam. The response of the temperature of the liquid in the tank will change following the step response of a classic first-order system. The temperature sensor, however, will not detect any noticeable change before the hot water travels the distance L, i.e. before time $\theta = L/v$ has elapsed. After that time the sensor will record a temperature response y_m that is identical to the response of the temperature in the tank, but delayed by θ, as shown in Figure 14.2.

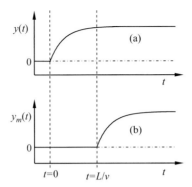

Figure 14.2 The response of the tank temperature and measured temperature to a step change in the steam flowrate.

In Figure 14.2a the response of the temperature of the tank $y(t)$ to a step change in the flowrate of steam is shown. The temperature $y_m(t)$, as recorded by the temperature sensor that is located at distance L in the exit pipe, is also shown in Figure 14.2b. Both responses are shown in deviation variables. We observe that the responses are identical but the measured response is delayed by time $\theta = L/v$:

$$y_m(t) = y(t-\theta) \qquad (14.1.4)$$

A similar process but with slightly different characteristics is shown in Figure 14.3. A stream of cold water with mass flowrate $\dot{M}$ and temperature T_{in} is mixed with saturated steam with mass flowrate $\dot{M}_{st}$ and temperature T_{st}. The mixing point is located at distance L upstream of the tank and the water has velocity v in the pipe. Any change in the mass flowrate of steam will not result in an immediate change in the incoming temperature of the water to the tank as the distance L needs to be traveled by the fluid first. The temperature at the mixing point T_0 can be related to the mass flowrate of water and steam and their temperatures

$$\dot{M}c_p T_{in} + \dot{M}_{st}\left(c_p T_{st} + \Delta \hat{h}_{vap,st}\right) = \left(\dot{M} + \dot{M}_{st}\right)c_p T_0 \qquad (14.1.5)$$

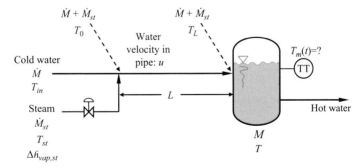

Figure 14.3 A process tank with input delay.

or, by assuming that $\dot{M}_{st} \ll \dot{M}$ in which case $\dot{M} + \dot{M}_{st} \approx \dot{M}$

$$T_0 = T_{in} + k\dot{M}_{st}, \quad k = \left[\frac{1}{\dot{M}}\left(T_{st} + \frac{\Delta h_{vap,st}}{c_p}\right)\right] \tag{14.1.6}$$

Based on the discussion for the system shown in Figure 14.1, we can conclude that the temperature at distance L downstream of the mixing point (entrance of the tank) T_L will be T_0, but delayed by $\theta = L/v$:

$$T_L(t) = T_0(t-\theta) = T_{in} + k\,\dot{M}_{st}(t-\theta) \tag{14.1.7}$$

The energy balance for the tank is written as follows

$$\frac{d(Mc_p T(t))}{dt} = \dot{M}c_p T_L(t) - \dot{M}c_p T(t) = \dot{M}c_p\left[T_{in} + k\dot{M}_{st}(t-\theta)\right] - \dot{M}c_p T(t) \tag{14.1.8}$$

which simplifies to

$$\frac{M}{\dot{M}}\frac{dT(t)}{dt} + T(t) = T_{in} + k\,\dot{M}_{st}(t-\theta) \tag{14.1.9}$$

Assuming that T_{in} is constant, and that y represents the tank temperature in deviation form and u represents the mass flowrate of steam in deviation form, the system takes the form

$$\tau \frac{dy(t)}{dt} + y(t) = k\,u(t-\theta) \tag{14.1.10}$$

In summary, we have seen that first-order, dead-time systems can be described either by Eqs. (14.1.3) and (14.1.4) when the dead time is attributed to the output, or by Eq. (14.1.10) when the dead time is attributed to the input. An nth-order linear system with measurement or output dead time has the following state-space description:

$$\begin{aligned}\frac{dx(t)}{dt} &= Ax(t) + bu(t) \\ y(t) &= cx(t) + du(t) \\ y_m(t) &= y(t-\theta)\end{aligned} \tag{14.1.11}$$

An nth-order linear system with input dead time has the following state-space description:

$$\begin{aligned}\frac{dx(t)}{dt} &= Ax(t) + bu(t-\theta) \\ y(t) &= cx(t) + du(t-\theta)\end{aligned} \tag{14.1.12}$$

We will now derive the transfer-function description of a first-order system with either an input or output dead time. In the case of output dead time, by taking the Laplace transform of Eqs. (14.1.3) and (14.1.4) under the assumption that the system is initially ($t \leq 0$) at zero, we obtain

$$Y(s) = \frac{k}{\tau s + 1} U(s) \tag{14.1.13}$$

$$Y_m(s) = e^{-\theta s} Y(s) \tag{14.1.14}$$

and the overall transfer function description becomes

$$Y_m(s) = \frac{k}{\tau s + 1} e^{-\theta s} U(s) \tag{14.1.15}$$

For the case of input dead time, Laplace transforming Eq. (14.1.10) under the assumption that the system is initially ($t \leq 0$) at zero, we obtain

$$Y(s) = \frac{k}{\tau s + 1} e^{-\theta s} U(s) \tag{14.1.16}$$

Comparing Eqs. (14.1.15) and (14.1.16) we see that the transfer function is the same irrespective of whether the dead time is associated to the input or the output. This result can be generalized to an nth-order linear, single-input–single-output system, Eqs. (14.1.11) or (14.1.12), to obtain the overall transfer function $G(s) = Y_m(s)/U(s)$ in the case of output dead time or $G(s) = Y(s)/U(s)$ for input dead time. In both cases, we find:

$$G(s) = G_0(s) e^{-\theta s} \tag{14.1.17}$$

where

$$G_0(s) = c(sI - A)^{-1} b + d \tag{14.1.18}$$

The role of dead time in the response of a dynamic system will be illustrated through a simple example.

Example 14.1 Step response of a first-order system with dead time

Consider the steam-heated tank shown in Figure 14.1, with the following numerical values of the parameters:

$$M = 1000 \text{ kg}, \dot{M} = 1000 \text{ kg/min}, T_{in} = 30 \text{ °C}, c_p = 4 \text{ kJ/(kg °C)},$$
$$\Delta \hat{h}_{vap,st} = 2000 \text{ kJ/kg}, \dot{M}_{st} = 100 \text{ kg/min}$$

At $t = 0$, the mass flowrate of the steam $\dot{M}_{st}$ is increased by 10%. Compare the temperature of the liquid content of the tank to the temperature reading of the thermometer that is located at distance $L = 10$ m downstream in the exit pipe. It is given that $v = 10$ m/min.

Solution

We first calculate the temperature in the tank under normal operating conditions. From Eq. (14.1.1) at steady state, we obtain

$$0 = \dot{M}c_p T_{in} + \dot{M}_{st,s} \Delta \hat{h}_{vap,st} - \dot{M}c_p T_s \Rightarrow T_s = T_{in} + \frac{\dot{M}_{st,s} \Delta \hat{h}_{vap,st}}{\dot{M}c_p} = 30 + \frac{100 \cdot 2000}{1000 \cdot 4} = 80 \text{ °C}$$

When the steam mass flowrate increases by 10% (i.e. a step increase by 10 kg/min), the variation of the temperature of the liquid in the tank (in deviation form) can be determined by solving Eq. (14.1.3) with $\tau = M/\dot{M} = 1000/1000 = 1$ min, $k = \Delta \hat{h}_{vap,st}/\dot{M}c_p = 2000/(100 \cdot 4) = 0.5$ °C/(kg/min). Using the step-response formula,

$$y(t) = T(t) - T_s = kM\left(1 - e^{-t/\tau}\right) \Rightarrow T(t) = T_s + kM\left(1 - e^{-t/\tau}\right) = 80 + 5\left(1 - e^{-t}\right)$$

The temperature measurement $T_m(t)$ will be equal to $T(t)$ but delayed by $\theta = L/v = 1$ min:

$$T_m(t) = T(t - \theta) = \begin{cases} 80, & 0 < t < 1 \\ 80 + 5\left(1 - e^{-(t-1)}\right), & t \geq 1 \end{cases}$$

The results are shown in Figure 14.4.

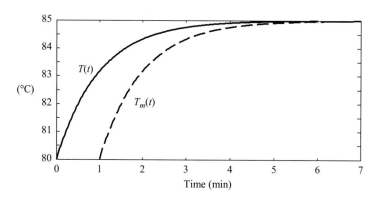

Figure 14.4 Comparison of the temperature in the tank and the delayed measurement.

14.1.2 Distributed Parameter Systems and Dead Time

The dead time observed in process systems is often associated with the dynamics of a distributed parameter system. In the present section, we will explain this physical origin of dead time. To this end, we consider a well-insulated pipe with cross-sectional area A and length L, as shown in Figure 14.5, in which a fluid with constant properties (density ρ and heat capacity c_p) and constant velocity v flows through the pipe. The temperature at the inlet of the pipe varies, and our aim is to derive an equation that can relate it to the temperature at the end of the pipe, i.e. we are seeking a relationship between the temperature at $z = 0$, $T_0(t)$, and the temperature at $z = L$, $T_L(t)$.

14.1 Introduction

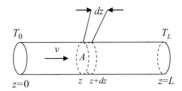

Figure 14.5 A fluid with variable inlet temperature flowing in a pipe.

Our analysis starts by writing an energy balance for a slice of the fluid of thickness dz which is located at point z along the length of the pipe. The energy balance is applied using the methods of Chapter 2 to obtain

$$\frac{d}{dt}\left(Adz \cdot \rho c_p T\right) = Av \cdot \rho c_p T(z) - Av \cdot \rho c_p T(z+dz) \tag{14.1.19}$$

By using the fact that the velocity, the properties of the fluid and the cross-sectional area of the pipe are constant, and by dividing by dz, we obtain

$$\frac{\partial T(z,t)}{\partial t} + v\frac{\partial T(z,t)}{\partial z} = 0 \tag{14.1.20}$$

which holds true for all t and all z. This is a linear partial differential equation (PDE) that describes the temperature dynamics of the fluid at all points in the pipe. It is a PDE and not an ODE, as the temperature varies with respect to time and axial distance along the pipe. The PDE (14.1.20) can be solved by using either classic calculus-based methods or Laplace-transform methods. In the former approach, one first finds the general solution of Eq. (14.1.20), which is

$$T(z,t) = \varphi\left(t - \frac{z}{v}\right) \tag{14.1.21}$$

where $\varphi(\cdot)$ is an arbitrary function. Applying the boundary condition at $z = 0$

$$T(0,t) = T_0(t) \tag{14.1.22}$$

it follows that $\varphi(t) = T_0(t)$ and thus the solution of the PDE is

$$T(z,t) = T_0\left(t - \frac{z}{v}\right) \tag{14.1.23}$$

The final conclusion is that the temperature at $z = L$, $T_L(t)$, is given by

$$T_L(t) = T_0\left(t - \frac{L}{v}\right) \tag{14.1.24}$$

The PDE (14.1.20) can also be solved with Laplace transforms, and the main product of the derivation is the calculation of the transfer function of the system. Suppose that the system is initially ($t \leq 0$) at steady state with the temperature being constant along the length of the pipe and equal to T_s. Defining the temperature deviation variable as

$$\bar{T}(z,t) = T(z,t) - T_s \qquad (14.1.25)$$

Eq. (14.1.20) becomes

$$\frac{\partial \bar{T}(z,t)}{\partial t} + v \frac{\partial \bar{T}(z,t)}{\partial z} = 0 \qquad (14.1.26)$$

Multiplying both sides of (14.1.26) by e^{-st} and integrating from $t = 0$ to $t = \infty$, we get

$$\int_0^\infty e^{-st} \frac{\partial \bar{T}(z,t)}{\partial t} dt + v \int_0^\infty e^{-st} \frac{\partial \bar{T}(z,t)}{\partial z} dt = 0$$

or

$$\left[s \int_0^\infty e^{-st} \bar{T}(z,t) dt - \bar{T}(z,0) \right] + v \frac{d}{dz} \left(\int_0^\infty e^{-st} \bar{T}(z,t) dt \right) = 0$$

and finally, defining the Laplace transform $\bar{T}(z,s) = \int_0^\infty e^{-st} \bar{T}(z,t) dt$ (with respect to the time variable) and noting that the initial temperature distribution is zero,

$$v \frac{d\bar{T}(z,s)}{dz} + s\bar{T}(z,s) = 0 \qquad (14.1.27)$$

It is important to note that, in the same way that the Laplace transform converts an ODE to an algebraic equation, the Laplace transform also converts a PDE to an ODE. Furthermore, it is interesting to note that the axial distance z along the pipe is the independent variable and s is simply a parameter of the ODE (14.1.27). Equation (14.1.27) is a linear, constant-coefficient, first-order ODE whose solution is

$$\bar{T}(z,s) = \bar{T}(0,s) \, e^{-\frac{s}{v} z}$$

and therefore

$$\bar{T}(L,s) = \bar{T}(0,s) \, e^{-\theta s} \qquad (14.1.28)$$

where $\theta = L/v$. But $\bar{T}(0,s)$ and $\bar{T}(L,s)$ are equal to the Laplace transforms of the inlet temperature $\bar{T}_0(t)$ and the outlet temperature $\bar{T}_L(t)$, respectively (in deviation form), so we can write

$$\bar{T}_L(s) = \bar{T}_0(s) \, e^{-\theta s} \qquad (14.1.29)$$

The above result is exactly the Laplace transform of Eq. (14.1.24) when written in deviation form. The foregoing derivation led to the same result that was found before, but it is useful in understanding the connection between the dead time and the nature of distributed systems. Moreover, from (14.1.29), we immediately obtain the transfer function of the system:

$$G(s) = \frac{\bar{T}_L(s)}{\bar{T}_0(s)} = e^{-\theta s} \tag{14.1.30}$$

An equally important result that helps to develop insight into the inherent characteristics of a system with dead time can be obtained from Eq. (14.1.20) by assuming that the pipe has been divided into N slices of thickness $\Delta z = L/N$. In this case, for the i-th slice of the pipe which has temperature T_i, Eq. (14.1.20) can be written approximately, in deviation form, as

$$\frac{d\bar{T}_i}{dt} + v \frac{\bar{T}_i - \bar{T}_{i-1}}{\frac{L}{N}} = 0$$

or

$$\left(\frac{\theta}{N}\right) \frac{d\bar{T}_i}{dt} + \bar{T}_i = \bar{T}_{i-1} \tag{14.1.31}$$

Taking the Laplace transform, we obtain the following approximate transfer function for the i-th slice of the pipe

$$\frac{\bar{T}_i(s)}{\bar{T}_{i-1}(s)} = \frac{1}{\left(\frac{\theta}{N}\right)s + 1} \tag{14.1.32}$$

Using the above, we can relate inlet temperature T_0 to the outlet temperature $T_N = T_L$ as follows:

$$\frac{\bar{T}_L(s)}{\bar{T}_0(s)} = \frac{\bar{T}_N(s)}{\bar{T}_{N-1}(s)} \frac{\bar{T}_{N-1}(s)}{\bar{T}_{N-2}(s)} \cdots \frac{\bar{T}_1(s)}{\bar{T}_0(s)} = \frac{1}{\left[\left(\frac{\theta}{N}\right)s + 1\right]^N} \tag{14.1.33}$$

The step response of N identical first-order systems in series with unit static gain and time constants equal to θ/N is shown in Figure 14.6, for an increasing N. In the limit as $N \to \infty$, we know from calculus that

$$\lim_{N \to \infty} \frac{1}{\left[\left(\frac{\theta}{N}\right)s + 1\right]^N} = e^{-\theta s} \tag{14.1.34}$$

Hence taking the limit as $N \to \infty$ in Eq. (14.1.33) gives

$$\frac{\bar{T}_L(s)}{\bar{T}_0(s)} = e^{-\theta s} \tag{14.1.35}$$

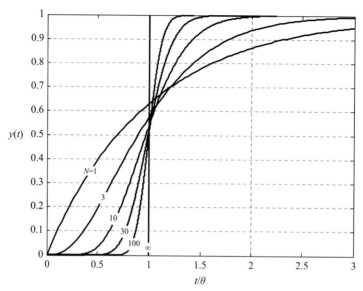

Figure 14.6 Comparison of the unit step response of a pure dead-time system and its approximation using N identical first-order systems in series.

which is, as expected, identical to Eq. (14.1.30) and justifies the step responses shown in Figure 14.6.

The conclusion from the foregoing discussion is that the presence of dead time could be the result of either the underlying distributed nature of a system, or of a large number of low-order systems connected in series, which is a common situation in process systems. Furthermore, Eq. (14.1.34) provides the means to approximate dead time by a rational function. Alternative forms of rational-function approximations of dead time will be presented in the sections that follow.

Before moving further, it is important to note that the dynamic analysis performed in the previous chapters is limited to systems without dead time. The study of systems with dead time can be achieved either by using special analysis methods or by using rational-function approximations of dead time, as was done in the pipe system. The latter approach is often easier, and it will be discussed further in the next section.

14.2 Approximation of Dead Time by Rational Transfer Functions

In the previous section we developed an approximation for the dead time, which is given by Eq. (14.1.34), and the resulting approximation of the unit step response of a pure dead time system shown in Figure 14.6. From this figure we can observe that the approximation of the step response is satisfactory provided that a very large number N (>100) of first-order systems in series can be used. In the present section, we would also like to consider the frequency response for evaluating the accuracy of rational approximations. For a pure dead

14.2 Approximation of Dead Time by Rational Transfer Functions

time with transfer function $G(s)=e^{-\theta s}$, the magnitude and the angle of $G(i\omega) = e^{-\theta i\omega}$ can be immediately obtained since it is already in polar form:

$$|G(i\omega)| = |e^{-\theta i \omega}| = 1 \qquad (14.2.1)$$
$$\arg\{G(i\omega)\} = \arg\{e^{-i\omega\theta}\} = -\omega\theta \qquad (14.2.2)$$

The Bode diagrams of a pure dead-time system are shown in Figure 14.7. Figure 14.8 depicts the Bode diagrams for the approximation of dead time given by (14.1.32). As can be seen

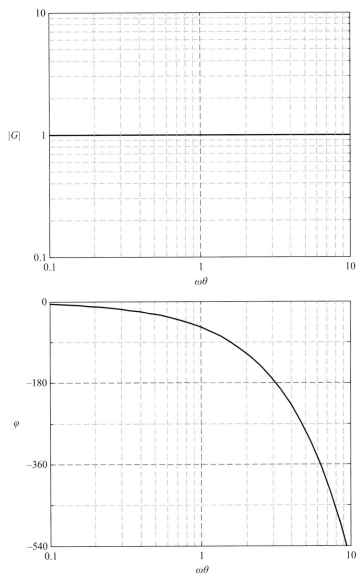

Figure 14.7 Magnitude (top) and phase angle (bottom) as a function of frequency for the pure dead-time system.

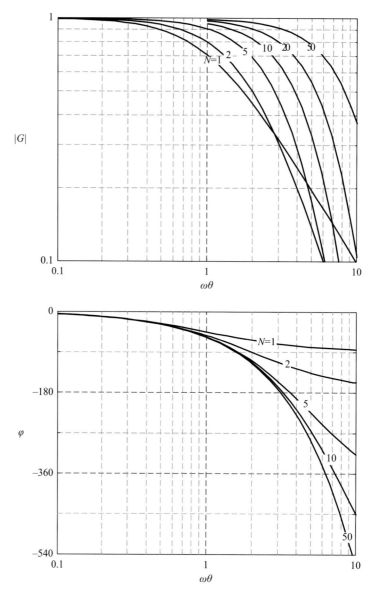

Figure 14.8 Magnitude (top) and phase angle (bottom) as a function of frequency for the dead-time approximation given by identical first-order systems in series.

from Figure 14.8, the approximation of the magnitude is satisfactory up to a frequency $\omega\theta \ll N$. In order to have an acceptable approximation of the phase angle over a fairly broad frequency range of interest, Figure 14.8 shows that N must be chosen to be fairly large (greater than 5). The conclusion from both the step-response and the frequency-response characteristics is that a fairly large value of N should be used in order to obtain reasonable accuracy. Needless to say, a large value of N will make manual calculations prohibitively complicated, and will necessitate the use of computer software.

14.2.1 Taylor and Padé Approximations

From basic engineering mathematics we know that an analytic function $f(s)$ can be expanded in power series in a neighborhood of the point $s = 0$ as follows:

$$f(s) = \sum_{n=0}^{\infty} \frac{f^{(n)}(0)}{n!} s^n = f(0) + \frac{f'(0)}{1!}s + \frac{f''(0)}{2!}s^2 + \ldots \quad (14.2.3)$$

and this is referred to as Taylor series, or Taylor–Maclaurin series when it is around 0. For the exponential function $e^{\theta s}$, the series expansion is:

$$e^{\theta s} = \sum_{n=0}^{\infty} \frac{f^{(n)}(0)}{n!} s^n = 1 + \theta s + \frac{1}{2}(\theta s)^2 + \frac{1}{6}(\theta s)^3 + \ldots + \frac{1}{k!}(\theta s)^k + \ldots \quad (14.2.4)$$

For the exponential function $e^{-\theta s}$, one can obtain similarly

$$e^{-\theta s} = \sum_{n=0}^{\infty} \frac{f^{(n)}(0)}{n!} s^n = 1 - \theta s + \frac{1}{2}(\theta s)^2 - \frac{1}{6}(\theta s)^3 + \ldots + \frac{1}{k!}(-\theta s)^k + \ldots \quad (14.2.5)$$

or, alternatively, one can invert the expansion of $e^{\theta s}$:

$$e^{-\theta s} = \frac{1}{e^{\theta s}} = \frac{1}{1 + \theta s + \frac{1}{2}(\theta s)^2 + \frac{1}{6}(\theta s)^3 + \ldots + \frac{1}{k!}(\theta s)^k + \ldots} \quad (14.2.6)$$

When the series of the denominator of Eq. (14.2.6) is truncated, this leads to rational approximations of dead time:

first-order Taylor approximation:

$$e^{-\theta s} \approx \frac{1}{1 + \theta s} \quad (14.2.7)$$

second-order Taylor approximation:

$$e^{-\theta s} \approx \frac{1}{1 + \theta s + \frac{1}{2}(\theta s)^2} \quad (14.2.8)$$

nth-order Taylor approximation:

$$e^{-\theta s} \approx \frac{1}{1 + \theta s + \frac{1}{2}(\theta s)^2 + \frac{1}{6}(\theta s)^3 + \ldots + \frac{1}{n!}(\theta s)^n} \quad (14.2.9)$$

The so-called Padé approximations are also based on a Taylor series expansion, but the construction of the approximation follows a slightly different rationale. The idea will be illustrated through a simple example.

Example 14.2 Analytic derivation of a Padé approximation

Derive a rational approximation of the function $e^{-\theta s}$ that is the ratio of a first-degree polynomial over a first-degree polynomial, i.e. of the form:

$$e^{-\theta s} \approx \frac{\beta_0 + \beta_1 \theta s}{1 + \alpha_1 \theta s} \tag{14.2.10}$$

Determine the coefficients β_0, β_1 and α_1 by matching the coefficients of the Taylor–Maclaurin series of right-hand side and left-hand side of Eq. (14.2.10) up to second-order terms.

Solution

The series expansion of the right-hand side $f(s) = \dfrac{\beta_0 + \beta_1 \theta s}{1 + \alpha_1 \theta s}$ is calculated as follows:

$$f(s) = \frac{\beta_0 + \beta_1 (\theta s)}{1 + \alpha_1 (\theta s)}\bigg|_{(\theta s)=0} + \frac{d}{d(\theta s)}\left(\frac{\beta_0 + \beta_1 (\theta s)}{1 + \alpha_1 (\theta s)}\right)\bigg|_{(\theta s)=0} (\theta s) + \frac{d^2}{d(\theta s)^2}\left(\frac{\beta_0 + \beta_1 (\theta s)}{1 + \alpha_1 (\theta s)}\right)\bigg|_{(\theta s)=0} \frac{(\theta s)^2}{2} + \cdots$$

$$= \beta_0 + (\beta_1 - \beta_0 \alpha_1)(\theta s) - \alpha_1 (\beta_1 - \beta_0 \alpha_1)(\theta s)^2 + \cdots$$

Matching the coefficients of the above series expansion to the one of $e^{-\theta s}$ given by Eq. (14.2.5) leads to the following equations for the unknown coefficients β_0, β_1 and α_1:

$$\beta_0 = +1$$

$$\beta_1 - \beta_0 \alpha_1 = -1$$

$$\alpha_1 (\beta_1 - \beta_0 \alpha_1) = -\frac{1}{2}$$

from which we obtain $\beta_0 = 1$, $\beta_1 = -1/2$, $\alpha_1 = 1/2$. The result is the so-called first-order Padé approximation:

$$e^{-\theta s} = \frac{1 - \dfrac{\theta}{2} s}{1 + \dfrac{\theta}{2} s} \tag{14.2.11}$$

The idea of Example 14.2 can be directly generalized. One can seek a rational approximation which is the ratio of an m-th-degree polynomial over an n-th-degree polynomial, i.e. of the form:

$$e^{-\theta s} \approx \frac{\beta_0 + \beta_1 \theta s + \beta_2 (\theta s)^2 + \ldots + \beta_m (\theta s)^m}{1 + \alpha_1 \theta s + \alpha_2 (\theta s)^2 + \ldots + \alpha_n (\theta s)^n} \tag{14.2.12}$$

The coefficients β_0, β_1, β_2, ..., β_m and α_1, α_2, ..., α_n can be determined by matching the coefficients of the Taylor–Maclaurin series of right-hand side and left-hand side of

(14.2.12) up to $(m+n)$th-order terms. The result is called an $[m, n]$ Padé approximation and it is given by

$$e^{-\theta s} \approx \frac{P_{[m,n]}(\theta s)}{Q_{[m,n]}(\theta s)} \qquad (14.2.13)$$

where

$$P_{[m,n]}(\theta s) = \sum_{k=0}^{m} \frac{(m+n-k)!m!}{(m+n)!k!(m-k)!}(-\theta s)^k \qquad (14.2.14)$$

$$Q_{[m,n]}(\theta s) = \sum_{k=0}^{n} \frac{(m+n-k)!n!}{(m+n)!k!(n-k)!}(\theta s)^k \qquad (14.2.15)$$

The most commonly used type of Padé approximation is with $m = n$, in which case

$$P_{[n,n]}(\theta s) = Q_{[n,n]}(-\theta s) \qquad (14.2.16)$$

The case $m = 0$ coincides with the Taylor approximation given by (14.2.9). Table 14.1 gives some representative low-order Padé approximations.

Table 14.1 Low-order Padé approximations of the dead time function $e^{-\theta s}$

n	Padé $[n, n]$	Padé $[n-1, n]$	Taylor = Padé $[0, n]$
1	$\dfrac{1 - \dfrac{\theta s}{2}}{1 + \dfrac{\theta s}{2}}$	$\dfrac{1}{1 + \theta s}$	$\dfrac{1}{1 + \theta s}$
2	$\dfrac{1 - \dfrac{\theta s}{2} + \dfrac{\theta^2 s^2}{12}}{1 + \dfrac{\theta s}{2} + \dfrac{\theta^2 s^2}{12}}$	$\dfrac{1 - \dfrac{\theta s}{3}}{1 + \dfrac{2\theta s}{3} + \dfrac{\theta^2 s^2}{6}}$	$\dfrac{1}{1 + \theta s + \dfrac{\theta^2 s^2}{2}}$
3	$\dfrac{1 - \dfrac{\theta s}{2} + \dfrac{\theta^2 s^2}{10} - \dfrac{\theta^3 s^3}{120}}{1 + \dfrac{\theta s}{2} + \dfrac{\theta^2 s^2}{10} + \dfrac{\theta^3 s^3}{120}}$	$\dfrac{1 - \dfrac{2\theta s}{5} + \dfrac{\theta^2 s^2}{20}}{1 + \dfrac{3\theta s}{5} + \dfrac{3\theta^2 s^2}{20} + \dfrac{\theta^3 s^3}{60}}$	$\dfrac{1}{1 + \theta s + \dfrac{\theta^2 s^2}{2} + \dfrac{\theta^3 s^3}{6}}$

Finally, it should be noted that the N tanks in series approximation derived in the previous section is a first-order Taylor approximation applied N times:

$$e^{-\theta s} = \left(e^{-\frac{\theta}{N}s}\right)^N \approx \left(\frac{1}{\left(\dfrac{\theta}{N}\right)s + 1}\right)^N = \frac{1}{\left[\left(\dfrac{\theta}{N}\right)s + 1\right]^N} \qquad (14.2.17)$$

The important question that remains to be answered is related to the accuracy of these approximations. To answer this question, we need a criterion that can be used to judge accuracy. The obvious criterion is the quality of the approximation achieved for the time response to a typical input variation, such as the step input. Another less-obvious criterion is the quality of the approximation achieved for the frequency-response characteristics. As we will see in the next subsection, an approximation that is of high quality in step response may be poor in frequency response or vice versa; however, both criteria have their own significance. Approximations will also be used later in this chapter, in order to calculate closed-loop stability limits.

14.2.2 Accuracy of the Taylor and Padé Approximations

In Figure 14.9 the unit step responses of the pure dead-time transfer function $e^{-\theta s}$ and the first, second and third $[n,n]$ Padé approximations are compared. All $[n, n]$ Padé approximations have numerator and denominator polynomials of equal order and, as a direct result, their unit step response at $t = 0+$ is different than zero. The low-order $[n,n]$ Padé approximations have also n right-half-plane zeros, which explains their peculiar step response. They also have an equal number of symmetrically located poles (with respect to the imaginary axis) in the complex plane.

In Figure 14.10 the phase angle of the pure dead time transfer function $e^{-\theta s}$ and the first, second and third $[n,n]$ Padé approximations are compared. The [1,1] approximation is unsatisfactory while the [2,2] approximation is satisfactory up to $-180°$ and the [3,3] approximation is satisfactory up to $-360°$. The magnitude plots are not shown as all $[n,n]$ Padé approximations have magnitude that is equal to 1 at all frequencies and are perfect approximations

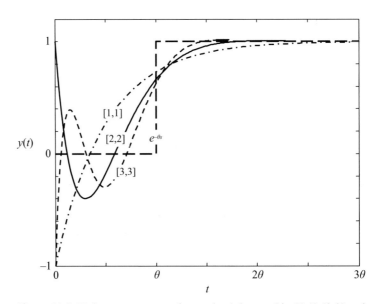

Figure 14.9 Unit step response of pure dead time and its [1,1], [2,2] and [3,3] Padé approximations.

14.2 Approximation of Dead Time by Rational Transfer Functions

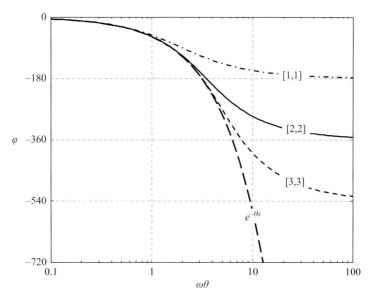

Figure 14.10 Phase angle of pure dead time and its [1,1], [2,2] and [3,3] Padé approximations.

of the magnitude of the pure dead-time transfer function. The main strength of $[n,n]$ Padé approximations is their ability to accurately describe frequency response even at low n.

In order to address the question of whether large-order approximations can offer a remedy to the poor approximation of the unit step response, we compare the unit step responses in Figure 14.11 for the cases of [5,5] and [50,50] Padé approximations. We observe that despite the improvement of the approximation the qualitative characteristics are remarkably different from the characteristics of the $e^{-\theta s}$, for reasons that have already been discussed.

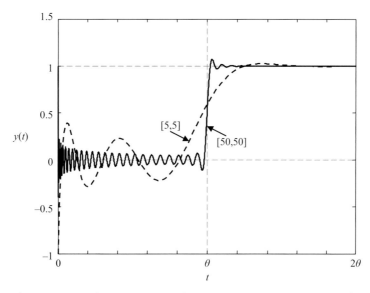

Figure 14.11 Unit step response of the [5,5] and [50,50] Padé approximations.

Taylor and Padé approximations are also used when dead time is present in series with other dynamic phenomena, as in the example of the heating tank (see Section 14.1.1), whose dynamic behavior follows Eq. (14.1.16). This is a first-order plus dead time (FOPDT) system with transfer function

$$G(s) = \frac{k}{\tau s + 1} e^{-\theta s} \qquad (14.2.18)$$

When a Taylor or Padé approximation is applied for the factor $e^{-\theta s}$, the overall approximate transfer function will be rational, and its accuracy will depend on the relative magnitude of time constant τ and dead time θ. When $\tau \ll \theta$, dead time will dominate the dynamic response, whereas when $\tau > \theta$, the first-order dynamics will dominate. In the latter case, the accuracy of the approximation may be significantly improved, and even low-order approximations may be sufficiently accurate.

Figure 14.12 depicts the unit step response of a FOPDT system with $k_p = 1$, $\theta = 1$ and $\tau = 2$ and its approximation using both low-order ([1,1]) and high-order ([10,10]) Padé approximations. In this figure the initial part of the response has been enlarged and included as an inset. The exact unit step response of the FOPDT system is not visible as it almost coincides, for $t/\theta > 1$, with the response of the [10,10] Padé approximation. It is seen from this figure that the approximation of the unit step response of a FOPDT system is significantly better than that of a pure time-delay system and may be considered as adequate even when the [1,1] Padé approximation is used. The accuracy of the approximation is improved for higher-order systems (second order, third order, etc.).

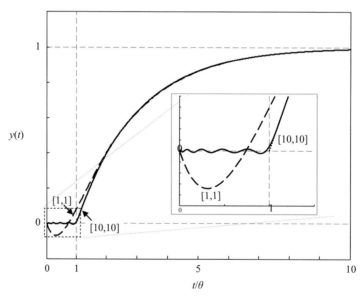

Figure 14.12 Unit step response of the FOPDT system $e^{-s}/(2s+1)$ and its [1,1] and [10,10] Padé approximations.

To summarize, we have examined alternative rational approximations of dead time. The first conclusion is that high-quality approximation of frequency response does not imply high-quality approximation of step response, and vice versa. The [n,n] Padé approximations are the best choice for frequency response even for low values of n, whereas step response is more meaningfully approximated by Taylor or tanks-in-series of high order. The second conclusion is that when the dead time is overshadowed by slower dynamics, the accuracy in step response is improved for all approximations.

14.2.3 Reverse Taylor Approximation

So far, we have considered approximating dead time by a rational transfer function. It is also conceivable to do the reverse approximation: given a higher-order process without dead time, one may wish to approximate it with, for example, a first-order plus dead time (FOPDT) function. This may be done by using the same principle of a Taylor approximation, but in a reverse manner. Consider, for example, a higher-order process consisting of a series of first-order processes:

$$G(s) = \frac{k}{(\tau_{dom} s + 1)\prod_{i=1}^{m}(\tau_i s + 1)}, \quad \text{where} \quad \tau_{dom} \gg \sum_{i=1}^{m} \tau_i \quad (14.2.19)$$

In other words, τ_{dom} is the dominant time constant, which means that the speed of the response will be governed by the decay of $e^{-t/\tau_{dom}}$, whereas the smaller time constants τ_i will contribute only in short time. Applying the first-order Taylor approximation in a reverse manner

$$\frac{1}{\tau_i s + 1} \approx e^{-\tau_i s} \quad (14.2.20)$$

one ends up with a first-order plus dead-time approximate transfer function:

$$G(s) \approx \frac{k}{\tau_{dom} s + 1} e^{-\left(\sum_{i=1}^{m} \tau_i\right)s} \quad (14.2.21)$$

with dead time equal to the sum of the small time constants.

Example 14.3 Approximation of a higher-order rational transfer function by first-order plus dead time (FOPDT)

Derive an FOPDT approximation for the transfer function

$$G(s) = \frac{1}{(5s+1)(s+1)(0.5s+1)}$$

Solution

The dominant time constant is $\tau_{dom}=5$. Applying Eq. (14.2.21), we find:

$$G_{FOPDT}(s) \approx \frac{1}{5s+1}e^{-1.5s}$$

Figure 14.13 compares the unit step responses of the exact and approximate transfer functions. We see that the approximation is very accurate, except for a small initial time period, corresponding to the sum of the small time constants.

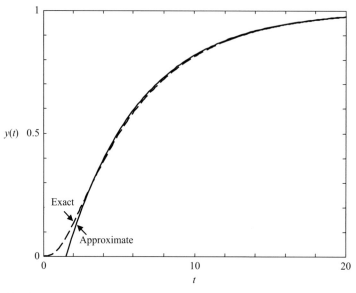

Figure 14.13 Comparison of the unit step response of the system $1/(5s+1)(s+1)(0.5s+1)$ and its FOPDT approximation $e^{-1.5s}/(5s+1)$.

14.3 Parameter Estimation for FOPDT Systems

Low-order models with dead time are often used to fit dynamic response data when a detailed first-principles model is unavailable. The most commonly used empirical dynamic models for chemical process systems are the following:

integrator plus dead time (IPDT)

$$G(s) = \frac{k_p}{s}e^{-\theta s} \quad (14.3.1)$$

first-order plus dead time (FOPDT)

$$G(s) = \frac{k_p}{\tau s+1}e^{-\theta s} \quad (14.3.2)$$

second-order plus dead time (SOPDT)

$$G(s) = \frac{k_p}{(\tau_1 s + 1)(\tau_2 s + 1)} e^{-\theta s} \quad (14.3.3)$$

The FOPDT model is by far the most popular empirical model, since its step response can approximately capture the sigmoidal shape of the step response observed in many process systems. In this section we will present some of the most commonly used methods to determine the parameters of a FOPDT system.

The response of a FOPDT model to a step input of magnitude M is given by

$$y(t) = \begin{cases} 0, & t < \theta \\ k_p M \left(1 - e^{-\left(\frac{t-\theta}{\tau}\right)} \right), & t \geq \theta \end{cases} \quad (14.3.4)$$

i.e. no response is obtained for time lower than the dead time, followed by the response of a classic first-order system delayed by θ, as shown in Figure 14.14. When a step response experiment is performed, one can try to determine the values of the parameters (k_p, τ, θ) of the FOPDT model, to obtain the best fit. The process static gain is obtained easily as the ratio of the new output steady state divided by the input steady state M, as can be seen from Eq. (14.3.4). The situation can become complicated when significant measurement noise is present in which case determining, among other things, when the process has reached a new steady state can be challenging.

To estimate the dead time and the time constant of the process, there are several options. The simplest is to visually identify the dead time and then use the slope of the response at

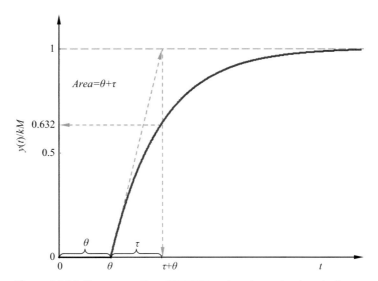

Figure 14.14 Response of an FOPDT system to a step input of magnitude M, applied at time $t = 0$.

$t = \theta$ to estimate the time constant, or alternatively the time at which the output covers 63.2% of the distance (see Figure 14.14). However, this could also be rather problematic in the presence of noise in the data, especially the estimation of the slope could be very inaccurate.

The sum of dead time plus time constant can estimated more reliably (even in the presence of noise) using the fact that (see Figure 14.14)

$$\tau + \theta = \int_0^\infty \left(1 - \frac{y(t)}{k_p M}\right) dt \qquad (14.3.5)$$

where the integral can be calculated numerically using, for instance, the trapezoidal rule. Thus, if it is possible to visually identify the dead time, the above result will immediately give the time constant. Otherwise, one could use (14.3.5) to estimate the sum $(\tau + \theta)$ and then τ from the integral (see also Problem 14.6):

$$\int_0^{\tau+\theta} \left(\frac{y(t)}{kM}\right) dt = \frac{\tau}{e} \qquad (14.3.6)$$

Finally, another possibility would be to use the values of the output at two or more time points $t_1, t_2, \ldots$, all larger than the apparent dead time θ, and then fit τ and θ to the set of equations

$$\theta - \ln\left(1 - \frac{y(t_i)}{y(\infty)}\right) \cdot \tau = t_i \qquad (14.3.7)$$

which immediately follow from (14.3.4).

Example 14.4 Approximation of a higher-order system by an FOPDT using step response data

In Table 14.2 the unit step response data of a higher-order system are given. Based on these data, estimate the parameters of an FOPDT model.

Solution

Initially we apply Eq. (14.3.5):

$$\tau + \theta = \int_0^\infty \left(1 - \frac{y(t)}{k_p M}\right) dt = \int_0^\infty (1 - y(t)) dt \approx \sum_{i=1}^{11} (t_{i+1} - t_i) \cdot \frac{(1 - y(t_{i+1})) + (1 - y(t_i))}{2}$$

$$= (1 - 0) \cdot \frac{(1 - 0.0190) + (1 - 0.0000)}{2} + (2 - 1) \cdot \frac{(1 - 0.1429) + (1 - 0.0190)}{2} + \ldots$$

$$= 0.9905 + 0.9191 + \ldots \approx 4$$

The details of the calculation are also shown in the last column of Table 14.2.

By visual inspection, $\theta \approx 1.5$ s, which can be combined with the above integration result to give $\tau \approx 4 - \theta = 2.5$ s, and thus we obtain the approximation shown in Figure 14.15.

As an alternative, we apply Eq. (14.3.7) with $t_1 = 3$ s and $t_2 = 6$ s. This gives $\tau = 2.01$ s and $\theta = 2.16$ s, and we obtain the approximation shown in Figure 14.15.

We see that neither one of the two fits is very accurate, and this is because the data came from a system that is not truly FOPDT.

Table 14.2 Unit step response data for Example 14.4

i	Time t(s)	$y(t)$	Integral approximation
1	0	0.0000	0.9905
2	1	0.0190	0.9191
3	2	0.1429	0.7522
4	3	0.3528	0.5404
5	4	0.5665	0.3492
6	5	0.7350	0.2081
7	6	0.8488	0.1165
8	7	0.9182	0.0621
9	8	0.9576	0.0318
10	9	0.9788	0.0158
11	10	0.9897	0.0211
12	15	1.0000	
			sum=4.0068

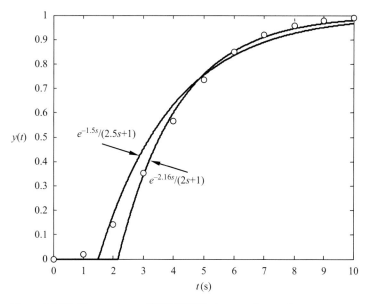

Figure 14.15 Comparison of FOPDT fits to experimental step response data – Example 14.3.

14.4 Feedback Control of Systems with Dead Time – Closed-Loop Stability Analysis

In this section, we will study the problem of feedback control of a system with dead time as shown in Figure 14.16: a controller $G_c(s)$ is placed around a process system whose transfer function is $G_0(s)e^{-\theta s}$, with dead time θ. There are two main issues to be studied: (i) stability analysis and (ii) calculation of the closed-loop response. The transfer function description of the feedback control system of Figure 14.16 is given by

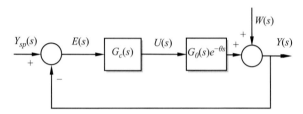

Figure 14.16 Block-diagram representation of feedback control system.

$$Y(s) = \frac{G_c(s)G_0(s)e^{-\theta s}}{1+G_c(s)G_0(s)e^{-\theta s}} Y_{sp}(s) + \frac{1}{1+G_c(s)G_0(s)e^{-\theta s}} W(s) \qquad (14.4.1)$$

When $G_0(s)$ and $G_c(s)$ are rational functions

$$G_0(s) = \frac{N(s)}{D(s)} \qquad (14.4.2)$$

$$G_c(s) = \frac{N_c(s)}{D_c(s)} \qquad (14.4.3)$$

with $N(s)$, $D(s)$, $N_c(s)$, $D_c(s)$ polynomials, (14.4.1) can be written as

$$Y(s) = \frac{N_c(s)N(s)e^{-\theta s}}{D_c(s)D(s)+N_c(s)N(s)e^{-\theta s}} Y_{sp}(s) + \frac{D_c(s)D(s)}{D_c(s)D(s)+N_c(s)N(s)e^{-\theta s}} W(s) \qquad (14.4.4)$$

We observe that the common denominator of the closed-loop transfer functions is $D_c(s)D(s)+N_c(s)N(s)e^{-\theta s}$, and it is not a polynomial. This creates some complications in the analysis of the closed-loop system.

The poles of the closed-loop system are the roots of the characteristic equation

$$D_c(s)D(s)+N_c(s)N(s)e^{-\theta s} = 0 \qquad (14.4.5)$$

and this equation has infinitely many roots. There is no general formula or general numerical method to calculate all the roots. Therefore, inversion of Laplace transforms arising from Eq.(14.4.4) is problematic. Also, Routh's criterion (see Chapter 12) cannot be used for

14.4 Feedback Control of Systems with Dead Time – Closed-Loop Stability Analysis

stability analysis here, because it is only applicable to polynomials. Because of these difficulties, special methods need to be used for closed-loop stability analysis and closed-loop response calculation. These will be presented in the remainder of this section and in the next section.

14.4.1 Approximate Stability Analysis using [n,n] Padé Approximations

One way to bypass the aforementioned mathematical difficulties is to use rational approximations of $e^{-\theta s}$ to derive approximate conditions for stability. The approximations that are most suitable for this purpose are Padé approximations with equal order of numerator and denominator ($m = n$).

When $e^{-\theta s}$ is approximated by a [n,n] Padé approximation, substituting (14.2.13) with (14.2.16) into (14.4.4), we obtain that

$$Y(s) \approx \frac{N_c(s)N(s)\dfrac{Q_{[n,n]}(-\theta s)}{Q_{[n,n]}(\theta s)}}{D_c(s)D(s) + N_c(s)N(s)\dfrac{Q_{[n,n]}(-\theta s)}{Q_{[n,n]}(\theta s)}} Y_{sp}(s)$$

$$+ \frac{D_c(s)D(s)}{D_c(s)D(s) + N_c(s)N(s)\dfrac{Q_{[n,n]}(-\theta s)}{Q_{[n,n]}(\theta s)}} W(s)$$

(14.4.6)

and the closed-loop transfer functions are approximated by rational functions. The poles of the approximate closed-loop transfer functions are the roots of the polynomial equation

$$D_c(s)D(s)Q_{[n,n]}(\theta s) + N_c(s)N(s)Q_{[n,n]}(-\theta s) = 0 \quad (14.4.7)$$

and stability can now be investigated using Routh's criterion. The procedure is best presented through a specific example.

Example 14.5 Approximate closed-loop stability analysis of proportional control of a FOPDT system

Consider a FOPDT system with transfer function (14.3.2), which is controlled by a P controller with transfer function $G_c(s) = k_c$. Derive closed-loop stability conditions using [1,1] and [2,2] Padé approximations of the dead time.

Solution

For the given problem, the transfer function description of the closed-loop system (14.4.1) takes the form

$$Y(s) = \frac{\frac{k_c k_p}{\tau s+1}e^{-\theta s}}{1+\frac{k_c k_p}{\tau s+1}e^{-\theta s}} Y_{sp}(s) + \frac{1}{1+\frac{k_c k_p}{\tau s+1}e^{-\theta s}} W(s)$$

$$= \frac{k_c k_p e^{-\theta s}}{\tau s+1+k_c k_p e^{-\theta s}} Y_{sp}(s) + \frac{\tau s+1}{\tau s+1+k_c k_p e^{-\theta s}} W(s) \quad (14.4.8)$$

When the [1,1] Padé approximation is used to approximate $e^{-\theta s}$ in (14.4.8), the characteristic equation becomes

$$1 + \frac{k_c k_p}{\tau s+1} e^{-\theta s} \approx 1 + \frac{k_c k_p}{\tau s+1} \cdot \frac{1-\frac{\theta}{2}s}{1+\frac{\theta}{2}s} = 0$$

or, after rearrangement,

$$\tau \frac{\theta}{2} s^2 + \left[\tau + \frac{\theta}{2}(1-k_c k_p)\right] s + (1+k_c k_p) = 0 \quad (14.4.9)$$

Using Routh's criterion it follows that the closed-loop system is stable if and only if the following two conditions hold true:

$$1 + k_c k_p > 0 \quad (14.4.10)$$
$$2\tau + \theta(1-k_c k_p) > 0 \quad (14.4.11)$$

These conditions can also be written as

$$-1 < k_c k_p < 1 + 2\left(\frac{\tau}{\theta}\right) \quad (14.4.12)$$

When the [2,2] Padé approximation is used and the same procedure is followed, then the following condition for stability is obtained

$$-1 < k_c k_p < \sqrt{\left[1+3\left(\frac{\tau}{\theta}\right)\right]^2 + 12\left(\frac{\tau}{\theta}\right)^2} - 3\left(\frac{\tau}{\theta}\right) \quad (14.4.13)$$

Table 14.3 summarizes the results for the approximate stability limits derived through [1,1] and [2,2] Padé approximations, and compares them with the exact stability limits, which will be derived in the next subsection. Both Padé approximations provide the exact value of the lower limit of stability (−1). The error in approximating the upper limit of stability is of the order of 20%–30% for the [1,1] Padé approximation. The [2,2] Padé approximation is very accurate with an error of around 1%. The percentage error is shown in Figure 14.17 as a function of τ/θ.

14.4 Feedback Control of Systems with Dead Time – Closed-Loop Stability Analysis

Table 14.3 Approximate stability limits for a FOPDT system under P control via low-order Padé approximations – comparison with exact stability limits

τ/θ	Stability range from [1,1] Padé	Stability range from [2,2] Padé	Exact stability range
0.01	$-1 < k_c k_p < 1.02$	$-1 < k_c k_p < 1.0006$	$-1 < k_c k_p < 1.0005$
0.10	$-1 < k_c k_p < 1.20$	$-1 < k_c k_p < 1.0454$	$-1 < k_c k_p < 1.0402$
0.20	$-1 < k_c k_p < 1.40$	$-1 < k_c k_p < 1.1436$	$-1 < k_c k_p < 1.1321$
0.50	$-1 < k_c k_p < 2.00$	$-1 < k_c k_p < 1.5414$	$-1 < k_c k_p < 1.5198$
1.00	$-1 < k_c k_p < 3.00$	$-1 < k_c k_p < 2.2915$	$-1 < k_c k_p < 2.2618$
2.00	$-1 < k_c k_p < 5.00$	$-1 < k_c k_p < 3.8489$	$-1 < k_c k_p < 3.8069$
5.00	$-1 < k_c k_p < 11.00$	$-1 < k_c k_p < 8.5797$	$-1 < k_c k_p < 8.5024$
10.00	$-1 < k_c k_p < 21.00$	$-1 < k_c k_p < 16.4866$	$-1 < k_c k_p < 16.3506$
50.00	$-1 < k_c k_p < 101.00$	$-1 < k_c k_p < 79.7847$	$-1 < k_c k_p < 79.1776$
100.00	$-1 < k_c k_p < 201.00$	$-1 < k_c k_p < 158.9128$	$-1 < k_c k_p < 157.7168$

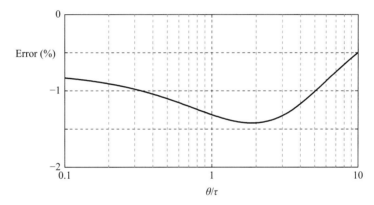

Figure 14.17 Relative percentage error in approximating the upper stability limit for an FOPDT system under P control, with a [2,2] Padé approximation

The high accuracy of the [2,2] Padé approximation makes it particularly useful for performing approximate stability analysis, and determining the critical values of the controller gain that brings the system onto the verge of instability.

Finally, from Table 14.3, we can observe the strong effect of the size of the dead time on the size of the stability range. For small values of the dead time ($\theta \to 0$), controller gain becomes unbounded from above ($-1 < k_c k_p < +\infty$), while when the dead time is significantly larger than the process time constant ($\theta \gg \tau$), the stability range becomes very limited ($-1 < k_c k_p < 1$).

14.4.2 Calculation of the Exact Stability Limits

In the previous subsection, we used Padé approximations to estimate the closed-loop stability limits. It turns out that the exact stability limits can also be calculated, even though the calculation generally requires more effort. The calculation starts with the observation that, at the limit of stability, a closed-loop pole must be zero or pure imaginary. Therefore, at the limit of stability, the characteristic equation has to be satisfied for $s = i\omega$ (see also Section 12.4 of Chapter 12).

For the particular example of P control of a FOPDT system that was considered in the previous subsection, the characteristic equation of the closed-loop system (14.4.8) is

$$1 + \frac{k_c k_p e^{-\theta s}}{\tau s + 1} = 0 \tag{14.4.14}$$

to which, substituting $s = i\omega$, we obtain

$$1 + \frac{k_c k_p}{1 + i\tau\omega} e^{-i\theta\omega} = 0$$

or

$$\frac{k_c k_p}{1 + i\tau\omega} e^{-i\theta\omega} = -1 \tag{14.4.15}$$

The above is clearly satisfied for $\omega = 0$ and $k_c k_p = -1$. This is the lower limit of the stability range that we have found using Padé approximations, and it is also the exact value of the lower limit.

The upper limit will correspond to $k_c k_p > 0$. To calculate it, we can convert both sides of Eq. (14.4.15) to polar form:

$$\frac{k_c k_p}{\sqrt{\tau^2 \omega^2 + 1}} e^{-i(\theta\omega + \tan^{-1}(\tau\omega))} = e^{-i\pi}$$

Thus, if we can find a pair of $k_c k_p$ and ω such that

$$\frac{k_c k_p}{\sqrt{\tau^2 \omega^2 + 1}} = 1 \tag{14.4.16a}$$

$$\theta\omega + \tan^{-1}(\tau\omega) = \pi \tag{14.4.16b}$$

the closed-loop system will have an imaginary pole at $i\omega$.

Equations (14.4.16a, b) can be solved numerically. We first compute the solution $\omega = \omega_p$ of (14.4.16b), and then substitute it to (14.4.16a) to obtain the upper limit of stability: $(k_c k_p)_{max} = \sqrt{\tau^2 \omega_p^2 + 1}$. In this way, we obtain the results of the right-hand column of Table 14.3.

It is important to note at this point that, when trying to match the polar forms of the two sides of Eq. (14.4.15), the arguments are not uniquely defined: any multiple of 2π can be added to the arguments and the equality still holds. So, matching the arguments in (14.4.15) will generally lead to the conclusion that $\theta\omega + \tan^{-1}(\tau\omega) =$ (odd multiple of π), and the question will be which one corresponds to the limit of stability. The reason that Eq. (14.4.16b) involves the correct multiple of π is that it gives rise to the smallest solution for ω, which in turn gives the smallest positive value of $k_c k_p$ such that the closed-loop system has an imaginary pole.

The method of the example can be generalized. The poles of the closed-loop system (14.4.1) are the roots of the characteristic equation:

$$1+G_c(s)G_0(s)e^{-\theta s}=0 \tag{14.4.17}$$

At the verge of instability, one of the poles will be either zero or pure imaginary, i.e. one of the roots of Eq. (14.4.17) will be $i\omega$, so we will have:

$$1+G_c(i\omega)G_0(i\omega)e^{-i\theta\omega}=0 \tag{14.4.18}$$

or

$$G_c(i\omega)G_0(i\omega)e^{-i\theta\omega}=-1=e^{-i\pi} \tag{14.4.19}$$

Conditions (14.4.18) or (14.4.19) will be satisfied when

$$\left|G_c(i\omega)G_0(i\omega)\right|=1 \tag{14.4.20}$$

and

$$\arg\left\{G_c(i\omega)G_0(i\omega)\right\}-\theta\omega=-\pi \tag{14.4.21}$$

Of course, there are complications arising from the presence of multiple (actually infinite) possibilities for the closed-loop system having a pole at $i\omega$, as we saw in the FOPDT example. There are two ways to resolve this issue. One possibility (brute-force) is to use Padé approximations as a guide, the other (mathematically rigorous) is to use the Bode or Nyquist stability criteria, which will be discussed in Chapter 17.

Example 14.6 Closed-loop stability analysis of proportional control of an IPDT system

Consider an integrator plus dead time (IPDT) system with transfer function (14.3.1), which is controlled by a P controller with transfer function $G_c(s) = k_c$. Derive closed-loop stability conditions using [1,1] and [2,2] Padé approximations of the dead time, as well as the exact stability conditions.

Solution

For the given problem, the transfer-function description of the closed-loop system (14.4.1) takes the form

$$Y(s) = \frac{\frac{k_c k_p}{s} e^{-\theta s}}{1 + \frac{k_c k_p}{s} e^{-\theta s}} Y_{sp}(s) + \frac{1}{1 + \frac{k_c k_p}{s} e^{-\theta s}} W(s)$$

$$= \frac{k_c k_p e^{-\theta s}}{s + k_c k_p e^{-\theta s}} Y_{sp}(s) + \frac{s}{s + k_c k_p e^{-\theta s}} W(s)$$

Using the [1,1] Padé approximation we have the characteristic equation

$$1 + \frac{k_c k_p}{s} \cdot \frac{1 - \frac{\theta}{2} s}{1 + \frac{\theta}{2} s} = 0$$

or

$$\frac{\theta}{2} s^2 + \left[1 - k_c k_p \frac{\theta}{2}\right] s + k_c k_p = 0$$

The Routh criterion gives the following conditions for closed-loop stability:

$$k_c k_p > 0$$
$$2 - k_c k_p \theta > 0$$

These can be combined as

$$0 < k_c k_p < \frac{2}{\theta}$$

When the [2,2] Padé approximation is used then the characteristic equation of the closed-loop system becomes

$$1 + \frac{k_c k_p}{s} \cdot \frac{\frac{\theta^2 s^2}{12} - \frac{\theta s}{2} + 1}{\frac{\theta^2 s^2}{12} + \frac{\theta s}{2} + 1} = 0$$

or

$$\frac{\theta^2}{12} s^3 + \left(\frac{\theta}{2} + k_c k_p \frac{\theta^2}{12}\right) s^2 + \left(1 - k_c k_p \frac{\theta}{2}\right) s + k_c k_p$$

Application of the Routh criterion gives the following conditions for closed-loop stability:

$$0 < k_c k_p < \frac{\sqrt{21} - 3}{\theta} \approx \frac{1.58}{\theta}$$

The exact characteristic equation of the closed-loop system is formed by using a P controller with gain k_c and an IPDT model is the following:

$$1 + \frac{k_c k_p}{s} e^{-\theta s} = 0 \quad \Leftrightarrow \quad s + k_c k_p e^{-\theta s} = 0$$

The characteristic equation is satisfied for $s = 0$ and $k_c k_p = 0$. This corresponds to the lower limit of the stability range and it agrees with the results found using Padé approximations.

The upper limit of the stability range will be for $k_c k_p > 0$. To calculate it, we can substitute $s = i\omega$ to the characteristic equation and convert it to polar form:

$$1 + \frac{k_c k_p}{i\omega} e^{-i\omega\theta} = 0 \iff \frac{k_c k_p}{i\omega} e^{-i\theta\omega} = -1 \iff \frac{k_c k_p}{\omega} e^{-i\left(\frac{\pi}{2}+\theta\omega\right)} = e^{-i\pi}$$

The above is satisfied when $\dfrac{k_c k_p}{\omega} = 1$ and $-\dfrac{\pi}{2} - \theta\omega = -\pi$, i.e. when $k_c k_p = \omega = \dfrac{\pi}{2\theta}$. This is the smallest positive value of $k_c k_p$ such that gives rise to an imaginary pole.

Thus, we can conclude that the exact stability range is $0 < k_c k_p < \dfrac{\pi}{2\theta} \approx \dfrac{1.5708}{\theta}$.

Comparing with the approximate stability ranges from Padé approximations, we see that the [1,1] approximation overestimates the upper limit of stability by a factor of $(4/\pi) \approx 1.27$, whereas the [2,2] approximation by a factor of $(2/\pi)(\sqrt{21} - 3) \approx 1.0075$.

14.5 Calculation of Closed-Loop Response for Systems involving Dead Time

In this section we will present briefly a general approach for calculating the closed-loop response for systems involving dead time. The general approach is similar to that followed in Chapter 13, but now the closed-loop system dynamics will be more complex due to the presence of dead time. We consider a system with input dead time described by a state-space model of the form

$$\frac{dx(t)}{dt} = Ax(t) + bu(t - \theta) \tag{14.5.1}$$
$$y(t) = cx(t)$$

which is controlled by a proportional feedback controller:

$$u(t) = k_c\left(y_{sp}(t) - y(t)\right) = k_c\left(y_{sp}(t) - k_c cx(t)\right) \tag{14.5.2}$$

To derive the state-space description of the closed-loop system, we substitute (14.5.2) into (14.5.1) to obtain

$$\frac{dx(t)}{dt} = Ax(t) - bk_c cx(t - \theta) + bk_c y_{sp}(t - \theta) \tag{14.5.3}$$
$$y(t) = cx(t)$$

We note that the closed-loop system has a "delayed" input ($y_{sp}(t - \theta)$) and a term proportional to the "delayed" state vector ($x(t - \theta)$), along with a term that depends on the current value of the state vector ($x(t)$). This is clearly not a system of ordinary differential equations (ODEs), it is in fact a different class of dynamical systems called differential delay systems (DDEs). In what follows we will outline a methodology for solving DDEs known as the Myshkys Method of Steps.

Equation (14.5.3) is usually accompanied by an auxiliary condition, stated in terms of a function revealing the state of the system for a period prior to the initial time. The auxiliary function is sometimes called the "history" function for the system. The need for the "history" function stems from the fact that knowing the input and the initial state of a system DDE is not enough in order to determine its solution. We will denote the "history" function by $x_o(t)$ and, in addition to the input function we will consider that

$$x(t) = x_o(t), \quad -\theta \leq t \leq 0 \tag{14.5.4}$$

is also known. If the "history" function is known then it can be substituted into the DDE and, as the input function and the initial state are known, we can determine the solution over the interval $[0, \theta]$. We then proceed by calculating, in the same way, the solution over the interval $[\theta, 2\theta]$, etc.

The methodology just described can best be demonstrated by considering the FOPDT system as an example (which corresponds to $A = -1/\tau$, $b = k_p/\tau$, $c = 1$):

$$\tau \frac{dy(t)}{dt} = -y(t) + k_p u(t - \theta) \tag{14.5.5}$$

When this system is under P control,

$$u(t) = k_c \left(y_{sp}(t) - y(t) \right) \tag{14.5.6}$$

the closed-loop dynamics follows the DDE

$$\tau \frac{dy(t)}{dt} = -y(t) - k_p k_c y(t - \theta) + k_p k_c y_{sp}(t - \theta) \tag{14.5.7}$$

If we assume that the system is initially at zero, i.e. $y(t) = y_o(t) = 0$ for $-\theta \leq t \leq 0$ and that the set point undergoes a unit step change at $t = 0$, then we can solve iteratively the DDE given by (14.5.7) as follows:

on the interval $[0, \theta]$:

$$\tau \frac{dy(t)}{dt} = -y(t) - k_p k_c \underbrace{y(t-\theta)}_{=0} + k_p k_c \underbrace{y_{sp}(t-\theta)}_{=0}, \quad y(0) = 0,$$

resulting in $y(t) = 0$, $t \in [0, \theta]$;

on the interval $[\theta, 2\theta]$:

$$\tau \frac{dy(t)}{dt} = -y(t) - k_p k_c \underbrace{y(t-\theta)}_{=0} + k_p k_c \underbrace{y_{sp}(t-\theta)}_{=1}, \quad y(\theta) = 0,$$

$$= -y(t) + k_p k_c$$

resulting in $y(t) = k_p k_c \left(1 - e^{-(t-\theta)/\tau}\right)$, $t \in [\theta, 2\theta]$;

on the interval $[2\theta, 3\theta]$:

$$\tau \frac{dy(t)}{dt} = -y(t) - k_p k_c \underbrace{y(t-\theta)}_{=k_p k_c (1-e^{-(t-2\theta)/\tau})} + k_p k_c \underbrace{y_{sp}(t-\theta)}_{=1}, \quad y(2\theta) = k_p k_c \left(1 - e^{-\theta/\tau}\right)$$

$$= -y(t) - (k_p k_c)^2 \left(1 - e^{-(t-2\theta)/\tau}\right) + k_p k_c$$

resulting in $y(t) = k_p k_c \left(1 - e^{-(t-\theta)/\tau}\right) - (k_p k_c)^2 \left(1 - \left(1 + \frac{t-2\theta}{\tau}\right) e^{-(t-2\theta)/\tau}\right)$, $t \in [2\theta, 3\theta]$;
etc.

Apparently, the calculation becomes too complex to handle manually as time progresses. Fortunately, computer software such as MATLAB and Maple are available to perform these calculations numerically or symbolically.

The same calculation may be performed using the transfer function description of the closed-loop system. For an FOPDT system under P control, the closed-loop transfer function with respect to the set point is given by

$$\frac{Y(s)}{Y_{sp}(s)} = \frac{\dfrac{k_c k_p}{\tau s + 1} e^{-\theta s}}{1 + \dfrac{k_c k_p}{\tau s + 1} e^{-\theta s}} \tag{14.5.8}$$

When the set point undergoes a unit step change, the Laplace transform of the output is given by

$$Y(s) = \frac{\dfrac{k_c k_p}{\tau s + 1} e^{-\theta s}}{1 + \dfrac{k_c k_p}{\tau s + 1} e^{-\theta s}} \cdot \frac{1}{s} \tag{14.5.9}$$

At this point, we cannot move forward in the usual way of Laplace-transform inversion, because of the presence of the dead time. However, it is possible to use the following power series expansion

$$\frac{z}{1+z} = z - z^2 + z^3 - z^4 + \cdots \tag{14.5.10}$$

to write Eq. (14.5.9) as follows

$$Y(s) = \left[\frac{k_c k_p}{\tau s + 1} e^{-\theta s} - \left(\frac{k_c k_p}{\tau s + 1}\right)^2 e^{-2\theta s} + \left(\frac{k_c k_p}{\tau s + 1}\right)^3 e^{-3\theta s} - \left(\frac{k_c k_p}{\tau s + 1}\right)^4 e^{-4\theta s} + \cdots\right] \cdot \frac{1}{s} \tag{14.5.11}$$

We then make use of the following results for the inverse Laplace transform

$$y_1(t) = \mathcal{L}^{-1}\left\{\frac{1}{\tau s+1} \cdot \frac{1}{s}\right\} = 1 - e^{-\frac{t}{\tau}} \tag{14.5.12a}$$

$$y_2(t) = \mathcal{L}^{-1}\left\{\left(\frac{1}{\tau s+1}\right)^2 \cdot \frac{1}{s}\right\} = 1 - \left(1 + \frac{t}{\tau}\right)e^{-\frac{t}{\tau}} \tag{14.5.12b}$$

$$y_3(t) = \mathcal{L}^{-1}\left\{\left(\frac{1}{\tau s+1}\right)^3 \cdot \frac{1}{s}\right\} = 1 - \left(1 + \frac{t}{\tau} + \frac{t^2}{2\tau^2}\right)e^{-\frac{t}{\tau}} \tag{14.5.12c}$$

$$y_\nu(t) = \mathcal{L}^{-1}\left\{\left(\frac{1}{\tau s+1}\right)^\nu \cdot \frac{1}{s}\right\} = 1 - \left[\sum_{\ell=0}^{\nu-1}\frac{1}{\ell!}\left(\frac{t}{\tau}\right)^\ell\right]e^{-\frac{t}{\tau}} \tag{14.5.12d}$$

Using Eqs. (14.5.12a–d) we can obtain the complete unit step response of the closed-loop system

$$y(t) = \sum_{\nu=1}^{N}(-1)^{\nu-1}(kk_c)^\nu y_\nu(t-\nu\theta), \quad N\theta < t < (N+1)\theta \tag{14.5.13}$$

which holds for every $N \geq 1$. Applying Eq. (14.5.13) in the interval $[2\theta, 3\theta]$ (i.e. for $N = 2$) results in

$$y(t) = \sum_{\nu=1}^{2}(-1)^{\nu-1}(kk_c)^\nu y_\nu(t-\nu\theta)$$
$$= (-1)^0(kk_c)^1 y_1(t-\theta) + (-1)^1(kk_c)^2 y_2(t-2\theta)$$
$$= kk_c\left(1 - e^{-(t-\theta)/\tau}\right) - (kk_c)^2\left\{1 - \left[1 + \left(\frac{t-2\theta}{\tau}\right)\right]e^{-\frac{(t-2\theta)}{\tau}}\right\}$$

which agrees with the results obtained by using the state-space description.

The foregoing series-expansion method can be generalized. Considering the general feedback loop of Figure 14.16, the closed-loop transfer functions in (14.4.1) can be expanded in the form of power series in $G_c(s)G_0(s)e^{-\theta s}$ as follows:

$$Y(s) = \left\{\sum_{\nu=1}^{\infty}(-1)^{\nu-1}[G_c(s)G_0(s)]^\nu e^{-\nu\theta s}\right\}Y_{sp}(s) + \left\{\sum_{\nu=0}^{\infty}(-1)^\nu[G_c(s)G_0(s)]^\nu e^{-\nu\theta s}\right\}W(s) \tag{14.5.14}$$

For a specific change in set point or disturbance, the resulting series can be inverted term by term. Each term involving a power $[G_c(s)G_0(s)]^\nu e^{-\nu\theta s}$ will become active after time $\nu\theta$, therefore over a finite time horizon, the response will have a finite number of terms. The application of Eq. (14.5.14) necessitates the use of symbolic software like Maple, and this will be demonstrated in the next section.

14.5.1 Using Approximations of the Open-Loop Transfer Function to Calculate the Closed-Loop Response

In the previous section, we have seen that low-order [n,n] Padé approximations of the dead time of the open-loop transfer function are accurate for calculating closed-loop stability limits. Even $n=2$ was sufficient for the FOPDT example that was considered.

14.5 Calculation of Closed-Loop Response for Systems involving Dead Time

It is therefore meaningful to try to use approximations of the open-loop transfer function, for the purpose of calculating the closed-loop response to, e.g., a step input.

Out of the dead-time approximations that we have seen, the ones that seem more promising are the [n,n] Padé approximations, since we already know that they are capable of calculating the stability limits. Taylor approximations are generally inaccurate, unless a high order is used.

Example 14.7 Comparison of alternative approximations of the open-loop transfer function in predicting the closed-loop response

For the following transfer functions

$$G(s) = \frac{1}{(5s+1)(s+1)(0.5s+1)}$$

$$G_{FOPDT}(s) = \frac{1}{5s+1} e^{-1.5s}$$

$$G_{[2,2]}(s) = \frac{1}{5s+1} \cdot \frac{1 - 0.75s + 0.1875s^2}{1 + 0.75s + 0.1875s^2}$$

calculate the closed-loop response to a unit step change in the set point, under P control and for the following values of the controller gain:

$$k_c = 1, \quad k_c = 5, \quad k_c = 7$$

Solution

In Example 14.3, we saw the first two transfer functions: the second one (FOPDT) was obtained by reverse Taylor approximations on the first. Their open-loop step responses are quite close to each other (see Figure 14.13). The third transfer function is the [2,2] Padé approximation of the FOPDT transfer function, and its open-loop step response is also quite close to the others. The closed-loop simulation of the three systems under P control is shown in Figure 14.18 for the different values of k_c. When $k_c = 1$, all three closed-loop step responses are reasonably close to each other. When the gain is increased to $k_c = 5$, the closed-loop step responses of FOPDT and its [2,2] Padé approximation are very close to each other for times beyond the dead time, but the one from the first transfer function significantly deviates from the others. When the gain is increased to $k_c = 7$, the closed-loop step responses of FOPDT and its [2,2] Padé approximation are both unstable, whereas the first transfer function is stable.

In all cases, the closed-loop responses from the Padé approximation are in excellent agreement with those of the FOPDT system for times larger than the dead time, but give meaningless oscillations over the time period before the dead time. Overall, the Padé approximation is successful in the closed-loop response, but because of the availability of software tools to numerically calculate the exact closed-loop response, there is little incentive to use it.

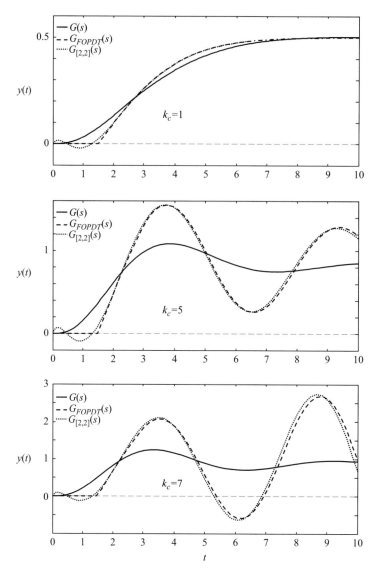

Figure 14.18 Comparison of the unit step response of alternative approximations – Example 14.7.

The main lesson to be learned in this example is the inadequacy of the reverse Taylor approximation in predicting the closed-loop response under high controller gains. A seemingly reasonable approximation of the open-loop step response is not necessarily capable of predicting the closed-loop response.

14.6 Software Tools

14.6.1 Handling of Time-Delay Systems in MATLAB

The command [num, den]=pade(theta, n) generates the coefficients of a [n,n] Padé approximation of $e^{-\theta s}$. For instance, the commands

```
[num,den]=pade(1,2);
G=tf(num,den);
step(G)
```

generates the unit step response of the [2,2] Padé approximation of e^{-s} shown in Figure 14.9.

A transfer function of a system with numerator polynomial num and denominator polynomial den that has a delay time theta is defined by the more general form of the tf command

```
G=tf(num,den,'InputDelay',theta)
```

In a similar way we can define the state space form of a system involving a delay time

```
sys=ss(A,b,c,d,'InputDelay',theta)
```

Needless to say that once we have defined the transfer function description or the state-space description in MATLAB, we can then use all available commands such as the bode, step, series, etc. The following commands are used, for instance, to generate the exact closed-loop response of the FOPDT system under P control shown in Figure 14.18:

```
kp=1;tau=5;theta=1.5;
G=tf(kp,[tau 1],'InputDelay',theta)
kc=1;
Gc=tf(kc,1)
Gcl=feedback(series(Gc,G),1);
[y,t]=step(Gcl,10);
plot(t,y)
```

14.6.2 Handling of Time-Delay Systems in Maple

The command pade(f,s,[m, n]) of the numapprox package of Maple can be used to generate the [m,n] Padé approximation of the function $f(s)$. In particular, to get the [8,8] Padé approximation of e^{-s},

```
> with(numapprox):
> pade(exp(-s),s,[8,8]);
```

$$\frac{\dfrac{1}{518918400}s^8 - \dfrac{1}{7207200}s^7 + \dfrac{1}{250920}s^6 - \dfrac{1}{9360}s^5 + \dfrac{1}{624}s^4 - \dfrac{1}{60}s^3 + \dfrac{7}{60}s^2 - \dfrac{1}{2}s + 1}{\dfrac{1}{518918400}s^8 + \dfrac{1}{7207200}s^7 + \dfrac{1}{250920}s^6 - \dfrac{1}{9360}s^5 + \dfrac{1}{624}s^4 - \dfrac{1}{60}s^3 + \dfrac{7}{60}s^2 + \dfrac{1}{2}s + 1}$$

Maple offers the advantage that the symbolic calculation of the closed-loop response via the method of steps presented in Section 14.5 can be performed efficiently. Calculation of the unit step response for P control of the FOPDT system over a time horizon of four dead times is performed as follows:

```
> with(inttrans):
> G0:=kp/(tau*s+1):G:=G0*exp(-theta*s);Gc:=kc;
```

$$G := \frac{kp\, c^{-\theta s}}{s\tau + 1}$$

$$Gc := kc$$

```
> NDT:=4:
> for i to NDT do y_0[i]:= invlaplace((G0*Gc)^i/s, s,t):
y_d[i]:=subs([t=t-i*theta],y_0[i]): end do:
> y:= sum((-1)^(v-1)*y_d[v]*Heaviside(t-v*theta), v=1.. NDT);
```

$$y := kpkc\left(1 - e^{-\frac{t-\theta}{\tau}}\right)\text{Heaviside}(t-\theta)$$

$$-kp^2kc^2\left(1 - \frac{e^{-\frac{t-\theta}{\tau}}(t-2\theta+\tau)}{\tau}\right)\text{Heaviside}(t-2\theta)$$

$$+\frac{1}{2}kp^3kc^3\left(2 - \frac{e^{-\frac{t-3\theta}{\tau}}\left((t-3\theta)^2 + 2(t-3\theta)\tau + 2\tau^2\right)}{\tau^2}\right)\text{Heaviside}(t-3\theta)$$

$$-\frac{1}{6}kp^4kc^4\left(6 - \frac{e^{-\frac{t-4\theta}{\tau}}\left((t-4\theta)^3 + 3(t-4\theta)^2\tau + 6(t-4\theta)\tau^2 + 6\tau^3\right)}{\tau^3}\right)\text{Heaviside}(t-4\theta)$$

Once the response is calculated symbolically, one can substitute specific values of the parameters and generate the plots of the exact closed-loop response that is shown in Figure 14.18.

LEARNING SUMMARY

- Many chemical process systems involve dead time in either the input or the output of the plant and they have the following state-space description:

$$\frac{dx(t)}{dt} = Ax(t) + bu(t)$$
$$y(t) = cx(t) + du(t) \quad \text{(dead time in the output)}$$
$$y_m(t) = y(t - \theta)$$

$$\frac{dx(t)}{dt} = Ax(t) + bu(t - \theta)$$
$$y(t) = cx(t) + du(t - \theta) \quad \text{(dead time in the input)}$$

- The transfer-function description of the above systems is

$$Y(s) = G_0(s) \, e^{-\theta s} \, U(s)$$

where

$$G_0(s) = c(sI - A)^{-1} b + d$$

- The following approximation is often used for the exponential factor $e^{-\theta s}$:

$$e^{-\theta s} \approx \frac{1}{\left[\left(\frac{\theta}{N}\right)s + 1\right]^N}, \text{ for large } N$$

This approximation is accurate only when N is large (>10).
- Taylor and Padé approximations of dead time are commonly used in process control. They are summarized in Table 14.1.
- Padé approximations of low order are quite accurate in approximating the frequency response of $e^{-\theta s}$. They can also provide accurate approximations in the study of the closed-loop dynamics of higher-order systems with dead time.
- Stability analysis of closed-loop systems involving dead time can be performed approximately, using Padé approximations and the Routh criterion (see Chapter 12). Exact stability conditions can also be derived for simple low-order systems.
- Despite the complexity of the analytic form of closed-loop responses, simulation of closed-loop systems involving dead time can be performed accurately and efficiently in MATLAB or Maple.

TERMS AND CONCEPTS

Dead time. The time that elapses between an input change and a noticeable effect on the system output or its measurement.

Padé approximation. A rational approximation that is used to approximate dead time.

FURTHER READING

Further analysis of dead time systems can be found in the following textbooks

Luyben, W. L. and Luyben, M. L., *Essentials of Process Control*. New York: McGraw Hill, 1997.

Ogata, K., *Modern Control Engineering, Pearson International Edition*, 5th edn, 2008.

Ogunnaike, B. and Ray, H., *Process Dynamics, Modelling and Control*. New York: Oxford University Press, 1994.

Stephanopoulos, G., *Chemical Process Control, An Introduction to Theory and Practice*. New Jersey: Prentice Hall, 1984.

PROBLEMS

14.1 Consider the pipe shown in Figure P14.1 with length L and diameter D and cross-sectional area A. The fluid that flows through the pipe at constant velocity v is cooled by natural convection to the environment (the temperature of the environment is T_∞ and the overall heat-transfer coefficient is U and is constant). The heat capacity c_p and density ρ of the fluid are assumed constant.

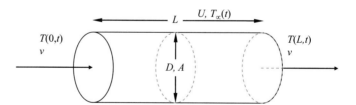

Figure P14.1

(a) Perform an energy balance at a shell located at distance z from the entrance of the pipe and show that the dynamics of the system is described by the following PDE:

$$\frac{\partial \bar{T}(z,t)}{\partial t} + v\frac{\partial \bar{T}(z,t)}{\partial z} = 4\frac{U}{\rho c_p D}\left(\bar{T}_\infty(t) - \bar{T}(z,t)\right)$$

(b) Use appropriate boundary and initial conditions to derive the transfer-function model that relates variation in the inlet temperature $T(0,t)$ to the outlet temperature $T(L,t)$ (the temperature of the environment is assumed constant):

$$\bar{T}(L,s) = k \cdot e^{-\theta s} \cdot \bar{T}(0,s)$$

where $k = e^{-\theta/\tau}$, $\theta = L/v$, $\tau = \rho c_p D / \pi U$.

(c) Consider the case of a pipe that is 1 m long and 0.05 m in diameter, in which an incoming stream with velocity 1 m/s is fed at temperature 100 °C. The heat-transfer coefficient is 1 kW/(m² °C) and the temperature of the environment is 0 °C. Derive the response of the heat-exchanger outlet temperature to a unit step change in the temperature of the feed.

14.2 Consider the tubular reactor shown in Figure P14.2 with length L and diameter D. The concentration of the reactant A at the inlet of the reactor is $c(0,t)$ and the reactor operation is isothermal. A first-order reaction is taking place in the reactor and the rate of consumption of A is $k \cdot c$. The density ρ of the reacting mixture and the cross-sectional area of the reactor are constant.

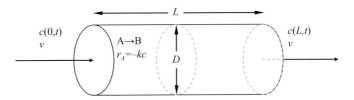

Figure P14.2

(a) Perform a mass balance of A at a shell located at distance z from the entrance of the reactor and show that the dynamics of the system is described by the following PDE:

$$\frac{\partial \bar{c}(z,t)}{\partial t} + v \frac{\partial \bar{c}(z,t)}{\partial z} = -k\bar{c}(z,t)$$

(b) Use appropriate boundary and initial conditions to derive the transfer function model that relates variation in the inlet concentration $c(0,t)$ of A and the outlet concentration $c(L,t)$ of A

$$\bar{c}(L,s) = K \cdot e^{-\theta s} \cdot \bar{c}(0,s)$$

where $K = e^{-\theta k}$, $\theta = L/v$.

(c) Consider the case of a tubular reactor that is 10 m long and 0.05 m in diameter, in which an incoming stream with velocity 1 m/s is fed with concentration of A $c(0,t) = 10$ kmol/m³. The reaction rate constant is $k = 1$ s⁻¹. Derive the response of the concentration of A to a unit step change in the inlet concentration of A at distances $L = 1$, 5 and 10 m from the reactor entrance.

14.3 Repeat Problem 14.2 to show that the dynamics of the product concentration is given by

$$\bar{c}_B(L,s) = (1-K) \cdot e^{-\theta s} \cdot \bar{c}(0,s)$$

14.4 Consider the tubular reactor studied in Problem 14.2 with $L = 1$ m. The outlet of the tubular reactor is fed to a CSTR with volume 0.02 m³ that is also operated isothermally, as shown in Figure P14.4.
   (a) Derive the overall transfer function between the inlet concentration of A to the tubular reactor and the outlet concentration from the CSTR and show that it is of the FOPDT form.
   (b) Derive the response of the outlet concentration of A to a unit step change of the inlet concentration of A.

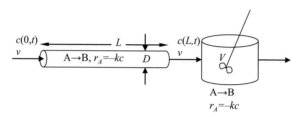

Figure P14.4

14.5 Consider the double-pipe heat exchanger shown in Figure P14.5, where the fluid that flows through the inner pipe at constant velocity v is heated by saturated steam condensing outside the pipe. The temperature of the fluid entering the pipe varies with time. The steam temperature varies with time, but not with position in the exchanger. Use Figure P14.5 and the following notation:
$T(z,t)$ is the fluid temperature
$T_{st}(t)$ is the saturated steam temperature
v is the fluid velocity
ρ is the (constant) fluid density of the fluid
c_p is the (constant) heat capacity of the fluid
A is the (constant) cross-sectional area for flow inside the pipe
U is the (constant) overall heat-transfer coefficient based on the inside area.

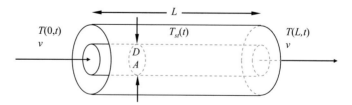

Figure P14.5

(a) Perform an energy balance at a shell located at distance z from the entrance of the inside pipe and show that the dynamics of the system is described by the following PDE:

$$\frac{\partial T(z,t)}{\partial t} + v\frac{\partial T(z,t)}{\partial z} = 4\frac{U}{\rho c_p D}\left(T_{st}(t) - T(z,t)\right)$$

(b) Use appropriate boundary and initial conditions to derive the transfer function model that relates variation in the inlet temperature $T(0,t)$ and the steam temperature $T_{st}(t)$ to the outlet temperature $T(L,t)$:

$$T(L,s) = k \cdot e^{-\theta s} \cdot T(0,s) + \left(\frac{1 - ke^{-\theta s}}{\tau s + 1}\right) \cdot T_{st}(s)$$

where $k = e^{-\theta/\tau}$, $\theta = L/v$, $\tau = \dfrac{\rho c_p D}{\pi U}$.

(c) Consider the case of a pipe that is 1 m long and 2 in in diameter, in which an incoming stream with velocity 1 m/s is fed at temperature 20 °C. The overall heat transfer coefficient is 1 kW/(m² °C) and the steam temperature is 160 °C. Derive the response of the heat-exchanger outlet temperature to a unit step change in either feed or steam temperature.

14.6 A method for estimating τ and θ of an FOPDT model from noisy experimental data is shown in Figure P14.6. The method is based on the following properties of a FOPDT model

$$\int_0^\infty \left(1 - \frac{y(t)}{kM}\right)dt = \tau + \theta, \quad \int_0^{\tau+\theta}\left(\frac{y(t)}{kM}\right)dt = \frac{\tau}{e}$$

Use the method, known as the area method of Åstrom and Hägglund, to rework Example 14.4.

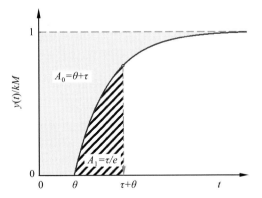

Figure P14.6

14.7 The dynamics of the reboiler of a distillation column is described by the transfer function:

$$\frac{H(s)}{F_{st}(s)} = \frac{1 - \gamma e^{-\theta s}}{s}$$

where H is the level of the liquid in the reboiler, F_{st} is the flowrate of the heating steam and $\gamma > 1$, $\theta > 0$ are constant parameters.

(a) Calculate the poles and the zeros of the system. Is the reboiler input–output stable?
(b) Calculate the response of the liquid level to a unit step change in the steam input and graph your result. What happens in the limit as $t \to \infty$?
(c) Someone suggested constructing a rational approximation of the transfer function, using a first-order Padé approximation of the exponential $e^{-\theta s}$. Repeat the calculations of question (b) for the approximate transfer function resulting from this approximation. In a common graph, compare the approximate step response to the exact step response that you calculated in question (b).

Use the following numerical values of process parameters: $\gamma = 1.5$, $\theta = 1$.

14.8 Suppose that the liquid level of the reboiler of Problem 14.7 is controlled by a P controller by manipulating the steam flowrate. Determine the range of values of the controller gain for closed-loop stability. Use a first-order or second-order Padé approximation of the exponential term in the transfer function. Can you find the exact stability range?

14.9 Use Eqs. (14.4.20) and (14.4.21) to derive the exact stability conditions for the case of P control ($G_c(s) = k_c$) of a pure dead-time system shown in Figure P14.9.

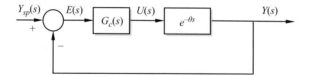

Figure P14.9

Describe qualitatively and quantitatively the unit step response of the closed-loop system.

14.10 Use the approximation given by Eq. (14.1.34)

$$e^{-\theta s} \approx \frac{1}{\left[\left(\dfrac{\theta}{N}\right)s + 1\right]^N}, \text{ for large } N$$

to derive approximate stability conditions for the system studied in Problem 14.9. Use several values of N in your analysis, e.g. $N = 3, 6, 10, \ldots$ What is the minimum value of N for the approximation to be considered acceptable?

14.11 For the control system of Problem 14.9, the following controller has been proposed

$$G_c(s) = \frac{1}{\lambda s + 1 - e^{-\theta s}}$$

(a) What is the transfer function of the closed-loop system?
(b) What is the range of values of λ for which stability is guaranteed?
(c) Give a rational function approximation of $G_c(s)$.
(d) Describe qualitatively and quantitatively the unit step response of the closed-loop system.

14.12 Consider the system shown in Figure P14.12. When a P controller is used the system becomes unstable when $k_c > 2$. When a PI controller is used with $\tau_I = 1$ min the closed-loop system becomes unstable when $k_c > 0.9$. Based on these data calculate the values of k and θ.

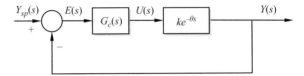

Figure P14.12

14.13 A system is known to exhibit first-order plus deadtime (FOPDT) system behavior. Several experiments have been performed to determine the characteristics of the system and a static gain of 2.5 was estimated. It was observed that the closed-loop system under P control exhibits sustained oscillations, with a period of 6.28 min ($\approx 2\pi$ min), when the P gain is equal to 0.5. Given this information, derive equations to calculate the parameters of the FOPDT model (k, τ, θ) for the process under study.

14.14 Consider an unstable FOPDT system with transfer function

$$G(s) = \frac{k_p}{\tau s - 1} e^{-\theta s}$$

which is controlled by a P controller with transfer function $G_c(s) = k_c$. Derive closed-loop stability conditions using [1,1] and [2,2] Padé approximations, as well as exact closed-loop stability conditions. Is it possible to achieve closed-loop stability when $\theta > \tau$?

COMPUTATIONAL PROBLEM

14.15 In the system shown in Figure P14.15, the heated tanks of equal and constant volume have been placed in series. The tanks are heated by steam that is condensing inside a coil immersed in the liquid. The flowrate of the steam in the second and third tanks

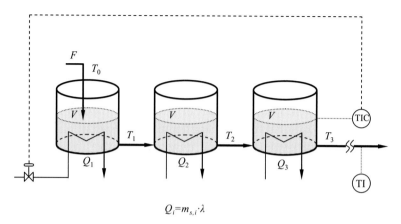

Figure P14.15

is constant. The flowrate of the steam to the first tank is varied by a P controller in order to control the temperature of the third tank. The closed-loop system has been operated successfully for some time. However, recently, due to the malfunctioning of the temperature sensor you are proposing its replacement. An engineer in your department suggests that you can save some money by using a temperature sensor that measures the temperature of the effluent steam.

You are not convinced that this is a good idea but before expressing your opinion you have decided to perform an analysis of the problem to quantify the potential loss in control quality. To this end you need to do the following.

(a) Derive the transfer function model of the existing system and calculate the controller gain, which is half of the value that destabilizes the system (the dynamics of the temperature sensor can be neglected).
(b) Derive the transfer function of the system with the sensor located at the effluent pipe.
(c) Determine the value of the P controller gain that destabilizes the system.
(d) Compare the steady-state error for the existing system and for the proposed system for a unit step change in the set point.
(e) Validate your results in MATLAB.
(f) Prepare simulations of the closed-loop response of the existing system and the proposed system when the proportional gain is half the value that destabilizes the system.
(g) Give a list of arguments to support your case against the solution suggested by your colleague.

Use the following notation and numerical values:

F: volumetric flowrate of the liquid $=1$ m³/min
V: volume of the liquid in the tanks $=1$ m³

T_0 : temperature of the feed stream = 10 °C
T_1 : temperature in the first tank = 40 °C at steady state
T_2 : temperature in the second tank = 60 °C at steady state
T_3 : temperature in the third tank = 80 °C at steady state
Q_i : heat transfer rate at tank i [kJ/min]
$m_{s,i}$: mass flowrate of steam in tank i [kg/min]
λ : latent heat of steam (constant) = 2000 kJ/kg steam
ρ : density of the liquid (constant) = 1000 kg/m³
C_p : specific heat capacity of the liquid (constant) = 2 kJ/(kg °C)
u : velocity of liquid in the effluent pipe = 60 m/min
L : distance of the location of the temperature sensor from the third tank = 30 m.

15 Parametric Analysis of Closed-Loop Dynamics – Root-Locus Diagrams

The aim of this chapter is to introduce a general methodology for analyzing the effect of controller parameters on the dynamic characteristics of the closed-loop system. This is accomplished through the root-locus diagrams, which depict the location of the closed-loop eigenvalues or poles in the complex plane, as one controller parameter varies from zero to infinity. Basic properties of the root-locus diagrams are discussed, as well as computer tools for generating root-locus diagrams and related information.

STUDY OBJECTIVES

After studying this chapter you should be able to do the following.

- "Read" a given root-locus diagram: extract qualitative and quantitative information from the diagram, regarding the dynamic characteristics of a closed-loop system as a function of a controller parameter.
- Sketch a root-locus diagram for a given simple process model and a given controller type.
- Use MATLAB or Maple to generate accurate root-locus diagrams, as well as related quantitative information on the closed-loop poles.

15.1 What is a Root-Locus Diagram? Some Examples

15.1.1 Introduction

The poles of a dynamic system provide valuable information about the dynamic behavior of the system: they determine stability or instability, the presence or absence of inherent tendency to oscillate, as well as the speed of response. When a feedback controller is applied to a process, the poles of the closed-loop system depend on the choice of controller parameters; a good choice of controller parameters can lead to a stable and fast-responding system without excessive oscillations, whereas a poor choice can lead to unacceptable performance. Therefore, in the design of a feedback controller, a key question is: *how do closed-loop poles vary as a function of controller parameters?* If the answer to this question can be found and appropriately recorded, this can be a valuable guide for the selection of controller parameters.

The root-locus diagram provides a very useful way of recording and depicting the dependence of the poles of the closed-loop system as a function of *one* controller parameter, when all the other parameters are fixed. Root-locus diagrams are commonly used to study the dependence of the closed-loop poles on the gain k_c of a P, PI, PD or PID controller, when integral time and/or derivative time are fixed. Alternatively, one can use a root-locus diagram to study the effect of the integral time or the derivative time when the other parameters are fixed.

Root-locus diagrams were first introduced by W. R. Evans in 1950. Since then, they have been widely used by engineers for the design of feedback control systems. In the 1950s, graphing the root locus was a long and painful procedure because, at that time, the engineer's only computational tool was the slide rule. Today, with the available software tools, accurate root-locus calculations can be performed fast and easily, but still it is very important to have a solid understanding of the mathematical principles on which root-locus diagrams are based.

Evans' theory, in its original form, had a predominantly geometric perspective, because in the 1950s root-locus diagrams were obtained via manual drafting on graph paper, using ruler, protractor, French curves, etc., as computer graphics were unavailable at that time.

The present chapter will give a modern introduction to root-locus theory, balanced in algebraic and geometric perspectives, and with the availability of computer tools (calculation of roots of polynomials, computer graphics, etc.) as a given. We will start with a simple example of a third-order process controlled with a P or PI or PID controller, to study the effect of the controller gain on the location of the closed-loop poles. We would like to see what a root-locus diagram looks like and what information it provides.

15.1.2 Motivating Examples

Example 15.1 Proportional control of a third-order process

Consider the feedback control system depicted in Figure 15.1, where $\tau_1 > 0$, $\tau_2 > 0$, $\tau_3 > 0$, $kk_c > 0$. The closed-loop transfer function is

$$\frac{Y(s)}{Y_{sp}(s)} = \frac{kk_c}{(\tau_1 s + 1)(\tau_2 s + 1)(\tau_3 s + 1) + kk_c}$$

and its poles are the roots of the denominator polynomial

$$Q(s) = (\tau_1 s + 1)(\tau_2 s + 1)(\tau_3 s + 1) + kk_c$$
$$= \tau_1 \tau_2 \tau_3 s^3 + (\tau_1 \tau_2 + \tau_2 \tau_3 + \tau_3 \tau_1) s^2 + (\tau_1 + \tau_2 + \tau_3) s + (1 + kk_c)$$

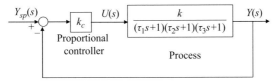

Figure 15.1 Feedback control loop consisting of a third-order process and a P controller.

We would like to study how the roots of the above polynomial depend on the choice of the controller gain k_c.

To make the example more concrete, let's use specific numbers for the process parameters: $\tau_1 = 1$, $\tau_2 = \dfrac{1}{2}$, $\tau_3 = \dfrac{1}{3}$, $k = \dfrac{1}{3}$. Then, the process transfer function becomes

$$\frac{Y(s)}{U(s)} = \frac{\frac{1}{3}}{(s+1)\left(\frac{s}{2}+1\right)\left(\frac{s}{3}+1\right)} = \frac{2}{(s+1)(s+2)(s+3)}$$

the closed-loop transfer function becomes

$$\frac{Y(s)}{Y_{sp}(s)} = \frac{\frac{1}{3}k_c}{(s+1)\left(\frac{s}{2}+1\right)\left(\frac{s}{3}+1\right)+\frac{1}{3}k_c} = \frac{2k_c}{(s+1)(s+2)(s+3)+2k_c}$$

and the poles of the closed-loop system are the roots of the polynomial

$$Q(s) = (s+1)(s+2)(s+3) + 2k_c = s^3 + 6s^2 + 11s + (6+2k_c)$$

For different values of the controller gain k_c, the roots of the above polynomial will be at different locations, and this can have a major impact on the dynamic characteristics of the closed-loop system. Figures 15.2 and 15.3 show the root locations and the corresponding unit step response of the closed-loop system for $k_c = 3, 30, 100$. We see that for $k_c = 3$, all roots are to the left of the imaginary axis, and therefore the response is stable and well behaved. When $k_c = 30$, there is a pair of roots on the imaginary axis, hence the response oscillates with constant amplitude. When $k_c = 100$, there is a pair of roots to the right of the imaginary axis, leading to explosively growing oscillations.

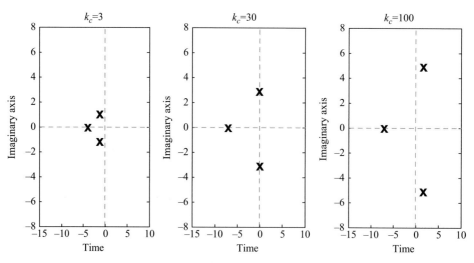

Figure 15.2 Location of the roots in the complex plane for $k_c = 3, 30, 100$.

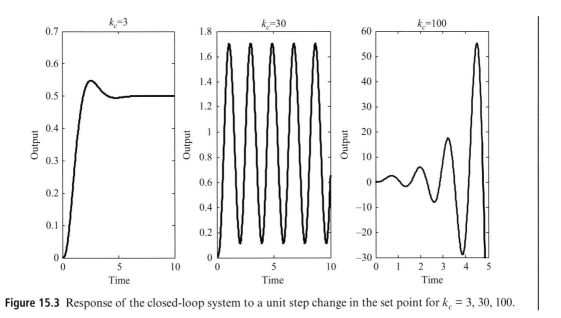

Figure 15.3 Response of the closed-loop system to a unit step change in the set point for $k_c = 3, 30, 100$.

Table 15.1 gives the roots ρ_1, ρ_2 and ρ_3 of the denominator polynomial $Q(s)$ of the closed-loop transfer function for representative values of the controller gain k_c.

Table 15.1 Roots of denominator polynomial versus k_c under P control

k_c	ρ_1	ρ_2	ρ_3
0	−1.0000	−2.0000	−3.0000
0.0100	−1.0102	−1.9800	−3.0099
0.0200	−1.0206	−1.9599	−3.0194
0.0500	−1.0544	−1.8990	−3.0467
0.1000	−1.1211	−1.7909	−3.0880
0.19245	−1.42265	−1.42265	−3.1547
0.2000	−1.4201 + 0.0932i	−1.4201 − 0.0932i	−3.1597
0.5000	−1.3376 + 0.5623i	−1.3376 − 0.5623i	−3.3247
1.0000	−1.2393 + 0.8579i	−1.2393 − 0.8579i	−3.5214
2.0000	−1.1018 + 1.1917i	−1.1018 − 1.1917i	−3.7963
3.0000	−1.0000 + 1.4142i	−1.0000 − 1.4142i	−4.0000
5.0000	−0.8455 + 1.7316i	−0.8455 − 1.7316i	−4.3089
10.0000	−0.5814 + 2.2443i	−0.5814 − 2.2443i	−4.8371
12.0000	−0.5000 + 2.3979i	−0.5000 − 2.3979i	−5.0000
20.0000	−0.2413 + 2.8773i	−0.2413 − 2.8773i	−5.5174
30.0000	0.0000 + 3.3166i	0.0000 − 3.3166i	−6.0000
60.0000	0.5000 + 4.2131i	0.5000 − 4.2131i	−7.0000
100.0000	0.9525 + 5.0152i	0.9525 − 5.0152i	−7.9050
200.0000	1.7066 + 6.3417i	1.7066 − 6.3417i	−9.4133

From Table 15.1, we see that:

→ For $k_c = 0$ (extreme case, where the controller is off), the roots are equal to the process poles: $p_1 = -1$, $p_2 = -2$, $p_3 = -3$. This also follows from the formula of $Q(s)$.
→ When k_c is very small, the roots are very close to the process poles.
→ As k_c increases, the first and second roots move towards each other, whereas the third root moves away from the other two, increasing in magnitude.
→ There is a critical value $k_c = 0.19245$ at which the first and second roots become equal (double root $p_b = -1.42265$). When k_c exceeds this critical value, the double root becomes a pair of complex conjugate roots, which move away from each other as k_c increases. At the same time, the third root keeps moving away from the other two, staying real and increasing in magnitude.
→ For $k_c < 30$, all roots have negative real parts, therefore the closed-loop system is stable. For $k_c = 30$ the first two roots are pure imaginary $\pm 3.3166i$, and for $k_c > 30$ the real part of the complex roots is positive, hence the system is unstable for $k_c \geq 30$.
→ If we were to use extremely large values of k_c ($\rightarrow +\infty$), we would see the magnitudes of all the roots becoming extremely large ($|p_1|, |p_2|, |p_3| \rightarrow +\infty$), but with each one of the roots moving towards infinity in a different direction.

The data from a table such as Table 15.1 can be depicted in the form of a diagram. By placing the roots in the complex plane for a very large number of values of k_c, we will obtain continuous curves that the roots trace in the complex plane as k_c varies from 0 to $+\infty$. This is called a **root-locus diagram**: it is the locus of the roots of the denominator polynomial of the closed-loop transfer function (i.e. the closed-loop poles) as the controller gain k_c varies from 0 to $+\infty$.

In our example, the root-locus diagram is given in Figure 15.4.

The root-locus diagram contains:

(a) on the real axis, the intervals $(-\infty, -3]$ and $[-2, -1]$.
(b) two symmetric branches off the real axis, that emerge from the point $p_b = -1.42265$ of the interval $[-2, -1]$, where the double root is located.

The root locus "starts" ($k_c = 0$) at the process poles $-1, -2$ and -3. It "ends" ($k_c = +\infty$) at infinity, but in three different directions. It turns out that the angles that these three directions form with the positive real axis are equal to $\dfrac{\pi}{3}$, π, $\dfrac{5\pi}{3}$.

By examining a root locus diagram, and knowing where the locus "starts" ($k_c = 0$), where it "ends" ($k_c = +\infty$) and the paths that it follows, we can immediately see how the dynamic characteristics of the closed-loop system change as k_c varies. The root-locus diagram is usually combined with a table such as Table 15.1, which gives accurate quantitative information.

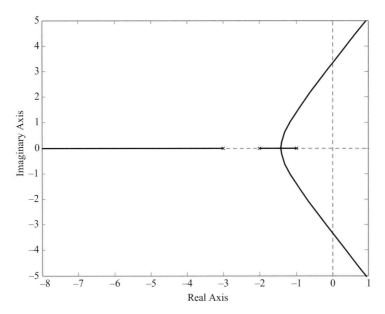

Figure 15.4 Root-locus diagram for third-order process under P control. In the diagram, the symbol x denotes a process pole.

It may also be combined with response plots for representative values of k_c. Figure 15.5 gives the unit step response of the closed-loop system for k_c = 5, 12 and 24, all in the stable range. They show that increasing k_c has a destabilizing effect on the closed-loop system, making it more and more oscillatory. Also, the presence of offset is clearly visible in the responses, which is a function of k_c.

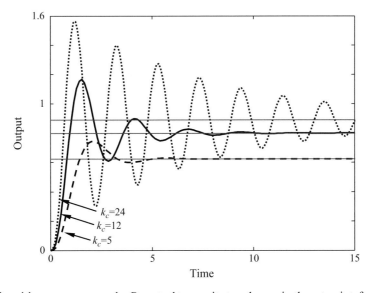

Figure 15.5 Closed-loop response under P control to a unit step change in the set point, for k_c = 5, 12, 24.

Example 15.2 PI and PID control of a third-order process

Consider the same control system, the only difference being that the controller is now dynamic, like PI, PD, PID, etc. Again, we would like to study the effect of the controller gain k_c on the closed-loop poles, with all the other controller parameters being fixed. In the feedback control system of Figure 15.6, $\tau_1 > 0$, $\tau_2 > 0$, $\tau_3 > 0$, $kk_c > 0$ and

$$G_c(s) = 1 + \frac{1}{\tau_I s} = \frac{s + \frac{1}{\tau_I}}{s}, \text{ if the controller is PI,}$$

$$G_c(s) = 1 + \frac{1}{\tau_I s} + \tau_D s = \frac{\tau_D s^2 + s + \frac{1}{\tau_I}}{s}, \text{ if the controller is ideal PID, etc.}$$

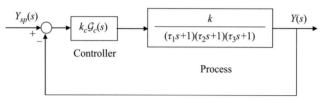

Figure 15.6 Feedback control loop consisting of a third-order process and a dynamic controller, like PI or PD or PID.

This problem is of the same type as in Example 15.1: the block diagram of Figure 15.6 is equivalent to the one of Figure 15.7.

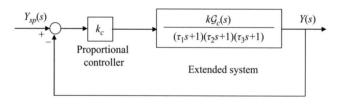

Figure 15.7 Equivalent feedback control loop involving an extended system under P control.

It is like having P control of a fictitious process, with transfer function $\dfrac{kG_c(s)}{(\tau_1 s + 1)(\tau_2 s + 1)(\tau_3 s + 1)}$, which includes the extra factors coming from the fixed dynamic part of the controller:

$$\frac{kG_c(s)}{(\tau_1 s + 1)(\tau_2 s + 1)(\tau_3 s + 1)} = \frac{k\left(s + \frac{1}{\tau_I}\right)}{s(\tau_1 s + 1)(\tau_2 s + 1)(\tau_3 s + 1)}, \text{ if the controller is PI,}$$

$$\frac{kG_c(s)}{(\tau_1 s + 1)(\tau_2 s + 1)(\tau_3 s + 1)} = \frac{k\left(\tau_D s^2 + s + \frac{1}{\tau_I}\right)}{s(\tau_1 s + 1)(\tau_2 s + 1)(\tau_3 s + 1)}, \text{ if the controller is PID, etc.}$$

PI Control

Let's use the same numbers for the process parameters as in Example 15.1:

$$\tau_1 = 1, \quad \tau_2 = \frac{1}{2}, \quad \tau_3 = \frac{1}{3}, \quad k = \frac{1}{3}$$

Then the transfer function of the closed-loop system becomes:

$$\frac{Y(s)}{Y_{sp}(s)} = \frac{k_c \dfrac{\dfrac{1}{3}\left(s+\dfrac{1}{\tau_I}\right)}{s(s+1)\left(\dfrac{s}{2}+1\right)\left(\dfrac{s}{3}+1\right)}}{1+k_c \dfrac{\dfrac{1}{3}\left(s+\dfrac{1}{\tau_I}\right)}{s(s+1)\left(\dfrac{s}{2}+1\right)\left(\dfrac{s}{3}+1\right)}} = \frac{2k_c\left(s+\dfrac{1}{\tau_I}\right)}{s(s+1)(s+2)(s+3)+2k_c\left(s+\dfrac{1}{\tau_I}\right)}$$

and the poles of the closed-loop transfer function are the roots of the polynomial

$$s(s+1)(s+2)(s+3)+2k_c\left(s+\frac{1}{\tau_I}\right) = s^4 + 6s^3 + 11s^2 + (6+2k_c)s + \frac{2k_c}{\tau_I}$$

Let's see now how the presence of integral action affects the root-locus diagram. Before we proceed, we must note here that the extended system is of fourth order, with four poles, at the points 0, –1, –2, –3, and one zero at $-\dfrac{1}{\tau_I}$. The polynomial is of fourth degree, and we want to trace the path of four roots.

Suppose now that we choose $\tau_I = 1\frac{1}{2}$, which is a medium-size value of integral time for the process under consideration. Table 15.2 lists the roots for different values of k_c, and Figure 15.8 gives the corresponding root-locus diagram.

Table 15.2 Roots of denominator polynomial versus k_c, under PI control with $\tau_I = 1\frac{1}{2}$

k_c	Roots			
0	–3.0000	–2.0000	–1.0000	0.0000
0.0100	–3.0077	–1.9867	–1.0034	–0.0022
0.0200	–3.0152	–1.9735	–1.0068	–0.0045
0.0500	–3.0370	–1.9342	–1.0177	–0.0112
0.1000	–3.0706	–1.8691	–1.0379	–0.0224
0.2000	–3.1307	–1.7351	–1.0891	–0.0451
0.3708	–3.2177	–1.3489	–1.3489	–0.0844
0.5000	–3.2747	–1.3054 +0.2688i	–1.3054 –0.2688i	–0.1146
1.0000	–3.4529	–1.1579 +0.5733i	–1.1579 –0.5733i	–0.2313
2.0000	–3.7088	–0.9365 +0.9176i	–0.9365 –0.9176i	–0.4183
5.0000	–4.1973	–0.6096 +1.5332i	–0.6096 –1.5332i	–0.5835
10.0000	–4.7089	–0.3312 +2.0962i	–0.3312 –2.0962i	–0.6287

Table 15.2 (continued)

k_c	Roots			
19.5934	−5.3518	0.0000 + 2.7443i	0.0000 − 2.7443i	−0.6482
30.0000	−5.8487	0.2517 + 3.2220i	0.2517 − 3.2220i	−0.6548
50.0000	−6.5529	0.6063 + 3.8802i	0.6063 − 3.8802i	−0.6596
100.0000	−7.7337	1.1984 + 4.9559i	1.1984 − 4.9559i	−0.6632
200.0000	−9.2327	1.9488 + 6.2960i	1.9488 − 6.2960i	−0.6649

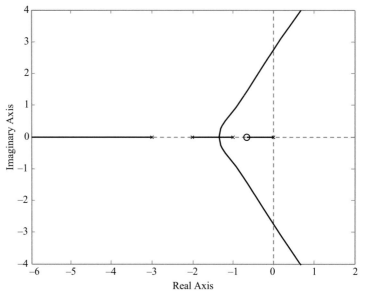

Figure 15.8 Root-locus diagram for third-order process under PI control with $\tau_I = 1\frac{1}{2}$. In the diagram, the symbol × denotes a process pole, and ○ denotes a process zero.

From the table and the graph, we observe the following.

→ For $k_c = 0$, the roots are equal to the poles of the extended system:
$$p_1 = 0, \ p_2 = -1, \ p_3 = -2, \ p_4 = -3$$

→ As k_c increases, the second and third roots move towards each other, whereas the other two roots move to the left.

→ There is a critical value $k_c = 0.3708$ at which the second and third roots become equal (double root $p_b = -1.3489$). When k_c exceeds this critical value, the double root becomes a pair of complex conjugate roots, which move away from each other as k_c increases.

→ For $k_c < 19.5934$, all roots have negative real parts, which means that the closed-loop system is stable, whereas for $k_c > 19.5934$, the real part of the complex roots becomes positive, hence the system becomes unstable.

→ As $k_c \to +\infty$, one of the roots tends to the zero of the extended system's transfer function $-\dfrac{1}{\tau_I} = -\dfrac{2}{3} = -0.6667$, whereas the other three roots tend to infinity in different directions, forming angles $\dfrac{\pi}{3}, \pi, \dfrac{5\pi}{3}$ with the positive real axis.

The root locus diagram contains:

(a) on the real axis, the intervals $(-\infty, -3]$, $[-2, -1]$ and $\left(-\frac{2}{3}, 0\right)$
(b) two symmetric branches off the real axis, that emerge from the point $\rho_b = -1.3489$ of the interval $[-2, -1]$, where the double root is located.

The root locus starts ($k_c = 0$) at the poles of the extended system: $0, -1, -2$ and -3; it ends ($k_c = +\infty$) at the zero $-\frac{1}{\tau_I} = -\frac{2}{3}$ and at infinity, forming angles $\frac{\pi}{3}$, π, $\frac{5\pi}{3}$.

Figure 15.9 gives the unit step response of the closed-loop system for $\tau_I = 1\frac{1}{2}$ and $k_c = 1, 2, 5$ and 10, all in the stable range. They show that increasing k_c has a destabilizing effect, making the system more and more oscillatory. Comparing with Figure 15.5 for P control, we now see that lower values of the gain must be used for a reasonably damped response, and the responses are somewhat slower. The main advantage is that now, because of the presence of integral action, the offset is zero.

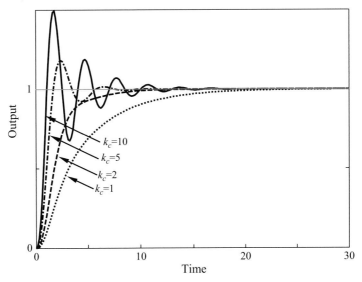

Figure 15.9 Closed-loop response under PI control with $\tau_I = 1\frac{1}{2}$ to a unit step change in the set point, for $k_c = 1, 2, 5, 10$.

Let's examine now the case of large integral action (small integral time τ_I). Suppose for example that we choose $\tau_I = \frac{1}{4}$. Then the system's zero is located at $-\frac{1}{\tau_I} = -4$, which is to the left of all the poles.

Table 15.3 lists the roots for different values of k_c, and Figure 15.10 gives the corresponding root-locus diagram.

Comparing this root-locus diagram with the one of Figure 15.8 for $\tau_I = 1\frac{1}{2}$, we see similarities, but also essential differences. In both diagrams, the root locus

- starts ($k_c = 0$) at the poles of the extended system $0, -1, -2$ and -3
- ends ($k_c = +\infty$) at the zero $-\frac{1}{\tau_I}$ and at infinity, with angles $\frac{\pi}{3}$, π, $\frac{5\pi}{3}$.

Table 15.3 Roots of denominator polynomial versus k_c under PI control with $\tau_I = \frac{1}{4}$

k_c	Roots			
0	−3.0000	−2.0000	−1.0000	0.0000
0.0100	−2.9966	−2.0196	−0.9701	−0.0136
0.0200	−2.9932	−2.0385	−0.9404	−0.0279
0.0500	−2.9825	−2.0920	−0.8501	−0.0754
0.1000	−2.9630	−2.1733	−0.6814	−0.1823
0.13874	−2.9461	−2.2323	−0.41082	−0.41082
0.2000	−2.9150	−2.3225	−0.3813 + 0.3016i	−0.3813 − 0.3016i
0.37119	−2.68595	−2.68595	−0.3140 + 0.5595i	−0.3140 − 0.5595i
0.5000	−2.7272 − 0.2284i	−2.7272 + 0.2284i	−0.2728 + 0.6780i	−0.2728 − 0.6780i
1.0000	−2.8486 − 0.4449i	−2.8486 + 0.4449i	−0.1514 + 0.9693i	−0.1514 − 0.9693i
1.91823	−3.0000 − 0.6005i	−3.0000 + 0.6005i	0.0000 + 1.2804i	0.0000 − 1.2804i
2.0000	−3.0111 − 0.6096i	−3.0111 + 0.6096i	0.0111 + 1.3020i	0.0111 − 1.3020i
3.0000	−3.1281 − 0.6922i	−3.1281 + 0.6922i	0.1281 + 1.5238i	0.1281 − 1.5238i
5.0000	−3.3018 − 0.7803i	−3.3018 + 0.7803i	0.3018 + 1.8396i	0.3018 − 1.8396i
10.0000	−3.5919 − 0.8577i	−3.5919 + 0.8577i	0.5919 + 2.3486i	0.5919 − 2.3486i
20.0000	−3.9569 − 0.8462i	−3.9569 + 0.8462i	0.9569 + 2.9760i	0.9569 − 2.9760i
30.0000	−4.2120 − 0.7574i	−4.2120 + 0.7574i	1.2120 + 3.4111i	1.2120 − 3.4111i
50.0000	−4.5849 − 0.4031i	−4.5849 + 0.4031i	1.5849 + 4.0461i	1.5849 − 4.0461i
57.01101	−4.69107	−4.69107	1.6911 + 4.2268i	1.6911 − 4.2268i
60.0000	−4.4673	−5.0000	1.7336 + 4.2993i	1.7336 − 4.2993i
90.0000	−4.1955	−6.0000	2.0977 + 4.9195i	2.0977 − 4.9195i
100.0000	−4.1668	−6.2341	2.2005 + 5.0946i	2.2005 − 5.0946i
140.0000	−4.1061	−7.0000	2.5530 + 5.6963i	2.5530 − 5.6963i
210.0000	−4.0653	−8.0000	3.0326 + 6.5161i	3.0326 − 6.5161i

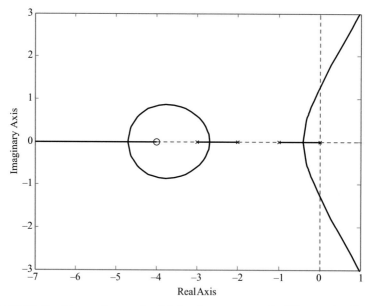

Figure 15.10 Root-locus diagram for third-order process under PI control with $\tau_I = \frac{1}{4}$.

The differences are for intermediate values of k_c. The root-locus diagram of Figure 15.10 has more branches off the real axis, including branches that form a circular shape. Also, the parts of the root locus that lie on the real axis are different. For example, the root locus of Figure 15.10 includes the interval [–2, –3] and does not include [–1, –2], whereas in the one of Figure 15.8, it is vice versa.

The only difference between the two systems is the location of the zero at $-\frac{1}{\tau_I}$: in Figure 15.8 the zero lies to the right of the process poles, whereas in Figure 15.10 it lies to the left of all poles. This is what made the two root-locus diagrams look so different.

It is also instructive to compare the step responses for the two different values of integral time. Figure 15.11 shows the unit step response of the closed-loop system for $\tau_I = ¼$ and $k_c = 0.05, 0.1, 0.2, 1$, all in the stable range. The responses clearly indicate that increasing k_c has a destabilizing effect, making the system more and more oscillatory, like in Figure 15.9 for $\tau_I = 1½$.

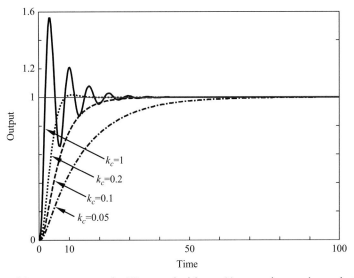

Figure 15.11 Closed-loop response under PI control with $\tau_I = ¼$ to a unit step change in the set point, for $k_c = 0.05, 0.1, 0.2, 1$.

But there is also a major difference: much lower values of the gain k_c had to be used in Figure 15.11 in order to get a reasonably damped response, and the responses are significantly slower. The reason is that the stability restriction does not allow using large enough gains, and this is what makes the response slower.

At this point, it will be useful to summarize some side conclusions that can be drawn from the present example regarding the effect of integral action. From the tables of the roots that were used for the construction of the root-locus diagram, the stability range is immediately obtained:

- in P control, where $\frac{1}{\tau_I} = 0$, the stability condition was found to be $0 < k_c < 30$
- in PI control with $\frac{1}{\tau_I} = \frac{2}{3}$, the stability condition was found to be $0 < k_c < 19.5934$
- in PI control with $\frac{1}{\tau_I} = 4$, the stability condition was found to be $0 < k_c < 1.91823$.

We saw therefore that as integral action increases, the stability range for the controller gain shrinks. *Integral action has a destabilizing effect on the closed-loop system*, which becomes stronger as the value of $\dfrac{1}{\tau_I}$ gets larger.

The use of very large integral action (very small integral time τ_I) is undesirable because it necessitates using very small gains (to prevent destabilization) and this significantly reduces the speed of achievable response of the closed-loop system.

On the other hand, using very small integral action (very large integral time τ_I) is also undesirable. In that case, the root locus will be like the one of Figure 15.8, and it will include a root in the interval $\left(-\dfrac{1}{\tau_I}, 0\right)$, which because $-\dfrac{1}{\tau_I}$ is too small, will necessarily be too close to 0. This means that the closed-loop system's response will involve an exponential that decays too slowly, therefore the response will be too slow.

To be able to come up with a "best" choice of integral time and controller gain, some trial-and-error simulations may be needed, but this can be done in a guided manner: the root tables give the numbers for the closed-loop poles to make a screening of alternative choices, and the root-locus diagrams give the "global picture" of the dynamic behavior of the closed-loop system.

PID Control

We keep the same numbers for the process parameters:

$$\tau_1 = 1, \quad \tau_2 = \frac{1}{2}, \quad \tau_3 = \frac{1}{3}, \quad k = \frac{1}{3}.$$

Then, the extended system's transfer function becomes

$$\frac{k\left(\tau_D s^2 + s + \dfrac{1}{\tau_I}\right)}{s(\tau_1 s+1)(\tau_2 s+1)(\tau_3 s+1)} = \frac{\dfrac{1}{3}\left(\tau_D s^2 + s + \dfrac{1}{\tau_I}\right)}{s(s+1)\left(\dfrac{s}{2}+1\right)\left(\dfrac{s}{3}+1\right)},$$

the closed-loop transfer function becomes

$$\frac{Y(s)}{Y_{sp}(s)} = \frac{k_c \dfrac{\dfrac{1}{3}\left(\tau_D s^2 + s + \dfrac{1}{\tau_I}\right)}{s(s+1)\left(\dfrac{s}{2}+1\right)\left(\dfrac{s}{3}+1\right)}}{1 + k_c \dfrac{\dfrac{1}{3}\left(\tau_D s^2 + s + \dfrac{1}{\tau_I}\right)}{s(s+1)\left(\dfrac{s}{2}+1\right)\left(\dfrac{s}{3}+1\right)}} = \frac{2k_c\left(\tau_D s^2 + s + \dfrac{1}{\tau_I}\right)}{s(s+1)(s+2)(s+3) + 2k_c\left(\tau_D s^2 + s + \dfrac{1}{\tau_I}\right)}$$

and the poles of the closed-loop transfer function are the roots of the polynomial

$$s(s+1)(s+2)(s+3) + 2k_c\left(\tau_D s^2 + s + \dfrac{1}{\tau_I}\right) = s^4 + 6s^3 + (11 + 2\tau_D k_c)s^2 + (6 + 2k_c)s + \dfrac{2k_c}{\tau_I}$$

The extended system is still of fourth order, with four poles, at the points 0, –1, –2, –3, as before, but now it has two zeros, which are the roots of the quadratic polynomial $\tau_D s^2 + s + \frac{1}{\tau_I}$. As in PI control, there will be four roots, therefore four branches of the root locus, starting from the poles of the extended system 0, – 1, – 2 and – 3. However, because of the presence of two zeros, two branches will terminate at them, and there will only be two branches terminating at infinity. As we will see, the directions along which the two roots will tend to infinity will now be different: they will be perpendicular to the real axis, forming angles $\frac{\pi}{2}$, $\frac{3\pi}{2}$.

Let's see now how the root-locus diagram changes when derivative action is used, in addition to proportional and integral. We will keep the same medium-size value of integral time $\tau_I = 1\frac{1}{2}$ and try out different values for the derivative time τ_D.

We first try the value of $\tau_D = \frac{1}{4}$. Table 15.4 gives the roots for representative values of the controller gain k_c and Figure 15.12 gives the corresponding root-locus diagram.

Comparing the root-locus diagram with the one of Figure 15.8 (PI control with the same value of integral time) we see major differences for high k_c. Now we have two zeros, $z_1 = -2 + \frac{2}{\sqrt{3}} = -0.84530$, $z_2 = -2 - \frac{2}{\sqrt{3}} = -3.15470$, at which two of the branches terminate. The other two branches terminate at infinity, at vertical directions passing through the point (–1).

Table 15.4 Roots of denominator polynomial versus k_c in PID control, with $\tau_I = 1\frac{1}{2}$ and $\tau_D = \frac{1}{4}$

k_c	Roots			
0	–3.0000	–2.0000	–1.0000	0.0000
0.0100	–3.0003	–1.9967	–1.0008	–0.0022
0.0200	–3.0006	–1.9933	–1.0017	–0.0045
0.0500	–3.0014	–1.9832	–1.0043	–0.0112
0.1000	–3.0027	–1.9661	–1.0088	–0.0224
0.2000	–3.0053	–1.9309	–1.0187	–0.0451
0.5000	–3.0125	–1.8147	–1.0575	–0.1153
1.0000	–3.0229	–1.5364	–1.2018	–0.2389
1.11278	–3.0250	–1.35364	–1.35364	–0.2677
2.0000	–3.0393	–1.2354 + 0.5147i	–1.2354 – 0.5147i	–0.4898
5.0000	–3.0696	–1.0847 + 1.2951i	–1.0847 – 1.2951i	–0.7610
10.0000	–3.0948	–1.0469 + 2.0529i	–1.0469 – 2.0529i	–0.8113
20.0000	–3.1168	–1.0267 + 3.0423i	–1.0267 – 3.0423i	–0.8299
50.0000	–3.1365	–1.0120 + 4.9290i	–1.0120 – 4.9290i	–0.8395
100.0000	–3.1450	–1.0063 + 7.0223i	–1.0063 – 7.0223i	–0.8424
200.0000	–3.1496	–1.0032 + 9.9661i	–1.0032 – 9.9661i	–0.8439

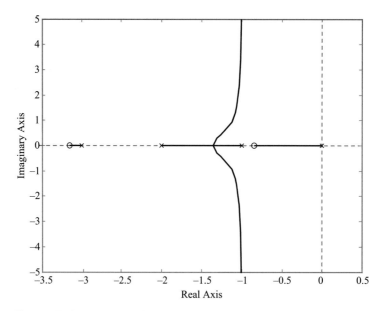

Figure 15.12 Root-locus diagram for PID control, with $\tau_I = 1\frac{1}{2}$ and $\tau_D = \frac{1}{4}$.

The root locus diagram contains:

(a) on the real axis, the intervals $(-3.1547, -3]$, $[-2, -1]$ and $(-0.8453, 0]$.
(b) two symmetric branches off the real axis, that emerge from the point $\rho_b = -1.3536$ of the interval $[-2, -1]$, where the double root is located.

What is really interesting is that now the entire root locus lies to the left of the imaginary axis, which means closed-loop stability for every positive k_c. The presence of *derivative action has a stabilizing effect on the closed-loop system.*

Figure 15.13 gives the unit step response of the closed loop system for $\tau_I = 1\frac{1}{2}$, $\tau_D = \frac{1}{4}$ and $k_c = 2, 5, 10$. Comparing with Figure 15.5 for P control and Figure 15.9 for PI control, we see that now much faster and well-damped responses are feasible. This is because of the stabilizing effect of the derivative action.

At this point one may start wondering whether complete stabilization is feasible for any value of τ_D. But the answer is negative.

If we draw the root locus diagram for $\tau_D = \frac{1}{8}$, for example, and for the same integral time, we see that the system becomes unstable for large values of the gain k_c (see Table 15.5 and Figure 15.14).

Despite the qualitative similarities between the root-locus diagrams for $\tau_D = \frac{1}{4}$ and $\tau_D = \frac{1}{8}$, there is a very essential difference: The vertical directions followed by the roots tending to infinity, in the first case pass from the point (-1), whereas in the second case they pass from the point $(+1)$. Thus, in the second case, two branches will necessarily cross the imaginary axis and move to the right half plane.

15.1 What is a Root-Locus Diagram? Some Examples

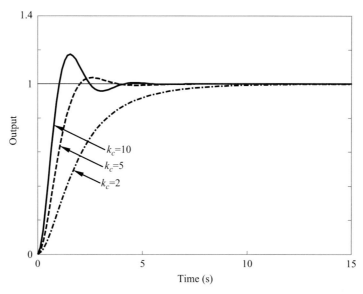

Figure 15.13 Closed-loop response under PID control with $\tau_I = 1\frac{1}{2}$ and $\tau_D = \frac{1}{4}$ to a unit step change in the set point, for $k_c = 2, 5, 10$.

Table 15.5 Roots of denominator polynomial versus k_c in PID control, with $\tau_I = 1\frac{1}{2}$ and $\tau_D = \frac{1}{8}$

k_c	Roots			
0	−3.0000	−2.0000	−1.0000	0.0000
0.0100	−3.0040	−1.9917	−1.0021	−0.0022
0.0200	−3.0080	−1.9834	−1.0042	−0.0045
0.0500	−3.0195	−1.9584	−1.0109	−0.0112
0.1000	−3.0379	−1.9169	−1.0228	−0.0224
0.2000	−3.0719	−1.8324	−1.0506	−0.0451
0.55624	−3.1719	−1.34988	−1.34988	−0.1283
1.0000	−3.2691	−1.2480 + 0.4225i	−1.2480 − 0.4225i	−0.2349
2.0000	−3.4325	−1.0605 + 0.7844i	−1.0605 − 0.7844i	−0.4465
5.0000	−3.7416	−0.8062 + 1.4521i	−0.8062 − 1.4521i	−0.6459
10.0000	−4.0513	−0.6267 + 2.0835i	−0.6267 − 2.0835i	−0.6952
20.0000	−4.4256	−0.4293 + 2.8694i	−0.4293 − 2.8694i	−0.7158
50.0000	−5.0046	−0.1342 + 4.2785i	−0.1342 − 4.2785i	−0.7270
73.8732	−5.2707	0.0000 + 5.0621i	0.0000 − 5.0621i	−0.7293
100.0000	−5.4801	0.1053 + 5.7701i	0.1053 − 5.7701i	−0.7305
200.0000	−5.9500	0.3412 + 7.8157i	0.3412 − 7.8157i	−0.7323
500.0000	−6.4900	0.6117 + 11.8196i	0.6117 − 11.8196i	−0.7333
1000.0000	−6.7935	0.7636 + 16.3379i	0.7636 − 16.3379i	−0.7337
10000.0000	−7.2055	0.9697 + 50.2013i	0.9697 − 50.2013i	−0.7340

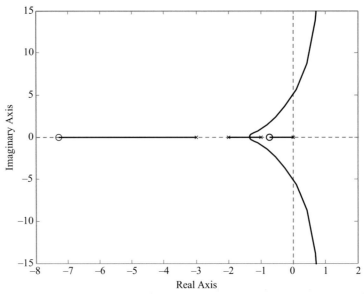

Figure 15.14 Root-locus diagram for PID control with $\tau_I = 1\frac{1}{2}$ and $\tau_D = \frac{1}{8}$.

But at any rate, derivative action has a stabilizing effect on the system. This can be seen from the enlargement of the stability range:

- in PI control, where $\tau_D = 0$, the stability condition for $\tau_I = 1\frac{1}{2}$ is $0 < k_c < 19.5934$
- in PID control with $\tau_D = \frac{1}{8}$, the stability condition for $\tau_I = 1\frac{1}{2}$ is $0 < k_c < 73.8732$.

It turns out that, for the specific value of integral time, complete stabilization – for all positive k_c – is achieved as long as $\tau_D \geq \frac{1}{6}$.

Figures 15.15 and 15.16 depict the root-locus diagrams for $\tau_D = \frac{3}{8}$ and $\tau_D = \frac{1}{2}$, and the same value of the integral time. As in the case of $\tau_D = \frac{1}{4}$, the closed-loop system is stable for all positive k_c. The differences between the three cases have to do with the nature and the location of the system zeros z_1, z_2, at which two of the branches terminate:

$\tau_D = \frac{1}{4} \leftrightarrow z_1 = -2 + \frac{2}{\sqrt{3}} = -0.84530$, $z_2 = -2 - \frac{2}{\sqrt{3}} = -3.15470$ (real simple)

$\tau_D = \frac{3}{8} \leftrightarrow z_1 = z_2 = -\frac{4}{3} = -1.3333$ (real double)

$\tau_D = \frac{1}{2} \leftrightarrow z_{1,2} = -1 \pm i\frac{1}{\sqrt{3}} = -1 \pm 0.57735i$ (complex conjugate)

15.1 What is a Root-Locus Diagram? Some Examples

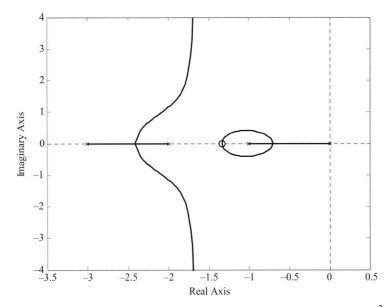

Figure 15.15 Root-locus diagram for PID control with $\tau_I = 1\frac{1}{2}$ and $\tau_D = \frac{3}{8}$.

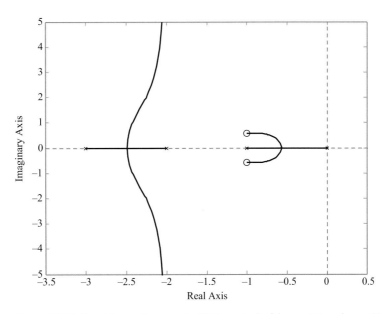

Figure 15.16 Root-locus diagram for PID control with $\tau_I = 1\frac{1}{2}$ and $\tau_D = \frac{1}{2}$.

15.2 Basic Properties of the Root Locus – Basic Rules for Sketching Root-Locus Diagrams

This section will give an outline of the basic mathematical properties of root locus. Root-locus theory as well as software tools (see Section 15.6 at the end of the chapter) *refer strictly to proportional control*. The understanding is that, after a math trick, using an appropriate extended or auxiliary system, the problem is transformed to a P control problem, as was done in the previous section for PI and PID control.

15.2.1 Basic Properties of the Root Locus

Consider a linear system with transfer function an irreducible fraction of two polynomials, of the form

$$G(s) = \frac{R(s)}{P(s)} = \frac{\kappa \cdot \prod_{j=1}^{m}(s-z_j)}{\prod_{i=1}^{n}(s-p_i)}, \text{ where } m \leq n,$$

i.e. a proper rational function with poles $p_1, \ldots, p_n$ and zeros $z_1, \ldots, z_m$.

Also, consider the feedback control system of Figure 15.17, where $G(s)$ is controlled with a P controller.

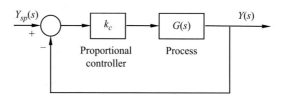

Figure 15.17 General feedback control loop with a P controller.

We would like to study the effect of the controller gain on the poles of the closed-loop system. These are the roots of the polynomial

$$Q(s) = P(s) + k_c R(s) = \prod_{i=1}^{n}(s-p_i) + k_c \kappa \prod_{j=1}^{m}(s-z_j)$$

In particular, we will study the *locus of the roots* ρ_k, $k = 1, \ldots, n$ of the above polynomial *in the complex plane*, as the *controller gain* k_c *varies from 0 to* $+\infty$.

(If k_c varies from 0 to $-\infty$, we can reverse the sign of the coefficient κ and apply the same procedure.)

The key mathematical properties of the root locus are outlined below. The derivation of some of these properties will be postponed to the next section.

15.2 Basic Properties of the Root Locus – Basic Rules for Sketching Root-Locus Diagrams

- The roots *for* $k_c = 0$ are equal to the poles $p_1, \ldots, p_n$ of the transfer function.
- For small k_c, all roots lie close to the poles $p_1, \ldots, p_n$ of the transfer function.
- For large k_c, m roots lie close to the zeros $z_1, \ldots, z_m$ of the transfer function, and the remaining $(n - m)$ roots tend to infinity.

Those roots that tend to infinity, asymptotically approach $(n - m)$ straight lines that pass through the point

$$\sigma = \frac{\sum_{i=1}^{n} p_i - \sum_{j=1}^{m} z_j}{n - m}$$

of the real axis, and form angles

$$\alpha_v = \begin{cases} \dfrac{(2v-1)\pi}{n-m}, & v = 1, \ldots, (n-m), \text{ if } \kappa > 0 \\[2ex] \dfrac{2(v-1)\pi}{n-m}, & v = 1, \ldots, (n-m), \text{ if } \kappa < 0 \end{cases}$$

with the positive real axis. The point σ is called the *center of gravity* of the root locus.

The *asymptotes* approached by the $(n - m)$ roots radiate from the center of gravity and divide the plane into $(n - m)$ equal angles of size $\dfrac{2\pi}{n - m}$.

– When $\kappa > 0$, the angles of the asymptotes with the positive real axis are:

for $n - m = 1$: $\alpha_1 = \pi$

for $n - m = 2$: $\alpha_1 = \dfrac{\pi}{2}$, $\alpha_2 = \dfrac{3\pi}{2}$

for $n - m = 3$: $\alpha_1 = \dfrac{\pi}{3}$, $\alpha_2 = \pi$, $\alpha_3 = \dfrac{5\pi}{3}$

for $n - m = 4$: $\alpha_1 = \dfrac{\pi}{4}$, $\alpha_2 = \dfrac{3\pi}{4}$, $\alpha_3 = \dfrac{5\pi}{4}$, $\alpha_4 = \dfrac{7\pi}{4}$

etc.

n−m=1   n−m=2   n−m=3   n−m=4

– When $\kappa < 0$, the angles of the asymptotes with the positive real axis are:

for $n - m = 1$: $\alpha_1 = 0$

for $n - m = 2$: $\alpha_1 = 0$, $\alpha_2 = \pi$

for $n - m = 3$: $\alpha_1 = 0$, $\alpha_2 = \dfrac{2\pi}{3}$, $\alpha_3 = \dfrac{4\pi}{3}$

for $n - m = 4$: $\alpha_1 = 0$, $\alpha_2 = \dfrac{\pi}{2}$, $\alpha_3 = \pi$, $\alpha_4 = \dfrac{3\pi}{2}$

etc.

- The roots *for intermediate* k_c must be computed numerically. However, it is possible to independently determine *the parts of the real axis that are contained in the root locus.*

In order to find out if a real number s is a root of the equation $Q(s) = P(s) + k_c R(s) = 0$, we can write it equivalently as

$$k_c = -\frac{P(s)}{R(s)} = -\frac{1}{\kappa} \frac{\prod_{i=1}^{n}(s-p_i)}{\prod_{j=1}^{m}(s-z_j)}$$

and examine under what condition the right-hand side $-\dfrac{1}{\kappa} \dfrac{\prod_{i=1}^{n}(s-p_i)}{\prod_{j=1}^{m}(s-z_j)}$ is a positive number;

this depends on the signs of κ and of the factors $(s - p_i)$ and $(s - z_j)$.

Notice that complex poles and complex zeros do not affect the sign, since they appear as conjugate pairs, therefore a product $(s-z)(s-\bar{z}) = (s - \mathrm{Re}\, z)^2 + (\mathrm{Im}\, z)^2$ is always positive.

- When $\kappa > 0$, the right-hand side will be a positive number as long as the total number of poles and zeros that lie to the right of s is odd.
- When $\kappa < 0$, the right-hand side will be a positive number as long as the total number of poles and zeros that lie to the right of s is even or zero.

The conclusion is therefore that:

Points of the real axis are inside the root locus if the total number of poles and zeros that lie to their right is $\begin{cases} \text{odd}, & \text{in case } \kappa > 0 \\ \text{even or zero}, & \text{in case } \kappa < 0 \end{cases}$.

Finally, an important property that must be noted is the *symmetry of the root locus with respect to the real axis*, as an immediate consequence of the fact that complex roots always appear in conjugate pairs.

15.2.1 Basic Rules for Sketching Root-Locus Diagrams

The theoretical properties outlined in the previous subsection can now be summarized. The summary is often referred to as the "basic sketching rules" for root-locus diagrams. Historically, these properties were the first steps for the construction of a root locus, before computer tools became available. But, even today, these rules have significant value. One can sketch root-locus diagrams for simple transfer functions and see the big picture, without having to rely on software. And more importantly, these basic rules provide consistency checks between theory and computer-generated root-locus diagrams, in terms of covering the proper ranges in the complex plane or in terms of accuracy of numerical interpolations in intermediate parts of the diagram.

15.2 Basic Properties of the Root Locus – Basic Rules for Sketching Root-Locus Diagrams

The basic sketching rules are as follows.

(1) The root locus diagram has as many branches as the order n of the process.
(2) The branches start ($k_c = 0$) from the process poles and terminate ($k_c = \infty$) at the process zeros or at infinity.
(3) Those branches that terminate at infinity, asymptotically approach $(n-m)$ straight lines that pass through the center of gravity $\sigma = \dfrac{\sum_{i=1}^{n} p_i - \sum_{j=1}^{m} z_j}{n-m}$

and form angles $\alpha_v = \begin{cases} \dfrac{(2v-1)\pi}{n-m}, & \text{if } \kappa > 0 \\[2mm] \dfrac{2(v-1)\pi}{n-m}, & \text{if } \kappa < 0 \end{cases}$, $v = 1, \ldots, (n-m)$

with the positive real axis.

(4) Points of the real axis are inside the root locus if the total number of poles and zeros that lie to their right is $\begin{cases} \text{odd}, & \text{in case } \kappa > 0 \\ \text{even or zero}, & \text{in case } \kappa < 0 \end{cases}$

(5) The branches that lie off the real axis are symmetric with respect to the real axis.

Example 15.3 Application of the basic rules to the root loci of Examples 15.1 and 15.2

The feedback control systems of Examples 15.1 and 15.2 will now be reexamined in light of the previous basic rules.

P Control

Consider the block diagram of Figure 15.1, with $\tau_1 > 0$, $\tau_2 > 0$, $\tau_3 > 0$, $k > 0$.

Here $G(s) = \dfrac{k}{(\tau_1 s + 1)(\tau_2 s + 1)(\tau_3 s + 1)} = \dfrac{\dfrac{k}{\tau_1 \tau_2 \tau_3}}{\left(s + \dfrac{1}{\tau_1}\right)\left(s + \dfrac{1}{\tau_2}\right)\left(s + \dfrac{1}{\tau_3}\right)}$, which has $n = 3$ poles, at $-\dfrac{1}{\tau_1}, -\dfrac{1}{\tau_2}, -\dfrac{1}{\tau_3}$ and no zeros ($m = 0$).

The coefficient $\kappa = \dfrac{k}{\tau_1 \tau_2 \tau_3} > 0$.

We can now apply the five basic rules, to see what general conclusions can be drawn, while at the same time comparing with Table 15.1 and Figure 15.4 (for $\tau_1 = 1$, $\tau_2 = \frac{1}{2}$, $\tau_3 = 1\frac{1}{3}$), to check for consistency.

(1) The root locus has $n = 3$ branches.
(2) The branches start from the poles $-\dfrac{1}{\tau_1}, -\dfrac{1}{\tau_2}, -\dfrac{1}{\tau_3}$.

Because there are no zeros, they all terminate at infinity.

(3) The $n - m = 3$ asymptotes

- all pass from the center of gravity

$$\sigma = \dfrac{\left(-\dfrac{1}{\tau_1} - \dfrac{1}{\tau_2} - \dfrac{1}{\tau_3}\right)}{3} = -\dfrac{1}{3}\left(\dfrac{1}{\tau_1} + \dfrac{1}{\tau_2} + \dfrac{1}{\tau_3}\right)$$

(for the specific values of the parameters, $\sigma = -2$)

- since $\kappa > 0$, they form angles $\dfrac{\pi}{3}, \pi, \dfrac{5\pi}{3}$ with the positive real axis.

(4) Because $\kappa > 0$, the parts of the real axis that are inside the root locus are those that have an odd number of poles to their right (for the specific values of the parameters, they are $(-\infty, -3]$ and $[-2, -1]$).

(5) The root-locus diagram is symmetric with respect to the real axis.

PI Control

Consider the block diagram of Figure 15.7, with $\mathcal{G}_c(s) = \dfrac{s + \dfrac{1}{\tau_I}}{s}$ and $\tau_I > 0, \tau_1 > 0, \tau_2 > 0, \tau_3 > 0$, $k > 0$. Here

$$G(s) = \dfrac{k\left(s + \dfrac{1}{\tau_I}\right)}{s(\tau_1 s + 1)(\tau_2 s + 1)(\tau_3 s + 1)} = \dfrac{\left(\dfrac{k}{\tau_1 \tau_2 \tau_3}\right)\left(s + \dfrac{1}{\tau_I}\right)}{s\left(s + \dfrac{1}{\tau_1}\right)\left(s + \dfrac{1}{\tau_2}\right)\left(s + \dfrac{1}{\tau_3}\right)}.$$

which has $n = 4$ poles at $0, -\dfrac{1}{\tau_1}, -\dfrac{1}{\tau_2}, -\dfrac{1}{\tau_3}$ and $m = 1$ zero at $-\dfrac{1}{\tau_I}$.

The coefficient $\kappa = \dfrac{k}{\tau_1 \tau_2 \tau_3} > 0$.

Applying the basic rules leads to the following conclusions.

(1) The root locus has $n = 4$ branches.
(2) The branches start from the poles $0, -\dfrac{1}{\tau_1}, -\dfrac{1}{\tau_2}, -\dfrac{1}{\tau_3}$. One branch terminates at the zero at $-\dfrac{1}{\tau_I}$ and the remaining branches at infinity.

(3) The $n - m = 3$ asymptotes
- all pass from the center of gravity
$$\sigma = \frac{\left(0 - \frac{1}{\tau_1} - \frac{1}{\tau_2} - \frac{1}{\tau_3}\right) - \left(-\frac{1}{\tau_I}\right)}{3} = -\frac{1}{3}\left(\frac{1}{\tau_1} + \frac{1}{\tau_2} + \frac{1}{\tau_3} - \frac{1}{\tau_I}\right)$$
(for the specific values of process parameters, $\sigma = -2 + \frac{1}{3\tau_I}$)

- since $\kappa > 0$, they form angles $\frac{\pi}{3}$, π, $\frac{5\pi}{3}$ with the positive real axis.

(4) Because $\kappa > 0$, the parts of the real axis that are inside the root locus are those that have an odd number of poles to their right (see Figures 15.8 and 15.10).
(5) The root-locus diagram is symmetric with respect to the real axis.

PID Control

Consider the block diagram of Figure 15.7, with $G_c(s) = 1 + \frac{1}{\tau_I s} + \tau_D s = \frac{\tau_D s^2 + s + \frac{1}{\tau_I}}{s}$ and $\tau_D > 0$, $\tau_I > 0$, $\tau_1 > 0$, $\tau_2 > 0$, $\tau_3 > 0$, $k > 0$. Here

$$G(s) = \frac{k\left(\tau_D s^2 + s + \frac{1}{\tau_I}\right)}{s(\tau_1 s + 1)(\tau_2 s + 1)(\tau_3 s + 1)} = \frac{\left(\frac{k\tau_D}{\tau_1 \tau_2 \tau_3}\right)\left(s^2 + \frac{1}{\tau_D}s + \frac{1}{\tau_D \tau_I}\right)}{s\left(s + \frac{1}{\tau_1}\right)\left(s + \frac{1}{\tau_2}\right)\left(s + \frac{1}{\tau_3}\right)},$$

which has $n = 4$ poles at 0, $-\frac{1}{\tau_1}$, $-\frac{1}{\tau_2}$, $-\frac{1}{\tau_3}$ and $m = 2$ zeros at the roots of the quadratic polynomial $s^2 + \frac{1}{\tau_D}s + \frac{1}{\tau_D \tau_I}$. The coefficient $\kappa = \frac{k\tau_D}{\tau_1 \tau_2 \tau_3} > 0$.

Applying the basic rules leads to the following conclusions.
(1) The root locus has $n = 4$ branches.
(2) The branches start from the poles 0, $-\frac{1}{\tau_1}$, $-\frac{1}{\tau_2}$, $-\frac{1}{\tau_3}$. Two branches terminate at the roots of the quadratic $s^2 + \frac{1}{\tau_D}s + \frac{1}{\tau_D \tau_I}$ and the remaining branches at infinity.
(3) The $n - m = 2$ asymptotes
- all pass from the center of gravity
$$\sigma = \frac{\left(0 - \frac{1}{\tau_1} - \frac{1}{\tau_2} - \frac{1}{\tau_3}\right) - \left(-\frac{1}{\tau_D}\right)}{2} = -\frac{1}{2}\left(\frac{1}{\tau_1} + \frac{1}{\tau_2} + \frac{1}{\tau_3} - \frac{1}{\tau_D}\right)$$

(for the specific values of process parameters, $\sigma = -3 + \dfrac{1}{2\tau_D}$, hence

$\tau_D = \dfrac{1}{8} \leftrightarrow \sigma = +1$, $\tau_D = \dfrac{1}{4} \leftrightarrow \sigma = -1$, $\tau_D = \dfrac{1}{2} \leftrightarrow \sigma = -2$, etc.)

– since $\kappa > 0$, they form angles $\dfrac{\pi}{2}$, $\dfrac{3\pi}{2}$ with the positive real axis.

(4) Because $\kappa > 0$, the parts of the real axis that are inside the root locus are those that have an odd number of poles to their right (see Figures 15.12 and 15.14–15.16).

(5) The root-locus diagram is symmetric with respect to the real axis.

15.3 Further Properties of the Root Locus – Additional Rules for Sketching Root-Locus Diagrams

In the previous section, five key theoretical properties (rules) of the root-locus diagram were outlined. The present section will provide more theoretical properties, leading to additional rules that enable more sophisticated predictions of the shape of the root-locus diagram, or more sophisticated consistency checks for a computer-generated root-locus diagram.

In particular, in this section, we will calculate (i) the so-called "angles of departure" and "angles of arrival" and (ii) the so-called "break-away" or "break-in" points. We start with two motivating examples.

Example 15.4 Proportional-derivative control of an underdamped second-order process

Consider the feedback control system of Figure 15.18, where the parameter τ_D of the controller is kept fixed. Also, it is assumed that $\tau_D > 0$, $\tau > 0$, $0 < \zeta < 1$, $k > 0$.

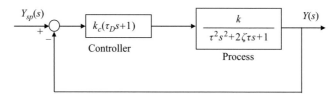

Figure 15.18 Feedback control loop consisting of an underdamped second-order process and a proportional-derivative controller.

In this system, there are $n = 2$ poles, at $p_{1,2} = \dfrac{-\zeta \pm i\sqrt{1-\zeta^2}}{\tau}$, from which the two branches of the root locus start, and $m = 1$ zero at $z_1 = -\dfrac{1}{\tau_D}$, to which one of the branches ends. The other branch ends at $-\infty$ (for $n - m = 1$ and $\kappa > 0$, the angle of the asymptote is $\alpha_1 = \pi$).

15.3 Further Properties of the Root Locus – Additional Rules for Sketching Root-Locus Diagrams

Figure 15.19 depicts the root-locus diagrams for $\tau = \zeta = \frac{1}{2}$ and two different values of derivative time, $\tau_D = \frac{1}{4}$ and $\tau_D = 1$. We see that, in the first case, the branches starting from the complex poles move father away from the real axis, but then turn around and move towards the real axis. In the second case, the branches starting from the complex poles initially move horizontally, and then towards the real axis.

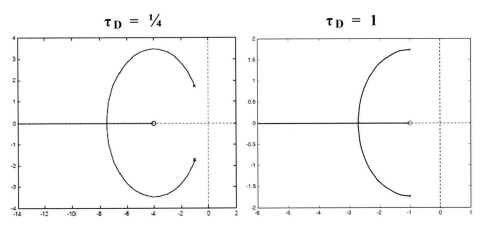

Figure 15.19 Root-locus diagram for underdamped second-order process under PD control with $\tau_D = \frac{1}{4}$ (left) and $\tau_D = 1$ (right).

The angles of the tangents of the root-locus curves at the complex poles are called "angles of departure," since they specify the directions along which the roots "depart" from the complex poles. Could we have predicted the angles of departure in this problem? As we will see, the answer is "yes."

From the root-locus locus diagrams of Figure 15.19 we also observe that, as k_c increases, the two roots get closer to the real axis and, at some value of k_c, they meet on the real axis. Under further increase of k_c, they separate, but staying in the real axis, with one of the roots going towards the zero at $-\frac{1}{\tau_D}$ and the other towards $-\infty$.

A point on the real axis where two branches meet is called a "break-away" point or "break-in" point, depending on whether the roots move "away" or "in" the real axis after they separate with increasing k_c. Would it be possible to independently calculate the location of the break-in point in this problem? As we will see, the answer is "yes."

Example 15.5 Proportional control of a third-order process, including cases of two equal or all three equal time constants

Consider again Example 15.1, where a third-order process is controlled with a P controller, as shown in Figure 15.1. When $\tau_1 = 1$, $\tau_2 = \frac{1}{2}$, $\tau_3 = \frac{1}{3}$, $k = \frac{1}{3}$, the root-locus diagram is given in Figure 15.4. From the diagram, as well as from the root table (Table 15.1), we see that there is a break-away point where two branches meet on the real axis: $\rho_b = -1.42265$, which corresponds to $k_c = 0.19245$. As k_c increases, the roots move away from the real axis.

It is interesting to also consider cases where two time constants or all three time constants are equal.

For example, for $\tau_1 = 1$, $\tau_2 = 1$, $\tau_3 = \frac{1}{2}$, $k = \frac{1}{2}$, i.e. for a process with transfer function $\dfrac{Y(s)}{U(s)} = \dfrac{1}{(s+1)^2(s+2)}$, which has a double pole at -1 and a simple pole at -2, the root-locus diagram is as shown in Figure 15.20.

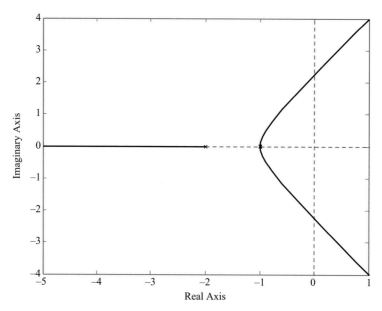

Figure 15.20 Root-locus diagram for third-order process with two equal time constants $\tau_1 = \tau_2 = 1$ and a smaller time constant $\tau_3 = \frac{1}{2}$, under P control.

From the real double pole, two symmetric branches emerge, lying outside the real axis.

In the case where all three time constants are equal, for example when $\tau_1 = \tau_2 = \tau_3 = 1$, $k = 1$, i.e. for a process transfer function $\dfrac{Y(s)}{U(s)} = \dfrac{1}{(s+1)^3}$, with a triple pole at -1, the root-locus diagram is as shown in Figure 15.21.

From the real triple pole, three branches emerge, which form angles $\dfrac{\pi}{3}$, π, $\dfrac{5\pi}{3}$ with the positive real axis.

Through this example, we see that from a multiple pole, multiple branches emerge, pointing in different directions in the complex plane. Depending on the system and the multiplicity of the pole, there are "angles of departure" from the multiple pole. As we will see, these angles can be calculated independently.

15.3 Further Properties of the Root Locus – Additional Rules for Sketching Root-Locus Diagrams

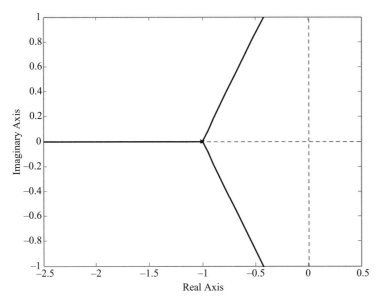

Figure 15.21 Root-locus diagram for third-order process with equal time constants $\tau_1 = \tau_2 = \tau_3 = 1$, under P control.

15.3.1 Small-Gain Approximate Formulas for the Roots – Angles of Departure

Consider a general feedback control loop, as described in Section 15.2.1.

When the controller gain k_c is small, all the roots lie close to the poles $p_1, \ldots, p_n$ of the transfer function. In particular we have the following:

→ Near every simple pole p_k, there is a simple root ρ_k given by the approximate formula

$$\rho_k \approx p_k - \gamma_k k_c \quad , \quad \text{for small } k_c$$

where $\displaystyle \gamma_k = \lim_{s \to p_k}\left[(s - p_k)G(s)\right] = \kappa \, \frac{\prod_{j=1}^{m}(p_k - z_j)}{\prod_{\substack{i=1 \\ (i \neq k)}}^{n}(p_k - p_i)}$.

- When the simple pole p_k is real, the root ρ_k for small k_c that lies close to it is also a real number. The location of the root ρ_k (to the right or to the left of the pole p_k) depends on the sign of the real number γ_k.
- When the simple pole p_k is complex, the root ρ_k for small k_c that lies close to it is also a complex number, since the number γ_k is complex. The straight line that connects the points ρ_k and p_k on the complex plane for small k_c, forms an angle of

$$\varphi_k = \arg(-\gamma_k) = \arg\left(-\kappa \, \frac{\prod_{j=1}^{m}(p_k - z_j)}{\prod_{\substack{i=1 \\ (i \neq k)}}^{n}(p_k - p_i)}\right)$$

with the real axis, which is called the *angle of departure* from the pole p_k. (In the above formula, arg(z) denotes the argument of the complex number z.)

→ Near every multiple pole p_k of multiplicity μ, there are μ simple roots p_{k_ℓ}, $\ell = 1, \ldots, \mu$, given by the approximate formula

$$p_{k_\ell} \approx p_k + e^{i\varphi_{k_\ell}} |\gamma_k|^{\frac{1}{\mu}} k_c^{\frac{1}{\mu}}, \quad \ell = 1, \ldots, \mu, \quad \text{for small } k_c$$

where $\gamma_k = \lim_{s \to p_k} \left[(s - p_k)^\mu G(s) \right]$ and $\varphi_{k_\ell} = \dfrac{\arg(-\gamma_k) + 2(\ell - 1)\pi}{\mu}$, $\ell = 1, \ldots, \mu$.

In the above, φ_{k_ℓ}, $\ell = 1, \ldots, \mu$ are the μ angles of departure from the pole p_k.

For a double real pole, the angles of departure are $0, \pi$ or $\dfrac{\pi}{2}, \dfrac{3\pi}{2}$

For a triple real pole, the angles of departure are $0, \dfrac{2\pi}{3}, \dfrac{4\pi}{3}$ or $\dfrac{\pi}{3}, \pi, \dfrac{5\pi}{3}$

etc., depending on the sign of γ_k.

Example 15.6 Proportional control of a third-order process: roots under small gain and angles of departure

Consider again the case of a third-order process controlled with a P controller, as shown in Figure 15.1, whose root locus was seen in Example 15.1 for the case of unequal time constants, and in Example 15.5 for the cases where two time constants or all three time constants are equal. Here we would like to calculate the roots for small k_c using the approximate formulas and calculate the angles of departure where appropriate.

For $\tau_1 = 1$, $\tau_2 = \dfrac{1}{2}$, $\tau_3 = \dfrac{1}{3}$, $k = \dfrac{1}{3}$ i.e. for transfer function $G(s) = \dfrac{2}{(s+1)(s+2)(s+3)}$, the partial fraction expansion is

$$G(s) = \dfrac{1}{s+1} - \dfrac{2}{s+2} + \dfrac{1}{s+3}$$

hence $\gamma_1 = 1$, $\gamma_2 = -2$, $\gamma_3 = 1$, and applying the approximate formula, it follows that

$p_1 \approx -1 - k_c$
$p_2 \approx -2 + 2k_c$, for small k_c.
$p_3 \approx -3 - k_c$

15.3 Further Properties of the Root Locus – Additional Rules for Sketching Root-Locus Diagrams

Comparing with Table 15.1, we see that the above formulas are quite accurate for k_c less than 0.1.

For $\tau_1 = 1$, $\tau_2 = 1$, $\tau_3 = \frac{1}{2}$, $k = \frac{1}{2}$ i.e. for transfer function $G(s) = \dfrac{2}{(s+1)^2(s+2)}$, which has a double pole at $p_1 = -1$ and a simple pole at $p_2 = -2$, the partial fraction expansion is

$$G(s) = \frac{1}{(s+1)^2} - \frac{1}{s+1} + \frac{1}{s+2}$$

hence $\gamma_1 = 1$ (coefficient of the quadratic term corresponding to the double pole) and $\gamma_2 = 1$, and applying the approximate formula, it follows that

$$p_{1_1} \approx -1 + e^{i\frac{\pi}{2}} k_c^{\frac{1}{2}} = -1 + i\sqrt{k_c}$$

$$p_{1_2} \approx -1 + e^{i\frac{3\pi}{2}} k_c^{\frac{1}{2}} = -1 - i\sqrt{k_c} \quad \text{, for small } k_c.$$

$$p_2 \approx -2 - k_c$$

Because the coefficient γ_1 (corresponding to the double pole) is positive, the angles of departure are $\dfrac{\pi}{2}$, $\dfrac{3\pi}{2}$, which is clear from the above formulas.

The complete root-locus diagram has been given in Figure 15.20. Zooming close to the double pole, we see that indeed the two branches emerging from the double pole are in vertical directions (see Figure 15.22).

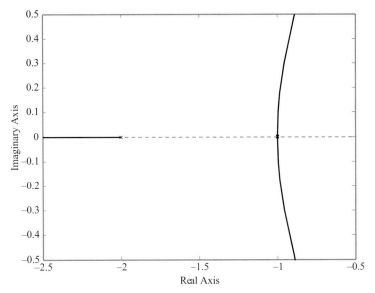

Figure 15.22 Detail of the root-locus diagram of Figure 15.20 (third-order process with time constants $\tau_1 = \tau_2 = 1$, $\tau_3 = \frac{1}{2}$, under P control).

Example 15.7 Proportional-derivative control of an underdamped second-order process: calculation of angles of departure

Consider the feedback control system of Example 15.4, where an underdamped second-order process is controlled with a proportional-derivative controller, as shown in Figure 15.18.

The transfer function $G(s) = \dfrac{k(\tau_D s + 1)}{\tau^2 s^2 + 2\zeta\tau s + 1}$ can be expanded in partial fractions as follows:

$$G(s) = \frac{k\tau_D}{2\tau^2}\left(\frac{1 + i\dfrac{\zeta - \dfrac{\tau}{\tau_D}}{\sqrt{1-\zeta^2}}}{s + \dfrac{\zeta}{\tau} - i\dfrac{\sqrt{1-\zeta^2}}{\tau}} + \frac{1 - i\dfrac{\zeta - \dfrac{\tau}{\tau_D}}{\sqrt{1-\zeta^2}}}{s + \dfrac{\zeta}{\tau} + i\dfrac{\sqrt{1-\zeta^2}}{\tau}}\right)$$

Therefore, for small k_c the roots are: $\rho_1 = \dfrac{-\zeta + i\sqrt{1-\zeta^2}}{\tau} - \gamma_1 k_c$, where $\gamma_1 = \dfrac{k\tau_D}{2\tau^2}\left(1 + i\dfrac{\zeta - \dfrac{\tau}{\tau_D}}{\sqrt{1-\zeta^2}}\right)$

$$\rho_2 = \frac{-\zeta - i\sqrt{1-\zeta^2}}{\tau} - \gamma_2 k_c, \text{ where } \gamma_2 = \frac{k\tau_D}{2\tau^2}\left(1 - i\frac{\zeta - \dfrac{\tau}{\tau_D}}{\sqrt{1-\zeta^2}}\right)$$

Because $k > 0$, $\tau_D > 0$, the corresponding angles of departure are given by

$$\varphi_1 = \arg(-\gamma_1) = \arg\left(-1 - i\frac{\zeta - \dfrac{\tau}{\tau_D}}{\sqrt{1-\zeta^2}}\right) \quad , \quad \varphi_2 = \arg(-\gamma_2) = \arg\left(-1 + i\frac{\zeta - \dfrac{\tau}{\tau_D}}{\sqrt{1-\zeta^2}}\right)$$

For the specific values of the process parameters $\tau = \zeta = \dfrac{1}{2}$ and

for $\tau_D = \dfrac{1}{4}$, $\varphi_1 = \arg(-1 + i\sqrt{3}) = \dfrac{2\pi}{3}$, $\varphi_2 = \arg(-1 - i\sqrt{3}) = \dfrac{4\pi}{3}$

for $\tau_D = 1$, $\varphi_1 = \arg(-1) = \pi$, $\varphi_2 = \arg(-1) = \pi$

and these are consistent with the root-locus diagrams of Figure 15.19.

15.3.2 Large-Gain Approximate Formulas for the Roots – Angles of Arrival

In the previous subsection, we gave general formulas for the roots under small k_c. These have led to formulas for the angles of departure from simple complex poles or from multiple poles. In the present subsection, we will give general formulas for the roots under large k_c ($\to +\infty$). These will lead to formulas for the angles of arrival to simple complex zeros or to multiple zeros.

Consider again the general feedback control loop, as described in Section 15.2.1.

When the controller gain k_c is large, m roots lie close to the zeros $z_1, \ldots, z_m$ of the transfer function, and $(n-m)$ roots tend to infinity. In particular:

→ Near every simple zero z_k, there is a simple root ρ_k given by the approximate formula

$$\rho_k \approx z_k - \frac{\delta_k}{k_c}, \quad \text{for large } k_c$$

where $\delta_k = \lim\limits_{s \to z_k} \left[(s - z_k) \cdot \frac{1}{G(s)} \right] = \dfrac{\prod\limits_{i=1}^{n}(z_k - p_i)}{\kappa \prod\limits_{\substack{j=1 \\ (j \ne k)}}^{m}(z_k - z_j)}.$

- When the simple zero z_k is real, the root ρ_k for large k_c that lies close to it is also a real number. The location of the root ρ_k (to the right or to the left of the zero z_k) depends on the sign of the real number δ_k.
- When the simple zero z_k is complex, the root ρ_k for large k_c that lies close to it is also a complex number, since the number δ_k is complex. The straight line that connects the points ρ_k and z_k on the complex plane for large k_c, forms an angle of

$$\psi_k = \arg(-\delta_k) = \arg\left(-\dfrac{\prod\limits_{i=1}^{n}(z_k - p_i)}{\kappa \prod\limits_{\substack{j=1 \\ (j \ne k)}}^{m}(z_k - z_j)} \right)$$

with the real axis, which is called the *angle of arrival* at the zero z_k. (Again, arg(z) denotes the argument of the complex number z.)

→ Near every multiple zero z_k of multiplicity μ, there are μ simple roots ρ_{k_ℓ}, $\ell = 1, \ldots, \mu$, given by the approximate formula

$$\rho_{k_\ell} \approx z_k + e^{i\psi_{k_\ell}} \frac{|\delta_k|^{\frac{1}{\mu}}}{k_c^{\frac{1}{\mu}}}, \quad \ell = 1, \ldots, \mu, \quad \text{for large } k_c$$

where $\delta_k = \lim\limits_{s \to z_k}\left[(s-z_k)^\mu \cdot \dfrac{1}{G(s)}\right]$

and $\psi_{k_\ell} = \dfrac{\arg(-\delta_k) + 2(\ell-1)\pi}{\mu}$, $\ell = 1, \ldots, \mu$ (angles of arrival)

→ Those $(n-m)$ roots whose magnitude tends to infinity for large k_c, can be calculated via the approximate formula

$$\rho_v \approx \dfrac{\sum\limits_{i=1}^{n} p_i - \sum\limits_{j=1}^{m} z_j}{n-m} + e^{i\alpha_v}\,|\kappa|^{\frac{1}{n-m}}\, k_c^{\frac{1}{n-m}}, \quad v = 1, \ldots, n-m, \quad \text{for large } k_c$$

where $\alpha_v = \dfrac{\arg(-\kappa) + 2(v-1)\pi}{n-m} = \begin{cases} \dfrac{(2v-1)\pi}{n-m}, & \text{if } \kappa > 0 \\ \dfrac{2(v-1)\pi}{n-m}, & \text{if } \kappa < 0 \end{cases}$, $v = 1, \ldots, (n-m)$

Note that the above formulas completely specify the asymptotes of the root locus, as stated in sketching rule number (3): the asymptotes all pass through the center of gravity

$$\sigma = \dfrac{\sum\limits_{i=1}^{n} p_i - \sum\limits_{j=1}^{m} z_j}{n-m}$$ and form angles α_v, $v = 1, \ldots, (n-m)$ as given above.

Example 15.8 P and PI control of a third-order process: roots under large gain and asymptotes

Consider again the case of a third-order process controlled with a P or PI controller, as in Examples 15.1 and 15.2, and for the specific values of process parameters $\tau_1 = 1$, $\tau_2 = \dfrac{1}{2}$, $\tau_3 = \dfrac{1}{3}$, $k = \dfrac{1}{3}$.

→ Under P control, $G(s) = \dfrac{2}{(s+1)(s+2)(s+3)}$, which has no zeros, therefore all three roots tend to infinity as k_c tends to infinity. Here $\sigma = -2$, $\kappa = 2 > 0$, $n - m = 3$, hence applying the general formulas gives:

$$\rho_1 \approx \sigma + e^{i\frac{\pi}{3}}(\kappa k_c)^{\frac{1}{3}} = -2 + \left(\dfrac{1}{2} + i\dfrac{\sqrt{3}}{2}\right)(2k_c)^{\frac{1}{3}}$$

$$\rho_2 \approx \sigma + e^{i\pi}(\kappa k_c)^{\frac{1}{3}} = -2 - (2k_c)^{\frac{1}{3}} \qquad \text{, for large } k_c$$

$$\rho_3 \approx \sigma + e^{i\frac{5\pi}{3}}(\kappa k_c)^{\frac{1}{3}} = -2 + \left(\dfrac{1}{2} - i\dfrac{\sqrt{3}}{2}\right)(2k_c)^{\frac{1}{3}}$$

15.3 Further Properties of the Root Locus – Additional Rules for Sketching Root-Locus Diagrams

The roots asymptotically approach three straight lines that radiate from the point -2 and form angles $\frac{\pi}{3}$, π, $\frac{5\pi}{3}$ with the positive real axis. The above approximate formulas have satisfactory accuracy for k_c of the order of 100 or larger. See Table 15.1.

→ Under PI control, $G(s) = \dfrac{2\left(s + \dfrac{1}{\tau_I}\right)}{s(s+1)(s+2)(s+3)}$, which has only one zero at $-\dfrac{1}{\tau_I}$.

One of the roots will get close to the zero for large k_c, according to the formula

$$\rho_1 \approx -\frac{1}{\tau_I} - \frac{\delta_1}{k_c},$$

where $\delta_1 = \lim\limits_{s \to -\frac{1}{\tau_I}} \left[\left(s + \dfrac{1}{\tau_I}\right)\dfrac{1}{G(s)}\right] = -\dfrac{1}{2}\dfrac{1}{\tau_I}\left(1 - \dfrac{1}{\tau_I}\right)\left(2 - \dfrac{1}{\tau_I}\right)\left(3 - \dfrac{1}{\tau_I}\right)$

The other three roots will therefore tend to infinity as k_c tends to infinity. Here $\sigma = -2 + \dfrac{1}{3\tau_I}$, $\kappa = 2 > 0$, $n - m = 3$, hence applying the general formulas leads to the following approximate expressions for these three roots:

$$\sigma + e^{i\frac{\pi}{3}}(\kappa k_c)^{\frac{1}{3}} = -2 + \frac{1}{3\tau_I} + \left(\frac{1}{2} + i\frac{\sqrt{3}}{2}\right)(2k_c)^{\frac{1}{3}}$$

$$\sigma + e^{i\pi}(\kappa k_c)^{\frac{1}{3}} = -2 + \frac{1}{3\tau_I} - (2k_c)^{\frac{1}{3}} \qquad \text{, for large } k_c$$

$$\sigma + e^{i\frac{5\pi}{3}}(\kappa k_c)^{\frac{1}{3}} = -2 + \frac{1}{3\tau_I} + \left(\frac{1}{2} + i\frac{\sqrt{3}}{2}\right)(2k_c)^{\frac{1}{3}}$$

Comparing the above with the large-gain roots under P control, we see that they are essentially the same, except that now the center of gravity is different.

Example 15.9 PID control of a third-order process, in the presence of a double zero or a pair of complex conjugate zeros: roots under large gain and angles of arrival

Consider again the case of a third-order process controlled with a PID controller, as in Example 15.2, and for the specific values of $\tau_1 = 1$, $\tau_2 = \dfrac{1}{2}$, $\tau_3 = \dfrac{1}{3}$, $k = \dfrac{1}{3}$. Here $G(s) = \dfrac{2\left(\tau_D s^2 + s + \dfrac{1}{\tau_I}\right)}{s(s+1)(s+2)(s+3)}$, which has two zeros at the roots of $\tau_D s^2 + s + \dfrac{1}{\tau_I}$.

Depending on the values of τ_I and τ_D, the zeros can be real simple, real double, or complex conjugate. We have already seen root locus diagrams for all three cases: Figures 15.12 and 15.14 for simple zeros, Figure 15.15 for double zero and Figure 15.16 for complex conjugate zeros. For all these cases, the roots for large k_c that are close to the zeros can be approximately calculated via the formulas and moreover, for the cases of double zero and complex zeros, the angles of arrival can be calculated.

Specifically, we have the following.

→ For $\tau_I = \dfrac{3}{2}$, $\tau_D = \dfrac{1}{4}$ ↔ $z_1 = -2 + \dfrac{2}{\sqrt{3}} = -0.84530$, $z_2 = -2 - \dfrac{2}{\sqrt{3}} = -3.15470$

(simple real zeros), the roots for large k_c that lie close to the zeros can be calculated from the approximate formulas:

$$p_1 \approx z_1 - \dfrac{\delta_1}{k_c}$$

$$p_2 \approx z_2 - \dfrac{\delta_2}{k_c}$$

where

$$\delta_1 = \lim_{s \to -2 + \frac{2}{\sqrt{3}}} \left[\left(s + 2 - \dfrac{2}{\sqrt{3}}\right) \dfrac{1}{G(s)} \right] = -\dfrac{2}{3}\left(1 - \dfrac{1}{\sqrt{3}}\right) = -0.28177$$

$$\delta_2 = \lim_{s \to -2 - \frac{2}{\sqrt{3}}} \left[\left(s + 2 + \dfrac{2}{\sqrt{3}}\right) \dfrac{1}{G(s)} \right] = -\dfrac{2}{3}\left(1 + \dfrac{1}{\sqrt{3}}\right) = -1.0516$$

Because $\delta_1 < 0$, $\delta_2 < 0$, the roots lie to the right of the zeros.

→ For $\tau_I = \dfrac{3}{2}$, $\tau_D = \dfrac{3}{8}$ ↔ $z_1 = z_2 = -\dfrac{4}{3}$ (double real zero), the roots for large k_c that lie close to the double zero may be calculated as follows:

We first calculate $\delta_1 = \lim_{s \to -\frac{4}{3}} \left[\left(s + \dfrac{4}{3}\right)^2 \dfrac{1}{G(s)} \right] = \dfrac{160}{243} = 0.65844$,

and because $\delta_1 > 0$, the roots are given by the approximate formulas

$$p_1 \approx -\dfrac{4}{3} + e^{i\frac{\pi}{2}} \dfrac{\sqrt{\delta_1}}{\sqrt{k_c}} = -1.3333 + i\dfrac{0.81144}{\sqrt{k_c}}$$

$$p_2 \approx -\dfrac{4}{3} + e^{i\frac{3\pi}{2}} \dfrac{\sqrt{\delta_1}}{\sqrt{k_c}} = -1.3333 - i\dfrac{0.81144}{\sqrt{k_c}}$$

The roots arrive to the double zero with angles $\dfrac{\pi}{2}$, $\dfrac{3\pi}{2}$ (angles of arrival).

The complete root locus diagram has been given in Figure 15.15. Zooming close to the double zero, we see that indeed the two branches approaching the double zero are in vertical directions (see Figure 15.23).

15.3 Further Properties of the Root Locus – Additional Rules for Sketching Root-Locus Diagrams

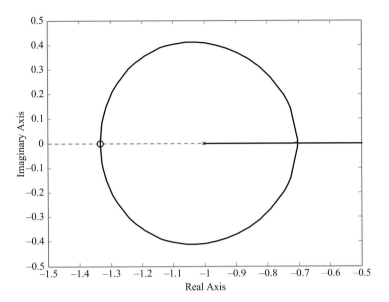

Figure 15.23 Detail of the root locus diagram of Figure 15.15 (third-order process under PID control with $\tau_I = 1\tfrac{1}{2}$ and $\tau_D = \tfrac{3}{8}$).

→ For $\tau_I = \dfrac{3}{2}$, $\tau_D = \dfrac{1}{2}$ ↔ $z_{1,2} = -1 \pm \dfrac{1}{\sqrt{3}} i$ (complex conjugate zeros), the roots for large k_c that lie close to the zeros can be calculated from the approximate formulas:

$$\rho_1 \approx \left(-1 + \dfrac{1}{\sqrt{3}} i\right) - \dfrac{\delta_1}{k_c}$$
$$\rho_2 \approx \left(-1 - \dfrac{1}{\sqrt{3}} i\right) - \dfrac{\delta_2}{k_c}$$

, where

$$\delta_1 = \lim_{s \to -1 + \tfrac{1}{\sqrt{3}} i} \left[\left(s + 1 - \dfrac{1}{\sqrt{3}} i\right) \dfrac{1}{G(s)}\right] = -\dfrac{4}{3} - \dfrac{2}{3\sqrt{3}} i$$
$$\delta_2 = \lim_{s \to -1 - \tfrac{1}{\sqrt{3}} i} \left[\left(s + 1 + \dfrac{1}{\sqrt{3}} i\right) \dfrac{1}{G(s)}\right] = -\dfrac{4}{3} + \dfrac{2}{3\sqrt{3}} i$$

The complete root-locus diagram has been given in Figure 15.16. Zooming close to the zeros (see Figure 15.24) seems to indicate that the approach towards the complex zeros is in approximately horizontal directions. However, zooming even closer to one of the complex zeros (see Figure 15.25) shows that the approach is not horizontal: there is a nonzero angle of arrival.

Exact calculation of the angles of arrival is possible through the formulas:

$$\psi_1 = \arg(-\delta_1) = \tan^{-1}\left(\dfrac{\tfrac{2}{3\sqrt{3}}}{\tfrac{4}{3}}\right) = \tan^{-1}\left(\dfrac{1}{2\sqrt{3}}\right) = 16.1°, \quad \psi_2 = 2\pi - \psi_1 = 360° - 16.1° = 343.9°$$

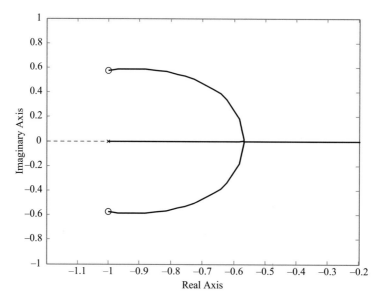

Figure 15.24 Detail of the root-locus diagram of Figure 15.16 (third-order process under PID control with $\tau_I = 1\frac{1}{2}$ and $\tau_D = \frac{1}{2}$).

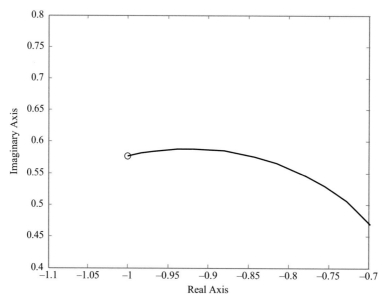

Figure 15.25 Detail of the root-locus diagram of Figure 15.16 (third-order process under PID control with $\tau_I = 1\frac{1}{2}$ and $\tau_D = \frac{1}{2}$), very close to one of the zeros.

15.3.3 Points of Intersection of Branches of the Root Locus – Break-away Points and Break-in Points

Root-locus diagrams often have branches that intersect. The point of intersection corresponds to a multiple root (at least double root). In the case where the point of intersection lies on the real axis, it is called a break-away or break-in point. If two roots are real and distinct for lower k_c, they meet on the real axis for some value of k_c, and then they separate and move away from the real axis for higher k_c; the meeting point is called the break-away point. In the reverse situation where two roots are complex conjugate for lower k_c, they meet on the real axis for some value of k_c, and then they separate and move along the real axis for higher k_c; the meeting point is called the break-in point. In what follows, we will see how to calculate the points of intersection (real or complex) of branches of the root locus.

Consider again the general feedback control loop, as described in Section 15.2.1.

A point of intersection s of two (or more) branches of the root locus will be (at least) double root the polynomial $Q(s) = P(s) + k_c R(s)$, hence the following will hold:

$$Q(s) = 0 \quad \Rightarrow \quad P(s) + k_c R(s) = 0$$

$$\frac{d}{ds}[Q(s)] = 0 \quad \Rightarrow \quad \frac{dP}{ds}(s) + k_c \frac{dR}{ds}(s) = 0$$

from which, eliminating k_c, it follows that

$$\frac{dP}{ds}(s) \cdot R(s) - P(s) \cdot \frac{dR}{ds}(s) = 0$$

The above equation is satisfied at all points of intersection of branches. It is a polynomial equation of degree $(n + m - 1)$, hence it has $(n + m - 1)$ roots, but not all of them are points of the root locus. A root s of the above equation is a point of the root locus if the corresponding gain $k_c = -\frac{P(s)}{R(s)}$ is a real and positive number.

For real roots of the above polynomial equation, one can easily check if it is inside the root locus by counting the total number of poles and zeros that lie to the right of it, per sketching rule number (4).

Finally it should be noted that the condition for intersection of branches of the root locus may also be written in alternative but equivalent forms, for example as

$$\frac{\frac{dP}{ds}(s)}{P(s)} = \frac{\frac{dR}{ds}(s)}{R(s)}$$

or as

$$\sum_{i=1}^{n} \frac{1}{s - p_i} = \sum_{j=1}^{m} \frac{1}{s - z_j}.$$

Example 15.10 P and PI control of a third-order process: calculation of the break-away point(s) of the root locus

Consider again the case of a third-order process controlled with a P or PI controller, as in Examples 15.1 and 15.2, and for the specific values of process parameters $\tau_1 = 1$, $\tau_2 = \frac{1}{2}$, $\tau_3 = \frac{1}{3}$, $k = \frac{1}{3}$. The root locus diagrams are given in Figure 15.4 for P control and in Figures 15.8 and 15.10 for PI control with $\tau_I = \frac{3}{2}$ and $\tau_I = \frac{1}{4}$, respectively.

P Control

At the break-away point, both $Q(s)$ and $\frac{dQ}{ds}(s)$ must vanish. We have:

$$Q(s) = (s+1)(s+2)(s+3) + 2k_c = s^3 + 6s^2 + 11s + 6 + 2k_c = 0$$

$$\frac{dQ(s)}{ds} = 3s^2 + 12s + 11 = 0$$

The last equation has two roots: $-2 + \frac{1}{\sqrt{3}} = -1.42265$ and $-2 - \frac{1}{\sqrt{3}} = -2.57735$. These are substituted to the first equation to check the corresponding value of the gain.

The first root corresponds to $k_c = \frac{1}{3\sqrt{3}} = 0.19245$ and is acceptable, but the second root corresponds to $k_c = -\frac{1}{3\sqrt{3}} = -0.19245 < 0$, therefore it is rejected.

So, there is one break-away point, $p_b = -2 + \frac{1}{\sqrt{3}} = -1.42265$, which corresponds to the value of the gain $k_{c_b} = \frac{1}{3\sqrt{3}} = 0.19245$. (See also Table 15.1 and Figure 15.4.)

PI Control

Again, at the break-away point, both $Q(s)$ and $\frac{dQ}{ds}(s)$ must vanish.

$$Q(s) = \left(s^4 + 6s^3 + 11s^2 + 6s\right) + 2k_c\left(s + \frac{1}{\tau_I}\right) = 0$$

$$\frac{dQ(s)}{ds} = 4s^3 + 18s^2 + 22s + 6 + 2k_c = 0$$

Eliminating k_c from the above equations, it follows that

$$\left(s^4 + 6s^3 + 11s^2 + 6s\right) - \left(4s^3 + 18s^2 + 22s + 6\right)\left(s + \frac{1}{\tau_I}\right) = 0$$

$$\Leftrightarrow 3s^4 + \left(12 + \frac{4}{\tau_I}\right)s^3 + \left(11 + \frac{18}{\tau_I}\right)s^2 + \frac{22}{\tau_I}s + \frac{6}{\tau_I} = 0$$

The above polynomial equation is of fourth degree, therefore it has four roots, but not all of them are in the root locus.

- For $\tau_I = \dfrac{3}{2}$, the roots are $-2.5619, -1.3489, -0.48903 \pm 0.38298i$, out of which only the second is acceptable (the first corresponds to negative gain, whereas the third and fourth correspond to complex gains). Hence there is one break-away point, $\rho_b = -1.3489$. This corresponds to the value of the gain $k_{c_b} = 0.3708$. (See also Table 15.2 and Figure 15.8.)

- For $\tau_I = \dfrac{1}{4}$, the roots are $-4.69107, -2.68595, -1.54549, -0.41082$, out of which only the third should be rejected (it corresponds to negative gain). Hence, the break-away points are: $\rho_{b_1} = -4.69107$, $\rho_{b_2} = -2.68595$, $\rho_{b_3} = -0.41082$. The corresponding gains are: $k_{c_{b_1}} = 57.01101$, $k_{c_{b_2}} = 0.37119$, $k_{c_{b_3}} = 0.13874$. (See also Table 15.3 and Figure 15.10.)

Example 15.11 PD control of an underdamped second-order process: calculation of the break-in point of the root locus

Consider the feedback control system of Example 15.4, where an underdamped second-order process is controlled with a PD controller. Here

$$Q(s) = \tau^2 s^2 + 2\zeta\tau s + 1 + k_c k(\tau_D s + 1) = 0$$

$$\frac{dQ(s)}{ds} = 2\tau^2 s + 2\zeta\tau + k_c k \tau_D = 0$$

Eliminating k_c from the above equations, it follows that

$$(\tau^2 s^2 + 2\zeta\tau s + 1) - (2\tau^2 s + 2\zeta\tau)\left(s + \frac{1}{\tau_D}\right) = 0$$

The above quadratic equation has two roots $-\dfrac{1}{\tau_D} \pm \sqrt{\dfrac{1}{\tau_D^2} + \dfrac{1}{\tau^2} - \dfrac{2\zeta}{\tau_D \tau}}$, out of which only the one with the $-\sqrt{\ }$ is acceptable (the other root corresponds to negative gain).

Hence there is one break-in point, $\rho_b = -\dfrac{1}{\tau_D} - \sqrt{\dfrac{1}{\tau_D^2} + \dfrac{1}{\tau^2} - \dfrac{2\zeta}{\tau_D \tau}}$.

For $\tau = \zeta = \dfrac{1}{2}$ and $\tau_D = \dfrac{1}{4}$, the break-in point is $\rho_b = -4 - 2\sqrt{3} = -7.46410$.

For $\tau = \zeta = \dfrac{1}{2}$ and $\tau_D = 1$, the break-in point is $\rho_b = -1 - \sqrt{3} = -2.73205$.

(See Figure 15.19.)

15.4 Calculation of the Points of Intersection of the Root Locus with the Imaginary Axis

Root-locus theory, as outlined in the previous two sections, does not include anything about the points of intersection of the root locus with the imaginary axis. However, these points of intersection are of major significance, since they are connected with closed-loop stability limits.

The points of intersection of the root locus with the imaginary axis can be identified from a table of roots as a function of the gain: it is where the real part of some of the roots changes sign. It is also possible to calculate these points independently of the root table or the root-locus diagram, using the following method.

Consider again the general feedback control loop, as described in Section 15.2.1.

A point of intersection with the imaginary axis will correspond to a pure imaginary root or a zero root of the polynomial $Q(s) = P(s) + k_c R(s)$. Therefore, substituting $s = i\omega$ into the equation $Q(s) = 0$, it follows that

$$P(i\omega) + k_c R(i\omega) = 0$$

The above equation implies two equations, since both the real part and the imaginary part of the left-hand side must vanish: $\text{Re}\{P(i\omega)\} + k_c \text{Re}\{R(i\omega)\} = 0$ and $\text{Im}\{P(i\omega)\} + k_c \text{Im}\{R(i\omega)\} = 0$
Eliminating k_c from the above equations, it follows that

$$\text{Re}\{P(i\omega)\}\text{Im}\{R(i\omega)\} - \text{Im}\{P(i\omega)\}\text{Re}\{R(i\omega)\} = 0$$

The above equation can be solved for ω, from which the corresponding value of the gain k_c can be immediately obtained. Acceptable are the positive solutions for ω, corresponding to a positive value of k_c.

Example 15.12 P, PI and PID control of a third-order process: calculation of point(s) of intersection with the imaginary axis

Consider again the case of a third-order process controlled with a P or PI controller, as in Examples 15.1 and 15.2.

P Control

The poles of the closed-loop system are the roots of the polynomial

$$Q(s) = (\tau_1 s + 1)(\tau_2 s + 1)(\tau_3 s + 1) + k_c k$$
$$= \tau_1 \tau_2 \tau_3 s^3 + (\tau_1 \tau_2 + \tau_2 \tau_3 + \tau_3 \tau_1)s^2 + (\tau_1 + \tau_2 + \tau_3)s + 1 + k_c k$$

In order for $s = \omega i$ to be a root of $Q(s)$, the following condition must hold:

$$-\tau_1 \tau_2 \tau_3 \omega^3 i - (\tau_1 \tau_2 + \tau_2 \tau_3 + \tau_3 \tau_1)\omega^2 + (\tau_1 + \tau_2 + \tau_3)\omega i + 1 + k_c k = 0$$

from which, setting the real and imaginary parts equal to zero leads to:

Real Part = 0 $\Rightarrow$ $-(\tau_1\tau_2 + \tau_2\tau_3 + \tau_3\tau_1)\omega^2 + 1 + k_c k = 0$

Imaginary Part = 0 $\Rightarrow$ $-\tau_1\tau_2\tau_3\omega^3 + (\tau_1 + \tau_2 + \tau_3)\omega = 0$ $\Rightarrow$ $\omega^2 = \dfrac{\tau_1 + \tau_2 + \tau_3}{\tau_1\tau_2\tau_3}$

Therefore, there are two points of intersection with the imaginary axis $\pm i\sqrt{\dfrac{\tau_1 + \tau_2 + \tau_3}{\tau_1\tau_2\tau_3}}$,

and the corresponding gain is $k_{c_u} = \dfrac{1}{k}\left\{(\tau_1\tau_2 + \tau_2\tau_3 + \tau_3\tau_1)\dfrac{\tau_1 + \tau_2 + \tau_3}{\tau_1\tau_2\tau_3} - 1\right\}$.

For the specific values of process parameters $\tau_1 = 1$, $\tau_2 = \dfrac{1}{2}$, $\tau_3 = \dfrac{1}{3}$, $k = \dfrac{1}{3}$, the points of intersection are $\pm i\sqrt{11} = \pm\, 3.3166\, i$ and the corresponding value of the gain is $k_{c_u} = 30$. This is what was found in Example 15.1. (See Table 15.1 and Figure 15.4.)

PI or PID Control

The poles of the closed-loop system are the roots of the polynomial

$$Q(s) = s(\tau_1 s + 1)(\tau_2 s + 1)(\tau_3 s + 1) + k_c k\left(\tau_D s^2 + s + \dfrac{1}{\tau_I}\right)$$

$$= \tau_1\tau_2\tau_3 s^4 + (\tau_1\tau_2 + \tau_2\tau_3 + \tau_3\tau_1)s^3 + (\tau_1 + \tau_2 + \tau_3)s^2 + s + k_c k\left(\tau_D s^2 + s + \dfrac{1}{\tau_I}\right)$$

In order for $s = \omega i$ to be a root of $Q(s)$, the following condition must hold:

$$\tau_1\tau_2\tau_3\omega^4 - (\tau_1\tau_2 + \tau_2\tau_3 + \tau_3\tau_1)\omega^3 i - (\tau_1 + \tau_2 + \tau_3)\omega^2 + \omega i + k_c k(-\tau_D\omega^2 + \omega i + \dfrac{1}{\tau_I}) = 0$$

from which, setting the real and imaginary parts equal to zero leads to:

Real Part = 0 $\Rightarrow$ $\tau_1\tau_2\tau_3\omega^4 - (\tau_1 + \tau_2 + \tau_3)\omega^2 + k_c k(-\tau_D\omega^2 + \dfrac{1}{\tau_I}) = 0$

Imaginary Part = 0 $\Rightarrow$ $-(\tau_1\tau_2 + \tau_2\tau_3 + \tau_3\tau_1)\omega^3 + \omega + k_c k\omega = 0$

$\Rightarrow -(\tau_1\tau_2 + \tau_2\tau_3 + \tau_3\tau_1)\omega^2 + 1 + k_c k = 0$

Eliminating $k_c k$ from the above equations leads to

$$\tau_1\tau_2\tau_3\omega^4 - (\tau_1 + \tau_2 + \tau_3)\omega^2 + \left((\tau_1\tau_2 + \tau_2\tau_3 + \tau_3\tau_1)\omega^2 - 1\right)(-\tau_D\omega^2 + \dfrac{1}{\tau_I}) = 0$$

which is a quadratic equation in ω^2, therefore easily solvable (with only positive roots being acceptable), and finally the corresponding value of the gain can be obtained from

$$k_{c_u} = \dfrac{1}{k}\left\{(\tau_1\tau_2 + \tau_2\tau_3 + \tau_3\tau_1)\omega^2 - 1\right\}$$

Now using the specific values of process parameters $\tau_1 = 1$, $\tau_2 = \frac{1}{2}$, $\tau_3 = \frac{1}{3}$, $k = \frac{1}{3}$, we can look at some representative cases for the controller parameters:

→ for $\tau_I = \frac{3}{2}$, $\tau_D = 0$ (PI control), there is one positive solution

$$\omega^2 = \frac{7 + \sqrt{65}}{2} = 7.5311 \Rightarrow \omega = 2.7443.$$

The points of intersection are $\pm 2.7443\, i$, and the corresponding gain is $k_{c_u} = 19.5934$. This is what was found in Example 15.2. (See Table 15.2 and Figure 15.8.)

→ For $\tau_I = \frac{3}{2}$, $\tau_D = \frac{1}{4}$, the solutions for ω^2 are both negative, therefore rejected, and the conclusion is that there is no point of intersection with the imaginary axis.
This is what was found in Example 15.2. (See Table 15.4 and Figure 15.12.)

→ For $\tau_I = \frac{1}{4}$, $\tau_D = \frac{1}{4}$, there are two positive solutions: $\omega^2 = \frac{29 \pm \sqrt{649}}{2}$

from which $\omega_1 = 1.3275$, $\omega_2 = 5.2190$, therefore there are four points of intersection $\pm 1.3275\, i$ and $\pm 5.2190\, i$, with corresponding gains $k_{c_{u1}} = 2.2868$ and $k_{c_{u2}} = 78.8132$.

The corresponding root-locus diagram is depicted in Figure 15.26.

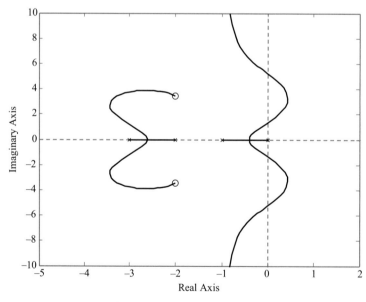

Figure 15.26 Root-locus diagram for third-order process under PID control with $\tau_I = \frac{1}{4}$, $\tau_D = \frac{1}{4}$.

From Figure 15.26, we observe that there are indeed four points of intersection. Despite the location of the center of gravity $\sigma = -1$, which implies that the asymptotes lie in the left half of the complex plane, the closed-loop system is not stable for all positive values of k_c. The closed-loop system will be stable for $0 < k_c < k_{c_{u_1}} = 2.2868$ and for $k_c > k_{c_{u_2}} = 78.8132$.

Remark: For the specific third-order process considered in this example, all the gains corresponding to a point of intersection with the imaginary axis are at the same time limiting values for the stability range, therefore they could have been alternatively calculated through the Routh array. However, in general, a point of intersection with the imaginary axis does not necessarily correspond to a stability limit, since there might be other roots that are already to the right of the imaginary axis.

15.5 Root Locus with Respect to Other Controller Parameters

In the previous sections, the root locus was always for varying gain k_c, with all the other controller parameters fixed. If one wants to study the effect of varying integral time τ_I with all the other controller parameters fixed, or the effect of varying derivative time τ_D with all the other controller parameters fixed, this is possible with a simple mathematical trick. This will be explained in the present section.

Consider a general feedback control loop, as in Section 15.2.1, except that the controller is now PI (see Figure 15.27).

The poles of the closed-loop system are the roots of the polynomial

$$Q(s) = s\, P(s) + k_c\left(s + \frac{1}{\tau_I}\right) R(s)$$

When the gain k_c is fixed but the integral time is variable, one can rearrange the above polynomial as follows:

$$Q(s) = \{s\, P(s) + k_c s\, R(s)\} + \left(\frac{1}{\tau_I}\right)\{k_c R(s)\}$$

With this rearrangement, we see that the problem of calculating the root locus under varying $\dfrac{1}{\tau_I}$ is mathematically identical to the one studied in Section 15.2: we have a linear combination of two polynomials and we want to study the effect of varying the coefficient. In particular, defining $P^*(s) = sP(s) + k_c s R(s)$, $R^*(s) = k_c R(s)$ and the auxiliary transfer function $G^*(s) = \dfrac{R^*(s)}{P^*(s)} = \dfrac{k_c R(s)}{sP(s) + k_c s R(s)}$ we see that, when $G^*(s)$ is controlled by a proportional

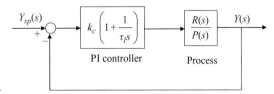

Figure 15.27 General PI control loop.

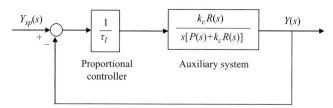

Figure 15.28 P control loop having the same roots as the one of Figure 15.27.

controller with gain $\frac{1}{\tau_I}$, the corresponding closed-loop poles are exactly the roots of the same polynomial Q(s). Figure 15.28 depicts the P control loop for $G^*(s)$.

By considering the auxiliary system, the original problem of constructing a root locus with respect to $\frac{1}{\tau_I}$ under PI control, has been recast to an equivalent P control problem.

Example 15.13 PI control of a third-order process: root locus for fixed controller gain and varying integral time

Consider again the third-order process studied in the previous examples, but now we wish to study the effect of varying integral action under fixed controller gain $k_c > 0$. The feedback control loop is depicted in Figure 15.29.

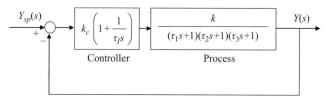

Figure 15.29 Feedback control loop consisting of a third-order process and a PI controller.

The poles of the closed-loop system are the roots of the polynomial

$$s(\tau_1 s+1)(\tau_2 s+1)(\tau_3 s+1) + kk_c\left(s+\frac{1}{\tau_I}\right) = s\left[(\tau_1 s+1)(\tau_2 s+1)(\tau_3 s+1) + kk_c\right] + \left(\frac{1}{\tau_I}\right)kk_c$$

and we would like to study the effect of the parameter $\frac{1}{\tau_I}$ on the roots.

Considering the auxiliary system with transfer function

$$G^*(s) = \frac{kk_c}{s\left[(\tau_1 s+1)(\tau_2 s+1)(\tau_3 s+1) + kk_c\right]}$$

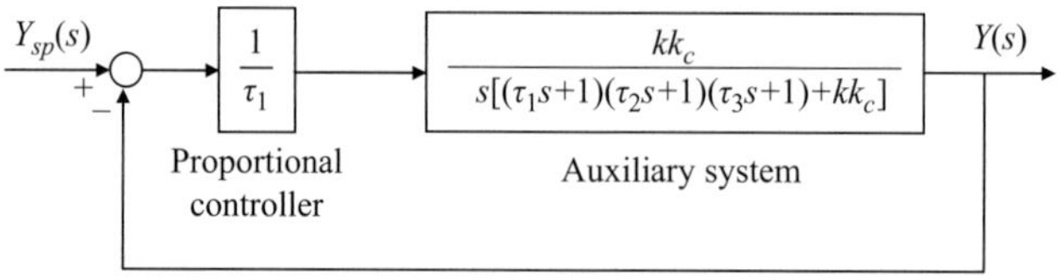

Figure 15.30 P control loop having the same roots as the one of Figure 15.29.

15.5 Root Locus with Respect to Other Controller Parameters

we see that when it is controlled with a P controller with gain $\frac{1}{\tau_I}$, the corresponding closed-loop poles are exactly the roots of the same polynomial.

Figure 15.30 depicts the equivalent P–control problem for the auxiliary system. We observe that the auxiliary system is of fourth order with

- no zeros
- three poles at the roots of the polynomial $(\tau_1 s + 1)(\tau_2 s + 1)(\tau_3 s + 1) + kk_c$ and one pole at 0.

A study of the locus of the roots as a function of $\frac{1}{\tau_I}$ will show the effect of varying size of integral action when the magnitude of the proportional action is preselected.

To make the example more concrete, we will use specific values of the process parameters, $\tau_1 = 1, \tau_2 = \frac{1}{2}, \tau_3 = \frac{1}{3}, k = \frac{1}{3}$, the same values as in the previous examples, and we will study two representative cases.

(i) Very small controller gain $k_c = \frac{4}{27} = 0.1481$, for which the poles of the auxiliary system are all real:
$$p_1 = -\frac{13 + \sqrt{33}}{6} = -3.1241, \quad p_2 = -\frac{5}{3} = -1.6667, \quad p_3 = -\frac{13 - \sqrt{33}}{6} = -1.2092$$

(ii) Medium size gain $k_c = 3$, for which two of the poles of the auxiliary system are complex:
$$p_1 = -4, \quad p_2 = -1 + i\sqrt{2}, \quad p_3 = -1 - i\sqrt{2}$$

Table 15.6 Roots of denominator polynomial versus $1/\tau_I$, under PI control with $k_c = 4/27$

$1/\tau_I$	Roots			
0	−3.1241	−1.6667	−1.2092	0.0000
0.0100	−3.1238	−1.6693	−1.2065	−0.0005
0.0200	−3.1234	−1.6719	−1.2037	−0.0009
0.0500	−3.1224	−1.6796	−1.1956	−0.0024
0.1000	−3.1207	−1.6920	−1.1826	−0.0047
0.2000	−3.1172	−1.7151	−1.1581	−0.0096
0.5000	−3.1066	−1.7759	−1.0929	−0.0246
1.0000	−3.0882	−1.8603	−1.0000	−0.0516
2.0000	−3.0475	−2.0000	−0.8362	−0.1163
3.77220	−2.9563	−2.2180	−0.41286	−0.41286
5.0000	−2.8621	−2.3813	−0.3783 + 0.2725i	−0.3783 − 0.2725i
6.00748	−2.64656	−2.64656	−0.3534 + 0.3595i	−0.3534 − 0.3595i
10.0000	−2.7260 − 0.4390i	−2.7260 + 0.4390i	−0.2740 + 0.5600i	−0.2740 − 0.5600i
20.0000	−2.8618 − 0.7417i	−2.8618 + 0.7417i	−0.1382 + 0.8117i	−0.1382 − 0.8117i
35.24177	−3.0000 − 0.9750i	−3.0000 + 0.9750i	0.0000 + 1.0244i	0.0000 − 1.0244i
50.0000	−3.0992 − 1.1229i	−3.0992 + 1.1229i	0.0992 + 1.1634i	0.0992 − 1.1634i
100.0000	−3.3309 − 1.4359i	−3.3309 + 1.4359i	0.3309 + 1.4638i	0.3309 − 1.4638i
200.0000	−3.6161 − 1.7869i	−3.6161 + 1.7869i	0.6161 + 1.8063i	0.6161 − 1.8063i
500.0000	−4.0910 − 2.3312i	−4.0910 + 2.3312i	1.0910 + 2.3435i	1.0910 − 2.3435i
1000.0000	−4.5386 − 2.8212i	−4.5386 + 2.8212i	1.5386 + 2.8298i	1.5386 − 2.8298i

Tables 15.6 and 15.7 provide the roots as a function of $\frac{1}{\tau_I}$, and Figures 15.31 and 15.32 show the corresponding root-locus diagrams, for the above two cases.

We observe that for very low gain ($k_c = \frac{4}{27} = 0.1481$), the system can take large size integral action without getting destabilized (as long as $\frac{1}{\tau_I} < 35.242$).

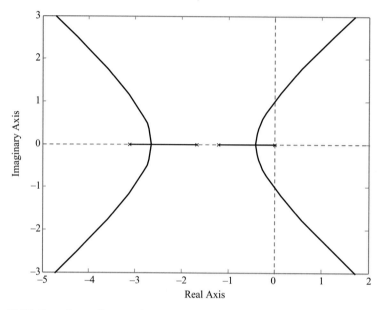

Figure 15.31 Root-locus diagram for third-order process under PI control with $k_c = 4/27$.

Table 15.7 Roots of denominator polynomial versus $1/\tau_I$, under PI control with $k_c = 3$

$1/\tau_I$	Roots			
0	−4.0000	−1.0000 − 1.4142i	−1.0000 + 1.4142i	0.0000
0.0100	−3.9986	−0.9982 − 1.4110i	−0.9982 + 1.4110i	−0.0050
0.0200	−3.9973	−0.9963 − 1.4078i	−0.9963 + 1.4078i	−0.0101
0.0500	−3.9931	−0.9906 − 1.3980i	−0.9906 + 1.3980i	−0.0256
0.1000	−3.9862	−0.9807 − 1.3813i	−0.9807 + 1.3813i	−0.0525
0.2000	−3.9721	−0.9587 − 1.3469i	−0.9587 + 1.3469i	−0.1105
0.5000	−3.9278	−0.8685 − 1.2345i	−0.8685 + 1.2345i	−0.3352
1.0000	−3.8455	−0.5773 − 1.1077i	−0.5773 + 1.1077i	−1.0000
2.0000	−3.6289	−0.1855 − 1.2724i	−0.1855 + 1.2724i	−2.0000
3.0000	−3.0000	0.0000 − 1.4142i	0.0000 + 1.4142i	−3.0000
5.0000	−3.2283 − 0.9322i	0.2283 − 1.6140i	0.2283 + 1.6140i	−3.2283 + 0.9322i
10.0000	−3.5576 − 1.5016i	0.5576 − 1.9269i	0.5576 + 1.9269i	−3.5576 + 1.5016i
20.0000	−3.9321 − 2.0121i	0.9321 − 2.2982i	0.9321 + 2.2982i	−3.9321 + 2.0121i
50.0000	−4.5276 − 2.7241i	1.5276 − 2.9003i	1.5276 + 2.9003i	−4.5276 + 2.7241i
100.0000	−5.0760 − 3.3344i	2.0760 − 3.4579i	2.0760 + 3.4579i	−5.0760 + 3.3344i
200.0000	−5.7291 − 4.0349i	2.7291 − 4.1219i	2.7291 + 4.1219i	−5.7291 + 4.0349i

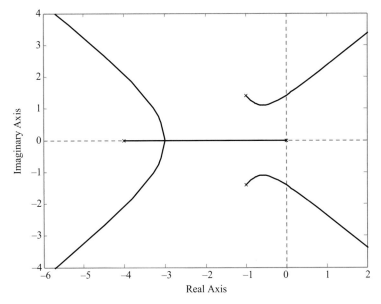

Figure 15.32 Root-locus diagram for third-order process under PI control with $k_c = 3$.

However, for larger gain ($k_c = 3$), the range of integral action for closed-loop stability is significantly narrower $\left(\dfrac{1}{\tau_I} < 3\right)$.

Finally, we also observe essential differences in the shape of the root-locus diagrams relative to what we have seen in the previous examples. Here $G^*(s)$ has $n - m = 4$, and consequently there are four asymptotes. The asymptotes form angles of $\dfrac{\pi}{4}, \dfrac{3\pi}{4}, \dfrac{5\pi}{4}, \dfrac{7\pi}{4}$ with the positive real axis.

15.6 Software Tools

15.6.1 Root-Locus Diagrams using MATLAB

In this section we will illustrate the use of MATLAB to calculate the roots as a function of the controller gain, plot the root locus and find points on the root-locus diagram.

Consider again the problem of Example 15.1. After entering the process transfer function,

```
» G=tf(2,[1 6 11 6]);
```

the `rlocus` command may be used to either build a root table for a given array of values of the gain (like Table 15.1)

```
» Kc=[0 0.01 0.02 0.05 0.1 0.19245 0.2 0.5 1 2 5 10 20 30 60];
» r=rlocus(G, Kc);
» gain_vs_closed_loop_poles=[Kc' r']
```

or to draw a root-locus diagram

```
» rlocus(G)
```

This gives the diagram of Figure 15.4. In case we want to set our own ranges for the axes (to see a detail of the diagram), this is done as follows:

» axis([-2.5 -0.5 -1 1])

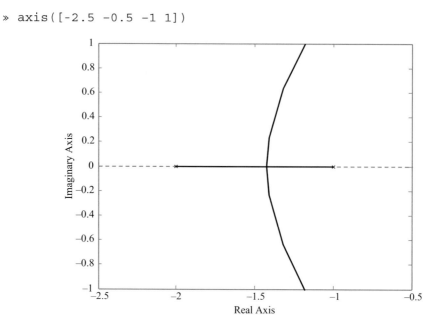

Also, it is possible to select a point on the diagram, and MATLAB will provide the corresponding value of the gain and all the poles of the closed-loop system. This is done with the rlocfind command as follows:

» [Kc_sel,roots] = rlocfind(G)
Select a point in the graphics window

After selecting the point of interest by clicking on the graph, MATLAB immediately returns our selected point, the corresponding value of the gain and the corresponding roots (closed-loop poles):

```
selected_point =
  -0.9502 + 1.5062i
Kc_sel =
  3.5189
roots =
  -4.0899
  -0.9551 + 1.5085i
  -0.9551 - 1.5085i
```

For the selected point on the root-locus diagram, we can further calculate the closed-loop transfer function and the response to a step change in the set point:

» Gcl=feedback(Kc_sel*G,1);
» step(Gcl)

When a PI or PID controller is used, and we want to study the effect of the controller gain (Example 15.2), the same commands can be used, but applied to the extended transfer function, which includes the fixed factor of the controller. For example, for PI control with $\tau_I = 1\frac{1}{2}$, the root locus is obtained with the following commands:

```
» G=tf(2,[1 6 11 6])
» tauI=1.5;
» Gc_fixed_part=tf([1 1/tauI],[1 0]);
» Gext=series(Gc_fixed_part,G)
» rlocus(Gext)
```

and this gives Figure 15.8. Then, using the rlocfind command, we can select a point on the root-locus diagram, obtain the corresponding value of the controller gain, and calculate the response to a step change in the set point:

```
» [Kc_sel,roots] = rlocfind(Gext)
  Select a point in the graphics window
```

and, after selecting a point of interest on the graph,

```
» Gcl=feedback(Kc_sel*Gext,1);
» step(Gcl)
```

By repeating this procedure, we can search for a good choice of gain by trial and error. Finally, it should be noted that, when the process model is given in state-space form, all the root-locus calculations may be performed directly from the matrices A, b, c, d of the state-space description – there is no need to compute the transfer function.

→ In Example 15.1 (P control), one can use the following commands:

```
A=[-1 0 0; 1 -2 0; 0 1 -3]; b=[2; 0; 0]; c=[0 0 1]; d=[0];
sys=ss(A,b,c,d);
rlocus(sys)
[Kc_sel,eigenvalues] = rlocfind(sys)
ACl=A-Kc_sel*b*c; bCl=Kc_sel*b;
sysCl=ss(ACl,bCl,c,d);
step(sysCl)
```

→ In Example 15.2, if we use PI control with $\tau_I = 1\frac{1}{2}$,

```
A=[-1 0 0; 1 -2 0; 0 1 -3]; b=[2; 0; 0]; c=[0 0 1];
tauI=1.5; zerovect=zeros(3,1);
Aext=[0 -c;zerovect A]; bext=[0;b]; cext=[-1/tauI c]; dext=[0];
sys=ss(Aext,bext,cext,dext;
rlocus(sys)
[Kc_sel,eigenvalues] = rlocfind(sys)
ACl=Aext-Kc_sel*bext*cext; bCl=[1;Kc_sel*b]; cCl=[0 c]; dCl=[0];
sysCl=ss(ACl,bCl,cCl,dCl);
step(sysCl)
```

15.6.2 Root-Locus Diagrams using Maple

Root-locus diagrams can be generated using the `RootLocusPlot` command of the `DynamicSystems` package of Maple, as follows.

Consider, for example, the third-order process of Example 15.1, which is under P control. We first define the dynamic system:

```
> with(DynamicSystems):
> G:=2/((s+1)*(s+2)*(s+3));
```

$$G(s) := \frac{2}{(s+1)(s+2)(s+3)}$$

```
> sys:=TransferFunction(G):
```

Then, the command

```
> RootLocusPlot(sys);
```

will generate the root-locus diagram of the P control loop. One also has the option of specifying the range of gains over which the points of the root-locus diagram will be calculated, as well as the ranges in the complex plane within which the root locus will be plotted. For example, one may set the range of gains $0 \leq k_c \leq 100$ and plot ranges $-8 \leq \text{Re}(s) \leq 1$, $-5 \leq \text{Im}(s) \leq 1$:

```
> RootLocusPlot(sys,0..100,view=[-8..1,-5..5]);
```

This will generate the following diagram.

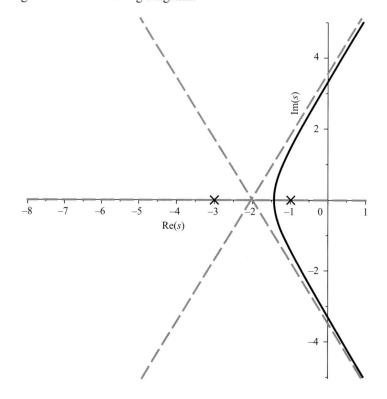

This is the same diagram as the MATLAB-generated diagram of Figure 15.4. The only difference is that it includes the asymptotes of the root locus. If it is not desirable to graph the asymptotes, one can add `plotasymptotes=false` in the arguments of the `RootLocusPlot` command.

Maple also has the capability of generating a root-locus diagram with respect to any symbolic parameter that appears linearly in the open-loop transfer function $G(s)G_c(s)$, but not necessarily multiplicatively. This is done by using the `RootContourPlot` command of the `DynamicSystems` package of Maple.

Consider for example the problem of Example 15.13, which has the same process transfer function, but it is under PI control with fixed k_c and varying $\dfrac{1}{\tau_I}$.

For the case $k_c = 3$, and $\text{invtauI} = \dfrac{1}{\tau_I}$ being the variable inverse integral time, we first define our system $G(s)G_c(s)$, which depends on the symbolic parameter invtauI:

```
> Gc:=3*(s+invtauI)/s:
> GGc:=Gc*G:
> sysnew:= TransferFunction(GGc):
```

Then, the command

```
> RootContourPlot(sysnew, 0..500, view=[-6..2,-4..4]);
```

will generate the root-locus diagram of the system as the parameter invtauI varies from 0 to 500, within the rectangle $-6 \leq \text{Re}(s) \leq 2$, $-4 \leq \text{Im}(s) \leq 4$ in the complex plane.

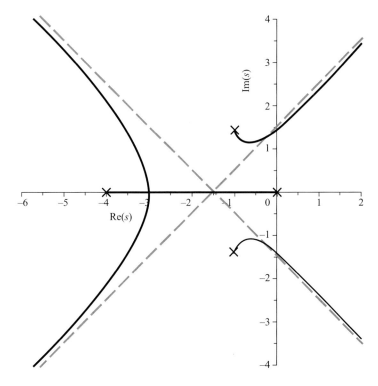

This is the same as the MATLAB-generated diagram of Figure 15.30, and, in addition, it has the asymptotes traced.

Finally, it should be noted that, when the system is given in state-space form, all the root-locus calculations of Maple remain unchanged, except that the system will need to be defined in terms of its A, b, c, d matrices.

LEARNING SUMMARY

- The root-locus diagram shows the "big picture" of how the dynamic characteristics of a closed-loop system change as one controller parameter varies from 0 to ∞. In particular, it depicts the locus in the complex plane traced by the closed-loop poles, i.e. the roots of the denominator of the closed-loop transfer functions.

- In this chapter we saw many examples of root-locus diagrams, to get a sense of what a root-locus diagram looks like, to know how to read it and to be able to relate the information that it provides to the dynamic characteristics of the closed-loop response.

- Root-locus diagrams can be generated through standard software packages such as MATLAB and Maple. However, for simple systems, it is also possible to do a sketch of the root locus guided by its theoretical properties. The basic theoretical properties are built for P control, where the controller gain is the variable parameter; however, through simple mathematical transformations, the effect of other variable parameters can be reformulated as P control problems.

- Some of the key theoretical properties include:
    - → the branches of the root locus start at the process poles
    - → the branches of the root locus terminate at the process zeros or at infinity
    - → those branches that terminate at infinity radiate from the center of gravity and form angles that can be calculated through the formulas of Section 15.2
    - → the parts of the real axis that form part of the root locus can be determined by counting the total number of process poles and zeros that lie to their right, as described in Section 15.2
    - → the root locus is symmetric with respect to the real axis.

- Additional theoretical properties allow us to calculate:
    - → the angles formed by the branches of the root-locus diagram, as they depart from the poles or as they approach the zeros
    - → the points at which two branches of the root locus meet on the real axis
    - → the points at which two branches of the root locus cross the imaginary axis.

TERMS AND CONCEPTS

Root locus. The root locus is the locus of the roots of the denominator polynomial of the closed-loop transfer functions (i.e. the closed-loop poles) as a controller parameter varies from 0 to $+\infty$.

Asymptotes. The asymptotes of the root-locus diagram are straight lines approached by the roots in the limit as the controller parameter tends to $+\infty$.

Center of gravity. The center of gravity of the root-locus diagram is the point of intersection of the asymptotes on the real axis.

Break-away/break-in points. These points of the root-locus diagram are the points of the real axis (if any) where branches of the root locus intersect.

Angles of departure/angles of arrival. These angles of the root-locus diagram are the angles formed by the root locus near a process pole/process zero.

FURTHER READING

Additional material on root-locus diagrams can be found in:
Dorf, C. D. and Bishop, R. H., *Modern Control Systems*, 12th edn. Prentice Hall, 2011.
Ogata, K., *Modern Control Engineering*, Pearson International Edition, 5th edn, 2008.

PROBLEMS

15.1 For a process with transfer function, $G(s) = \dfrac{k(\gamma s + 1)}{(\tau_1 s + 1)(\tau_2 s + 1)(\tau_3 s + 1)}$, with $\tau_1 > \tau_2 > 0$, $k > 0$, which is controlled with a P controller, sketch the root-locus diagram for the following cases:

(i) $\gamma > \tau_1 > \tau_2 > 0$
(ii) $\tau_1 > \gamma > \tau_2 > 0$
(iii) $\tau_1 > \tau_2 > \gamma > 0$
(iv) $\gamma = 0$
(v) $\gamma < 0$

Use the basic rules of Section 15.2 and check your result with MATLAB or Maple.

For which cases could the closed-loop system be destabilized for large k_c?

15.2 For a process with transfer function $G(s) = \dfrac{2(3s+1)}{(14s+3)(2s+1)}$, which is controlled with a PI controller, sketch the root-locus diagram for fixed integral time τ_I, for the following cases:

(i) $\tau_I > 14$ (larger than the process time constants)
(ii) $2 < \tau_I < 14$ (in between the process time constants)
(iii) $\tau_I < 2$ (smaller than the process time constants).

Use the basic rules of Section 15.2 and check your result with MATLAB or Maple.

15.3 For the control loop of Problem 15.2, and for a given value of the integral time τ_I, write a MATLAB program that
(a) generates the root-locus diagram
(b) computes and plots the response to a unit step change in the set point for a selected point on the root-locus diagram.

15.4 For the P control system of Example 15.1, suppose that we would like to draw the root-locus diagram for k_c varying from 0 to $-\infty$. For this purpose, you can simply invert the sign of your transfer function, i.e. use $-G(s) = -\dfrac{2}{(s+1)(s+2)(s+3)}$ as your auxiliary transfer function, apply the basic rules of Section 15.2 and finally check your result with MATLAB or Maple.

15.5 A second-order unstable process with transfer function $G(s) = \dfrac{(s+2)}{(s+3)(s-8)}$ is controlled with a PI controller, with k_c fixed and $1/\tau_1$ varying. Sketch the root-locus diagram for the following choices of controller gain:
(i) $k_c = 6$
(ii) $k_c = 18$.

After defining an appropriate auxiliary system following the method of Section 15.5, use the basic rules of Section 15.2 and check your result with MATLAB or Maple. Of the above choices of controller gain, which one is better? Why?

15.6 For the control loop of Problem 15.5, and for a given value of the controller gain k_c, write a MATLAB program that
(a) generates the root-locus diagram as $1/\tau_1$ varies from 0 to ∞
(b) computes and plots the response to a unit step change in the set point for a selected point on the root-locus diagram.

15.7 Use the the `RootContourPlot` command of the `DynamicSystems` package of Maple to directly plot the root-locus diagrams for Problem 15.5, without defining an auxiliary system. Compare with your results in Problem 15.5.

15.8 For controlling an unstable process with transfer function $G(s) = \dfrac{6(s+2)}{s(s-3)(s+4)}$, consider the following options:
(i) P control,
(ii) ideal PD control with $\tau_D = 1/5$.

For each of the above options, sketch the root-locus diagram using the basic rules of Section 15.2 and check your result with MATLAB or Maple.
From your root-locus diagrams, what conclusions can you draw regarding the stability of the closed-loop system? What is the effect of derivative action on closed-loop stability?

15.9 A P controller is used to control a third-order process with transfer function $G(s) = \dfrac{k}{(\tau_1 s+1)(\tau_2 s+1)(\tau_3 s+1)}$, where $\tau_1 > 0$, $\tau_2 > 0$, $\tau_3 > 0$, $k > 0$ are constant parameters.
(a) Calculate the points of intersection of the root locus with the imaginary axis and the corresponding value of k_c.
(b) Calculate the break-away point of the root locus.
(c) Derive approximate formulas for small k_c and for large k_c.

15.10 In Figure 15.22, we have seen the root-locus diagram of a third-order process under P control where the two larger time constants are equal. Consider now the case where the two smaller time constants are equal, e.g. constants $\tau_1 = 1$, $\tau_2 = \tau_3 = \frac{1}{2}$, $k = \frac{1}{4}$.

(a) Calculate the roots for small k_c and for large k_c.
(b) Calculate the angles of departure from the double pole.
(c) Sketch the root-locus diagram using the basic rules of Section 15.2 and check your result with MATLAB or Maple.

15.11 For the control system of Example 15.13 and for $k_c = 3$, calculate: (a) the angles of departure from the complex poles and (b) the break-away point. Compare your results with Table 15.7 and Figure 15.32.

15.12 For the following process transfer functions:

$$G_1(s) = \dfrac{1}{s(s^2+4s+5)}$$

$$G_2(s) = \dfrac{1}{s(s^2+6s+25)}$$

(a) calculate the break-away points (if any),
(b) calculate the angles of departure from the complex poles,
(c) draw the root-locus diagrams,
(d) comment on the similarities and differences between the two root-locus diagrams, as well as on the significance of the break-away point calculation.

16 Optimal Selection of Controller Parameters

The aim of this chapter is to introduce a general methodology of controller synthesis that is based on optimization theory. By considering a controller with predefined structure, such as the classical P, PI or PID controller, the number of tunable parameters and their effect on closed-loop performance is predictable. By defining a performance criterion, or an objective function, we then can use classical mathematical optimization tools to determine the optimal values of the controller parameters. The performance criteria available are presented in this chapter and analyzed.

STUDY OBJECTIVES

After studying this chapter you should be able to do the following.

- Describe the closed-loop performance criteria commonly used in process control.
- Design optimal P, PI and PID controllers for low-order systems using analytical equations.
- Use MATLAB and Maple to optimize the parameters of classical P, PI or PID controllers and complex performance criteria.

16.1 Control Performance Criteria

16.1.1 Introduction

In Chapter 13 we discussed the response of a heating tank under PI control and we derived the equation that describes the closed-loop system. The responses of the closed-loop system under PI control, for three different values of the proportional gain, are shown in Figure 16.1. The issue that we will be discussing first is which one of the responses is optimal or simply better than the others.

To answer this question, we must first choose a criterion to evaluate and then rank the alternatives. This criterion should express our opinion on what is a perfect or good response. Qualitative as well as quantitative criteria are employed in control theory to judge the quality of the closed-loop responses achieved using specific values for the controller parameters.

Performance criteria can vary considerably depending on the type of process and the preferences of the control engineer. Designing a control system for a race car or fighter aircraft is

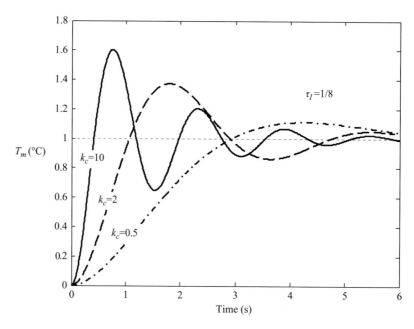

Figure 16.1 Typical responses of a heating tank under PI control and for a unit step change in the set point.

significantly different from that of a passenger car or aircraft. Performance is the overriding objective in the first case and comfort and reliability in the second. Process-control practitioners use quantitative criteria that are related to the speed of the closed-loop response, the degree and rate of damping of the oscillations or the sensitivity to model uncertainty. Such specifications determine the values of the adjustable parameters of the controller. If, for instance, a second-order system is controlled by a P controller, then setting either the overshoot or the damping ratio automatically determines the value of the proportional gain. Additional specifications can be achieved by using a PI or PID controller: a specific overshoot and rise time can be achieved simultaneously by using a PI controller that has two adjustable parameters. This approach works well when either a parametric description (mathematical model/transfer function) or a nonparametric description (point(s) of the Bode plot) is available for the process being controlled. The specifications set are normally achieved using a trial and error approach.

When a mathematical model is available, then we can devise more elaborate methodologies for controller design. One systematic methodology is to use mathematical optimization theory so as to optimize an analytic objective function or criterion over the controller parameters. The search space, i.e. the range over which the controller parameters can be varied, is restricted by the need to ensure stability and the limitations imposed by the equipment (valve saturation). Although optimization methodologies that consider many objectives are available, in this chapter we will consider criteria that can be expressed as a single objective function of the controller parameters. These criteria must transform all information included in the response of the closed-loop system (a function of time) to a single number, in order for a mathematical optimization methodology to be applicable.

16.1.2 Criteria of Performance

In order to formulate a systematic procedure to design controllers it is highly desirable to define criteria that measure how good the closed-loop response of a system is. These criteria must be functions of the controller parameters, must distinguish clearly a "good" design from a "bad" design and must be general enough to be practical. As most controllers act against the error signal, it comes as no surprise that available criteria of performance are defined with respect to the error signal. In Figure 16.2a the error signal of the closed-loop system response shown in Figure 16.1 is shown for the case where $k_c = 10$ and $\tau_I = 1/8$. A successful controller must keep the error as small as possible and the ideal controller achieves $e(t) = 0$, for every t. This is an unachievable objective, and in real systems we allow the controlled variable to deviate from the set point by an acceptable amount. We can define, as a first attempt, a criterion of performance as the integral of the error (IE), i.e.

$$J = \int_0^\infty e(t)\, dt \tag{16.1.1}$$

which is the shaded area in Figure 16.2a. However, as can be seen from the same figure, the error can be negative or positive and the positive shaded areas cancel with the negative shaded areas. The error can be arbitrarily large and at the same time the criterion given by Eq. (16.1.1) can be small. We can, therefore, dismiss this criterion from further consideration.

In order to devise a criterion that does not have the drawback of the integral of the error, we may use either the integral of the absolute value of the error (IAE) or the integral of the squared error (ISE)

$$J_{IAE} = \int_0^\infty |e(t)|\, dt \tag{16.1.2}$$

$$J_{ISE} = \int_0^\infty [e(t)]^2\, dt \tag{16.1.3}$$

These criteria are also depicted by the shaded areas in Figure 16.2b and 16.2c for the same values of controller parameters. The absolute value that appears in the IAE criterion makes it difficult to evaluate analytically as the sign of the error must be determined beforehand. The IAE criterion weighs all errors equally, whereas ISE puts more weight on larger errors and less weight on smaller errors. ISE is the most widely used criterion of performance due to the fact that it can, in many cases, be calculated analytically.

An important issue about the three criteria that have been defined so far is the fact that the upper limit of the integration is infinite. However, for the cases where the error becomes (and remains) zero after a finite time t_1 (note that for $t > t_1 = 10$ the error becomes zero in Figures 16.1 and 16.2) then integrals may be written as

$$J = \int_0^\infty f(e(t))\, dt = \int_0^{t_1} f(e(t))\, dt + \int_{t_1}^\infty f(e(t))\, dt \approx \int_0^{t_1} f(e(t))\, dt \tag{16.1.4}$$

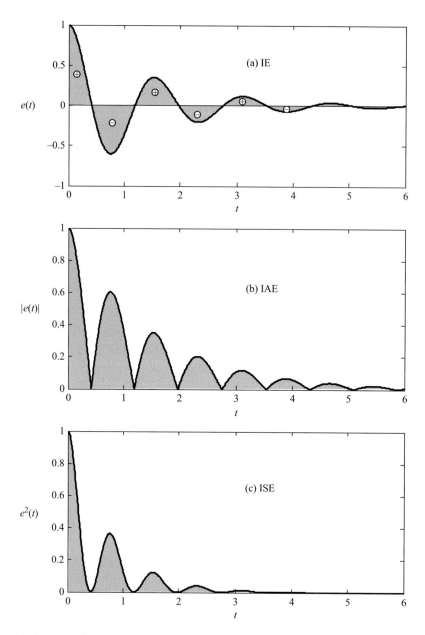

Figure 16.2 (a) Error, (b) absolute value of the error and (c) square of the error as a function of time for the heating tank under PI control with $k_c = 10$ and $\tau_I = 1/8$.

i.e. the infinite-time calculation can be simplified to a finite-time calculation, provided that the error becomes zero after some time and a judicious selection of t_1 can be made beforehand. This makes the numerical calculation of the performance criteria possible. However, it should be kept in mind that the theoretical (analytical) determination of the performance criteria is more easily done by using infinite time in the calculation.

An additional consideration in defining performance criteria is that errors that appear at early times are unavoidable and could be deemphasized in the performance criteria. Persistent errors are, on the other hand, important and could be penalized more heavily. A criterion that is employed to consider the time of the occurrence of a specific error is the integral of time-weighted absolute or squared error (ITAE or ITSE):

$$J_{\text{ITAE}} = \int_0^\infty t|e(t)|\,dt \qquad (16.1.5)$$

$$J_{\text{ITSE}} = \int_0^\infty t[e(t)]^2\,dt \qquad (16.1.6)$$

The two criteria ISE and ITSE are shown as shaded areas in Figure 16.3, from which it is apparent that errors occurring at early times are, in the case of ITSE, deemphasized.

All criteria that have been defined so far are functions of the error signal. Many alternative performance criteria have been proposed, including criteria that are based on the derivative of the error (rate of change of the error signal) or the deviation of the control signal from its asymptotic value, for example

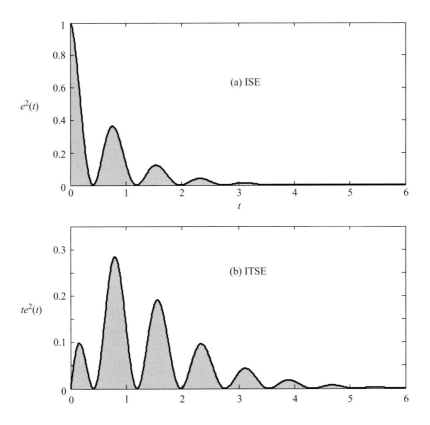

Figure 16.3 (a) Squared error and (b) time-weighted squared error as a function of time for the heating tank under PI control with $k_c = 10$ and $\tau_I = 1/8$.

$$J_{IADE} = \int_0^\infty \left| \frac{de}{dt}(t) \right| dt \qquad (16.1.7)$$

$$J_{ISDE} = \int_0^\infty \left[\frac{de}{dt}(t) \right]^2 dt \qquad (16.1.8)$$

or

$$J_{IAC} = \int_0^\infty |u(\infty) - u(t)| \, dt \qquad (16.1.9)$$

$$J_{ISC} = \int_0^\infty [u(\infty) - u(t)]^2 \, dt \qquad (16.1.10)$$

Here, $u(\infty)$ denotes the final value of the input after the output returns to set point. Also, it is possible to define time-weighted versions of the above criteria.

All these alternative criteria are used to penalize the aggressive behavior of the input signal that can generate a high-performance output response (as measured by the error criterion). The following mixed or combined criterion can be defined

$$J = \int_0^\infty [e(t)]^2 \, dt + \rho \int_0^\infty [u(\infty) - u(t)]^2 \, dt = J_{ISE} + \rho J_{ISC} \qquad (16.1.11)$$

This combined criterion can be used as a compromise between reducing the error and preventing overly aggressive input variation. Here, ρ is an adjustable positive constant that is used to weigh the contribution of the two terms to the overall objective. Choosing $\rho = 0$, for instance, results in the ISE criterion, whereas $\rho \to \infty$ corresponds to the ISC criterion. In a particular application, the choice of the weight coefficients will need to be done by trial and error, based on the time responses of input and output, as well as the physical constraints of the manipulated input.

Example 16.1 An integrating process controlled by a proportional controller

Consider the feedback control system consisting of an integrating process and a P controller with proportional gain k_c, as shown in Figure 16.4.

(a) Calculate IAE, ISE, ITAE, and ITSE for a unit step change of the set point. What is the optimal value of k_c that minimizes each of these criteria?
(b) Find the optimal value of k_c that minimizes the performance criterion J given by Eq, (16.1.11), for a unit step change of the set point.

Solution

(a) Based on Figure 16.4 we can derive the following transfer function between the set point and the error

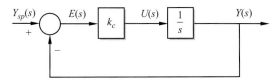

Figure 16.4 A closed-loop system consisting of an integrating system and a P controller.

$$\frac{E(s)}{Y_{sp}(s)} = \frac{s}{s+k_c}$$

For a unit step change of the set point $Y_{sp}(s) = 1/s$,

$$E(s) = \frac{1}{s+k_c} \Rightarrow e(t) = e^{-k_c t}$$

Having the analytic expression for $e(t)$ we can now determine the following performance criteria

$$J_{IAE} = \int_0^\infty |e(t)| dt = \int_0^\infty e^{-k_c t} dt = -\frac{1}{k_c}\left[e^{-k_c t}\right]_0^\infty = \frac{1}{k_c}$$

$$J_{ISE} = \int_0^\infty [e(t)]^2 d\tau = \int_0^\infty e^{-2k_c t} dt = -\frac{1}{2k_c}\left[e^{-2k_c t}\right]_0^\infty = \frac{1}{2k_c}$$

$$J_{ITAE} = \int_0^\infty t|e(t)| dt = \int_0^\infty e^{-k_c t} t\, dt = -\frac{1}{k_c^2}\left[(k_c t + 1)e^{-k_c t}\right]_0^\infty = \frac{1}{k_c^2}$$

$$J_{ITSE} = \int_0^\infty t[e(t)]^2 dt = \int_0^\infty t e^{-2k_c t} dt = -\frac{1}{4k_c^2}\left[(2k_c t + 1)e^{-2k_c t}\right]_0^\infty = \frac{1}{4k_c^2}$$

We note that in all cases the criterion is minimized as $k_c \to +\infty$.

(b) To determine the performance criterion given by Eq. (16.1.11) we first determine the control signal

$$u(t) = k_c e(t) = k_c e^{-k_c t}$$

from which we note that $u(\infty) = 0$, and

$$J_{ISC} = \int_0^\infty [u(\infty) - u(t)]^2 dt = k_c^2 \int_0^\infty e^{-2k_c t} dt = -\frac{k_c}{2}\left[e^{-2k_c t}\right]_0^\infty = \frac{k_c}{2}$$

Then the criterion can be calculated

$$J = J_{ISE} + \rho J_{ISC} = \int_0^\infty [e(t)]^2 \, dt + \rho \int_0^\infty [u(\infty) - u(t)]^2 \, dt = \frac{1}{2k_c} + \rho \frac{k_c}{2}$$

To find the value of k_c that minimizes $J(k_c)$, we calculate the first derivative of J with respect to k_c and then set the derivative equal to zero

$$\frac{dJ(k_c)}{dk_c} = -\frac{1}{2}\left(\frac{1}{k_c^2} - \rho\right) = 0 \Rightarrow k_c = \frac{1}{\sqrt{\rho}}$$

In Figure 16.5 the two integrals associated with the error and the control input deviation are plotted as a function of the parameter k_c. The overall objective function is also plotted as a function of k_c for three values of ρ and the minimum values are indicated. We observe that by increasing the parameter k_c the ISE is reduced while at the same time the control effort is increased. When both contributions are considered, a clear minimum is observed for reasonable values of $\rho > 0$.

It is important to note that the optimal value of k_c depends not only on the parameter ρ but also on the type of set-point variation assumed. A unit step change in the set point is normally considered but other changes are possible including pulse-type variations of the set point or the disturbance. It is important to note that the optimal controller parameters depend strongly on the performance criterion selected and that different performance criteria result in different optimal values for the controller parameters. An extreme case was presented in Example 16.1 in which the ISE optimal controller corresponds to $k_c \to +\infty$ while an optimal controller according to the ISC criterion given by Eq. (16.1.10) corresponds to $k_c = 0$.

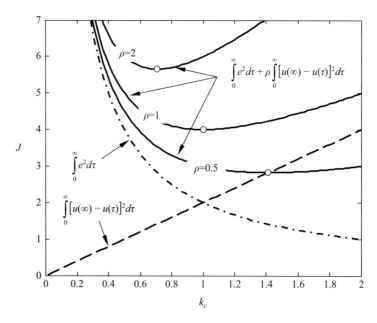

Figure 16.5 Performance criteria evaluated in Example 16.1 as a function of the controller gain.

16.2 Analytic Calculation of Quadratic Criteria for a Stable System and a Step Input

Consider the situation where

- the closed-loop system is stable
- the set point and/or a disturbance undergoes a step change and
- the closed-loop response has zero steady-state error (zero offset).

This means that $y(t) \to y_{sp}$ as $t \to \infty$, which implies that $e(t) = y_{sp} - y(t) = y(\infty) - y(t)$, and the error-based performance criteria can be written as

$$J_{IAE} = \int_0^\infty |y(\infty) - y(t)| \, dt \tag{16.2.1}$$

$$J_{ISE} = \int_0^\infty [y(\infty) - y(t)]^2 \, dt \tag{16.2.2}$$

etc., where $y(t)$ is the step response of the closed-loop system. It is important to point out here the significance of closed-loop stability and zero steady-state error conditions: unless they are satisfied, error-based criteria become infinite and thus they cannot serve as measures of performance. Also, it is important to observe that in the above form, J_{IAE} and J_{ISE} are of exactly the same form as the control signal criteria J_{IAC} and J_{ISC} given by (16.1.9) and (16.1.10).

In general, performance criteria are calculated numerically. However, there is one important exception: criteria involving the integral of the square of the deviation, i.e. J_{ISE} and J_{ISC}. These are called *quadratic criteria* and can be calculated analytically as a function of the controller parameters. This is a definite advantage, because having a formula available for the criterion can facilitate subsequent optimization.

16.2.1 Calculation of Quadratic Criteria using the State-Space Description of the Closed-Loop System

Suppose that the closed-loop system is stable (all eigenvalues of the A matrix lie in the left half complex plane) and that it has a state-space description of the form

$$\frac{dx(t)}{dt} = Ax(t) + bv(t) \tag{16.2.3}$$

$$z(t) = cx(t) + dv(t)$$

with input being $v = y_{sp}$ (the set point) or $v = w$ (a disturbance input), and output being $z = y$ (controlled output) or $z = u$ (control signal).

We also consider the quadratic performance index

$$J = \int_0^\infty (z(\infty) - z(t))^2 \, dt \tag{16.2.4}$$

which becomes J_{ISE} when $z = y$ and J_{ISC} when $z = u$. When the input undergoes a step change of size M, we apply the result from Table 6.3 of Chapter 6 to calculate the response of the system (16.2.3):

$$z(t) = \left(c(e^{At} - I)A^{-1}b + d\right)M \quad (16.2.5a)$$

$$z(\infty) = \left(-cA^{-1}b + d\right)M \quad (16.2.5b)$$

We substitute Eqs. (16.2.5a, b) into Eq. (16.2.4) to obtain

$$J = \int_0^\infty \left(-cA^{-1}b + d - c(e^{At} - I)A^{-1}b - d\right)^2 M^2 \, dt$$

$$= \int_0^\infty \left(-ce^{At}A^{-1}b\right)^2 M^2 \, dt = \int_0^\infty \left(-ce^{At}A^{-1}b\right)^T \left(-ce^{At}A^{-1}b\right) M^2 \, dt$$

$$= \left(A^{-1}b\right)^T \left[\int_0^\infty \left(e^{At}\right)^T c^T c\left(e^{At}\right) dt\right] \left(A^{-1}b\right) M^2$$

We define the following matrix

$$P = \int_0^\infty \left(e^{At}\right)^T c^T c\left(e^{At}\right) dt \quad (16.2.6)$$

for which we can observe that

$$A^T P + PA = \int_0^\infty \left[A^T \left(e^{At}\right)^T c^T c\left(e^{At}\right) + \left(e^{At}\right)^T c^T c\left(e^{At}\right) A\right] dt = \int_0^\infty \left\{\frac{d}{dt}\left[\left(e^{At}\right)^T c^T c\left(e^{At}\right)\right]\right\} dt$$

$$= \left(e^{At}\right)^T c^T c\left(e^{At}\right)\Big|_0^\infty$$

or, because the system is stable,

$$A^T P + PA = -c^T c \quad (16.2.7)$$

To conclude, if a matrix P can be found by solving Eq. (16.2.7) then the performance criterion J given by Eq. (16.2.4) can be readily evaluated:

$$J = \left(A^{-1}b\right)^T P\left(A^{-1}b\right) M^2 \quad (16.2.8)$$

Equation (16.2.7) is known as a *Lyapunov equation*. In general, a Lyapunov matrix equation is of the form

$$A^T P + PA = -Q \quad (16.2.9)$$

where A and Q are given square matrices and P is an unknown square matrix, all of the same size. If the matrix A has all its eigenvalues in the left half complex plane, then the corresponding Lyapunov equation is guaranteed to have a solution and the solution is unique. Solving a Lyapunov equation reduces to solving a set of n^2 linear algebraic equations with n^2 unknowns, where n is the size of the matrices. In our case, $Q = c^T c$ is a symmetric matrix, and this implies that the solution P is also a symmetric matrix. This property reduces the number of algebraic equations to be solved to $n(n + 1)/2$. The solution of Lyapunov equations, even though conceptually simple, involves tedious algebra, and for this reason one generally needs

16.2 Analytic Calculation of Quadratic Criteria for a Stable System and a Step Input

to use software to perform the calculations. In MATLAB the command `lyap`, and in Maple the command `LyapunovSolve`, calculate the solution of Eq. (16.2.9).

In what follows, an example will be presented to illustrate the steps involved in the calculation of a quadratic criterion through the solution of a Lyapunov equation. It will be seen in the example that the steps are simple but the algebra is very heavy, and the importance of software will become evident. In the last section of this chapter, the same example will be considered and the same calculations will be performed using Maple.

Example 16.2 ISE criterion for the heating tank under PI control and a step change in the set point

In Chapter 13 we derived the following state-space description of a heating tank under PI control (see Eq. (13.1.21)):

$$\frac{d}{dt}\begin{bmatrix} \bar{e}_I \\ \bar{T} \\ \bar{T}_m \end{bmatrix} = \begin{bmatrix} 0 & 0 & -1 \\ \frac{\Delta h_{vap,st}}{V\rho c_p}k_v\frac{k_c}{\tau_I} & -\frac{F}{V} - \frac{\Delta h_{vap,st}}{V\rho c_p}k_v k_c & 0 \\ 0 & \frac{1}{\tau_m} & -\frac{1}{\tau_m} \end{bmatrix} \begin{bmatrix} \bar{e}_I \\ \bar{T} \\ \bar{T}_m \end{bmatrix} + \begin{bmatrix} 1 \\ \frac{\Delta h_{vap,st}}{V\rho c_p}k_v k_c \\ 0 \end{bmatrix} \bar{T}_{sp}$$

We consider the actual temperature T as the output of the closed-loop system

$$y = \bar{T} = \begin{bmatrix} 0 & 1 & 0 \end{bmatrix} \begin{bmatrix} \bar{e}_I \\ \bar{T} \\ \bar{T}_m \end{bmatrix}$$

and the following values for the process parameters

$$\frac{F}{V} = 1, \quad \frac{\Delta h_{vap,st}}{V\rho c_p} = 1, \quad k_v = \frac{1}{5}, \quad \tau_m = \frac{1}{10}$$

Derive a formula for the ISE criterion

$$J_{ISE} = \int_0^\infty [T(\infty) - T(t)]^2 dt$$

as a function of the PI controller parameters, when the set point T_{sp} undergoes a unit step change.

Solution

By substituting the numerical values, the A matrix becomes

$$A = \begin{bmatrix} 0 & 0 & -1 \\ \dfrac{1}{5}\dfrac{k_c}{\tau_I} & -1 & -\dfrac{k_c}{5} \\ 0 & 10 & -10 \end{bmatrix}$$

and Eq. (16.2.7) becomes

$$\begin{bmatrix} 0 & 0 & -1 \\ \dfrac{1}{5}\dfrac{k_c}{\tau_I} & -1 & -\dfrac{k_c}{5} \\ 0 & 10 & -10 \end{bmatrix}^T P + P \begin{bmatrix} 0 & 0 & -1 \\ \dfrac{1}{5}\dfrac{k_c}{\tau_I} & -1 & -\dfrac{k_c}{5} \\ 0 & 10 & -10 \end{bmatrix} = -\begin{bmatrix} 0 & 1 & 0 \end{bmatrix}^T \begin{bmatrix} 0 & 1 & 0 \end{bmatrix}$$

or

$$\begin{bmatrix} 0 & \dfrac{1}{5}\dfrac{k_c}{\tau_I} & 0 \\ 0 & -1 & 10 \\ -1 & -\dfrac{k_c}{5} & -10 \end{bmatrix} \begin{bmatrix} p_{11} & p_{12} & p_{13} \\ p_{12} & p_{22} & p_{23} \\ p_{13} & p_{23} & p_{33} \end{bmatrix} + \begin{bmatrix} p_{11} & p_{12} & p_{13} \\ p_{12} & p_{22} & p_{23} \\ p_{13} & p_{23} & p_{33} \end{bmatrix} \begin{bmatrix} 0 & 0 & -1 \\ \dfrac{1}{5}\dfrac{k_c}{\tau_I} & -1 & -\dfrac{k_c}{5} \\ 0 & 10 & -10 \end{bmatrix} = -\begin{bmatrix} 0 & 0 & 0 \\ 0 & 1 & 0 \\ 0 & 0 & 0 \end{bmatrix}$$

where in the above, we have already substituted $p_{21} = p_{12}, p_{31} = p_{13}, p_{32} = p_{23}$ since the solution P must be a symmetric matrix. By performing the algebra and taking into consideration that both sides of the equation are symmetric matrices, we obtain the following equations

element 1,1: $\quad 2\dfrac{1}{5}\dfrac{k_c}{\tau_I} p_{12} = 0 \quad \Rightarrow \quad p_{12} = 0$ \hfill (a)

element 1,2: $\quad \dfrac{1}{5}\dfrac{k_c}{\tau_I} p_{22} + 10 p_{13} = 0$ \hfill (b)

element 1,3: $\quad \dfrac{1}{5}\dfrac{k_c}{\tau_I} p_{23} - p_{11} - 10 p_{13} = 0$ \hfill (c)

element 2,2: $\quad -2 p_{22} + 20 p_{23} = -1$ \hfill (d)

element 2,3: $\quad -11 p_{23} + 10 p_{33} - \dfrac{k_c}{5} p_{22} = 0$ \hfill (e)

element 3,3: $\quad -2 p_{13} - \dfrac{2 k_c}{5} p_{23} - 20 p_{33} = 0$ \hfill (f)

We solve the linear equations (b)–(f) (using Gauss elimination) to finally obtain

16.2 Analytic Calculation of Quadratic Criteria for a Stable System and a Step Input

$$P = \frac{1}{2} \frac{1}{\left[11(5+k_c) - \dfrac{k_c}{\tau_I}\right]} \begin{bmatrix} \left(11 + \dfrac{1}{50}\dfrac{k_c}{\tau_I}\right)\left(\dfrac{k_c}{\tau_I}\right) & 0 & -\left(\dfrac{11}{10} + \dfrac{k_c}{50}\right)\left(\dfrac{k_c}{\tau_I}\right) \\ 0 & (55+k_c) & -k_c + \dfrac{1}{10}\dfrac{k_c}{\tau_I} \\ -\left(\dfrac{11}{10} + \dfrac{k_c}{50}\right)\left(\dfrac{k_c}{\tau_I}\right) & -k_c + \dfrac{1}{10}\dfrac{k_c}{\tau_I} & \dfrac{k_c^2}{50} + \dfrac{11}{100}\dfrac{k_c}{\tau_I} \end{bmatrix}$$

We then calculate the inverse of the A matrix

$$A^{-1} = \begin{bmatrix} 0 & 0 & -1 \\ \dfrac{1}{5}\dfrac{k_c}{\tau_I} & -1 & -\dfrac{1}{5}k_c \\ 0 & 10 & -10 \end{bmatrix}^{-1} = \frac{1}{-2\left(\dfrac{k_c}{\tau_I}\right)} \begin{bmatrix} 10+2k_c & -10 & -1 \\ 2\left(\dfrac{k_c}{\tau_I}\right) & 0 & -\dfrac{1}{5}\left(\dfrac{k_c}{\tau_I}\right) \\ 2\left(\dfrac{k_c}{\tau_I}\right) & 0 & 0 \end{bmatrix}$$

We finaly calculate the product $A^{-1}b$

$$A^{-1}b = \frac{1}{-2\left(\dfrac{k_c}{\tau_I}\right)} \begin{bmatrix} 10+2k_c & -10 & -1 \\ 2\left(\dfrac{k_c}{\tau_I}\right) & 0 & -\dfrac{1}{5}\left(\dfrac{k_c}{\tau_I}\right) \\ 2\left(\dfrac{k_c}{\tau_I}\right) & 0 & 0 \end{bmatrix} \begin{bmatrix} 1 \\ \dfrac{1}{5}k_c \\ 0 \end{bmatrix} = \frac{1}{-\left(\dfrac{k_c}{\tau_I}\right)} \begin{bmatrix} 5 \\ \left(\dfrac{k_c}{\tau_I}\right) \\ \left(\dfrac{k_c}{\tau_I}\right) \end{bmatrix}$$

and then we substitute into Eq. (16.2.8) to obtain

$$J_{ISE} = \frac{1}{2} \frac{\left[\dfrac{31}{100}\left(\dfrac{k_c}{\tau_I}\right)^2 + \left(\dfrac{k_c^2}{50} - \dfrac{6}{5}k_c - \dfrac{89}{2}\right)\left(\dfrac{k_c}{\tau_I}\right) + 275\right]}{\left[11(5+k_c) - \dfrac{k_c}{\tau_I}\right]\left(\dfrac{k_c}{\tau_I}\right)}$$

16.2.2 Calculation of Quadratic Criteria using the Closed-Loop Transfer Function for Low-Order Systems

In this section we summarize some analytical results for quadratic criteria of the form (16.2.4) in terms of the transfer function description of system (16.2.3), when the input undergoes a unit step change. These are the Newton–Gould–Kaiser (NGK) formulas. They are given in Table 16.1 for systems up to fourth order. The NGK formulas may be derived from the state-space results of the previous subsection or alternatively via the Parseval theorem of Laplace transforms (see Eq. (A.3.13) of Appendix A) and analytical evaluation of the resulting integral along the imaginary axis.

As an example, we consider the heating tank under PI control considered in Example 16.2. In Chapter 12, Example 12.2, we have derived the following transfer function

$$\bar{T}_m(s) = \frac{k_p k_v \left(k_c s + \dfrac{k_c}{\tau_I}\right)}{\tau_m \tau_p s^3 + (\tau_m + \tau_p)s^2 + (1 + k_p k_v k_c)s + k_p k_v \dfrac{k_c}{\tau_I}} \bar{T}_{sp}(s) \qquad (16.2.10a)$$

and

$$\bar{T}_m(s) = \frac{1}{(\tau_m s + 1)} \bar{T}(s) \qquad (16.2.10b)$$

Combining Eqs. (16.2.10a) and (16.2.10b) we obtain

$$\bar{T}(s) = \frac{k_p k_v \left(k_c s + \dfrac{k_c}{\tau_I}\right)(\tau_m s + 1)}{\tau_m \tau_p s^3 + (\tau_m + \tau_p)s^2 + (1 + k_p k_v k_c)s + k_p k_v \dfrac{k_c}{\tau_I}} \bar{T}_{sp}(s) \qquad (16.2.11)$$

Substituting the numerical values of the parameters ($k_p k_v = 1/5$, $\tau_p = 1$, $\tau_m = 1/10$) the transfer function becomes

$$\bar{T}(s) = \frac{\dfrac{1}{5}\left(k_c s^2 + \left(10 k_c + \dfrac{k_c}{\tau_I}\right)s + 10 \dfrac{k_c}{\tau_I}\right)}{s^3 + 11 s^2 + (10 + 2k_c)s + 2\dfrac{k_c}{\tau_I}} \bar{T}_{sp}(s) \qquad (16.2.12)$$

In the notation of Table 16.1 the above transfer function has

$$b_0 = 0,\ b_1 = \frac{k_c}{5},\ b_2 = 2k_c + \frac{1}{5}\frac{k_c}{\tau_I},\ b_3 = 2\frac{k_c}{\tau_I},\ a_1 = 11,\ a_2 = 10 + 2k_c,\ a_3 = 2\frac{k_c}{\tau_I}$$

from which, using Table 16.1, we obtain

16.2 Analytic Calculation of Quadratic Criteria for a Stable System and a Step Input

Table 16.1 NGK formulas for closed-loop systems up to fourth order

System order	Closed-loop transfer function $Z(s)/V(s)$
1	$\dfrac{b_0 s + b_1}{s + a_1}$

$$J = \frac{1}{2}\frac{\varphi_0^2}{a_1} \quad \text{where} \quad \varphi_0 = \frac{b_1}{a_1} - b_0$$

2	$\dfrac{b_0 s^2 + b_1 s + b_2}{s^2 + a_1 s + a_2}$

$$J = \frac{1}{2}\frac{\varphi_0^2 a_2 + \varphi_1^2}{a_1 a_2} \quad \text{where} \quad \varphi_0 = \frac{b_2}{a_2} - b_0, \quad \varphi_1 = \frac{b_2}{a_2} a_1 - b_1$$

3	$\dfrac{b_0 s^3 + b_1 s^2 + b_2 s + b_3}{s^3 + a_1 s^2 + a_2 s + a_3}$

$$J = \frac{1}{2}\frac{\varphi_0^2 a_2 a_3 + (\varphi_1^2 - 2\varphi_0 \varphi_2) a_3 + \varphi_2^2 a_1}{(a_1 a_2 - a_3) a_3}$$

$$\text{where} \quad \varphi_0 = \frac{b_3}{a_3} - b_0, \quad \varphi_1 = \frac{b_3}{a_3} a_1 - b_1, \quad \varphi_2 = \frac{b_3}{a_3} a_2 - b_2$$

4	$\dfrac{b_0 s^4 + b_1 s^3 + b_2 s^2 + b_3 s + b_4}{s^4 + a_1 s^3 + a_2 s^2 + a_3 s + a_4}$

$$J = \frac{1}{2}\frac{\varphi_0^2 (a_2 a_3 - a_1 a_4) a_4 + (\varphi_1^2 - 2\varphi_0 \varphi_2) a_3 a_4 + (\varphi_2^2 - 2\varphi_1 \varphi_3) a_1 a_4 + \varphi_3^2 (a_1 a_2 - a_3)}{(a_1 a_2 a_3 - a_3^2 - a_1^2 a_4) a_4}$$

$$\text{where} \quad \varphi_0 = \frac{b_4}{a_4} - b_0, \quad \varphi_1 = \frac{b_4}{a_4} a_1 - b_1, \quad \varphi_2 = \frac{b_4}{a_4} a_2 - b_2, \quad \varphi_3 = \frac{b_4}{a_4} a_3 - b_3$$

$$\varphi_0 = \frac{b_3}{a_3} - b_0 = 1$$

$$\varphi_1 = \frac{b_3}{a_3}a_1 - b_1 = 11 - \frac{k_c}{5}$$

$$\varphi_2 = \frac{b_3}{a_3}a_2 - b_2 = (10 + 2k_c) - \left(2k_c + \frac{1}{5}\frac{k_c}{\tau_I}\right) = 10 - \frac{1}{5}\frac{k_c}{\tau_I}$$

and finally

$$J_{ISE} = \frac{1}{2}\frac{\varphi_0^2 a_2 a_3 + (\varphi_1^2 - 2\varphi_0\varphi_2)a_3 + \varphi_2^2 a_1}{(a_1 a_2 - a_3)a_3}$$

$$= \frac{1}{2}\frac{2(10+2k_c)\frac{k_c}{\tau_I} + 2\left[\left(11 - \frac{k_c}{5}\right)^2 - 2\left(10 - \frac{1}{5}\frac{k_c}{\tau_I}\right)\right]\frac{k_c}{\tau_I} + 11\left(10 - \frac{1}{5}\frac{k_c}{\tau_I}\right)^2}{2\left[11(10+2k_c) - 2\frac{k_c}{\tau_I}\right]\frac{k_c}{\tau_I}}$$

After performing the necessary algebraic simplifications in the numerator we obtain

$$J_{ISE} = \frac{1}{2}\frac{\left[\frac{31}{100}\left(\frac{k_c}{\tau_I}\right)^2 + \left(\frac{k_c^2}{50} - \frac{6}{5}k_c - \frac{89}{2}\right)\left(\frac{k_c}{\tau_I}\right) + 275\right]}{\left[11(5+k_c) - \frac{k_c}{\tau_I}\right]\left(\frac{k_c}{\tau_I}\right)} \tag{16.2.13}$$

which agrees with the results of Example 16.2.

16.2.3 Performance Indices Based on the Variability of the Manipulated Variable

Most of the performance indices that were defined in Section 16.1 are based on the deviation of the output of the system from the desired operating point or set point. The choice of any of these output performance indices is self-evident because the performance of a closed-loop system and its process controller can reasonably be evaluated based on how well the output follows a predefined variation (such as a step change in the set point). However, any closed-loop system must take into consideration the necessary input variation in order to achieve the observed output variation. There are a number of reasons why input variation can be important, especially for chemical process systems. First of all, final control elements, such as control valves, have physical limitations that cannot be exceeded by any control system: the opening of a control valve can never exceed 100% even if the control signal drives the valve outside the permissible range. This particular practical issue has not been taken into consideration in our study so far. A second reason is that an aggressive variation of the control signal may cause excessive wear to the final control element and finally affect its reliability and increase maintenance costs. For these reasons a performance criterion that is related to the variability of the control signal like IAC and ISC (Eqs. (16.1.9) and (16.1.10)) plays a very significant role in evaluating a control system.

Returning now to the heating tank considered in Example 16.2, let's calculate the integral of the squared control signal (ISC) criterion

$$J_{ISC} = \int_0^\infty \left[u(\infty) - u(t)\right]^2 dt \qquad (16.2.14)$$

For a PI controller the control signal is defined as

$$u = k_c e + \frac{k_c}{\tau_I} \overline{e}_I = k_c \left(\overline{T}_{sp} - \overline{T}_m\right) + \frac{k_c}{\tau_I} \overline{e}_I = \begin{bmatrix} \frac{k_c}{\tau_I} & 0 & -k_c \end{bmatrix} \begin{bmatrix} \overline{e}_I \\ \overline{T} \\ \overline{T}_m \end{bmatrix} + k_c \overline{T}_{sp} \qquad (16.2.15)$$

We therefore have expressed the control signal as a linear combination of the state vector and external signal to the closed-loop system. Following the steps of Example 16.2 it can be found that the ISC performance index is given by the following expression (the details are left as an exercise for the reader)

$$J_{ISC} = \frac{1}{2} \frac{\left[\frac{1}{2}\left(\frac{k_c}{\tau_I}\right)^3 + \frac{1331}{4}\left(\frac{k_c}{\tau_I}\right)^2 + \left(k_c^3 + \frac{111}{2}k_c^2 - 580k_c - \frac{3275}{2}\right)\left(\frac{k_c}{\tau_I}\right) + 6875\right]}{\left[11(5 + k_c) - \left(\frac{k_c}{\tau_I}\right)\right]\left(\frac{k_c}{\tau_I}\right)} \qquad (16.2.16)$$

16.3 Calculation of Optimal Controller Parameters for Quadratic Criteria

In the previous sections, we presented analytical techniques for expressing the performance criteria that are the integrals of the error signal (J_{ISE} or ISE) or the control signal (J_{ISC} or ISC) as a closed-form expression that relates directly these criteria with the adjustable parameters of the controller. In the case of a P controller we have only one parameter over which we optimize the performance criterion: the controller gain k_c. As we have seen in Example 16.1, the performance criterion J is, in this case, a real-valued function of the controller gain, i.e. $J = J(k_c)$. To determine the optimal value of the controller gain we have used the optimality condition

$$\frac{dJ(k_c)}{dk_c} = 0 \qquad (16.3.1)$$

Applying this condition results in a polynomial equation that, when solved, gives the optimal value of the controller gain k_c^{opt}. In Example 16.1 we have presented the application of this technique. It is noted that condition (16.3.1) is a necessary but not sufficient condition, and the examination of the second derivative is required in order to ensure that k_c^{opt} corresponds to a local minimum (as Eq. (16.3.1) is also satisfied at a local maximum or saddle point).

When a PI controller is considered, then the performance index is a real valued function of both the controller gain k_c and the integral time τ_I, i.e. $J = J(k_c, \tau_I)$, as has been shown for the heating tank (see Eqs. (16.2.13) and (16.2.16)). In this case the necessary optimality conditions are

$$\frac{\partial J(k_c, \tau_I)}{\partial k_c} = 0 \qquad (16.3.2a)$$

$$\frac{\partial J(k_c,\tau_I)}{\partial \tau_I}=0 \qquad (16.3.2b)$$

which must be satisfied simultaneously. Again the first-order optimality conditions have transformed a minimization problem to a problem of solving a system of nonlinear algebraic equations in the controller parameters k_c and τ_I. Second-order conditions need to be applied in order to validate that the solution obtained corresponds to a local minimum.

For the ISE criterion of Eq. (16.2.13), the partial derivatives are given by

$$\frac{\partial J_{ISE}}{\partial k_c}=\frac{1}{2}\frac{\left(-\dfrac{1}{50}k_c^2+\dfrac{1231}{20}\right)\left(\dfrac{k_c}{\tau_I}\right)^2+\left(\dfrac{11}{50}k_c^3+\dfrac{11}{5}k_c^2-\dfrac{1111}{2}k_c+550\right)\left(\dfrac{k_c}{\tau_I}\right)-6050k_c-15125}{\left[11(5+k_c)-\left(\dfrac{k_c}{\tau_I}\right)\right]^2\left(\dfrac{k_c^2}{\tau_I}\right)}$$

$$\frac{\partial J_{ISE}}{\partial \tau_I}=\frac{1}{2}\frac{-\left(\dfrac{1}{50}k_c^2+\dfrac{221}{100}k_c+\dfrac{1231}{20}\right)\left(\dfrac{k_c}{\tau_I}\right)^2-550\left(\dfrac{k_c}{\tau_I}\right)+3025k_c+15125}{\left[11(5+k_c)-\left(\dfrac{k_c}{\tau_I}\right)\right]^2 k_c}$$

Setting the derivatives equal to zero leads to a system of two equations with two unknowns

$$\left(-\frac{1}{50}k_c^2+\frac{1231}{20}\right)\left(\frac{k_c}{\tau_I}\right)^2+\left(\frac{11}{50}k_c^3+\frac{11}{5}k_c^2-\frac{1111}{2}k_c+550\right)\left(\frac{k_c}{\tau_I}\right)-6050k_c-15125=0$$

$$-\left(\frac{1}{50}k_c^2+\frac{221}{100}k_c+\frac{1231}{20}\right)\left(\frac{k_c}{\tau_I}\right)^2-550\left(\frac{k_c}{\tau_I}\right)+3025k_c+15125=0$$

The simultaneous solution of these two equations is feasible, and includes the solution k_c = 55 and $(1/\tau_I)$ = 10/21 at which J_{ISE} assumes its minimum value, but manual calculation is very tedious. In general, because the criteria are rational functions of k_c and $1/\tau_I$, setting the time derivatives equal to zero leads to high-degree polynomial equations in k_c and $1/\tau_I$, which need to be solved numerically. Using derivatives to calculate the optimal values of the controller parameters is generally impractical, except for very simple systems.

An alternative approach is the direct numerical minimization of the performance criterion. Before doing this, it is useful to investigate the functions we seek to minimize in order to obtain some insight. To this end, the contours of the ISE and ISC are shown in Figure 16.6. We recall that a contour line for a function of two variables is a curve where the function has the same particular value. The gradient of the function, which in our case is a vector with elements the partial derivatives given by Eqs. (16.3.2a, b), is always perpendicular to the contour lines pointing in the direction of the maximum local rate of increase of the function. The negative of the gradients corresponds to the direction of

16.3 Calculation of Optimal Controller Parameters for Quadratic Criteria

the maximum rate of local decrease of a function. When the gradient vector is zero no direction can be found in which the function is decreasing, i.e. a local minimum has been attained.

The points where the minima are located for the ISE and ISC criteria can be approximately identified in the contour plots of Figure 16.6. For the ISE criterion the minimum is at $k_c = 55$ and $(1/\tau_I) = 10/21 \approx 0.48$, whereas for the ISC criterion the minimum is at $k_c \approx 4.6$ and $(1/\tau_I) \approx 0.97$. It should be noted once more that the optimal controller parameters are different for different criteria.

In general, the minimum of a performance criterion can be found by using numerical optimization algorithms. Both MATLAB and Maple have appropriate and efficient optimization commands; their use will be illustrated in the last section of this chapter.

The responses of the closed-loop system using the optimal controller parameters are shown in Figure 16.7. The response obtained under the ISE-optimal PI controller is aggressive at early times when the error is significant. The output quickly reaches and overshoots the set point (with mild oscillations) and then moves slowly towards its asymptotic value. The control signal follows a similar trend. When the ISC-optimal PI controller is used the response is more sluggish and does not exhibit overshoot or oscillations. The control signal is almost constant and equal to its asymptotic value as this is the aim of the specific performance metric.

We remember that each PI controller (ISE-optimal or ISC-optimal) is optimal according to the specific criterion. An ISE-optimal controller will tend to drive the output as close as possible to the set point and then have a slow response (as error has already become small). An ISC-optimal controller will tend to bring the control signal close to its asymptotic value as fast as possible and then respond slowly. There is no point in trying to compare the two responses and trying to find which one is best. Every optimal design is optimal according to the performance criterion used and, as a result, the designer needs to choose a performance criterion that meets the demands of the particular case study or his/her preferences about what is an optimal control performance (which clearly involves subjective elements). If the optimum of a performance criterion does not meet these demands or preferences, it needs to be replaced by a more meaningful criterion.

16.3.1 Optimizing a Combined Error and Input-Deviation Criterion

Based on the previous discussion, it appears that neither the ISE criterion nor the ISC criterion can capture all the demands of an engineer for a high-quality response. The ISE leads to a fast but overly aggressive response, whereas the ISC leads to a tame but too-slow response. Intuitively, it is expected that the best response should come from a compromise between these two criteria. For this reason, it is meaningful to optimize a combined criterion consisting of a linear combination of the individual ones, as defined previously by Eq. (16.1.11):

560 Optimal Selection of Controller Parameters

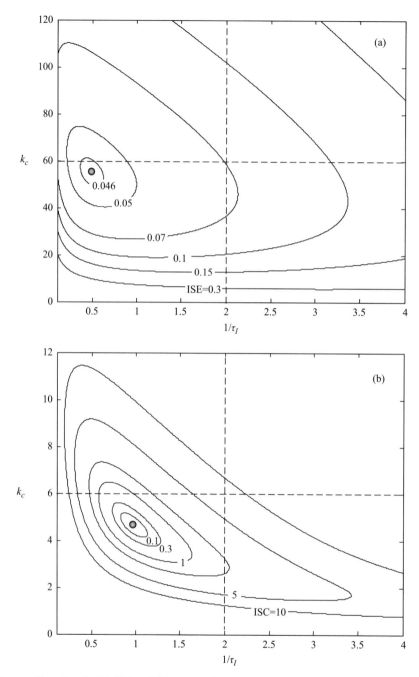

Figure 16.6 Contour lines for the (a) ISE and (b) ISC performance criteria in the parameter space (k_c, $1/\tau_I$).

16.3 Calculation of Optimal Controller Parameters for Quadratic Criteria

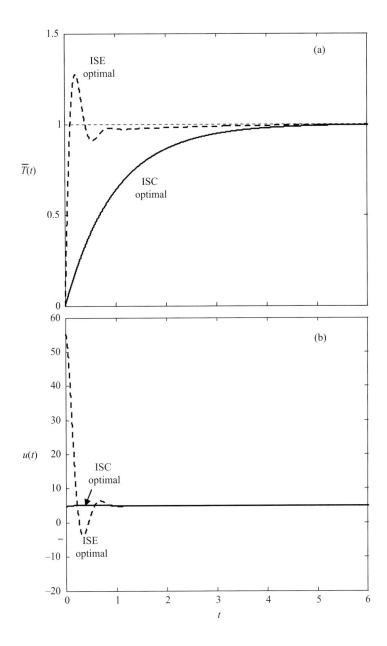

Figure 16.7 Response of the (a) heating tank temperature and (b) control input to a unit step change in the setpoint temperature using an ISE-optimal PI controller (dashed lines) and an ISC-optimal PI controller (continuous line).

$$J = J_{ISE}(k_c, \tau_I) + \rho\, J_{ISC}(k_c, \tau_I) \tag{16.3.3}$$

where ρ is a nonnegative weight coefficient that adjusts the relative contribution of the two specific criteria to the combined one. This approach has been followed in Example 16.1 for the case of an integrating system with a P controller. Here the combined index is a function of two controller parameters, and also depends on the weight coefficient ρ.

In the example of PI control of a heater, we have already derived formulas for ISE and ISC in Eqs. (16.2.13) and (16.2.16), therefore we already have a formula for the combined criterion J given by Eq. (16.3.3). Table 16.2 summarizes the corresponding optimization results, for various values of the weight coefficient ρ. The second and third columns give the values of controller gain and integral time that minimize J, whereas the fourth and fifth columns give the corresponding values of ISE and ISC.

The case $\rho = 0$ corresponds to minimizing ISE, whereas $\rho = \infty$ corresponds to minimizing ISC. We see from Table 16.2 that ISE and ISC are two conflicting objectives: improving one deteriorates the other. Increasing ρ makes ISE higher, which corresponds to slower response with larger error, but at the same time it makes ISC lower, which corresponds to smaller deviations of the control signal from its final value.

Another observation from Table 16.2 is that for $\rho = 10^{-3}$ or less, ISE is close to being minimal, whereas for $\rho = 10^0$ or higher, ISC is close to being minimal. A fair compromise between the two criteria will be reached when ρ is of the order of 10^{-2} or 10^{-3}. Figure 16.8 compares the optimal responses for $\rho = 10^{-6}$ (close to ISE optimal), $\rho = 10^{-3}$ and $\rho = 10^0$ (close to ISC optimal). We see that the compromise response ($\rho = 10^{-3}$) is quite fast, with settling time about the same as the ISE-optimal response, and without the spike in the control signal of the ISE-optimal response. Thus, the combined criterion J with this choice of $\rho = 10^{-3}$ has given a meaningful optimal compromise.

In a particular application, one would need to calculate step responses for several values of ρ within the appropriate order-of-magnitude range, and take into account the manipulated input constraints, and/or other constraints, for the final selection of ρ.

Table 16.2 Results of the numerical optimization of the combined performance index

ρ	k_c	τ_I	ISE	ISC
0	55.00000	2.10000	0.04545	244.58874
10^{-6}	54.77159	2.09346	0.04545	242.91206
10^{-5}	52.88182	2.03971	0.04553	229.22362
10^{-4}	42.06084	1.74581	0.04892	157.17499
10^{-3}	22.32010	1.28928	0.08639	53.83539
10^{-2}	9.61316	1.08105	0.22553	8.48121
10^{-1}	5.42784	1.03546	0.42325	0.36043
10^0	4.72937	1.02928	0.49081	0.01145
10^1	4.65238	1.02862	0.49951	0.00670
∞	4.64373	1.02855	0.50050	0.00665

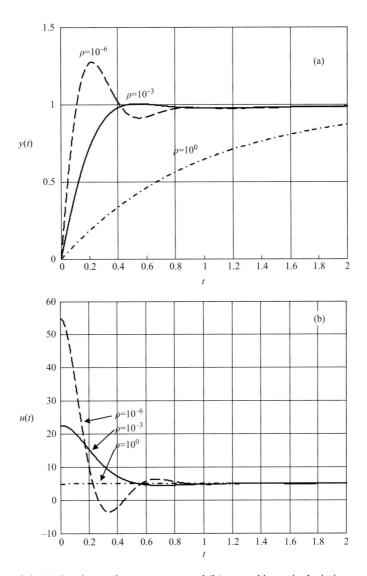

Figure 16.8 Response of the (a) heating tank temperature and (b) control input in deviation variables to a unit step change in the set-point temperature using the combined objective with $\rho = 10^{-6}$ (dashed lines), $\rho = 10^{-3}$ (continuous line) and $\rho = 10^0$ (dash-dotted line).

16.4 Software Tools

16.4.1 Optimizing PI Controller Parameters for General State-Space Systems using the Combined ISE/ISC Criterion

In this section we will demonstrate the steps that can be followed in order to construct a MATLAB m-file that can be used to determine the optimal PI controller parameters (k_c and τ_I or k_c and $k_I = k_c/\tau_I$) for any system that has the following open-loop state-space description

564   Optimal Selection of Controller Parameters

$$\frac{dx}{dt} = Ax + bu$$
$$y = cx$$
(16.4.1)

In this description the c vector corresponds to the selection of the controlled variable in the PI control law. As we have shown in Chapter 13, the closed-loop system formed using a PI controller has the following state-space description

$$\frac{d}{dt}\begin{bmatrix} e_I \\ x \end{bmatrix} = A_{CL}\begin{bmatrix} e_I \\ x \end{bmatrix} + b_{CL} y_{sp}$$
(16.4.2a)

where

$$A_{CL} = \begin{bmatrix} 0 & -c \\ b\dfrac{k_c}{\tau_I} & A - bk_c c \end{bmatrix}, \quad b_{CL} = \begin{bmatrix} 1 \\ bk_c \end{bmatrix}$$
(16.4.2b)

The performance criterion (see the general definition in Eq. (16.2.4)) is not necessarily defined in terms of the error of the output variable $y = cx$, but it could be defined in terms of an output y' that may be different from y, e.g. it could be the actual physical controlled variable instead of its measurement y. This is the case in the heating tank example under consideration. The closed-loop system is thus complemented with the following generalized output variable definition

$$z = y' = \begin{bmatrix} 0 & c' \end{bmatrix}\begin{bmatrix} e_I \\ x \end{bmatrix}$$
(16.4.3)

where c', as noted above, could be different from c. For the particular case of the ISC criterion, z is the control signal

$$z = u = \begin{bmatrix} \dfrac{k_c}{\tau_I} & -k_c c \end{bmatrix}\begin{bmatrix} e_I \\ x \end{bmatrix}$$
(16.4.4)

If the closed-loop system is stable, i.e.

$$\max\left\{\operatorname{Re} \operatorname{eig}\begin{bmatrix} 0 & -c \\ b\dfrac{k_c}{\tau_I} & A - bk_c c \end{bmatrix}\right\} < 0$$
(16.4.5)

then we can solve the following two Lyapunov functions (see Eq. (16.2.7)):

$$A_{CL}^T P_Y + P_Y A_{CL} = -\begin{bmatrix} 0 & c' \end{bmatrix}^T \begin{bmatrix} 0 & c' \end{bmatrix}$$
(16.4.6a)

$$A_{CL}^T P_U + P_U A_{CL} = -\begin{bmatrix} \dfrac{k_c}{\tau_I} & -k_c c \end{bmatrix}^T \begin{bmatrix} \dfrac{k_c}{\tau_I} & -k_c c \end{bmatrix}$$
(16.4.6b)

Finally the combined ISE/ISC criterion can be evaluated using Eq. (16.2.8):

$$J = J_{ISE} + \rho J_{ISC} = \left(A_{CL}^{-1} b_{CL}\right)^T P_Y \left(A_{CL}^{-1} b_{CL}\right) M^2 + \rho \left(A_{CL}^{-1} b_{CL}\right)^T P_U \left(A_{CL}^{-1} b_{CL}\right) M^2$$

or

$$J = \left(A_{CL}^{-1}b_{CL}\right)^T \left(P_Y + \rho P_U\right) \left(A_{CL}^{-1}b_{CL}\right) M^2 \qquad (16.4.7)$$

The following m-file implements the steps of determining the optimal PI controller parameters as described above

```
function [J,JISE,JISC]=ISE_ISC(k,A,b,c,cp,rho)

kc=k(1); tauI= k(2);
% form the closed loop system
ACL =[0              -c;
      b*kc/tauI      A-b*kc*c];   % Eq (16.4.2.b)
bCL  =[1             b'*kc]'    ; % Eq (16.4.2.b)
cCLY =[0             cp]        ; % Eq (16.4.3)
cCLU =[kc/tauI       -kc*c]     ; % Eq (16.4.4)

if (max(real(eig(ACL)))<0)
%    stable system Eq (16.4.5)
    PY   = lyap(ACL',cCLY'*cCLY);
    JISE = (inv(ACL)*bCL)'*PY*(inv(ACL)*bCL);

    PU   = lyap(ACL',cCLU'*cCLU);
    JISC = (inv(ACL)*bCL)'*PU*(inv(ACL)*bCL);

    J    = JISE + rho*JISC;
else
%    unstable system, return a large value
    JISE=1E12;JISC=1E12;J=1E12;
end
```

The user needs to supply the matrices of the open-loop system (A, b, c, cp), the weight coefficient rho together with the PI controller parameters as a vector k where k(1)=k_c and k(2)=τ_I. Figure 16.6 is then generated using the following script file:

```
clear all
F     = 1 ;  % m3/s
V     = 1 ;  % m3
taum  = 1/10; %s
Dhv   = 2000; % kJ/kg
kv    = 1/5;
dens  = 1000; % kg/m3
Cp= 2; % kJ/(kg deg)

A = [-F/V 0;1/taum -1/taum];
b = [Dhv*kv/(V*dens*Cp); 0];
c = [0 1];
```

```
cp= [1 0];
rho=0.001

kc=linspace(0.1,120,300);
tauI=linspace(0.1,4,200);

for i=1:length(kc)
    for j=1:length(tauI)
        [J(i,j),JISE(i,j),JISC(i,j)]= ...
            ISE_ISC([kc(i);tauI(j)],A,b,c,cp,rho);
    end
end
figure(1)
contour(1./tauI,kc,JISE,[0.046 0.05 0.07 0.1 0.15 0.3])
axis([0.1 4 0 120])

figure(2)
contour(1./tauI,kc,JISC,[0.1 0.3 1 2 5 10])
axis([0.1 4 0 12])

figure(3)
clf
contour(1./tauI,kc,J,[0.15 0.2 0.3 0.4 0.5 0.7])
axis([0.1 4 0 80])
```

Apart from generating the contour lines of the objective function, one can use directly the unconstrained optimization routines available in MATLAB to find the optimal controller parameters for $\rho = 0.001$:

```
>> fminsearch(@ISE_ISC,[1;1], optimset('Display','Iter'),
A,b,c,cp,0.001)
```

Iteration	Func-count	min f(x)	Procedure
0	1	2.48983	
1	3	2.36798	initial simplex
2	5	2.11635	expand
3	7	1.82643	expand
...			
67	123	0.140223	contract inside
68	125	0.140223	contract inside
69	127	0.140223	contract outside

Optimization terminated:

ans =

    22.3200
     1.2893

The fminsearch is a derivative-free, unconstrained optimization method that is particularly suited for the problem that we study in this chapter. The initial values (second argument of the fminsearch command) that have been used are $[k_c, \tau_I] = [1, 1]$. Any numerical optimization technique needs an initial guess for the optimization parameters and all derivative-based optimization techniques are sensitive to the initial guesses supplied by the user. A derivative-free technique, such as the one implemented in fminsearch, is less sensitive to the initial guesses. A judicious choice of initial values can be obtained by trial and error using simulation or by several classical controller synthesis methodologies such as the root-locus methodology.

In what follows we will demonstrate a numerical approach, which is more general since it is also applicable to nonquadratic criteria that cannot be calculated analytically. To this end, we consider the following criterion

$$J = \int_0^\infty \left[|e(t)| + \rho |u(\infty) - u(t)| \right] dt = J_{IAE} + \rho J_{IAC} \tag{16.4.8}$$

To evaluate this criterion for a closed-loop system formed using a PI controller, the following m-file can be used

```
function J=simIAE(k,A,b,c,cp,rho)

kc=k(1); tauI=k(2);
% form the closed loop system
ACL = [0              -c;
       b*kc/tauI      A-b*kc*c];   % Eq (16.4.2.b)
bCL  = [1             b'*kc]'      ; % Eq (16.4.2.b)
cCLY = [0             cp]          ; % Eq (16.4.3)
cCLU = [kc/tauI       -kc*c]       ; % Eq (16.4.4)
cCL  = [cCLY;cCLU]                 ;
dCL  = [0;            kc]          ;
sys=ss(ACL,bCL,cCL,dCL);

% calculate simulation time
maxT=ceil(10*(1/min(abs(real(eig(ACL))))));
T=linspace(0,maxT,1e3);

% simulate the unit step response
[z,t]=step(sys,T);

% calculate the asymptotic value of the outputs
zinf=-cCL*inv(ACL)*bCL+dCL;

% calculate deviations from the asymptotic values
eY=zinf(1)*ones(length(t),1)-z(:,1);
eU=zinf(2)*ones(length(t),1)-z(:,2);
```

Optimal Selection of Controller Parameters

```
% evaluate the objective by numerical integration
JIAEe = trapz(t,abs(eY));
JIAEu = trapz(t,abs(eU));
J     = JIAEe + rho*JIAEu;
```

Observe that after forming the closed-loop system in state space the unit step response of the closed-loop system is calculated. The final time of the simulation is based on the eigenvalue of the closed-loop system whose real part has the smallest absolute value, as discussed in Chapter 7. The final time of the simulation is set equal to five times the dominant time constant of the closed-loop system. After calculating the step response, the criterion is calculated by numerical integration using the trapezoidal rule as has been implemented in MATLAB. `fminsearch` can be used to obtain the results shown in Figure 16.9.

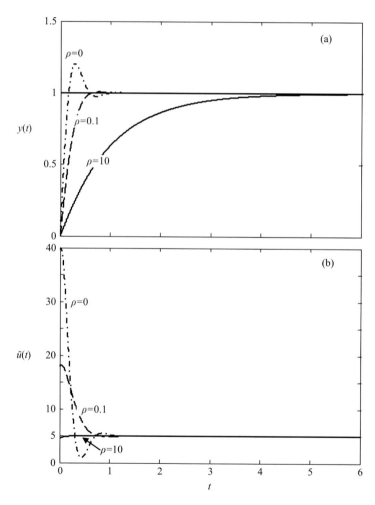

Figure 16.9 Closed-loop response to a unit step change in the set point using a PI controller that is optimal according to criterion (16.4.8) using $\rho = 0$, 0.1 and 10.

16.4.2 Symbolic Calculation of Quadratic Criteria and Subsequent Optimization

Symbolic calculation of the solution of the matrix equation (16.2.7) and subsequent substitution into (16.2.8) is straightforward using Maple. The same calculations that were performed manually in Example 16.2, can be performed painlessly with Maple, as follows:

```
> with(LinearAlgebra):
> A:=Matrix(3,[[0,0,-1],[(1/5)*kc/tauI,-1,-(1/5)*kc],[0,10,-10]]):
> b:=Vector(3,[1,(1/5)*kc,0]):
> c:=Vector[row](3,[0,1,0]):
```

The matrix equation (16.2.7) is formed and converted to a system of linear algebraic equations that are solved symbolically, leading to an exact expression for the matrix P:

```
> P:=Matrix(3,[[P11,P12,P13],[P21,P22,P23],[P31,P32,P33]]):
> R:=Multiply(Transpose(A),P)+Multiply(P,A)+Multiply(Transpose
  (c),c):
> res:=solve({R[1,1]=0,R[1,2]=0,R[1,3]=0,R[2,1]=0,R[2,2]=0,R[2,3
  ]=0,R[3,1]=0,R[3,2]=0,R[3,3]=0},{P11,P12,P13,P21,P22,P23,P31,P32
  ,P33}): assign(res);
> print(P);
```

$$\begin{bmatrix} \dfrac{1}{100}\dfrac{kc(kc+550\,taul)}{(11\,kc\,taul-kc+55\,taul)\,taul} & 0 & -\dfrac{1}{100}\dfrac{kc(kc+55\,taul)}{11\,kc\,taul-kc+55\,taul} \\ 0 & \dfrac{1}{2}\dfrac{taul(kc+55)}{11\,kc\,taul-kc+55\,taul} & -\dfrac{1}{20}\dfrac{kc(10\,taul-1)}{11\,kc\,taul-kc+55\,taul} \\ -\dfrac{1}{100}\dfrac{kc(kc+55)}{11\,kc\,taul-kc+55\,taul} & -\dfrac{1}{20}\dfrac{kc(10\,taul-1)}{11\,kc\,taul-kc+55\,taul} & \dfrac{1}{200}\dfrac{kc(2kc\,taul+11)}{11\,kc\,taul-kc+55\,taul} \end{bmatrix}$$

This is exactly the result for the matrix P that was obtained manually in Example 16.2, in a slightly rearranged form. Next, $A^{-1}b$ is calculated and finally ISE from (16.2.8):

```
> invA_1b:=simplify(Multiply(MatrixInverse(A),b));
```

$$invA\_1b := \begin{bmatrix} -\dfrac{5\,taul}{kc} \\ -1 \\ -1 \end{bmatrix}$$

```
> ISE:=normal(Multiply(Multiply(Transpose(A_1b),P),A_1b));
```

$$ISE := \dfrac{1}{200}\dfrac{2kc^3\,taul-120kc^2\,taul+31kc^2+4450kc\,taul+27500\,taul^2}{kc(11\,kc\,taul-kc+55\,taul)}$$

This is exactly the result of (16.2.13) in a slightly rearranged form.

Exactly the same calculation is used for ISC. The only difference is that the c-matrix from (16.4.4) needs to be used. The result obtained by Maple is

$$ISC := \frac{1}{8} \frac{4kc^4 taul^2 + 222 kc^3 taul^2 - 2320 kc^2 taul^2 + 2kc^2 + 1331 kc^2 taul - 6550 kc\, taul^2 + 27500\, taul^3}{kctaul(11kctaul - kc + 55 taul)}$$

This is exactly the result of (16.2.16) in a slightly rearranged form.

Once ISE or ISC have been calculated in terms of a formula, one may try to calculate the minimum by calculating the partial derivatives and setting them equal to zero. Maple can calculate the partial derivatives using the `diff` command, and then try to solve the resulting system of algebraic equations using the `solve` command. The equations are polynomial in k_c and τ_I and they might be exactly solvable when the problem is not too complex.

When this method is applied to the ISE formula of Eq. (16.2.13), one obtains:

```
> ISE:=(1/2)*((31/100)*(kc/tauI)^2+(kc^2/50-(6/5)*kc+(89/2))*(kc/
  tauI)+275)/((55+11*kc-kc/tauI)*(kc/tauI)):
> dISE_dkc:=diff(ISE,kc): dISE_dtauI:=diff(ISE,tauI):
> solve({dISE_dkc=0,dISE_dtauI=0},{kc,tauI});
```

$$\left\{kc = 35, taul = -\frac{133}{110}\right\}, \left\{kc = 55, taul = \frac{21}{10}\right\}$$

Maple finds two solutions; both are roots of the partial derivatives of ISE. The first solution corresponds to an unstable closed-loop system and should be rejected. The second one is the optimal solution.

Note, however, that in general, there may not be an exact solution (in terms of rational numbers and radicals), when the expression for the objective function is more complex. Numerical calculation will then be necessary. One option is to numerically solve the algebraic equations resulting from setting the partial derivatives equal to zero. The other option, which is usually more effective, is to numerically minimize the objective function subject to the stability constraints, using Maple's Nonlinear Programming package:

```
> with(Optimization):
> Minimize(ISE,{55+11*kc-(kc/tauI)>=0,kc>=0,tauI>=0});
```

$$\left[0.0454545454545454489, [kc = 54.9999999999844, taul = 2.09999999992497]\right]$$

LEARNING SUMMARY

- The most commonly used closed-loop performance criteria in process control are the integral of the absolute value of the error (IAE) and the integral of the square error (ISE):

$$J_{IAE} = \int_0^\infty |e(t)|\, dt$$

$$J_{ISE} = \int_0^\infty [e(t)]^2 \, dt$$

- Other important criteria are defined with respect to the input variation or the deviation of the control signal from its asymptotic value, for example:

$$J_{IAC} = \int_0^\infty |u(\infty) - u(t)| \, dt$$

$$J_{ISC} = \int_0^\infty [u(\infty) - u(t)]^2 \, dt$$

- The following mixed or combined criteria are also used:

$$J = \int_0^\infty [e(t)]^2 \, dt + \rho \int_0^\infty [u(\infty) - u(t)]^2 \, dt = J_{ISE} + \rho J_{ISC}$$

$$J = \int_0^\infty |e(t)| \, dt + \rho \int_0^\infty |u(\infty) - u(t)| \, dt = J_{IAE} + \rho J_{IAC}$$

- For low-order closed-loop systems and quadratic criteria such as ISE and ISC, analytic expressions have been derived from state space (via the Lyapunov Equation) or from the transfer function (Table 16.1).
- For nonquadratic criteria such as IAE and IAC, numerical optimization can be used effectively to determine the optimal controller parameters in both MATLAB and Maple.

TERMS AND CONCEPTS

Integral Absolute Error (IAE). A closed-loop performance criterion that is based on the integral of the absolute value of the deviation of the controlled variable from its set point. **IAC** is defined in a similar way in terms of the deviation of the manipulated variable (control signal) from its asymptotic value.

Integral Squared Error (ISE). A closed-loop performance criterion that is based on the integral of the squared deviation of the controlled variable from its set point. **ISC** is defined in a similar way in terms of the deviation of the manipulated variable (control signal) from its asymptotic value.

FURTHER READING

Further analytical results on the calculation of quadratic performance indexes can be found in the following book

Newton, G. C., Gould, L. A. and Kaiser, J. F., *Analytical Design of Linear Feedback Controls*. New York: Wiley, 1957.

PROBLEMS

16.1 Consider the liquid storage system of Example 13.1 under PI control, and suppose that the set point undergoes a unit step change.
  (a) Use MATLAB or Maple to calculate and do a contour plot of the ISE in the parameter space k_c versus $1/\tau_I$. What are the optimal values of k_c and τ_I?
  (b) Use MATLAB or Maple to calculate and do a contour plot of the ISC in the parameter space k_c versus $1/\tau_I$. What are the optimal values of k_c and τ_I?
  (c) Minimize the combined criterion ISE+ρISC for various values of ρ and tabulate your results. What would be reasonable values for ρ?

16.2 Consider a process with transfer function

$$G(s) = \frac{-s+4}{s^2 + 3s + 2}$$

which is controlled with a PI controller with transfer function

$$G_c(s) = k_c\left(1 + \frac{1}{\tau_I s}\right), \quad \tau_I > 0$$

  (a) Derive the necessary and sufficient conditions for closed-loop stability.
  (b) For a unit step change in the set point, calculate the ISE as a function of the controller gain k_c and the integral time τ_I.
  (c) Use MATLAB or Maple to do a contour plot of the ISE in the parameter space k_c versus $1/\tau_I$.
  (d) Calculate the optimal values of k_c and τ_I.
  (e) Simulate and plot the step response of the closed-loop system for the optimal controller parameters.

16.3 Repeat Problem 16.2 using the ISC criterion.

16.4 Given the results of Problems 16.2 and 16.3 for the ISE and ISC criteria, minimize the combined criterion ISE+ρISC for various values of ρ and tabulate your results. What would be reasonable values for ρ?

16.5 Use criterion (16.4.8) and MATLAB or Maple to repeat Problem 16.1.

16.6 Use criterion (16.4.8) and MATLAB or Maple to repeat Problem 16.4. Follow a numerical approach to calculate and minimize the performance criterion.

16.7 A second-order process is controlled with a PI controller, as shown in Figure P16.7. Given the values of the process parameters $k = 1$, $\tau = 1$, $\zeta = 1/2$ and that the PI controller has integral time $\tau_I = 1$,

  (a) Determine the range of values of the controller gain for which the closed-loop system is stable.
  (b) Calculate the ISE as a function of the controller gain for a unit step change in the set point Y_{sp}. What is the optimal value of the controller gain?
  (c) Calculate the ISE as a function of the controller gain for a unit step change in the disturbance W. What is the optimal value of the controller gain?

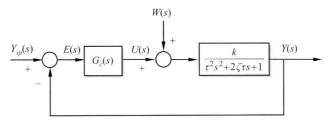

Figure P16.7

(d) Is the optimal gain for set point changes the same as the optimal gain for disturbance changes? If not, what would be a reasonable compromise?

16.8 For the system shown in Figure P16.8, where a third-order system is controlled by a PI controller with gain k_c, of the same sign with k, and integral time $\tau_I > 0$, suppose the set point undergoes a unit step change.

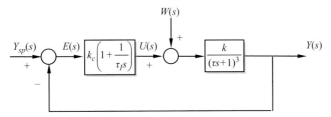

Figure P16.8

(a) Derive the necessary and sufficient conditions for closed-loop stability.
(b) Derive analytic formulas for the ISE and ISC criteria as a function of the controller gain and integral time using Maple.
(c) Show that the minimum for the ISE criterion is achieved when $\tau_I = (11/2)\tau$ and $k_c = (11/4)/k$.
(d) Use numerical optimization in MATLAB to verify the results obtained in (c) for $k = 1$ and $\tau = 1$.
(e) Determine the optimal values of the PI controller parameters using the combined performance criterion given by Eq. (16.1.11) when $\rho = k^2/100$, $\rho = k^2/10$, $\rho = k^2$ and $+\infty$. What are the corresponding values of the ISE and ISC criteria?
(f) What is the effect of the weight parameter ρ on the closed-loop responses? (Use MATLAB to plot the response of the output and control input.)

16.9 For the system shown in Figure P16.9, where a first-order system is controlled by a PI controller with gain k_c, of the same sign with k and integral time $\tau_I > 0$, suppose that the set point undergoes a unit step change.

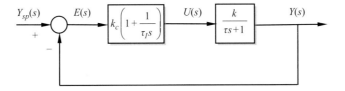

Figure P16.9

(a) Derive analytic formulas for the ISE and ISC criteria as a function of the controller gain and integral time.

(b) Show that the minimum for the combined criterion ISE+ρISC is achieved when

$$\tau_I = \tau, \quad k_c = \frac{1}{k}\sqrt{1+\frac{k^2}{\rho}}$$

(c) Determine the closed-loop transfer function for the optimal values of the controller parameters determined above. Are there any cancellations?

(d) Calculate the closed-loop response of the system analytically and sketch the process output y and the control input u as a function of time.

(e) What is the effect of the weight parameter ρ on the closed-loop responses? What happens when $\rho \to 0$ and when $\rho \to \infty$?

(f) What are your conclusions for the case when a first-order system is controlled by a PI controller?

16.10 Consider the system shown in Figure P16.10, where τ_1, τ_2, τ_3, τ_I and $k_c k > 0$.

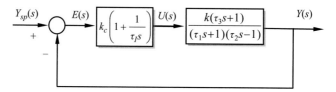

Figure P16.10

(a) Derive the necessary and sufficient conditions for closed-loop stability.
(b) Derive an analytic formula for the ISE criterion as a function of the controller gain and integral time for a unit step change in the set point.
(c) Determine the optimal values of the controller parameters and simulate the corresponding closed-loop response.

16.11 Consider the system shown in Figure P16.11, where $\tau_I > 0$ and $k_c > 0$.

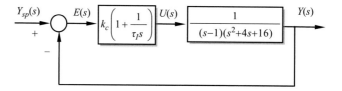

Figure P16.11

(a) Derive the necessary and sufficient conditions for closed-loop stability.
(b) Derive an analytic formula for the ISE criterion as a function of the controller gain and integral time for a unit step change in the set point.
(c) Determine the optimal values of the controller parameters and simulate the corresponding closed-loop response.

17 Bode and Nyquist Stability Criteria – Gain and Phase Margins

STUDY OBJECTIVES

After studying this chapter you should be able to do the following.

- Apply the Bode stability criterion to study the stability of closed-loop systems.
- Calculate and interpret the phase and gain margins, which are quantitative measures of relative stability.
- Apply the Nyquist stability criterion to study the stability of general closed-loop systems.
- Use software tools, like MATLAB or Maple, to calculate gain and phase margins.

17.1 Introduction

17.1.1 Closed-Loop Stability Analysis Based on the Frequency Response Characteristics of the Loop Transfer Function

The notion of stability was introduced in Chapters 7 and 8, and relatively simple, workable criteria for closed-loop stability analysis were introduced in Chapter 12. In this chapter, alternative criteria for investigating the stability of the closed-loop system that are based upon the frequency response characteristics of the loop transfer function will be presented. By the term "loop transfer function" we refer to the product of the controller transfer function and the overall process transfer function (the latter includes the final control element, the physical process and the sensor). We will be using the notation $L(s)$ to denote the loop transfer function, i.e.

$$L(s) = G_c(s)\underbrace{G_v(s)G_p(s)G_m(s)}_{G(s)} = G_c(s)G(s) \tag{17.1.1}$$

where $G_v(s)$ is the transfer function of the final control element (a control valve in most cases), $G_p(s)$ is the transfer function of the process and $G_m(s)$ is the transfer function of the measuring device (sensor). In this chapter we will present the Bode and the Nyquist stability criteria that are based on the frequency response characteristics of the loop transfer function

$L(s)$. Their distinctive advantages lie in the fact that they can be used to study the stability of dead-time systems without the need to use approximations and, additionally, they offer measures of relative stability of a closed-loop system. The latter have proved useful in controller analysis and synthesis.

A system with loop transfer function $L(s)$ is closed-loop stable if all roots of the characteristic equation $1 + L(s) = 0$ lie in the left half of the complex plane (as these are the poles of the closed-loop system). The Nyquist stability criterion is based on a well-known result in complex variable theory known as "the principle of the argument," which will be presented in Section 17.3 in an informal way. Before presenting the Nyquist criterion, we will present a result derived from the Nyquist criterion that is known as the Bode stability criterion. The Bode stability criterion applies to cases where the loop transfer function has no poles with positive real parts and it follows from the more general Nyquist criterion. It makes use of the properties of the Bode plots of the loop transfer function. The reader is referred to Chapter 9, where the construction of Bode plots has been presented in detail.

17.2 The Bode Stability Criterion

17.2.1 The Bode Criterion

Consider the closed-loop system shown in Figure 17.1 whose loop transfer function is

$$L(s) = G(s)G_c(s) \tag{17.2.1}$$

where $G(s)$ represents the transfer function of the entire instrumented process, including sensor and final control element, and $G_c(s)$ represents the controller transfer function. The following assumptions will be made:

(1) that $L(s)$ is strictly proper, i.e. $\lim_{s \to +\infty} L(s) = 0$
(2) that all the poles of $L(s)$ have negative real parts, except possibly a simple pole at $s = 0$.

The Bode stability criterion applies to three special classes of loop transfer functions. Whenever $L(s)$ belongs to one of these classes, an immediate conclusion can be drawn regarding closed-loop stability. Otherwise, one needs to resort to the more complicated but general Nyquist criterion, which will be discussed in Sections 17.3 and 17.4.

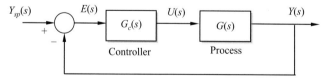

Figure 17.1 Block-diagram representation of a feedback control system.

The Bode Stability Criterion

Suppose that the loop transfer function satisfies assumptions (1) and (2) stated above.

(a) If it so happens that the magnitude of the loop transfer function is below 1, i.e. $|L(i\omega)| < 1$, for all $\omega > 0$, then the closed-loop system is stable.
(b) If it so happens that the phase of the loop transfer function is above $-180°$, i.e. $\arg\{L(i\omega)\} > -180°$, for all $\omega > 0$, then the closed-loop system is stable.
(c) If it so happens that there is a unique frequency ω_p (called the phase crossover frequency), where the phase crosses $-180°$ in the sense that

- $\arg\{L(i\omega)\} > -180°$ for $0 < \omega < \omega_p$
- $\arg\{L(i\omega)\} = -180°$ for $\omega = \omega_p$ (17.2.2)
- $\arg\{L(i\omega)\} < -180°$ for $\omega > \omega_p$

then the closed-loop system is stable if $|L(i\omega_p)| < 1$, unstable if $|L(i\omega_p)| > 1$.
(See Figure 17.2 for a graphical illustration of this case.)

It should be emphasized here that the Bode stability criterion applies only to those classes of functions identified as cases (a), (b) and (c), which also satisfy assumptions (1) and (2). For example, for a process with a positive pole, the Bode criterion is not applicable, or in the presence of multiple phase crossovers, it is not applicable either. From a mathematical point of view, the classes of loop transfer functions for which the Bode criterion is applicable are rather limited; however, in actuality, these classes are quite common. This is the reason for the popularity of the Bode stability criterion. The Nyquist criterion, which is more complicated but very general, may be used when the Bode criterion is not applicable.

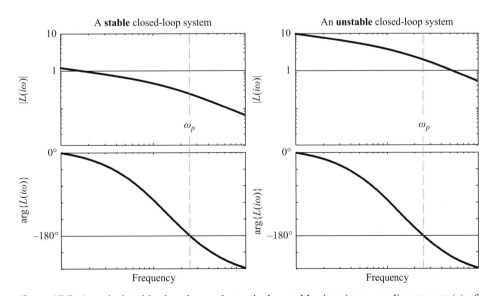

Figure 17.2 A typical stable situation and a typical unstable situation according to part (c) of the Bode stability criterion.

In order to apply the Bode stability criterion we need to construct the Bode diagrams of the loop transfer function (following the methodology for constructing Bode diagrams that has been presented in Chapter 9).

The steps in applying the Bode stability criterion are best demonstrated through an example. Consider the example of the heating tank shown in Figures 10.1 and 10.9. In Example 12.3, the stability of the closed-loop system formed by using a PI controller is studied using the Routh stability criterion (Figure 12.5 summarizes the results).

Initially we form the loop transfer function using a PI controller:

$$L(s) = G_c(s)G(s) = k_c\left(1 + \frac{1}{\tau_I s}\right)\frac{k_p k_v}{(\tau_m s + 1)(\tau_p s + 1)} \quad (17.2.3)$$

We use the numerical values for the model parameters of Example 12.3 ($\tau_p = 1$, $\tau_m = 0.1$, $k_v = 0.2$ and $k_p = 1$) and we assume that the controller parameters k_c and τ_I are positive.

$$L(s) = k_c\left(\frac{\tau_I s + 1}{\tau_I s}\right)\frac{0.2}{(0.1s+1)(s+1)} \quad (17.2.4)$$

The magnitude and the phase of the loop transfer function are

$$|L(i\omega)| = 0.2\frac{k_c}{\tau_I \omega}\sqrt{\frac{\tau_I^2 \omega^2 + 1}{\left(\frac{\omega^2}{100}+1\right)(\omega^2+1)}} \quad (17.2.5)$$

$$\arg\{L(i\omega)\} = \tan^{-1}(\omega\tau_I) - \tan^{-1}(\omega) - \tan^{-1}(0.1\omega) - 90° \quad (17.2.6)$$

Following the results of Example 12.3, we choose $\tau_I = 1/12.1$. Figure 17.3 depicts the Bode diagrams of the loop transfer function, for three different values of the controller gain: $k_c = 10$, 50 and 100. Note that the value of k_c only affects the magnitude plot. From Figure 17.3b, we see that there is a unique frequency at which the argument of $L(i\omega)$ is equal to $-180°$, which is approximately $\omega_p = 10.5$. The exact value is $\omega_p = \sqrt{110} = 10.4882$, and can be determined by solving the algebraic equation $\arg\{L(i\omega_p)\} = -180°$, i.e.

$$\tan^{-1}(\omega_p/12.1) - \tan^{-1}(\omega_p) - \tan^{-1}(0.1\omega_p) - 90° = -180° \quad (17.2.7)$$

or

$$\tan^{-1}(\omega_p/12.1) - \tan^{-1}(\omega_p) - \tan^{-1}(0.1\omega_p) = -90° \quad (17.2.8)$$

When the controller gain is selected to be $k_c = 10$, then

$$|L(i\omega_p)| = 0.2 < 1 \quad (17.2.9)$$

and part (c) of the Bode stability criterion implies that the closed-loop system is stable. When the controller gain is $k_c = 100$, then

$$|L(i\omega_p)| = 2 > 1 \qquad (17.2.10)$$

therefore an unstable closed-loop system is obtained. The critical value of the controller gain is $k_c = 50$ for which the magnitude of the loop transfer function is equal to 1. We therefore conclude that for $\tau_I = 1/12.1$ the system is stable if $k_c < 50$. This is exactly the same result that has been obtained in Chapter 12 using Routh's criterion.

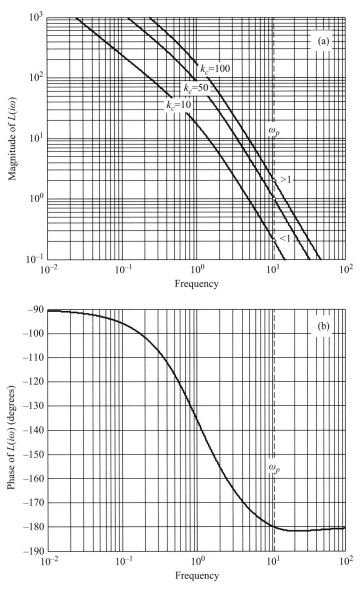

Figure 17.3 Bode diagrams of the loop transfer function for the heating tank under PI control: (a) magnitude plot and (b) phase plot.

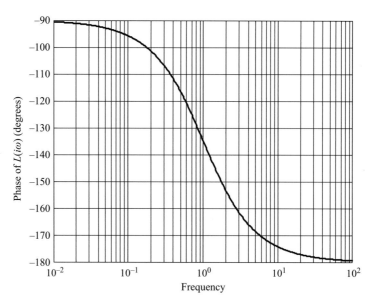

Figure 17.4 Phase plot of the heating tank under PI control and $\tau_I = 1/10$.

In Figure 17.4 the phase plot of the heating tank for $\tau_I = 1/10$ is shown. For this selection of integral time, the phase is always larger than $-180°$ and according to case (b) of the Bode criterion, the closed-loop system will be stable irrespective of the value selected for the controller gain, i.e. we have closed-loop stability for every $k_c > 0$.

Example 17.1 Stability of a third-order system under PI control

A PI controller is used to control a third-order system with transfer function

$$G(s) = \frac{k}{(\tau_1 s + 1)(\tau_2 s + 1)(\tau_3 s + 1)}$$

If $\tau_1 = 1$, $\tau_2 = 1/2$, $\tau_3 = 1/3$ and $\tau_I = 1/3$, derive conditions for closed-loop stability in terms of the product of the controller gain and process gain $k_c k$. Assume that $k_c k > 0$.

Solution

Initially we form the loop transfer function and calculate its magnitude and argument:

$$L(s) = G_c(s)G(s) = k_c\left(1 + \frac{1}{\tau_I s}\right)\frac{k}{(\tau_1 s + 1)(\tau_2 s + 1)(\tau_3 s + 1)}$$

$$|L(i\omega)| = k_c k \sqrt{\frac{\tau_I^2 \omega^2 + 1}{\tau_I^2 \omega^2 (\tau_1^2 \omega^2 + 1)(\tau_2^2 \omega^2 + 1)(\tau_3^2 \omega^2 + 1)}}$$

$$\arg\{L(i\omega)\} = \tan^{-1}(\tau_I \omega) - \tan^{-1}(\tau_1 \omega) - \tan^{-1}(\tau_2 \omega) - \tan^{-1}(\tau_3 \omega) - 90°$$

We substitute the numerical values of the time constants to obtain

$$\frac{|L(i\omega)|}{k_c k} = \sqrt{\frac{1}{\frac{\omega^2}{9}(\omega^2+1)\left(\frac{\omega^2}{4}+1\right)}}$$

$$\arg\{L(i\omega)\} = -\tan^{-1}(\omega) - \tan^{-1}(\omega/2) - 90°$$

The Bode diagrams of the loop transfer function are presented in Figure 17.5. We note that the phase crossover frequency is approximately $\omega_p \approx 1.4$. By using the exact expression for the

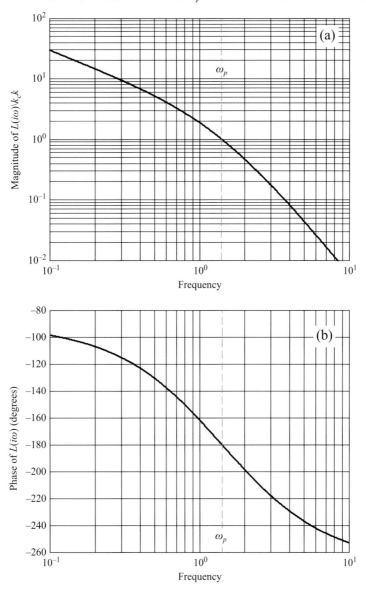

Figure 17.5 Bode diagrams of a third-order system with PI control: (a) magnitude plot and (b) phase plot.

argument of the loop transfer function and solving the equation $\arg\{L(i\omega)\} = -180°$ we obtain the exact value for the phase crossover frequency, which is $\omega_p = \sqrt{2}$. For this frequency it follows that

$$\frac{|L(i\omega_p)|}{k_c k} = 1$$

which agrees with the result shown in Figure 17.5a. The closed-loop system is, therefore, stable if $k_c k < 1$ and unstable if $k_c k > 1$.

Example 17.2 Stability of a third-order system under P control

Consider the case of a third-order system consisting of three identical first-order systems in series, with time constant $\tau/3$ each. The system is controlled with a P controller with controller gain k_c. Derive conditions for closed-loop stability in terms of the product of the controller gain and process gain $k_c k$.

Generalize for an arbitrary nth-order system of the form $G(s) = \dfrac{k}{\left(\dfrac{\tau}{n}s+1\right)^n}$ with $n > 2$.

Solution

The loop transfer function for the P controller and the third-order system is:

$$L(s) = G_c(s)G(s) = \frac{k_c k}{\left(\dfrac{\tau}{3}s+1\right)^3}$$

The sign of $k_c k$ will affect the phase of $L(s)$, therefore we must distinguish two cases: $k_c k > 0$ and $k_c k < 0$.

Consider first the case $k_c k > 0$. The magnitude and the phase of the loop transfer function are

$$|L(i\omega)| = \frac{k_c k}{\left(\dfrac{\tau^2\omega^2}{3^2}+1\right)^{3/2}} \quad \text{and} \quad \arg\{L(i\omega)\} = -3\tan^{-1}\left(\frac{\tau\omega}{3}\right)$$

We observe that the phase is a strictly decreasing function of ω. It tends to 0° as $\omega \to 0$ and to $-270°$ as $\omega \to \infty$. There is therefore a unique phase crossover frequency ω_p, which is the solution of

$$-3\tan^{-1}\left(\frac{\tau\omega_p}{3}\right) = -180° \Rightarrow \frac{\tau\omega_p}{3} = \tan\frac{180°}{3} = \tan\frac{\pi}{3} = \sqrt{3}$$

Applying part (c) of the Bode stability criterion, we conclude that the closed-loop system will be stable if and only if

$$|L(i\omega_p)| < 1 \Rightarrow \frac{k_c k}{\left(\dfrac{\tau^2\omega_p^2}{3^2}+1\right)^{3/2}} < 1 \Rightarrow k_c k < [(\sqrt{3})^2+1]^{3/2} = 2^3 = 8$$

For the case $k_c k < 0$, the magnitude and the phase of the loop transfer function are

$$|L(i\omega)| = \frac{|k_c k|}{\left(\frac{\tau^2 \omega^2}{3^2} + 1\right)^{3/2}} \text{ and } \arg\{L(i\omega)\} = -3\tan^{-1}\left(\frac{\omega\tau}{3}\right) - 180°$$

which implies that the entire phase diagram is below −180°. Therefore parts (b) and (c) of the Bode criterion are not applicable. Applying part (a) gives $\frac{|k_c k|}{\left(\frac{\tau^2 \omega^2}{3^2} + 1\right)^{3/2}} < 1$ for all ω, which is satisfied when $|k_c k| < 1$, i.e. when $k_c k > -1$.

Combining the cases of positive and negative $k_c k$, we obtain the entire stability range:

$$-1 < k_c k < 8.$$

The foregoing analysis easily generalizes for an nth-order system with $n > 2$. We have

$$L(s) = G_c(s)G(s) = \frac{k_c k}{\left(\frac{\tau}{n} s + 1\right)^n}$$

When $k_c k > 0$, the magnitude and the phase of the loop transfer function are

$$|L(i\omega)| = \frac{k_c k}{\left(\frac{\tau^2 \omega^2}{n^2} + 1\right)^{n/2}} \text{ and } \arg\{L(i\omega)\} = -n\tan^{-1}\left(\frac{\tau\omega}{n}\right)$$

There is a unique phase crossover frequency ω_p, which is the solution of

$$-n\tan^{-1}\left(\frac{\tau\omega_p}{n}\right) = -180° \Rightarrow \frac{\tau\omega_p}{n} = \tan\frac{180°}{n} = \tan\frac{\pi}{n}$$

Applying part (c) of the Bode stability criterion, we conclude that the closed-loop system will be stable if and only if

$$|L(i\omega_p)| < 1 \Rightarrow \frac{k_c k}{\left(\frac{\tau^2 \omega^2}{n^2} + 1\right)^{n/2}} < 1 \Rightarrow k_c k < \left[\left(\tan\frac{\pi}{n}\right)^2 + 1\right]^{n/2}$$

For $n = 4$, $k_c k < 2^2$ while for $n = 6$, $k_c k < (4/3)^3$. As $n \to \infty$, $k_c k < 1$ to guarantee stability (note that as as $n \to \infty$, $\tan(\pi/n) \approx \pi/n \to 0$).

When $k_c k < 0$, applying part (a) of the Bode criterion gives $\frac{k_c k}{\left(\frac{\tau^2 \omega^2}{n^2} + 1\right)^{n/2}} < 1$ for all ω, which is satisfied when $|k_c k| < 1$, i.e. when $k_c k > -1$.

17.2.2 The Bode Stability Criterion Applied to Dead-Time Systems

Dead-time systems have been discussed in Chapter 14, and the complexities introduced by dead time in the analysis of open- or closed-loop systems have been discussed. The stability analysis of systems involving dead time can be performed approximately by using Padé approximations. However, no approximation is necessary in order to study the stability of systems with dead time when the Bode stability criterion is used. In this case the loop transfer function is the following

$$L(s) = G_c(s)G(s)\, e^{-\theta s} \qquad (17.2.11a)$$

where θ is the dead time. As $\left| e^{-\theta i\omega} \right| = 1$ at all frequencies, the presence of the dead time does not affect the magnitude of the loop transfer function:

$$|L(i\omega)| = |G_c(i\omega)G(i\omega)| \qquad (17.2.11b)$$

However, as $\arg\left\{ e^{-\theta i\omega} \right\} = -\theta\omega$, it follows that

$$\arg\{L(i\omega)\} = \arg\{G_c(i\omega)G(i\omega)\} - \theta\omega \qquad (17.2.12)$$

The presence of dead time drops the phase of the loop transfer function, and necessarily $\lim_{\omega \to \infty} \arg\{L(i\omega)\} = -\infty$, which excludes the possibility of applying part (b) of the Bode stability criterion. However, part (c) or part (a) may be applicable. This will be illustrated in the following example.

Example 17.3 Stability analysis of P control of a FOPDT process

Consider a FOPDT process: $G(s) = \dfrac{k e^{-\theta s}}{\tau s + 1}$, $\tau > 0$, $\theta > 0$, which is controlled by a P controller: $G_c(s) = k_c$. Use the Bode stability criterion to derive conditions for closed-loop stability.

Solution

The loop transfer function of a FOPDT system and a P controller is the following:

$$L(s) = G_c(s)G(s) = \dfrac{k_c k\, e^{-\theta s}}{\tau s + 1}$$

The sign of $k_c k$ will affect the phase of $L(s)$, therefore we must distinguish two cases: $k_c k > 0$ and $k_c k < 0$.

Consider first the case $k_c k > 0$. The magnitude and the phase of the loop transfer function are then given by

$$|L(i\omega)| = \dfrac{k_c k}{\sqrt{1 + \tau^2 \omega^2}}$$

$$\arg\{L(i\omega)\} = -\tan^{-1}(\tau\omega) - \theta\omega$$

We observe that the phase is a strictly decreasing function of ω. It tends to 0 as $\omega \to 0$, and to $-\infty$ as $\omega \to \infty$. There is therefore a unique phase crossover frequency ω_p, which is the solution of

$$-\tan^{-1}(\tau\omega_p) - \theta\omega_p = -\pi$$

or

$$\tan^{-1}(\tau\omega_p) + \frac{\theta}{\tau}(\tau\omega_p) = \pi$$

This equation can be solved numerically for $\tau\omega_p$, given a specific value of $\frac{\theta}{\tau}$. Once the crossover frequency is calculated, we know from part (c) of the Bode stability criterion that the closed-loop system will be stable if and only if $|L(i\omega_p)|<1$, i.e. when

$$\frac{k_c k}{\sqrt{1+\tau^2\omega_p^2}} < 1 \quad \Rightarrow \quad k_c k < \sqrt{1+(\tau\omega_p)^2}$$

The results for phase crossover frequency and upper stability limit of $k_c k$ are given in Table 17.1 for different values of θ/τ.

Table 17.1 Phase crossover frequency and upper limit of stability range for P control of a FOPDT process

τ/θ	$\tau\omega_p$	$(k_c k)_{max}$
0.01	0.0311	1.0005
0.10	0.2863	1.0402
0.20	0.5307	1.1321
0.50	1.1445	1.5198
1.00	2.0288	2.2618
2.00	3.6732	3.8069
5.00	8.4434	8.5024
10.00	16.3199	16.3506
50.00	79.1713	79.1776
100.00	157.7137	157.7168

For the case $k_c k < 0$, the magnitude and the phase of the loop transfer function are

$$|L(i\omega)| = \frac{|k_c k|}{\sqrt{1+\tau^2\omega^2}} \quad \text{and} \quad \arg\{L(i\omega)\} = -\tan^{-1}(\tau\omega) - \theta\omega - \pi$$

which implies that the entire phase diagram is below $-180°$. Therefore parts (b) and (c) of the Bode criterion are not applicable. Applying part (a) gives $\frac{|k_c k|}{\sqrt{1+\tau^2\omega^2}} < 1$ for all ω, which is satisfied when $|k_c k| < 1$, i.e. when $k_c k > -1$.

The reader is reminded that the problem of Example 17.3 was studied in Chapter 14 (Section 14.2), where Padé approximations were used and subsequently the exact stability limits were calculated by substituting $s = i\omega$ in the characteristic equation. It should be stressed that the brute-force substitution of $s = i\omega$ may be used to calculate the values of the controller gain for a system to be on the verge of instability, but the Bode stability criterion provides precise mathematical conditions for closed-loop stability.

One important observation from the results of the stability analysis of Example 17.3 is that, as the dead time increases, the phase crossover frequency decreases and the stability range shrinks. This is typical behavior in systems with dead time. Figure 17.6 depicts the phase plot of the loop transfer function of the heating tank example for several values of a hypothetical dead time. We note that for $\theta = 0.025$ the phase crossover frequency is $\omega_p \approx 5$. At this frequency we note from Figure 17.3a that the magnitude of the loop transfer function for $k_c = 10$ is close to 1. This shows that a small dead time can result in a marginally stable system even if the system was away from instability in the absence of dead time. When dead time is increased to $\theta = 0.1$ the phase crossover frequency is $\omega_p \approx 3$ and at this frequency the magnitude of the loop transfer function for $k_c = 10$ is greater than 1, i.e. the system is closed-loop unstable.

The example demonstrates the advantage of the Bode stability criterion when applied to dead-time systems as it does not rely on any approximation while it retains its simplicity. This simple example also shows the destabilizing effect that dead time has on process systems. It was shown that a stable system can easily become marginally stable or even unstable if a relatively small dead time is present in the system and has been ignored. This is an alarming observation, as any model of a real process will always be an approximation of the real system. If the dynamics ignored intentionally (simplification) or unintentionally (modeling error) can cause instability then we need to take some precautionary measures. In classical

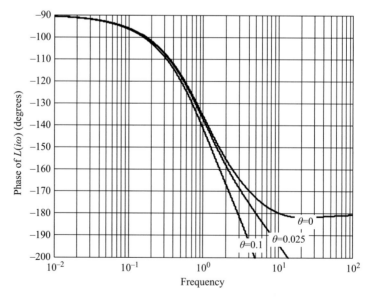

Figure 17.6 Phase plot of the heating tank with PI control and dead time in the loop transfer function.

control theory, in addition to absolute stability, which is normally studied using Routh's criterion, root-locus analysis or Nyquist criterion, relative stability is also evaluated and studied using two indicators known as gain and phase margins. Their definition is presented in the section that follows.

17.2.3 Relative Stability – Phase and Gain Margins

The notion of relative stability is used to answer questions relative to the conditions under which a system that is deemed to be stable can become unstable when model uncertainty and process variation are present. Absolute stability is studied with classical criteria such as the Routh's criterion or the root locus. Although these criteria can be used to study unambiguously the stability of a given system, they offer no information on how close a system is to instability. As model uncertainty or changes occurring in a system as a result of its operation are unavoidable, the possibility of a system becoming unstable is real. The process-control engineer needs to evaluate the "robustness" of a system to model uncertainty or process variation and to give quantitative information on how much "variation" on the model structure or parameters can be tolerated before the system becomes unstable. In addition, the engineer needs to incorporate elements of robustness into the controller synthesis methodology so as to generate control systems whose performance is relatively unaffected by model–process mismatch.

Two of the most well-known classical robustness indicators are the phase and gain margin. We have already defined the phase crossover frequency ω_p as the frequency at which the phase of the loop transfer function becomes equal to $-180°$. The gain crossover frequency ω_g is defined as the frequency at which the magnitude of the loop transfer function becomes equal to 1, i.e.

$$|L(i\omega_g)| = 1 \tag{17.2.13}$$

These frequencies are important in the definition of the phase and gain margin, which are measures of robustness in the classical control theory.

The phase margin (PM) is defined as the phase shift of the loop transfer function at ω_g that will cause the system to become marginally stable, i.e. will cause the phase at ω_g to become $-180°$. This can be interpreted as the distance in the phase diagram from $-180°$, i.e.

$$PM = \arg\{L(i\omega_g)\} - (-180°) \tag{17.2.14}$$

This is shown in Figure 17.7.

Let A be the magnitude of the loop transfer function at the phase crossover frequency, i.e.

$$A = |L(i\omega_p)| \tag{17.2.15}$$

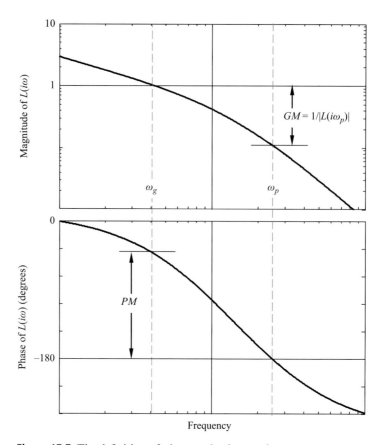

Figure 17.7 The definition of phase and gain margins.

Then, if the loop transfer function is multiplied by the gain factor $1/A$, the magnitude at ω_p will become equal to 1, i.e. the system will be on the verge of instability. This is exactly the definition of the gain margin (GM) (see also Figure 17.7):

$$GM = \frac{1}{|L(i\omega_p)|} \tag{17.2.16}$$

It is usually proposed in the literature that controllers should be designed to achieve phase margins between 30° and 45° (or larger) and gain margins between 1.4 and 1.8 (or larger). These are choices that combine reasonable robustness with acceptable dynamic performance. Many also propose designing controllers to achieve simultaneously prespecified values for both the PM and the GM. If a controller with two parameters is to be designed, such as a PI controller, then the resulting problem is well defined and the solution requires the simultaneous consideration of a set of nonlinear algebraic equations.

Example 17.4 PM and GM of a third-order system under PI control

Consider the third-order system with PI control studied in Example 17.1. For the numerical values of the parameters studied in Example 17.1 using $k_c k = 0.5$, calculate the values of the PM and the GM of the system.

Solution

In Example 17.1 we have developed the following equations for calculating the magnitude and the argument of the loop transfer function

$$|L(i\omega)| = k_c k \sqrt{\frac{1}{\frac{\omega^2}{9}(\omega^2+1)\left(\frac{\omega^2}{4}+1\right)}}$$

$$\arg\{L(i\omega)\} = -\tan^{-1}(\omega) - \tan^{-1}\left(\frac{\omega}{2}\right) - 90°$$

We note that the phase does not depend on the controller gain and remains unaltered when the controller gain is varied. We, therefore, conclude that the phase crossover frequency that has been found in Example 17.1 ($\omega_p = \sqrt{2}$) remains constant as we vary the gain of the controller. We also note from the magnitude of the loop transfer function that for $\omega = \omega_p$ the magnitude equals the product of the controller and the process gain, i.e.

$$|L(i\omega_p)| = k_c k$$

It immediately follows that the gain margin is equal to the inverse of the product of the controller and process gain, $GM = 1/k_c k$. We may conclude that by selecting a value for the product $k_c k$ we directly set the GM of the process. By selecting $k_c k = 1/2$, we set $GM = 2$.

In order to calculate the PM we need first to determine the gain crossover frequency. This is achieved by setting the magnitude equal to 1:

$$k_c k \sqrt{\frac{1}{\frac{\omega_g^2}{9}(\omega_g^2+1)\left(\frac{\omega_g^2}{4}+1\right)}} = 1$$

from which it follows that

$$\omega_g^6 + 5\omega_g^4 + 4\omega_g^2 - (6k_c k)^2 = 0$$

This polynomial equation can be solved easily in MATLAB or Maple to obtain $\omega_g = 0.9693$ and

$$\arg\{L(i\omega_g)\} = -\tan^{-1}(\omega_g) - \tan^{-1}\left(\frac{\omega_g}{2}\right) - 90° = -159.96° \approx -160°$$

The phase margin is, therefore, $PM = 20°$.

17.2.4 Selecting Controller Parameters for PI Controllers to Achieve Prespecified Phase and Gain Margins for FOPDT Systems

When a PI controller is used to control a process system then the controller parameters that achieve prespecified values of the phase and gain margin can be determined by solving a set of nonlinear equations. To demonstrate the idea we consider the case of a FOPDT system controlled by a PI controller with positive k_c and τ_I:

$$L(s) = G_c(s)G(s) = k_c\left(\frac{1+\tau_I s}{\tau_I s}\right)\frac{ke^{-\theta s}}{\tau s+1} \qquad (17.2.17)$$

The magnitude and the phase of the loop transfer function are given by

$$|L(i\omega)| = \frac{k_c k}{\omega \tau_I}\sqrt{\frac{1+\omega^2 \tau_I^2}{1+\omega^2 \tau^2}} \qquad (17.2.18)$$

$$\arg\{L(i\omega)\} = -\frac{\pi}{2} - \omega\theta - \tan^{-1}(\omega\tau) + \tan^{-1}(\omega\tau_I) \qquad (17.2.19)$$

If we seek to find the values of the controller parameters k_c and τ_I so as to achieve given specifications on the phase and gain margins, then we need to satisfy the following equations on the magnitude and the phase of the loop transfer function:

$$\frac{1}{GM} = \frac{k_c k}{\omega_p \tau_I}\sqrt{\frac{1+\omega_p^2 \tau_I^2}{1+\omega_p^2 \tau^2}} \qquad (17.2.20)$$

$$PM = \frac{\pi}{2} - \omega_g\theta - \tan^{-1}(\omega_g\tau) + \tan^{-1}(\omega_g\tau_I) \qquad (17.2.21)$$

The gain and the phase crossover frequencies are unknown but can be determined by their definition

$$1 = \frac{k_c k}{\omega_g \tau_I}\sqrt{\frac{1+\omega_g^2 \tau_I^2}{1+\omega_g^2 \tau^2}} \qquad (17.2.22)$$

$$0 = \frac{\pi}{2} - \omega_p\theta - \tan^{-1}(\omega_p\tau) + \tan^{-1}(\omega_p\tau_I) \qquad (17.2.23)$$

Equations (17.2.20)–(17.2.23) form a set of four nonlinear equations in four unknowns k_c, τ_I, ω_g, ω_p, which can be solved numerically in MATLAB or Maple.

In the case of small dead time ($\theta \ll \tau$), it is possible to derive an approximate solution. Using the approximation $\tan^{-1}(x) = (\pi/2) - 1/x$, which is accurate for $x \gg 1$, Eqs. (17.2.21) and (17.2.23) can be written approximately as

$$PM = \frac{\pi}{2} - \omega_g\theta + \frac{1}{\omega_g\tau} - \frac{1}{\omega_g\tau_I} \qquad (17.2.24)$$

$$0 = \frac{\pi}{2} - \omega_p\theta + \frac{1}{\omega_p\tau} - \frac{1}{\omega_p\tau_I} \qquad (17.2.25)$$

Moreover, because $\omega_p \tau$ and $\omega_g \tau$ are large, combining Eqs. (17.2.20) and (17.2.22) leads to the approximation $\omega_p = GM \cdot \omega_g$. This allows us to rewrite Eq. (17.2.24) as

$$PM = \frac{\pi}{2} - \frac{\omega_p}{GM}\theta + \frac{GM}{\omega_p \tau} - \frac{GM}{\omega_p \tau_I} \quad (17.2.26)$$

Multiplying Eq. (17.2.25) by the GM, subtracting it from Eq. (17.2.26) and solving the resulting expression for ω_p, we find:

$$\omega_p = \frac{GM\left(PM + (GM-1)\dfrac{\pi}{2}\right)}{GM^2 - 1} \cdot \frac{1}{\theta} \quad (17.2.27)$$

Given the process parameters τ and θ and the design specifications PM and GM, we calculate ω_p from Eq. (17.2.27) and then $\omega_g = \omega_p/GM$. We then solve Eq. (17.2.25) for τ_I:

$$\frac{1}{\tau_I} = \frac{1}{\tau} + \omega_p\left(\frac{\pi}{2} - \omega_p\theta\right) \quad (17.2.28)$$

We finally use Eq. (17.2.20) to determine the controller gain

$$k_c k = \frac{\omega_p \tau_I}{GM}\sqrt{\frac{1+\omega_p^2 \tau^2}{1+\omega_p^2 \tau_I^2}} \quad (17.2.29)$$

As a specific example, consider the following FOPDT system

$$G(s) = \frac{1}{s+1}e^{-0.1s} \quad (17.2.30)$$

with desirable $PM = 30°$ (or $\pi/6$) and $GM = 1.8$. Applying Eq. (17.2.27) we obtain $\omega_p = 14.3055$ and $\omega_g = 7.9475$. The controller parameters follow immediately $k_c = 7.7966$ and $\tau_I = 0.3326$. The actual PM and GM values achieved are only slightly different from the specifications: $PM = 29.67°$ and $GM = 1.804$. Their calculation is left as an exercise to the reader.

Example 17.5 Ultimate gain and ultimate period for FOPDT systems – the Ziegler–Nichols tuning rules

Some empirical tuning rules for PID controllers that are appropriate for FOPDT systems are based on implementing a P controller and increasing the controller gain until sustained oscillations are observed, i.e. until the closed-loop system is on the verge of instability. The corresponding value of the controller is called the critical or ultimate gain k_u whereas the corresponding period of oscillations is called the ultimate period P_u. A well-known PID controller tuning recipe based on the ultimate gain and ultimate period is given in Table 17.2 (Ziegler–Nichols tuning rules).

(a) Use the approximation $\tan^{-1}(x) = (\pi/2) - 1/x$ to derive analytic expressions for the ultimate gain and ultimate period of a FOPDT system with $\theta \ll \tau$.

(b) Use your results from the previous question and Table 17.2 to tune a PI controller for the system given by (17.2.30) and calculate the gain and phase margins achieved.

Table 17.2 The Ziegler–Nichols tuning rules for P, PI and PID controllers

Controller type	k_c	τ_I	τ_D
P	$k_u/2$		
PI	$k_u/2.2$	$P_u/1.2$	
PID	$k_u/1.7$	$P_u/2$	$P_u/8$

Solution

(a) The loop transfer function of a FOPDT system and a P controller is the following:

$$L(s) = G_c(s)G(s) = k_c \frac{ke^{-\theta s}}{\tau s + 1}$$

The magnitude and the phase of the loop transfer function are given by

$$|L(i\omega)| = \frac{k_c k}{\sqrt{1+\omega^2\tau^2}}$$

$$\arg\{L(i\omega)\} = -\omega\theta - \tan^{-1}(\omega\tau)$$

On the verge of instability, the phase is $-180°$ or

$$-\omega_p\theta - \tan^{-1}(\omega_p\tau) = -\pi$$

This is a nonlinear equation and we may use the approximation $\tan^{-1}(x) = (\pi/2) - 1/x$ to obtain a polynomial equation

$$\frac{\pi}{2} - \omega_p\theta + \frac{1}{\omega_p\tau} = 0$$

or

$$\frac{2}{\pi}\left(\frac{\theta}{\tau}\right)(\omega_p\tau)^2 - (\omega_p\tau) - \frac{2}{\pi} = 0$$

The solution is

$$\omega_p\tau = \frac{\pi}{4}\frac{\tau}{\theta} + \sqrt{\left(\frac{\pi}{4}\frac{\tau}{\theta}\right)^2 + \frac{\tau}{\theta}}$$

On the verge of instability, the magnitude of the loop transfer function must be equal to 1 at $\omega = \omega_p$, hence the ultimate gain is

$$k_u = \frac{1}{k}\sqrt{1+(\omega_p \tau)^2}$$

The ultimate period is given by

$$P_u = \frac{2\pi}{\omega_p}$$

(b) When the FOPDT system given by (17.2.30) is considered ($k = 1$, $\tau = 1$, $\theta = 0.1$), $\omega_p \tau = 16.32$ and $k_u = 16.35$, $P_u = 0.385$. Calculating the PI controller parameters using Table 17.2, we find: $k_c = k_u/2.2 = 7.4324$ and $\tau_I = P_u/1.2 = 0.3208$. The Bode diagram of the loop transfer function is shown in Figure 17.8. From the figure we observe that $PM \approx 30°$ and $GM \approx 2$ (the exact values are $PM = 30.32°$ and $GM \approx 1.8783$), which are considered satisfactory. The application of the Ziegler–Nichols tuning rules gives reasonable results when the ratio of the dead time to the time constant is less than about 0.2.

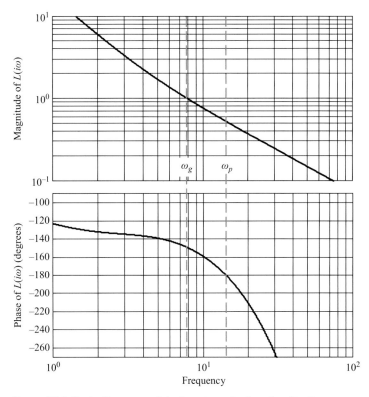

Figure 17.8 Bode diagrams of the loop transfer function for the process $e^{-0.1s}/(s + 1)$ controlled by a PI controller tuned by the Ziegler–Nichols tuning rule.

17.3 The Nyquist Stability Criterion

The Nyquist stability criterion is a comprehensive stability criterion with a strong theoretical basis and a wide range of application. It is based on the use of polar diagrams and the principle of the argument, which will be presented briefly and informally in the section that follows.

17.3.1 The Principle of the Argument

Let $F(s)$ be a single-valued function of s, i.e. there is one and only one point in the $F(s)$-plane corresponding to each value s, that has a finite number of poles. Let us also choose a closed (but otherwise arbitrary) trajectory in the s-plane that does not go through any poles of $F(s)$ and traverse this trajectory in the clockwise (CW) direction. According to the principle of the argument, the trajectory in the $F(s)$-plane, which is also closed, encircles the origin N times, and N is given by the following equation:

$$N = Z - P \qquad (17.3.1)$$

where Z is the number of zeros of $F(s)$ encircled by the trajectory chosen in the s-plane, and P is the number of poles of $F(s)$ encircled by the trajectory chosen in the s-plane when traversed in the CW direction. N can be positive, zero or negative. N is negative when the origin is encircled but the trajectory in the $F(s)$-space has a counterclockwise (CCW) direction (opposite to that in the s-space). The proof of this theorem can be found in many textbooks on complex functions.

To demonstrate the validity of the theorem and also to facilitate understanding let us choose the trajectory in the s-plane to be the unit cycle in the complex plane. Let us also choose the following function

$$F_1(s) = 1 + 2s \qquad (17.3.2)$$

where $F_1(s)$ has one zero at $z = -1/2$ and no pole. The s-plane trajectory and the $F_1(s)$-plane trajectory are shown in Figure 17.9. As can be seen from Figure 17.9, the zero of $F_1(s)$ at

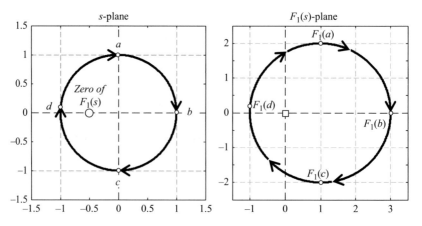

Figure 17.9 Example of the determination of N.

$z = -1/2$ is encircled by the closed trajectory in the s-plane when traversed in the CW direction. Applying Eq. (17.3.1) with $Z = 1$ (one zero is encircled in the s-plane trajectory) and $P = 0$ (no pole is encircled in the s-plane trajectory) results in $N = 1$, i.e. the origin must be encircled once in the $F_1(s)$-plane by the trajectory when traversed in the CW direction. This agrees well with the trajectory in the $F_1(s)$-space shown in Figure 17.9. We also consider the following functions:

$$F_2(s) = \frac{1+2s}{s}, \quad F_3(s) = \frac{s}{s+0.5}, \quad F_4(s) = \frac{s}{s+1.25}, \quad F_5(s) = \frac{s+1.25}{s+0.5} \quad (17.3.3)$$

The trajectories in the $F(s)$-space are shown in Figure 17.10. Note that $N = 1 - 1 = 0$ for the cases of $F_2(s)$ and $F_3(s)$, $N = 1 - 0 = 1$ for $F_4(s)$ (the origin is encircled once and the directions of traversing the trajectories in the s-space and $F_4(s)$-space are the same) and $N = 0 - 1 = -1 < 0$ for $F_5(s)$ (the origin is encircled once and the traversing of the trajectories in the s-space and $F_4(s)$-space are in the opposite directions).

17.3.2 The Nyquist Stability Criterion

Nyquist was the first to note that the principle of the argument can be used to study the stability of linear control systems. He proposed defining a trajectory in the s-plane that covers the whole right half complex plane (RHP) excluding any poles or zeros of $F(s)$ that fall on the imaginary axis as shown in Figure 17.11. When $F(s) = 1 + L(s)$ then we count the number of encirclements of the origin in the F-space. When attention is restricted to systems that do not have any zeros in the RHP, then the number of encirclements of the origin is equal to the poles of the system in the RHP. These are clearly unstable poles and the system is unstable. As the origin of the F-plane (point $(0, 0)$) corresponds to the point $(-1, 0)$ in the L-plane the Nyquist stability criterion can be expressed as follows.

> If $L(s)$ has no zeros on the right half complex plane then the closed-loop system is stable if the plot of $L(s)$ that corresponds to the Nyquist path in the s-space encircles the $(-1, 0)$ point as many times as the number of poles of $L(s)$ in the right half of the s-plane.

For open-loop systems with no poles or zeros in the RHP the Nyquist criterion simplifies as there must be no encirclement of the $(-1, 0)$ point for stability. If the system has no zeros in the RHP but has N unstable poles then there must be N encirclements of the $(-1, 0)$ point and the trajectory of $L(s)$ must be traversed in the CCW direction. When the Nyquist plot passes over the $(-1, 0)$ point then the criterion is inconclusive.

The definition of the phase margin and the gain margin on a polar plot are shown in Figure 17.12. Their interpretation is exactly the same as in the case of the Bode plots. As the frequency is not denoted on a Nyquist plot most engineers prefer the use of Bode plots.

Figure 17.10
Examples of the determination of N.

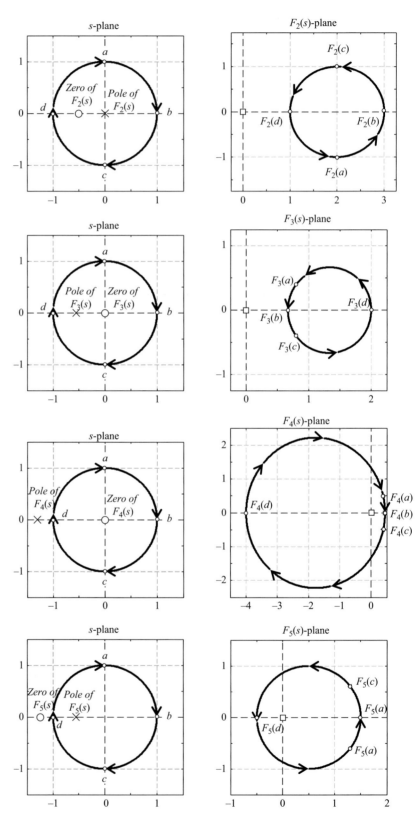

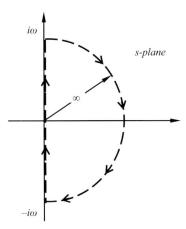

Figure 17.11 The Nyquist path in the s-plane.

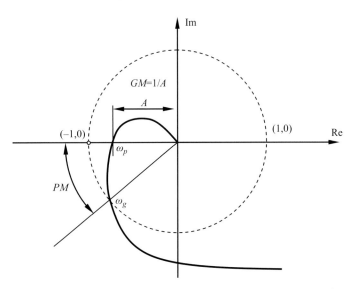

Figure 17.12 The phase margin and the gain margin on a Nyquist (polar) plot.

The Nyquist plots of common transfer functions have been presented and their construction demonstrated in Chapter 9 together with the Bode plots for studying the dynamics of a system in the frequency domain. The reader must review the relevant material.

17.4 Example Applications of the Nyquist Criterion

In order to apply the Nyquist criterion we need to be able to sketch the Nyquist plot of $L(s)$ quickly and determine the conditions that need to be satisfied to guarantee stability of the closed-loop system. To achieve that we need to make the substitution $s = i\omega$ and study first the behavior of the $L(i\omega)$ for $\omega \to 0$ and $\omega \to +\infty$. Then we establish the values of ω for which the polar plot crosses the real axis and the imaginary axis. We finally determine a few locus points in the vicinity of the critical point $(-1, 0)$. It is needless to say that MATLAB and Maple have very well-developed capabilities for plotting the Nyquist plot, but we insist that the reader must always try to generate a brief sketch of the Nyquist plot before using software to produce the exact plot. Before giving some examples on how to apply the Nyquist criterion, we note that only the portion of the locus from $\omega \to 0^+$ to $\omega \to +\infty$ needs to be constructed, as the portion from $\omega \to -\infty$ to $\omega \to 0^-$ is its mirror image about the real axis (i.e. $L(-i\omega)$ is the conjugate of the $L(i\omega)$), as has been presented in Chapter 9. Several Nyquist plots of low-order systems have been presented in Chapter 9 and the reader must review them before moving forward.

17.4.1 Stability analysis of a third-order system under proportional control

When a third-order system with a pole at $s = 0$ is controlled by a proportional controller with controller gain k_c, the following open-loop transfer function is obtained

$$L(s) = \frac{kk_c}{s(s+p_1)(s+p_2)}, \quad p_1, p_2 > 0 \tag{17.4.1}$$

We substitute $s = i\omega$ to obtain

$$\frac{L(i\omega)}{kk_c} = \frac{-(p_1 + p_2)}{(\omega^2 + p_1^2)(\omega^2 + p_2^2)} - i\frac{(p_1 p_2 - \omega^2)}{\omega(\omega^2 + p_1^2)(\omega^2 + p_2^2)} \tag{17.4.2}$$

or

$$\frac{L(i\omega)}{kk_c} = \frac{-a(\omega) + i\beta(\omega)}{[a(\omega)]^2 + [\beta(\omega)]^2} \tag{17.4.3}$$

where

$$a(\omega) = (p_1 + p_2)\omega^2 \tag{17.4.4}$$

$$\beta(\omega) = \omega(\omega^2 - p_1 p_2) \tag{17.4.5}$$

We observe that

$$\left.\frac{L(i\omega)}{kk_c}\right|_{\omega \to 0} = \left.\frac{-a(\omega) + i\beta(\omega)}{[a(\omega)]^2 + [\beta(\omega)]^2}\right|_{\omega \to 0} = -\frac{p_1 + p_2}{(p_1 p_2)^2} - i\infty \tag{17.4.6}$$

and

$$\left.\frac{L(i\omega)}{kk_c}\right|_{\omega \to +\infty} = \left.\frac{-a(\omega) + i\beta(\omega)}{[a(\omega)]^2 + [\beta(\omega)]^2}\right|_{\omega \to +\infty} \approx \frac{1}{(i\omega)^3} = 0 \angle -270° \tag{17.4.7}$$

To find the point where the Nyquist plot intersects the real axis, we set the imaginary part equal to zero to obtain

$$\beta(\omega) = 0 \tag{17.4.8}$$

or $\omega = +\sqrt{p_1 p_2}$. The real part at this frequency can be calculated

$$\text{Re}\left[\frac{L(i\omega)}{kk_c}\right] = -\frac{1}{a(\omega)} = -\frac{1}{(p_1 + p_2)p_1 p_2} \tag{17.4.9}$$

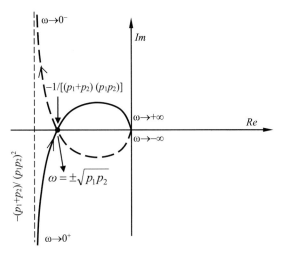

Figure 17.13 The quantitative Nyquist plot of $L(s)/kk_c = 1/s(s+p_1)(s+p_2)$.

A qualitative Nyquist plot is presented in Figure 17.13. The stability depends on the exact point on the negative real axis that the trajectory crosses the axis. For a stable system, the $(-1,0)$ point must lie on the left of the crossing point on the negative real axis, which, using Equation (17.4.9), results in the following stability condition

$$kk_c < (p_1 + p_2)p_1 p_2 \tag{17.4.10}$$

Example 17.6 Stability of a third-order system under proportional control

Consider the closed-loop system shown in Figure 17.14, where a P controller with gain $k_c > 0$ is used to control a third-order system. What are the values of the controller gain for which the closed-loop system is stable? Generate the Nyquist plot and the unit step response of the closed-loop system for $k_c = 1, 3, 6$ and 9.

Solution

We note that the system is of the general form of Eq. (17.4.1) with $p_1 = 1$ and $p_2 = 2$ and that the loop transfer function is

$$L(s) = k_c G(s) = k_c \frac{1}{s(s+1)(s+2)}$$

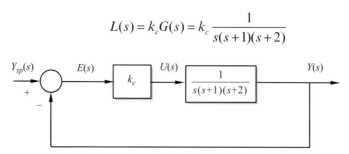

Figure 17.14 A third-order system with proportional feedback control system.

Repeating the steps between Eqs. (17.4.1) and (17.4.9) we find that the negative real axis is crossed when $L(i\omega)$ obtains the value

$$\text{Re}[G(i\omega)] = -\frac{k_c}{(p_1 + p_2)p_1 p_2} = -\frac{k_c}{6}$$

For the closed-loop system to be stable this value must be greater that -1, i.e. $k_c < 6$. The Nyquist plot for $k_c = 1, 3, 6$ and 9 and the corresponding unit step responses are shown in Figures 17.15 and 17.16. It is important to note that for $k_c < 6$ the unit step responses are stable while for $k_c > 6$ are unstable. Sustained oscillations are observed for $k_c = 6$.

Example 17.7 Stability of a third-order system with dead time under proportional control

Consider the closed-loop system shown in Figure 17.17 where a P controller with gain $k_c > 0$ is used to control a third-order system with dead time. Assume that $k_c = 1$ has been selected. Generate the Nyquist plot and the unit step response of the closed-loop system for $\theta = 0, 1, 2$ and 3. What is the effect of dead time to the stability of the closed-loop system?

Solution

We have seen that dead time only adds a phase and does not affect the magnitude of the polar plot of a transfer function. We use the results from the previous example and generate the Nyquist plot of the system for increasing values of the dead time. The results are presented in Figure 17.18. We note that, as the dead time increases, points on the Nyquist plot with given magnitude increase their phase angle and this has a detrimental effect on stability. In the present case study the system is unstable for $\theta = 3$ because for this value of the dead time the Nyquist plot encircles the $(-1, 0)$ point. Apart from this fact this example also demonstrates that the

Nyquist criterion can be used to derive the exact conditions of stability for systems involving dead time (it is recalled that in order to use Routh's criterion we first have to approximate the dead time by using Padé or any similar approximation).

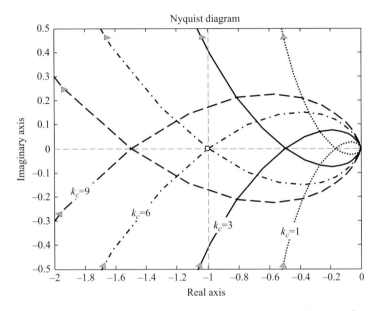

Figure 17.15 The Nyquist plot of the $k_c/s(s + p_1)(s + p_2)$ system for different values of k_c (generated in MATLAB).

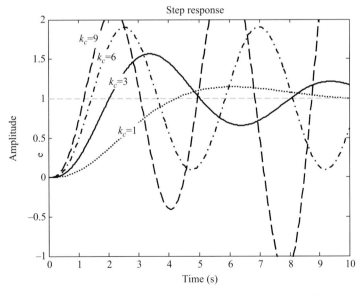

Figure 17.16 Unit step response of the closed-loop system with loop transfer function $k_c/s(s + p_1)(s + p_2)$ for different values of k_c.

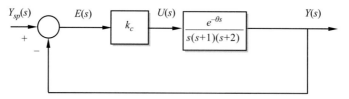

Figure 17.17 A third-order system with proportional feedback control system.

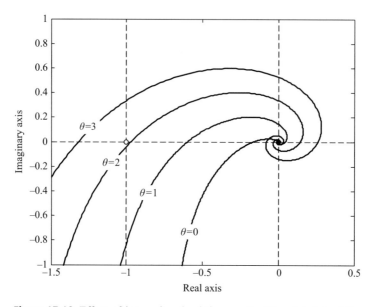

Figure 17.18 Effect of increasing dead time on the Nyquist plot of the third-order system with dead time $e^{-\theta s}/s(s+1)(s+2)$.

17.4.2 Stability Analysis of an Open-Loop Unstable System

In this section we consider the feedback control of a first-order but unstable system that has a pole in the right half complex plane:

$$G(s) = \frac{k}{\tau s - 1}, \quad k, \tau > 0 \tag{17.4.11}$$

We restrict attention to a P controller with gain k_c as shown in Figure 17.19. Our aim is to derive, using the Nyquist criterion, the conditions under which the closed-loop system is stable. To this end, we substitute $s = i\omega$ into Eq. (17.4.11) to obtain:

$$k_c G(i\omega) = \frac{k_c k}{i\omega\tau - 1} \cdot \frac{-i\omega\tau - 1}{-i\omega\tau - 1} = -k_c k \left(\frac{1}{(\omega\tau)^2 + 1} + i \frac{\omega\tau}{(\omega\tau)^2 + 1} \right) \tag{17.4.12}$$

The real and imaginary parts are

$$\text{Re}[G(i\omega)] = -k_c k \left(\frac{1}{(\omega \tau)^2 + 1} \right) \qquad (17.4.13)$$

$$\text{Im}[G(i\omega)] = -k_c k \left(\frac{\omega \tau}{(\omega \tau)^2 + 1} \right) \qquad (17.4.14)$$

We note that for $\omega \to 0^+$ the plot starts from the $(-k_c k, 0)$ point and ends for $\omega \to +\infty$ at the $(0, 0)$ point, forming a semicircle for the intermediate frequencies with center at $(-k_c k/2, 0)$ point with radius $k_c k/2$. A brief sketch is shown in Figure 17.20. The open-loop system has a pole on the right half complex plane. For the closed-loop system to be stable there must exactly one encirclement of the $(-1, 0)$ point as the direction of traversal is opposite (CCW) to the traversal of the Nyquist path (CW). For this condition to be satisfied the $(-1, 0)$ point must be on the right of the $(-k_c k, 0)$ point, i.e. $-k_c k < -1$, or $k_c > 1/k$.

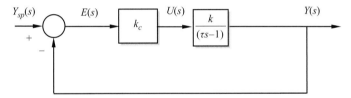

Figure 17.19 An open-loop unstable system with proportional feedback control system.

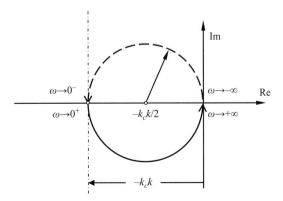

Figure 17.20 Qualitative Nyquist diagram for the first-order unstable system $k/(\tau s - 1)$ under P control with gain k_c.

17.5 Software Tools

17.5.1 MATLAB Commands for Bode and Nyquist Plots and Stability Margins

The commands available in MATLAB for constructing the Bode and Nyquist diagrams were presented in detail in Chapter 9. MATLAB also offers the opportunity to determine the phase and gain margins using the command `margin`. To demonstrate the use of the `margin` command we consider the FOPDT system given by Eq. (17.2.30). For this system the Ziegler–Nichols tunings for a PI controller have been calculated in Example 17.5. The commands that follow form the process and controller transfer function and then the loop transfer function. Then the margin command is used to determine the phase and gain margins and the phase and gain crossover frequencies:

```
» clear all
» k=1; tau=1; theta=0.1;
» sys_p=tf(k,[tau 1],'iodelay',theta);
» zeta=(4/pi)*(theta/tau);
» omega_p=(1+sqrt(1+(4/pi)*zeta))/zeta;
» ku=(1/k)*sqrt(1+(omega_p*tau)^2);
» Pu=2*pi/omega_p;
» kc=ku/2.2;
» tau_I=Pu/1.2;
» sys_PI=tf(kc*[tau_I 1],[tau_I 0]);
» sys_OL=series(sys_PI,sys_p);
» margin(sys_OL)
```

When the `margin` command is invoked without left-hand arguments then MATLAB generates the Bode diagram of the loop transfer function and denotes the stability margins and the crossover frequencies, as shown in Figure 17.21 (compare with Figure 17.8). Note that MATLAB reports the GM in dB.

When the margin command is invoked with left-hand arguments then the stability margins and the crossover frequencies can be obtained:

```
» [GM,PM,wp,wg]=margin(sys_OL)

GM =
    1.8783

PM =
    30.3189

wp =
    14.2549

wg =
    7.9239
```

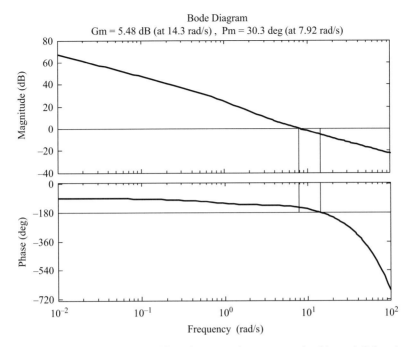

Figure 17.21 Result of invoking the `margin` command without left-hand arguments.

17.5.2 Maple Commands for Bode and Nyquist Plots and Stability Margins

The commands available in Maple for constructing the Bode and Nyquist diagrams were presented in detail in Chapter 9. In addition, the `DynamicSystems` package of Maple also has the special commands `GainMargin` and `PhaseMargin` that compute the gain and phase margin, respectively.

Consider again Example 17.4, where gain and phase margins were computed manually. The same calculations can be performed with Maple as follows:

```
> with(DynamicSystems):
> kkc:=1/2:tau1:=1:tau2:=1/2:tau3:=1/3:tauI:=1/3:
> sys:=TransferFunction(kkc*(s+1/tauI)/(s*(tau1*s+1)*
(tau2*s+1)*(tau3*s+1))):
> g:=GainMargin(sys,decibels=false):GM:=g[1];wp:=g[2];
```

$$GM := 1.999999998$$
$$wp := 1.414213562$$

```
> plots[display](plot([[wp,1],[wp,1/GM]]),MagnitudePlot
(sys,decibels=false));
```

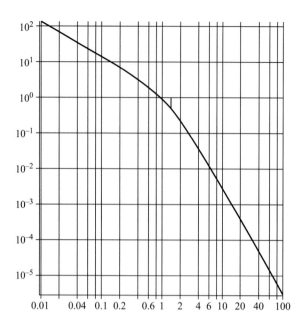

```
> p:=PhaseMargin(sys):PM := p[1];wg:=p[2];
            PM := 20.03808683
            wg := 0.9692600573

> plots[display](plot([[wg,-180],[wg,PM-180]]), PhasePlot(sys));
```

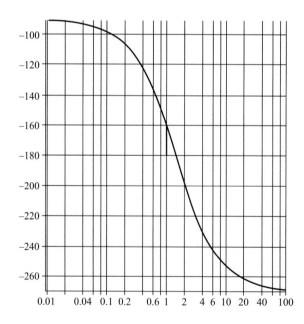

LEARNING SUMMARY

- The loop transfer function is the product of the process transfer function (including the dynamics of the final control element and the sensor) and the transfer function of the controller.
- The Bode stability criterion, which applies to open-loop stable processes, states that the closed-loop system is stable if the magnitude of the loop transfer function is less than 1 at the frequency at which its phase equals $-180°$. The closed-loop system is also stable if the phase is above $-180°$ or the magnitude less than 1 at all frequencies.
- The phase crossover frequency is the frequency at which the argument of the loop transfer function equals $-180°$.
- The gain crossover frequency is the frequency at which the magnitude of the loop transfer function is equal to 1.
- The gain margin is the inverse of the magnitude of the loop transfer function at the phase crossover frequency.
- The phase margin is the distance in degrees of the argument of the loop transfer function from $-180°$ at the gain crossover frequency.
- Experience has shown that a gain margin in the range 1.4 to 1.8 or larger and a phase margin between $30°$ and $45°$ or larger give satisfactory performance and robustness to the closed-loop system.

TERMS AND CONCEPTS

Bode stability criterion. This applies to open-loop stable processes, and states that the closed-loop system is stable if the magnitude of the loop transfer function is less than 1 at the frequency at which its phase equals $-180°$. The system is also stable if the phase is less than $-180°$ or the magnitude less than 1 at all frequencies.

Gain crossover frequency. The frequency at which the magnitude of the loop transfer function is equal to 1.

Gain margin. The inverse of the magnitude of the loop transfer function at the phase crossover frequency.

Nyquist stability criterion. It states that a closed-loop system is stable if and only if the contour in the $L(s)$-plane does not encircle the $(-1, 0)$ point when the number of poles of $L(s)$ in the right half s-plane is zero. If $L(s)$ has P poles in the right half s-plane, then the number of counterclockwise encirclements of the $(-1, 0)$ point must be equal to P for a stable system.

Phase crossover frequency. The frequency at which the argument of the loop transfer function equals $-180°$.

Phase margin. The distance of the argument of the loop transfer function in degrees from $-180°$ at the gain crossover frequency.

Principle of the argument (or Cauchy's theorem). This states that if a contour encircles Z zeros and P poles of $L(s)$ traversing clockwise in the s-plane, the corresponding contour in the $L(s)$-plane encircles the origin $N = Z - P$ times clockwise.

FURTHER READING
Additional information can be found in

Ogata, K., *Modern Control Engineering*, Pearson International Edition, 5th edn, 2008.

Ogunnaike, B. and Ray, H., *Process Dynamics, Modelling and Control*. New York: Oxford University Press, 1994.

The derivation of approximate solutions for determining PI controller parameters that achieve specified PM and GM (Section 17.2.4) has been adapted from the following article

Ho, W. K., Hang, C. C. and Cao, L. S., Tuning of PID controllers based on gain and phase margin specifications, *Automatica*, **31**(3), 497–502, 1995.

PROBLEMS

17.1 A process with transfer function

$$G(s) = \frac{2}{s(s+1)^2}$$

is controlled with a P controller with positive gain k_c.

(a) For what value of k_c is the gain margin 2?
(b) For what value of k_c is the phase margin 30°?
(c) Use MATLAB or Maple to calculate and plot the response of the closed-loop system to a unit step change in the set point for the values of k_c found in (a) and (b).
(d) If there is an unmodeled dead time θ in the process transfer function, what value of the dead time in each case will drive the system to the verge of instability?

Solve (a), (b) and (d) algebraically and then validate your results using MATLAB.

17.2 A process with transfer function

$$G(s) = \frac{1}{(s+1)^3}$$

is controlled with a P controller with positive gain k_c.

(a) For what value of k_c is the gain margin 2?
(b) For what value of k_c is the phase margin 30°?
(c) Use MATLAB or Maple to calculate and plot the response of the closed-loop system to a unit step change in the set point for the values of k_c found in (a) and (b).
(d) If there is an unmodeled dead time θ in the process transfer function, what value of the dead time in each case will drive the system to the verge of instability?

Solve (a), (b) and (d) algebraically and then validate your results using MATLAB.

17.3 Consider an integrating process with dead time

$$G(s) = \frac{k}{s} e^{-\theta s}$$

that is controlled with a P controller with gain k_c, of the same sign with k.
(a) Find the value of k_c for which the gain margin assumes a prespecified value for the GM.
(b) Find the value of k_c for which the phase margin assumes a prespecified value for the PM.
(c) If there is an unmodeled dead time θ in the process transfer function, what value of the dead time in each case will drive the system to the verge of instability?

17.4 Consider an integrating process with dead time

$$G(s) = \frac{k}{s} e^{-\theta s}$$

that is controlled with a P controller with gain k_c, of the same sign with k.
(a) Calculate the ultimate gain and the ultimate period as a function of k and θ.
(b) If you use Ziegler–Nichols tuning rule to select the gain k_c of the P controller, calculate the phase and gain margin.
(c) Use MATLAB to simulate the closed-loop response to a unit step change in the set point when $k = 1$ and $\theta = 0.1$.

17.5 A FOPDT process with $k = 0.4$, $\tau = 6$ and $\theta = 1$ is controlled with a P controller with positive gain k_c.
(a) For what value of k_c is the gain margin 1.6?
(b) For what value of k_c is the phase margin 30°?
Use MATLAB to do the above calculations.

17.6 Consider a process with transfer function

$$G(s) = \frac{k e^{-\theta s}}{\tau^2 s^2 + 2\zeta\tau s + 1}$$

with $\tau = 1$, $\zeta = 1/2$, $k = 1$ and $\theta = 0.1$, that is controlled with a P controller with positive gain k_c.
(a) For what value of k_c is the gain margin 1.8?
(b) For what value of k_c is the phase margin 40°?
Use MATLAB to do the above calculations.

17.7 A FOPDT process is controlled by a PI controller with gain k_c of the same sign as the process static gain and positive integral time τ_I. Using the conditions of the Bode stability criterion, derive a relationship between k_c and τ_I that defines the limit of stability. In a diagram of kk_c versus τ/τ_I, plot this relationship for various values of θ/τ (e.g. 0.1,

0.2, 0.5, 1, 5). Identify the regions of stability and instability and comment on the effect of dead time on the size of the stability region.

17.8 Consider an integrating process with dead time

$$G(s) = \frac{k}{s(\tau s + 1)} e^{-\theta s}$$

that is controlled with a P controller with gain k_c, of the same sign as k. Use the Bode stability criterion to derive conditions for closed-loop stability.

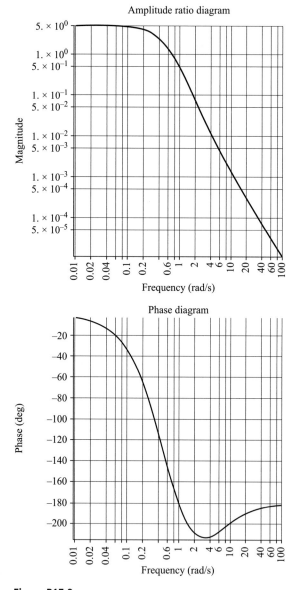

Figure P17.9

17.9 Figure P17.9 depicts the Bode diagrams of an unknown process.
  (a) If this process is controlled with a P controller with positive gain k_c, determine the range of values of the controller gain for which the closed-loop system is stable. What will be the offset (as a function of k_c) for a unit step change in the set point?
  (b) If this process is controlled with a P controller with gain $k_c = 1$, determine the phase and gain margins.
  (c) If this process is controlled with an ideal PD controller with positive gain k_c and derivative time $\tau_D = 25$, sketch the Bode diagrams of the corresponding loop transfer function and then determine the range of values of the controller gain k_c for closed-loop stability.

17.10 The frequency response of a process has been studied experimentally. By fitting the data, the Bode plot (magnitude plot) shown in Figure P17.10 was obtained.

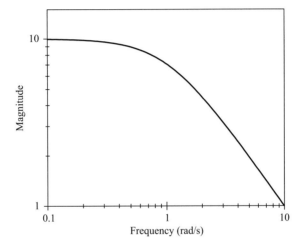

Figure P17.10

Moreover, a number of closed-loop tests have been performed under P control. It was observed that for $0 < k_c < 0.5$, the closed-loop system was stable, with damped oscillations, whereas for $k_c = 0.5$, the oscillations were sustained. If it is known that the process dynamics can be adequately described through an approximate FOPDT transfer function, use the given information to determine the values of the parameters k, τ and θ.

17.11 Consider a FOPDT process with known steady-state gain k, for which we have determined experimentally the ultimate gain k_u and the ultimate period P_u. Show that the time constant and the dead time can be calculated through the equations

$$\tau = \frac{P_u}{2\pi}\sqrt{(k_u k)^2 - 1}$$

$$\theta = \frac{P_u}{2\pi}\left[\pi - \tan^{-1}\left(\frac{2\pi\tau}{P_u}\right)\right]$$

17.12 Repeat the calculations of Section 17.2.4 using the approximation $\tan^{-1}(x) \approx \pi/2 - \pi/(4x)$, which has been proposed by Ho et al. (*Automatica*, **31**(3), 497–502, 1995). Compare your results with the approximation used in Section 17.2.4. Comment on the accuracy of the two approximations, including the accuracy in determining the exact values of the phase and gain crossover frequencies and the PM and GM achieved.

17.13 Use Eqs. (17.2.20), (17.2.22), (17.2.24) and (17.2.25) in the form obtained with the approximation of the previous problem to show that, given any PI controller parameters k_c and τ_I for a FOPDT system and PI controller, then the phase and gain margin can be approximated by the following equations:

$$GM = \frac{\pi}{4}\left(\frac{1}{k_c k}\right)\left(\frac{\tau}{\theta}\right)\left(1 + \sqrt{1 + \frac{4}{\pi}\left(\frac{\theta}{\tau} - \frac{\theta}{\tau_I}\right)}\right)$$

$$PM = \frac{\pi}{2} - k_c k\left(\frac{\theta}{\tau}\right) + \pi\left(\frac{1}{k_c k}\right)\left(1 - \frac{\tau}{\tau_I}\right)$$

17.14 Consider a FOPDT process whose dynamics is dominated by the dead time, i.e. $\theta > \tau$. Consider the cases where $\theta = \tau$, $\theta = 3\tau$ and $\theta = 5\tau$. Use the Ziegler–Nichols rules to tune PI controllers. Calculate the phase and gain margins and the step response to setpoint changes. What are your conclusions about the success of Ziegler–Nichols tuning in FOPDT systems whose dynamics is dominated by dead time?

17.15 Consider a FOPDT process whose dynamics is dominated by the dead time, i.e. $\theta > \tau$. Chien and Fruehauf (*Chemical Engineering Progress*, **86**(10), 33, 1990) proposed that a PI controller can be tuned using $\tau_I = \tau$ and $k_c k = \tau/(\lambda + \theta)$, where λ is an adjustable time constant for the closed-loop system. Calculate the PM and GM of a FOPDT system controlled by a PI controller tuned according to this rule. Express λ as a fraction of τ. Use MATLAB to determine the closed-loop response to a unit step change in the set point. What are your conclusions?

17.16 Consider an unstable first-order plus dead time (UFOPDT) process:

$$G(s) = \frac{k}{\tau s - 1} e^{-\theta s}, \quad \tau, \theta > 0$$

with specific values of the parameters $k = 1$, $\tau = 1$ and $\theta = 0.1$. If the process is to be controlled with a P controller, we want to find the range of values of k_c that result in a closed-loop stable system.

(a) Use [1,1] and [2,2] Padé approximations of the dead time to do approximate stability analysis.
(b) Use MATLAB to draw the Nyquist plots of the loop transfer function for $k_c = 2$, 10, 20, 40 and 100. What are your conclusions from the Nyquist stability criterion in each case? Note that you have an open-loop unstable system with one pole in the RHP and a counterclockwise encirclement of (–1, 0) is necessary for stability.
(c) Calculate the exact limits of stability for the given system.
(d) Can you generalize your conclusions for a general UFOPDT system?

18 Multi-Input–Multi-Output Linear Systems

The aim of this chapter is to introduce the basic elements of systems with multiple inputs and multiple outputs, or MIMO systems. MIMO systems are very common in industrial process control systems. Interaction is an inherent characteristic of a MIMO feedback control system that needs to be taken into consideration when designing the controllers. We will show that many characteristics of the SISO systems can be generalized to MIMO systems, including the conditions for stability. Understanding the basic features of MIMO system dynamics is essential in the study of complex process control systems.

STUDY OBJECTIVES

After studying this chapter you should be able to do the following.

- Describe the closed-loop and open-loop MIMO system in state space.
- Derive the transfer-function description of a MIMO system when the state-space description is known.
- Perform stability analysis of the closed-loop system using the state-space description or the transfer-function matrix description of a MIMO system.
- Develop a decoupling precompensator to simplify the controller synthesis for MIMO systems.
- Use MATLAB and Maple to simulate and analyze a MIMO system.

18.1 Introduction

Multi-input–multi-output (MIMO) linear dynamic systems have already been introduced in Chapter 6 (Section 6.9), in both state-space and transfer-function descriptions. We will start with a brief review before we proceed to the analysis of MIMO feedback control systems.

18.1.1 A Mixing Tank Example

Consider the tank shown in Figure 18.1 where two liquid streams of the same component are mixed. One stream (the hot stream) has a high, constant temperature T_H and variable volumetric flowrate F_H, and the other stream (the cold stream) has a low, constant temperature T_C and variable flowrate F_C. The exit stream has flowrate F and temperature T, while the height of the liquid level in the tank is h. The tank has a constant cross-sectional area A and the exit volumetric flowrate is proportional to the height of the liquid level in the tank. The mass and energy balances for the tank can be written as follows

$$\frac{d}{dt}(\rho A h) = F_H \rho + F_C \rho - F \rho \tag{18.1.1}$$

$$\frac{d}{dt}(\rho C_p A h T) = F_H \rho C_p T_H + F_C \rho C_p T_C - F \rho C_p T \tag{18.1.2}$$

If we assume that the density and the specific heat capacity of the liquid are weak functions of the temperature and that $F = h/R$ we can write

$$A\frac{dh}{dt} = F_H + F_C - \frac{h}{R} \tag{18.1.3}$$

$$Ah\frac{dT}{dt} = F_H T_H + F_C T_C - F_H T - F_C T \tag{18.1.4}$$

F_H and F_C are the inputs, and h, T are the states of the system.

Note that Eqs. (18.1.3) and (18.1.4) are nonlinear, but can be linearized in the vicinity of a steady state as follows:

$$A\frac{d\bar{h}}{dt} = \bar{F}_H + \bar{F}_C - \frac{\bar{h}}{R} \tag{18.1.5}$$

$$Ah_s\frac{d\bar{T}}{dt} = (T_H - T_s)\bar{F}_H + (T_C - T_s)\bar{F}_C - (F_{H,s} + F_{C,s})\bar{T} \tag{18.1.6}$$

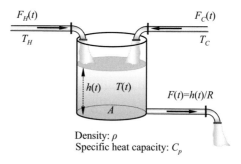

Figure 18.1 A mixing tank system.

where

$$\begin{aligned} \bar{F}_H &= F_H - F_{H,s} \\ \bar{F}_C &= F_C - F_{C,s} \\ \bar{T} &= T - T_s \\ \bar{h} &= h - h_s \end{aligned} \quad (18.1.7)$$

are the deviation variables. Noting that $F_{H,s} + F_{c,s} = F_s = h_s/R$, and converting the state equations to matrix form, we obtain

$$\frac{d}{dt}\begin{bmatrix} \bar{h} \\ \bar{T} \end{bmatrix} = \begin{bmatrix} -\dfrac{1}{AR} & 0 \\ 0 & -\dfrac{1}{AR} \end{bmatrix} \begin{bmatrix} \bar{h} \\ \bar{T} \end{bmatrix} + \begin{bmatrix} \dfrac{1}{A} & \dfrac{1}{A} \\ \dfrac{T_H - T_s}{Ah_s} & \dfrac{T_C - T_s}{Ah_s} \end{bmatrix} \begin{bmatrix} \bar{F}_H \\ \bar{F}_C \end{bmatrix} \quad (18.1.8)$$

18.1.2 State-Space Description of MIMO Systems

Considering the height of the liquid in the tank and its temperature as the output variables, and defining the vectors of state, input and output variables

$$x = \begin{bmatrix} \bar{h} \\ \bar{T} \end{bmatrix}, \quad u = \begin{bmatrix} \bar{F}_H \\ \bar{F}_C \end{bmatrix}, \quad y = \begin{bmatrix} \bar{h} \\ \bar{T} \end{bmatrix} \quad (18.1.9)$$

Eq. (18.1.8) can be written as

$$\begin{aligned} \frac{dx}{dt} &= Ax + Bu \\ y &= Cx + Du \end{aligned} \quad (18.1.10)$$

where

$$A = \begin{bmatrix} -\dfrac{1}{AR} & 0 \\ 0 & -\dfrac{1}{AR} \end{bmatrix}, \; B = \begin{bmatrix} \dfrac{1}{A} & \dfrac{1}{A} \\ \dfrac{T_H - T_s}{Ah_s} & \dfrac{T_C - T_s}{Ah_s} \end{bmatrix}, \; C = \begin{bmatrix} 1 & 0 \\ 0 & 1 \end{bmatrix}, \; D = \begin{bmatrix} 0 & 0 \\ 0 & 0 \end{bmatrix} \quad (18.1.11)$$

In the more general setting where we have n state variables, m input variables and p output variables, the state-space description of the system is given by Eq. (18.1.10), where

$$x = \begin{bmatrix} x_1 \\ x_2 \\ \vdots \\ x_n \end{bmatrix}, \quad u = \begin{bmatrix} u_1 \\ u_2 \\ \vdots \\ u_m \end{bmatrix}, \quad y = \begin{bmatrix} y_1 \\ y_2 \\ \vdots \\ y_p \end{bmatrix} \quad (18.1.12)$$

$$A = \begin{bmatrix} a_{11} & a_{12} & \cdots & a_{1n} \\ a_{21} & a_{22} & \cdots & a_{2n} \\ \vdots & \vdots & \ddots & \vdots \\ a_{n1} & a_{n2} & \cdots & a_{nn} \end{bmatrix}, \quad B = \begin{bmatrix} b_{11} & b_{12} & \cdots & b_{1m} \\ b_{21} & b_{22} & \cdots & b_{2m} \\ \vdots & \vdots & \ddots & \vdots \\ b_{n1} & b_{n2} & \cdots & b_{nm} \end{bmatrix}$$

(18.1.13)

$$C = \begin{bmatrix} c_{11} & c_{12} & \cdots & c_{1n} \\ c_{21} & c_{22} & \cdots & c_{2n} \\ \vdots & \vdots & \ddots & \vdots \\ c_{p1} & c_{p2} & \cdots & c_{pn} \end{bmatrix}, \quad D = \begin{bmatrix} d_{11} & d_{12} & \cdots & d_{1m} \\ d_{21} & d_{22} & \cdots & d_{2m} \\ \vdots & \vdots & \ddots & \vdots \\ d_{p1} & d_{p2} & \cdots & d_{pm} \end{bmatrix}$$

Under steady-state conditions, $dx/dt = 0$ and Eqs. (18.1.10) give

$$0 = Ax_s + Bu_s \qquad (18.1.14a)$$
$$y_s = Cx_s + Du_s \qquad (18.1.14b)$$

If the matrix A is invertible, it follows from (18.1.14a) that

$$x_s = -A^{-1}Bu_s \qquad (18.1.15)$$

Eliminating the state vector from (18.1.14b) we obtain

$$y_s = K u_s \qquad (18.1.16)$$

where

$$K = -CA^{-1}B + D \qquad (18.1.17)$$

is the static or steady-state gain matrix.

Example 18.1 A liquid storage system of two interacting tanks

In Figure 18.2 a two-tank liquid storage system with interaction is shown. There are two incoming liquid streams, one for each tank, the flowrates of which are manipulated through control valves. The outflow of each tank is proportional to the liquid level in the tank, whereas the flow through the interconnecting pipe is proportional to the difference of liquid levels. Derive the state-space description of the system for the parameter values $A_1 = 1$, $A_2 = 0.5$, $R_1 = 1$, $R_2 = 2$, $R_{12} = 0.5$.

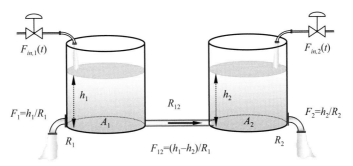

Figure 18.2 A liquid storage system of two interacting tanks.

Solution

Assuming constant density, mass balances for each tank give the following state equations:

$$A_1 \frac{dh_1}{dt} = F_{in,1} - \frac{h_1}{R_1} - \frac{h_1 - h_2}{R_{12}}$$

$$A_2 \frac{dh_2}{dt} = F_{in,2} - \frac{h_2}{R_2} + \frac{h_1 - h_2}{R_{12}}$$

or

$$\frac{dh_1}{dt} = -\left(\frac{1}{A_1 R_1} + \frac{1}{A_1 R_{12}}\right) h_1 + \frac{1}{A_1 R_{12}} h_2 + \frac{1}{A_1} F_{in,1}$$

$$\frac{dh_2}{dt} = \frac{1}{A_2 R_{12}} h_1 - \left(\frac{1}{A_2 R_2} + \frac{1}{A_2 R_{12}}\right) h_2 + \frac{1}{A_2} F_{in,2}$$

The state-space description of the system under consideration is given by Eq. (18.1.10), where

$$A = \begin{bmatrix} -\left(\dfrac{1}{A_1 R_1} + \dfrac{1}{A_1 R_{12}}\right) & \dfrac{1}{A_1 R_{12}} \\ \dfrac{1}{A_2 R_{12}} & -\left(\dfrac{1}{A_2 R_2} + \dfrac{1}{A_2 R_{12}}\right) \end{bmatrix} = \begin{bmatrix} -3 & 2 \\ 4 & -5 \end{bmatrix}, \quad B = \begin{bmatrix} \dfrac{1}{A_1} & 0 \\ 0 & \dfrac{1}{A_2} \end{bmatrix} = \begin{bmatrix} 1 & 0 \\ 0 & 2 \end{bmatrix}$$

Considering the state variables as output variables, we have

$$C = \begin{bmatrix} 1 & 0 \\ 0 & 1 \end{bmatrix}, \quad D = \begin{bmatrix} 0 & 0 \\ 0 & 0 \end{bmatrix}$$

18.1.3 Matrix Transfer-Function Description of MIMO Systems

By taking the Laplace transform of Eq. (18.1.10) under the assumption of zero initial conditions we obtain

$$(sI - A)X(s) = BU(s)$$
$$Y(s) = CX(s) + DU(s)$$

or

$$Y(s) = G(s)U(s) \tag{18.1.18}$$

where

$$G(s) = C(sI - A)^{-1} B + D = \begin{bmatrix} G_{11}(s) & G_{12}(s) & \cdots & G_{1m}(s) \\ G_{21}(s) & G_{22}(s) & \cdots & G_{2m}(s) \\ \vdots & \vdots & \ddots & \\ G_{p1}(s) & G_{p2}(s) & \cdots & G_{pm}(s) \end{bmatrix} \tag{18.1.19}$$

is the p-by-m transfer-function matrix with elements

$$G_{ij}(s) = c_i(sI - A)^{-1} b_j + d_{ij} \tag{18.1.20}$$

where c_i is the i-th row of the C matrix and b_j is the j-th column of the B matrix.

From Eqs. (18.1.17) and (18.1.19) we conclude that the steady-state gain matrix of the system is equal to the transfer-function matrix evaluated at $s = 0$: $K = G(0)$.

For the case of the mixing tank shown in Figure 18.1, using the A, B, C, D matrices from Eq. (18.1.11), we have that

$$G(s) = C(sI - A)^{-1} B + D = \begin{bmatrix} 1 & 0 \\ 0 & 1 \end{bmatrix} \begin{bmatrix} s + \dfrac{1}{AR} & 0 \\ 0 & s + \dfrac{1}{AR} \end{bmatrix}^{-1} \begin{bmatrix} \dfrac{1}{A} & \dfrac{1}{A} \\ \dfrac{T_H - T_s}{Ah_s} & \dfrac{T_C - T_s}{Ah_s} \end{bmatrix} + \begin{bmatrix} 0 & 0 \\ 0 & 0 \end{bmatrix}$$

$$= \begin{bmatrix} \dfrac{1}{s + \dfrac{1}{AR}} & 0 \\ 0 & \dfrac{1}{s + \dfrac{1}{AR}} \end{bmatrix} \begin{bmatrix} \dfrac{1}{A} & \dfrac{1}{A} \\ \dfrac{T_H - T_s}{Ah_s} & \dfrac{T_C - T_s}{Ah_s} \end{bmatrix}$$

or

$$G(s) = \begin{bmatrix} G_{11}(s) & G_{12}(s) \\ G_{21}(s) & G_{22}(s) \end{bmatrix} = \begin{bmatrix} \dfrac{R}{ARs + 1} & \dfrac{R}{ARs + 1} \\ \dfrac{\dfrac{R}{h_s}(T_H - T_s)}{ARs + 1} & \dfrac{\dfrac{R}{h_s}(T_C - T_s)}{ARs + 1} \end{bmatrix} \tag{18.1.21}$$

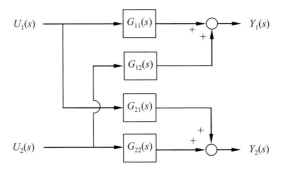

Figure 18.3 Block diagram of a two-input–two-output system.

which is the transfer function matrix of the mixing tank that relates the two inputs with the two outputs. The block-diagram description of a two-input–two-output system is shown in Figure 18.3. We also observe that the steady-state gain matrix of the mixing tank is

$$K = G(0) = \begin{bmatrix} R & R \\ \dfrac{R}{h_s}(T_H - T_s) & \dfrac{R}{h_s}(T_C - T_s) \end{bmatrix}$$

Example 18.2 Transfer function of a system of two interacting tanks

Consider the two-tank liquid storage system with interaction shown in Figure 18.2. In Example 18.1, the state-space description of the system has been derived. Use the results of the Example 18.1 to derive the transfer function matrix of the system.

Solution

Using Eq. (18.1.19) and the results of Example 18.1 we obtain

$$G(s) = C(sI - A)^{-1} B + D = \begin{bmatrix} 1 & 0 \\ 0 & 1 \end{bmatrix} \begin{bmatrix} s+3 & -2 \\ -4 & s+5 \end{bmatrix}^{-1} \begin{bmatrix} 1 & 0 \\ 0 & 2 \end{bmatrix} + \begin{bmatrix} 0 & 0 \\ 0 & 0 \end{bmatrix}$$

$$= \begin{bmatrix} \dfrac{s+5}{(s+1)(s+7)} & \dfrac{4}{(s+1)(s+7)} \\ \dfrac{4}{(s+1)(s+7)} & \dfrac{2(s+3)}{(s+1)(s+7)} \end{bmatrix}$$

18.2 Dynamic Response of MIMO Linear Systems

In Chapter 6 (Section 6.9) we have derived analytic expressions of the time response of a MIMO linear system of the form (18.1.10). The results are summarized below.

The response of the state vector is given by Eq. (6.9.17), and the response of the outputs by Eq. (6.9.18), and these are repeated below.

$$x(t) = e^{At}x(0) + \sum_{j=1}^{m} \int_0^t e^{A(t-t')} b_j u_j(t') dt' \qquad (18.2.1)$$

$$y_i(t) = c_i e^{At}x(0) + \sum_{j=1}^{m} \left(\int_0^t c_i e^{A(t-t')} b_j u_j(t') dt' + d_{ij} u_j(t) \right), \quad i = 1, \ldots, p \qquad (18.2.2)$$

As in single-input–single-output systems, we can distinguish:
(1) unforced response, i.e. the response of the system state under zero input

$$x(t) = e^{At}x(0) \qquad (18.2.3)$$

(2) forced response under zero initial condition
 – for the state vector

$$x(t) = \sum_{j=1}^{m} \int_0^t e^{A(t-t')} b_j u_j(t') dt' \qquad (18.2.4)$$

 – for the outputs

$$y_i(t) = \sum_{j=1}^{m} \left(\int_0^t c_i e^{A(t-t')} b_j u_j(t') dt' + d_{ij} u_j(t) \right), \quad i = 1, 2, \ldots, p \qquad (18.2.5)$$

Because the system is linear, the effects of each input are additive. This means that the effect of each input may be calculated separately and the results can be added up.

When the matrix transfer function description is used, the responses of the outputs may be calculated, again by superimposing the effects of each input:

$$y_i(t) = \sum_{j=1}^{m} \mathcal{L}^{-1}\{G_{ij}(s) U_j(s)\}, \quad i = 1, 2, \ldots, p \qquad (18.2.6)$$

In view of Eq. (18.1.20) and the convolution theorem of the Laplace transform, expressions (18.2.5) and (18.2.6) are equivalent.

Example 18.3 Response of the system of two interacting tanks to a simultaneous change in its two inputs

In the two-tank system studied in Examples 18.1 and 18.2, a unit step change is applied to both inputs simultaneously. What is the response of the two outputs?

Solution

From Example 18.2 we have that

$$\begin{bmatrix} Y_1(s) \\ Y_2(s) \end{bmatrix} = \begin{bmatrix} \dfrac{s+5}{(s+1)(s+7)} & \dfrac{4}{(s+1)(s+7)} \\ \dfrac{4}{(s+1)(s+7)} & \dfrac{2(s+3)}{(s+1)(s+7)} \end{bmatrix} \begin{bmatrix} U_1(s) \\ U_2(s) \end{bmatrix}$$

We apply Eq. (18.2.6) in order to calculate $y_1(t)$:

$$y_1(t) = \mathcal{L}^{-1}\{G_{11}(s)U_1(s)\} + \mathcal{L}^{-1}\{G_{12}(s)U_2(s)\}$$

$$= \mathcal{L}^{-1}\left\{\frac{s+5}{(s+1)(s+7)} \cdot \frac{1}{s}\right\} + \mathcal{L}^{-1}\left\{\frac{4}{(s+1)(s+7)} \cdot \frac{1}{s}\right\}$$

$$= \left(\frac{5}{7} - \frac{2}{3}e^{-t} - \frac{1}{21}e^{-7t}\right) + \left(\frac{4}{7} - \frac{2}{3}e^{-t} + \frac{2}{21}e^{-7t}\right)$$

$$= \frac{9}{7} - \frac{4}{3}e^{-t} + \frac{1}{21}e^{-7t}$$

We also find in exactly the same way that

$$y_2(t) = \frac{10}{7} - \frac{4}{3}e^{-t} - \frac{2}{21}e^{-7t}$$

18.2.1 Poles, Zeros and Stability of MIMO Linear Systems

The notion of asymptotic stability of linear systems has been defined in Chapter 7 as the property of the unforced response to return to zero as $t \to \infty$. It was shown that, for any linear system (SISO or MIMO), asymptotic stability is equivalent to all eigenvalues of the matrix A having negative real parts.

In Chapter 8, the notion of input–output stability was defined for SISO systems in the sense that, for every bounded input, the output is bounded. When it comes to MIMO systems, input–output stability can be defined by imposing the requirement that for every bounded input u_j, $j = 1 \ldots m$, every output y_i, $i = 1 \ldots p$, is bounded. In other words, all the individual transfer functions $G_{ij}(s)$, $j = 1 \ldots m$, $i = 1 \ldots p$ (entries of the transfer-function matrix $G(s)$) are required to be input–output stable for the overall MIMO system to be input–output stable. Note that, from Eq. (18.1.20),

$$G_{ij}(s) = c_i \frac{\text{Adj}(sI - A)}{\det(sI - A)} b_j + d_{ij} \tag{18.2.7}$$

which shows that all the poles of every $G_{ij}(s)$ are eigenvalues of the matrix A. Hence asymptotic stability implies input–output stability of the MIMO linear system.

It is also possible to define poles and zeros of a transfer-function matrix $G(s)$, in analogy to their definition for the scalar transfer function of a SISO linear system.

A number p is called a *pole* of the transfer-function matrix $G(s)$ if at least one of its entries $G_{ij}(s)$ becomes infinite at $s = p$. All the poles are eigenvalues of the A matrix of the state-space description from which $G(s)$ was generated.

For a square transfer-function matrix $G(s)$, i.e. one that relates equal number of inputs and outputs, a number z is called a *zero* of $G(s)$ if its determinant $\det[G(s)] = 0$ at $s = z$. Equivalently, z is a zero of $G(s)$ if it is a pole of its matrix inverse $[G(s)]^{-1}$.

Consider, for example, the following transfer-function matrix

$$G(s) = \begin{bmatrix} \dfrac{s-1}{s} & \dfrac{2}{s} \\ \dfrac{1}{s} & \dfrac{1}{s} \end{bmatrix}$$

At $s = 0$, all four entries of $G(s)$ become infinite, whereas everywhere else they are all finite. Therefore, $G(s)$ has one pole, at $s = 0$. To find the zeros, we calculate the determinant $\det[G(s)] = [(s-1)-2]/s^2 = (s-3)/s^2$. Hence $z = 3$ is the zero of $G(s)$. It is important to observe here that no individual entry of $G(s)$ vanishes at $s = 3$. Also, note that the 11-element of $G(s)$ has a zero at $s = 1$, but this is not a zero of the system.

We also consider the transfer-function matrix of the mixing tank example that is given by Eq. (18.1.21). At $s = -1/(AR)$, all four entries of $G(s)$ become infinite, whereas everywhere else they are finite. Therefore, $G(s)$ has one pole, at $-1/(AR)$. To find the zeros, we calculate the determinant of $G(s)$:

$$\det G(s) = \left(\frac{R}{ARs+1}\right)^2 \left(\frac{T_C - T_H}{h_s}\right)$$

which is nonzero for all s, therefore $G(s)$ does not have any zeros.

A MIMO linear system with an equal number of inputs and outputs that does not have RHP zeros or dead-time elements is called a minimum-phase system. Consider for example the following transfer-function matrix

$$G(s) = \begin{bmatrix} \dfrac{-1}{s+2} & \dfrac{2}{s+1} \\ \dfrac{2}{s+2} & \dfrac{7(-s+1)}{(s+1)(s+2)} \end{bmatrix}$$

Calculating $\det(G(s))$, we find:

$$\det(G(s)) = \frac{7(s-1)}{(s+1)(s+2)^2} - \frac{4}{(s+1)(s+2)} = \frac{3}{(s+1)(s+2)^2}(s-5)$$

We immediately see that the system has an RHP zero at $s = +5$ and therefore the system is a nonminimum-phase system.

18.3 Feedback Control of MIMO Systems: State-Space versus Transfer-Function Description of the Closed-Loop System

In this section, we will introduce feedback control of MIMO process systems. A standing assumption will be that the number of manipulated inputs is equal to the number of controlled outputs. We will first derive the state-space description of the closed-loop system, and then the transfer-function description.

Consider, for example, the mixing tank of Figure 18.1 whose state-space description is given by Eq. (18.1.8), the objective being to control the liquid level h to the set point h_{sp} and the temperature T to the set point T_{sp}. Suppose that we use two proportional controllers, one to control the level h by manipulating the volumetric flowrate F_H of the hot stream, and the other to control the temperature T by manipulating the volumetric flowrate F_C of the cold stream:

$$\overline{F}_H = k_{c,1}(\overline{h}_{sp} - \overline{h}) \tag{18.3.1}$$

$$\overline{F}_C = k_{c,2}(\overline{T}_{sp} - \overline{T}) \tag{18.3.2}$$

The feedback control system is shown in Figure 18.4. To obtain a state-space description of the closed-loop system, we substitute Eqs. (18.3.1) and (18.3.2) into (18.1.8) to obtain

$$\begin{bmatrix} \dfrac{d\overline{h}}{dt} \\ \dfrac{d\overline{T}}{dt} \end{bmatrix} = \begin{bmatrix} -\dfrac{1}{AR} & 0 \\ 0 & -\dfrac{1}{AR} \end{bmatrix} \begin{bmatrix} \overline{h} \\ \overline{T} \end{bmatrix} + \begin{bmatrix} \dfrac{1}{A} & \dfrac{1}{A} \\ \dfrac{T_H - T_s}{Ah_s} & \dfrac{T_C - T_s}{Ah_s} \end{bmatrix} \begin{bmatrix} k_{c,1}(\overline{h}_{sp} - \overline{h}) \\ k_{c,2}(\overline{T}_{sp} - \overline{T}) \end{bmatrix}$$

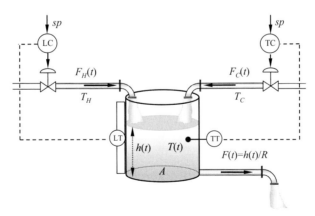

Figure 18.4 A mixing-tank system controlled by two P controllers.

or

$$\begin{bmatrix} \dfrac{d\bar{h}}{dt} \\ \dfrac{d\bar{T}}{dt} \end{bmatrix} = \begin{bmatrix} -\dfrac{1}{AR} & 0 \\ 0 & -\dfrac{1}{AR} \end{bmatrix} \begin{bmatrix} \bar{h} \\ \bar{T} \end{bmatrix} + \begin{bmatrix} \dfrac{1}{A} & \dfrac{1}{A} \\ \dfrac{T_H - T_s}{Ah_s} & \dfrac{T_C - T_s}{Ah_s} \end{bmatrix} \begin{bmatrix} k_{c,1} & 0 \\ 0 & k_{c,2} \end{bmatrix} \left(\begin{bmatrix} \bar{h}_{sp} \\ \bar{T}_{sp} \end{bmatrix} - \begin{bmatrix} \bar{h} \\ \bar{T} \end{bmatrix} \right)$$

and finally

$$\begin{bmatrix} \dfrac{d\bar{h}}{dt} \\ \dfrac{d\bar{T}}{dt} \end{bmatrix} = \begin{bmatrix} -\dfrac{1}{AR} - \dfrac{1}{A}k_{c,1} & -\dfrac{1}{A}k_{c,2} \\ -\dfrac{T_H - T_s}{Ah_s}k_{c,1} & -\dfrac{1}{AR} - \dfrac{T_C - T_s}{Ah_s}k_{c,2} \end{bmatrix} \begin{bmatrix} \bar{h} \\ \bar{T} \end{bmatrix} + \begin{bmatrix} \dfrac{1}{A}k_{c,1} & \dfrac{1}{A}k_{c,2} \\ \dfrac{T_H - T_s}{Ah_s}k_{c,1} & \dfrac{T_C - T_s}{Ah_s}k_{c,2} \end{bmatrix} \begin{bmatrix} \bar{h}_{sp} \\ \bar{T}_{sp} \end{bmatrix}$$

In the general case, where a proportional controller (in vector form):

$$u = K_c(y_{sp} - y) \tag{18.3.3}$$

is used to control a process with state-space description

$$\dfrac{dx}{dt} = Ax + Bu + B_w w$$
$$y = Cx \tag{18.3.4}$$

where, above, u represents the manipulated input vector and w represents the disturbance input vector. Then by substituting the control law, we obtain the closed-loop system

$$\dfrac{dx}{dt} = (A - BK_cC)x + BK_c y_{sp} + B_w w$$
$$y = Cx \tag{18.3.5}$$

If instead of a static controller (18.3.3), a dynamic controller is used, such as a multivariable PI controller, with a state-space description of the form

$$\dfrac{dx_c}{dt} = A_c x_c + B_c(y_{sp} - y)$$
$$u = C_c x_c + D_c(y_{sp} - y) \tag{18.3.6}$$

where the subscript c denotes the controller, then we can combine the state-space equations (18.3.4) and (18.3.6) as follows. We first substitute $y = Cx$ in the controller equation (18.3.6) and then u into the process equations (18.3.4) to obtain

$$\dfrac{dx}{dt} = (A - BD_cC)x + BC_c x_c + BD_c y_{sp} + B_w w$$
$$\dfrac{dx_c}{dt} = -B_c Cx + A_c x_c + B_c y_{sp}$$
$$y = Cx$$

The state-space description of the closed-loop system is then

$$\frac{d}{dt}\begin{bmatrix} x \\ x_c \end{bmatrix} = \begin{bmatrix} (A - BD_cC) & BC_c \\ -B_cC & A_c \end{bmatrix}\begin{bmatrix} x \\ x_c \end{bmatrix} + \begin{bmatrix} BD_c \\ B_c \end{bmatrix} y_{sp} + \begin{bmatrix} B_w \\ 0 \end{bmatrix} w \quad (18.3.7)$$

$$y = \begin{bmatrix} C & 0 \end{bmatrix}\begin{bmatrix} x \\ x_c \end{bmatrix}$$

When the process system is given in transfer-function matrix form

$$Y(s) = G(s)U(s) + G_w(s)W(s) \quad (18.3.8)$$

where the manipulated input vector u is of the same size as the output vector y, and w represents the vector of disturbances, and it is controlled by a controller described by

$$U(s) = G_c(s)\,(Y_{sp}(s) - Y(s)) \quad (18.3.9)$$

then the closed-loop system description is obtained by substituting Eq. (18.3.9) into (18.3.8) to eliminate $U(s)$:

$$Y(s) = G(s)G_c(s)\,(Y_{sp}(s) - Y(s)) + G_w(s)W(s)$$

from which we can solve for $Y(s)$ and obtain:

$$Y(s) = [I + G(s)G_c(s)]^{-1}\,G(s)G_c(s)Y_{sp}(s) + [I + G(s)G_c(s)]^{-1}\,G_w(s)W(s) \quad (18.3.10)$$

It is important to note at this point that the calculation of the closed-loop transfer-function matrices through Eq. (18.3.10) is extremely demanding in terms of the matrix algebra involved. However, the calculation can be efficiently performed using MATLAB or Maple, as will be seen later in this chapter.

Example 18.4 Closed-loop transfer-function matrix of a system of two interacting tanks

The system of two interacting tanks has been studied in Examples 18.1 and 18.2. More specifically, its state-space description was derived in Example 18.1 and its transfer function matrix was derived in Example 18.2. Suppose that a P controller with gain k_{c1} is used to control the liquid level h_1 in the first tank by manipulating the inlet flowrate $F_{in,1}$:

$$\bar{F}_{in,1} = k_{c1}(\bar{h}_{1,sp} - \bar{h}_1)$$

and a proportional controller with gain k_{c2} is used to control the liquid level h_2 in the second tank by manipulating the inlet flowrate $F_{in,2}$:

$$\bar{F}_{in,2} = k_{c2}(\bar{h}_{2,sp} - \bar{h}_2)$$

626 Multi-Input–Multi-Output Linear Systems

resulting in the following controller transfer-function matrix:

$$G_c(s) = \begin{bmatrix} k_{c1} & 0 \\ 0 & k_{c2} \end{bmatrix}$$

Derive the closed-loop system in transfer-function form and in state-space form. Assume that there are no external disturbances affecting the system.

Solution

We will derive the closed-loop transfer function using Eq. (18.3.10) and the results of Example 18.2. We first calculate $G(s)G_c(s)$:

$$G(s)G_c(s) = \begin{bmatrix} \dfrac{s+5}{(s+1)(s+7)} & \dfrac{4}{(s+1)(s+7)} \\ \dfrac{4}{(s+1)(s+7)} & \dfrac{2(s+3)}{(s+1)(s+7)} \end{bmatrix} \begin{bmatrix} k_{c1} & 0 \\ 0 & k_{c2} \end{bmatrix} = \begin{bmatrix} \dfrac{k_{c1}(s+5)}{(s+1)(s+7)} & \dfrac{4k_{c2}}{(s+1)(s+7)} \\ \dfrac{4k_{c1}}{(s+1)(s+7)} & \dfrac{2k_{c2}(s+3)}{(s+1)(s+7)} \end{bmatrix}$$

Then we calculate the matrix $(I + G(s)G_c(s))$ and its inverse, and finally obtain the closed-loop transfer-function matrix (after some lengthy algebra, which is omitted):

$$[I + G(s)G_c(s)]^{-1} G(s)G_c(s) = \dfrac{1}{s^2 + \beta s + \gamma} \begin{bmatrix} k_{c1}s + (5k_{c1} + 2k_{c1}k_{c2}) & 4k_{c2} \\ 4k_{c1} & 2k_{c2}s + (6k_{c2} + 2k_{c1}k_{c2}) \end{bmatrix}$$

where $\beta = 8 + k_{c1} + 2k_{c2}$ and $\gamma = 7 + 5k_{c1} + 6k_{c2} + 2k_{c1}k_{c2}$.

The state-space description of the process has been derived in Example 18.1 and is given by

$$\dfrac{d}{dt}\begin{bmatrix} \bar{h}_1 \\ \bar{h}_2 \end{bmatrix} = \begin{bmatrix} -3 & 2 \\ 4 & -5 \end{bmatrix}\begin{bmatrix} \bar{h}_1 \\ \bar{h}_2 \end{bmatrix} + \begin{bmatrix} 1 & 0 \\ 0 & 2 \end{bmatrix}\begin{bmatrix} F_{u1} \\ F_{u2} \end{bmatrix}$$

Using the equations of the proportional controllers and Eq. (18.3.5) we have

$$BK_c = \begin{bmatrix} 1 & 0 \\ 0 & 2 \end{bmatrix}\begin{bmatrix} k_{c1} & 0 \\ 0 & k_{c2} \end{bmatrix} = \begin{bmatrix} k_{c1} & 0 \\ 0 & 2k_{c2} \end{bmatrix}$$

$$A - BK_c C = \begin{bmatrix} -3 & 2 \\ 4 & -5 \end{bmatrix} - \begin{bmatrix} k_{c1} & 0 \\ 0 & 2k_{c2} \end{bmatrix}\begin{bmatrix} 1 & 0 \\ 0 & 1 \end{bmatrix} = \begin{bmatrix} -(3+k_{c1}) & 2 \\ 4 & -(5+2k_{c2}) \end{bmatrix}$$

The state-space description of the closed-loop system is therefore:

$$\dfrac{d}{dt}\begin{bmatrix} \bar{h}_1 \\ \bar{h}_2 \end{bmatrix} = \begin{bmatrix} -(3+k_{c1}) & 2 \\ 4 & -(5+2k_{c2}) \end{bmatrix}\begin{bmatrix} \bar{h}_1 \\ \bar{h}_2 \end{bmatrix} + \begin{bmatrix} k_{c1} & 0 \\ 0 & 2k_{c2} \end{bmatrix}\begin{bmatrix} \bar{h}_{1,sp} \\ \bar{h}_{2,sp} \end{bmatrix}$$

To investigate stability of the closed-loop system, we calculate the characteristic polynomial of the matrix $(A - BK_cC)$:

$$\det[\lambda I - (A - BK_cC)] = \det \begin{bmatrix} \lambda + (3 + k_{c1}) & -2 \\ -4 & \lambda + (5 + 2k_{c2}) \end{bmatrix}$$

$$= \lambda^2 + (8 + k_{c1} + 2k_{c2})\lambda + (7 + 5k_{c1} + 6k_{c2} + 2k_{c1}k_{c2})$$

From the Routh criterion it follows that $8 + k_{c1} + 2k_{c2} > 0$ and $7 + 5k_{c1} + 6k_{c2} + 2k_{c1}k_{c2} > 0$ are the stability conditions for the closed-loop system. Note that the characteristic polynomial of the closed-loop system matrix $(A - BK_cC)$ has exactly the same roots as the common denominator polynomial of the closed-loop transfer function matrix (which are the poles of the closed-loop system). The location of these roots in the complex plane determines the stability of the closed-loop system.

18.4 Interaction in MIMO Systems

Figure 18.5 shows a feedback control system for a two-input–two-output process, where controller $G_{c1}(s)$ is used to control y_1 by manipulating u_1 and controller $G_{c2}(s)$ is used to control y_2 by manipulating u_2, i.e. the controller transfer-function matrix is

$$G_c = \begin{bmatrix} G_{c1} & 0 \\ 0 & G_{c2} \end{bmatrix} \tag{18.4.1}$$

With this diagonal controller, the transfer-function matrix of the closed-loop system $G_{CL} = [I + GG_c]^{-1} GG_c$ takes the form:

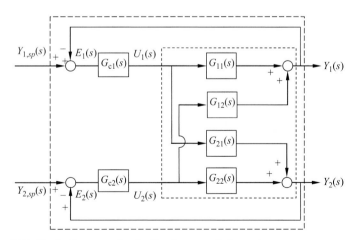

Figure 18.5 Block diagram of a feedback-controlled two-input–two-output system.

$$G_{CL} = \left(\begin{bmatrix} 1 & 0 \\ 0 & 1 \end{bmatrix} + \begin{bmatrix} G_{11} & G_{12} \\ G_{21} & G_{22} \end{bmatrix} \begin{bmatrix} G_{c1} & 0 \\ 0 & G_{c2} \end{bmatrix} \right)^{-1} \begin{bmatrix} G_{11} & G_{12} \\ G_{21} & G_{22} \end{bmatrix} \begin{bmatrix} G_{c1} & 0 \\ 0 & G_{c2} \end{bmatrix}$$

$$= \frac{1}{D} \begin{bmatrix} 1+G_{22}G_{c2} & -G_{12}G_{c2} \\ -G_{21}G_{c1} & 1+G_{11}G_{c1} \end{bmatrix} \begin{bmatrix} G_{11}G_{c1} & G_{12}G_{c2} \\ G_{21}G_{c1} & G_{22}G_{c2} \end{bmatrix}$$

or

$$G_{CL} = \frac{1}{D} \begin{bmatrix} G_{11}G_{c1} + (G_{11}G_{22} - G_{12}G_{21})G_{c1}G_{c2} & G_{12}G_{c2} \\ G_{21}G_{c1} & G_{22}G_{c2} + (G_{11}G_{22} - G_{12}G_{21})G_{c1}G_{c2} \end{bmatrix} \quad (18.4.2)$$

where

$$D = \det(I + GG_c) = (1+G_{11}G_{c1})(1+G_{22}G_{c2}) - G_{12}G_{21}G_{c1}G_{c2} \quad (18.4.3)$$

With the above result, we can now relate the system outputs to their set points via

$$\begin{bmatrix} Y_1 \\ Y_2 \end{bmatrix} = G_{CL} \begin{bmatrix} Y_{1,sp} \\ Y_{2,sp} \end{bmatrix}$$

i.e.

$$Y_1 = \frac{[G_{11}G_{c1} + (G_{11}G_{22} - G_{12}G_{21})G_{c1}G_{c2}]}{D} Y_{1,sp} + \frac{G_{12}G_{c2}}{D} Y_{2,sp} \quad (18.4.4a)$$

$$Y_2 = \frac{G_{21}G_{c1}}{D} Y_{1,sp} + \frac{[G_{22}G_{c2} + (G_{11}G_{22} - G_{12}G_{21})G_{c1}G_{c2}]}{D} Y_{2,sp} \quad (18.4.4b)$$

It is apparent from Eqs. (18.4.4a, b) that both outputs depend on both inputs. If the goal is to change the set point of any one of the controlled outputs, then both outputs will be affected. When one controller tries to change the set point of one of the outputs, this will affect the operation of the other controller that tries to maintain the other output at its set point. This characteristic of the MIMO systems is called interaction, and it complicates the analysis of the feedback control system and the design of successful controllers.

The same conclusions can be drawn by using the state-space description of the closed-loop system. To this end, we consider a system with two inputs u_1 and u_2, and two outputs y_1 and y_2, described by

$$\frac{dx}{dt} = Ax + b_1 u_1 + b_2 u_2$$
$$y_1 = c_1 x \quad (18.4.5)$$
$$y_2 = c_2 x$$

which is controlled by two proportional controllers

$$u_1 = k_{c,1}(y_{1,sp} - y_1) \quad (18.4.6a)$$

$$u_2 = k_{c,2}(y_{2,sp} - y_2) \quad (18.4.6b)$$

Then, the closed-loop system is described by

$$\frac{dx}{dt} = \left(A - b_1 k_{c,1} c_1 - b_2 k_{c,2} c_2\right) x + b_1 k_{c,1} y_{1,sp} + b_2 k_{c,2} y_{2,sp}$$

$$y_1 = c_1 x \tag{18.4.7}$$

$$y_2 = c_2 x$$

Again, we see that any change in one of the set points, $y_{1,sp}$ or $y_{2,sp}$, will affect the state vector and, consequently, both outputs.

In the exceptional case where the off-diagonal elements of the process transfer function are zero, i.e. $G_{21} = G_{21} = 0$ then the closed-loop transfer-function description (18.4.4a, b) simplifies to the following two noninteracting SISO transfer functions

$$Y_1 = \frac{G_{11} G_{c1}}{1 + G_{11} G_{c1}} Y_{1,sp} \tag{18.4.8a}$$

$$Y_2 = \frac{G_{22} G_{c2}}{1 + G_{22} G_{c2}} Y_{2,sp} \tag{18.4.8b}$$

In this uncommon situation, the SISO control system analysis and design methods that have been developed in the previous chapters are applicable.

In order to stress further the importance of interaction present in the MIMO systems we consider the system of the two interacting tanks shown in Figure 18.2. More specifically we consider the closed-loop system formed using two proportional SISO controllers as shown in Figure 18.6. The volumetric flowrate of the feed stream to each tank is used to control the liquid level in the same tank. The parameters of the system (A_1, A_2, R_1 and R_2) assume the specific values given in Figure 18.6 (the cross-sectional areas are given in m² and the resistances in (m³/min)/m). There is an important difference between the physical systems shown in Figures 18.2 and 18.6. In the system shown in Figure 18.6, a manual valve has been placed in the pipe that connects the two tanks. In this way, we can vary the resistance to flow by opening or closing the valve.

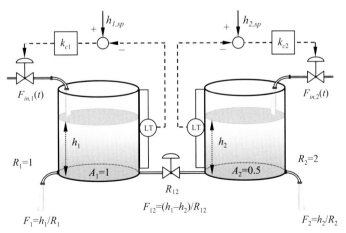

Figure 18.6 Control of the levels of two interacting liquid storage tanks.

We initially consider the case where we close the valve in which case $R_{12} \to \infty$. The model of the process that has been developed in Example 18.1 is valid, and when $R_{12} \to \infty$ the A matrix becomes

$$A = \begin{bmatrix} -\left(\dfrac{1}{A_1 R_1} + \dfrac{1}{A_1 R_{12}}\right) & \dfrac{1}{A_1 R_{12}} \\ \dfrac{1}{A_2 R_{12}} & -\left(\dfrac{1}{A_2 R_2} + \dfrac{1}{A_2 R_{12}}\right) \end{bmatrix} \to \begin{bmatrix} -\dfrac{1}{A_1 R_1} & 0 \\ 0 & -\dfrac{1}{A_2 R_2} \end{bmatrix} = \begin{bmatrix} -1 & 0 \\ 0 & -1 \end{bmatrix}$$

i.e. when the valve in the interconnecting pipe is closed, the interaction is eliminated.

The transfer-function matrix of the process, in the absence of interaction, can be calculated as follows:

$$G(s) = C(sI - A)^{-1} B + D = \begin{bmatrix} 1 & 0 \\ 0 & 1 \end{bmatrix} \begin{bmatrix} s+1 & 0 \\ 0 & s+1 \end{bmatrix}^{-1} \begin{bmatrix} 1 & 0 \\ 0 & 2 \end{bmatrix} + \begin{bmatrix} 0 & 0 \\ 0 & 0 \end{bmatrix}$$

$$= \begin{bmatrix} \dfrac{1}{s+1} & 0 \\ 0 & \dfrac{2}{s+1} \end{bmatrix} \qquad (18.4.9)$$

Based on this diagonal transfer function matrix, two SISO proportional controllers with $k_{c1} = k_{c2} = 10$ can be used to control each output independently. The responses of the closed-loop system to unit step changes in the set points are shown in Figure 18.7. We observe that the desired set-point changes are achieved, with some offset, as expected.

We now consider the case where the manually controlled valve in the interconnecting pipe is opened and the resistance to flow becomes $R_{12} = 0.5$. The corresponding process transfer function has been derived in Example 18.2. If the controllers remain the same as in the previous case, then the responses of the closed-loop system to unit step changes in the set points are as shown in Figure 18.8. We observe that the quality of the response deteriorates when interaction is present. Not only is the offset increased but also, due to interaction, both output variables are affected when a change in any of the set points takes place, and they deviate significantly from their set points. We therefore conclude that although the controller is successful in the absence of interaction, it becomes particularly unsuccessful when interaction is present. A more elaborate controller design method is necessary in order to build an effective controller that can alleviate the effects of interaction.

A final remark concerning the use of SISO controllers for MIMO systems, usually referred to as decentralized controllers, is related to the options available for pairing controlled and manipulated variables. For the case of two controlled variables (y_1 and y_2) and two manipulated variables (u_1 and u_2) there are two potential pairings available:

option 1: $u_1 \to y_1$ and $u_2 \to y_2$
option 2: $u_1 \to y_2$ and $u_2 \to y_1$

18.4 Interaction in MIMO Systems

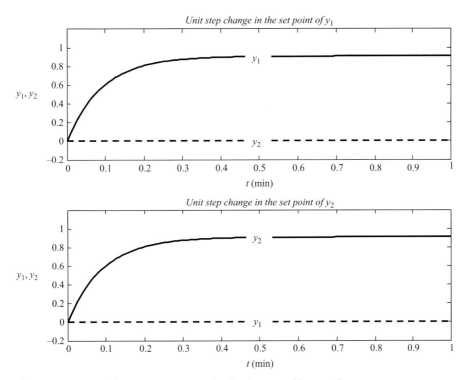

Figure 18.7 Closed-loop step responses in the absence of interaction.

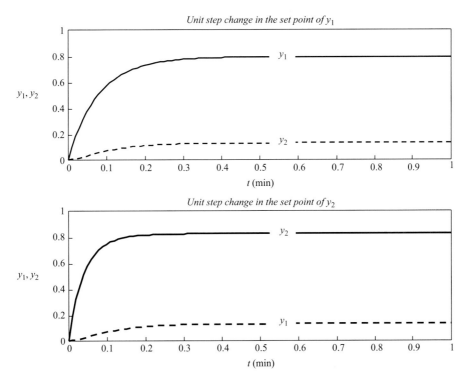

Figure 18.8 Closed-loop step responses in the presence of interaction.

The control engineer needs to choose the best pairing as a part of the controller design problem. The number of alternatives increases as the potential manipulated and controlled variables increase. In fact, the number of potential pairings for the case on n controlled and n manipulated variables is equal to $n!$ (n factorial). In simple words, this means that for a five-input–five-output system there are 120 potential alternative control structures. This creates a combinatorial problem that can be particularly difficult to solve. This problem is bypassed if a full non-diagonal controller is selected (such a controller is usually referred to as a centralized controller). However, in this case, a more elaborate controller design method will be needed, as the centralized controller generally involves a larger number of parameters to be selected.

18.5 Decoupling in MIMO Systems

The discussion of the previous section suggests that the use of decentralized controllers for MIMO process systems is desirable from the point of view of simplicity. However, because of interaction, a decentralized control system may suffer from loss of performance. There is a simple idea that can be useful in many cases to bypass the problem of interaction. The idea is based on the design of a precompensator $P(s)$ that is located between the controllers and the process, as shown in Figure 18.9. The role of the precompensator is to "remove" the interaction present between the process input and output variables. If the transfer-function matrix of the precompensator is $P(s)$, then the transfer-function matrix between the control signals and the process outputs will be:

$$Y(s) = G(s)P(s)U(s) \qquad (18.5.1)$$

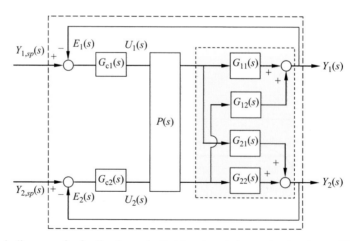

Figure 18.9 Block diagram of a feedback-controlled two-input–two-output system with precompensator.

18.5 Decoupling in MIMO Systems

The elements of the precompensator can be selected so that the off-diagonal elements of $G(s)P(s)$ are zero. For a two-input–two-output system, we have:

$$GP = \begin{bmatrix} G_{11} & G_{12} \\ G_{21} & G_{22} \end{bmatrix} \begin{bmatrix} P_{11} & P_{12} \\ P_{21} & P_{22} \end{bmatrix} = \begin{bmatrix} G_{11}P_{11} + G_{12}P_{21} & G_{11}P_{12} + G_{12}P_{22} \\ G_{21}P_{11} + G_{22}P_{21} & G_{21}P_{12} + G_{22}P_{22} \end{bmatrix} \quad (18.5.2)$$

If we specify the off-diagonal elements to be zero, we obtain the following two equations:

$$G_{11}P_{12} + G_{12}P_{22} = 0$$
$$G_{21}G_{11} + G_{22}P_{21} = 0 \quad (18.5.3)$$

To determine the precompensator, we need two additional specifications. A common choice is to select $P_{11} = P_{22} = 1$, and in this case

$$P_{12} = -\frac{G_{12}}{G_{11}}$$
$$P_{21} = -\frac{G_{21}}{G_{22}} \quad (18.5.4)$$

and

$$P = \begin{bmatrix} 1 & -\dfrac{G_{12}}{G_{11}} \\ -\dfrac{G_{21}}{G_{22}} & 1 \end{bmatrix} \quad (18.5.5)$$

To illustrate the idea, we consider the system of two liquid storage tanks that was studied in Examples 18.1 and 18.2. In Example 18.2, it was shown that the transfer-function matrix of the process is

$$G(s) = \begin{bmatrix} \dfrac{s+5}{(s+1)(s+7)} & \dfrac{4}{(s+1)(s+7)} \\ \dfrac{4}{(s+1)(s+7)} & \dfrac{2(s+3)}{(s+1)(s+7)} \end{bmatrix} \quad (18.5.6)$$

The precompensator that is appropriate for this system is calculated using Eq. (18.5.5) as follows

$$P = \begin{bmatrix} 1 & -\dfrac{\frac{4}{(s+1)(s+7)}}{\frac{s+5}{(s+1)(s+7)}} \\ -\dfrac{\frac{4}{(s+1)(s+7)}}{\frac{2(s+3)}{(s+1)(s+7)}} & 1 \end{bmatrix} = \begin{bmatrix} 1 & -\dfrac{4}{s+5} \\ -\dfrac{2}{s+3} & 1 \end{bmatrix} \quad (18.5.7)$$

The product $G(s)P(s)$ can also be calculated:

$$GP = \begin{bmatrix} \dfrac{s+5}{(s+1)(s+7)} & \dfrac{4}{(s+1)(s+7)} \\ \dfrac{4}{(s+1)(s+7)} & \dfrac{2(s+3)}{(s+1)(s+7)} \end{bmatrix} \begin{bmatrix} 1 & -\dfrac{4}{s+5} \\ -\dfrac{2}{s+3} & 1 \end{bmatrix}$$

$$= \begin{bmatrix} \dfrac{s+5}{(s+1)(s+7)} - \dfrac{8}{(s+1)(s+7)(s+3)} & 0 \\ 0 & \dfrac{2(s+3)}{(s+1)(s+7)} - \dfrac{16}{(s+1)(s+7)(s+5)} \end{bmatrix} \quad (18.5.8)$$

and it is indeed a diagonal transfer-function matrix. The design of the decentralized controllers can now be performed using the methods that have been developed in the previous chapters for SISO systems. It is interesting to note that if the controller is decentralized (diagonal transfer function matrix), then the same holds true for the product $G(s)P(s)G_c(s)$.

18.6 Software Tools

18.6.1 MIMO Systems Analysis using MATLAB

MATLAB handles MIMO systems in exactly the same way as it handles SISO systems. Therefore, all commands that we have used in the previous chapters can be used for MIMO systems as well. To demonstrate this fact, we will use the commands that we have introduced in Chapters 6, 8 and 11 to define a MIMO system and derive the state-space description and the transfer-function description of both the open-loop and the closed-loop system. We first build the A, B, C and D matrices of the two-tank system that we have studied in Examples 18.1–18.4:

```
>> A1=1;
>> A2=0.5;
>> R1=1;
>> R2=2;
>> R12=0.5
>> A=[-(1/(A1*R1)+1/(A1*R12))  1/(A1*R12);
      1/(A2*R12)  -(1/(A2*R2)+1/(A2*R12))];
>> B=[1/A1 0;0 1/A2];
>> C=eye(2);
>> D=zeros(2,2);
```

We then create the state-space description and transfer-function description

```
>> sys_ss=ss(A,B,C,D);
>> sys_tf=tf(sys_ss)
```

```
sys_tf =

  From input 1 to output ...
          s + 5
  1:   -------------
       s^2 + 8 s + 7

            4
  2:   -------------
       s^2 + 8 s + 7

  From input 2 to output ...
            4
  1:   -------------
       s^2 + 8 s + 7

          2 s + 6
  2:   -------------
       s^2 + 8 s + 7
```

Continuous-time transfer function.

We then define the state-space description of a two-input–two-output proportional controller with unity gain in both loops:

```
>> sys_C_ss=ss([],[],[],eye(2))

sys_C_ss =

  d =
        u1   u2
   y1   1    0
   y2   0    1
```

Static gain.

We then form the closed-loop system

```
>> sys_OL_ss=series(sys_C_ss,sys_ss);
>> sys_CL_ss=feedback(sys_OL_ss,eye(2))

sys_CL_ss =

  a =
        x1   x2
   x1   -4   2
   x2   4    -7
```

```
b =
         u1   u2
    x1    1    0
    x2    0    2

c =
         x1   x2
    y1    1    0
    y2    0    1

d =
         u1   u2
    y1    0    0
    y2    0    0
```

Continuous-time state-space model.

```
>> sys_CL_tf=tf(sys_CL_ss)

sys_CL_tf =

  From input 1 to output ...
           s + 7
   1:  ---------------
        s^2 + 11 s + 20

             4
   2:  ---------------
        s^2 + 11 s + 20

  From input 2 to output ...
             4
   1:  ---------------
        s^2 + 11 s + 20

           2 s + 8
   2:  ---------------
        s^2 + 11 s + 20
```

Continuous-time transfer function.

These results agree with the results derived in Examples 18.1–18.4. Asymptotic stability of the closed-loop system can be checked through its eigenvalues:

```
>> eig(sys_CL_ss.a)

ans =

    -2.2984
    -8.7016
```

18.6.2 MIMO System Analysis using Maple

MIMO linear systems can be handled with Maple in essentially the same way as SISO systems. The same type of calculations presented in Chapters 6 and 8 for open-loop dynamics, and in Chapters 11–13 for closed-loop dynamics, can be performed for MIMO systems using the `LinearAlgebra` package of Maple. Consider, for example, the two-tank liquid storage system studied in Examples 18.1, 18.3 and 18.4. After entering the process parameters and the matrices that define the system in state-space form,

```
> with(LinearAlgebra):
> A1:=1:A2:=1/2:R1:=1:R12:=1/2:R2:=2:
> A:=Matrix([[-1/(A1*R1)-1/(A1*R12),1/(A1*R12)],
  [1/(A2*R12),-1/(A2*R2)-1/(A2*R12)]]):
> B:=Matrix([[1/A1,0],[0,1/A2]]):
> C:=Matrix([[1,0],[0,1]]):
```

one can calculate the transfer-function matrix as

```
G:=Multiply(C,Multiply(MatrixInverse(s*IdentityMatrix(2)-A),B));
```

$$G := \begin{bmatrix} \dfrac{s+5}{s^2+8s+7} & \dfrac{4}{s^2+8s+7} \\ \dfrac{4}{s^2+8s+7} & \dfrac{2(s+3)}{s^2+8s+7} \end{bmatrix}$$

which is exactly the result of Example 18.2.

Considering now two decentralized proportional controllers as in Example 18.4, the closed-loop system can be obtained in state-space form by using Eq. (18.3.5) as follows. First, the proportional controller is defined in matrix form, and then the A and B matrices of the closed-loop system are defined:

```
> Kc:=Matrix([[kc1,0],[0,kc2]]):
> A_cl:=A-Multiply(B,Multiply(Kc,C)); B_cl:=Multiply(B,Kc);
```

$$A\_cl := \begin{bmatrix} -3-kc1 & 2 \\ 4 & -5-2kc2 \end{bmatrix}$$

$$B\_cl := \begin{bmatrix} kc1 & 0 \\ 0 & 2kc2 \end{bmatrix}$$

The characteristic polynomial of the closed-loop system is then immediately obtained

```
> CharacteristicPolynomial(A_cl,lambda);
```

$$\lambda^2 + (8+2kc2+kc1)\lambda + 2kc1\ kc2 + 5kc1 + 6kc2 + 7$$

as well as the closed-loop transfer-function matrix

```
> G_cl:=Multiply(C,Multiply(MatrixInverse(s*IdentityMatrix(2) -
A_cl),B_cl));
```

$$G\_cl := \begin{bmatrix} \dfrac{(s+5+2kc2)kc1}{2kc1\,kc2+kc1s+2kc2s+s^2+5kc1+6kc2+8s+7} & \dfrac{4kc2}{2kc1\,kc2+kc1s+2kc2s+s^2+5kc1+6kc2+8s+7} \\ \dfrac{4kc1}{2kc1\,kc2+kc1s+2kc2s+s^2+5kc1+6kc2+8s+7} & \dfrac{2(s+3+kc1)kc2}{2kc1\,kc2+kc1s+2kc2s+s^2+5kc1+6kc2+8s+7} \end{bmatrix}$$

which are exactly the results of Example 18.4. Alternatively, the transfer function matrix can be calculated by applying the formula of Eq. (18.3.10):

```
> G_cl:=Multiply(MatrixInverse(IdentityMatrix(2)+Multiply(G,Kc)),
Multiply(G,Kc));
```

leading to exactly the same result.

LEARNING SUMMARY

- The state-space description of MIMO systems is given by

$$\frac{dx}{dt} = Ax + Bu$$

$$y = Cx + Du$$

- The corresponding transfer-function matrix is given by

$$G(s) = C(sI - A)^{-1} B + D$$

- The poles of the transfer-function matrix are those values of s at which at least one of the entries of $G(s)$ becomes infinite. All poles are eigenvalues of the A matrix of the state-space description. The zeros of the transfer-function matrix are those values of s at which $\det[G(s)]$ is zero.
- When a controller with state-space representation

$$\frac{dx_c}{dt} = A_c x_c + B_c (y_{sp} - y)$$

$$u = C_c x_c + D_c (y_{sp} - y)$$

is applied, then the state-space description of the closed-loop system is

$$\frac{d}{dt}\begin{bmatrix} x \\ x_c \end{bmatrix} = \begin{bmatrix} (A - BD_c C) & BC_c \\ -B_c C & A_c \end{bmatrix} \begin{bmatrix} x \\ x_c \end{bmatrix} + \begin{bmatrix} BD_c \\ B_c \end{bmatrix} y_{sp}$$

$$y = \begin{bmatrix} C & 0 \end{bmatrix} \begin{bmatrix} x \\ x_c \end{bmatrix}$$

- A major challenge in designing feedback controllers for MIMO process systems is the presence of interactions between individual control loops.
- Decoupling precompensators can be used to eliminate interaction and simplify the controller design problem.

TERMS AND CONCEPTS

Decoupling. A controller design methodology that is based on the idea of first eliminating interaction, and then designing controllers independently for each SISO subsystem.

Multi-input–multi-output (MIMO) system. A system with many inputs and/or many outputs.

Poles of a MIMO system. Those values of s at which at least one of the entries of the transfer-function matrix $G(s)$ becomes infinite. All poles are eigenvalues of the A matrix of the state-space description from which $G(s)$ was generated.

Zeros of a MIMO system. Those values of s for which the transfer-function matrix $G(s)$ loses rank (or $\det[G(s)] = 0$).

FURTHER READING

A comprehensive discussion of the design methodologies and control design challenges for MIMO systems can be found in the following textbooks

Albertos, P. and Sala, A., *Multivariable Control Systems: An Engineering Approach*. London: Springer, 2004.

Skogestad, S. and Postlethwaite, I., *Multivariable Feedback Control*. New York: Wiley, 1996.

PROBLEMS

18.1 Consider the two-tank system shown in Figure P18.1. Derive the state-space description and the transfer-function matrix of the system.

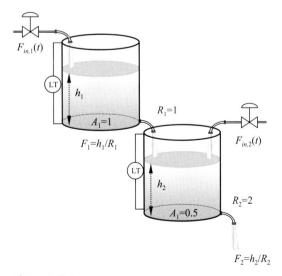

Figure P18.1

18.2 Consider the two-tank system shown in Figure P18.1. Calculate the response of the system (in deviation form) to a unit step change in $F_{in,1}$. Also, calculate the response to a unit step change in $F_{in,2}$.

18.3 The levels h_1 and h_2 of the two-tank system of Figure P18.1 need to be controlled to set point by manipulating the feed flowrates $F_{in,1}$ and $F_{in,2}$. Derive the state-space description and the transfer-function description of the closed-loop system formed when decentralized proportional controllers are used for this purpose.

18.4 Derive the necessary and sufficient conditions for closed-loop stability for the system of Problem 18.3. Pick an arbitrary pair of controller gains that satisfies the stability conditions and simulate the closed-loop response to a unit step change in the set point of the liquid level in the first tank.

18.5 A two-input–two-output process of the form (18.4.5) is controlled with two PI controllers: one controls y_1 by manipulating u_1, and the other controls y_2 by manipulating u_2. Derive the resulting closed-loop system in state-space form.

18.6 Apply your result in Problem 18.5 to the two-tank interacting system of Example 18.1. Simulate the closed-loop response to a unit step change in the set point of the liquid level of the first tank, when $k_{c1} = k_{c2} = 10$ and $\tau_{I1} = \tau_{I2} = 0.5$. Is the closed-loop system stable for these values of the controller parameters?

18.7 Consider the well-stirred mixing tank shown in Figure P18.7. The tank is fed with two inlet streams with variable volumetric flowrates $F_1(t)$ and $F_2(t)$ but constant concentrations c_1 and c_2 of an active compound. The outlet flowrate is $F(t)$ and the concentration $c(t)$. The outlet flowrate is proportional to the square root of the liquid level in the tank. Use the following steady-state information and parameter values: $V_s = 1$ m³, $F_{1s} = 0.05$ m³/s, $F_{2s} = 0.05$ m³/s, $c_1 = 1$ kmol/m³, $c_2 = 3$ kmol/m³, $c_s = 2$ kmol/m³, $A = 1$ m².
(a) Derive the nonlinear state-space model of the process.
(b) Linearize the nonlinear process model around the given steady-state conditions.
(c) Derive the transfer-function matrix of the process, with the outlet flowrate and concentration as outputs and the two inlet flowrates as inputs.
(d) Determine the poles and the zeros of the process transfer-function matrix.
(e) Derive the closed-loop system in state-space form and in transfer-function form, when decentralized proportional controllers are used.
(f) Choose any values of the controller gains for which the closed-loop system is stable and calculate the closed-loop responses to step changes by 0.01 units in the set points of the outlet flowrate and outlet concentration.

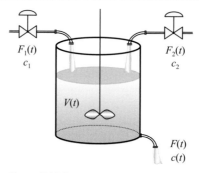

Figure P18.7

19 Synthesis of Model-Based Feedback Controllers

The aim of this chapter is to introduce key ideas and methods of model-based feedback control. A controller, which may or may not be PID, is built on the basis of a mathematical model of the process. The model-based controller is initially given in transfer-function form and subsequently in state-space form.

STUDY OBJECTIVES

After studying this chapter, you should be able to do the following.

- Apply the controller synthesis formula to derive the transfer function of the controller on the basis of a mathematical model of the process.
- Identify in which cases the model-based controller is PID and in which it is not.
- Understand the predictive capability of the model-based controller in the presence of process dead time.
- Evaluate the effect of modeling errors on closed-loop stability.
- Derive the model-based controller in state-space form.

19.1 Introduction

19.1.1 The Idea of Advanced Control

The term "**advanced control**" refers to control algorithms that are not necessarily PID or any of its variants. A control algorithm is considered to be "advanced":

- *either in the sense that* the mathematical function of the controller is intelligently chosen, so that the control algorithm is tailor-made for a given process. For this purpose, a dynamic model of the given process is used, and the resulting control algorithm is called "model-based";
- *or in the sense that* it uses more on-line information, in addition to the measurement of the controlled process output. For example, cascade and feedforward/feedback control systems utilize additional process measurements.

In the present chapter we will study model-based feedback controllers; in the next chapter we will study cascade and feedforward/feedback control systems.

In the previous chapters, the understanding was that the mathematical function of the controller is fixed (PID or any of its variants), and the objective was to make good choices of the controller parameters (controller gain, integral time, derivative time). Model-based control has a fundamentally different philosophy. The model of the process is considered known and the aim is to select the mathematical function of the controller (not necessarily PID), including its parameters, so as to achieve high performance of the control system.

Model-based control algorithms began being used from the 1970s when chemical plants started being equipped with computer hardware. Computers allowed the implementation of essentially any kind of control algorithm that the engineer had designed, not necessarily as simple as PID. However, the freedom of being able to use more sophisticated algorithms has created the need for appropriate theory that could guide building the control algorithm. In this chapter, we will give a brief elementary introduction to model-based control. As we will see, in some special cases, the mathematical form of the model-based controller will be PID, or similar to PID. However, in general, the model-based controller will be non-PID.

19.1.2 The General Framework

Consider the classical feedback control system shown in Figure 19.1 where

- $G(s)$, the process transfer function, is understood to include the actuator and the sensor dynamics (see Figure 11.1)
- $D(s)$ is understood to represent the overall effect of process disturbances on the controlled output $Y(s)$.

The objective is to select the controller transfer function $G_c(s)$, without restricting ourselves to the PID function. To be able to do that, we must first set the following specifications.

Closed-loop stability: What kind of controller transfer functions $G_c(s)$ guarantee closed-loop stability?

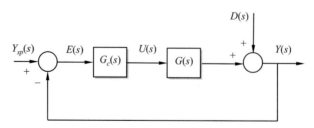

Figure 19.1 The classical structure of a feedback control system.

Zero offset: What kind of controller transfer functions $G_c(s)$ guarantee zero offset for step changes in the set point or the disturbance?

Dynamic performance: Out of those controller transfer functions that provide closed-loop stability and zero offset, which one leads to the best performance of the control system?

To provide answers to the above issues, the starting point will be the transfer-function description of the closed-loop system:

$$Y(s) = \frac{G(s)G_c(s)}{1+G(s)G_c(s)} Y_{sp}(s) + \frac{1}{1+G(s)G_c(s)} D(s) \qquad (19.1.1)$$

We will make the assumptions that

(1) the process $G(s)$ is a stable system (all poles have strictly negative real parts)
(2) the steady-state gain of the process is nonzero ($G(0) \neq 0$).

The question will be what choice of $G_c(s)$ meets the above specifications. To facilitate our analysis, we define the quantity $Q(s)$ as follows

$$Q(s) = \frac{G_c(s)}{1+G(s)G_c(s)} \qquad (19.1.2)$$

and then Eq. (19.1.1) takes the simpler form:

$$Y(s) = G(s)Q(s)Y_{sp}(s) + (1 - G(s)Q(s))D(s) \qquad (19.1.3)$$

If we can select a $Q(s)$ such that the closed-loop system (19.1.3) meets the specifications of stability, zero offset and optimal performance, then we can solve Eq. (19.1.2) to determine the controller transfer function $G_c(s)$:

$$G_c(s) = \frac{Q(s)}{1 - G(s)Q(s)} \qquad (19.1.4)$$

It is interesting to note at this point that with $G_c(s)$ given by Eq. (19.1.4), the block diagram of Figure 19.1 may be alternatively represented as shown in Figure 19.2. The two block diagrams are completely equivalent in the sense that they give rise to the same closed-loop transfer functions, the only difference being the way that the controller is described. However, Figure 19.2 reveals some interesting aspects of the controller representation (19.1.4). The controller incorporates an on-line simulation of the process model $G(s)$, concurrently with the process, the output of the model simulation is subtracted from the process output measurement, and the difference is fed back. In this way, it is the effect of disturbances that is fed back, and the control system, through $Q(s)$, must apply the appropriate corrections.

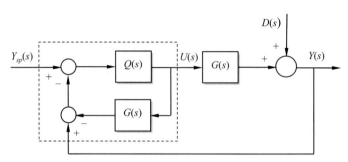

Figure 19.2 Alternative representation of the feedback control system when the controller is given by Eq. (19.1.4).

19.1.3 The Class of Stabilizing Controllers

We will now derive the working conditions under which the closed-loop system in Figures 19.1 or 19.2 will be stable. We recall that we have restricted ourselves to open-loop stable processes. As $G(s)$ is stable, we may conclude that both closed-loop transfer functions $G(s)Q(s)$ and $(1-G(s)Q(s))$ in Eq. (19.1.3) will be stable as long as the function $Q(s)$ is a stable function. The conclusion from the above simple observation is a very important one.

> *Any controller transfer function of the form given by Eq. (19.1.4), with $Q(s)$ stable, guarantees stability of the closed-loop system.*

It is important to note that, unlike the PID controller whose free parameters (k_c, τ_I, τ_D) are numbers, Eq. (19.1.4) defines a large class of controllers whose free "parameter" is an arbitrary stable function $Q(s)$. By design, this type of controller guarantees closed-loop stability for a process with transfer function $G(s)$. In the literature, Eq. (19.1.4) is referred to as the *Q-parameterization* of stabilizing controllers for a given stable process $G(s)$. We now proceed by adding additional constraints to the members of this class of controllers, so as to come up with the most satisfactory controller within this class.

19.1.4 The Class of Stabilizing Controllers that Achieve Zero Offset

For step changes in the set point or the disturbance, applying the final value theorem to (19.1.3) leads to:

$$\lim_{t\to\infty} y(t) = G(0)Q(0)[\lim_{t\to\infty} y_{sp}(t)] + (1 - G(0)Q(0))[\lim_{t\to\infty} d(t)] \tag{19.1.5}$$

For zero offset, we must have $\lim_{t\to\infty} y(t) = \lim_{t\to\infty} y_{sp}(t)$. This will hold true as long as $G(0)Q(0)=1$, or equivalently $Q(0) = \dfrac{1}{G(0)}$.

Thus, so far we know that every controller transfer function of the form given by Eq. (19.1.4) with $Q(s)$ stable and $Q(0)=1/G(0)$, guarantees stability of the closed-loop system and achieves zero offset.

19.1.5 Optimal Dynamic Performance

Out of the class of controller transfer functions that guarantee closed-loop stability and zero offset, we would like to select the one that has the best possible dynamic performance. One meaningful criterion of performance that can form the basis of selection is the minimization of the integral of the squared error (ISE):

$$ISE = \int_0^\infty [e(t)]^2 \, dt \tag{19.1.6}$$

Considering unit step changes in the set point and/or the disturbance, and using Eq. (19.1.3), we have

$$E(s) = (1 - Q(s)G(s))\frac{1}{s} \tag{19.1.7}$$

and

$$e(t) = \mathcal{L}^{-1}\left\{(1 - Q(s)G(s))\frac{1}{s}\right\} \tag{19.1.8}$$

Thus, the problem may be posed as that of minimizing the ISE criterion defined by Eqs. (19.1.6) and (19.1.8), over the entire class of stable functions $Q(s)$. This is a challenging mathematical problem. But it has been solved and the answer is quite easy to compute. To be able to write down a formula for the solution to the optimization problem, we must first factorize the stable transfer function $G(s)$, along the lines of the factorization that we have used to calculate frequency response and sketch the Bode diagrams: the transfer function is a product of gain, lag elements, lead elements, all-pass elements, and/or dead time. In particular, the factorization needed here is of the following form:

$$G(s) = G^+(s)\, G^-(s) \tag{19.1.9}$$

where

- $G^+(s)$ contains all the first-order, second-order all-pass and dead-time factors, if any, otherwise it is equal to 1. Hence it has unity amplitude ratio: $|G^+(i\omega)| = 1$ at all frequencies;
- $G^-(s)$ contains the gain and all lag and/or lead factors. Hence it has all its poles and all its zeros with negative real parts, and no dead time.

$G^-(s)$ is called the *minimum-phase factor* of $G(s)$, since it excludes those elements that generate excess phase (i.e. right-half-plane zeros and dead time), and its amplitude ratio agrees with $G(s)$. With the above factorization of the process transfer function, the solution of the ISE optimization problem can be written as follows:

$$Q^{opt}(s) = \frac{1}{G^-(s)} \qquad (19.1.10)$$

Note that, because $G^-(s)$ has all its zeros with negative real parts, its inverse is stable, therefore it is a choice of $Q(s)$ that leads to closed-loop stability. Also, because $G^+(0) = 1$, it follows that $Q(0) = 1/G^-(0) = 1/G(0)$, leading to zero offset for step changes in set point or disturbance. Under the optimal choice of $Q(s)$ that gives rise to minimal ISE, the corresponding controller transfer function is obtained by substituting Eq. (19.1.10) into Eq. (19.1.4) to obtain:

$$G_c(s) = \frac{1}{1 - G^+(s)} \cdot \frac{1}{G^-(s)} \qquad (19.1.11)$$

and the resulting transfer-function description of the closed-loop system is:

$$Y^{opt}(s) = G^+(s) Y_{sp}(s) + (1 - G^+(s)) D(s) \qquad (19.1.12)$$

Example 19.1 A stable process $G(s)$ without right-half-plane zeros and without dead time

When a stable process has no zeros with positive real part and no dead time, it is called a "*minimum-phase process*", since its factorization does not contain any elements that generate excess phase. Calculate the optimal $Q(s)$ and the corresponding optimal feedback controller $G_c(s)$ for a minimum-phase process. Is ISE-optimal performance feasible? If not, could it be approximately achieved?

Solution

For a minimum-phase process, $G^+(s) = 1$ and $G^-(s) = G(s)$. Therefore, ISE is optimized by choosing $Q(s)$ to be equal to:

$$Q^{opt}(s) = \frac{1}{G^-(s)} = \frac{1}{G(s)}$$

and (19.1.12) gives the optimal closed-loop system's transfer-function description: $Y(s) = Y_{sp}(s)$ for every s, which means *perfect control* (y identically equal to y_{sp}). However, when we calculate the corresponding controller transfer function,

$$G_c^{opt}(s) = \frac{1}{1 - G^+(s)} \cdot \frac{1}{G^-(s)} = \frac{1}{1-1} \cdot \frac{1}{G(s)} = \frac{1}{0} \cdot \frac{1}{G(s)}!$$

Of course, the above infinite controller cannot be realized, and perfect control is not possible to achieve. However, perfect control is a theoretical limit which, even though it is unattainable, any control system could try to get reasonably close to it. We must, therefore, back off from the idea of achieving perfect control and accept a nearly optimal, but physically realizable, controller.

In this direction, one may try using a $Q(s)$ that is an approximate inverse of $G(s)$, instead of the exact inverse. For example, one might try a filtered inverse

$$Q(s) = Q^{opt}(s)F(s) = \frac{1}{G(s)} \cdot F(s) \tag{19.1.13}$$

where

$$F(s) = \frac{1}{\lambda s + 1} \tag{19.1.14}$$

is a first-order filter, and $\lambda > 0$ is a small time constant. This leads to the controller

$$G_c(s) = \frac{Q(s)}{1 - G(s)Q(s)} = \frac{\frac{1}{G(s)} \cdot F(s)}{1 - G(s) \cdot \frac{1}{G(s)} \cdot F(s)} = \frac{F(s)}{1 - F(s)} \cdot \frac{1}{G(s)} \tag{19.1.15}$$

hence, for a first-order filter,

$$G_c(s) = \frac{\frac{1}{\lambda s + 1}}{1 - \frac{1}{\lambda s + 1}} \cdot \frac{1}{G(s)} = \frac{1}{\lambda s} \cdot \frac{1}{G(s)} \tag{19.1.16}$$

which is finite and results in the closed-loop transfer functions:

$$Y(s) = F(s)Y_{sp}(s) + (1 - F(s))D(s) = \frac{1}{\lambda s + 1}Y_{sp}(s) + \frac{\lambda s}{\lambda s + 1}D(s) \tag{19.1.17}$$

Equation (19.1.17) does not correspond to perfect control but, in the limit as $\lambda \to 0$, then $Y(s) \to Y_{sp}(s)$ and the response tends to the theoretical limit of perfect control.

Example 19.2 A stable process $G(s)$ with a right-half-plane zero and without dead time

Consider the following nonminimum phase-transfer function that has a positive zero:

$$G(s) = \frac{1 - s}{(s + 2)(s + 3)(s + 4)}$$

Calculate the optimal $Q(s)$ and the corresponding optimal feedback controller $G_c(s)$. Is ISE-optimal performance feasible? If not, could it be approximately achieved?

Solution
The factorization is as follows: $G^+(s) = \dfrac{1-s}{s+1}$ and $G^-(s) = \dfrac{s+1}{(s+2)(s+3)(s+4)}$.

Therefore, ISE is optimized by choosing $Q(s)$ to be equal to:

$$Q^{opt}(s) = \frac{1}{G^-(s)} = \frac{(s+2)(s+3)(s+4)}{s+1}$$

and the corresponding controller transfer function is

$$G_c^{opt}(s) = \frac{1}{1-G^+(s)} \cdot \frac{1}{G^-(s)} = \frac{1}{1-\frac{1-s}{s+1}} \cdot \frac{(s+2)(s+3)(s+4)}{s+1} = \frac{(s+2)(s+3)(s+4)}{2s}$$

leading to the following optimal closed-loop system:

$$Y^{opt}(s) = G^+(s)Y_{sp}(s) + (1-G^+(s))D(s) = \frac{1-s}{s+1}Y_{sp}(s) + \frac{2s}{s+1}D(s)$$

The above expression describes the optimal closed-loop response, which is not even close to being perfect control. It is not possible to achieve anything better than the above optimal closed-loop response, which represents an upper limit of performance. It is the presence of the positive zero in the process transfer function that has imposed limitations on achievable dynamic performance. But even this nonperfect optimal response is not really achievable. Notice that both $Q^{opt}(s)$ and $G_c^{opt}(s)$ are improper transfer functions that cannot be exactly realized. The situation is very similar to PD and PID controller transfer functions, which are not proper: a filter must be applied in series with the ideal PD or PID transfer function, to make the overall transfer function realizable. Here, the degree of the numerator transfer function is three and that of the denominator is one. Therefore, we need to add a second-order transfer function in the denominator to make the controller implementable:

$$Q(s) = Q^{opt}(s)F(s) = \frac{(s+2)(s+3)(s+4)}{s+1} \cdot \frac{1}{(\lambda s+1)^2}$$

where λ is a small filter time constant.

19.2 Nearly Optimal Model-Based Controller Synthesis

Motivated by the foregoing examples, we will now present a model-based controller synthesis methodology that is trying to approximate the (perfect or not) optimal response. This methodology was originally developed in the late 1980s by M. Morari and coworkers, and has had many industrial applications. Consider a general process model of the form:

$$G(s) = \frac{N(s)}{D(s)}e^{-\theta s} \qquad (19.2.1)$$

where $\theta \geq 0$ is the dead time, $N(s)$ is the numerator polynomial and $D(s)$ is the denominator polynomial. We denote by $r \geq 0$ the relative order, defined as the difference between the degree of the denominator polynomial and the degree of the numerator polynomial

$$r = \deg\{D(s)\} - \deg\{N(s)\} \tag{19.2.2}$$

Also, consider the factorization of the process model:

$$G(s) = G^+(s)\, G^-(s) \tag{19.2.3}$$

where

- $G^+(s)$ contains all right-half-plane zeros and dead time present in $G(s)$ and $|G^+(i\omega)| = 1$ at all frequencies
- $G^-(s)$ contains all left-half-plane poles and zeros.

Then, one could use the optimal $Q(s)$ in series with an r-th-order filter

$$Q(s) = Q^{opt}(s) F(s) = \frac{F(s)}{G^-(s)} \tag{19.2.4}$$

where

$$F(s) = \frac{1}{(\lambda s + 1)^r} \tag{19.2.5}$$

and where $\lambda > 0$ is a small time constant. In this way, $Q(s)$ is a stable and proper rational function for which, in the limit as $\lambda \to 0$, $Q(s) \to Q^{opt}(s) = \frac{1}{G^-(s)}$. With this choice of approximately optimal $Q(s)$, the corresponding controller transfer function is:

$$G_c(s) = \frac{F(s)}{1 - G^+(s) F(s)} \cdot \frac{1}{G^-(s)} = \frac{1}{(\lambda s + 1)^r - G^+(s)} \cdot \frac{1}{G^-(s)} \tag{19.2.6}$$

and the resulting closed-loop system's transfer-function description is:

$$Y(s) = G^+(s) F(s) Y_{sp}(s) + \left(1 - G^+(s) F(s)\right) D(s)$$
$$= \frac{G^+(s)}{(\lambda s + 1)^r} Y_{sp}(s) + \left(1 - \frac{G^+(s)}{(\lambda s + 1)^r}\right) D(s) \tag{19.2.7}$$

which tends to the optimal closed-loop system $Y^{opt}(s) = G^+(s) Y_{sp}(s) + \left(1 - G^+(s)\right) D(s)$ in the limit as $\lambda \to 0$. Formula (19.2.6) is often called the *model-based controller synthesis formula*. It is an easy-to-use formula that directly provides the transfer function of a nearly optimal feedback controller, given the process model.

650 Synthesis of Model-Based Feedback Controllers

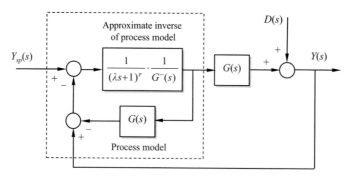

Figure 19.3 Internal Model Controller (IMC) structure of a nearly optimal feedback control system.

Remembering that the controller transfer function (19.2.6) has been generated by the representation (19.1.4) with $Q(s)$ given by (19.2.4) and (19.2.5), the corresponding feedback control system can be equivalently represented as shown in Figure 19.3. The feedback controller (19.2.6) is also referred to as the *Internal Model Controller* or IMC. The controller includes a copy of the process model, as well as an approximate inverse of the process model, connected together as shown in Figure 19.3.

19.3 Controller Synthesis for Low-Order Models

19.3.1 First-Order Process

We first consider the first-order process

$$G(s) = \frac{k}{\tau s + 1}, \quad \tau > 0 \tag{19.3.1}$$

For this system, the factorization is

$$G^+(s) = 1, \quad G^-(s) = G(s) \tag{19.3.2}$$

the relative order is $r = 1$, and the controller transfer function can be calculated from Eq. (19.2.6) as follows:

$$G_c(s) = \frac{1}{(\lambda s + 1)^r - G^+(s)} \cdot \frac{1}{G^-(s)} = \frac{1}{\lambda s} \cdot \frac{\tau s + 1}{k} = \frac{1}{k} \cdot \frac{\tau}{\lambda} \left(1 + \frac{1}{\tau s} \right) \tag{19.3.3}$$

This is the transfer function of a PI controller with controller gain and integral time given by

$$k_c = \frac{1}{k} \cdot \frac{\tau}{\lambda}, \quad \tau_I = \tau \tag{19.3.4}$$

We observe that the integral time is equal to the time constant of the system, and the controller gain is equal to the inverse of the process gain multiplied by the ratio of the system time constant and the filter time constant λ. The resulting closed-loop system follows:

$$Y(s) = \frac{1}{(\lambda s + 1)} Y_{sp}(s) + \frac{\lambda s}{(\lambda s + 1)} D(s) \quad (19.3.5)$$

In the limit as $\lambda \to 0$, the system output becomes $Y(s) \to Y_{sp}(s)$, i.e. perfect control is achieved in the limit as the controller gain becomes infinite.

19.3.2 Second-Order Process

We now consider the second-order process

$$G(s) = \frac{k}{\tau^2 s^2 + 2\zeta\tau s + 1}, \quad \tau > 0, \ \zeta > 0 \quad (19.3.6)$$

For this system, the factorization is

$$G^+(s) = 1, \ G^-(s) = G(s) \quad (19.3.7)$$

the relative order is $r = 2$, and the controller transfer function can be calculated from Eq. (19.2.6) as follows:

$$G_c(s) = \frac{1}{\left[(\lambda s + 1)^2 - 1\right]} \cdot \frac{(\tau^2 s^2 + 2\zeta\tau s + 1)}{k} = \frac{1}{2\lambda s\left(\frac{\lambda}{2}s + 1\right)} \cdot \frac{(\tau^2 s^2 + 2\zeta\tau s + 1)}{k}$$

$$= \frac{1}{2\lambda k} \cdot \left(\tau^2 s + 2\zeta\tau + \frac{1}{s}\right) \cdot \frac{1}{\left(\frac{\lambda}{2}s + 1\right)} \quad (19.3.8)$$

$$= \frac{\zeta\tau}{\lambda k} \cdot \left(1 + \frac{1}{2\zeta\tau} \cdot \frac{1}{s} + \frac{\tau}{2\zeta}s\right) \cdot \frac{1}{\left(\frac{\lambda}{2}s + 1\right)}$$

This is a PID controller in series with a first-order filter and where the controller gain, integral time and derivative time are given by

$$k_c = \frac{\zeta}{k} \cdot \frac{\tau}{\lambda}, \ \tau_I = 2\zeta\tau, \ \tau_D = \frac{\tau}{2\zeta} \quad (19.3.9)$$

The resulting closed-loop system follows:

$$Y(s) = \frac{1}{(\lambda s + 1)^2} Y_{sp}(s) + \frac{(\lambda s + 2)\lambda s}{(\lambda s + 1)^2} D(s) \quad (19.3.10)$$

In the limit as $\lambda \to 0$, the system output becomes $Y(s) \to Y_{sp}(s)$, i.e. perfect control is again achieved in the limit as the controller gain becomes infinite.

We will now consider a representative example of a nonminimum-phase system that has a positive zero.

19.3.3 A Second-Order Process with a Positive Zero

We now consider the second-order process with a positive zero

$$G(s) = k \frac{(-\tau_0 s + 1)}{(\tau_1 s + 1)(\tau_2 s + 1)}, \quad \tau_0 > 0, \ \tau_1 > 0, \ \tau_2 > 0 \tag{19.3.11}$$

For this system, the factorization is

$$G^+(s) = \frac{-\tau_0 s + 1}{\tau_0 s + 1}, \quad G^-(s) = k \frac{(\tau_0 s + 1)}{(\tau_1 s + 1)(\tau_2 s + 1)} \tag{19.3.12}$$

the relative order is $r = 1$, and the controller transfer function is calculated from Eq. (19.2.6):

$$G_c(s) = \frac{(\tau_1 s + 1)(\tau_2 s + 1)}{\left(\lambda \tau_0 s^2 + (2\tau_0 + \lambda)s\right)k}$$

$$= \frac{\tau_1 + \tau_2}{(2\tau_0 + \lambda)k} \cdot \left(1 + \frac{1}{(\tau_1 + \tau_2)} \cdot \frac{1}{s} + \frac{\tau_1 \tau_2}{\tau_1 + \tau_2} s\right) \cdot \frac{1}{\left(\frac{\lambda \tau_0}{2\tau_0 + \lambda} s + 1\right)} \tag{19.3.13}$$

This is again a PID controller in series with a first-order filter, where the controller gain, integral time and derivative time are given by

$$k_c = \frac{\tau_1 + \tau_2}{(2\tau_0 + \lambda)k}, \quad \tau_I = \tau_1 + \tau_2, \quad \tau_D = \frac{\tau_1 \tau_2}{\tau_1 + \tau_2} \tag{19.3.14}$$

The resulting closed-loop system follows:

$$Y(s) = \frac{1}{(\lambda s + 1)}\left(\frac{-\tau_0 s + 1}{\tau_0 s + 1}\right) Y_{sp}(s) + \frac{(\lambda \tau_0 s + 2\tau_0 + \lambda)s}{(\lambda s + 1)(\tau_0 s + 1)} D(s) \tag{19.3.15}$$

In the limit as $\lambda \to 0$, the system output becomes

$$Y(s) = \left(\frac{-\tau_0 s + 1}{\tau_0 s + 1}\right) Y_{sp}(s) + \frac{2\tau_0 s}{(\tau_0 s + 1)} D(s) \tag{19.3.16}$$

and clearly $Y(s) \neq Y_{sp}(s)$, i.e. perfect control cannot be achieved due to the presence of the positive zero.

For comparison purposes we note that, if the zero is negative,

$$G(s) = k \frac{(\tau_0 s + 1)}{(\tau_1 s + 1)(\tau_2 s + 1)}, \quad \tau_0 > 0, \ \tau_1 > 0, \ \tau_2 > 0 \tag{19.3.17}$$

then the controller has the following transfer function

$$G_c(s) = \frac{1}{\lambda s} \cdot \frac{(\tau_1 s + 1)(\tau_2 s + 1)}{k(\tau_0 s + 1)} = \frac{\tau_1 + \tau_2}{\lambda k} \cdot \left(1 + \frac{1}{(\tau_1 + \tau_2)} \cdot \frac{1}{s} + \frac{\tau_1 \tau_2}{\tau_1 + \tau_2} s\right) \cdot \frac{1}{(\tau_0 s + 1)} \quad (19.3.18)$$

which is again a PID controller. The resulting closed-loop response is given by

$$Y(s) = \frac{1}{(\lambda s + 1)} Y_{sp}(s) + \frac{\lambda s}{(\lambda s + 1)} D(s) \quad (19.3.19)$$

and perfect control is achieved in the limit as $\lambda \to 0$.

In all the cases considered so far, a PI or PID controller in series with a first-order filter have been obtained. This is not always the case; this happens only for first- and second-order systems. In general, the model-based controller is not PID.

19.3.4 A Third-Order Process

We now consider the third-order process

$$G(s) = \frac{k}{(\tau_1 s + 1)(\tau_2 s + 1)(\tau_3 s + 1)}, \quad \tau_1 > 0, \ \tau_2 > 0, \ \tau_3 > 0 \quad (19.3.20)$$

For this system, the factorization is

$$G^+(s) = 1, \ G^-(s) = G(s) \quad (19.3.21)$$

the relative order is $r = 3$, and the controller transfer function is calculated from Eq. (19.2.6):

$$G_c(s) = \frac{1}{k} \cdot \frac{(\tau_1 s + 1)(\tau_2 s + 1)(\tau_3 s + 1)}{(\lambda^3 s^3 + 3\lambda^2 s^2 + 3\lambda s)} \quad (19.3.22)$$

The structure of the controller does not resemble that of a classical PID controller or any of its variants. The closed-loop system is described by

$$Y(s) = \frac{1}{(\lambda s + 1)^3} Y_{sp}(s) + \frac{(\lambda^3 s^3 + 3\lambda^2 s^2 + 3\lambda s)}{(\lambda s + 1)^3} D(s) \quad (19.3.23)$$

Another case of an important system that does not result in a PID controller follows.

19.3.5 First-Order System with Dead Time

We now consider the case of a first-order plus dead time (FOPDT) system

$$G(s) = \frac{k}{\tau s + 1} e^{-\theta s}, \quad \tau > 0, \ \theta > 0 \quad (19.3.24)$$

For this system, the factorization is

$$G^+(s) = e^{-\theta s}, \quad G^-(s) = \frac{k}{\tau s + 1} \tag{19.3.25}$$

the relative order is $r = 1$, and the controller transfer function can be calculated from Eq. (19.2.6):

$$G_c(s) = \frac{1}{\lambda s + 1 - e^{-\theta s}} \cdot \frac{\tau s + 1}{k} \tag{19.3.26}$$

which does not have the structure of a PID controller. The resulting closed-loop system follows:

$$Y(s) = \frac{e^{-\theta s}}{\lambda s + 1} Y_{sp}(s) + \left(1 - \frac{e^{-\theta s}}{\lambda s + 1}\right) D(s) \tag{19.3.27}$$

In the limit as $\lambda \to 0$, the system output becomes

$$Y(s) = e^{-\theta s} Y_{sp}(s) + (1 - e^{-\theta s}) D(s) \tag{19.3.28}$$

which are the ISE-optimal transfer functions. The closed-loop system inherits the dead time of the process. Dead time cannot be eliminated by the controller, and, similarly to a right-half-plane zero, it poses limits on achievable closed-loop performance.

When the dead time θ is small relative to the time constant τ, it is possible to do a first-order Padé approximation of the dead time

$$G(s) = \frac{k}{\tau s + 1} e^{-\theta s} \approx \frac{k}{\tau s + 1} \left(\frac{-\frac{\theta}{2} s + 1}{\frac{\theta}{2} s + 1} \right) \tag{19.3.29}$$

in which case using the results of Section 19.3.3, we may obtain a PID controller with a filter by setting $\tau_0 = \tau_2 = \theta/2$:

$$G_c(s) = \frac{\tau + \frac{\theta}{2}}{(\theta + \lambda)k} \cdot \left(1 + \frac{1}{\left(\tau + \frac{\theta}{2}\right)} \cdot \frac{1}{s} + \frac{\tau \theta}{2\tau + \theta} s\right) \cdot \frac{1}{\left(\frac{\lambda \theta}{2(\theta + \lambda)} s + 1\right)} \tag{19.3.30}$$

When θ is extremely small compared to τ, we might even totally neglect the dead time θ and then Eqs. (19.3.26) or (19.3.30) simplify to a PI controller:

$$G_c(s) = \frac{\tau}{\lambda k} \cdot \left(1 + \frac{1}{\tau s}\right) \tag{19.3.31}$$

The same result would have been obtained by using the results of Section 19.3.1.

Example 19.3 Control of a first-order plus dead time (FOPDT) process – comparison of the model-based controller with its approximations

Consider the FOPDT system

$$G(s) = \frac{e^{-\theta s}}{s+1}$$

For a unit step change in the set point, compare the performance of the model-based controller (19.3.26) to its PID and PI approximations given by Eqs. (19.3.30) and (19.3.31), for the following cases: (a) $\theta = 0.01$, (b) $\theta = 0.1$ and (c) $\theta = 1$.

In all cases, use a filter time constant of $\lambda = 0.1$.

Solution

To solve the problem, we use MATLAB to perform the calculations:

```
% process and filter parameters
K=1; tau=1; theta=0.01; lambda=0.1;
G=tf(K,[tau 1],'InputDelay',theta);

% model-based controller
Q=tf([tau 1],[lambda*K K]);
Gc=feedback(Q,-G);
GCL=feedback(series(Gc,G),1);

% PID approximation of model based controller
GcPID=tf(conv([tau 1],[theta/2 1]),[K*lambda*theta/2 K*(lambda+theta) 0]);
GCLPID=feedback(series(GcPID,G),1);

% PI approximation of model based controller
GcPI=tf([tau 1],[K*lambda 0]);
GCLPI=feedback(series(GcPI,G),1);

% simulate step response
step(GCL)
hold on
step(GCLPID)
hold on
step(GCLPI)
```

The results are shown in Figure 19.4. We see that, for $\theta = 0.01$, which is two orders of magnitude smaller than the process time constant, all three controllers perform equally well, with the responses being essentially on top of each other, and following the closed-loop transfer function in Eq. (19.3.27). For $\theta = 0.1$, which is one order of magnitude smaller than the process time constant, the PI approximation is not performing well, but the other two controllers are essentially on top

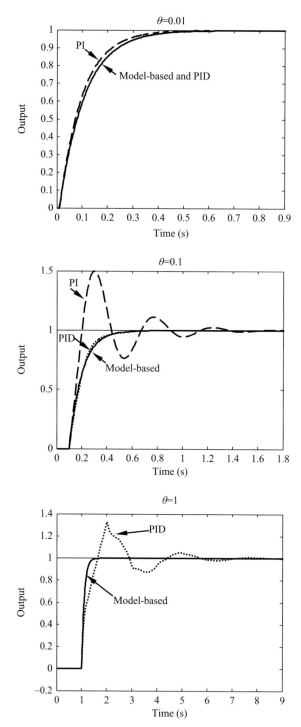

Figure 19.4 Closed-loop response to unit step change in the set point for the FOPDT system for the three controllers and for $\theta = 0.01$, $\theta = 0.1$ and $\theta = 1$.

of each other and follow the response given by the closed-loop transfer function in Eq. (19.3.27). For $\theta = 1$, which is equal to the process time constant, only the model-based controller (19.3.26) follows the response given by the closed-loop transfer function in Eq. (19.3.27). The PID approximation manages to bring the system to the set point but with large overshoot and oscillations, whereas the PI approximation (not shown in the figure) yields an unstable closed-loop response.

An important conclusion from this example is that, in the presence of small dead time, the model-based controller can be effectively approximated by PID or even PI, but when the dead time is of the same order of magnitude as, or larger than, the process time constant, the non-PID controller (19.3.26) will be the right choice.

19.4 The Smith Predictor for Processes with Large Dead Time

The so-called "Smith predictor" was developed by O. J. M. Smith in the 1950s, as a modification of a PI or PID controller to enhance the stability and performance of the closed-loop system in the presence of dead time. Over the years, the Smith predictor has found numerous applications and it has become an industrial standard for the control of processes with large dead times. The idea is very intuitive and will be outlined below.

Consider a process with dead time, with transfer-function description

$$G(s) = G_0(s)e^{-\theta s} \qquad (19.4.1)$$

where $G_0(s)$ is a rational function (ratio of two polynomials). This is also shown in Figure 19.5a, where the dead time is indicated as being in series with the rest of the transfer function. The measurement y is delayed; it is "behind" the intermediate variable z, which we wish that we could measure, but we can't. However, because the process input is known, it is possible to estimate "how far behind" y is from z.

Combining the transfer-function description of the process

$$Y(s) = G_0(s)e^{-\theta s}U(s)$$

with the transfer function description for the intermediate variable

$$Z(s) = G_0(s)U(s)$$

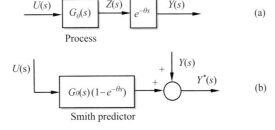

Figure 19.5 (a) A process with dead time. (b) The Smith predictor.

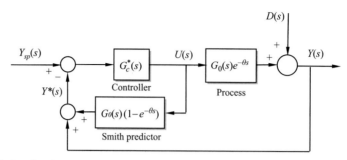

Figure 19.6 Feedback control system for a process with dead time using the Smith predictor.

we can immediately derive their difference:

$$Z(s) - Y(s) = G_0(s)\left(1 - e^{-\theta s}\right)U(s)$$

This difference can be calculated on-line and added to the measurement signal $Y(s)$, to generate an on-line estimate of the intermediate signal $Z(s)$:

$$Y^*(s) = G_0(s)\left(1 - e^{-\theta s}\right)U(s) + Y(s) \tag{19.4.2}$$

This is the *Smith predictor*. The signal $Y^*(s)$ is calculated on-line from the process input and output signals and the process model, and it represents an estimate of $Z(s)$.

The signal y^* also represents a *prediction of the output y, θ time units ahead*.
Because $y(t) = z(t - \theta)$ and $y^*(t) = z(t)$, it follows that $y^*(t) = y(t + \theta)$.
That is why it is called a "predictor."

Consider now a *stable* process with transfer function of the form (19.4.1). If a Smith predictor is used to calculate the predicted output y^*, this can be fed to a standard PI- or PID-type controller, which will control y^* to the set point, and therefore y to the set point. This is depicted in Figure 19.6.

Simplifying the block diagram of Figure 19.6, we find the transfer-function description of the closed-loop system:

$$Y(s) = \frac{G_c^*(s)G_0(s)}{1 + G_c^*(s)G_0(s)} e^{-\theta s} Y_{sp}(s) + \left(1 - \frac{G_c^*(s)G_0(s)}{1 + G_c^*(s)G_0(s)} e^{-\theta s}\right) D(s) \tag{19.4.3}$$

An interesting observation from Eq. (19.4.3) is that, with $G_c^*(s)$ and $G_0(s)$ being ratios of polynomials, closed-loop stability analysis and optimization/tuning of the controller $G_c^*(s)$ become straightforward, using methods from Chapters 12 and 16. *The use of the Smith predictor untangles the dead time from the control loop.*

It will also be interesting to apply what we learned in the present chapter to the feedback control system based on the Smith predictor. Using the synthesis formula (19.2.6) to select the controller $G_c^*(s)$ based on $G_0(s)$,

$$G_c^*(s) = \frac{1}{(\lambda s + 1)^r - G_0^+(s)} \cdot \frac{1}{G_0^-(s)} \tag{19.4.4}$$

the closed-loop system becomes:

$$Y(s) = \frac{G_0^+(s)}{(\lambda s + 1)^r} e^{-\theta s} Y_{sp}(s) + \left(1 - \frac{G_0^+(s)}{(\lambda s + 1)^r} e^{-\theta s}\right) D(s) \quad (19.4.5)$$

But this is identical to Eq. (19.2.7) because the nonminimum-phase factor of $G_0(s)e^{-\theta s}$ is exactly $G_0^+(s) \cdot e^{-\theta s}$, the product of the nonminimum-phase factor of $G_0(s)$ times the dead time.

The conclusion is therefore that the feedback control structure based on the Smith predictor is completely equivalent to IMC, when its controller $G_c^*(s)$ is selected using the synthesis formula for the dead-time-free part of the transfer function $G_0(s)$.

One of the most popular industrial applications of model-based feedback control is for FOPDT processes that have significant dead time. The reason is that empirical FOPDT models can be easily fit to experimental data, and the model can be directly embedded into the IMC or the Smith predictor.

Example 19.4 Model-based control of a first-order plus dead time (FOPDT) process

Consider the FOPDT system

$$G(s) = \frac{k \, e^{-\theta s}}{\tau s + 1}$$

Compare the IMC control structure to the control structure based on the Smith predictor. In the latter, use a PI controller with integral time equal to the process time constant.

Solution

For the IMC structure, the Q-system is obtained by applying Eq. (19.2.4): $Q(s) = \frac{1}{k} \frac{\tau s + 1}{\lambda s + 1}$

For the structure based on the Smith predictor, the controller is PI with integral time $\tau_I = \tau$, i.e. $G_c^*(s) = k_c \left(1 + \frac{1}{\tau s}\right)$.

The two control structures are depicted in Figure 19.7.

For the IMC structure, the closed-loop system is described by Eq. (19.3.27). For the control structure based on the Smith predictor, the closed-loop system is described by

$$Y(s) = \frac{k_c k}{\tau s + k_c k} e^{-\theta s} Y_{sp}(s) + \left(1 - \frac{k_c k}{\tau s + k_c k} e^{-\theta s}\right) D(s) \quad (19.4.6)$$

When k_c and λ are related through $k_c k = \frac{\tau}{\lambda}$, the closed-loop transfer functions (19.3.27) and (19.4.6) become identical and the two control structures are completely equivalent.

The above example shows two alternative implementations of the nonrational controller transfer function (19.3.26). They are equivalent, and it is up to the user to pick either one of them, depending on whether he/she prefers to tune a filter time constant λ or a PI controller gain k_c.

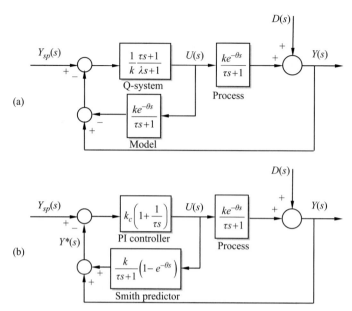

Figure 19.7 Feedback control system for a FOPDT process using (a) IMC structure and (b) Smith-predictor and a PI controller with integral time equal to the time constant.

19.5 Effect of Modeling Error

Models are never perfect representations of real physical systems and model–process mismatch is inevitable. Therefore, there is a key question regarding the effect of modeling error on the performance of a model-based controller. To this end, consider a process with transfer function $G(s)$ and suppose that the model has transfer function $\tilde{G}(s)$. When model-based control is implemented based on the available model, we obtain the block-diagram representation shown in Figure 19.8.

Simplifying the block diagram, we obtain the transfer-function description of the closed-loop system:

$$Y(s) = \frac{G(s)Q(s)}{1+\left(G(s)-\tilde{G}(s)\right)Q(s)} Y_{sp}(s) + \frac{1-\tilde{G}(s)Q(s)}{1+\left(G(s)-\tilde{G}(s)\right)Q(s)} D(s) \qquad (19.5.1)$$

In the absence of modeling error $G(s) = \tilde{G}(s)$, and Eq. (19.5.1) simplifies to Eq. (19.1.3). However, in the presence of modeling error, the denominator of the above closed-loop transfer function will be $\neq 1$, and this will affect the dynamic behavior of the closed-loop system. The closed-loop system may even be unstable despite the stability of $Q(s)$, when the modeling

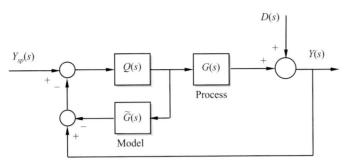

Figure 19.8 Model-based control structure in the presence of modeling error.

error is significant. In the presence of the modeling error, one may still want to select $Q(s)$ as the filtered inverse of the minimum-phase factor of the model, but then important questions arise regarding sensitivity to modeling errors.

Given some bound on the modeling error, what conditions could guarantee stability of the closed-loop system?

What conditions could guarantee a certain level of performance of the closed-loop system?

These types of questions are the subject of "robust control theory" a very important branch of control theory, which is normally covered in graduate-level control classes. An important qualitative conclusion from robust control theory is that, the closer we get to optimality, the more sensitive the closed-loop system becomes to modeling error. This implies that accurate modeling permits using lighter filtering (smaller λ), whereas inaccurate modeling necessitates using heavier filtering (larger λ). In practice, λ is tuned by simulation of various modeling-error scenarios and/or trial and error in the plant.

Example 19.5 Model-based control of a first-order plus dead time (FOPDT) process, with possible error in the dead time

A process follows the FOPDT transfer function:

$$\tilde{G}(s) = \frac{k\,e^{-\theta s}}{\tau s + 1}$$

The steady-state gain k and the time constant τ are accurately known, $k = 1$ and $\tau = 1$, but the dead time θ is uncertain and can take any value within the interval $\theta \in [0.6, 1]$.

Suppose that this process is controlled with a model-based controller using the above values of k and τ, and the mid-range value of dead time $\theta = 0.8$, and implemented with an IMC structure.

When the set point undergoes a unit step change, calculate the closed-loop responses for representative values of dead times within the uncertainty range, and evaluate their proximity to the response predicted by the model through the transfer function in (19.3.27). Try three different values of the filter time constant $\lambda = 0.6$, $\lambda = 0.2$ and $\lambda = 0.1$.

Solution

To solve the problem, we may use MATLAB to perform the calculations:

```
K=1;tau=1;theta=0.8;theta_actual=0.6;lambda=0.6;
G_model=tf(K,[tau 1],'InputDelay',theta)
G_actual=tf(K,[tau 1],'InputDelay',theta_actual)

Q=tf([tau 1],[lambda*K K])
Gc=feedback(Q,-G_model);

GCL=feedback(series(Gc,G_model),1);
GCL_actual=feedback(series(Gc,G_actual),1);

step(GCL)
hold on
step(GCL_actual)
```

We observe that, as the modeling error increases, control quality decreases. The worst-case scenario is when the modeling error is maximal.

For $\lambda = 0.6$ (which is a fairly large value for the filter parameter) and for actual process dead time of 0.6 (which corresponds to maximal dead-time error of 0.2), there is a fairly good agreement between the actual closed-loop response and the one predicted by the model, as seen in Figure 19.9a. For process dead times that are closer to the model's dead time of $\theta = 0.8$, the agreement is even better.

For $\lambda = 0.2$ and, again, for an actual process dead time of 0.6 (maximal dead-time error of 0.2), the actual closed-loop response is poor compared with the one predicted by the model, as seen in Figure 19.9b. When the filter parameter is further reduced to $\lambda = 0.1$ and the maximal dead-time error of 0.2 is considered, the response is unstable.

An important conclusion from the simulation results in this example is that:

- for large λ, responses are stable and well-behaved over the entire uncertainty range, but they are too slow,
- for small λ, the response may become unstable for large enough error in θ.

There is a range of intermediate values of λ giving stability and good performance over the entire uncertainty range.

In the specific problem under consideration, a value of λ around 0.3–0.4 is a good choice.

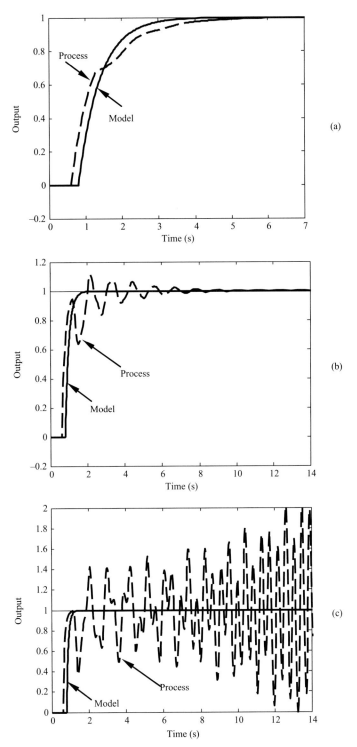

Figure 19.9 Comparison of the closed-loop response for actual process dead time of 0.6 to the one predicted by the model, when (a) $\lambda = 0.6$, (b) $\lambda = 0.2$ and (c) $\lambda = 0.1$.

19.5.1 A Simple Sufficient Condition for Robust Stability

Even though robust control theory is beyond the scope of this book, we can still give a simple sufficient condition for the closed-loop system of Figure 19.8 to be stable under all possible process transfer functions that lie within a certain error band. This condition is a special case of the so-called Doyle–Stein theorem of robust stability, but can be proved independently using the Bode stability criterion.

Consider the feedback control system of Figure 19.8 and assume that the transfer functions $G(s)$, $\tilde{G}(s)$ and $Q(s)$ are stable. Also, suppose that an upper bound of the modeling error is available in the following form:

$$\left| \frac{G(i\omega) - \tilde{G}(i\omega)}{\tilde{G}(i\omega)} \right| < \ell(\omega) \quad \text{for all } \omega > 0 \tag{19.5.2}$$

The closed-loop system (19.5.1) will be stable for all $G(s)$ that satisfy (19.5.2), if the following condition is met:

$$\left| \tilde{G}(i\omega) Q(i\omega) \right| < \frac{1}{\ell(\omega)} \quad \text{for all } \omega > 0 \tag{19.5.3}$$

To prove this result, one can first observe that because of the assumption of stability of $G(s)$, $\tilde{G}(s)$ and $Q(s)$, the transfer functions in (19.5.1) will be stable if the function

$$\frac{1}{1 + (G(s) - \tilde{G}(s))Q(s)} = 1 - \frac{(G(s) - \tilde{G}(s))Q(s)}{1 + (G(s) - \tilde{G}(s))Q(s)} \tag{19.5.4}$$

is stable; or, equivalently, if the feedback loop of Figure 19.10 is stable.

Using part (a) of the Bode stability criterion (see Chapter 17, Section 17.2), we conclude that a sufficient condition for this loop to be stable is that

$$\left| (G(i\omega) - \tilde{G}(i\omega)) Q(i\omega) \right| < 1 \quad \text{for all } \omega > 0 \tag{19.5.5}$$

which is true, as an immediate consequence of Eqs. (19.5.2) and (19.5.3).

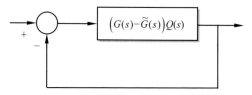

Figure 19.10 Feedback loop for the proof of the robust stability condition.

Example 19.6 Robust stability condition for the model-based controller in the presence of uncertainty in dead time

A process with dead time has been modeled by the transfer function

$$\tilde{G}(s) = G_0(s)e^{-\theta s}$$

where the dead-time-free factor $G_0(s)$ is an accurately known rational function, but the dead time is uncertain. The true process transfer function is

$$G(s) = G_0(s)e^{-(\theta + \Delta\theta)s}$$

where $\Delta\theta$ is the uncertainty in dead time.
Assuming that:

(i) $G_0(s)$ is a minimum-phase stable function of relative order 1,
(ii) an upper bound $\Delta\theta_{max}$ on the uncertainty is known, i.e. that $|\Delta\theta| \le \Delta\theta_{max}$,

derive a robust stability condition, which guarantees closed-loop stability for all possible process dead times within the given range.
  Specialize your results to the specific system of Example 19.5. What is the smallest value of the filter time constant that guarantees stability over the entire range of possible dead times?

Solution

Because of assumption (i), the model-based controller will have $Q(s) = \dfrac{1}{\lambda s + 1} \cdot \dfrac{1}{G_0(s)}$, and the robust stability condition (19.5.3) becomes:

$$\left|\frac{e^{-\theta i \omega}}{\lambda i \omega + 1}\right| < \frac{1}{\ell(\omega)} \Leftrightarrow \frac{1}{\sqrt{\lambda^2 \omega^2 + 1}} < \frac{1}{\ell(\omega)} \Leftrightarrow \sqrt{\lambda^2 \omega^2 + 1} > \ell(\omega) \quad \text{for all } \omega > 0$$

It remains for us to calculate $\ell(\omega)$ using assumption (ii). We have:

$$\frac{G(s) - \tilde{G}(s)}{\tilde{G}(s)} = \frac{G_0(s)e^{-(\theta + \Delta\theta)s} - G_0(s)e^{-\theta s}}{G_0(s)e^{-\theta s}} = e^{-\Delta\theta s} - 1$$

$$\Rightarrow \frac{G(i\omega) - \tilde{G}(i\omega)}{\tilde{G}(i\omega)} = e^{-i\omega\Delta\theta} - 1 = \cos(\omega\Delta\theta) - i\sin(\omega\Delta\theta) - 1$$

$$= -2\sin\left(\frac{\omega\Delta\theta}{2}\right)\left\{\sin\left(\frac{\omega\Delta\theta}{2}\right) + i\cos\left(\frac{\omega\Delta\theta}{2}\right)\right\}$$

$$\Rightarrow \left|\frac{G(i\omega) - \tilde{G}(i\omega)}{\tilde{G}(i\omega)}\right| = 2\left|\sin\left(\frac{\omega\Delta\theta}{2}\right)\right|$$

For $\Delta\theta$ ranging from $-\Delta\theta_{max}$ to $+\Delta\theta_{max}$, the above expression is bounded by the function:

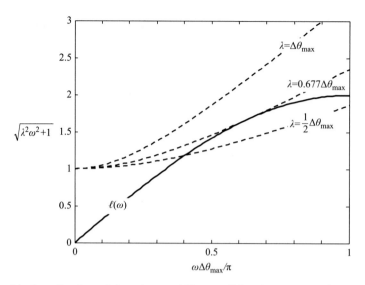

Figure 19.11 Graphical application of the robust stability condition for systems with uncertain dead time: plots of $\ell(\omega)$ and $\sqrt{\lambda^2\omega^2+1}$ for $\lambda = 0.5\Delta\theta_{max}$, $0.677\Delta\theta_{max}$ and $\Delta\theta_{max}$.

$$\ell(\omega) = \begin{cases} 2\sin\left(\dfrac{\omega\,\Delta\theta_{max}}{2}\right), & 0 \leq \omega\,\Delta\theta_{max} \leq \pi \\ 2, & \omega\,\Delta\theta_{max} \geq \pi \end{cases}$$

Figure 19.11 depicts the uncertainty function $\ell(\omega)$ along with $\sqrt{\lambda^2\omega^2+1}$, for various values of filter time constant λ.

We see from the graph that the robust stability condition will be satisfied when $\lambda \geq 0.677\,\Delta\theta_{max}$. For the specific process of Example 19.5, $\Delta\theta_{max} = 0.2$, and thus the robust stability condition specializes to $\lambda \geq 0.1354$. This is consistent with the simulation results of Figure 19.9: when $\lambda = 0.2$, the closed-loop system is stable over the entire uncertainty range, whereas when $\lambda = 0.1$, the closed-loop system becomes unstable near the ends of the uncertainty range.

It should be emphasized here that the robust stability condition (19.5.3) is a *sufficient* condition, therefore it is natural to suspect that it might be conservative. The exact robust stability condition can be found through brute-force trial-and-error simulations. In the specific example, the robust stability condition is $\lambda \gtrapprox 0.127$, and one can verify that for $\lambda = 0.1273239545$ and $\Delta\theta = +0.2$, the closed-loop system has a pair of imaginary poles at $\pm 7.853981634i$, whereas all other poles have negative real parts.

Example 19.7 Robust stability condition for the model-based controller in the presence of uncertainty in the time constant

A process with dead time has been modeled by the transfer function

$$\tilde{G}(s) = \frac{k}{\tau s + 1}e^{-\theta s}$$

where the parameters k and θ are accurately known, but the time constant τ is uncertain. The true process transfer function is

$$G(s) = \frac{k}{(\tau + \Delta\tau)s + 1} e^{-\theta s}$$

where $\Delta\tau$ is the uncertainty in the time constant.

Assuming that an upper bound $\Delta\tau_{max}$ on the uncertainty is known, i.e. $|\Delta\tau| \leq \Delta\tau_{max}$, with $\Delta\tau_{max} < \tau$, derive a robust stability condition that guarantees closed-loop stability for all possible process time constants within the given range.

Specialize your result for the specific values of the parameters $k = 1$, $\theta = 0.8$, $\tau = 1$ and $\Delta\tau_{max} = 0.6$.

Solution

The model-based controller will have $Q(s) = \dfrac{1}{\lambda s + 1} \cdot \dfrac{\tau s + 1}{k}$ and the robust stability condition (19.5.3) becomes:

$$\left| \frac{e^{-\theta i\omega}}{\lambda i\omega + 1} \right| < \frac{1}{\ell(\omega)} \Leftrightarrow \frac{1}{\sqrt{\lambda^2 \omega^2 + 1}} < \frac{1}{\ell(\omega)} \Leftrightarrow \sqrt{\lambda^2 \omega^2 + 1} > \ell(\omega) \quad \text{for all } \omega > 0$$

It remains for us to calculate $\ell(\omega)$. We have:

$$\frac{G(s) - \tilde{G}(s)}{\tilde{G}(s)} = \frac{\dfrac{k}{(\tau + \Delta\tau)s + 1} e^{-\theta s} - \dfrac{k}{\tau s + 1} e^{-\theta s}}{\dfrac{k}{\tau s + 1} e^{-\theta s}} = \frac{-\Delta\tau s}{(\tau + \Delta\tau)s + 1}$$

$$\Rightarrow \left| \frac{G(i\omega) - \tilde{G}(i\omega)}{\tilde{G}(i\omega)} \right| = \frac{|\Delta\tau|\omega}{\sqrt{(\tau + \Delta\tau)^2 \omega^2 + 1}} = \frac{\dfrac{|\Delta\tau|}{\tau}(\tau\omega)}{\sqrt{\left(1 + \dfrac{\Delta\tau}{\tau}\right)^2 (\tau\omega)^2 + 1}}$$

For $\Delta\tau$ ranging from $-\Delta\tau_{max}$ to $+\Delta\tau_{max}$, it is possible to prove that the above expression is bounded by the function:

$$\ell(\omega) = \frac{\dfrac{\Delta\tau_{max}}{\tau}(\tau\omega)}{\sqrt{\left(1 - \dfrac{\Delta\tau_{max}}{\tau}\right)^2 (\tau\omega)^2 + 1}}$$

which corresponds to the lower end of the uncertainty range, $\Delta\tau = -\Delta\tau_{max}$.

- When $\dfrac{\Delta\tau_{max}}{\tau} \leq \dfrac{1}{2}$, i.e. for time-constant error less than 50%, one can prove that $\ell(\omega) < 1$ for all ω, which implies that the robust stability condition $\sqrt{\lambda^2 \omega^2 + 1} > \ell(\omega)$ is satisfied for any value of the filter constant λ.
- When $\dfrac{1}{2} < \dfrac{\Delta\tau_{max}}{\tau} < 1$, i.e. for time-constant error between 50% and 100%, one can prove that the robust stability condition is satisfied for $\dfrac{\lambda}{\tau} > 2\dfrac{\Delta\tau_{max}}{\tau} - 1$

We see that the effect of error in the time constant is much milder than error in the dead time: only a huge time-constant error, larger than 50%, can destabilize the system.

Figure 19.12 shows the application of the robust stability condition for $\Delta\tau_{max}/\tau = 0.6$.

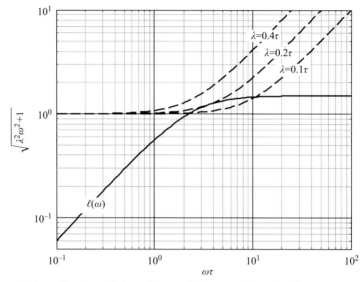

Figure 19.12 Graphical application of the robust stability condition for time-constant uncertainty with $\Delta\tau_{max}/\tau = 0.6$: plots of $\ell(\omega)$ and $\sqrt{\lambda^2\omega^2 + 1}$ for $\lambda = 0.1\tau$, 0.2τ and 0.4τ.

It is possible to consider other scenarios of model uncertainty, such as combined errors in dead time and time constant, unmodeled higher-order dynamics, error in a zero, etc. These will be left as exercises. For the specific system of Examples 19.6 and 19.7, with combined uncertainty in dead time $|\Delta\theta| \leq 0.2$ and time constant $|\Delta\tau| \leq 0.6$, the robust stability condition turns out to be $\lambda \geq 0.283$.

19.6 State-Space Form of the Model-Based Controller

The model-based controllers derived in the previous sections may also be described in state-space form. The state-space form of the controller is actually what will be simulated on-line, therefore it is what will be used in the application of the model-based controller in practice. In this section, we will derive a state-space representation of the controller (19.2.6).

19.6.1 Definition of Process Dynamics, Relative Order and Minimum-Phase Factor in State Space

Consider a stable linear dynamic system, possibly containing dead time, of the form

$$\frac{dx}{dt}(t) = Ax(t) + bu(t)$$
$$y(t) = cx(t-\theta)$$
(19.6.1)

The transfer function of this system is

$$G(s) = e^{-\theta s} c(sI - A)^{-1} b = e^{-\theta s} \frac{\beta_1 s^{n-1} + \cdots + \beta_{n-1} s + \beta_n}{s^n + a_1 s^{n-1} + \cdots + a_{n-1} s + a_n} \qquad (19.6.2)$$

consisting of a ratio of two polynomials times (possibly) an exponential, where

$$s^n + a_1 s^{n-1} + \cdots + a_{n-1} s + a_n = \det(sI - A) \qquad (19.6.3)$$

is the characteristic polynomial of A, and

$$[\beta_1 \ \beta_2 \ \cdots \ \beta_n] = [cb \ \ cAb + a_1 cb \ \cdots \ cA^{n-1}b + a_1 cA^{n-2}b + \cdots + a_{n-1}cb] \qquad (19.6.4)$$

(See Chapter 8, Eqs. (8.2.3) and (8.2.5).)

Moreover (see Chapter 8, Section 8.3), the system is said to have:

- relative order $r = 1$, if $cb \neq 0$,
- relative order $r = 2$, if $cb = 0$ but $cAb \neq 0$,
- relative order $r = 3$, if $cb = cAb = 0$ but $cA^2 b \neq 0$, etc.

In general, the relative order r is the smallest positive integer for which $cA^{r-1}b \neq 0$, and it equals the difference of degrees of denominator and numerator polynomials.

In the case where the polynomial $\beta_1 s^{n-1} + \cdots + \beta_{n-1} s + \beta_n$ has roots with positive real parts, one must factorize the rational part of the transfer function

$$\frac{\beta_1 s^{n-1} + \cdots + \beta_{n-1} s + \beta_n}{s^n + a_1 s^{n-1} + \cdots + a_{n-1} s + a_n} = G_{\text{all-pass}}(s) \cdot \frac{\beta_1^- s^{n-1} + \cdots + \beta_{n-1}^- s + \beta_n^-}{s^n + a_1 s^{n-1} + \cdots + a_{n-1} s + a_n} \qquad (19.6.5)$$

where $G_{\text{all-pass}}(s)$ will involve first- and/or second-order all-pass elements, corresponding to the roots with positive real parts, and the polynomial $\beta_1^- s^{n-1} + \cdots + \beta_{n-1}^- s + \beta_n^-$ will have all its roots with negative real parts. In this way, the nonminimum-phase factor of $G(s)$ will be the product of $e^{-\theta s}$ times $G_{\text{all-pass}}(s)$, whereas the minimum-phase factor will be

$$G^-(s) = \frac{\beta_1^- s^{n-1} + \cdots + \beta_{n-1}^- s + \beta_n^-}{s^n + a_1 s^{n-1} + \cdots + a_{n-1} s + a_n} \qquad (19.6.6)$$

The above minimum-phase factor appears in the controller formula (19.2.6), and it will be useful to identify its state-space matrices before attempting to put (19.2.6) in state-space form.

Notice that Eq. (19.6.6) is of the same form as the rational part of (19.6.2), with the denominator polynomials being the same. The numerator polynomials are, in general, different, but they have to be of the same degree, therefore the relative orders are the same. Also, an equation of the form (19.6.4) will have to hold, as it relates the coefficients of the numerator polynomial of the function (19.6.6) to the A, b, c matrices of the state-space description of the system represented by (19.6.6). If we want to keep the same A matrix so that the denominators match, and also keep the same b matrix as in (19.6.1), we can have an equation (19.6.4) satisfied, but with a different c matrix:

$$[\beta_1^- \ \beta_2^- \ \cdots \ \beta_n^-] = [c^-b \ \ c^-Ab + a_1c^-b \ \cdots \ c^-A^{n-1}b + a_1c^-A^{n-2}b + \cdots + a_{n-1}c^-b]$$
$$= c^-[b \ \ Ab + a_1b \ \cdots \ A^{n-1}b + a_1A^{n-2}b + \cdots + a_{n-1}b]$$

from which, solving for c^-, we find:

$$c^- = [\beta_1^- \ \beta_2^- \ \cdots \ \beta_n^-][b \ \ Ab + a_1b \ \cdots \ A^{n-1}b + a_1A^{n-2}b + \cdots + a_{n-1}b]^{-1} \tag{19.6.7}$$

The conclusion is therefore that

$$G^-(s) = c^-(sI - A)^{-1}b \tag{19.6.8}$$

where c^- is given by Eq. (19.6.7), i.e. that the state-space matrices of the minimum-phase factor are A, b and c^-.

19.6.2 Rearrangements of the Block Diagram of the Model-Based Controller and Conversion to State Space

The model-based controller (19.2.6) can be represented as a feedback connection between the Q-system

$$Q(s) = \frac{1}{(\lambda s + 1)^r} \cdot \frac{1}{G^-(s)} = \frac{1}{(\lambda s + 1)^r} \cdot \frac{1}{c^-(sI - A)^{-1}b} \tag{19.6.9}$$

and the process model transfer function

$$G(s) = e^{-\theta s}c(sI - A)^{-1}b \tag{19.6.10}$$

as shown in the top and middle block diagrams of Figure 19.13.

Moreover, the Q-system can be represented as a feedback loop, as shown in the bottom block diagram of Figure 19.13. To prove that this is indeed the case, we calculate the transfer function of the inner loop

$$\frac{U(s)}{V(s)} = \frac{\dfrac{1}{\lambda^r c^- A^{r-1} b}}{1 + \dfrac{c^-(\lambda A + I)^r (sI - A)^{-1} b}{\lambda^r c^- A^{r-1} b}} = \frac{1}{\lambda^r c^- A^{r-1} b + c^-(\lambda A + I)^r (sI - A)^{-1} b}$$

and, using the following matrix identity (the proof is left as an exercise)

$$(\lambda s + 1)^r c^- (sI - A)^{-1} b - c^-(\lambda A + I)^r (sI - A)^{-1} b = \lambda^r c^- A^{r-1} b \tag{19.6.11}$$

we conclude that indeed $\dfrac{U(s)}{V(s)} = \dfrac{1}{(\lambda s + 1)^r c^- (sI - A)^{-1} b} = Q(s)$.

An important observation that can be made from the block diagram is that $X_I(s) = X_M(s)$, since they are outputs of the same transfer function, driven by the same input. This allows us to rearrange the blocks as shown in Figure 19.14. In the same figure, the key ingredients of the model-based controller are identified.

19.6 State-Space Form of the Model-Based Controller

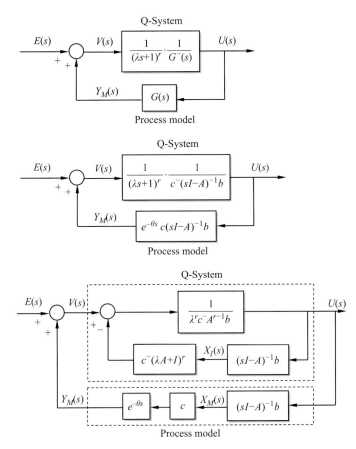

Figure 19.13 Steps for converting the model-based controller to state-space form.

The model-based controller simulates the state $x_M(t)$ of the process model. This is used to calculate the process output $y_M(t)$, which is added to the error signal and fed back, along with the model state, with appropriate constant gains.

We are now ready to convert the block diagram of Figure 19.14 to state space. Combining

$y_M(t) = c x_M(t - \theta)$ [calculated process output]

$v(t) = e(t) + y_M(t)$ [addition of calculated output to error signal]

$u(t) = \dfrac{1}{\lambda^r c^- A^{r-1} b}\left(v(t) - c^-(\lambda A + I)^r x_M(t)\right)$ [internal feedback]

$\dfrac{dx_M}{dt}(t) = A x_M(t) + b u(t)$ [simulation of process model states]

leads to the overall state-space description:

$$\dfrac{dx_M}{dt}(t) = A x_M(t) + \dfrac{b}{\lambda^r c^- A^{r-1} b}\left(e(t) + c x_M(t-\theta) - c^-(\lambda A + I)^r x_M(t)\right)$$

$$u(t) = \dfrac{1}{\lambda^r c^- A^{r-1} b}\left(e(t) + c x_M(t-\theta) - c^-(\lambda A + I)^r x_M(t)\right)$$

(19.6.12)

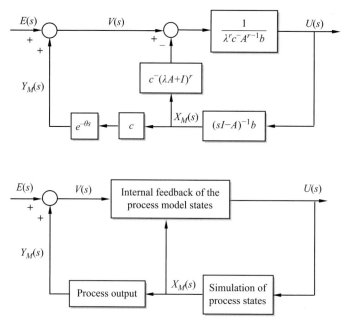

Figure 19.14 Rearrangement of the controller block diagram of Figure 19.13 – identification of the key components of the model-based controller.

19.6.3 Special Cases and Specific Examples

In the special case of a minimum-phase process, $c^- = c$ and $\theta = 0$, and the model-based controller (19.6.12) simplifies to:

$$\frac{dx_M}{dt}(t) = Ax_M(t) + \frac{b}{\lambda^r cA^{r-1}b}\left(e(t) - c\left[(\lambda A + I)^r - I\right]x_M(t)\right)$$

$$u(t) = \frac{1}{\lambda^r cA^{r-1}b}\left(e(t) - c\left[(\lambda A + I)^r - I\right]x_M(t)\right)$$

(19.6.13)

When the relative order $r = 1$, this takes the form

$$\frac{dx_M}{dt}(t) = Ax_M(t) + \frac{b}{\lambda cb}\left(e(t) - \lambda cAx_M(t)\right)$$

$$u(t) = \frac{1}{\lambda cb}\left(e(t) - \lambda cAx_M(t)\right)$$

(19.6.14)

when the relative order $r = 2$, it takes the form

$$\frac{dx_M}{dt}(t) = Ax_M(t) + \frac{b}{\lambda^2 cAb}\left(e(t) - (\lambda^2 cA^2 + 2\lambda cA)x_M(t)\right)$$

$$u(t) = \frac{1}{\lambda^2 cAb}\left(e(t) - (\lambda^2 cA^2 + 2\lambda cA)x_M(t)\right)$$

(19.6.15)

etc.

We can now apply the general state-space representation formula (19.6.12), or its special cases (19.6.13)–(19.6.15) to some of the specific examples considered earlier in Section 19.3. For a first-order process with transfer function (19.3.1) or state-space description

$$\frac{dx}{dt}(t) = -\frac{1}{\tau}x(t) + \frac{k}{\tau}u(t)$$
$$y(t) = x(t)$$
(19.6.16)

which is minimum-phase with relative order $r = 1$, application of (19.6.14) gives the model-based controller in state-space form:

$$\frac{dx_M}{dt}(t) = \frac{1}{\lambda}e(t)$$
$$u(t) = \frac{1}{k}\left(x_M(t) + \frac{\tau}{\lambda}e(t)\right)$$
(19.6.17)

The controller has the transfer function given by Eq. (19.3.3). One can also see that the state of the controller (19.6.17) equals $1/\lambda$ times the integral of the error, and its output has both integral and proportional actions, therefore the controller is PI.

For a second-order process with transfer function (19.3.6) or state-space description

$$\frac{dx_1}{dt}(t) = x_2(t)$$
$$\frac{dx_2}{dt}(t) = -\frac{1}{\tau^2}x_1(t) - \frac{2\zeta}{\tau}x_2(t) + \frac{k}{\tau^2}u(t)$$
$$y(t) = x_1(t)$$
(19.6.18)

which is minimum phase with relative order $r = 2$, application of Eqs. (19.6.15) gives the model-based controller in state-space form:

$$\frac{dx_{M1}}{dt}(t) = x_{M2}(t)$$
$$\frac{dx_{M2}}{dt}(t) = -\frac{2}{\lambda}x_{M2}(t) + \frac{1}{\lambda^2}e(t)$$
$$u(t) = \frac{1}{k}\left(x_{M1}(t) + 2\zeta\tau x_{M2}(t) - \frac{2\tau^2}{\lambda}x_{M2}(t) + \frac{\tau^2}{\lambda^2}e(t)\right)$$
(19.6.19)

The controller has the transfer function given by Eq. (19.3.8). One can also see that the controller's second state is the filtered error with filter time constant $\lambda/2$, and its first state is exactly the integral of the first state, i.e. the integral of the filtered error. The overall controller is PID with filter.

Consider now the second-order process with a right-half-plane (RHP) zero described by the transfer-function model (19.3.11) or the state-space model

$$\frac{dx_1}{dt}(t) = x_2(t)$$
$$\frac{dx_2}{dt}(t) = -\frac{1}{\tau_1\tau_2}x_1(t) - \frac{\tau_1+\tau_2}{\tau_1\tau_2}x_2(t) + \frac{k}{\tau_1\tau_2}u(t)$$
$$y(t) = x_1(t) - \tau_0 x_2(t)$$
(19.6.20)

This is non-minimum phase with relative order $r = 1$. Its minimum-phase factor is given in (19.3.12), and can be described by exactly the same states as in (19.6.20), but a different output equation:

$$y^-(t) = x_1(t) + \tau_0 x_2(t) \qquad (19.6.21)$$

i.e. it can be described by the same A and b matrices, but with a different c matrix. Applying the general formula (19.6.12) for $r = 1$, $c = [1\ -\tau_0]$, $c^- = [1\ -\tau_0]$, and the corresponding A, b matrices of the state equations of (19.6.20), we obtain the model-based controller in state-space form:

$$\frac{dx_{M1}}{dt}(t) = x_{M2}(t)$$

$$\frac{dx_{M2}}{dt}(t) = -\left(\frac{2}{\lambda} + \frac{1}{\tau_0}\right) x_{M2}(t) + \frac{1}{\lambda \tau_0} e(t) \qquad (19.6.22)$$

$$u(t) = \frac{1}{k}\left(x_{M1}(t) + (\tau_1 + \tau_2) x_{M2}(t) - \tau_1 \tau_2 \left(\frac{2}{\lambda} + \frac{1}{\tau_0}\right) x_{M2}(t) + \frac{\tau_1 \tau_2}{\lambda \tau_0} e(t)\right)$$

The controller has the transfer function given by (19.3.13). One can also see that the controller's second state is the filtered error with filter time constant $\lambda \tau_0/(2\tau_0 + \lambda)$ and its first state is exactly the integral of the first state, i.e. the integral of the filtered error. The overall controller is PID with filter.

Last but not least, consider the FOPDT system described by the transfer-function model (19.3.24) or the state-space model

$$\frac{dx}{dt}(t) = -\frac{1}{\tau} x(t) + \frac{k}{\tau} u(t) \qquad (19.6.23)$$
$$y(t) = x(t - \theta)$$

The system is nonminimum phase due to the presence of dead time, but its rational part does not have any RHP zero, and has relative order $r = 1$. Applying the general formula (19.6.12) for $r = 1$ and $c^- = c$, we obtain the model-based controller in state-space form:

$$\frac{dx_M}{dt}(t) = \frac{1}{\lambda}\left(x_M(t - \theta) - x_M(t) + e(t)\right)$$

$$u(t) = \frac{1}{k}\left(\frac{\tau}{\lambda} x_M(t - \theta) + \left(1 - \frac{\tau}{\lambda}\right) x_M(t) + \frac{\tau}{\lambda} e(t)\right) \qquad (19.6.24)$$

which has the transfer function given by Eq. (19.3.26).

19.7 Model-Based Controller Synthesis for MIMO Systems

We will now show that the controller synthesis method developed in this chapter can be generalized to square MIMO systems. To this end we consider again the general feedback control structure shown in Figure 19.1 with $G(s)$ and $G_c(s)$ being n-by-n (square) transfer-function matrices. In this case we can also define the following Q-parameterization of the feedback controller (see Eq. (19.1.4) for the SISO analog):

$$G_c(s) = Q(s)[I - G(s)Q(s)]^{-1} \tag{19.7.1}$$

It is noted that the closed-loop transfer-function description of the MIMO system is

$$Y(s) = [I + G(s)G_c(s)]^{-1} G(s)G_c(s) Y_{sp}(s) + [I + G(s)G_c(s)]^{-1} D(s) \tag{19.7.2}$$

Our aim is to derive the closed-loop transfer function in terms of $Q(s)$. We first multiply both sides of Eq. (19.7.1) with $G(s)$ from the left and then add the unity matrix to obtain

$$I + G(s)G_c(s) = [I - G(s)Q(s)][I - G(s)Q(s)]^{-1} + G(s)Q(s)[I - G(s)Q(s)]^{-1}$$

It then follows that

$$I + G(s)G_c(s) = [I - G(s)Q(s)]^{-1}$$

and finally

$$[I + G(s)G_c(s)]^{-1} = [I - G(s)Q(s)] \tag{19.7.3}$$

$$[I + G(s)G_c(s)]^{-1} G(s)G_c(s) = G(s)Q(s) \tag{19.7.4}$$

We substitute Eqs. (19.7.3) and (19.7.4) into (19.7.2) to obtain

$$Y(s) = G(s)Q(s)Y_{sp}(s) + [I - G(s)Q(s)]D(s) \tag{19.7.5}$$

It is interesting to compare Eq. (19.7.5), which holds true for MIMO square systems, with Eq. (19.1.3), which holds true for SISO systems, to note the striking similarity. Thus, we see that similarly to SISO systems, it is possible to define a Q-parameterization for MIMO systems through Eq. (19.7.1), leading to the closed-loop system (19.7.5). Provided $Q(s)$ is stable, the corresponding controller $G_c(s)$ will lead to closed-loop stability. Moreover, the controller can achieve zero offset if $Q(0)$ is selected to satisfy:

$$Q(0) = [G(0)]^{-1} \tag{19.7.6}$$

If $G(s)$ is a minimum-phase MIMO transfer-function matrix (rational and with all its zeros having negative real parts), the optimal choice of $Q(s)$ is the inverse system:

$$Q^{opt}(s) = [G(s)]^{-1} \tag{19.7.7}$$

which will be stable and satisfy (19.7.6). However, the inverse involves improper transfer functions that cannot be exactly realized, and leads to an infinite $G_c(s)$. To overcome this difficulty, we can use a filter $F(s)$ in series with the process inverse:

$$Q(s) = [G(s)]^{-1} F(s) \tag{19.7.8}$$

For example, $F(s)$ can be chosen as a diagonal filter

$$F(s) = \begin{bmatrix} \dfrac{1}{(\lambda_1 s + 1)^{r_1}} & 0 & \cdots & 0 \\ 0 & \dfrac{1}{(\lambda_2 s + 1)^{r_2}} & 0 & \vdots \\ \vdots & 0 & \ddots & 0 \\ 0 & \cdots & 0 & \dfrac{1}{(\lambda_n s + 1)^{r_n}} \end{bmatrix} \tag{19.7.9}$$

where $\lambda_1, \lambda_2, \ldots, \lambda_n$ are small time constants affecting the speed of the closed-loop response, and the integers $r_1, r_2, \ldots, r_n$ are such that all elements of $Q(s)$ are proper.

An important result for the closed-loop system response is obtained when Eq. (19.7.8) is substituted into Eq. (19.7.5):

$$Y(s) = F(s) Y_{sp}(s) + [I - F(s)] D(s) \tag{19.7.10}$$

The result is the generalization of Eq. (19.1.17), which holds for SISO systems. We immediately observe from Eq. (19.7.10) that, for minimum-phase square MIMO systems, perfect control can be achieved at the limit as $\lambda_i \to 0$, $\forall i$. It can also be observed that a diagonal closed-loop system is obtained, i.e. the multivariable model-based controller is a generalization of the decoupling controller studied in Chapter 18. It is also interesting to note that the transfer function matrix of the controller is given by

$$G_c(s) = [G(s)]^{-1} F(s) [I - F(s)]^{-1} \tag{19.7.11}$$

which, for the diagonal filter of Eq. (19.7.9), takes the form:

$$G_c(s) = [G(s)]^{-1} \begin{bmatrix} \dfrac{1}{(\lambda_1 s + 1)^{r_1} - 1} & 0 & \cdots & 0 \\ 0 & \dfrac{1}{(\lambda_2 s + 1)^{r_2} - 1} & 0 & \vdots \\ \vdots & 0 & \ddots & 0 \\ 0 & \cdots & 0 & \dfrac{1}{(\lambda_n s + 1)^{r_n} - 1} \end{bmatrix} \tag{19.7.12}$$

Example 19.8 Model-based control of the two-tank system

In Example 18.2 the transfer-function matrix of the system of two interacting tanks was derived. Use the transfer-function matrix of Example 18.2 to derive a multivariable model-based controller.

Solution

Based on the results of Example 18.2 we first calculate the inverse of the transfer-function matrix of the open-loop process

$$[G(s)]^{-1} = \begin{bmatrix} \dfrac{s+5}{(s+1)(s+7)} & \dfrac{4}{(s+1)(s+7)} \\ \dfrac{4}{(s+1)(s+7)} & \dfrac{2(s+3)}{(s+1)(s+7)} \end{bmatrix}^{-1} = \dfrac{1}{2}\begin{bmatrix} 2(s+3) & -4 \\ -4 & (s+5) \end{bmatrix}$$

Using a first-order diagonal filter (of the form (19.7.9) with $r_1 = r_2 = 1$),

$$Q(s) = [G(s)]^{-1} F(s) = \dfrac{1}{2}\begin{bmatrix} 2(s+3) & -4 \\ -4 & (s+5) \end{bmatrix}\begin{bmatrix} \dfrac{1}{\lambda_1 s+1} & 0 \\ 0 & \dfrac{1}{\lambda_2 s+1} \end{bmatrix} = \begin{bmatrix} \dfrac{s+3}{\lambda_1 s+1} & \dfrac{-2}{\lambda_2 s+1} \\ \dfrac{-2}{\lambda_1 s+1} & \dfrac{0.5(s+5)}{\lambda_2 s+1} \end{bmatrix}$$

is proper, and the controller transfer-function matrix is obtained from Eq. (19.7.12):

$$G_c(s) = \dfrac{1}{2}\begin{bmatrix} 2(s+3) & -4 \\ -4 & (s+5) \end{bmatrix}\begin{bmatrix} \dfrac{1}{\lambda_1 s} & 0 \\ 0 & \dfrac{1}{\lambda_2 s} \end{bmatrix} = \begin{bmatrix} \dfrac{(s+3)}{\lambda_1 s} & \dfrac{-2}{\lambda_2 s} \\ \dfrac{-2}{\lambda_1 s} & \dfrac{s+5}{2\lambda_2 s} \end{bmatrix}$$

The controller has proportional and integral actions:

$$G_c(s) = \begin{bmatrix} \dfrac{1}{\lambda_1} & 0 \\ 0 & \dfrac{1}{2\lambda_2} \end{bmatrix} + \begin{bmatrix} \dfrac{3}{\lambda_1} & \dfrac{-2}{\lambda_2} \\ \dfrac{-2}{\lambda_1} & \dfrac{5}{2\lambda_2} \end{bmatrix}\dfrac{1}{s}$$

i.e. it is a multivariable generalization of PI control.

LEARNING SUMMARY

- Model-based control involves selecting a controller transfer function tailored to a given process model.
- First, a class of stabilizing controllers is defined, depending on the process model and an adjustable stable function $Q(s)$. Subsequently, $Q(s)$ is selected for zero offset and nearly optimal performance.
- The controller synthesis formula (19.2.6) provides the transfer function of the model-based controller. To apply the formula, one must first calculate the relative order of the system and factorize the process transfer function as the product of an all-pass and a minimum-phase factor.
- The model-based controller (19.2.6) is also called an "internal model controller" since it includes a copy of the process model. It also includes an approximate inverse of the process model, connected with the process model through a feedback connection.
- For simple low-order process models, the model-based controller turns out to be PID. For higher-order models or models that include dead time, the model-based controller is not PID.
- For processes with dead time, the model-based controller can be interpreted as performing a model-based prediction of the future behavior of the process output. This is is referred to as the "Smith predictor."
- For model-based controllers, a key question is the effect of modeling error on the stability and performance of the control system. The Doyle–Stein theorem on robust stability gives a sufficient condition for the closed-loop system to be stable for all process transfer functions within a prespecified error band.
- Model-based controllers may also be described in state-space form, if the process model is given in state-space form. The state-space form is useful in the on-line implementation of the model-based controller.
- Model-based control can be generalized to MIMO processes.

TERMS AND CONCEPTS

Model-based controller or internal model controller. This includes a copy of the process model in its transfer-function or state-space description.

Relative order of a transfer function. The difference between the degree of denominator polynomial and numerator polynomial.

Smith predictor. Involves a model-based on-line calculation that predicts the future behavior of the process output, for processes with dead time.

Robust stability. Refers to the property of a control system of being stable for all process transfer functions within a prespecified error band.

FURTHER READING

Additional information on model-based control and robust control can be found in

Morari, M. and Zafiriou, E., *Robust Process Control*. Englewood Cliffs, NJ: Prentice Hall, 1989.

Ogunnaike, B. and Ray, H., *Process Dynamics, Modelling and Control*. New York: Oxford University Press, 1994.

The state space descriptions of model-based controllers are based on the results in

Kravaris, C., Daoutidis, P. and Wright, R. A., Output feedback control of nonminimum-phase nonlinear processes, *Chemical Engineering Science*, **49**(13), 2107–22, 1994.

PROBLEMS

19.1 Apply the controller synthesis formula (19.2.6) for the following process transfer functions:

(a) $G(s) = \dfrac{k(1+\tau_0 s)}{(\tau_1 s+1)(\tau_2 s+1)(\tau_3 s+1)}$, $\tau_0 > 0$, $\tau_1 > 0$, $\tau_2 > 0$, $\tau_3 > 0$

(b) $G(s) = \dfrac{k(1-\tau_0 s)}{(\tau_1 s+1)(\tau_2 s+1)(\tau_3 s+1)}$, $\tau_0 > 0$, $\tau_1 > 0$, $\tau_2 > 0$, $\tau_3 > 0$

(c) $G(s) = \dfrac{k}{(\tau_1 s+1)(\tau_2 s+1)} e^{-\theta s}$, $\tau_1 > 0$, $\tau_2 > 0$, $\theta > 0$

For the last transfer function, under the assumption of small dead time, apply a first-order Padé approximation to derive an approximate rational controller transfer function. In all cases, determine if the controller transfer function is PID.

19.2 For model-based control of an *unstable process*, it is possible to use the control structure in Figure P19.2a.

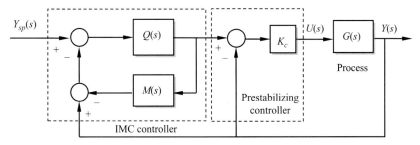

Figure P19.2a

- The process is prestabilized with a proportional controller in an inner loop, and then a model-based controller is built around it.
- The controller gain K_c of the inner loop assumes an arbitrary value within the range of values that guarantee stability.

- The model $M(s)$ in the IMC controller is the transfer function of the inner loop, and $Q(s)$ is chosen accordingly.

We would like you to follow this approach to build a model-based controller for the unstable first-order process:

$$G(s) = \frac{K}{\tau s - 1}, \quad \tau > 0$$

(a) Select appropriate K_c, $M(s)$ and $Q(s)$ and then calculate the overall controller (Figure P19.2b).

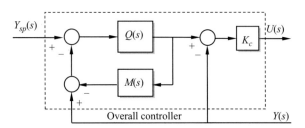

Figure P19.2b

Is it a standard PI or PID? If not, what are the key similarities and differences?

(b) Calculate the transfer function of the closed-loop system under the controller you built in part (a).

19.3 In an isothermal chemical reactor, where series-parallel Van de Vusse reactions take place, the transfer function between the outlet concentration measurement and the feed flowrate is given by

$$\frac{\bar{C}_{R_m}(s)}{\bar{F}(s)} = \frac{115.5 - s}{(s+70)(s+84)} e^{-0.5s}$$

The outlet concentration is to be controlled by manipulating the feed flowrate, using a Smith predictor and a PI controller.

(a) Derive necessary and sufficient conditions for closed-loop stability.
(b) Calculate the ISE-optimal PI controller parameters (gain and integral time).
(c) For the optimal controller parameters found in the previous question, simulate the closed-loop response to a unit step change in the set point.

Would you get a better closed-loop response if a model-based PID controller is used instead of an optimal PI?

19.4 For the FOPDT process of Example 19.7 ($k = 1$, $\theta = 0.8$, $\tau = 1$, $\Delta \tau_{max} = 0.6$), simulate the closed-loop response for the worst-case scenario ($\Delta \tau = -\Delta \tau_{max}$), for the following values of the filter time constant: $\lambda = 0.4$, $\lambda = 0.2$ and $\lambda = 0.1$. What do you observe?

19.5 For an FOPDT process with uncertainty in both dead time and time constant, and in particular with $k = 1$, $\tau = 1$, $\Delta \tau_{max} = 0.6$, $\theta = 0.8$, $\Delta \theta_{max} = 0.2$, do a graphical application of the Doyle–Stein theorem to determine the range of values of the filter time constant λ that guarantee robust stability.

You can assume that the worst-case scenario is for $\Delta \tau = -0.6$ and $\Delta \theta = 0.2$.

19.6 Apply the Doyle–Stein theorem to investigate the effect of error in a process zero on closed-loop stability.

In particular, for the following cases of second-order plus dead time process:

(a) $G(s) = k \dfrac{(\tau_0 + \Delta \tau_0)s + 1}{(\tau_1 s + 1)(\tau_2 s + 1)} e^{-\theta s}$, $\tau_0 > 0$, $\tau_1 > 0$, $\tau_2 > 0$, $\theta > 0$, $|\Delta \tau_0| \leq \Delta \tau_{0_{max}}$

(b) $G(s) = k \dfrac{-(\tau_0 + \Delta \tau_0)s + 1}{(\tau_1 s + 1)(\tau_2 s + 1)} e^{-\theta s}$, $\tau_0 > 0$, $\tau_1 > 0$, $\tau_2 > 0$, $\theta > 0$, $|\Delta \tau_0| \leq \Delta \tau_{0_{max}}$

find the range of values of the filter time constant λ that guarantee robust stability.

19.7 A process has transfer function of the form $G(s) = \dfrac{k}{(\tau s + 1)(\varepsilon s + 1)^2}$, $\tau > 0$, $\varepsilon > 0$, but because the time constant ε is much smaller than the time constant τ, the fast dynamics was neglected and the process was modeled by $\tilde{G}(s) = \dfrac{k}{\tau s + 1}$.

If we use a PI controller (19.3.3) to control it, what will be the range of values of the filter time constant λ that guarantee stability? It is given that $\dfrac{\varepsilon}{\tau} \leq 0.1$.

19.8 Prove the matrix identity (19.6.11).

Hint: First prove that, for every positive integer k,

$$s^k c^{\mathsf{T}} (sI - A)^{-1} b - c^{\mathsf{T}} A^k (sI - A)^{-1} b = s^{k-1} c^{\mathsf{T}} b + \cdots + s c^{\mathsf{T}} A^{k-2} b + c^{\mathsf{T}} A^{k-1} b.$$

Apply the above for $k = 1, \ldots, r$, where r is the relative order, and combine.

19.9 Derive model-based controllers in state-space form for the following process models:

(a) $\begin{cases} \dfrac{dx_1}{dt}(t) = x_2(t) \\ \dfrac{dx_2}{dt}(t) = -\dfrac{1}{\tau_1 \tau_2} x_1(t) - \dfrac{\tau_1 + \tau_2}{\tau_1 \tau_2} x_2(t) + \dfrac{k}{\tau_1 \tau_2} u(t) \\ y(t) = x_1(t) + \tau_0 x_2(t) \end{cases}$

where $\tau_0 > 0$, $\tau_1 > 0$, $\tau_2 > 0$;

(b) $\begin{cases} \dfrac{dx_1}{dt}(t) = x_2(t) \\ \dfrac{dx_2}{dt}(t) = -\dfrac{1}{\tau^2} x_1(t) - \dfrac{2\zeta}{\tau} x_2(t) + \dfrac{k}{\tau^2} u(t) \\ y(t) = x_1(t - \theta) \end{cases}$

where $\tau > 0$, $\zeta > 0$, $\theta > 0$.

19.10 A system of two noninteracting tanks, with separate feed streams in each tank, has transfer-function matrix

$$G(s) = \begin{bmatrix} \dfrac{1}{s+2} & 0 \\ \dfrac{2}{(s+1)(s+2)} & \dfrac{1}{s+1} \end{bmatrix}$$

Derive a multivariable model-based controller for this system.

20 Cascade, Ratio and Feedforward Control

The aim of this chapter is to introduce the basic ideas of cascade, ratio and feedforward control. These are advanced control techniques, in the sense of incorporating additional on-line information in the control system, for the improvement of disturbance rejection. We will study how to build a cascade control system and a feedforward or combined feedforward/feedback control system.

STUDY OBJECTIVES

After studying this chapter you should be able to do the following.

- Identify the benefits of applying cascade control and/or feedforward control.
- Explain the main differences between feedback and feedforward control.
- Design model-based cascade controllers.
- Design model-based feedforward controllers and combined feedforward/feedback controllers.

20.1 Introduction

Up to now, we have presented and analyzed the feedback control structure. We have seen that feedback action can be particularly successful in controlling process systems, and we have also pointed out the inherent limitations in achievable control quality arising from the presence of dead time or right-half-plane (RHP) zeros. A natural question to ask is whether there is a fundamentally different control strategy that can be used to further improve control performance. To answer this question, we need to step back and think about the chain of events that cause a feedback controller to act. When a disturbance acts on the process, it causes process variables to deviate from their nominal or desirable values. The measuring device detects the deviation from the set point and passes this information to the feedback controller, which takes corrective action. It is clear that a feedback controller acts after the disturbance has caused noticeable changes to the controlled variable. In this chapter, we will study how additional information that detects the presence of disturbances, whenever available, can be used in order to apply corrective action before the disturbance has a significant effect on the controlled variable. To this end, we will introduce the ideas of feedforward and cascade control, which are two very common strategies for improving the disturbance rejection capabilities of a control system. In cascade control, an additional measurement of a secondary output is utilized, which is strongly and quickly affected by the presence of a disturbance. In feedforward and ratio control, a disturbance is directly measured, and correction is applied before the effect is sensed by the controlled output.

20.2 Cascade Control

20.2.1 The Cascade Control Structure and its Advantages

The cascade control structure involves two feedback loops, one inside the other, and it is very common in practice. We will present the idea and its potential advantages through an example that we have already seen, albeit in a different context, in Chapters 10 and 11. We consider the liquid storage tank shown in Figure 10.10 (redrawn in Figure 20.1 for convenience). The system has a major disturbance, which is the changes in the volumetric flowrate of the feed stream $F_d(t)$. A feedback control system is used to control the liquid level in the tank by manipulating the other feed stream to the tank $F_{in}(t)$. To this end, a control valve has been installed for manipulating the volumetric flowrate $F_{in}(t)$ by adjusting the valve opening. This feedback control system is effective in rejecting an unexpected change (disturbance) in the volumetric flowrate of the unregulated stream $F_d(t)$, or changes in the set point of the liquid level in the tank. However, there is an additional disturbance that necessitates corrective action. Because the flowrate through the valve depends not only on the valve opening but also on the pressure drop across the valve, the valve will not deliver the proper flowrate if the upstream pressure changes. Such a disturbance will eventually cause a deviation in the liquid level in the tank, which will be handled by the level controller. However, it can be handled in a much more effective manner using a cascade control system.

A cascade control system for this problem is shown in Figure 10.14, which is redrawn in Figure 20.2 for convenience. A flow control loop is installed in the feed line of the manipulated stream, which continuously measures and regulates the volumetric flowrate to the value $F_{sp}(t)$ demanded by the level controller.

It is important to note that the level control loop needs to adjust the volumetric flowrate $F_{in}(t)$ in order to keep the liquid level at the set point. But this is done indirectly by adjusting the valve position, and the relation between valve position and flowrate may be affected by the presence of

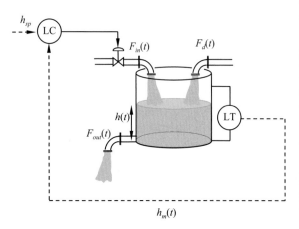

Figure 20.1 A liquid-level control example with classical feedback control structure.

20.2 Cascade Control

disturbances in upstream pressure. The role of the inner or secondary feedback loop is to ensure that the proper flowrate is, in effect, applied to the process. The outer or primary control loop is "in command" and determines the proper value of the flowrate, which then becomes the set point of the inner or secondary loop. From the point of view of the initial feedback control problem, nothing has changed apart from the fact that now the actual flowrate is adjusted instead of the valve opening, whose effect can be uncertain.

The installation of a cascade control system, such as the one shown in Figure 20.2, necessitates the use of additional measuring devices and increases the cost of installing the control system. The justification for the additional inner feedback loop in the cascade control system comes from the fact that the inner loop reacts to the disturbance before its effect becomes visible to the important process variables that need to be controlled. As a result, it offers improved control by reducing or even eliminating the effect of the internal disturbance. In addition, if the external loop is, for any reason, taken out of service, then the operator can supply the set point of a physically meaningful variable (volumetric flowrate instead of valve opening).

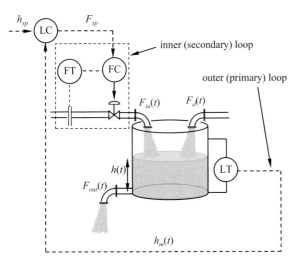

Figure 20.2 A liquid-level control example with a cascade system installed.

We can also use quantitative arguments to evaluate potential advantages of cascade control. To this end we consider the block diagram of a more-general cascade control system shown in Figure 20.3. The variables appearing in this block diagram must be elaborated further. The inner loop's disturbance W_i represents fluctuations in upstream pressure, whereas the outer loop's disturbance W_o represents fluctuations in the feed flowrate F_d. The inner or secondary loop control loop applies necessary corrections to the valve opening to counteract the disturbance W_i, before it affects the level of

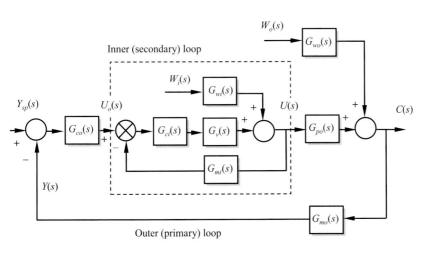

Figure 20.3 Block-diagram representation of a cascade control system.

the tank (the controlled process variable). The block diagram may be simplified by using the methods of Chapter 11. We first simplify the inner loop, and the result is presented in Figure 20.4. The transfer functions of the two new blocks that appear in Figure 20.4 are given by:

$$G_i(s) = \frac{G_v(s)G_{ci}(s)}{1+G_v(s)G_{ci}(s)G_{mi}(s)}, \quad G'_{wi}(s) = \frac{G_{wi}(s)}{1+G_v(s)G_{ci}(s)G_{mi}(s)} \quad (20.2.1)$$

It is interesting to compare, quantitatively and qualitatively, the cascade control system of Figures 20.3 and 20.4 to the case of single-loop feedback, where only the outer feedback loop is present, as shown in Figure 20.5. We observe that *for the rejection of W_i in single-loop control*, $G_{co}(s)$ tries to reduce the effect of the disturbance on the output, which is $G_{po}(s)G_{wi}(s)W_i(s)$ while in cascade control, $G_{co}(s)$ tries to reduce the effect of the disturbance on the output, which is

$$G_{po}(s)G'_{wi}(s)W_i(s) = G_{po}(s)G_{wi}(s) \cdot \frac{1}{1+G_v(s)G_{ci}(s)G_{mi}(s)} \cdot W_i(s) \quad (20.2.2)$$

We can choose $G_{ci}(s)$ so that the sensitivity function $1/(1+G_v G_{ci} G_{mi})$ is small, and in this way $G_{co}(s)$ has to deal with a smaller effective disturbance than that in single-loop feedback

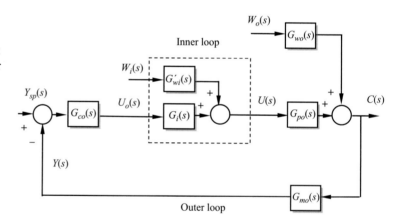

Figure 20.4 Block-diagram representation of cascade control system after simplifying the inner loop.

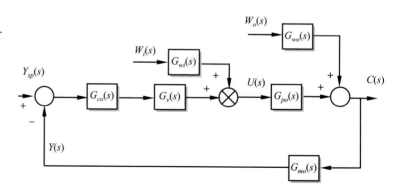

Figure 20.5 Block-diagram representation of the single-loop feedback control system.

control. This is the quantitative justification of why cascade control can perform better than single-loop feedback control.

In a similar way we observe that, *for the rejection of W_o in both cases*, $G_{co}(s)$ tries to reduce the same effect of the disturbance on the output, which is $G_{wo}(s)W_o(s)$ in both cases. We therefore conclude that, as far as W_o is considered, single-loop control and cascade control will have similar performance.

In industrial applications where the flowrate is manipulated through a standard control valve, it is very common to have an inner flow control loop to properly set the flowrate, as requested by the primary control loop. The outer or primary control loop can be associated with temperature, pressure, liquid level or composition control. Two classical examples of cascade control are shown in Figures 20.6 and 20.7.

In the cascade system shown in Figure 20.6, a CSTR is considered where an exothermic reaction is taking place. The major disturbances of the system are changes in temperature of the feed stream or feed composition, which affect the reaction rate. To keep the temperature

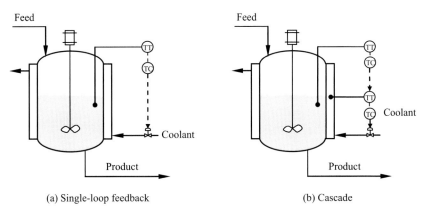

Figure 20.6 (a) Single-loop feedback and (b) cascade temperature control of a CSTR.

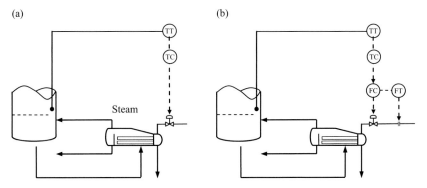

Figure 20.7 (a) Single-loop and (b) cascade temperature control structures at the bottom of a distillation column.

of the reacting mixture at set point, a feedback control structure (shown in Figure 20.6a) is commonly employed, that manipulates the flowrate of the cooling medium. However, because the temperature of the coolant could also vary, a cascade system can be installed, with an inner loop that controls the jacket temperature by manipulating the coolant flowrate. The set point for the jacket temperature is now determined by the outer or primary feedback controller. A similar cascade control system is shown in Figure 20.7 where the lower part and the reboiler of a distillation column is considered. A single-loop feedback structure is shown in Figure 20.7a, where the opening of the valve in the steam line is adjusted by a feedback controller to regulate the temperature in the lower trays of the column. In the cascade control structure of Figure 20.7b, an inner feedback loop is installed that directly measures and controls the steam flowrate, at the set point determined by the outer temperature control loop. In this way, the temperature of one of the lower trays is effectively controlled, with the inner loop absorbing the disturbances in the steam line.

In industrial applications, one can often find "triple" cascades, with three feedback loops, one inside the other. For example, the cascaded control loops of Figure 20.7b could be inside a composition control loop that gives the set point for the temperature.

When is cascade control most useful? Generally speaking, using additional online information is helpful, as long as it is properly used. But the benefit could be minor or major, depending on the situation. For cascade control to bring a significant benefit,

- the inner loop must have the potential to alleviate or even eliminate the effect of one (or more) significant process disturbance,
- the inner loop must be significantly faster that the outer loop,
- the output of the inner loop must have a direct and significant effect on the output of the outer loop.

In industrial applications, PI controllers are the most commonly used for the inner loop, and are usually designed/tuned first (i.e. the outer loop is designed with the inner loop in operation). The outer loop can be P, PI, PID, model-based, etc., depending on the needs of the particular application.

20.2.2 Model-Based Cascade Control

Consider a process consisting of two parts in series, with outputs Y_1 and Y_2 being measurable and with disturbance(s) affecting the first part (see also Figure 20.8). When the first part, denoted by $G_1(s)$, is fast and is followed by a slow and possibly delayed second part, denoted by $G_2(s)$, there is strong motivation to develop a cascade control structure to reduce the effect of the disturbance(s) before affecting the slow but more important output Y_2. The situation is depicted in the Figure 20.8. In what follows, we would like to compare single-loop feedback control to cascade control, and build model-based controllers for each case. The analysis will allow us to compare the achievable performance of single-loop feedback to cascade control and thus demonstrate, in a quantitative manner, the potential benefits of cascade control.

In Figure 20.9 we show a classical single-loop feedback structure (Figure 20.9a) for the process shown in Figure 20.8. The corresponding cascade control structure is also shown in Figure 20.9b. We begin by first analyzing the single-loop feedback control structure. We note that the closed-loop transfer function can be expressed as

$$Y_2(s) = \frac{G(s)G_c(s)}{1+G(s)G_c(s)} Y_{sp}(s) + \frac{G_w(s)}{1+G(s)G_c(s)} W(s) \qquad (20.2.3)$$

where $G(s) = G_1(s)G_2(s)$ is the overall process transfer function and $G_w(s) = G_2(s)$ is the transfer function for the process disturbance.

Expressing the controller in Q-parameterized form:

$$G_c(s) = \frac{Q(s)}{1-G(s)Q(s)} = \frac{Q(s)}{1-G_1(s)G_2(s)Q(s)} \qquad (20.2.4)$$

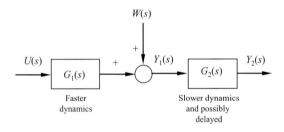

Figure 20.8 Open-loop process considered for applying cascade control.

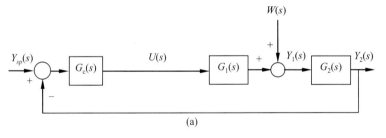

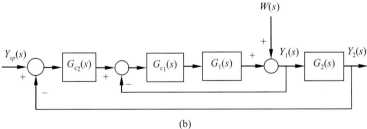

Figure 20.9 (a) Single-loop feedback and (b) cascade control structure for the system shown in Figure 20.8.

the transfer-function description of closed-loop system takes the form

$$Y_2(s) = Q(s)G_1(s)G_2(s)Y_{sp}(s) + (1 - Q(s)G_1(s)G_2(s))G_2(s)W(s) \qquad (20.2.5)$$

The "parameter" $Q(s)$ will be chosen as an approximate inverse of the minimum-phase factor of $G(s) = G_1(s)G_2(s)$, as has been discussed in Chapter 19.

Following the same approach for the case of cascade control we obtain

$$\begin{aligned}Y_2(s) &= Q_1(s)Q_2(s)G_1(s)G_2(s)Y_{sp}(s) \\ &+ (1 - Q_1(s)Q_2(s)G_1(s)G_2(s))G_2(s)(1 - Q_1(s)G_1(s))W(s)\end{aligned} \qquad (20.2.6)$$

where the controllers have been expressed in the following Q-parameterized form:

$$G_{c1}(s) = \frac{Q_1(s)}{1 - G_1(s)Q_1(s)}, \quad G_{c2}(s) = \frac{Q_2(s)}{1 - G_1(s)G_2(s)Q_1(s)Q_2(s)} \qquad (20.2.7)$$

Following the results of Chapter 19, the "parameter" $Q_1(s)$ can be chosen as an approximate inverse of $G_1(s)$, and $Q_2(s)$ so that the product $Q(s) = Q_1(s)Q_2(s)$ equals the single-loop choice of $Q(s)$, i.e. an approximate inverse of the minimum-phase factor of $G(s) = G_1(s)G_2(s)$. We can then compare the closed-loop transfer function between the output and the disturbance:

$$\left.\frac{Y_2(s)}{W(s)}\right)_{single\text{-}loop} = (1 - Q(s)G_1(s)G_2(s))G_2(s) \qquad (20.2.8a)$$

$$\left.\frac{Y_2(s)}{W(s)}\right)_{cascade} = (1 - Q_1(s)Q_2(s)G_1(s)G_2(s))G_2(s)(1 - Q_1(s)G_1(s)) \qquad (20.2.8b)$$

We note that in cascade control there is an extra factor of $(1 - Q_1(s)G_1(s))$ which, for $G_1(s)$ being minimum-phase and $Q_1(s)$ an approximate inverse of $G_1(s)$, it is a nearly zero transfer function. We therefore conclude that, in cascade control, the outer controller has to deal with a much smaller effective disturbance, since the disturbance undergoes a major reduction through the inner loop. The design approach implied by the aforementioned analysis is best demonstrated through examples.

Example 20.1 Cascade control of a second-order process

For $G_1(s) = \dfrac{k_1}{\tau_1 s + 1}$ and $G_2(s) = \dfrac{k_2}{\tau_2 s + 1}$ calculate the model-based feedback controllers for a single-loop and a cascade control configuration.

Solution

Both G_1 and G_2 are minimum-phase, so is their product $G = G_1G_2$. G_1 and G_2 have relative order 1, therefore their product $G = G_1G_2$ has relative order 2.

20.2 Cascade Control

Single-loop feedback control: $Q(s)$ will be a filtered inverse of the overall transfer function:

$$Q(s) = \frac{(\tau_1 s + 1)(\tau_2 s + 1)}{k_1 k_2} \cdot \frac{1}{(\lambda s + 1)^2}$$

and this gives the feedback controller

$$G_c(s) = \frac{(\tau_1 s + 1)(\tau_2 s + 1)}{k_1 k_2} \cdot \frac{1}{(\lambda s + 1)^2 - 1} = \frac{(\tau_1 + \tau_2)}{2\lambda k_1 k_2}\left(1 + \frac{1}{(\tau_1 + \tau_2)s} + \frac{\tau_1 \tau_2}{(\tau_1 + \tau_2)}s\right)\frac{1}{\frac{\lambda}{2}s+1}$$

which is a filtered PID controller. The resulting closed-loop transfer function with respect to the disturbance is:

$$\left.\frac{Y_2(s)}{W(s)}\right|_{single\text{-}loop} = (1 - Q(s)G_1(s)G_2(s))G_2(s) = \left(1 - \frac{1}{(\lambda s + 1)^2}\right)\cdot\frac{k_2}{\tau_2 s + 1}$$

Cascade control: $Q_1(s)$ will be a filtered inverse of $G_1(s)$:

$$Q_1(s) = \frac{(\tau_1 s + 1)}{k_1} \cdot \frac{1}{(\lambda s + 1)}$$

and $Q_2(s)$ can be chosen so that the product $Q_1(s)Q_2(s)$ equals the single-loop choice of $Q(s)$, i.e.

$$Q_2(s) = \frac{(\tau_2 s + 1)}{k_2} \cdot \frac{1}{(\lambda s + 1)}$$

These give the feedback controllers

$$G_{c1}(s) = \frac{(\tau_1 s + 1)}{k_1} \cdot \frac{1}{\lambda s} = \frac{\tau_1}{\lambda k_1}\left(1 + \frac{1}{\tau_1 s}\right)$$

$$G_{c2}(s) = \frac{(\tau_2 s + 1)}{k_2} \cdot \frac{\lambda s + 1}{(\lambda s + 1)^2 - 1} = \frac{(\tau_2 + \lambda)}{2\lambda k_2}\left(1 + \frac{1}{(\tau_2 + \lambda)s} + \frac{\tau_2 \lambda}{(\tau_2 + \lambda)}s\right)\frac{1}{\frac{\lambda}{2}s+1}$$

and the resulting closed-loop transfer function with respect to the disturbance is:

$$\left.\frac{Y_2(s)}{W(s)}\right|_{cascade} = (1 - Q_1(s)Q_2(s)G_1(s)G_2(s))G_2(s)(1 - Q_1(s)G_1(s))$$

$$= \left(1 - \frac{1}{(\lambda s + 1)^2}\right)\cdot\frac{k_2}{\tau_2 s + 1}\cdot\frac{\lambda s}{\lambda s + 1}$$

The inner loop's controller $G_{c1}(s)$ is a PI, whereas the outer $G_{c2}(s)$ is filtered PID. Comparing the closed-loop transfer functions with respect to the disturbance, we see that, in cascade control, the presence of the factor $\lambda s/(\lambda s + 1)$ will result in a major reduction of the effect of the disturbance on the output when the filter time constant λ is very small.

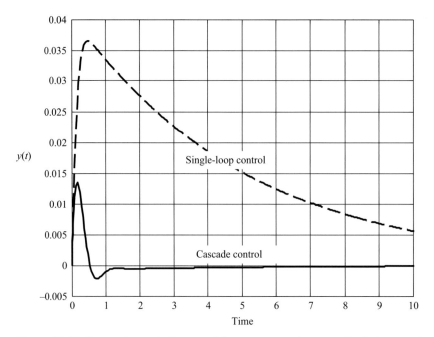

Figure 20.10 Comparison of the closed-loop response between single-loop and cascade control to a unit step change in disturbance – Example 20.1.

Simulation of closed-loop response to a step disturbance: For specific values of the process parameters $k_1 = k_2 = 1$, $\tau_1 = 1$, $\tau_2 = 5$ and for filter time constant $\lambda = 0.1$, the output response to a unit step change in the disturbance is depicted in Figure 20.10.

Example 20.2 Cascade control of a second-order process with dead time

For $G_1(s) = \dfrac{k_1}{\tau_1 s + 1}$ and $G_2(s) = \dfrac{k_2}{\tau_2 s + 1} e^{-\theta s}$ calculate the model-based feedback controllers for a single-loop and a cascade control configuration.

Solution

G_1 is minimum-phase, but G_2 is not, because of the presence of dead time. The product $G = G_1 G_2$ is also nonminimum-phase because of the dead time. G_1 and G_2 have relative order 1, therefore their product $G = G_1 G_2$ has relative order 2.

Single-loop feedback control: $Q(s)$ will be a filtered inverse of the minimum-phase factor of the overall transfer function

$$Q(s) = \frac{(\tau_1 s + 1)(\tau_2 s + 1)}{k_1 k_2} \cdot \frac{1}{(\lambda s + 1)^2}$$

and this gives the feedback controller

$$G_c(s) = \frac{(\tau_1 s + 1)(\tau_2 s + 1)}{k_1 k_2} \cdot \frac{1}{(\lambda s + 1)^2 - e^{-\theta s}}$$

which is not PID; it involves dead time compensation. The resulting closed-loop transfer function with respect to the disturbance is:

$$\left.\frac{Y_2(s)}{W(s)}\right|_{single\text{-}loop} = \left(1 - Q(s)G_1(s)G_2(s)\right)G_2(s) = \left(1 - \frac{e^{-\theta s}}{(\lambda s + 1)^2}\right) \cdot \frac{k_2}{\tau_2 s + 1} e^{-\theta s}$$

Cascade control: $Q_1(s)$ will be a filtered inverse of $G_1(s)$:

$$Q_1(s) = \frac{(\tau_1 s + 1)}{k_1} \cdot \frac{1}{(\lambda s + 1)}$$

and $Q_2(s)$ so that the product $Q_1(s)Q_2(s)$ equals the single-loop choice of $Q(s)$, i.e.

$$Q_2(s) = \frac{(\tau_2 s + 1)}{k_2} \cdot \frac{1}{(\lambda s + 1)}$$

These give the feedback controllers

$$G_{c1}(s) = \frac{(\tau_1 s + 1)}{k_1} \cdot \frac{1}{\lambda s} = \frac{\tau_1}{\lambda k_1}\left(1 + \frac{1}{\tau_1 s}\right)$$

$$G_{c2}(s) = \frac{(\tau_2 s + 1)}{k_2} \cdot \frac{\lambda s + 1}{(\lambda s + 1)^2 - e^{-\theta s}}$$

and the resulting closed-loop transfer function with respect to the disturbance is:

$$\left.\frac{Y_2(s)}{W(s)}\right|_{cascade} = \left(1 - Q_1(s)Q_2(s)G_1(s)G_2(s)\right)G_2(s)(1 - Q_1(s)G_1(s))$$

$$= \left(1 - \frac{e^{-\theta s}}{(\lambda s + 1)^2}\right) \cdot \frac{k_2}{\tau_2 s + 1} \cdot e^{-\theta s} \cdot \frac{\lambda s}{\lambda s + 1}$$

The inner loop's controller $G_{c1}(s)$ is a PI, whereas the outer $G_{c2}(s)$ is non-PID.

Comparison: Comparing the closed-loop transfer functions with respect to the disturbance, we see again the presence of the additional factor $\lambda s/(\lambda s + 1)$ in the cascade control case, which will result in a major reduction of the effect of the disturbance on the output when the filter time constant λ is very small.

Simulation of closed-loop response to a step disturbance: For specific values of the process parameters $k_1 = k_2 = 1$, $\tau_1 = 1$, $\tau_2 = 5$, $\theta = 5$ and for filter time constant $\lambda = 0.1$, the output response to a unit step change in the disturbance is depicted in Figure 20.11.

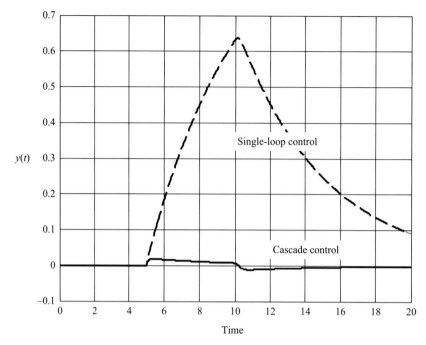

Figure 20.11 Comparison of closed-loop response between single-loop and cascade control to a unit step change in disturbance – Example 20.2.

20.3 Ratio Control

In many applications, proper operation of a process is achieved by adjusting the ratio of the flowrates of two feed streams. Usually, only one of the flows is manipulated, and the other one, usually called the wild stream, is measured on-line. A classical example of ratio control is in industrial burners. Fuel (e.g. methane, propane) and oxidant (usually air or oxygen-enriched air) must be fed in a constant ratio (determined by the combustion stoichiometry) in order to achieve and maintain an ideal combustion flame. The necessary stoichiometry of air must be guaranteed, otherwise system safety may be compromised or emissions can increase beyond limits. A simplified ratio control scheme is shown in Figure 20.12. Given the desired ratio of the two flowrates $r\ (= F_a/F_f)$, the ratio control scheme involves continuous measurement of the flowrate of the combustion gas F_f, multiplication of the measured value by r and adjustment of the oxidant flowrate $F_a(t)$ to be equal to $r \cdot F_f(t)$. Accurate adjustment of the oxidant flowrate is accomplished through an inner flow control feedback loop, with set point $SP = r \cdot F_f$.

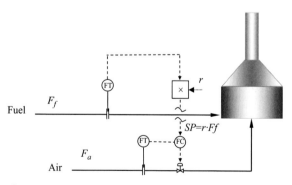

Figure 20.12 Ratio control of an industrial burner.

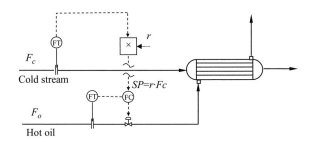

Figure 20.13 Ratio control of a heat exchanger.

A similar ratio control scheme applied to a heat exchanger is shown in Figure 20.13. A cold process stream is heated to the desired temperature by a hot oil (utility) stream. The flowrate of the cold process stream may vary to satisfy other operational objectives. The flowrate of the process stream is constantly monitored and the measurement is multiplied by a prespecified ratio between the flowrate of the process and utility stream. The result is supplied to a flow control loop that manipulates the flowrate of the heating oil. It is important to note that the signals involved in the determination of the set point for the hot oil are absolute and not deviation variables (i.e. the desired ratio is the ratio between the actual values of the two flowrates and not the ratio of the deviation from their nominal or steady-state values).

Mathematically, a ratio controller can be seen as a simple multiplication operation: the desirable ratio is simply multiplied by the measured uncontrolled flow to obtain the set point for the flow of the controlled stream. There is, however, a new feature involved in a ratio controller: the ratio controller uses the measurement of a disturbance in order to react in advance in a feedforward manner, before the disturbance affects the important process variables. The strategy of feedforward control is a much more general idea, fundamentally different from the feedback control, that is applied extensively in process control systems and is analyzed in the section that follows.

20.4 Feedforward Control

20.4.1 The Feedforward Control Structure

Feedforward control is a general strategy for alleviating the effect of disturbances on important process variables. This is feasible when the disturbance(s) can be measured reliably and a relatively accurate process model is available. The basic idea is to use the online information on the disturbance together with the process model, so as to be able to predict the effect of the disturbance on the system. We can then calculate the necessary change of the manipulated input, in order to keep the controlled output variable unaffected. Feedforward control is based on a fundamentally different idea compared to feedback control. When we apply feedback control, we wait until the effect of a disturbance is visible on the controlled process output and then apply corrective action. In feedforward control, we use the measurement of the disturbance, so as to act before it affects the process. It must be emphasized that feedforward control relies heavily on the accuracy of the process model, which cannot be a perfect match to the real system. The inevitable process–model mismatch could be a source of degradation in the performance of any feedforward controller, as well as the effect of other, unmeasured disturbances. For this reason, feedforward control is usually combined with feedback control, to reduce the adverse effect of model inaccuracies and the effect of unmeasured disturbances.

To put the above ideas in context, we consider the steam boiler shown in Figure 20.14, where water is fed to a boiler and hot oil is used to heat and vaporize the water and produce high-pressure steam. The demand of steam by downstream processes determines the flow-rate of steam that exits the boiler (downstream flow control loops set its flowrate), which means that the exiting steam flowrate is actually an input to the boiler, whose changes represent disturbances to the boiler's operation. As the demand for steam changes, the feed-water flowrate will also need to change. For this reason, level control is very important: if the level is too low, the heat exchanger coils may be exposed to steam, whereas if the level is too high, the steam may be mixed with water. Figure 20.14a shows a feedback controller, where the liquid level is measured and controlled by manipulating the feed-water flowrate. When there is an increase in steam flowrate, this causes the liquid level to decrease. But this decrease is detected by the level sensor and is fed to the feedback controller, which increases the flowrate of the water fed to the boiler.

A feedforward control system for the boiler is shown in Figure 20.14b. In the feedforward control system, the mass flowrate of the steam is continuously measured and the feed-water mass flowrate is continuously adjusted so as to match the steam flowrate. By keeping the inlet and outlet mass flowrates equal at all times, the water level will necessarily be kept constant. This is a clear advantage, because disturbances are handled ahead of time, without waiting till they affect the process.

The main disadvantage of the feedforward control system of Figure 20.14b is that it is vulnerable to metering errors. Suppose, for instance, that there is a systematic positive error in the measurement of the steam flowrate (i.e. the measured value is greater than the actual value of the flowrate). In this case, more water will be fed to the boiler than is actually removed in the form of steam, and the liquid level will start to rise. It will keep rising, and

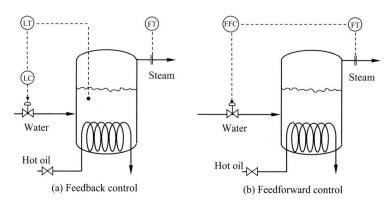

Figure 20.14 (a) Feedback and (b) feedforward control of an industrial steam boiler.

the boiler will eventually be filled completely, and the steam line will contain liquid, and thus fail. One practical way to increase the reliability of the feedforward control system is to combine it with feedback control. In this way, we benefit from both schemes at the expense of having to design a more complex control system. A combined feedback–feedforward control scheme for the steam boiler system is shown in Figure 20.15.

We will now analyze a general feedforward control system as shown in Figure 20.16. In this figure the basic elements of a feedforward system are shown. The disturbance $W(s)$ is measured by a measuring device with transfer function $G_m(s)$ and the measurement signal is fed to the feedforward controller, which has a transfer function $G_{FF}(s)$. The disturbance acts directly on the output $Y(s)$ through the transfer function $G_w(s)$ while the manipulated variable $U(s)$ drives a control valve that affects the process $G_p(s)$. The transfer-function description of the open-loop system that relates the disturbance and the manipulated variable with the process output is:

$$Y(s) = G_p(s)G_v(s)U(s) + G_w(s)W(s) \qquad (20.4.1)$$

To simplify the analysis, we will assume that the set point is unchanged and equal to zero. Under this assumption, the action of the feedforward controller is given by

$$U(s) = -G_{FF}(s)G_m(s)W(s) \qquad (20.4.2)$$

Substituting Eq. (20.4.2) into Eq. (20.4.1) we obtain

$$Y(s) = \left(-G_p(s)G_v(s)G_{FF}(s)G_m(s) + G_w(s)\right)W(s) \qquad (20.4.3)$$

The goal is to achieve $Y(s) = Y_{sp}(s) = 0$, and this can be obtained by simply choosing

$$G_{FF}(s) = \frac{G_w(s)}{G_p(s)G_v(s)G_m(s)} \qquad (20.4.4)$$

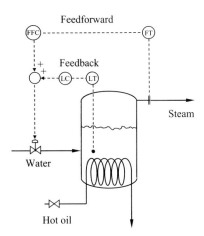

Figure 20.15 Combined feedback–feedforward control of an industrial steam boiler.

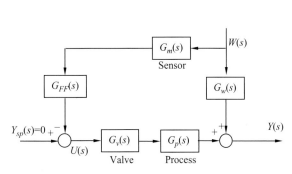

Figure 20.16 Block diagram of a feedforward control system.

The fact that the ideal feedforward controller involves the inversion of system dynamics comes as no surprise. However, we also know that the inversion of the open-loop dynamics is not always feasible, as in some cases it may lead to an unstable or improper controller transfer function.

Example 20.3 Feedforward control of a stirred tank heater

Consider the simple tank heater shown in Figure 20.17. Assume constant properties (heat capacity c_p and density ρ) and constant liquid level in the tank (i.e. constant volume V), as well as constant and equal inlet and outlet flowrates F. The liquid in the tank is heated at temperature T with an internal coil with heat transfer rate Q, which is the manipulated variable. The temperature of the inlet stream T_{in} varies significantly and the goal is to determine the transfer function of a feedforward controller $G_{FF}(s)$ that will eliminate the effect of inlet temperature changes on the outlet temperature. Repeat the analysis when the temperature sensor of the inlet temperature has first-order dynamics with time constant τ_m (and unity static gain).

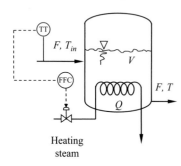

Figure 20.17 The tank heater example.

Solution

The unsteady-state energy balance for the heater has been developed in Chapter 2 and is given by Eq. (2.3.14):

$$V\rho c_p \frac{dT}{dt} = F\rho c_p (T_{in} - T) + Q$$

Introducing deviation variables and rearranging, we have

$$\frac{V}{F}\frac{d\bar{T}(t)}{dt} + \bar{T}(t) = \bar{T}_{in}(t) + \frac{\bar{Q}(t)}{F\rho c_p}$$

In order to achieve $\bar{T}(t) = 0$ at all times, the necessary heating rate is given by

$$\bar{Q}(t) = -F\rho c_p \bar{T}_{in}(t)$$

The same conclusion can be reached from transfer-function description of the process:

$$\bar{T}(s) = G_p(s)\bar{Q}(s) + G_w(s)\bar{T}_{in}(s) = \frac{\frac{1}{F\rho c_p}}{\frac{V}{F}s+1}\bar{Q}(s) + \frac{1}{\frac{V}{F}s+1}\bar{T}_{in}(s)$$

In order for $\bar{T}(s) = 0$, we must have:

$$\bar{Q}(s) = -F\rho c_p \bar{T}_{in}(s)$$

We see that the feedforward controller's transfer function is constant:

$$G_{FF}(s) = F\rho c_p$$

This is like the proportional controller in feedback control.

In the above analysis, the underlying assumption is that both the valve and the thermometer for the inlet temperature are very fast (negligible time constants). But when the temperature sensor exhibits first-order dynamics with time constant τ_m, i.e.

$$\frac{T_m(s)}{T_{in}(s)} = G_m(s) = \frac{1}{\tau_m s + 1}$$

substituting into Eq. (20.4.4), we obtain

$$G_{FF}(s) = \frac{\dfrac{1}{\left(\dfrac{V}{F}s+1\right)}}{\dfrac{1}{F\rho c_p}\cdot\dfrac{1}{\left(\dfrac{V}{F}s+1\right)(\tau_m s+1)}} = F\rho c_p (1+\tau_m s)$$

We immediately observe that including the sensor dynamics results in a feedforward controller that has the transfer function of an ideal proportional-derivative controller, which is improper. A low-pass filter needs to be added to make it proper:

$$G_{FF}(s) = F\rho c_p \frac{1+\tau_m s}{1+\tau_f s}$$

where τ_f is the filter time constant.

20.4.2 Combination of Feedforward with Feedback Control

To take advantage of the benefits of feedforward control and feedback control, it is possible to combine them as shown in Figure 20.18. Feedforward control can be used to eliminate or reduce the effect of some measurable disturbances, but there are unmeasurable disturbances and modeling errors, and these can be handled through feedback control.

Simplifying the block diagram of the combined feedforward and feedback control structure of Figure 20.18, we find:

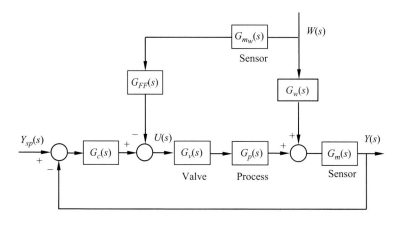

Figure 20.18 The general structure of a feedforward plus feedback controller.

$$Y(s) = \frac{G_m(s)G_p(s)G_v(s)G_c(s)}{1+G_m(s)G_p(s)G_v(s)G_c(s)} Y_{sp}(s) + \frac{G_m(s)[G_w(s)-G_p(s)G_v(s)G_{FF}(s)G_{m_w}(s)]}{1+G_m(s)G_p(s)G_v(s)G_c(s)} W(s)$$

(20.4.5)

From the last equation we immediately observe that:

- for changes in set point, the control system acts as pure feedback,
- for changes in the disturbance, it is possible to select $G_{FF}(s)$ to make the numerator $G_w(s) - G_p(s)G_v(s)G_{FF}(s)G_{m_w}(s)$ approximately vanish, while retaining the corrective and stabilizing action of feedback.

The feedforward and feedback controllers could be designed or tuned independently. However, if optimal disturbance rejection is sought, one could follow the methods given in the following section.

20.5 Model-Based Feedforward Control

In this section, we discuss how to build model-based feedforward controllers that best reject the effect of a particular measured disturbance. While feedback control can correct for all possible disturbances, feedforward control focuses on rejecting a particular disturbance that is measured on-line.

The starting point of our discussion will be disturbance rejection in feedback control. The problem of optimally selecting a feedforward controller will then be discussed, and finally we will see how to combine feedback control and feedforward control. We will consider the process shown in Figure 20.19, where the disturbance transfer function $G_w(s)$ describes the effect of a particular disturbance w. For simplicity of exposition of the results, we will assume that the actuator and sensor elements are either very fast, or their dynamics has been included in the process transfer functions.

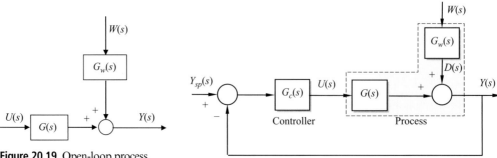

Figure 20.19 Open-loop process considered for model-based control.

Figure 20.20 Standard feedback control structure.

20.5.1 Disturbance Rejection in Model-Based Feedback Control

Figure 20.20 depicts the standard feedback control loop for the process under consideration, which has the following transfer function description:

$$Y(s) = \frac{G(s)G_c(s)}{1+G(s)G_c(s)} Y_{sp}(s) + \frac{1}{1+G(s)G_c(s)} G_w(s)W(s) \quad (20.5.1)$$

Comparing with the feedback control loop considered in Chapter 19 (see Figure 19.1), we see that here the disturbance does not directly affect the output, but goes through the transfer function $G_w(s)$ before it affects the output. The signal $D(s)$ describes the effect of the disturbance $W(s)$ on the output y. The analysis of Chapter 19 can be directly extended to the feedback control loop under consideration. Assuming that:

(1) the process transfer functions $G(s)$ and $G_w(s)$ are both stable,
(2) the steady-state gain of the process is nonzero ($G(0) \neq 0$)

and performing exactly the same Q-parameterization of the controller

$$G_c(s) = \frac{Q(s)}{1-G(s)Q(s)} \quad (20.5.2)$$

where $Q(s)$ is an arbitrary stable transfer function, Eq. (20.5.1) takes the simpler form:

$$Y(s) = G(s)Q(s)Y_{sp}(s) + (1-G(s)Q(s))G_w(s)W(s) \quad (20.5.3)$$

Based on the same analysis as in Chapter 19, we may conclude that every controller of the form given by Eq. (20.5.2) with $Q(s)$ stable and $Q(0) = 1/G(0)$, guarantees stability of the closed-loop system and zero offset.

So far, everything looks exactly the same as in Chapter 19. The main difference is in terms of the optimality properties of the feedback controller.

- *If the process $G(s)$ is minimum-phase, there is no difference*: Choosing $Q(s)$ to be a proper filtered approximation of $1/G(s)$, makes $Q(s)G(s) \approx 1$ and $(1 - G(s)Q(s)) \approx 0$, hence the response will be nearly optimal for step changes in either y_{sp} or w.
- *But if the process $G(s)$ is nonminimum-phase, we do have an issue*: Optimizing with respect to step changes in y_{sp} does not imply optimality with respect to w, and vice versa. The *feedback controller cannot be optimal in both* servo and regulatory operation, unless $G_w(s) = 1$.

General formulas of optimal $Q(s)$ for step disturbances w with transfer function $G_w(s)$ are available in the literature (see, for example, Morari and Zafiriou's 1989 book – details are given in the Further Reading section at the end of this chapter). But, in general, such a $Q(s)$ will not be optimal with respect to other disturbances or set-point changes.

The point of view in feedback control is that the controller should protect the process from all kinds of disturbances, and all sizes and all kinds of G_w. This is the point of view that was followed in Chapter 19, where the overall effect of all disturbances was considered lumped together in the quantity $D(s)$ that directly affects the output. If it so happens that the overall effect of disturbances changes in a stepwise manner, the choice $Q(s) = 1/\bar{G}(s)$ will be disturbance optimal; however, in general, it will only be set-point optimal.

20.5.2 Optimal Feedforward Control

In Figure 20.21 a complete feedforward control system is shown that is capable of handling both setpoint tracking and disturbance rejection. This is a generalization of the control system shown in Figure 20.16, where only disturbance rejection was considered. The feedforward controller shown in Figure 20.21 consists of two transfer functions:

- $G_{ST}(s)$ for set-point tracking, to guide the system so that the output tracks $y_{sp}(t)$,
- $G_{FF}(s)$ with feedforward action, to correct for the effect of the disturbance.

Simplifying the block diagram, we find:

$$Y(s) = G(s)G_{ST}(s)Y_{sp}(s) + \left(G_w(s) - G(s)G_{FF}(s)\right)W(s) \tag{20.5.4}$$

Assuming that the process transfer functions $G(s)$ and $G_w(s)$ are stable, the overall transfer functions in Eq. (20.5.4) will be stable as long as the controller transfer functions $G_{ST}(s)$ and $G_{FF}(s)$ are stable.

Ideally, one would want to achieve *perfect control*: complete elimination of the disturbance and perfect tracking of the set point. This would imply choosing the controller transfer functions as follows:

$$G_{ST}(s) = \frac{1}{G(s)} \quad , \quad G_{FF}(s) = \frac{G_w(s)}{G(s)} \tag{20.5.5}$$

Whenever the transfer functions in Eqs. (20.5.5) are stable and causal (i.e. not involving predictions $e^{+\theta s}$), they can be implemented, either exactly or approximately (exactly when the degree of the numerator is not larger than the degree of the denominator, otherwise approximately by adding an appropriate filter in series). In general, however, the transfer functions of (20.5.5) will not necessarily be stable and causal, in which case the best possible response might not be even close to ideality. For general $G(s)$ and $G_w(s)$, the ISE-optimal choices are (see Morari and Zafiriou's book for a proof):

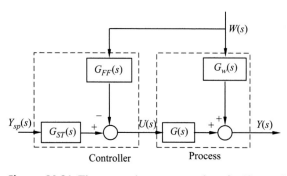

Figure 20.21 The general structure of a feedforward controller.

20.5 Model-Based Feedforward Control

$$G_{ST}^{opt}(s) = \frac{1}{G^-(s)} \quad , \quad G_{FF}^{opt}(s) = \frac{s}{G^-(s)} \left\{ \frac{G_w(s)}{s\, G^+(s)} \right\}_* \tag{20.5.6}$$

where the operator $\{\ \}_*$ denotes that, after a partial fraction expansion of the operand, all terms involving unstable poles and/or predictions of the form $e^{+\theta s}$ are omitted. With these optimal choices, the resulting overall transfer-function description (20.5.4) takes the form:

$$Y^{opt}(s) = G^+(s) Y_{sp}(s) + \left[G_w(s) - s\, G^+(s) \left\{ \frac{G_w(s)}{s\, G^+(s)} \right\}_* \right] W(s) \tag{20.5.7}$$

A general feedforward control system has the flexibility of using a separate controller for set-point tracking and a separate one for rejecting the measured disturbance, and these can be chosen completely independently of each other. Therefore, it is possible to achieve optimality for both set-point tracking and disturbance rejection (of course keeping in mind that only measured disturbances can be rejected by feedforward control).

It will be instructive to compare the overall transfer-function description in feedforward and in feedback control: Eq. (20.5.4) versus Eq. (20.5.3). The first observation is that the transfer functions with respect to the set point are identical, with $G_{ST}(s)$ in feedforward control being $Q(s)$ in feedback control. Moreover, their ISE-optimal choices are exactly the same: $1/G^-(s)$. Thus, the immediate conclusion is that *feedforward and feedback control perform exactly the same in terms of set-point tracking.*

Comparing the transfer functions with respect to the measured disturbance, we see $G_w(s) - G(s) G_{FF}(s)$ in feedforward control versus $(1 - G(s) Q(s)) G_w(s)$ in feedback control. These can be in principle nullified – exactly or approximately – by appropriate choice of $G_{FF}(s)$ or $Q(s)$, whenever $G(s)$ is minimum-phase. However, in general, there will be differences in performance, and this will be examined in the examples that will be given below.

Example 20.4 Optimal feedforward control of a second-order process with a positive zero

Consider a second-order process with a positive zero, and a disturbance transfer function that is first order:

$$G(s) = \frac{k(-\tau_0 s + 1)}{(\tau_1 s + 1)(\tau_2 s + 1)} \quad , \quad G_w(s) = \frac{k_w}{\tau_w s + 1} \quad , \quad \tau_0 > 0,\ \tau_1 > 0,\ \tau_2 > 0,\ \tau_w > 0$$

Derive a feedforward controller $G_{FF}(s)$ that is nearly ISE-optimal for step changes in the disturbance w. Compare its performance with the feedback controller derived in Chapter 19 (Section 19.3.3).

Solution

For this system, the process transfer function $G(s)$ is factorized as follows

$$G^+(s) = \frac{-\tau_0 s + 1}{\tau_0 s + 1}, \quad G^-(s) = k \frac{(\tau_0 s + 1)}{(\tau_1 s + 1)(\tau_2 s + 1)}$$

From formula (20.5.6), we have $G_{FF}^{opt}(s) = \dfrac{s}{G^-(s)} \left\{ \dfrac{G_w(s)}{s\, G^+(s)} \right\}_*$.

In the problem under consideration,

$$\dfrac{G_w(s)}{sG^+(s)} = \dfrac{k_w(\tau_0 s+1)}{s(\tau_w s+1)(-\tau_0 s+1)} = k_w \left(\dfrac{1}{s} + \dfrac{\dfrac{\tau_0-\tau_w}{\tau_0+\tau_w} \tau_w}{\tau_w s+1} + \dfrac{\dfrac{2\tau_0^2}{\tau_0+\tau_w}}{-\tau_0 s+1} \right)$$

$$\Rightarrow \left\{ \dfrac{G_w(s)}{sG^+(s)} \right\}_* = k_w \left(\dfrac{1}{s} + \dfrac{\dfrac{\tau_0-\tau_w}{\tau_0+\tau_w} \tau_w}{\tau_w s+1} \right) = k_w \dfrac{\dfrac{2\tau_0 \tau_w}{\tau_0+\tau_w} s+1}{s(\tau_w s+1)}$$

and thus finally

$$G_{FF}^{opt}(s) = \dfrac{s}{G^-(s)} \left\{ \dfrac{G_w(s)}{sG^+(s)} \right\}_* = \dfrac{s(\tau_1 s+1)(\tau_2 s+1)}{k(\tau_0 s+1)} \cdot k_w \dfrac{\dfrac{2\tau_0 \tau_w}{\tau_0+\tau_w} s+1}{s(\tau_w s+1)}$$

$$= \dfrac{k_w}{k} \cdot \dfrac{(\tau_1 s+1)(\tau_2 s+1)\left(\dfrac{2\tau_0 \tau_w}{\tau_0+\tau_w} s+1\right)}{(\tau_w s+1)(\tau_0 s+1)}$$

The above optimal feedforward controller has an improper transfer function. A realizable approximation is obtained by adding a first-order filter in series:

$$G_{FF}(s) = G_{FF}^{opt}(s) \dfrac{1}{\lambda s+1} = \dfrac{k_w}{k} \cdot \dfrac{(\tau_1 s+1)(\tau_2 s+1)\left(\dfrac{2\tau_0 \tau_w}{\tau_0+\tau_w} s+1\right)}{(\tau_w s+1)(\tau_0 s+1)(\lambda s+1)}$$

In Chapter 19, the following model-based feedback controller was derived (see Eq. (19.3.13)):

$$G_c(s) = \dfrac{1}{k} \dfrac{(\tau_1 s+1)(\tau_2 s+1)}{(\lambda \tau_0 s + 2\tau_0 + \lambda)s}$$

Figure 20.22 compares the closed-loop responses for a unit step change in the disturbance and zero set point, for the following values of process parameters:

$$\tau_0 = 1, \quad \tau_1 = \dfrac{1}{2}, \quad \tau_2 = \dfrac{1}{3}, \quad k = \dfrac{1}{6}, \quad k_w = \dfrac{1}{2}, \quad \tau_w = \dfrac{1}{2}$$

and for a filter time constant $\lambda = 0.1$.

We see that the feedforward controller eliminates the effect of the disturbance faster, even though the feedback controller can still do a reasonable job in disturbance rejection. Calculating ISE for each of the responses, we find $ISE_{FF\ controller} = 0.27$, whereas $ISE_{FB\ controller} = 0.36$.

In the next example, we will consider the case of a process transfer function $G(s)$ with dead time, where again the disturbance-optimal controller is different from the set-point-optimal controller, and the same issues arise.

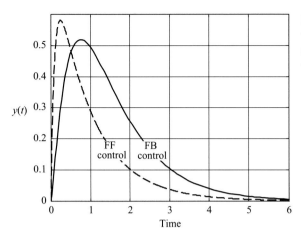

Figure 20.22 Comparison of the responses of the feedforward controller and the feedback controller to a unit step disturbance.

Example 20.5 Optimal feedforward control of a FOPDT process

Consider a process that is first-order plus dead time (FOPDT), and a disturbance transfer function that is also FOPDT:

$$G(s) = \frac{k}{\tau s + 1} e^{-\theta s}, \quad G_w(s) = \frac{k_w}{\tau_w s + 1} e^{-\theta_w s}, \quad \tau > 0, \ \tau_w > 0, \ \theta > 0, \ \theta_w > 0$$

Derive a feedforward controller $G_{FF}(s)$ that is nearly ISE-optimal for step changes in the disturbance w. Distinguish two cases: $\theta_w \geq \theta$ and $\theta_w < \theta$.

Also, calculate the corresponding optimal closed-loop transfer function $\dfrac{Y^{opt}(s)}{W(s)}$ and the optimal output response when the disturbance undergoes a unit step change and the set point is zero.

Compare the performance of the feedforward controller with the feedback controller derived in Chapter 19 (Section 19.3.5).

Solution

For this system, the process transfer function $G(s)$ is factorized as follows:

$$G^+(s) = e^{-\theta s}, \quad G^-(s) = \frac{k}{\tau s + 1}$$

Moreover,

$$\frac{G_w(s)}{s\, G^+(s)} = \frac{k_w e^{-\theta_w s}}{\tau_w s + 1} \frac{e^{\theta s}}{s} = \frac{k_w}{\tau_w s + 1} \frac{e^{(\theta - \theta_w)s}}{s}$$

Case 1: $\theta_w \geq \theta$

The ratio $\dfrac{G_w(s)}{s\, G^+(s)}$ does not involve prediction, therefore

$$\left\{ \frac{G_w(s)}{s\, G^+(s)} \right\}_* = \frac{G_w(s)}{s\, G^+(s)} = \frac{k_w}{\tau_w s + 1} \frac{e^{(\theta - \theta_w)s}}{s}$$

and so
$$G_{FF}^{opt}(s) = \frac{s}{G^-(s)}\left\{\frac{G_w(s)}{s\,G^+(s)}\right\}_* = \frac{s}{k}\cdot\frac{k_w}{\tau_w s+1}\cdot\frac{e^{(\theta-\theta_w)s}}{s} = \frac{k_w}{k}\frac{\tau s+1}{\tau_w s+1}e^{-(\theta_w-\theta)s}$$

and
$$\frac{Y^{opt}(s)}{W(s)} = G_w(s) - G(s)G_{FF}^{opt}(s) = 0$$

We see that in this case, *perfect control* is achieved under feedforward control: the effect of the disturbance is completely eliminated in no time. This is impossible under feedback control: in the presence of dead time θ_w in the disturbance (delayed detection) and dead time θ in the manipulated input (delayed controller action), nothing will happen until after $(\theta + \theta_w)$ time units.

Case 2: $\theta_w < \theta$

The ratio $\dfrac{G_w(s)}{s\,G^+(s)}$ involves the prediction $e^{(\theta-\theta_w)s}$, therefore it should be expanded and the terms involving predictions should be dropped. We find:

$$\frac{G_w(s)}{sG^+(s)} = \frac{k_w}{\tau_w s+1}\cdot\frac{e^{(\theta-\theta_w)s}}{s} = k_w\left(\frac{1}{s} - \frac{e^{-\frac{\theta-\theta_w}{\tau_w}}}{s+\frac{1}{\tau_w}} + (\text{terms involving prediction})\right)$$

$$\Rightarrow \left\{\frac{G_w(s)}{sG^+(s)}\right\}_* = k_w\left(\frac{1}{s} - \frac{e^{-\frac{\theta-\theta_w}{\tau_w}}}{s+\frac{1}{\tau_w}}\right) = \frac{k_w\left[\left(1-e^{-\frac{\theta-\theta_w}{\tau_w}}\right)\tau_w s+1\right]}{s(\tau_w s+1)}$$

and finally

$$G_{FF}^{opt}(s) = \frac{s}{G^-(s)}\left\{\frac{G_w(s)}{sG^+(s)}\right\}_* = \frac{s}{k}\cdot\frac{\tau s+1}{\tau_w s+1}\cdot\frac{k_w\left[\left(1-e^{-\frac{\theta-\theta_w}{\tau_w}}\right)\tau_w s+1\right]}{s(\tau_w s+1)}$$

$$= \frac{k_w}{k}\frac{\tau s+1}{\tau_w s+1}\left[\left(1-e^{-\frac{\theta-\theta_w}{\tau_w}}\right)\tau_w s+1\right]$$

The corresponding optimal closed-loop transfer function is

$$\frac{Y^{opt}(s)}{W(s)} = G_w(s) - G(s)G_{FF}^{opt}(s) = \frac{k_w}{\tau_w s+1}e^{-\theta_w s} - \frac{k_w\left[\left(1-e^{-\frac{\theta-\theta_w}{\tau_w}}\right)\tau_w s+1\right]}{\tau_w s+1}e^{-\theta s}$$

$$= k_w\frac{1-\left[\left(1-e^{-\frac{\theta-\theta_w}{\tau_w}}\right)\tau_w s+1\right]e^{-(\theta-\theta_w)s}}{\tau_w s+1}e^{-\theta_w s}$$

For a unit step change in the disturbance $W(s) = 1/s$ and zero set point, we invert the Laplace transform of $Y^{opt}(s)$ and find:

$$y^{opt}(t) = \begin{cases} 0 & , \text{ for } 0 \le t \le \theta_w \\ k_w \left(1 - e^{-\frac{t-\theta_w}{\tau_w}} \right) & , \text{ for } \theta_w \le t < \theta \\ 0 & , \text{ for } t > \theta \end{cases}$$

Figure 20.23 depicts the optimal response. We see that the effect of the disturbance is completely eliminated *after time θ*. Because the disturbance is measured, the controller acts immediately after the step change occurs, but because the manipulated input has dead time, elimination of the effect of the disturbance can happen *after the manipulated input's dead time*.

The derived optimal feedforward controller has an improper transfer function. A realizable approximation is obtained by adding a first-order filter in series:

$$G_{FF}(s) = G_{FF}^{opt}(s) \frac{1}{\lambda s + 1} = \frac{k_w}{k} \frac{\tau s + 1}{\tau_w s + 1} \frac{\left(1 - e^{-\frac{\theta - \theta_w}{\tau_w}}\right) \tau_w s + 1}{\lambda s + 1}$$

With the addition of the filter, elimination of the effect of the disturbance will not happen instantaneously at time θ, but after going through a fast exponential transient. This nearly optimal disturbance rejection is depicted in Figure 20.24 (solid line) along with the optimal response (dashed line).

It is instructive to compare the performance of the feedforward controller to the one of the model-based feedback controller derived in Chapter 19 (see Eq. (19.3.26)):

$$G_c(s) = \frac{1}{\lambda s + 1 - e^{-\theta s}} \cdot \frac{\tau s + 1}{k}$$

This controller gives the following closed-loop transfer function with respect to the disturbance:

$$\frac{Y(s)}{W(s)} = \left(1 - \frac{e^{-\theta s}}{\lambda s + 1}\right) \frac{k_w}{\tau_w s + 1} e^{-\theta_w s}$$

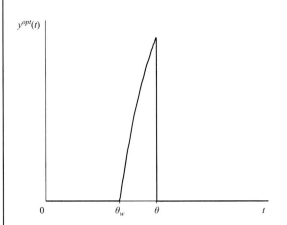

Figure 20.23 ISE-optimal response under feedforward control to a unit step disturbance.

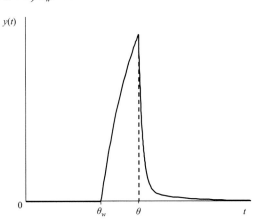

Figure 20.24 Response to a unit step disturbance under feedforward control (dashed line ISE-optimal, solid line nearly optimal).

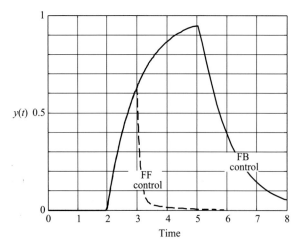

Figure 20.25 Response to a unit step disturbance under feedforward control and feedback control

The feedback controller will not be able to detect the disturbance till after time θ_w (disturbance dead time) and will not be able to act till after additional time θ (manipulated input dead time). The end result is that the elimination of the effect of the disturbance will start *after time* $\theta + \theta_w$.

Figure 20.25 compares the responses under feedforward control and feedback control for a unit step change in the disturbance and for the following values of process parameters:

$$k = k_w = 1, \quad \tau = \tau_w = 1, \quad \theta = 3, \quad \theta_w = 2$$

and for a filter time constant of $\lambda = 0.1$.

We see that, in feedforward control, the disturbance is eliminated after $\theta = 3$ time units and a short transient, but in feedback control we need to wait after $\theta_w + \theta = 5$ time units.

20.5.3 Comparison of Performance of Feedback and Feedforward Control in Disturbance Rejection

A direct comparison of the optimal closed-loop transfer functions with respect to the disturbance will allow us to evaluate the potential benefits of applying feedforward action using a disturbance measurement.

Under feedback control, choosing the set-point-optimal $Q(s) = 1/\bar{G}(s)$, from Eq. (20.5.3) we obtain

$$\left. \frac{Y^{sp-opt}(s)}{W(s)} \right|_{FB} = (1 - G(s)Q^{sp-opt}(s))G_w(s) = G_w(s) - G^+(s)G_w(s) \tag{20.5.8}$$

whereas under feedforward control, from Eq. (20.5.7),

$$\left. \frac{Y^{opt}(s)}{W(s)} \right|_{FF} = G_w(s) - G(s)G_{FF}^{opt}(s) = G_w(s) - s\, G^+(s) \left\{ \frac{G_w(s)}{s\, G^+(s)} \right\}_* \tag{20.5.9}$$

Some immediate conclusions may be drawn, comparing Eqs. (20.5.8) and (20.5.9).

(i) *When $G(s)$ is minimum-phase*, the optimal closed-loop transfer functions are zero in both FB and FF, and this can be accomplished at least approximately. The optimal closed-loop transfer function of FF can become exactly zero when the relative order of

$G_w(s)$ is larger or equal to the relative order of $G(s)$, in which case $G_{FF}^{opt}(s)$ is a proper transfer function. In this case, FF control outperforms FB.

(ii) *When $G(s)$ is nonminimum-phase, but $G_w(s)$ is minimum-phase*, the effect of the disturbance cannot be completely eliminated, neither in FB nor in FF. But because FB controllers are usually optimized for set-point tracking instead of disturbance rejection, performance in FF control is generally higher than FB. This was the case in Example 20.4.

(iii) *When both $G(s)$ and $G_w(s)$ are nonminimum-phase*, FF control may provide a very big improvement over FB *if there are cancellations of all-pass elements in the ratio* $\dfrac{G_w(s)}{G^+(s)}$ *inside $\{\ \}_*$*. If $G^+(s)$ is completely canceled, the optimal closed-loop transfer function of FF is zero (perfect rejection of w). In Example 20.5, complete cancelation happened in the case $\theta_w \geq \theta$, and partial cancelation when $\theta_w < \theta$.

In summary, FF generally outperforms FB, but the increase in performance can be very significant only in case (iii), when $G_w(s)$ inherits (some of or all of) the nonminimum-phase characteristics of $G(s)$. From an intuitive point of view, we can say that FF can give a big benefit in the case of delayed disturbances: if the disturbance dead time is larger than the manipulated input dead time, there is more than enough time to completely eliminate the disturbance, but even if it is smaller, the controller has the time to eliminate a significant part of the disturbance before it affects the output. From a practical standpoint, the use of feedforward measurements from an upstream unit can be extremely valuable in improving the operation of the downstream unit.

20.5.4 Model-Based Feedforward Plus Feedback Control System

Feedforward control can provide the best possible rejection of a particular measurable disturbance, but when used alone, it is generally inadequate because of the presence of other, unmeasured disturbances, modeling error and measurement error. However, when it is used together with feedback, the feedback loop will (i) ensure stability and (ii) provide adequate rejection of all kinds of disturbances, measured and unmeasured, despite measurement error or model–process mismatch. Thus, the combination of feedback with feedforward will provide superior performance in rejecting the measurable disturbance, while possessing all the valuable properties of feedback control. Figure 20.26 depicts a feedforward plus feedback control system that can handle both set-point tracking and disturbance rejection. Simplifying the block diagram, we obtain:

$$Y(s) = \frac{G(s)G_c(s)}{1+G(s)G_c(s)} Y_{sp}(s) + \frac{G_w(s)-G(s)G_{FF}(s)}{1+G(s)G_c(s)} W(s) \qquad (20.5.10)$$

From the transfer functions in Eq. (20.5.10), we see that set-point tracking is affected only by the feedback controller $G_c(s)$. However, both controllers are working synergistically for disturbance rejection:

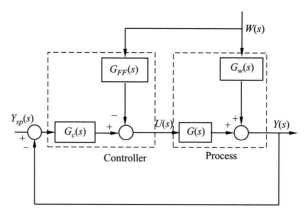

Figure 20.26 The general structure of feedforward plus feedback control.

- $G_c(s)$ is trying to make the factor $1/(1+G(s)G_c(s))$ small, whereas
- $G_{FF}(s)$ is trying to make the factor $(G_w(s) - G(s)G_{FF}(s))$ ideally zero, or at least small.

It is this double action, feedforward combined with feedback, that gives very successful disturbance rejection.

The individual feedback and feedforward controllers $G_c(s)$ and $G_{FF}(s)$ may be designed independently, with:

- $G_c(s) = \dfrac{Q(s)}{1 - G(s)Q(s)}$, with $Q(s)$ a proper filtered approximation of $\dfrac{1}{G^-(s)}$

- $G_{FF}(s) = \dfrac{s}{G^-(s)} \left\{ \dfrac{G_w(s)}{s\, G^+(s)} \right\}_*$ or a proper filtered approximation of it, if needed.

Example 20.6 Combined feedforward and feedback control of a second-order minimum-phase process

Consider a second-order minimum-phase process, with a disturbance transfer function that is first order:

$$G(s) = \frac{k}{(\tau_1 s + 1)(\tau_2 s + 1)}, \quad G_w(s) = \frac{k_w}{\tau_1 s + 1}, \quad \tau_1 > 0,\ \tau_2 > 0$$

Derive a feedback controller, a feedforward controller and a combined feedforward plus feedback controller, and the resulting closed-loop transfer functions, with respect to the disturbance. Compare their responses to a unit step change in the disturbance.

Solution

For this system, the relative order of $G(s)$ is 2, and the relative order of $G_w(s)$ is 1. Perfect cancelation of the disturbance is not possible since $G_w(s)/G(s)$ is not proper. The model-based feedback and feedforward controllers are:

$$G_c(s) = \frac{1}{(\lambda s+1)^2 - 1} \cdot \frac{(\tau_1 s+1)(\tau_2 s+1)}{k} = \frac{(\tau_1 s+1)(\tau_2 s+1)}{k\lambda s(\lambda s+2)} \quad (filtered\ PID)$$

$$G_{FF}(s) = \frac{1}{\lambda' s+1} \cdot \frac{\dfrac{k_w}{\tau_1 s+1}}{\dfrac{k}{(\tau_1 s+1)(\tau_2 s+1)}} = \frac{k_w}{k} \cdot \frac{\tau_2 s+1}{\lambda' s+1} \quad (gain + lead\ /\ lag)$$

The resulting closed-loop transfer functions with respect to the disturbance, under feedback only, feedforward only and feedforward/feedback are as follows:

$$\left.\frac{Y(s)}{W(s)}\right|_{FB} = \frac{\lambda s(\lambda s+2)}{(\lambda s+1)^2} \cdot \frac{k_w}{\tau_1 s+1}, \quad \left.\frac{Y(s)}{W(s)}\right|_{FF} = \frac{\lambda' s}{\lambda' s+1} \cdot \frac{k_w}{\tau_1 s+1}$$

$$\left.\frac{Y(s)}{W(s)}\right|_{FF+FB} = \frac{\lambda s(\lambda s+2)}{(\lambda s+1)^2} \cdot \frac{\lambda' s}{\lambda' s+1} \cdot \frac{k_w}{\tau_1 s+1}$$

In the feedforward plus feedback control system's closed-loop transfer function, we can identify two factors that reduce the effect of the disturbance: $\dfrac{\lambda s(\lambda s+2)}{(\lambda s+1)^2}$ from the action of the feedback controller and $\dfrac{\lambda' s}{\lambda' s+1}$ from the action of the feedforward. The overall effect makes feedforward plus feedback the most effective choice for disturbance rejection.

In Figure 20.27, the closed-loop responses for a unit step change in the disturbance and zero set point are compared, for the process parameter values $k_w = \tau_1 = 1$ and filter time constants $\lambda = \lambda' = 0.1$.

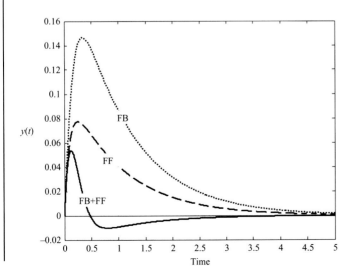

Figure 20.27 Comparison of feedback (FB), feedforward (FF) and combined feedback and feedforward (FB+FF) control system response to a unit step disturbance.

20.5.5 Q-Parameterization of a Combined Feedforward Plus Feedback Control System

Consider again the feedforward plus feedback control system of Figure 20.26, whose transfer-function description is given by Eq. (20.5.10) and assume that both $G(s)$ and $G_w(s)$ are stable. Rearranging (20.5.10) as

$$Y(s) = G(s)\frac{G_c(s)}{1+G(s)G_c(s)}Y_{sp}(s) + \left(G_w(s) - \frac{G_w(s)G_c(s)+G_{FF}(s)}{1+G(s)G_c(s)}G(s)\right)W(s)$$

and setting

$$Q(s) = \frac{G_c(s)}{1+G(s)G_c(s)}, \quad Q_w(s) = \frac{G_w(s)G_c(s)+G_{FF}(s)}{1+G(s)G_c(s)} \tag{20.5.11}$$

or equivalently,

$$G_c(s) = \frac{Q(s)}{1-G(s)Q(s)}, \quad G_{FF}(s) = \frac{Q_w(s)-G_w(s)Q(s)}{1-G(s)Q(s)} \tag{20.5.12}$$

the transfer-function description of the closed-loop system takes the form

$$Y(s) = G(s)Q(s)Y_{sp}(s) + (G_w(s) - Q_w(s)G(s))W(s) \tag{20.5.13}$$

The closed-loop system will be stable for all controllers $G_c(s)$, $G_{FF}(s)$ that are generated by stable $Q(s)$ and $Q_w(s)$ via Eq. (20.5.12). Also, the closed-loop system will have zero offset as long as $Q(0)=1/G(0)$ and $Q_w(0)=G_w(0)/G(0)$. Moreover, we observe that with the foregoing parameterization, the closed-loop transfer functions take a form identical to pure feedforward control (compare with Eq. (20.5.4)), with $Q(s)$ and $Q_w(s)$ in place of $G_{ST}(s)$ and $G_{FF}(s)$. Thus, applying Eq. (20.5.6) we obtain the optimal choices

$$Q^{opt}(s) = \frac{1}{G^-(s)}, \quad Q_w^{opt}(s) = \frac{s}{G^-(s)}\left\{\frac{G_w(s)}{s\,G^+(s)}\right\}_* \tag{20.5.14}$$

and the resulting transfer-function description of the closed-loop system becomes identical to Eq. (20.5.7).

The feedforward plus feedback control system with Q-parameterized controllers can be seen as a control system that includes internal models of $G(s)$ and $G_w(s)$. For this reason, it is often referred to as the *Feedforward–Feedback Internal Model Control* structure (FF+FB IMC). The structure of this equivalent system representation is depicted in the block diagram of Figure 20.28.

20.5 Model-Based Feedforward Control

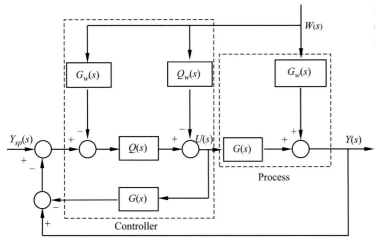

Figure 20.28 The general Q-parameterized combined feedforward plus feedback control structure.

Example 20.7 Feedforward plus feedback control of a second-order process with a positive zero

Consider the following second-order process with a positive zero, and a disturbance transfer function that is first order:

$$G(s) = \frac{1-s}{(s+2)(s+3)}, \qquad G_w(s) = \frac{1}{s+2}$$

Derive nearly optimal transfer functions for a feedforward plus feedback control system using the Q-parameterization approach.

Solution

For this system,

$$G^-(s) = \frac{s+1}{(s+2)(s+3)}, \quad G^+(s) = \frac{1-s}{1+s}, \quad G_w(s) = \frac{1}{s+2}$$

and we can calculate the optimal feedforward controllers

$$Q^{opt}(s) = \frac{1}{G^-(s)} = \frac{(s+2)(s+3)}{s+1}$$

$$Q_w^{opt}(s) = \frac{s}{G^-(s)} \left\{ \frac{G_w(s)}{sG^+(s)} \right\}_* = \frac{s(s+2)(s+3)}{1+s} \left\{ \frac{1+s}{s(s+2)(1-s)} \right\}_*$$

$$= \frac{s(s+2)(s+3)}{1+s} \left\{ \frac{\frac{1}{2}}{s} + \frac{\frac{1}{6}}{s+2} + \frac{\frac{2}{3}}{1-s} \right\}_* = \frac{s(s+2)(s+3)}{1+s} \left(\frac{\frac{1}{2}}{s} + \frac{\frac{1}{6}}{s+2} \right) = \frac{(s+3)\left(\frac{2}{3}s+1\right)}{1+s}$$

Hence, we can choose

$$Q(s) = \frac{1}{\lambda s + 1} Q^{opt}(s) = \frac{(s+2)(s+3)}{(\lambda s + 1)(s+1)}$$

$$Q_w(s) = \frac{1}{\lambda' s + 1} Q_w^{opt}(s) = \frac{(s+3)\left(\frac{2}{3}s+1\right)}{(\lambda' s + 1)(s+1)}$$

Also, for simplicity, take $\lambda = \lambda'$. This leads to

$$G_c(s) = \frac{(s+2)(s+3)}{s(\lambda s + \lambda + 2)} \quad (\textit{filtered PID})$$

$$G_{FF}(s) = \frac{\frac{2}{3}(s+3)}{\lambda s + \lambda + 2} \quad (\textit{gain + lead / lag})$$

Under the above controllers, the closed-loop transfer function with respect to the disturbance is:

$$\left.\frac{Y(s)}{W(s)}\right|_{FF+FB} = \left(1 - \frac{\frac{2}{3}s+1}{\lambda s + 1} \cdot \frac{1-s}{1+s}\right)\frac{1}{s+2}$$

In the limit as $\lambda \to 0$, one obtains the ISE-optimal closed-loop transfer function

$$\frac{Y^{opt}(s)}{W(s)} = \left(1 - \left(\frac{2}{3}s+1\right)\frac{1-s}{1+s}\right)\frac{1}{s+2} = \frac{\frac{2}{3}s}{1+s}$$

LEARNING SUMMARY

- Cascade control is commonly used in practice for improved disturbance rejection. Cascade control can improve control performance through an inner feedback loop that reduces the effect of a significant disturbance before it affects the controlled output.
- Cascade control is most commonly used when a flowrate needs to be manipulated, but it is disturbed by upstream or downstream pressure variations.
- Model-based design of cascade control systems can be achieved using the methods of Chapter 19, which are explored further in this chapter.
- Ratio control is a special case of feedforward control, which is used when the flowrates of two streams must be kept at a given fixed ratio. It involves measuring one flowrate and adjusting the other.
- Feedforward control uses measurements of disturbances to predict their effect on the controlled output, and act proactively to try to cancel their effect.

- Feedforward control is not used alone, but it is combined with feedback control. Feedforward and feedback actions are applied in parallel, and this results in combining their advantages.
- Feedforward controllers can be optimally designed based on a process model, in the same spirit as model-based feedback controllers in Chapter 19.

TERMS AND CONCEPTS

Cascade control. Involves two feedback loops, one inside the other, where the outer loop provides the set point of the inner loop.

Feedforward control. Involves measuring a disturbance and adjusting the manipulated input in a preventive manner, so that the controlled output remains unaffected.

Inner or secondary loop of a cascade control system. The interior feedback loop, which measures and controls an auxiliary output (secondary output).

Outer or primary loop of a cascade control system. The exterior feedback loop, which measures and controls the controlled output (primary output).

Ratio control. A special case of feedforward control, where one flowrate is measured (disturbance) and another is adjusted, so as to keep them at a fixed ratio.

FURTHER READING

Additional information on the design of model-based cascade and feedforward control systems can be found in

Morari, M. and Zafiriou, E., *Robust Process Control*. Englewood Cliffs, NJ: Prentice Hall, 1989.

PROBLEMS

20.1 Consider the level-control problem of the tanks in series system of Example 13.1. Suppose that, in addition to the level sensor for the second tank that measures the controlled variable, we also have a level sensor installed for the first tank, and we want to use this additional measurement to build a cascade control system for the tanks. The model equations for the tanks are given by

$$A_1 \frac{dh_1}{dt} = -\frac{h_1}{R_1} + k_v u + F_w$$

$$A_2 \frac{dh_2}{dt} = \frac{h_1}{R_1} - \frac{h_2}{R_2}$$

and the parameter values are $A_1 = R_1 = A_2 = R_2 = k_v = 1$. The primary output of the system (controlled variable) is h_2, whereas h_1 is a secondary output, and we assume that they are both measured with very fast level sensors (negligible time constants). The main disturbance of the system is the flowrate F_w, which can vary in size in an unpredictable manner.

(a) Draw the block diagram of a cascade control system, where the inner (secondary) loop regulates h_1 and the outer (primary) loop regulates h_2.
(b) Build a model-based PI controller for the inner loop and a model-based PID controller with filter for the outer loop.
(c) Calculate and plot the closed-loop response of the controlled output h_2 in deviation form, when F_w undergoes a unit step increase. Use $\lambda = 0.1$ as the filter parameter for both controllers.
(d) If single-loop feedback control is applied to the system, with the same PID controller as in the previous question, what will be the closed-loop response? In a common graph, compare it with the closed-loop response of the cascade control system of the previous question (c).

20.2 Repeat the previous problem for the case where, instead of h_2, the primary output is the outlet flowrate $F_2 = h_2/R_2$ and it is measured downstream with dead time $\theta = 2$.

20.3 Consider temperature control of the process of Problem 11.11 (Figure P11.11). The model equations for the tanks are given by:

$$V_1 \rho c_p \frac{dT_1}{dt} = F\rho c_p (T_0 - T_1) + Q$$

$$V_2 \rho c_p \frac{dT_2}{dt} = F\rho c_p (T_1 - T_2)$$

The heating rate Q is the manipulated input, the inlet temperature T_0 is the disturbance input, the outlet temperature T_2 is the controlled output, and V_1, V_2, F, ρ, c_p are constant parameters.

Suppose that manipulation of the heating rate Q can be accomplished very quickly and that both temperatures T_1 and T_2 can be measured on-line with very fast sensors. We would like to build a cascade control system with T_2 as the primary output and T_1 as secondary output.

(a) Draw the block diagram of a cascade control system, where the inner (secondary) loop regulates T_1 and the outer (primary) loop regulates T_2.
(b) Build a model-based PI controller for the inner loop and a model-based PID controller with filter for the outer loop.
(c) Calculate the closed-loop transfer function that relates T_2 to T_0. Compare this with the closed-loop transfer function under single-loop feedback control, with the same PID controller as in the cascade control system. Do you expect improved disturbance rejection under cascade control?

20.4 Consider the heater example that was studied in Chapters 12 and 13 from the point of view of stability under P and PI control, whereas in Chapter 16 optimal PI controller parameters were calculated for step changes in the set point. The model equations for the instrumented process are given by

$$\frac{dT}{dt} = \frac{F}{V}(T_0 - T) + \frac{\Delta h_{vap,st}}{V\rho c_p} k_v u$$

$$\frac{dT_m}{dt} = \frac{1}{\tau_m}(T - T_m)$$

and the following parameter values were used:

$$\frac{F}{V} = 1, \quad \frac{\Delta h_{vap,st}}{V\rho c_p} = 1, \quad k_v = \frac{1}{5}, \quad \tau_m = \frac{1}{10}$$

It was found in Chapter 16 that the PI controller parameter values $k_c = 55$ and $\tau_I = 2.1$ correspond to minimal ISE for step changes in the set point.

Suppose now that the set point is unchanged but there is a unit step disturbance in the temperature T_0 of the inlet stream.
(a) Calculate and plot the closed-loop response in deviation form under PI control, using the above set-point-optimal controller parameters.
(b) Assuming that the temperature T_0 is measured on line, derive a static feedforward controller for the heater and combine it with the PI controller of question (a). Calculate and plot the closed-loop response under this feedforward plus feedback controller and compare it with the response of the pure feedback controller of question (a).
(c) Repeat the previous question (b) using a dynamic feedforward controller, with filter parameter $\tau_f = 0.01$.

20.5 Consider a second-order process with transfer functions:

$$G(s) = \frac{k(-\tau_0 s + 1)}{(\tau_1 s + 1)(\tau_2 s + 1)}, \quad G_w(s) = \frac{k_w(-\tau_0 s + 1)}{(\tau_1 s + 1)(\tau_w s + 1)}, \quad \tau_0 > 0, \ \tau_1 > 0, \ \tau_2 > 0, \ \tau_w > 0$$

Apply formula (20.5.6) to derive a feedforward controller $G_{FF}(s)$ that is ISE-optimal for step changes in the disturbance. Will it be able to achieve perfect elimination of the disturbance? Compare your results with the results of Example 20.4, which has the same $G(s)$ but different $G_w(s)$. What is the key difference?

20.6 Consider the FOPDT process of Example 20.5. Combine the derived feedforward and feedback controllers in a feedforward plus feedback control structure.
(a) Calculate the closed-loop transfer function with respect to the disturbance.

(b) Simulate the response to a unit step change in the disturbance and compare it with the responses under pure feedback and pure feedforward control. Use the same parameter values as in Example 20.5.

20.7 Generalize the analysis of Example 20.7 for a second-order process with transfer functions:

$$G(s) = \frac{k(-\tau_0 s + 1)}{(\tau_1 s + 1)(\tau_2 s + 1)}, \quad G_w(s) = \frac{k_w}{\tau_w s + 1}, \quad \tau_0 > 0, \; \tau_1 > 0, \; \tau_2 > 0, \; \tau_w > 0$$

APPENDIX A
Laplace Transform

In this appendix, a short but comprehensive introduction to the Laplace transform will be presented. The material has been organized so as to serve both those readers who have already taken a course covering the Laplace transform and its application to the solution of differential equations, and as a crash course for the newcomer in the field.

A.1 Definition of the Laplace Transform

Let $f(t)$ be an integrable function defined on $[0, +\infty)$. The Laplace transform of $f(t)$ is defined as:

$$F(s) = \int_0^\infty e^{-st} f(t)\, dt \qquad (A.1.1)$$

where the parameter s in the integral is understood to be a complex number, therefore the integral defines a function of a complex variable s.

The notation $\mathcal{L}[f(t)]$ is also used to denote the Laplace transform of the function $f(t)$. Note that the lower-case symbol f is normally used to denote a function of t and the corresponding upper-case symbol F is used to denote its Laplace transform, function of s.

In dynamics applications, the variable t represents time. The variable s does not have a physical significance; it is a mathematical entity, with units of inverse time.

A very important observation from the definition of the Laplace transform is that the values of $f(t)$ for $t < 0$ do not affect the Laplace transform. For example, the Laplace transform of the Heaviside function (unit step function)

$$\mathcal{H}(t) = \begin{cases} 0, & \text{for } t < 0 \\ 1, & \text{for } t \geq 0 \end{cases} \qquad (A.1.2)$$

is exactly the same with the Laplace transform of the unity function $f(t) = 1$, $t \in \mathbb{R}$.

In math books, one can find conditions for the improper integral in (A.1.1) to converge. In particular, it is shown that if $f(t)$ is integrable and of exponential order, i.e. there exist $t_0 > 0$, $M > 0$ and $\alpha \in \mathbb{R}$ such that

$$|f(t)| \leq M e^{\alpha t} \ \forall t \geq t_0 \geq 0$$

then the improper integral $\int_0^\infty e^{-st} f(t)\, dt$ converges for all $s \in \mathbb{C}$ such that $Re\, s > \alpha$, and the function $F(s)$ that it defines is analytic in the open half plane $\{s \in \mathbb{C} / Re\, s > \alpha\}$.

In simple terms, a function will possess a Laplace transform as long as it does not grow faster than an exponential as $t \to \infty$. This is not a restrictive condition, as it is met by all commonly used functions.

A.2 Laplace Transforms of Elementary Functions

In Table A.1, the Laplace transform of some commonly encountered elementary functions are given. The derivation of these results is straightforward by direct application of the definition of the Laplace transform.

In addition to the standard elementary functions (constant, linear, power, exponential, sine, cosine), the table includes the pulse and impulse functions, which play a very important role in dynamics (see also Figure A.1).

The unit pulse function is defined as

$$\pi_\varepsilon(t) = \begin{cases} 0, & \text{for } t < 0 \\ \dfrac{1}{\varepsilon}, & 0 \leq t < \varepsilon \\ 0, & \text{for } t \geq \varepsilon \end{cases} \tag{A.2.1}$$

and can be expressed in terms of the Heaviside function as

$$\pi_\varepsilon(t) = \frac{1}{\varepsilon}\left(\mathcal{H}(t) - \mathcal{H}(t-\varepsilon)\right) \tag{A.2.2}$$

In other words, the unit pulse function is the superposition of a step of size $1/\varepsilon$ and a step of size $-1/\varepsilon$ delayed by ε.

As the width ε of the unit pulse $\pi_\varepsilon(t)$ goes to 0, its height $1/\varepsilon$ gets larger and larger, so the pulse gets "thinner" and "taller," while preserving the area under the curve (= 1). The so-called unit impulse function or Dirac delta function, is the limit of $\pi_\varepsilon(t)$ as $\varepsilon \to 0$ and is denoted by $\delta(t)$. It should be noted that $\delta(t)$ is not a function in a usual sense, it is a generalized function or distribution. The Dirac delta function is very useful in engineering calculations as an approximation of "thin and tall" pulses.

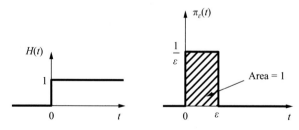

Figure A.1 Unit step function (left) and unit pulse function (right).

Table A.1 Laplace transforms of elementary functions

	$f(t)$	$F(s)$
Unity or unit step	1 or $\mathcal{H}(t)$	$\dfrac{1}{s}$
Unit ramp	t	$\dfrac{1}{s^2}$
Power	$t^n,\ n \in \mathbb{N}$	$\dfrac{n!}{s^{n+1}}$
Exponential	e^{-at}	$\dfrac{1}{s+a}$
Sine	$\sin(\omega t)$	$\dfrac{\omega}{s^2+\omega^2}$
Cosine	$\cos(\omega t)$	$\dfrac{s}{s^2+\omega^2}$
Unit pulse	$\pi_\varepsilon(t)$	$\dfrac{1-e^{-\varepsilon s}}{\varepsilon s}$
Unit impulse	$\delta(t)$	1

A.3 Properties of Laplace Transforms

If $\mathcal{L}[f(t)] = F(s)$, the following properties hold.

(1) *Linearity*:

$$\mathcal{L}[\lambda_1 f_1(t) + \lambda_2 f_2(t)] = \lambda_1 F_1(s) + \lambda_2 F_2(s) \qquad \forall \lambda_1, \lambda_2 \in \mathbb{R} \tag{A.3.1}$$

(2) *Transform variable translation*:

$$\mathcal{L}[e^{-at} f(t)] = F(s+a) \tag{A.3.2}$$

(3) *Transform differentiation*:

$$\mathcal{L}[t^n f(t)] = (-1)^n \frac{d^n}{ds^n} F(s) \tag{A.3.3}$$

(4) *Time-variable translation*: Given a function $f(t)$ and $t_0 > 0$, the translation of $f(t)$ by t_0 is defined as follows (see Figure A.2):

$$f^{transl}(t) = \begin{cases} 0, & \text{for } t < t_0 \\ f(t - t_0), & \text{for } t \geq t_0 \end{cases} \quad (A.3.4)$$

and the following holds true

$$\mathcal{L}\left[f^{transl}(t)\right] = e^{-t_0 s} F(s) \quad (A.3.5)$$

Note that, because the translated function may be expressed as

$$f^{transl}(t) = f(t - t_0) \, \mathcal{H}(t - t_0)$$

the time-variable translation property (A.3.5) may be written equivalently as

$$\mathcal{L}\left[f(t - t_0) \, \mathcal{H}(t - t_0)\right] = e^{-t_0 s} F(s) \quad (A.3.6)$$

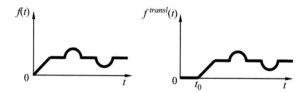

Figure A.2 Function $f(t)$ and its translation by t_0.

(5) *Differentiation with respect to time*: If $f(t)$ and its derivatives up to order $(n-1)$ are continuous functions of t, then

$$\mathcal{L}\left[\frac{df}{dt}(t)\right] = sF(s) - f(0) \quad (A.3.7a)$$

$$\mathcal{L}\left[\frac{d^2 f}{dt^2}(t)\right] = s^2 F(s) - sf(0) - \frac{df}{dt}(0) \quad (A.3.7b)$$

$$\vdots$$

$$\mathcal{L}\left[\frac{d^n f}{dt^n}(t)\right] = s^n F(s) - s^{n-1} f(0) - s^{n-2}\frac{df}{dt}(0) - \ldots - \frac{d^{n-1} f}{dt^{n-1}}(0) \quad (A.3.7c)$$

(6) *Integration with respect to time*:

$$\mathcal{L}\left[\int_0^t f(t')dt'\right] = \frac{1}{s} F(s) \quad (A.3.8)$$

(7) *Convolution theorem*:

$$\mathcal{L}\left[(f_1 * f_2)(t)\right] = F_1(s) F_2(s) \quad (A.3.9)$$

where

$$(f_1 * f_2)(t) = \int_0^t f_1(t-t') f_2(t') dt' = \int_0^t f_1(t') f_2(t-t') dt' \qquad (A.3.10)$$

is the convolution of $f_1(t)$ and $f_2(t)$.

(8) *Initial-value theorem*: If $F(s)$ tends to zero as $|s| \to \infty$, then

$$\lim_{t \to 0^+} f(t) = \lim_{|s| \to \infty} [sF(s)] \qquad (A.3.11)$$

(9) *Final-value theorem*: If $sF(s)$ is finite for every s with Re $s \geq 0$, then $\lim_{t \to \infty} f(t)$ exists and

$$\lim_{t \to \infty} f(t) = \lim_{s \to 0} [sF(s)] \qquad (A.3.12)$$

(10) *Parseval's theorem*:

$$\int_0^\infty [f(t)]^2 dt = \frac{1}{2\pi} \int_{-\infty}^\infty |F(i\omega)|^2 d\omega \qquad (A.3.13)$$

All properties are given here without proof and the reader is referred to any book on engineering mathematics for their proofs.

A.3.1 Significance of the Properties of Laplace Transforms

Properties (2) and (3) can be used to obtain the Laplace transforms of exponential times $f(t)$ or power times $f(t)$, if the Laplace transform of $f(t)$ is known. Table A.2 gives the Laplace transforms of some commonly used functions that are immediately obtained from Table A.1 by applying property (2) or property (3).

Properties (1) and (5) are of paramount importance since they enable the solution of linear differential equations with constant coefficients through Laplace transforms. The linearity property enables term-by-term transformation, multiplying by the appropriate coefficients. The differentiation property makes transforms of derivatives algebraic functions of s, and this implies that the Laplace transform of the linear differential equation is an algebraic equation. The solution of linear differential equations with constant coefficients will be discussed in detail in Section A.6.

Property (6) is, in a sense, symmetric to property (5): the transformed derivative involves multiplication by s, whereas the transformed integral division by s. Property (6) could be used for solving linear integral equations via Laplace transforms.

Property (4) tells us how the Laplace transforms are affected by time translation: the Laplace transform of the translated function, with translation time t_0, equals $e^{-t_0 s}$ times the Laplace transform of the original function. Physically, time translation occurs when a fluid travels through a pipe: the concentration at the exit of the pipe equals the concentration at the entrance, t_0 time units ago, where t_0 is the time needed for the fluid to travel through the

Table A.2 Laplace transform pairs of some commonly used functions

$f(t)$	$F(s)$
$t^n e^{-at}, n \in \mathbb{N}$	$\dfrac{n!}{(s+a)^{n+1}}$
$e^{-at} \sin(\omega t)$	$\dfrac{\omega}{(s+a)^2 + \omega^2}$
$e^{-at} \cos(\omega t)$	$\dfrac{s+a}{(s+a)^2 + \omega^2}$
$t \sin(\omega t)$	$\dfrac{2\omega s}{(s^2+\omega^2)^2}$
$t \cos(\omega t)$	$\dfrac{s^2 - \omega^2}{(s^2+\omega^2)^2}$

pipe. In dynamics, time translation is referred to as time delay or dead time. Because time delays are common in practice, property (4) is used quite frequently to relate the corresponding Laplace transforms.

Another application of property (4) is for the calculation of piecewise defined functions. For example, to calculate the Laplace transform of the unit pulse function, instead of directly applying the definition, one can "decompose" the pulse function according to (A.2.2) and do term-by-term transformation, applying (A.3.6) for the second term:

$$\mathcal{L}[\pi_\varepsilon(t)] = \frac{1}{\varepsilon}\mathcal{L}[\mathcal{H}(t)] - \frac{1}{\varepsilon}\mathcal{L}[\mathcal{H}(t-\varepsilon)] = \frac{1}{\varepsilon}\cdot\frac{1}{s} - \frac{1}{\varepsilon}\cdot\frac{1}{s}e^{-\varepsilon s} = \frac{1-e^{-\varepsilon s}}{\varepsilon s}$$

and thus the result of Table A.1 is obtained.

Property (7) is useful when we know the relation between Laplace transforms and we want to conclude how the corresponding time functions are related. Suppose for example that $G(s) = \dfrac{1}{s+2}F(s)$ and we want to know how the corresponding time functions $g(t)$ and $f(t)$ are related. Because $1/(s+2) = \mathcal{L}[e^{-2t}]$, it would be tempting to conclude that $g(t) = e^{-2t}\cdot f(t)$, but this is wrong! The correct answer is given by the convolution theorem: $g(t)$ equals the convolution of e^{-2t} and $f(t)$

$$g(t) = e^{-2t} * f(t) = \int_0^t e^{-2(t-t')}f(t')dt' = \int_0^t e^{-2t'}f(t-t')dt'$$

Properties (8) and (9) are useful when we know the Laplace transform $F(s)$ and we only need the initial value or the final value of the corresponding time function $f(t)$, and we do not want to go through the effort of calculating the whole function $f(t)$. Suppose, for example, that we are given the Laplace transform function

$$F(s) = \frac{2}{s(s+2)(s+4)}$$

$F(s)$ satisfies the assumption of the *initial-value theorem*, and we can calculate

$$\lim_{t \to 0^+} f(t) = \lim_{|s| \to \infty} [sF(s)] = \lim_{|s| \to \infty} \left[\frac{2}{(s+2)(s+4)} \right] = 0$$

Also, $sF(s)$ satisfies the assumption of the *final-value theorem*, since it can only become infinite at –2 and –4, which are negative numbers, and we can calculate

$$\lim_{t \to \infty} f(t) = \lim_{s \to 0} [sF(s)] = \lim_{s \to 0} \left[\frac{2}{(s+2)(s+4)} \right] = \frac{1}{4}$$

A.4 Inverse Laplace Transform

The problem of inverting the Laplace transform $F(s)$ is to find the function $f(t)$, $t > 0$ such that the following holds true:

$$\mathcal{L}[f(t)] = F(s) \tag{A.4.1}$$

If such a function can be found, then we say that the function $f(t)$ is the inverse Laplace transform of $F(s)$ and we use the following notation to denote this operation

$$f(t) = \mathcal{L}^{-1}[F(s)] \tag{A.4.2}$$

For simple functions $F(s)$, inversion may be achieved by directly using Tables A.1 and A.2. In the solution of differential equations, inversion of rational functions $F(s)$ arises in the last step of the solution procedure, and these are not always simple enough to be handled by table look-up. The inversion for this class of functions will be discussed in detail in the section that follows.

A.5 Calculation of the Inverse Laplace Transform of Rational Functions via Partial Fraction Expansion

A rational function is any function that can be written as the ratio of two polynomial functions, i.e. if it is of the form

$$F(s) = \frac{P(s)}{Q(s)} \tag{A.5.1}$$

where $P(s)$, $Q(s)$ are polynomial functions.

A rational function with $\deg P(s) < \deg Q(s)$ may be expanded in partial fractions, and this is the basis for inversion of the Laplace transform.

In order to calculate the partial fraction expansion and subsequently the inverse Laplace transform of (A.5.1), we need to consider two cases depending on whether the denominator polynomial $Q(s)$ has simple or repeated roots. It is recalled that, for any polynomial function $\varphi(s)$, the solutions of the equation $\varphi(s) = 0$ are called the roots of the polynomial φ. A number ρ is a root if and only if $\varphi(\rho) = 0$ and the polynomial $(s - \rho)$ divides $\varphi(s)$. It may happen that higher powers of $(s - \rho)$ divide $\varphi(s)$, in which case ρ is called a multiple root. Otherwise, ρ is called a simple root. The highest power m, such that $(s - \rho)^m$ divides $\varphi(s)$, is called the multiplicity of the root ρ.

A.5.1 Case 1: All Roots of Denominator Polynomial $Q(s)$ are Simple Roots

In this case $F(s)$ can be expanded in partial fractions as follows:

$$F(s) = \sum_j \frac{c_j}{s - \rho_j} \tag{A.5.2}$$

where the coefficients c_j can be calculated as follows

$$c_j = \left[(s - \rho_j) \frac{P(s)}{Q(s)} \right]_{s = \rho_j} \tag{A.5.3}$$

and ρ_j are the simple roots of the denominator polynomial $Q(s)$. Using the fact that

$$\mathcal{L}\left[e^{\rho t}\right] = \frac{1}{s - \rho}$$

we conclude that when all roots are simple roots, then the inverse Laplace of $F(s)$ is given by

$$f(t) = \mathcal{L}^{-1}\left[\sum_j \frac{c_j}{s - \rho_j}\right] = \sum_j c_j \mathcal{L}^{-1}\left[\frac{1}{s - \rho_j}\right] = \sum_j c_j e^{\rho_j t} \tag{A.5.4}$$

The case where $\rho = \alpha + i\beta$ is a complex root is of particular importance. It is important to note that, in this case, $\rho^* = \alpha - i\beta$ is also a root and ρ, ρ^* are called complex conjugate roots. The coefficients c that correspond to complex conjugate roots are also complex conjugates and the partial fraction expansion of $F(s)$ will be of the form

$$F(s) = \cdots + \underbrace{\frac{\overbrace{A + iB}^{c}}{s - \underbrace{(\alpha + i\beta)}_{\rho}}} + \underbrace{\frac{\overbrace{A - iB}^{c^*}}{s - \underbrace{(\alpha - i\beta)}_{\rho^*}}} \tag{A.5.5}$$

Inverting the Laplace transform we obtain

$$f(t) = \cdots + (A + iB)e^{(\alpha + i\beta)t} + (A - iB)e^{(\alpha - i\beta)t} + \cdots \tag{A.5.6}$$

The complex exponentials in (A.5.6) may be expanded using Euler's formula:

$$e^{(\alpha+i\beta)t} = e^{\alpha t}(\cos\beta t + i\sin\beta t)$$
$$e^{(\alpha-i\beta)t} = e^{\alpha t}(\cos\beta t - i\sin\beta t)$$

and, at the end, after simplifying the resulting expression, we finally obtain

$$f(t) = \cdots + 2Ae^{\alpha t}\cos(\beta t) - 2Be^{\alpha t}\sin(\beta t) + \cdots \qquad (A.5.7)$$

Example A.1 Inverse Laplace transform with simple real roots

Find the inverse Laplace transform of

$$F_1(s) = \frac{1}{s^2 + 3s + 2}$$

$$F_2(s) = \frac{1}{s(s^2 + 3s + 2)}$$

$$F_3(s) = \frac{s^2 + s + 1}{s(s^2 + 3s + 2)}$$

Solution

$F_1(s)$: We observe that the denominator polynomial has roots $\rho_1 = -1$, $\rho_2 = -2$. We then have that

$$F_1(s) = \frac{1}{s^2 + 3s + 2} = \frac{c_1}{s+1} + \frac{c_2}{s+2}$$

and

$$c_1 = (s+1)\frac{1}{s^2 + 3s + 2}\bigg|_{s=-1} = (s+1)\frac{1}{(s+1)(s+2)}\bigg|_{s=-1} = \frac{1}{(s+2)}\bigg|_{s=-1} = 1$$

$$c_2 = (s+2)\frac{1}{s^2 + 3s + 2}\bigg|_{s=-2} = (s+2)\frac{1}{(s+1)(s+2)}\bigg|_{s=-2} = \frac{1}{(s+1)}\bigg|_{s=-2} = -1$$

We therefore obtain

$$F_1(s) = \frac{1}{s+1} - \frac{1}{s+2}$$

and

$$f_1(t) = \mathcal{L}^{-1}[F_1(s)] = \mathcal{L}^{-1}\left[\frac{1}{s+1} - \frac{1}{s+2}\right] = \mathcal{L}^{-1}\left[\frac{1}{s+1}\right] - \mathcal{L}^{-1}\left[\frac{1}{s+2}\right] = e^{-t} - e^{-2t}$$

$F_2(s)$: We observe that the denominator polynomial has the roots $p_1 = -1$, $p_2 = -2$, as in the case of $F_1(s)$, and additionally it has the root $p_3 = 0$. We then have

$$F_2(s) = \frac{1}{s(s^2 + 3s + 2)} = \frac{1}{s(s+1)(s+2)} = \frac{c_1}{s+1} + \frac{c_2}{s+2} + \frac{c_3}{s}$$

and

$$c_1 = (s+1)\frac{1}{s(s+1)(s+2)}\bigg|_{s=-1} = \frac{1}{s(s+2)}\bigg|_{s=-1} = -1$$

$$c_2 = (s+2)\frac{1}{s(s+1)(s+2)}\bigg|_{s=-2} = \frac{1}{s(s+1)}\bigg|_{s=-2} = \frac{1}{2}$$

$$c_3 = s\frac{1}{s(s+1)(s+2)}\bigg|_{s=0} = \frac{1}{(s+1)(s+2)}\bigg|_{s=0} = \frac{1}{2}$$

The inverse Laplace transform is as follows

$$f_2(t) = \mathcal{L}^{-1}[F_2(s)] = \mathcal{L}^{-1}\left[\frac{1}{2}\frac{1}{s} - \frac{1}{s+1} + \frac{1}{2}\frac{1}{s+2}\right] = \frac{1}{2}\mathcal{L}^{-1}\left[\frac{1}{s}\right] - \mathcal{L}^{-1}\left[\frac{1}{s+1}\right] + \frac{1}{2}\mathcal{L}^{-1}\left[\frac{1}{s+2}\right]$$

$$= \frac{1}{2} - e^{-t} + \frac{1}{2}e^{-2t}$$

$F_3(s)$: We observe that the denominator polynomial has the same roots as in $F_2(s)$ but the numerator polynomial is different

$$F_3(s) = \frac{s^2 + s + 1}{s(s^2 + 3s + 2)} = \frac{s^2 + s + 1}{s(s+1)(s+2)} = \frac{c_1}{s+1} + \frac{c_2}{s+2} + \frac{c_3}{s}$$

and

$$c_1 = (s+1)\frac{(s^2 + s + 1)}{s(s+1)(s+2)}\bigg|_{s=-1} = \frac{s^2 + s + 1}{s(s+2)}\bigg|_{s=-1} = -1$$

$$c_2 = (s+2)\frac{(s^2 + s + 1)}{s(s+1)(s+2)}\bigg|_{s=-2} = \frac{(s^2 + s + 1)}{s(s+1)}\bigg|_{s=-2} = \frac{3}{2}$$

$$c_3 = s\frac{(s^2 + s + 1)}{s(s+1)(s+2)}\bigg|_{s=0} = \frac{(s^2 + s + 1)}{(s+1)(s+2)}\bigg|_{s=0} = \frac{1}{2}$$

The inverse Laplace transform is as follows

$$f_3(t) = \mathcal{L}^{-1}[F_3(s)] = \frac{1}{2}\mathcal{L}^{-1}\left[\frac{1}{s}\right] - \mathcal{L}^{-1}\left[\frac{1}{s+1}\right] + \frac{3}{2}\mathcal{L}^{-1}\left[\frac{1}{s+2}\right] = \frac{1}{2} - e^{-t} + \frac{3}{2}e^{-2t}$$

Example A.2 Inverse Laplace transform with simple imaginary roots

Find the inverse Laplace transform of

$$F_4(s) = \frac{s^2+2}{(s+1)(s^2+1)}$$

Solution

We note that the roots of the denominator polynomial are: a simple real root $\rho_1 = -1$, and two imaginary roots $\rho_2 = -i$, $\rho_3 = +i$. We therefore have that

$$F_4(s) = \frac{s^2+2}{(s+1)(s^2+1)} = \frac{c_1}{s+1} + \frac{c_2}{s+i} + \frac{c_3}{s-i}$$

where c_2 and c_3 are complex conjugates but we assume that we ignore that prior information. We calculate the constants

$$c_1 = \left.\frac{s^2+2}{(s^2+1)}\right|_{s=-1} = \frac{3}{2}$$

$$c_2 = (s+i)\left.\frac{s^2+2}{(s+1)(s^2+1)}\right|_{s=-i} = \left.\frac{s^2+2}{(s+1)(s-i)}\right|_{s=-i} = \frac{1}{(1-i)(-2i)} = -\frac{1}{2(1+i)}\frac{(1-i)}{(1-i)} = \frac{-1+i}{4}$$

$$c_3 = (s-i)\left.\frac{s^2+2}{(s+1)(s^2+1)}\right|_{s=+i} = \left.\frac{s^2+2}{(s+1)(s+i)}\right|_{s=+i} = \frac{1}{(i+1)(2i)} = \frac{1}{2(-1+i)}\frac{(-1-i)}{(-1-i)} = \frac{-1-i}{4}$$

We can now calculate the inverse Laplace transform as follows

$$f_4(t) = \mathcal{L}^{-1}[F_4(s)] = \frac{3}{2}\mathcal{L}^{-1}\left[\frac{1}{s+1}\right] - \left(\frac{1-i}{4}\right)\mathcal{L}^{-1}\left[\frac{1}{s+i}\right] - \left(\frac{1+i}{4}\right)\mathcal{L}^{-1}\left[\frac{1}{s-i}\right]$$

$$= \frac{3}{2}e^{-t} - \left(\frac{1-i}{4}\right)e^{-it} - \left(\frac{1+i}{4}\right)e^{it}$$

$$= \frac{3}{2}e^{-t} - \left(\frac{1-i}{4}\right)(\cos t - i\sin t) - \left(\frac{1+i}{4}\right)(\cos t + i\sin t)$$

$$= \frac{3}{2}e^{-t} - \left[\left(\frac{1-i}{4}\right) + \left(\frac{1+i}{4}\right)\right]\cos t + i\left[\left(\frac{1-i}{4}\right) - \left(\frac{1+i}{4}\right)\right]\sin t$$

$$= \frac{3}{2}e^{-t} - \frac{1}{2}\cos t + \frac{1}{2}\sin t$$

Example A.3 Inverse Laplace transform with simple complex conjugate roots

Find the inverse Laplace transform of

$$F_5(s) = \frac{2}{s(s^2 + 2s + 2)}$$

Solution

We note that the roots of the denominator polynomial are: a simple real root $p_1 = 0$, and two complex conjugate roots $p_2 = -1 + i$, $p_3 = -1 - i$. We therefore have that

$$F_5(s) = \frac{2}{s(s+1-i)(s+1+i)} = \frac{c_1}{s} + \frac{c_2}{s+1-i} + \frac{c_3}{s+1+i}$$

where c_2 and c_3 are complex conjugates but we ignore this information. We calculate the constants

$$c_1 = s \cdot \frac{2}{s(s+1-i)(s+1+i)} \bigg|_{s=0} = \frac{2}{(1-i)(1+i)} = 1$$

$$c_2 = (s+1-i) \frac{2}{s(s+1-i)(s+1+i)} \bigg|_{s=-1+i} = \frac{2}{s(s+1+i)} \bigg|_{s=-1+i} = -\frac{1-i}{2}$$

$$c_3 = (s+1+i) \frac{2}{s(s+1-i)(s+1+i)} \bigg|_{s=-1-i} = \frac{2}{s(s+1-i)} \bigg|_{s=-1-i} = -\frac{1+i}{2}$$

We can now calculate the inverse Laplace transform as follows

$$f_5(t) = \mathcal{L}^{-1}[F_5(s)] = \mathcal{L}^{-1}\left[\frac{1}{s}\right] - \left(\frac{1-i}{2}\right)\mathcal{L}^{-1}\left[\frac{1}{s+1-i}\right] - \left(\frac{1+i}{2}\right)\mathcal{L}^{-1}\left[\frac{1}{s+1+i}\right]$$

$$= 1 - \left(\frac{1-i}{2}\right)e^{(-1+i)t} - \left(\frac{1+i}{2}\right)e^{(-1-i)t}$$

$$= 1 - e^{-t}\left[\left(\frac{1-i}{2}\right)e^{it} + \left(\frac{1+i}{2}\right)e^{-it}\right]$$

$$= 1 - e^{-t}\left\{\left[\left(\frac{1-i}{2}\right) + \left(\frac{1+i}{2}\right)\right]\cos t + i\left[\left(\frac{1-i}{2}\right) - \left(\frac{1+i}{2}\right)\right]\sin t\right\}$$

$$= 1 - e^{-t}(\cos t + \sin t)$$

A.5.2 Case 2: Some roots of denominator polynomial Q(s) are multiple roots

In this case, if root p_j has multiplicity m_j, then $F(s)$ is expanded as follows:

$$F(s) = \cdots + \left(\frac{c_{j,1}}{(s-p_j)^{m_j}} + \frac{c_{j,2}}{(s-p_j)^{m_j-1}} + \cdots + \frac{c_{j,m_j}}{(s-p_j)} \right) + \cdots \qquad (A.5.8)$$

where

$$c_{j,1} = \left[(s-p_j)^{m_j} F(s) \right]_{s=p_j}$$

$$c_{j,2} = \frac{1}{1!} \left\{ \frac{d}{ds} \left[(s-p_j)^{m_j} F(s) \right] \right\}_{s=p_j}$$

$$c_{j,3} = \frac{1}{2!} \left\{ \frac{d^2}{ds^2} \left[(s-p_j)^{m_j} F(s) \right] \right\}_{s=p_j}$$

$$\vdots$$

$$c_{j,m_j} = \frac{1}{(m_j-1)!} \left\{ \frac{d^{m_j-1}}{ds^{m_j-1}} \left[(s-p_j)^{m_j} F(s) \right] \right\}_{s=p_j} \qquad (A.5.9)$$

The inverse Laplace transform has the following form

$$f(t) = \cdots + \left[c_{j,1} \frac{t^{m_j-1}}{(m_j-1)!} + c_{j,2} \frac{t^{m_j-2}}{(m_j-2)!} + \cdots + c_{j,m_j} \right] e^{p_j t} + \cdots \qquad (A.5.10)$$

Example A.4 Inverse Laplace transform with multiple real root

Find the inverse Laplace transform of

$$F_6(s) = \frac{1}{s(s^3 + 3s^2 + 3s + 1)}$$

Solution

We note that the roots of the denominator polynomial are: a simple real root $p_1 = 0$, and a multiple real root $p_2 = -1$ with multiplicity $m = 3$. We therefore have that

$$F_6(s) = \frac{1}{s(s^3 + 3s^2 + 3s + 1)} = \frac{1}{s(s+1)^3} = \frac{c_1}{s} + \frac{c_{2,1}}{(s+1)^3} + \frac{c_{2,2}}{(s+1)^2} + \frac{c_{2,3}}{(s+1)}$$

We calculate the constants

$$c_1 = s \cdot \frac{1}{s(s+1)^3}\bigg|_{s=0} = 1$$

$$c_{2,1} = (s+1)^3 \cdot \frac{1}{s(s+1)^3}\bigg|_{s=-1} = -1$$

$$c_{2,2} = \frac{1}{1!}\left\{\frac{d}{ds}\left[(s-p_j)^{m_j} F(s)\right]\right\}_{s=p_j} = \left\{\frac{d}{ds}\left[\frac{1}{s}\right]\right\}_{s=-1} = \left\{-\frac{1}{s^2}\right\}_{s=-1} = -1$$

$$c_{2,3} = \frac{1}{2!}\left\{\frac{d^2}{ds^2}\left[(s-p_j)^{m_j} F(s)\right]\right\}_{s=p_j} = \frac{1}{2}\left\{\frac{d^2}{ds^2}\left[\frac{1}{s}\right]\right\}_{s=-1} = -\frac{1}{2}\left\{\frac{d}{ds}\left[\frac{1}{s^2}\right]\right\}_{s=-1} = -1$$

We can now calculate the inverse Laplace transform as follows

$$f_6(t) = \mathcal{L}^{-1}[F_6(s)] = \mathcal{L}^{-1}\left[\frac{1}{s}\right] - \mathcal{L}^{-1}\left[\frac{1}{(s+1)^3}\right] - \mathcal{L}^{-1}\left[\frac{1}{(s+1)^2}\right] - \mathcal{L}^{-1}\left[\frac{1}{(s+1)}\right]$$

$$= 1 - \frac{t^2}{2}e^{-t} - te^{-t} - e^{-t}$$

$$= 1 - \left(\frac{t^2}{2} + t + 1\right)e^{-t}$$

A.6 Solution of Linear Ordinary Differential Equations using the Laplace Transform

The solution of linear ordinary differential equations using the Laplace transform method follows these steps.

Step 1: We take the Laplace transform of the linear differential equation, taking into consideration the initial conditions.

Step 2: We solve the resulting algebraic equation for the Laplace transform of the dependent variable.

Step 3: We find the inverse Laplace transform of the expression derived in step 2.

Example A.5 Solving linear ODEs with the Laplace transform method

Solve the following linear ordinary differential equations:

$$\frac{d^2y}{dt^2} + 3\frac{dy}{dt} + 2y = 2$$

$$y(0) = 0, \quad \frac{dy}{dt}(0) = 0$$

Solution

Step 1: We take the Laplace transform of the differential equation

$$\mathcal{L}\left[\frac{d^2y}{dt^2}+3\frac{dy}{dt}+2y\right]=\mathcal{L}[2]$$

$$\Rightarrow \mathcal{L}\left[\frac{d^2y}{dt^2}\right]+3\mathcal{L}\left[\frac{dy}{dt}\right]+2\mathcal{L}[y]=\mathcal{L}[2]$$

$$\Rightarrow \left[s^2Y(s)-sy(0)-\frac{dy}{dt}(0)\right]+3[sY(s)-y(0)]+2Y(s)=\frac{2}{s}$$

$$\Rightarrow (s^2+3s+2)Y(s)=\frac{2}{s}$$

Step 2: We solve for $Y(s)$ to obtain

$$Y(s)=\frac{2}{s(s^2+3s+2)}$$

Step 3: We observe that the right-hand side equals $2F_2(s)$ examined in Example A.1 and it follows that

$$y(t)=2f_2(t)=1-2e^{-t}+e^{-2t}$$

Example A.6 Solving linear ODEs with Laplace transform method

Solve the following linear ordinary differential equation:

$$\frac{dy^3}{dt^3}+2\frac{dy^2}{dt^2}-\frac{dy}{dt}-2y=4+e^{2t}$$

$$y(0)=1,\ \frac{dy}{dt}(0)=0,\ \frac{d^2y}{dt^2}(0)=-1$$

Solution

Step 1: We take the Laplace transform of the differential equation by using Eqs. (A.3.7a–c)

$$\mathcal{L}\left[\frac{dy^3}{dt^3}\right]=s^3Y(s)-s^2y(0)-s\frac{dy}{dt}(0)-\frac{d^2y}{dt^2}(0)=s^3Y(s)-s^2+1$$

$$\mathcal{L}\left[\frac{dy^2}{dt^2}\right]=s^2Y(s)-sy(0)-\frac{dy}{dt}(0)=s^2Y(s)-s$$

$$\mathcal{L}\left[\frac{dy}{dt}\right]=sY(s)-y(0)=sY(s)-1$$

or

$$[s^3Y(s)-s^2+1]+2[s^2Y(s)-s]-[sY(s)-1]-2Y(s)=\frac{4}{s}+\frac{1}{s-2}$$

Step 2: We solve for $Y(s)$ to obtain

$$(s^3+2s^2-s-2)Y(s)-(s^2+2s-2)=\frac{4(s-2)+s}{s(s-2)}$$

$$\Rightarrow (s^3+2s^2-s-2)Y(s)=\frac{4(s-2)+s}{s(s-2)}+\frac{s(s-2)}{s(s-2)}(s^2+2s-2)$$

or

$$Y(s)=\frac{s^4-6s^2+9s-8}{s(s-2)(s^3+2s^2-s-2)}=\frac{s^4-6s^2+9s-8}{s(s-2)(s+2)(s+1)(s-1)}$$

Step 3: Since all roots of the denominator polynomial are simple and real, we expand in partial fractions as follows:

$$Y(s)=\frac{s^4-6s^2+9s-8}{s(s-2)(s+1)(s+2)(s-1)}=\frac{-2}{s}+\frac{1/12}{s-2}+\frac{11/3}{s+1}+\frac{-17/12}{s+2}+\frac{2/3}{s-1}$$

and finally, inverting,

$$y(t)=-2+\frac{1}{12}e^{2t}+\frac{11}{3}e^{-t}-\frac{17}{12}e^{-2t}+\frac{2}{3}e^{t}$$

Example A.7 Solving linear ODEs with Laplace transform method

Solve the following system of linear ordinary differential equations

$$\frac{dx_1}{dt}=2x_1+3x_2+1$$

$$\frac{dx_2}{dt}=2x_1+x_2+1$$

$$x_1(0)=x_2(0)=0$$

Solution

Step 1: We take the Laplace transform of the differential equation by using Eq. (A.3.7a)

$$sX_1(s)-x_1(0)=2X_1(s)+3X_2(s)+\frac{1}{s}$$

$$sX_2(s)-x_2(0)=2X_1(s)+X_2(s)+\frac{1}{s}$$

Applying the initial conditions and rearranging,

$$(s-2)X_1(s) - 3X_2(s) = \frac{1}{s}$$

$$-2X_1(s) + (s-1)X_2(s) = \frac{1}{s}$$

Step 2: We solve the system of algebraic equations for $X_1(s)$ and $X_2(s)$:

$$X_1(s) = \frac{s+2}{s(s^2 - 3s - 4)} = \frac{s+2}{s(s+1)(s-4)}$$

$$X_2(s) = \frac{1}{s^2 - 3s - 4} = \frac{1}{(s+1)(s-4)}$$

Step 3: We expand in partial fractions as follows

$$X_1(s) = -\frac{1}{2}\frac{1}{s} + \frac{1}{5}\frac{1}{s+1} + \frac{3}{10}\frac{1}{s-4}$$

$$X_2(s) = -\frac{1}{5}\frac{1}{s+1} + \frac{1}{5}\frac{1}{s-4}$$

and finally, inverting,

$$x_1(t) = -\frac{1}{2} + \frac{1}{5}e^{-t} + \frac{3}{10}e^{4t}$$

$$x_2(t) = -\frac{1}{5}e^{-t} + \frac{1}{5}e^{4t}$$

A.7 Software Tools

A.7.1 Partial Fraction Expansion in MATLAB

Polynomial functions of order n are declared in MATLAB as vectors of length $n + 1$ whose elements are the coefficients of the polynomial in descending powers. If we consider $F_1(s)$ studied in Example A.1 we have that the following numerator and denominator polynomials

$$P(s) = 1 = 1 \cdot s^0$$

$$Q(s) = s^2 + 3s + 2 = \begin{bmatrix} 1 & 3 & 2 \end{bmatrix} \cdot \begin{bmatrix} s^2 \\ s^1 \\ s^0 \end{bmatrix}$$

These two polynomials can be defined in MATLAB by typing the following in the command window

```
» P=1;
» Q=[1 3 2];
```

We can calculate the roots of the polynomial $Q(s)$ through the following command

```
» roots(Q)

ans =

    -2
    -1
```

In order to calculate the roots and the corresponding coefficients in the partial fraction expansion we type the following

```
» [c,rho]=residue(P,Q)

c =

    -1
     1

rho =

    -2
    -1
```

which validates the result in Example A.1. We can multiply two polynomials by using the conv command available in MATLAB. To multiply the polynomials $(s + 1)$ and $(s^2 + 1)$ that appear in the denominator of $F_4(s)$ studied in Example A.2 we type the following in the command window

```
» Q=conv([1 1],[1 0 1])

Q =

     1     1     1     1
```

To find the partial fraction expansion we type

```
» P=[1 0 2];
» [c,rho]=residue(P,Q)

c =

     1.5000
    -0.2500 - 0.2500i
    -0.2500 + 0.2500i
```

```
rho =

   -1.0000
   -0.0000 + 1.0000i
   -0.0000 - 1.0000i
```

which is in agreement with the results found in Example A.2. The reader can use the help command to find how MATLAB handles the case of roots with multiplicity and validate the results found in Example A.4.

A.7.2 Laplace Transform, Inverse Laplace Transform and Partial Fraction Expansion in Maple

Laplace transforms and inverse Laplace transforms can be obtained by using the `laplace` and `invlaplace` commands of the `inttrans` package of Maple.

Suppose, for example, that we wish to calculate the Laplace transform of the function $f(t) = t\sin(\omega t)$. This is done as follows:

```
> with(inttrans):
> f(t):=t*sin(omega*t);
```

$f(t) := t \sin(\omega t)$

```
> F(s):=laplace(f(t),t,s);
```

$$F(s) := \frac{2s\omega}{(s^2 + \omega^2)^2}$$

and thus we obtain one of the results of Table A.2.

A similar procedure can be followed for calculating inverse Laplace transforms. Suppose for example that we wish to calculate the inverse Laplace transform of the function $F_1(s)$ studied in Example A.1. This is done as follows:

```
> with(inttrans):
> F1(s):=1/(s^2+3*s+2);
```

$$F1(s) := \frac{1}{s^2 + 3s + 2}$$

```
> f1(t):=invlaplace(F1(s),s,t);
```

$f1(t) := e^{-t} - e^{-2t}$

which validates the result in Example A.1. Likewise, for the function $F_4(s)$ of Example A.2,

```
> F4(s):=(s^2+2)/((s+1)*(s^2+1));
```

$$F4(s) := \frac{s^2+2}{(s+1)(s^2+1)}$$

```
> f4(t):=invlaplace(F4(s),s,t);
```

$$f4(t) := -\frac{1}{2}\cos(t) + \frac{1}{2}\sin(t) + \frac{3}{2}e^{-t}$$

which validates the result in Example A.2.

It is also possible to obtain partial fraction expansion with Maple, if we want to follow the steps of the manual inversion of the Laplace transform. For the same functions $F_1(s)$ and $F_4(s)$ as above,

```
> convert(F1(s),parfrac);
```

$$-\frac{1}{s+2} + \frac{1}{s+1}$$

which is readily invertible using the tables of Laplace transforms.
Likewise,

```
> convert(F4(s),parfrac);
```

$$\frac{1}{2}\frac{-s+1}{s^2+1} + \frac{3}{2(s+1)}$$

which is also readily invertible using the tables of Laplace transforms. Notice that the first term of the above partial fraction expansion has a second-order denominator, which combines the pair of imaginary roots of the denominator of $F_4(s)$. If a complex partial fraction expansion is desired, one can type

```
> convert(F4(s),parfrac,complex);
```

$$\frac{-0.25-0.25I}{s-1.I} + \frac{-0.25+0.25I}{s+1.I} + \frac{1.5}{s+1.}$$

As a last example, consider the function $F_6(s)$ of Example A.4, whose denominator has one simple and one triple root. Its partial fraction expansion is calculated in exactly the same way

```
> F6(s):=1/(s*(s^3+3*s^2+3*s+1)):
> convert(F6(s),parfrac);
```

$$\frac{1}{s} - \frac{1}{s+1} - \frac{1}{(s+1)^2} - \frac{1}{(s+1)^3}$$

validating the result of Example A.4.

PROBLEMS

A.1 For the following functions of time:

- $f(t) = 3 + 5t - t^2$ $\quad\Rightarrow\quad$ $F(s) = ?$
- $f(t) = e^{-2t} t^3$ $\quad\Rightarrow\quad$ $F(s) = ?$
- $f(t) = e^{-4t} \cos(2t)$ $\quad\Rightarrow\quad$ $F(s) = ?$
- $f(t) = -e^{-t} - \frac{t^2}{2} e^{-3t} + \sin t$ $\quad\Rightarrow\quad$ $F(s) = ?$
- $f(t) = 2\sin(t-1)$ $\quad\Rightarrow\quad$ $F(s) = ?$

A.2 For the following Laplace transforms:

- $F(s) = \dfrac{5}{s+3}$ $\quad\Rightarrow\quad$ $f(t) = ?$
- $F(s) = \dfrac{5e^{-2s}}{s+3}$ $\quad\Rightarrow\quad$ $f(t) = ?$
- $F(s) = \dfrac{5}{s(s+3)}$ $\quad\Rightarrow\quad$ $f(t) = ?$
- $F(s) = \dfrac{1}{s}\left(1 + \dfrac{1}{2s}\right)$ $\quad\Rightarrow\quad$ $\lim_{t \to 0+} f(t) = ?$
- $F(s) = \dfrac{1}{s(s+1)^2}$ $\quad\Rightarrow\quad$ $\lim_{t \to \infty} f(t) = ?$
- $F(s) = \dfrac{1}{s(s-1)^2}$ $\quad\Rightarrow\quad$ $\lim_{t \to \infty} f(t) = ?$

A.3 Calculate the inverse Laplace transforms of the following functions:

(a) $F(s) = \dfrac{4}{s^2(s^2 + 4s + 4)}$

(b) $F(s) = \dfrac{s+2}{s^2(s^2 + 2s + 2)}$

A.4 Apply the convolution theorem to calculate the inverse Laplace transform of

$$F(s) = \frac{1}{(s^2+1)^2}$$

A.5 Solve the following initial-value problem

$$\frac{dy}{dt} + y = \sin(t)$$

$$y(0) = 1$$

A.6 Solve the following initial-value problem

$$\frac{d^2y}{dt^2} + 2\frac{dy}{dt} + 5y = 1$$

$$y(0) = \frac{dy}{dt}(0) = 0$$

A.7 Solve the following initial-value problem

$$\frac{d^3y}{dt^3} + 6\frac{d^2y}{dt^2} + 11\frac{dy}{dt} + 6y = 1$$

$$y(0) = \frac{dy}{dt}(0) = \frac{d^2y}{dt^2}(0) = 0$$

A.8 Solve the following initial value problem

$$\frac{d^3y}{dt^3} + 6\frac{d^2y}{dt^2} + 12\frac{dy}{dt} + 8y = e^{-t}$$

$$y(0) = \frac{dy}{dt}(0) = \frac{d^2y}{dt^2}(0) = 0$$

A.9 Solve the following initial-value problem

$$\frac{d^3y}{dt^3} + 3\frac{d^2y}{dt^2} + 4\frac{dy}{dt} + 2y = 2$$

$$y(0) = \frac{dy}{dt}(0) = \frac{d^2y}{dt^2}(0) = 0$$

A.10 Solve the initial-value problem for the following system of first-order differential equations:

$$\frac{dx_1}{dt} = 2x_1 + 3x_2 + 1$$

$$\frac{dx_2}{dt} = 2x_1 + x_2 + e^t$$

$$x_1(0) = x_2(0) = 0$$

A.11 Solve the initial-value problem for the following system of first-order differential equations:

$$\frac{dx_1}{dt} = -3x_1 + 3x_2 + f(t)$$

$$\frac{dx_2}{dt} = x_1 - 5x_2$$

$$x_1(0) = x_2(0) = 0$$

where $f(t)$ is a given function.

A.12 Solve the initial-value problem for the following system of first-order differential equations:

$$\frac{dx_1}{dt} = -2x_1 + 2x_2 + f(t)$$

$$\frac{dx_2}{dt} = x_1 - 3x_2$$

$$x_1(0) = x_2(0) = 0$$

where $f(t)$ is a given function.

A.13 Calculate the Laplace transform of the function

$$u(t) = \begin{cases} 0, & t < 0 \\ at, & 0 \leq t < \dfrac{M}{a} \\ M, & t \geq \dfrac{M}{a} \end{cases}$$

What happens in the limit as $a \to \infty$?

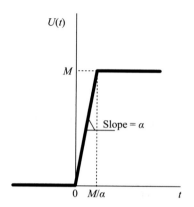

A.14 A triangular pulse of duration 2ε and unit magnitude is defined as follows:

$$f(t) = \begin{cases} 0, & \text{for } t < 0 \\ \dfrac{t}{\varepsilon^2}, & \text{for } 0 \le t < \varepsilon \\ \dfrac{2}{\varepsilon} - \dfrac{t}{\varepsilon^2}, & \text{for } \varepsilon \le t < 2\varepsilon \\ 0, & \text{for } t \ge 2\varepsilon \end{cases}$$

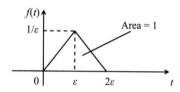

Calculate the Laplace transform of the above unit triangular pulse. What happens in the limit as $\varepsilon \to 0$?

Appendix B
Basic Matrix Theory

This appendix provides a brief summary of basic matrix theory that includes definitions and calculations useful in process control. As in the previous appendix, the material has been organized so as to serve both as a review for those readers who already had matrices in math classes, and as a crash course for a newcomer in the field.

B.1 Basic Notations and Definitions

A matrix is a rectangular array of elements arranged in rows and columns. The elements (or entries) of a matrix can be real or complex numbers, polynomials, functions, etc. A matrix with n rows and m columns is called an $n \times m$ matrix and can be written as

$$A = \begin{bmatrix} a_{11} & a_{12} & \cdots & a_{1m} \\ a_{21} & a_{22} & \cdots & a_{2m} \\ \vdots & \vdots & \ddots & \vdots \\ a_{n1} & a_{n2} & \cdots & a_{nm} \end{bmatrix} \qquad (B.1.1)$$

where a_{ij} denotes the element that lies in the i-th row and j-th column. Uppercase letters are normally (but not always) used to denote matrices, while vectors (a matrix with one column or one row) are usually (but not always) denoted by lowercase letters. Scalars can be considered as matrices with one column and one row.

The addition of two $n \times m$ matrices, A with elements a_{ij} and B with elements b_{ij}, denoted by $A + B$, is performed by adding the corresponding entries together

$$\begin{aligned} A + B &= \begin{bmatrix} a_{11} & a_{12} & \cdots & a_{1m} \\ a_{21} & a_{22} & \cdots & a_{2m} \\ \vdots & \vdots & \ddots & \vdots \\ a_{n1} & a_{n2} & \cdots & a_{nm} \end{bmatrix} + \begin{bmatrix} b_{11} & b_{12} & \cdots & b_{1m} \\ b_{21} & b_{22} & \cdots & b_{2m} \\ \vdots & \vdots & \ddots & \vdots \\ b_{n1} & b_{n2} & \cdots & b_{nm} \end{bmatrix} \\ &= \begin{bmatrix} a_{11}+b_{11} & a_{12}+b_{12} & \cdots & a_{1m}+b_{1m} \\ a_{21}+b_{21} & a_{22}+b_{22} & \cdots & a_{2m}+b_{2m} \\ \vdots & \vdots & \ddots & \vdots \\ a_{n1}+b_{n1} & a_{n2}+b_{n2} & \cdots & a_{nm}+b_{nm} \end{bmatrix} \end{aligned} \qquad (B.1.2)$$

Subtraction of matrices $A - B$ is defined similarly. Matrix addition is commutative, i.e. the matrix sum does not depend on the order of the summands:

$$A + B = B + A \tag{B.1.3}$$

and associative:
$$A + (B + C) = (A + B) + C \tag{B.1.4}$$

If A is an $n \times q$ matrix and B is a $q \times m$ matrix, then their multiplication AB is defined as an $n \times m$ matrix whose ij-th element is $\sum_{k=1}^{q} a_{ik} b_{kj}$, i.e. the inner product of the i-th row of A and the j-th column of B:

$$AB = \begin{bmatrix} \sum_{k=1}^{q} a_{1k} b_{k1} & \sum_{k=1}^{q} a_{1k} b_{k2} & \cdots & \sum_{k=1}^{q} a_{1k} b_{km} \\ \sum_{k=1}^{q} a_{2k} b_{k1} & \sum_{k=1}^{q} a_{2k} b_{k2} & \cdots & \sum_{k=1}^{q} a_{2k} b_{km} \\ \vdots & \vdots & \ddots & \vdots \\ \sum_{k=1}^{q} a_{nk} b_{k1} & \sum_{k=1}^{q} a_{nk} b_{k2} & \cdots & \sum_{k=1}^{q} a_{nk} b_{km} \end{bmatrix} \tag{B.1.5}$$

The matrix product is associative
$$A(BC) = (AB)C \tag{B.1.6}$$

and distributive over matrix addition
$$A(B + C) = AB + AC \tag{B.1.7}$$
$$(B + C)A = BA + CA \tag{B.1.8}$$

but is not commutative, i.e. in general
$$AB \ne BA \tag{B.1.9}$$

A square matrix is a matrix with the same number or columns and rows. An $n \times n$ matrix is of the form:

$$A = \begin{bmatrix} a_{11} & a_{12} & \cdots & a_{1n} \\ a_{21} & a_{22} & \cdots & a_{2n} \\ \vdots & \vdots & \ddots & \vdots \\ a_{n1} & a_{n2} & \cdots & a_{nn} \end{bmatrix} \tag{B.1.10}$$

The elements $a_{11}, a_{22}, \ldots, a_{nn}$ are called diagonal elements of the square matrix A. In other words, the diagonal elements are those a_{ij} for which $i = j$. The nondiagonal elements are the ones with $i \ne j$.

A diagonal matrix is a square matrix with all nondiagonal elements equal to 0. An upper triangular matrix is any matrix with all elements below the diagonal equal to 0 while a lower triangular matrix has all elements above the diagonal equal to 0.

The zero matrix is a matrix with all elements equal to 0 and it is denoted also by 0. The unit matrix, denoted by I, is a diagonal matrix with all diagonal elements equal to 1. The identity matrix has the property that

$$AI = IA = A \tag{B.1.11}$$

for every square matrix A.

The transpose of an $n \times m$ matrix A, denoted A^T, is defined as the $m \times n$ matrix whose i-th row is the i-th column of A, and whose j-th column is the j-th row of A. In other words, $[A^T]_{ij} = [A]_{ji}$ for all i and j. The following are some useful properties of the transpose

$$\left(A^T\right)^T = A \tag{B.1.12}$$

$$(A+B)^T = A^T + B^T \tag{B.1.13}$$

$$(AB)^T = B^T A^T \tag{B.1.14}$$

If a matrix has the property that $A = A^T$, it is called symmetric matrix. A symmetric matrix is necessarily a square matrix.

If p is a positive integer and A is a square matrix, the pth power of A is defined as:

$$A^p = \underbrace{AA \cdots A}_{p \text{ times}} \tag{B.1.15}$$

An immediate consequence of the above definition is that

$$A^{p+q} = A^p A^q = A^q A^p \tag{B.1.16}$$

for every positive integers p and q. By convention, the zeroth power of a square matrix is the identity matrix

$$A^0 = I \tag{B.1.17}$$

Thus, Eq. (B.1.16) holds even when p or q is zero.

Moreover, it is possible to define functions $f(A)$ of a square matrix argument, where f is an analytic scalar function, via the Taylor series expansion of f. For example, the Taylor series expansion of the scalar exponential function

$$e^x = 1 + x + \frac{x^2}{2} + \cdots + \frac{x^k}{k!} + \cdots = \sum_{k=0}^{\infty} \frac{x^k}{k!}$$

may be used to define a matrix exponential accordingly

$$e^A = I + A + \frac{A^2}{2} + \cdots + \frac{A^k}{k!} + \cdots = \sum_{k=0}^{\infty} \frac{A^k}{k!} \tag{B.1.18}$$

Similarly, it is possible to define other functions like $\sin(A)$, $\cos(A)$, etc.

Example B.1

Consider the following matrices:

$$A = \begin{bmatrix} 1 & 2 \\ 0 & 3 \end{bmatrix}, \quad B = \begin{bmatrix} 4 & 0 \\ 2 & 2 \end{bmatrix}, \quad C = \begin{bmatrix} 7 & 0 & 0 \\ 0 & 8 & 9 \end{bmatrix}$$

The matrices A and B are square 2×2 matrices while C is a 2×3 matrix and is not square. The sum of A and B is the following

$$A+B = \begin{bmatrix} a_{11}+b_{11} & a_{12}+b_{12} \\ a_{21}+b_{21} & a_{22}+b_{22} \end{bmatrix} = \begin{bmatrix} 1+4 & 2+0 \\ 0+2 & 3+2 \end{bmatrix} = \begin{bmatrix} 5 & 2 \\ 2 & 5 \end{bmatrix}$$

or

$$B+A = \begin{bmatrix} b_{11}+a_{11} & b_{12}+a_{12} \\ b_{21}+a_{21} & b_{22}+a_{22} \end{bmatrix} = \begin{bmatrix} 4+1 & 0+2 \\ 2+0 & 2+3 \end{bmatrix} = \begin{bmatrix} 5 & 2 \\ 2 & 5 \end{bmatrix}$$

As expected, $A + B$ and $B + A$ are equal (commutativity property of addition). The sum of either A or B with C is not defined as they do not have the same number of rows and columns.
The multiplication of A and B is defined as

$$AB = \begin{bmatrix} a_{11}b_{11}+a_{12}b_{21} & a_{11}b_{12}+a_{12}b_{22} \\ a_{21}b_{11}+a_{22}b_{21} & a_{21}b_{12}+a_{22}b_{22} \end{bmatrix} = \begin{bmatrix} 1\cdot 4+2\cdot 2 & 1\cdot 0+2\cdot 2 \\ 0\cdot 4+3\cdot 2 & 0\cdot 0+3\cdot 2 \end{bmatrix} = \begin{bmatrix} 8 & 4 \\ 6 & 6 \end{bmatrix}$$

or

$$BA = \begin{bmatrix} b_{11}a_{11}+b_{12}a_{21} & b_{11}a_{12}+b_{12}a_{22} \\ b_{21}a_{11}+b_{22}a_{21} & b_{21}a_{12}+b_{22}a_{22} \end{bmatrix} = \begin{bmatrix} 4\cdot 1+0\cdot 0 & 4\cdot 2+0\cdot 3 \\ 2\cdot 1+2\cdot 0 & 2\cdot 2+2\cdot 3 \end{bmatrix} = \begin{bmatrix} 4 & 8 \\ 2 & 10 \end{bmatrix}$$

It is observed that $AB \neq BA$ (the noncommutativity property of multiplication). Multiplication of C on the left by A or B is possible and gives

$$AC = \begin{bmatrix} 1 & 2 \\ 0 & 3 \end{bmatrix}\begin{bmatrix} 7 & 0 & 0 \\ 0 & 8 & 9 \end{bmatrix} = \begin{bmatrix} 1\cdot 7+2\cdot 0 & 1\cdot 0+2\cdot 8 & 1\cdot 0+2\cdot 9 \\ 0\cdot 7+3\cdot 0 & 0\cdot 0+3\cdot 8 & 0\cdot 0+3\cdot 9 \end{bmatrix} = \begin{bmatrix} 7 & 16 & 18 \\ 0 & 24 & 27 \end{bmatrix}$$

$$BC = \begin{bmatrix} 4 & 0 \\ 2 & 2 \end{bmatrix}\begin{bmatrix} 7 & 0 & 0 \\ 0 & 8 & 9 \end{bmatrix} = \begin{bmatrix} 28 & 0 & 0 \\ 14 & 16 & 18 \end{bmatrix}$$

However, the products CA and CB are not defined.
Matrix A is upper triangular and matrix B is lower triangular. Their transposes can be found as follows

$$A^T = \begin{bmatrix} 1 & 2 \\ 0 & 3 \end{bmatrix}^T = \begin{bmatrix} 1 & 0 \\ 2 & 3 \end{bmatrix}$$

$$B^T = \begin{bmatrix} 4 & 0 \\ 2 & 2 \end{bmatrix}^T = \begin{bmatrix} 4 & 2 \\ 0 & 2 \end{bmatrix}$$

The transpose of C is

$$C^T = \begin{bmatrix} 7 & 0 & 0 \\ 0 & 8 & 9 \end{bmatrix}^T = \begin{bmatrix} 7 & 0 \\ 0 & 8 \\ 0 & 9 \end{bmatrix}$$

For the matrix $D = A + B$, we have that

$$D^T = \begin{bmatrix} 5 & 2 \\ 2 & 5 \end{bmatrix}^T = \begin{bmatrix} 5 & 2 \\ 2 & 5 \end{bmatrix} = D$$

i.e. D is a symmetric matrix.

The square A^2 is defined as

$$A^2 = AA = \begin{bmatrix} 1 & 2 \\ 0 & 3 \end{bmatrix} \begin{bmatrix} 1 & 2 \\ 0 & 3 \end{bmatrix} = \begin{bmatrix} 1 & 8 \\ 0 & 9 \end{bmatrix}$$

and similarly the cube A^3 as

$$A^3 = AA^2 = \begin{bmatrix} 1 & 2 \\ 0 & 3 \end{bmatrix} \begin{bmatrix} 1 & 8 \\ 0 & 9 \end{bmatrix} = \begin{bmatrix} 1 & 26 \\ 0 & 27 \end{bmatrix}$$

The zeroth power of A is the 2×2 identity matrix:

$$A^0 = I = \begin{bmatrix} 1 & 0 \\ 0 & 1 \end{bmatrix}$$

The exponential of the matrix A is given by:

$$e^A = \sum_{k=0}^{\infty} \frac{1}{k!} \begin{bmatrix} 1 & 2 \\ 0 & 3 \end{bmatrix}^k = \sum_{k=0}^{\infty} \frac{1}{k!} \begin{bmatrix} 1 & 3^k - 1 \\ 0 & 3^k \end{bmatrix} = \begin{bmatrix} \sum_{k=0}^{\infty} \frac{1}{k!} & \sum_{k=0}^{\infty} \frac{3^k}{k!} - \sum_{k=0}^{\infty} \frac{1}{k!} \\ 0 & \sum_{k=0}^{\infty} \frac{3^k}{k!} \end{bmatrix} = \begin{bmatrix} e & e^3 - e \\ 0 & e^3 \end{bmatrix}$$

B.2 Determinant of a Square Matrix

The determinant of a square $n \times n$ matrix A is denoted by $\det(A)$ and is a scalar quantity. The value of the determinant can be calculated in terms of lower-order determinants through the Laplace expansion. This expansion is defined in terms of the minors or the cofactors of elements of the matrix. The minor m_{ij} of the element a_{ij} in a matrix A is the determinant of the array formed by deleting the i-th row and the j-th column from the original matrix. The cofactor c_{ij} corresponding to the element a_{ij} of A is

$$c_{ij} = (-1)^{i+j} m_{ij} \tag{B.2.1}$$

In terms of the Laplace expansion, the determinant of a matrix A is given by

$$\det(A) = \sum_{i=1}^{n} a_{ij} c_{ij}, \tag{B.2.2}$$

for any j, or

$$\det(A) = \sum_{j=1}^{n} a_{ij} c_{ij}, \tag{B.2.3}$$

for any i. The first of these is called an expansion along the j-th column and the second an expansion along the i-th row. All such expansions yield identical values. The Laplace expansion expresses an n-th-order determinant as a linear combination of $(n-1)$-th-order determinants. Each of the required $(n-1)$-th-order determinants can itself be expressed, by Laplace expansion, in terms of $(n-2)$-th-order determinants, and so on. This expansion together with the definition of the determinant for a scalar ($\det(a) = a$) is sufficient to determine the value of any determinant.

For a square 2×2 matrix A

$$A = \begin{bmatrix} a_{11} & a_{12} \\ a_{21} & a_{22} \end{bmatrix} \tag{B.2.4}$$

from Eq. (B.2.2), expanding along the first column ($j = 1$), we have that

$$\det(A) = \sum_{i=1}^{2} a_{i1} c_{i1} = a_{11} c_{11} + a_{21} c_{21}$$

The cofactors are calculated as follows

$$c_{11} = (-1)^{1+1} m_{11} = (-1)^{2} \det(a_{22}) = a_{22}$$
$$c_{21} = (-1)^{2+1} m_{21} = (-1)^{3} \det(a_{12}) = -a_{12}$$

and finally

$$\det(A) = \det\left(\begin{bmatrix} a_{11} & a_{12} \\ a_{21} & a_{22} \end{bmatrix}\right) = a_{11} a_{22} - a_{12} a_{21} \tag{B.2.5}$$

Similarly, for a 3×3 matrix A, expanding along the first column, we have

$$\det(A) = \det\left(\begin{bmatrix} a_{11} & a_{12} & a_{13} \\ a_{21} & a_{22} & a_{23} \\ a_{31} & a_{32} & a_{33} \end{bmatrix}\right)$$
$$= a_{11} c_{11} + a_{21} c_{21} + a_{31} c_{31}$$
$$= a_{11}(a_{22}a_{33} - a_{32}a_{23}) - a_{21}(a_{12}a_{33} - a_{32}a_{13}) + a_{31}(a_{12}a_{23} - a_{22}a_{13})$$
$$= a_{11}a_{22}a_{33} + a_{12}a_{23}a_{31} + a_{13}a_{32}a_{21} - a_{11}a_{23}a_{32} - a_{22}a_{13}a_{31} - a_{33}a_{12}a_{21} \tag{B.2.6}$$

The following are fundamental properties of the determinant:

$$\det(AB) = \det(A)\det(B) \tag{B.2.7}$$

$$\det(A) = \det(A^T) \tag{B.2.8}$$

If A is diagonal or upper/lower triangular then it is equal to the product of the diagonal elements:

$$\det(A) = \prod_{i=1}^{n} a_{ii} \tag{B.2.9}$$

Finally, if A is a square $n \times n$ matrix and c is a scalar, then:

$$\det(cA) = c^n \det(A) \tag{B.2.10}$$

B.3 Matrix Inversion

The inverse of a square $n \times n$ matrix A, denoted by A^{-1}, is a square $n \times n$ matrix with the following property

$$AA^{-1} = A^{-1}A = I \qquad (B.3.1)$$

The inverse of a matrix A exists if and only if $\det(A) \neq 0$. If the determinant is zero the matrix is said to be singular, and no inverse exists. The inverse of a matrix can be calculated through Cramer's rule

$$A^{-1} = \frac{1}{\det(A)} \operatorname{Adj}(A) \qquad (B.3.2)$$

The matrix $\operatorname{Adj}(A)$ is called the adjoint or adjugate matrix of A and is defined as the transpose of the matrix of cofactors of the elements of A:

$$\operatorname{Adj}(A) = \begin{bmatrix} c_{11} & c_{12} & \cdots & c_{1n} \\ c_{21} & c_{22} & \cdots & c_{2n} \\ \vdots & \vdots & \ddots & \vdots \\ c_{n1} & c_{n2} & \cdots & c_{nn} \end{bmatrix}^T = \begin{bmatrix} c_{11} & c_{21} & \cdots & c_{n1} \\ c_{12} & c_{22} & \cdots & c_{n2} \\ \vdots & \vdots & \ddots & \vdots \\ c_{1n} & c_{2n} & \cdots & c_{nn} \end{bmatrix} \qquad (B.3.3)$$

For a 2×2 matrix we have that

$$\operatorname{Adj}(A) = \begin{bmatrix} c_{11} & c_{12} \\ c_{21} & c_{22} \end{bmatrix}^T = \begin{bmatrix} c_{11} & c_{21} \\ c_{12} & c_{22} \end{bmatrix} = \begin{bmatrix} m_{11} & -m_{21} \\ -m_{12} & m_{22} \end{bmatrix} = \begin{bmatrix} a_{22} & -a_{12} \\ -a_{21} & a_{11} \end{bmatrix} \qquad (B.3.4)$$

and the determinant $\det(A)$ is given by Eq. (B.2.5). Therefore

$$A^{-1} = \frac{1}{\det(A)} \operatorname{Adj}(A) = \frac{1}{a_{11}a_{22} - a_{21}a_{12}} \begin{bmatrix} a_{22} & -a_{12} \\ -a_{21} & a_{11} \end{bmatrix} \qquad (B.3.5)$$

In a similar way, for a 3×3 matrix A, the adjugate matrix is found to be

$$\operatorname{Adj}(A) = \begin{bmatrix} a_{22}a_{33} - a_{32}a_{23} & -a_{12}a_{33} + a_{32}a_{13} & a_{12}a_{23} - a_{22}a_{13} \\ -a_{21}a_{33} + a_{31}a_{23} & a_{11}a_{33} - a_{31}a_{13} & -a_{11}a_{23} + a_{21}a_{13} \\ a_{21}a_{32} - a_{31}a_{22} & -a_{11}a_{32} + a_{31}a_{12} & a_{11}a_{22} - a_{21}a_{12} \end{bmatrix} \qquad (B.3.6)$$

and the determinant $\det(A)$ is given by Eq. (B.2.6), from which the inverse A^{-1} is immediately obtained via Eq. (B.3.2).

Fundamental properties of the matrix inverse are the following

$$(AB)^{-1} = B^{-1}A^{-1} \qquad (B.3.7)$$

$$\left(A^{-1}\right)^{-1} = A \qquad (B.3.8)$$

$$\left(A^{-1}\right)^T = \left(A^T\right)^{-1} \qquad (B.3.9)$$

$$(\lambda A)^{-1} = \frac{1}{\lambda} A^{-1} \qquad (B.3.10)$$

$$\det(A)\det(A^{-1}) = 1 \qquad (B.3.11)$$

Example B.2
Consider the matrix

$$A = \begin{bmatrix} 1 & 3 \\ 4 & 2 \end{bmatrix}$$

Its determinant, its adjugate and is inverse may be calculated by directly applying Eqs. (B.2.5), (B.3.4) and (B.3.5):

$$\det(A) = \det\left(\begin{bmatrix} 1 & 3 \\ 4 & 2 \end{bmatrix}\right) = 1 \cdot 2 - 3 \cdot 4 = -10$$

$$\mathrm{Adj}(A) = \mathrm{Adj}\left(\begin{bmatrix} 1 & 3 \\ 4 & 2 \end{bmatrix}\right) = \begin{bmatrix} 2 & -3 \\ -4 & 1 \end{bmatrix}$$

$$A^{-1} = \frac{1}{\det(A)} \mathrm{Adj}(A) = \begin{bmatrix} -\frac{1}{5} & \frac{3}{10} \\ \frac{2}{5} & -\frac{1}{10} \end{bmatrix}$$

B.4 Eigenvalues

The scalar λ is called an eigenvalue of a square $n \times n$ matrix A if there exists a nonzero column vector u that satisfies

$$Au = \lambda u \qquad (B.4.1)$$

When λ is an eigenvalue, any $u \neq 0$ satisfying Eq. (B.4.1) is called an eigenvector.

If for a specific value of λ, $\det(\lambda I - A) \neq 0$, then the linear system (B.4.1) has only the trivial solution $u = 0$, hence λ is not an eigenvalue. On the other hand, if $\det(\lambda I - A)$ vanishes, then (B.4.1) admits a nonzero solution. The eigenvalues are therefore the roots of the equation:

$$\det(\lambda I - A) = 0 \qquad (B.4.2)$$

The polynomial $\det(\lambda I - A)$ is called the characteristic polynomial of the $n \times n$ matrix A and Eq. (B.4.2) is called the characteristic equation of the matrix A. The characteristic polynomial is an n-th-degree polynomial in λ, with the coefficient of λ^n equal to 1 and with its n roots equal to the n eigenvalues $\lambda_1, \lambda_2, \ldots, \lambda_n$ of the matrix A. The characteristic polynomial can therefore be written in the form

$$\det(\lambda I - A) = (\lambda - \lambda_1)(\lambda - \lambda_2) \cdots (\lambda - \lambda_n) \tag{B.4.3}$$

This equation implies $\det(-A) = (-\lambda_1)(-\lambda_2)\cdots(-\lambda_n) = (-1)^n \lambda_1 \lambda_2 \cdots \lambda_n$ or $\det(A) = \lambda_1 \lambda_2 \cdots \lambda_n$, i.e. the determinant of an $n \times n$ matrix equals the product of its eigenvalues. Therefore, a square matrix is invertible if and only if all its eigenvalues are nonzero.

Other important properties of the eigenvalues include the following.

- The eigenvalues of a diagonal or triangular matrix are equal to its diagonal elements a_{ii}.
- The eigenvalues of the transpose A^T are the same as the eigenvalues of A.
- If λ_i, $i = 1, \ldots, n$ are the eigenvalues of a matrix A, then λ_i^p are the eigenvalues of the p-th power A^p, and e^{λ_i} are the eigenvalues of the exponential e^A.

For a square 2×2 matrix A, the characteristic polynomial is

$$\det(\lambda I - A) = \det\left(\lambda \begin{bmatrix} 1 & 0 \\ 0 & 1 \end{bmatrix} - \begin{bmatrix} a_{11} & a_{12} \\ a_{21} & a_{22} \end{bmatrix}\right) = \det\left(\begin{bmatrix} \lambda - a_{11} & -a_{12} \\ -a_{21} & \lambda - a_{22} \end{bmatrix}\right)$$

from which

$$\det(\lambda I - A) = \lambda^2 - (a_{11} + a_{22})\lambda + (a_{11}a_{22} - a_{12}a_{21}) \tag{B.4.4}$$

The eigenvalues are the roots of the polynomial (B.4.4). We observe that the constant term of (B.4.4) equals $\det(A)$ (see Eq. (B.2.5)). Hence, if $\det(A) = 0$, then $\lambda = 0$ is an eigenvalue of A.

For a square 3×3 matrix A, the characteristic polynomial is

$$\det(\lambda I - A) = \det\left(\begin{bmatrix} \lambda - a_{11} & -a_{12} & -a_{13} \\ -a_{21} & \lambda - a_{22} & -a_{23} \\ -a_{31} & -a_{32} & \lambda - a_{33} \end{bmatrix}\right) =$$

$$\lambda^3 - (a_{11} + a_{22} + a_{33})\lambda^2 + [(a_{11}a_{22} - a_{12}a_{21}) + (a_{22}a_{33} - a_{23}a_{32}) + (a_{33}a_{11} - a_{31}a_{13})]\lambda$$
$$-(a_{11}a_{22}a_{33} + a_{12}a_{23}a_{31} + a_{13}a_{32}a_{21} - a_{11}a_{23}a_{32} - a_{22}a_{13}a_{31} - a_{33}a_{12}a_{21}) \tag{B.4.5}$$

The eigenvalues are the three roots of the polynomial (B.4.5).

Example B.3

Consider the matrix

$$A = \begin{bmatrix} 1 & 3 \\ 4 & 2 \end{bmatrix}$$

The characteristic polynomial is

$$\det(\lambda I - A) = \det\left(\lambda \begin{bmatrix} 1 & 0 \\ 0 & 1 \end{bmatrix} - \begin{bmatrix} 1 & 3 \\ 4 & 2 \end{bmatrix}\right) = \det\left(\begin{bmatrix} \lambda - 1 & -3 \\ -4 & \lambda - 2 \end{bmatrix}\right)$$
$$= (\lambda - 1)(\lambda - 2) - (-4)(-3) = \lambda^2 - 3\lambda - 10$$

We could have obtained the same result by using Eq. (B.4.4). Note that $\det(A) = -10$. The characteristic equation is

$$\lambda^2 - 3\lambda - 10 = 0 \Rightarrow (\lambda - 5)(\lambda + 2) = 0$$

and the eigenvalues are $\lambda_1 = 5$ and $\lambda_2 = -2$.

Example B.4

Consider the matrix

$$A = \begin{bmatrix} 1 & 0 & 2 \\ 1 & 0 & 1 \\ 2 & 0 & 4 \end{bmatrix}$$

The characteristic polynomial is

$$\det(\lambda I - A) = \det\left(\lambda \begin{bmatrix} 1 & 0 & 0 \\ 0 & 1 & 0 \\ 0 & 0 & 1 \end{bmatrix} - \begin{bmatrix} 1 & 0 & 2 \\ 1 & 0 & 1 \\ 2 & 0 & 4 \end{bmatrix}\right) = \det\left(\begin{bmatrix} \lambda-1 & 0 & -2 \\ -1 & \lambda & -1 \\ -2 & 0 & \lambda-4 \end{bmatrix}\right)$$

$$= \lambda\{(\lambda-1)(\lambda-4) - (-2)\cdot(-2)\} = \lambda^3 - 5\lambda^2$$

We could have obtained the same result by using Eq. (B.4.5). The characteristic equation is therefore

$$\lambda^2(\lambda - 5) = 0$$

and it has one double root $\lambda_{1,2} = 0$ (double eigenvalue) and and one simple root $\lambda_3 = 5$ (simple eigenvalue). Because of the presence of the zero eigenvalue, we can immediately conclude that the matrix A is singular (does not have an inverse).

B.5 The Cayley–Hamilton Theorem and the Resolvent Identity

The Cayley–Hamilton theorem states the following: Every square $n \times n$ matrix A satisfies its characteristic equation. In other words, if the characteristic equation of a square $n \times n$ matrix A is

$$\lambda^n + a_1\lambda^{n-1} + \cdots + a_{n-1}\lambda + a_n = 0 \tag{B.5.1}$$

then the following holds true

$$A^n + a_1 A^{n-1} + \cdots + a_{n-1}A + a_n I = 0 \tag{B.5.2}$$

Basic Matrix Theory

The Cayley–Hamilton theorem is useful in calculating the inverse of a matrix, if its characteristic polynomial has already been calculated. Indeed, multiplying Eq. (B.5.2) by A^{-1}, we obtain $A^{n-1} + a_1 A^{n-2} + \cdots + a_{n-1} I + a_n A^{-1} = 0$, hence

$$A^{-1} = -\frac{1}{a_n}\left(A^{n-1} + a_1 A^{n-2} + \cdots + a_{n-1} I\right) \tag{B.5.3}$$

Another important byproduct of the Cayley–Hamilton theorem is the resolvent identity. The matrix $(\lambda I - A)^{-1}$ is called the resolvent matrix, and plays a very important role in the analysis of dynamic systems. The Cayley–Hamilton theorem leads to an explicit formula for the resolvent matrix. In particular, if the characteristic polynomial of a square $n \times n$ matrix A has already been calculated and is equal to $\lambda^n + a_1 \lambda^{n-1} + \cdots + a_{n-1} \lambda + a_n$, then the adjugate of $(\lambda I - A)$ is given by the formula:

$$\operatorname{Adj}(\lambda I - A) = I \lambda^{n-1} + (A + a_1 I)\lambda^{n-2} + \cdots + \left(A^{n-1} + a_1 A^{n-2} + \cdots + a_{n-2} A + a_{n-1} I\right) \tag{B.5.4}$$

hence the resolvent matrix can be expressed as

$$\begin{aligned}(\lambda I - A)^{-1} &= \frac{\operatorname{Adj}(\lambda I - A)}{\det(\lambda I - A)} \\ &= \frac{I \lambda^{n-1} + (A + a_1 I)\lambda^{n-2} + \cdots + \left(A^{n-1} + a_1 A^{n-2} + \cdots + a_{n-2} A + a_{n-1} I\right)}{\lambda^n + a_1 \lambda^{n-1} + a_2 \lambda^{n-2} + \cdots + a_{n-1}\lambda + a_n}\end{aligned} \tag{B.5.5}$$

The last equation is called the resolvent identity.

Example B.5

We have seen in Example B.3 that the matrix

$$A = \begin{bmatrix} 1 & 3 \\ 4 & 2 \end{bmatrix}$$

has the following characteristic polynomial

$$\det(\lambda I - A) = \lambda^2 - 3\lambda - 10$$

Based on the Cayley–Hamilton theorem, the following must hold true

$$A^2 - 3A - 10I = 0$$

To verify the validity of the previous relation, we perform the calculation

$$\begin{aligned}A^2 - 3A - 10I &= \begin{bmatrix} 1 & 3 \\ 4 & 2 \end{bmatrix}\begin{bmatrix} 1 & 3 \\ 4 & 2 \end{bmatrix} - 3\begin{bmatrix} 1 & 3 \\ 4 & 2 \end{bmatrix} - 10\begin{bmatrix} 1 & 0 \\ 0 & 1 \end{bmatrix} \\ &= \begin{bmatrix} 13 & 9 \\ 12 & 16 \end{bmatrix} - \begin{bmatrix} 3 & 9 \\ 12 & 6 \end{bmatrix} - \begin{bmatrix} 10 & 0 \\ 0 & 10 \end{bmatrix} = \begin{bmatrix} 13 & 9 \\ 12 & 16 \end{bmatrix} - \begin{bmatrix} 13 & 9 \\ 12 & 16 \end{bmatrix} = \begin{bmatrix} 0 & 0 \\ 0 & 0 \end{bmatrix}\end{aligned}$$

which shows that it is indeed true. We can now use the result of Cayley–Hamilton theorem to calculate the inverse of A:

$$A^2 - 3A - 10I = 0 \quad \Rightarrow \quad A - 3I - 10A^{-1} = 0 \quad \Rightarrow \quad A^{-1} = \frac{1}{10}A - \frac{3}{10}I$$

Hence

$$A^{-1} = \frac{1}{10}\begin{bmatrix} 1 & 3 \\ 4 & 2 \end{bmatrix} - \frac{3}{10}\begin{bmatrix} 1 & 0 \\ 0 & 1 \end{bmatrix} = \begin{bmatrix} -\frac{1}{5} & \frac{3}{10} \\ \frac{2}{5} & -\frac{1}{10} \end{bmatrix}$$

which is exactly the result found in Example B.2.

Example B.6

We have seen in Example B.4 that the matrix

$$A = \begin{bmatrix} 1 & 0 & 2 \\ 1 & 0 & 1 \\ 2 & 0 & 4 \end{bmatrix}$$

has the following characteristic polynomial

$$\det(\lambda I - A) = \lambda^3 - 5\lambda^2 = \lambda^2(\lambda - 5)$$

Applying Eq. (B.5.4) for a 3 × 3 matrix, we can calculate

$$\text{Adj}(\lambda I - A) = I\lambda^2 + (A + a_1 I)\lambda + (A^2 + a_1 A + a_2 I) = I\lambda^2 + (A - 5I)\lambda + (A^2 - 5A)$$

and substituting A,

$$\text{Adj}(\lambda I - A) = \begin{bmatrix} 1 & 0 & 0 \\ 0 & 1 & 0 \\ 0 & 0 & 1 \end{bmatrix}\lambda^2 + \begin{bmatrix} -4 & 0 & 2 \\ 1 & -5 & 1 \\ 2 & 0 & -1 \end{bmatrix}\lambda + \begin{bmatrix} 0 & 0 & 0 \\ -2 & 0 & 1 \\ 0 & 0 & 0 \end{bmatrix}$$

$$= \begin{bmatrix} \lambda^2 - 4\lambda & 0 & 2\lambda \\ \lambda - 2 & \lambda^2 - 5\lambda & \lambda + 1 \\ 2\lambda & 0 & \lambda^2 - \lambda \end{bmatrix}$$

Thus finally the resolvent matrix is given by

$$(\lambda I - A)^{-1} = \frac{\text{Adj}(\lambda I - A)}{\det(\lambda I - A)} = \begin{bmatrix} \frac{\lambda - 4}{\lambda(\lambda - 5)} & 0 & \frac{2}{\lambda(\lambda - 5)} \\ \frac{\lambda - 2}{\lambda^2(\lambda - 5)} & \frac{1}{\lambda} & \frac{\lambda + 1}{\lambda^2(\lambda - 5)} \\ \frac{2}{\lambda(\lambda - 5)} & 0 & \frac{\lambda - 1}{\lambda(\lambda - 5)} \end{bmatrix}$$

B.6 Differentiation and Integration of Matrices

Let

$$F(t) = \begin{bmatrix} F_{11}(t) & F_{12}(t) & \cdots & F_{1m}(t) \\ F_{21}(t) & F_{22}(t) & \cdots & F_{2m}(t) \\ \vdots & \vdots & \vdots & \vdots \\ F_{n1}(t) & F_{n2}(t) & \cdots & F_{nm}(t) \end{bmatrix} \qquad (B.6.1)$$

be a matrix function of a real variable t. The derivative of $F(t)$ is obtained by differentiating each element of $F(t)$:

$$\frac{dF}{dt}(t) = \begin{bmatrix} \dfrac{dF_{11}}{dt}(t) & \dfrac{dF_{12}}{dt}(t) & \cdots & \dfrac{dF_{1m}}{dt}(t) \\ \dfrac{dF_{21}}{dt}(t) & \dfrac{dF_{22}}{dt}(t) & \cdots & \dfrac{dF_{2m}}{dt}(t) \\ \vdots & \vdots & \vdots & \vdots \\ \dfrac{dF_{n1}}{dt}(t) & \dfrac{dF_{n2}}{dt}(t) & \cdots & \dfrac{dF_{nm}}{dt}(t) \end{bmatrix} \qquad (B.6.2)$$

The integral of $F(t)$, either definite or indefinite, is obtained by integrating each element of $F(t)$:

$$\int F(t)\,dt = \begin{bmatrix} \int F_{11}(t)\,dt & \int F_{12}(t)\,dt & \cdots & \int F_{1m}(t)\,dt \\ \int F_{21}(t)\,dt & \int F_{22}(t)\,dt & \cdots & \int F_{2m}(t)\,dt \\ \vdots & \vdots & & \vdots \\ \int F_{n1}(t)\,dt & \int F_{n2}(t)\,dt & \cdots & \int F_{nm}(t)\,dt \end{bmatrix} \qquad (B.6.3)$$

Example B.7

Given the matrix function $F(t) = \begin{bmatrix} t^2+1 & e^{2t} \\ \sin t & 45 \end{bmatrix}$, we wish to calculate its derivative $\dfrac{dF}{dt}(t)$ and its integral $\int_0^t F(\xi)\,d\xi$.

Differentiating element by element, we find

$$\frac{dF}{dt}(t) = \begin{bmatrix} \dfrac{d}{dt}(t^2+1) & \dfrac{d}{dt}(e^{2t}) \\ \dfrac{d}{dt}(\sin t) & \dfrac{d}{dt}(45) \end{bmatrix} = \begin{bmatrix} 2t & 2e^{2t} \\ \cos t & 0 \end{bmatrix}$$

and integrating element by element,

$$\int_0^t F(\xi)d\xi = \begin{bmatrix} \int_0^t (\xi^2+1)d\xi & \int_0^t e^{2\xi}d\xi \\ \int_0^t \sin\xi\, d\xi & \int_0^t 45 d\xi \end{bmatrix} = \begin{bmatrix} \frac{1}{3}t^3 + t & \frac{1}{2}(e^{2t}-1) \\ 1-\cos t & 45t \end{bmatrix}$$

Table B.1 Main matrix operations in MATLAB

Matrix operation	in MATLAB
Matrix addition, subtraction	A+B, A-B
Matrix multiplication	A*B
$n \times n$ Identity matrix	eye(n)
Determinant	det(A)
Matrix inverse	inv(A)
Eigenvalues	eig(A)
Characteristic polynomial	poly(A)
Exponential matrix e^A	expm(A)

B.7 Software Tools

B.7.1 Matrix Calculations in MATLAB

In Table B.1 the main matrix operations in MATLAB are summarized. Calculations in Example B.1 can be performed with MATLAB as follows:

```
» A=[1 2;0 3];
» B=[4 0;2 2];
» C=[7 0 0;0 8 9];
» A+B
ans =
       5  2
       2  5
» A*B
ans =
       8  4
       6  6
» A*C
```

```
ans =
      7  16  18
      0  24  27
» A^3
ans =
      1   26
      0   27
» expm(A)
ans =
      2.7183  17.3673
           0  20.0855
» [exp(1) exp(3)-exp(1);0 exp(3)]
ans =
      2.7183  17.3673
           0  20.0855
```

The transpose of matrix A is calculated as follows:

```
» A'
ans =
      1  0
      2  3
```

Calculations in Examples B.2, B.3 and B.5 can be performed as follows:

```
» A=[1 3;4 2];
» det(A)
ans =
      -10
» inv(A)
ans =
      -0.2000   0.3000
       0.4000  -0.1000
» eig(A)
ans =
      -2
       5
» p=poly(A)
p =
      1   -3   -10
» p(1)*A^2+p(2)*A+p(3)*eye(2)
ans =
      0  0
      0  0
```

Table B.2 Main matrix operations in Maple

Matrix operation	in Maple (`LinearAlgebra` package)
Matrix addition, subtraction	`A+B, A-B`
Matrix multiplication	`A.B` or `Multiply(A,B)`
$n \times n$ identity matrix	`IdentityMatrix(n)`
Determinant	`Determinant(A)`
Matrix inverse	`MatrixInverse(A)`
Eigenvalues	`Eigenvalues(A)`
Characteristic polynomial	`CharacteristicPolynomial(A,lambda)`
Exponential matrix e^A	`MatrixExponential(A)`

B.7.2 Matrix Calculations in Maple

The basic operations of addition, subtraction and multiplication are performed in Maple as `A+B`, `A-B` and `A.B`, respectively. Further matrix operations and calculations may be performed using the `LinearAlgebra` package of Maple. Table B.2 summarizes the main matrix operations in Maple.

Calculations in Example B.1 can be performed with Maple as follows:

```
> with(LinearAlgebra):
> A:=Matrix([[1,2],[0,3]]):
> B:=Matrix([[4,0],[2,2]]):
> C:=Matrix([[7,0,0],[0,8,9]]):
> A+B;
```

$$\begin{bmatrix} 5 & 2 \\ 2 & 5 \end{bmatrix}$$

```
> A.B;
```

$$\begin{bmatrix} 8 & 4 \\ 6 & 6 \end{bmatrix}$$

```
> Multiply(A,B);
```

$$\begin{bmatrix} 8 & 4 \\ 6 & 6 \end{bmatrix}$$

```
> A.C;
```

$$\begin{bmatrix} 7 & 16 & 18 \\ 0 & 24 & 27 \end{bmatrix}$$

```
> Multiply(A,C);
```

$$\begin{bmatrix} 7 & 16 & 18 \\ 0 & 24 & 27 \end{bmatrix}$$

```
> Transpose(A);
```

$$\begin{bmatrix} 1 & 0 \\ 2 & 3 \end{bmatrix}$$

> A^3;

$$\begin{bmatrix} 1 & 26 \\ 0 & 27 \end{bmatrix}$$

> MatrixExponential(A);

$$\begin{bmatrix} e & e^3 - e \\ 0 & e^3 \end{bmatrix}$$

Calculations in Examples B.2 and B.3 can be performed as follows:

> with(LinearAlgebra):
> A:=Matrix([[1,3],[4,2]]):
> Determinant(A);

$$-10$$

> MatrixInverse(A);

$$\begin{bmatrix} -\dfrac{1}{5} & \dfrac{3}{10} \\ \dfrac{2}{5} & -\dfrac{1}{10} \end{bmatrix}$$

> CharacteristicPolynomial(A,lambda);

$$\lambda^2 - 3\lambda - 10$$

> Eigenvalues(A);

$$\begin{bmatrix} 5 \\ -2 \end{bmatrix}$$

Finally, the characteristic polynomial and the resolvent matrix $(\lambda I - A)^{-1}$ of Examples B.4 and B.6 can be calculated as follows:

> with(LinearAlgebra):
> A:=Matrix([[1,0,2],[1,0,1],[2,0,4]]):
> CharacteristicPolynomial(A,lambda);

$$\lambda^3 - 5\lambda^2$$

> MatrixInverse(lambda*IdentityMatrix(3)-A);

$$\begin{bmatrix} \dfrac{\lambda-4}{\lambda(\lambda-5)} & 0 & \dfrac{2}{\lambda(\lambda-5)} \\ \dfrac{\lambda-2}{\lambda^2(\lambda-5)} & \dfrac{1}{\lambda} & \dfrac{\lambda-1}{\lambda^2(\lambda-5)} \\ \dfrac{2}{\lambda(\lambda-5)} & 0 & \dfrac{\lambda-1}{\lambda(\lambda-5)} \end{bmatrix}$$

Index

absolute pressure, 330
advanced control, 641
all-pass element, 312, 669
all-pass filter, 81
amplitude ratio, 78, 159, 297
asymptotes, 503
asymptotic stability
   and eigenvalues, 221
   and forced response, 224
   and ultimate response, 225
   bioreactor example, 233
   CSTR example, 237
   definition, 221
   of nonlinear systems, 231
   theorem of Lyapunov, 233
automatic control, 1

backward rectangle approximation, 84
ball in a valley, 220
band-pass filter, 81
band-stop filter, 81
block diagram, 118, 120
   heating tank example, 333
   multi-loop systems, 356
   representation, 351
   simplification, 350
   simplification rules, 353
Bode diagrams, 298
   and straight-line approximation, 303
   first-order systems, 79, 299
   high-frequency, 311
   higher-order systems, 302
   integrating systems, 300
   lead–lag element, 306
   low-frequency, 311
   nonminimum phase element, 310
   second-order systems, 299
Bode plots, 298
bounded-input–bounded-output stability, 273
   and asymptotic response, 275
   conditions, 274

bounded-input–bounded-state property, 224

cascade control
   "triple" cascade systems, 688
   advantages, 685
   benefits, 688
   examples, 684
   introduction, 684
   model-based, 688
Cayley–Hamilton theorem, 752
center of gravity, 503
characteristic polynomial, 216, 253
closed loop, 328
   characteristic polynomial, 385
   poles, 385
   stability, 385
closed-loop response
   systems with dead time, 467
closed-loop system, 5
   and P control in state space, 415
   and PI control in state space, 417
   general case in state space, 419
closed-loop transfer function, 352
conservation laws
   energy, 22
   general form, 20
   mass of a component, 22
   total mass, 21
conserved quantities, 20
control actions
   general comments, 395
control laws
   introduction, 338
   proportional controller, 339
   proportional-derivative controller, 340
   proportional-integral controller, 340
   proportional-integral-derivative controller, 342
   state space, 341
control room, 6

control valve, 331
control volume, 20
controlled variable, 5
controller, 2, 328
controller bias, 339
controller gain, 339
convolution integral, 83
convolution theorem, 722
CSTR
   nonlinear closed loop, 427

damped-oscillation frequency, 152
damping factor, 146
dead time
   accuracy of approximations, 452
   and distributed parameter systems, 442
   Bode stability criterion, 583
   derivation of Padé approximation, 450
   low-order Padé approximations, 451
   n first-order systems in series, 445
   Padé approximation, 449
   reverse Taylor approximation, 455
   stability, 460
   step response of a first-order system, 441
   Taylor approximation, 449
decay ratio, 154
decibels, 303
delayed measurement, 438
denominator polynomial, 259
derivative time, 342
deviation variables, 59, 63, 176
diagonal elements, 744
differential delay systems, 468
differential pressure, 330
differential pressure cell, 330
discrete time model
   general SISO system, 193
   general state space, 422
distributed control system, 7

Index

disturbance rejection, 379
disturbance variables, 5
dynamic output response
    standard inputs, 190
dynamic response
    forced response, 187
    to an arbitrary input, 191
    to standard inputs, 188
    unforced response, 187
    using the matrix exponential
        function, 195

eigenvalues and dynamic response, 218
equal-percentage valve, 332
equilibrium point, 221
error, 5, 8
Euclidian norm, 232

Faraday's law, 35
feedback control
    disturbance rejection, 701
feedback control loop
    block diagram, 377
    parts, 351
    steady-state analysis, 378
    steady-state error, 378, 380
feedback control loop examples
    flow control, 337
    heating tank, 332
    liquid level, 335
    pressure control, 337
feedback control system, 5
feedforward control
    comparison with feedback, 708
    disadvantages, 696
    examples, 696
    FOPDT system, 705
    ideal form, 698
    introduction, 695
    model-based, 700
    optimal, 702
feedforward–feedback control, 699
    model-based, 709
    Q-parameterization, 712
filter factor, 87
filtering, 87
final control element, 2, 328, 351
final-value theorem, 723
first order plus dead time
    approximation of higher-order
        systems, 455

Bode stability criterion, 584
exact stability limits under P
    control, 463
feedforward control, 705
general model, 441
model-based control, 661
parameter estimation, 456
stability analysis with P control, 461
step response, 441
first-order system
    Bode diagrams, 79
    CSTR, 57
    definition, 55
    discrete, 82
    estimation of parameters, 70
    frequency response, 78
    general form, 99
    general model, 58
    mercury thermometer, 55
    sinusoidal response, 78
    solution by integrating factor, 60
    solution by Laplace transform, 61
    transfer function, 64
first-order system response
    to impulse input, 75
    to pulse input, 73
    to ramp input, 75
    to sinusoidal input, 77
    to step input, 68
first-order systems
    in parallel, 119
    in series, 115
    interacting, 122
forced response, 63
forward rectangle approximation,
    84
fractional opening of a valve, 332
frequency response, 298
    first-order system, 78

gain margin, 588

heat of reaction, 31
heating tank
    P control, 360
    PI control, 363
    with dead time, 437
heating tank example
    combined performance criterion,
        562
    discrete PI control, 424

eigenvalue analysis, 412, 415
ISE performance criterion, 551
stability under P control, 393
stability under PI control, 386, 391
stability under PID control, 396
state-space model, 409
state-space under PI control, 413
steady-state analysis, 412
Heaviside function, 719
higher-order systems
    approximation by FOPDT, 455
high-pass filter, 81

improper transfer function, 257
information flow, 4
inherent characteristics of a valve, 332
initial-value theorem, 723
input–output model, 254
integral action, 384
integral time, 340
integrating process
    P control, 364
integrating system, 90
integrating system plus dead time
    stability analysis, 465
internal model controller, 650
inventory rate equation, 20
inverse Laplace transform, 725
inverse response, 272

Kirchhoff's law, 40

Laplace expansion, 747
Laplace transform, 719
    definition, 719
    elementary functions, 720
    properties, 721
    solving ODE, 732
lead–lag element, 306
level sensors, 330
linear approximation of a function, 90
linear partial differential equation, 443
linear time-invariant systems, 175
linear valve, 332
linearization, 91, 94, 228, 230
    general first-order system, 100
locally asymptotically stable, 231
locus of the roots, 502
loop transfer function, 353, 576
low-pass filter, 81, 87
    equation, 82

762   Index

Luedeking and Piret model, 34
Lyapunov equation, 550
Lyapunov theorem, 233

manipulated variable, 2, 5
Maple command
   `BodePlot`, 320
   `dsolve`, 45
   `GainMargin`, 605
   `ImpulseResponse`, 285
   `int`, 104
   `invlaplace`, 105, 737
   `laplace`, 105
   `matrix`, 205
   `MatrixExponential`, 205
   `pade`, 473
   `parfrac`, 738
   `PhaseMargin`, 606
   `plot`, 165
   `residue`, 284
   `RootLocusPlot`, 534
   `vectror`, 205
   `ZeroPoleGain`, 283
mathematical models
   distributed, 19
   empirical, 19
   general comments, 18
   lumped, 19
   mechanistic, 19
   steps in developing, 19
MATLAB command
   `bode`, 319
   `c2d`, 204
   `conv`, 736
   `dcgain`, 280, 400
   `expm`, 202
   `feedback`, 367
   `fminsearch`, 566
   `impulse`, 281
   `linspace`, 202
   `margin`, 604
   `nyquist`, 319
   `ode45`, 43
   `pole`, 280
   `residue`, 280, 736
   `rlocfind`, 532
   `rlocus`, 531
   `roots`, 736
   `series`, 281, 367
   `ss`, 203, 473
   `ss2tf`, 203
   `step`, 203
   `tf`, 281, 473
   `tf2zp`, 280
   `zero`, 280
matrix
   addition, 743
   adjoint, 749
   commands in Maple, 758
   commands in MATLAB, 756
   definition, 743
   determinant, 747
   diagonal, 744
   differentiation, 755
   eigenvalues, 750
   identity, 744
   integration, 755
   inversion, 749
   square, 744
   subtraction, 743
   symmetric, 745
   transpose, 745
   zero, 744
matrix exponential, 745
matrix exponential function, 180
   definition, 180
   properties, 180, 181
maximum overshoot, 154
measured variables, 5
measurement, 8
mercury thermometer
   selection of reference temperature, 62
minimum-phase factor, 645, 669
model-based control, 642
   cascade control, 688
   controller synthesis formula, 649
   dynamic performance, 645
   factorization of the transfer function, 645
   feedforward, 700
   feedforward and feedback, 709
   filter selection, 676
   in state space, 668
   introduction, 642
   ISE optimal, 645
   large dead time, 657
   low-order models, 650
   MIMO systems, 674
   modeling error, 660
   nearly optimal controllers, 648
   specifications, 642
   zero offset, 644
modelling example
   bioreactor, 32
   blending process, 25
   CSTR isothermal, 28
   CSTR, nonisothermal, 30
   electrical circuit, 39
   flash distillation, 37
   heating tank, 26
   integrating system, 88
   interacting tanks, 122
   liquid storage tank, 23
   mechanical systems, 41
   mercury thermometer, 55
   mixing tank, 614
   n tanks in series, 174
   solid-oxide fuel cell, 34
   spring with an attached object, 145
   tank with input bypass, 97
   tanks is parallel, 119
   three tanks in series, 173
   two tanks in series, 116
Monod equation, 33
multi-input–multi-output systems, 197
   state-space description, 197
   transfer function description, 201
multiple input–multiple output
   closed loop in state space, 625
   closed-loop transfer function, 625
   decoupling, 632
   dynamic response, 620
   interaction, 627
   introduction, 613
   poles, 621
   stability, 621
   state space, 615
   steady-state gain matrix, 616
   transfer-function matrix, 618
   zeros, 621
Myshkys method of steps, 468

natural frequency, 147
Nernst potential, 36
Newton's second law, 42
noise, 87
nonlinear systems
   closed-loop system in state space, 426
nonminimum phase element, 310
numerator polynomial, 259
Nyquist criterion for stability, 595

Nyquist plots, 298, 312
  first-order systems, 313
  higher-order systems, 317
  integrator, 315
  lead system, 316
  second-order systems, 314

offset, 379, 380, 381, 382
optimal PID controller parameters
  combined criteria, 559
  quadratic criteria, 557

Padé approximation
  and step response, 470
Parseval's theorem, 723
partial fraction expansion, 725
partial fractions, 216
peak time, 154
pendulum system, 220
performance criteria
  based on input variation, 556
  calculation for infinite and finite time, 544
  calculation in Laplace domain, 554
  combined, 546
  derivative of the error, absolute, 546
  derivative of the error, squared, 546
  input variation, 546
  integral of the absolute error, 543
  integral of the error, 543
  integral of the squared error, 543
  integral of the time-weighted squared error, 545
  integral of time-weighted absolute error, 545
  integrating process, example, 546
  introduction, 541
  Lyapunov equation, 550
  quadratic criteria, calculation, 549
  symbolic calculation, 569
phase angle, 298
phase margin, 587
polar plot, 298, 312
poles, 259, 484
  and eigenvalues, 259
pressure sensors, 330
principle of the argument, 594
Process and Instrumentation Diagrams, 8
process control, 1

process dynamics
  typical responses, 10
Process Flow Diagram, 8
proper transfer function, 257
proportional-derivative controller
  ideal, 342
  real, 343
  real, state-space, 343
proportional-integral controller
  phase and gain margin specification, 590
proportional-integral-derivative controller
  ideal, 342
  real, 344
  summary, 344
  Ziegler–Nichols tuning, 592

Q-parameterization, 644

rate of cell growth, 32
ratio control
  examples, 694
  introduction, 694
rational function, 253
RC circuit, 40
reaction rate, 22
recursive calculation, 84
relative order, 257, 649
relative stability, 587
resistance thermometers detectors, 329
resolvent identity, 753
resolvent matrix, 177, 216, 753
resonance, 160
rise time, 154
robust stability, 664
  conditions, 664
root-locus diagram, 485, 488
  angles of arrival, 508, 515
  angles of departure, 508, 512
  approximate sketching, 504
  basic properties, 502
  break-away points, 508
  controller parameters, 527
  points of intersection, 521, 524
  third-order system and P control, 505
  third-order system and PI control, 506
  third-order system and PID control, 507

Routh array, 389
Routh criterion, 389

sampling period, 83
  choice and accuracy, 225
second-order system
  as two first-order systems in series, 147
  Bode diagram, 159
  estimation of the parameters, 158
  example, 145
  frequency response, 159
  general equation, 146
  general form, 161
  step response, 148
  step response, characteristics, 154
  step response, critically damped, 151
  step response, overdamped, 150
  step response, underdamped, 152
  transfer function, 146
sensitivity coefficient, 381, 384
sensitivity function, 381, 384
sensor, 6, 328, 351
set point, 2, 5
set point tracking, 378
settling time, 154
single-input–single-output systems, 175
Smith predictor, 657
specific enthalpy, 23
specifications
  in model based control, 642
stability, 10
  Bode criterion, 576, 577
  criteria in the frequency domain, 575
  Nyquist criterion, 594
stability limits, 394
stabilizing controllers, 644
standard inputs
  pulse, 66
  ramp, 66
  sinusoidal, 67
  step, 65
Standard ISA-S5, 8
state space
  with input delay, 440
  with output delay, 440
state-space description, 171
state-space equations
  general solution, 182

state-space model, 117
static gain, 176, 260
steady-state gain, 58, 69
step response
    and zeros of the transfer function, 268
stoichiometric number, 22
summing point, 120
superposition principle, 61
surge tank, 88
symbols
    in P&ID diagrams, 9
system eigenvalues, 217

take-off point, 121
temperature sensors
    models, 330
thermocouples, 329
thermoelectric circuit, 329
third-order process
    P control, 485
    PI control, 490
    PID control, 490
time constant, 58
transducer, 9
transfer function, 252

transient behavior, 10
transmitter, 328

ultimate gain, 591
ultimate period, 591
unforced response, 63

vapor–liquid equilibrium, 37

zero-order hold, 84
zeros, 259
Ziegler–Nichols tuning rules, 591